EMOTION, WELL-BEING, AND RESILIENCE

Theoretical Perspectives and Practical Applications

EMOTION, WELL-BEING, AND RESILIENCE

Theoretical Perspectives and Practical Applications

Edited by

Rabindra Kumar Pradhan, PhD

Updesh Kumar, PhD

AAP | APPLE ACADEMIC PRESS

First edition published 2021

Apple Academic Press Inc.
1265 Goldenrod Circle, NE,
Palm Bay, FL 32905 USA

4164 Lakeshore Road, Burlington,
ON, L7L 1A4 Canada

CRC Press
6000 Broken Sound Parkway NW,
Suite 300, Boca Raton, FL 33487-2742 USA

2 Park Square, Milton Park,
Abingdon, Oxon, OX14 4RN UK

First issued in paperback 2021

Library and Archives Canada Cataloguing in Publication

Title: Emotion, well-being, and resilience : theoretical perspectives and practical applications / edited by Rabindra Kumar Pradhan, PhD, Updesh Kumar, PhD.

Names: Pradhan, Rabindra Kumar, editor. | Kumar, Updesh, editor.

Description: Includes bibliographical references and index.

Identifiers: Canadiana (print) 20200316818 | Canadiana (ebook) 20200317105 | ISBN 9781771888905 (hardcover) | ISBN 9781003057802 (ebook)

Subjects: LCSH: Resilience (Personality trait) | LCSH: Emotions. | LCSH: Well-being. | LCSH: Mental health.

Classification: LCC BF698.35.R47 E46 2021 | DDC 155.2/4—dc23

Library of Congress Cataloging-in-Publication Data

Names: Pradhan, Rabindra Kumar, editor. | Kumar, Updesh, editor.

Title: Emotion, well-being, and resilience : theoretical perspectives and practical applications / edited by Rabindra Kumar Pradhan, PhD, Updesh Kumar, PhD.

Description: Palm Bay, FL : Apple Academic Press, 2021. | Includes bibliographical references and index. | Summary: "This important new volume discusses the role of emotion, resilience, and well-being in many contexts of human life, including home, school, and workplace. Leading researchers and academicians from around the world and from various fields-such as health, education, information and technology, military, and manufacturing-explore the theoretical and practical implications of many studies in this area. They present new concepts, models, and knowledge for practical applications that address challenges to well-being. The volume also considers the roles of several other influencing factors, such as emotional intelligence, performance, productivity, and employee's health and happiness. The book's editors state that, "At this juncture of human and technological development, when artificial intelligence and automation are slowly taking over the world, holding on to the study of emotions, well-being, and resilience has become imperative, as these influence sustainable performances and growth by individuals as well as organizations." This volume addresses the important issues of how globalization has affected employee mental health and poses workplace-related challenges in both developed and emerging economies. It considers the growing state of social apathy, aggression, and violence in many places across the globe along with impact of the frequent natural and man-made disasters that have been continuously disturbing and destabilizing human life in society and community. Key features of Emotion, Well-Being, and Resilience: Theoretical Perspectives and Practical Applications: Presents empirical findings from many studies on well-being and resilience with practical applications Provides a comprehensive understanding of the theoretical concepts and models related to emotion, well-being, and resilience Reviews findings in the field of resilience and well-being studies in the context of extreme environments, such as the military Addresses a mental health and well-being in many contexts, including workplace, disease and chronic illness, motherhood, suicide risk, caring for the disabled, aging, school (both teachers and students), community, sports, etc. With contributions from over 70 authors from six continents, this volume will be exceptionally valuable for research scholars and teachers in psychology, human resource managers, and mental health practitioners working in the field of emotion, resilience, and well-being"-- Provided by publisher.

Identifiers: LCCN 2020038294 (print) | LCCN 2020038295 (ebook) | ISBN 9781771888905 (hbk) | ISBN 9781003057802 (ebk)

Subjects: LCSH: Resilience (Personality trait) | Well-being. | Emotions.

Classification: LCC BF698.35.R47 E66 2021 (print) | LCC BF698.35.R47 (ebook) | DDC 158.1--dc23

LC record available at https://lccn.loc.gov/2020038294

LC ebook record available at https://lccn.loc.gov/2020038295

ISBN: 978-1-77188-890-5 (hbk)
ISBN: 978-1-77463-759-3 (pbk)
ISBN: 978-1-00305-780-2 (ebk)

About the Editors

Rabindra Kumar Pradhan, PhD, is currently working as an Associate Professor of Psychology and Human Resources in the Department of Humanities and Social Sciences, Indian Institute of Technology Kharagpur, India. His research areas are focused on employee health and well-being, resilience, emotional intelligence, emotional labor, employee performance, employee health and well-being, workplace spirituality, organizational citizenship behavior, HR flexibility, positive leadership, and HR sustainability. He has more than 20 years of research and teaching experience and has published numerous papers in reputed peer-reviewed journals, such as *Journal of Knowledge Management, The Qualitative Report, Benchmarking: An International Journal, International Journal of Entrepreneurship and Innovation Management, International Journal of Productivity and Performance Management, International Journal of Work, Organization and Emotion, Journal of Workplace Learning, International Journal of Human Resources Development and Management, Psychological Studies, Journal of Indian Academy of Applied Psychology, Global Business Review, Vision-The Journal of Business Perspectives, Psychologist-Manager, Cogent Business & Management, Management and Labor Studies*, and *Journal of Health Management, etc.* He has also published 20 international book chapters and four edited books in the area of emotional intelligence, human resources management, and training and development.

He is a member of the International Association of Applied Psychology, Academy of Management, National Academy of Psychology, India, Indian Academy of Applied Psychology, and many others. He is a recipient of many awards; including an ISTD Best Trainer Award, 2012; Best Teacher Award in HRD, 2013; and Best Research Paper Award-ICOM-2012, Abu Dhabi, etc.

Dr. Pradhan has published four edited books: *Human Resources Management in India: Emerging Issues* and *Challenges* by Rabindra Kumar Pradhan and C. K. Podder (New Century Publications, New Delhi, 2013); *Emotional Intelligence: Optimizing Human Performance at Work* by Rabindra Kumar Pradhan and N. K. S. Thingujam (LAP Lambart Publisher, Germany, 2012; ISBN:978-3-8473-7530-2); *Value Creation through Human Development: The Emerging Dimensions* by Uday Kumar Haldar, Rabindra Kumar Pradhan, Arun Roy Chaudhury, and Juthika Sarkar (Published by Indian Society for Training & Development [ISTD], 2012, Kolkata); *Emotional Intelligence: Perspectives in Organization* by Rabindra Kumar Pradhan and Purnima Mathur (Academic Excellence, New Delhi, 2008).

Updesh Kumar, PhD, is a Scientist "G" (R) and the former head of the Mental Health Division, Follow-Up Division, and Technical Coordination Division at the Defense Institute of Psychological Research (DIPR), R & D Organization (DRDO), Ministry of Defense, Delhi, India. After obtaining his doctorate degree in the area of suicidal behavior from Punjab University, Chandigarh, India, he gained more than 30 years of experience as a Scientist in an R&D organization. He specializes in the area of military psychology, personality assessment, suicidal behavior, and health psychology. Dr. Kumar has been a psychological assessor (Psychologist) on various Services Selection boards for eight years for the selection of officers in the Indian Armed Forces and was responsible for monitoring the entire selection system of the Indian Armed Forces for more than 22 years. He has been the project director of a mega project titled *Comprehensive Soldier Fitness Program: Resilience Building* for the Indian Armed Forces.

Dr. Kumar has edited 12 volumes that include *Suicidal Behaviour: Assessment of People-at-Risk* (Sage, 2010); *Countering Terrorism: Psychosocial Strategies* (Sage, 2012); *Understanding Suicide Terrorism: Psychosocial Dynamics* (Sage, 2014); *Suicidal Behaviour: Underlying Dynamics* (Routledge, UK, 2015); *Positive Psychology: Applications in Work, Health and Well-being* (Pearson Education, 2015); *The Wiley Handbook of Personality Assessment* (John Wiley & Sons, 2016); *The Routledge International Handbook of Psychosocial Resilience* (Routledge, 2016); *Handbook of Suicidal Behavior* (Springer Nature, 2017); and most recently *The Routledge International Handbook of Military Psychology and Mental Health* (Taylor & Francis Group, 2019). He is also the past editor of *the Journal of Indian Academy of Applied Psychology* (JIAAP). Dr. Kumar has authored manuals on *Suicide and Fratricide: Dynamics and Management* for defense personnel; *Managing Emotions in Daily Life & at Work Place for the general population; Overcoming Obsolescence & Becoming Creative in R&D Environment for R&D organizations; and Self-Help Techniques in Military Settings.*

In addition to completing many important research projects of the Indian Armed Forces, Dr. Kumar has authored more than 65 other academic publications in the form of research papers, journal articles, and book chapters and has represented his institute at both the national and international level. He is a certified psychologist by The British Psychological Society with level 'A' and level 'B' Certificates of Competence in Occupational Testing. He has been conferred with several awards, which include the DRDO's Best Popular Science Communication Award–2009 by Hon'ble Defense Minister of India; DRDO Technology Group Award in 2001 and 2009; Laboratory Scientist of the Year Award–2012–2013; and the prestigious DRDO's Scientist of the Year Award–2013 by the Government of India.

Contents

Contributors ... *xi*

Abbreviations ... *xxv*

About This Book ... *xxvii*

Foreword ... *xxix*

Acknowledgments ... *xxxi*

Preface ... *xxxiii*

PART I: Emotion, Well-Being and Resilience in Clinical and Counseling Context ... 1

1. **Stress, Resilience, and Coping ... 3**
   Rachel N. Ward, Abbie J. Brady, Rebekah Jazdzewski, and Matthew M. Yalch

2. **Resilience and Reducing Self-Destructive Behaviors ... 15**
   Bruce Bongar, Kristen J. Vescera, Paola Mendoza-Rivera, and Rachel Bacigalupi

3. **Resilience and Mental Health ... 33**
   Amanda Wallick, Anna Feinman, Jacie T. Brown, and Lisa M. Brown

4. **Cherishing and Psychological Well-Being: Some Empirical Observations ... 51**
   Pratishtha Bhattacharyya and Rabindra Kumar Pradhan

5. **Social Support and Emotion Regulation as Predictors of Well-Being ... 61**
   Meera Padhy and Padiri Ruth Angiel

6. **The Role of Perceived Hope in Harmony in Life among Czech and South African Adults ... 77**
   Alena Slezackova, Tharina Guse, and Tomas Prosek

7. **Difficulties in Emotion Regulation as a Barrier to Resilience: Therapeutic Implications ... 95**
   Parwinder Singh

8. **Assessment of Coping with Stress: A Short Version of a Coping Strategy Indicator (CSI) in Peruvian Adults ... 109**
   César Merino-Soto and Arturo Juárez-García

9. **Are You Satisfied with Your Care? A Study to Examine the Predictors of Indian Cancer Patients' Satisfaction with Their Medical Consultation ... 129**
   Shweta Chawak and Mahati Chittem

10. Self-Efficacy and Resilience: Correlates of Happiness and Well-Being Among Adolescents with Disability 145
Swati Patra

11. Resilience, Coping, and Chronic Illnesses in the Workplace 161
Manisha Sawhney and Julijana Nevland

12. Resilience and Suicide Risk 175
Renata Sargon, Marissa N. Eusebio, Rachel Bacigalupi, Kevin Rodriguez, and Bruce Bongar

PART II: Resilience and Well-Being in Home, School, Community, and Sports Environments 193

13. Systemic Analysis of Family Resilience 195
Alexander V. Makhnach and Anna I. Laktionova

14. Resilience and Successful Aging in the Elderly: Exploring the Indian Scenario 209
Shubhangi Goswami and Rabindra Kumar Pradhan

15. Making of a Resilient Student: Tracing Challenges and Interventions 229
Samridhi Ahuja, Ayesha Saha, and Updesh Kumar

16. Teachers' Resilience: An Empirical Investigation Based on Demographic Profiles 243
G. N. Sumathi, R. Arthi, and Feba Thomas

17. Community Resilience and Place Identity 255
Jhuma Mukhopadhyay, and Shubhangi Goswami

18. Resilience and Well-Being in Sports 267
Mohita Junnarkar, Sanjeev P. Sahni, and Shivjot Gill

PART III: Emotion, Resilience, and Well-Being in Work and Organizational Context 281

19. Positive Emotions at Workplace 283
Archana and Updesh Kumar

20. Navigating Through the Thunderstorm: Understanding and Developing Leadership Resilience at Workplace 297
Surendra Kumar Sia and K. Alphonsa Jose

21. Promoting Workplace Health: A Positive Occupational Health Psychology Perspective 309
Aishwarya Jaiswal, Yogesh Kumar Arya, and Tushar Singh

22. Linkage Among Work-Life Balance, Psychological Capital, and Life Satisfaction: Testing a Spillover-Crossover Model Among Indian Dual-Working Couples 323
Nidhi Bansal and Upasna A. Agarwal

23. Validation of Resilience at Work (RAW) Scale: An Insight from an Indian Sample.................. 339
Pooja Garg and Han Ki-Soon

24. Can Meaningful Work Mitigate the Impact of Technostress Creators? Evidence from India.. 359
Monalisa Mahapatra, Surya Prakash Pati, and Rabindra Kumar Pradhan

25. Antecedents and Consequences of Employee Well-Being: A Theoretical Framework................ 379
Lopamudra Hati, and Rabindra Kumar Pradhan

PART IV: Resilience in Extreme Environment...395

26. Psychological Flexibility: An Overview and Avenues for Application in the Military 397
Swati Mukherjee and Updesh Kumar

27. Psychological Resources of Resilience to Terrorist Threat.. 409
Nadezhda N. Kazymova, Nadezhda V. Tarabrina, and Julia V. Bykhovets

28. Terrorism, Emotion, and Well-Being 425
Catalina Udani, and Jonathan Matusitz

29. Post-Traumatic Growth: Positive Transformation beyond Resilience... 439
N. D. S. Naga Seema

30. Post-Traumatic Growth Following the Transition to Motherhood: Investigating the Role of Social Support... 451
Jillian M. Millar and Julie Ann Pooley

31. The Utility of Adventure Programming for the Development of Resilience and Related Constructs: Examples from Adolescent, Corporate, and Clinical Contexts 465
Conrad S. Zygmont, A. Vernon Naidoo, and Pieter Snyman

Index.. 487

Contributors

Upasna A. Agarwal, PhD, is an Associate Professor in Organization Behavior and Human Resource Management at NITIE. An MBA and Masters Labor Law from Symbiosis, Pune, she is a PhD from Indian Institute of Technology, Mumbai. Upasna launched her academic career at S.P. Jain Institute of Management and Research, and has seven years of corporate experience prior to that. In her teaching she has designed and conducted modules on various aspects of personal and team effectiveness for middle and senior managers of public- and private-sector organizations and NGOs. Her areas of research interest are work engagement, cross generational differences, psychological contract, trust, organizational conflict, power politics and bullying, employee well-being, and innovation.

Samridhi Ahuja is a RCI certified counselor trained in child and adolescent guidance and counseling from NIPCCD, New Delhi. She has pursued her Bachelors in Psychology from Gargi College and Masters in Psychology from North Campus, University of Delhi. Having previously practiced at NIPCCD, she is passionate about working with children with developmental disorders and groups with special needs. Currently, she is pursuing her PhD in the area of clinical psychology. Her areas of interest also include Indian psychology, developmental psychology, suicidology, and counseling. As a Research Scholar at the Defence Institute of Psychological Research, Defence R&D Organisation, Ministry of Defence, Government of India, she is associated with the major research project on "Comprehensive Soldier Fitness Program: Resilience Building," under the directorship of Dr. Updesh Kumar, Scientist "G". Apart from this, she has presented papers in various national and international conferences and has original publications in peer-reviewed journals.

Padiri Ruth Angiel, PhD, has completed her Doctoral studies under the guidance of Dr. Meera Padhy at the University of Hyderabad. She is currently working as Assistant Professor in the Department of Psychology, School of Social and Behavioral Sciences at the Central University of Karnataka. Although her primary research area is Health Psychology, her research interests include child development, adjustment during adolescence, challenges of young adults, and women's mental health. She has facilitated workshops with various groups of adolescents and has worked in the rehabilitation center for adolescents. She has also presented papers in various national seminars and has publications in esteemed journals.

Archana, PhD, is a Scientist "E" at Defence Institute of Psychological Research and is associated with R&D activities in the area of mental health and well-being of soldiers deployed in peace and field locations. She has authored two manuals on *Stress Management* and *Psychological Well-being of Soldiers at High Altitude*. In addition, she has made valuable contributions by organizing many workshops and seminars for the soldiers in the Indian Army. She has also been actively involved in delivering lectures on stress management and other related areas in active field environments. She is the co-editor of a volume on *Positive Psychology: Applications in Work, Health and Well-being* published by Pearson Education and has made significant contributions in the field of research through publications. She has been a recipient of DRDO National Technology Oration Award in the year 2014.

R. Arthi is a Research Scholar at VIT University. She completed her graduate BE (ECE) and postgraduate (MBA) at colleges affiliated to Anna University. She is pursuing her doctoral work at VIT University.

She has around 2 years of experience in academics as Assistant Professor and 4 years of Research experience. Her research interest lies in the areas of organizational behavior, especially on leadership and emotional intelligence. She is also keen on exploring other areas of research on analytics, marketing, and behavioral finance. She has published around eight papers in national and international journals, attended around 15 workshops, and 10 conferences. She served as a member of various events like workshops, international conferences, industry-academia conclave, and seminars. She is very much interested in playing leadership roles and get in connect with people.

Yogesh Kumar Arya, PhD, is currently working as an Associate Professor at the Department of Psychology, Banaras Hindu University (BHU), Varanasi, India. He has more than 18 years of teaching experience in central universities such as the University of Allahabad and Banaras Hindu University. He has completed his PhD in Psychology from Banaras Hindu University and is actively engaged in teaching and researches in the areas of health and well-being, cognition, neuropsychology, and forensic psychology. Dr Arya is currently associated with three research projects awarded by the Education Research and Innovation Council (ERIC), NCERT, New Delhi, UGC and DBT, India. Dr Arya has published more than a dozen of research papers as well as book chapters. He has conducted several workshops on research methodologies and has organized several national and international conferences.

Rachel Bacigalupi is a second-year PhD student studying Clinical Psychology at Palo Alto University (PAU). Prior to attending PAU, she earned a Bachelor of Sciences in Psychology at the University of New Orleans. She is currently a member of Dr. Bongar's Clinical Crises and Emergencies (CCER) research group. While working with members of the CCER, she has contributed to projects studying culturally informed practice, suicide, and global mental health care. Her research and clinical interests include suicide, serious mental illness, cultural sensitivity, and treatment utilization.

Pratishtha Bhattacharyya is a Doctoral Research Scholar and pursuing her PhD at the Indian Institute of Technology Kharagpur in the Department of Humanities and Social Sciences. Her specializations include positive organizational behavior, positive psychology, and health psychology. She is working on the concept of cherishing as her PhD topic. She has published papers in national and international journals. She has also presented papers in many national and international seminars and conferences. Recently, she has joined as Assistant Professor, Xavier University, and Bhubaneswar, India.

Nidhi Bansal is a PhD scholar in Organization Behavior and Human Resource Management at NITIE, Mumbai, India. A graduate from Kota University, she holds her graduation degree in BSc (Biology) and done MBA (HR and Marketing) from Rajasthan technical University. She is also NET and JRF qualified. Her research interests include: work-life interface, gender role ideology, prosocial workplace behaviors like work engagement, innovative workplace behaviors, etc.

Bruce Bongar, PhD, ABPP, FAPM, CPsychol, CSci is the Calvin Distinguished Professor of Psychology at Palo Alto University. Early in his career, Dr Bongar developed strong interests in identifying the risk factors associated with suicidal behavior and other clinical emergencies, and he has remained a very prolific contributor to the scholarly literature and research in this area. He is a past president of the Section on Clinical Emergencies and Crises of Division 12 (Clinical Psychology) of the American Psychological Association (APA). The American Association of Suicidology recognized Dr. Bongar for his work by the early career achievement Edwin S. Shneidman Award (1993) for outstanding contributions to research in suicidology and the Louis I. Dublin Award (2004) for significant lifetime career contributions to research in suicidology. In 2008, Professor Bongar was presented with the

Florence Halpern award from the Division of Clinical Psychology of the APA for his distinguished contributions to the profession of clinical psychology. In 2010 he also was awarded the Career Achievement Award from the Section on Clinical Emergencies and Crises of the Division of Clinical Psychology of APA for his work on suicide and life-threatening behaviors. In 2016, Professor Bongar's research and scholarly work on suicide was recognized by a lifetime achievement award in professional psychology by the California Psychological Association.

Abbie J. Brady is a first-year student in the trauma track of the PhD program after earning her BA in Psychology at Ashland University in Ohio. During her undergraduate career, she served at a transitional living center and as a research assistant at Ashland University. She is interested in working with veterans and active duty military to treat combat-related PTSD and treatment effects. Research interests include studying military populations and the development and treatment of phantom limb syndrome.

Jacie T. Brown is a second-year student in the Clinical Psychology PhD program at PAU. She is currently interested in the Forensic and Trauma Areas of Emphasis. She received her Bachelor of Science in Psychology and Criminal Justice from Ferris State University. She has forensic experience working for a rehabilitation and reentry program specializing in sex offenders. Jacie is currently a student therapist at the Gronowski Clinic. Her primary research interests include the psychology of terrorism, firearm access and mental health, clinical emergencies, physical bravery, as well as risk assessment and management.

Lisa M. Brown, PhD, ABPP, is a Professor, Director of the Trauma Program and Director of the Risk and Resilience Research Lab at Palo Alto University. Her clinical and research focus is on trauma and resilience, global mental health, aging, and vulnerable populations. As a researcher, she is actively involved in developing and evaluating mental health programs used nationally and internationally, crafting recommendations aimed at protecting individuals and communities during catastrophic events, facilitating participation of key stakeholders, and improving access to resources and services. From 2007 to 2014, Dr Brown served as the Assistant Clinical Director of Disaster Behavioral Health Services, Florida Department of Health where she helped write the state disaster behavioral health response plan, develop regional disaster behavioral health teams, and conduct program evaluations of SAMHSA and FEMA crisis counseling programs. From 2008 to 2011, Dr Brown served on the Disaster Mental Health Subcommittee of the National Bio-defense Science Board Federal Advisory Committee, US Department of Health and Human Services where she contributed to the development of a national behavioral health response to disasters, terrorism, and pandemics. Her research experience and collaborative relationships with first responder groups and long-term care organizations led to the development of the second edition of the Psychological First Aid Field Guide for Nursing Home Residents. Dr. Brown is a Fellow of the American Psychological Association Division 20 and the Gerontological Society of America. She is the recipient of Fulbright Specialist awards with the University of the West Indies, Mona, Jamaica (2014) and with Massey University, Palmerston North, New Zealand (2015).

Julia V. Bykhovets, PhD, is a senior researcher in the Laboratory of Psychology of Human Development in Normal and Post-traumatic Conditions, Institute of Psychology at the Russian Academy of Sciences (Moscow, Russia) and also a member of the International Society for Traumatic Stress. Her areas of expertise are post-traumatic stress disorder, sensitivity to a terroristic threat, personality psychology, and family psychology. In her doctoral dissertation, she has investigated the impact of a terrorist threat on indirect victims.

Shweta Chawak, PhD, is a Research Scholar at the Department of Liberal Arts, Indian Institute of Technology Hyderabad, India. Her areas of interest include communication with cancer patients in Indian healthcare settings and illness perceptions specific to chronic diseases. As part of her PhD research, she aims to develop an intervention to improve cancer communication during consultation in the Indian context. Ms. Chawak was awarded with the Certificate of Excellence for the outstanding doctoral research in liberal arts for the year 2018 by the Indian Institute of Technology Hyderabad, India. In 2015, she was awarded a silver medal for her MPhil degree. Her research was on psycho-oncology and focused on understanding patient's perceptions on cancer consultation in the Indian setting. Her papers have been published in peer-reviewed journals such as the Journal of Cancer Education and Diabetes & Metabolic Syndrome: Clinical Research & Reviews.

Mahati Chittem, PhD, is an Associate Professor of Health & Medical Psychology in the Department of Liberal Arts at the Indian Institute of Technology Hyderabad, India. Her area of expertise includes *Psycho-oncology* (with a focus on pathways to care, communication, and end-of-life issues), *Patient adherence* among chronically ill, and *Health Behaviors* (diet, exercise, and safe sex). Dr. Chittem has been an active contributor to the field of psycho-oncology in India. She has published several research articles and book chapters on these topics. Her research has brought to light various psycho-social issues faced by chronically ill patients within the Indian context. In 2014, she received Young Investigator Award at the Asia-Pacific Psycho-oncology Network's (APPON) for her contributions to psycho-oncology in India. Dr. Chittem is a founding member of Hyderabad Psycho-oncology Working Group (HPWG) India and Association of Health Psychologists India. She is also an active member of the International Psycho-oncology Society (IPOS) and Asia-Pacific Psycho-oncology Network (APPON). Besides supervising MPhil and Doctoral scholars, she is also a consultant for various clinics and health care start-ups within India.

Marissa N. Eusebio is in her second-year of the clinical psychology PhD program at Palo Alto University in the duel adult and child areas of emphases. She will begin an external practicum at the Northern California VA in the Cognitive Rehabilitation Clinic. Marissa is predominantly interested in working with military and veteran populations in various capacities. She studies risk factors for veteran suicide, physical bravery in veterans, and currently works in a research lab at the San Francisco VA studying PTSD. She is also working on a project studying suicide in gender and sexual minorities. Marissa received a BS in Psychology with an emphasis in neuroscience and behavior from the University of California, San Diego.

Anna Feinman is a first-year student in the Clinical Psychology PhD program at Palo Alto University (PAU) where she is looking forward to completing the Forensic Area of Emphasis. She received her Bachelor of Arts in Psychology and Political Science from the University of California, Santa Barbara and her Masters in Marriage and Family Therapy from the University of Southern California. Anna completed her MFT Internship at Ventura County Behavioral Health working with SMI adolescents and their families. Her primary research interests include the psychology of terrorism and murder-suicide, firearm access and mental health, and police training.

Pooja Garg, PhD, is an Associate Professor in the Department of Humanities and Social Sciences, at the Indian Institute of Technology Roorkee, India. She received her PhD in Psychology from the Indian Institute of Technology Roorkee, India. Her area of specialization is organizational behavior, human resource management, and positive psychology. Dr. Pooja Garg has published more than 25 research articles in the journals of national and international repute, such as: *The International Journal of Human Resource Management, Asia-Pacific Journal of Business Administration, Journal of Organizational Change Management, International Journal of Society Systems Science, Journal of Management Development, and International Journal of Educational Management*. She is the winner of Outstanding Paper Award at the Emerald Literati Network Awards for Excellence 2007.

Shivjot Gill is a Doctoral Research Fellow at the Jindal Institute of Behavioral Sciences, O.P. Jindal Global University. She is a former international and national level athlete in shooting sport and is currently writing her dissertation on "Mental Training and its effect on Psychological well-being of Target Sports Athletes." Her research interests incorporate exploring the effect of performance stressors on psycho-social well-being, resilience, and motivation on athlete from varied age groups and backgrounds.

Shubhangi Goswami is currently a Research Scholar in the Department of Humanities and Social Sciences, Indian Institute of Technology Kharagpur. Her broad area of research is positive psychology. She is a Rehabilitation Council of India certified Clinical Psychologist. After completing MPhil in 2014, she worked for a research project, worked as a guest lecturer in Jogamaya Devi College (Kolkata), and also practiced in a private clinic in Kolkata for a year before joining the PhD program in IIT Kharagpur. She was awarded the UGC-NET Lectureship in the year 2012. She has presented scientific papers in several national and international conferences and has two research publications peer-reviewed journals.

Tharina Guse, PhD, is a Professor and Head of the Department of Psychology at the University of Pretoria, South Africa. She is currently serving on the Professional Board for Psychology of the Health Professions Council of South Africa. Professor Guse has presented several scientific papers at conferences and has also presented workshops on hypnosis and psychotherapy to psychologists locally and internationally. Her research interests include psychological well-being, psychological strengths (particularly hope and gratitude), positive psychology interventions, and the application of hypnosis for mental health promotion. She was acknowledged as an established researcher by the National Research Foundation of South Africa in 2017.

Han Ki-Soon, PhD, is a Professor in the Department of Creative HRD, School of Social Sciences, Incheon National University, Incheon, South Korea. She received her PhD from the University of Nebraska, USA. Her major area of research is gifted education. She is acting as Vice Dean of the Graduate School at the Department of Creative HRD and has published research articles in the journal of international repute.

Lopamudra Hati is currently working as a PhD Research Scholar under the supervision of Dr. Rabindra Kumar Pradhan in the Department of Humanities and Social Sciences, Indian Institute of Technology Kharagpur, India. Her research interests focus on antecedents and consequences of Employee well-being and tools and techniques to measure employee well-being. She has published papers in national and international journals. Many times, she has presented papers in reputed international conferences.

Aishwarya Jaiswal is a UGC-JRF and is currently pursuing research under the supervision of Dr. Yogesh Kumar Arya, at the Department of Psychology, Banaras Hindu University (BHU), Varanasi. She has completed her Bachelors and Masters in Psychology from BHU and has received Prof. M.M. Sinha Endowment Scholarship during her Masters. She has presented papers in national and international conferences; and has attended training workshops on psychological testing and assessment, research methodology, statistical computing packages (Excel, SPSS and NVivo), etc. She is an associate member of the Society of Industrial and Organizational Psychology, India.

Rebekah Jazdzewski, BA, is currently working on her doctorate in clinical psychology at Palo Alto University. Rebekah earned her bachelor's degree from Point Loma Nazarene University where she studied coping mechanisms in under-served populations. Currently, Rebekah is working with Native American tribes to study suicide prevention and resilience. Rebekah also serves as a student therapist at the Gronowski Center, working as a therapist in community mental health.

K. Alphonsa Jose, PhD, is an Assistant Professor in the School of Business Studies and Social Sciences,

CHRIST (Deemed to be University) Bangalore. She was a Maulana Azad National Senior Research Fellow in the Department of Applied Psychology, Pondicherry University. She has cleared UGC-NET and holds a PG diploma in Industrial Psychology. Her research in the area of environmental psychology attempts to identify the antecedents of building eco-friendly houses. She has authored more than seven research articles in various peer-reviewed national and international journals, as well as in edited books. She presented research papers in various national and international forums and received the best paper award.

Arturo Juarez-Garcia, PhD, is Professor at the School of Psychology, Psychological and Research Services Unit, National Autonomous University of Mexico (UNAM), Mexico. He has received several academic awards as an outstanding student. He received a doctorate degree in psychology in 2005 from UNAM. He works at the National Institute for Workers' Health (INSAT, Havana, Cuba), and the Center for Occupational and Environmental Health, School of Medicine, UCLA, in the United States of America. He has published many research papers in refereed journals, conference proceedings, and books.

Mohita Junnarkar, PhD, is currently working as an Assistant Professor at the Jindal Institute of Behavioral Sciences, O.P. Jindal Global University. She has completed her PhD in Psychology from IIT Bombay in 2013. She has about 6 years of research experience and 3 years of teaching experience. She has completed postdoctoral fellowship in the area of positive psychology focusing on assessment and validation of psychological scales from IIT Delhi in 2014. Her key teaching areas are applied positive psychology, psychometrics, adolescent development, and cognitive psychology and her research areas are eyewitness testimony, psychometrics, and adolescent development. Currently, she is working in the area of eyewitness testimony, she is investigating "how can eyewitness memory accuracy be enhanced and what is difference in neuronal pathways for correct identification and misidentification by eyewitness?" In the field of adolescent development, she is seeking answers to "how to identify the early signs of languishing mental health of adolescent at home and in school and how to communicate effectively with adolescents on day-to-day basis so that they flourish?" Apart from seeking answers to these questions, she is also interested to explore the role of spirituality and wisdom in adolescent's mental health and how smaller nuances of spirituality can be reached out to adolescents. She has published about 14 research papers in national and international peer-reviewed indexed journals in the area of psychometric validation of positive psychology constructs and has a book chapter and two books to her credit. Currently, she is also serving as a reviewer for the Journal of the Indian Academy of Applied Psychology.

Nadezhda N. Kazymova is a junior researcher in the Laboratory of Psychology of Human Development in Normal and Post-traumatic Conditions, Institute of psychology at Russian Academy of Science (Moscow, Russia). Her research interests include different types of traumatic situations, traumatic experience, personal resources and psychological resilience and well-being in extreme conditions of life.

Updesh Kumar, PhD, is Scientist "G" (R) and Former Head of the Mental Health Division at the Defence Institute of Psychological Research (DIPR) for the Indian Government. His research and publications mainly focuses on personnel selection, military mental health, suicidal behavior, personality assessment, dynamics of psychological resilience, and counter-terrorism. His publications include *Suicidal Behavior: Assessment of People-at-Risk (2010)*; *Countering Terrorism: Psychosocial Strategies (2012)*; *Understanding Suicide Terrorism: Psychosocial Dynamics (2014)*; *Suicidal Behavior: Underlying Dynamics (2015)*; *Positive Psychology: Applications in Work, Health, and Well-being (2015)*, *The Wiley Handbook of Personality Assessment (2016)*;*The Routledge International Handbook* *of Psychosocial Resilience (2016)*; *Handbook of Suicidal Behavior (2017)*, and most recently *The Routledge International Handbook of Military Psychology and Mental Health* (2019). Among many achievements, he was conferred the DRDO's Best Popular Science Communication Award-2009 by the Honorable Defense Minister of India and the prestigious DRDO's Scientist of the Year Award-2013 by the Indian government.

Anna I. Laktionova, PhD, is a senior researcher at the Institute of Psychology, Russian Academy of Sciences, Moscow. In 2003–2005, she worked in Moscow State Psychology and Pedagogy University in the Research Laboratory of Psychological and Social Problems of Child Abandonment and Neglect Prevention. 2002–till present time, she is a Director of Counseling Center in the NGO "Institute of Psychology and Psychotherapy," Moscow. He is an author of two books and more than 40 chapters and papers in books and professional journals.

Alexander V. Makhnach, PhD, is a senior researcher at the Institute of Psychology, Russian Academy of Sciences, Moscow. In 1994–1995, he worked on the development of the International program: The Challenge of Foster Care under the auspice of the Russian Ministry of Education that designed a system of foster care in Russia. From 1993 until the present he is working as a rector at the NGO Institute of Psychology and Psychotherapy, Moscow. In 2000–2005, he was on the board of experts and then as deputy director for methodology at the "Assistance to Russian Orphans' Program." In 2008–2012, he was a principal researcher of the program of the psychological evaluation of foster parents' applicants under the patronage of the Russian Ministry of Education and Science. His areas of expertise are orphan-hood, social adaptation, adoptive family, family stress factors and resources, and family resilience. He is an author of five books and more than 100 chapters and papers in books and professional journals.

Monalisa Mahapatra is Former Fellow Program in Management Student (Information Systems), Indian Institute of Management Kozhikode, Kerala, India. She has presented papers in reputed international conference and has also published in international journals.

Jonathan Matusitz (PhD, University of Oklahoma, 2006) is currently a Tenured Associate Professor in the Nicholson School of Communication at the University of Central Florida (UCF). His research focuses on the role of communication in terrorism, symbolism in terrorism, the globalization of culture, and health communication. His research methodologies include qualitative interviewing, content analysis, semiotics, and theoretical analysis. On top of having more than 140 academic publications and over 150 conference presentations, he taught at a NATO-affiliated military base in Belgium in 2010. Originally from Belgium himself, he moved to the United States in 2000. His first book on terrorism, titled *Terrorism & Communication: A Critical Introduction*, was published by SAGE in 2013. His second book, *Symbolism in Terrorism: Motivation, Communication, and Behavior*, was published by Rowman & Littlefield in 2014. In 2011, Dr. Matusitz's research was cited by the US Supreme Court. Since 2014, he has collaborated with a representative of the Defence Research & Development Organisation (DRDO), Ministry of Defence, in India to examine suicide terrorism.

Paola Mendoza-Rivera, BA, is a undergraduate student at Paul Alto University, USA. Paola has a BA in General Psychology and a BS in Criminal Justice from the University of New Haven in West Haven, CT and is now working toward a PhD in Clinical Psychology at Palo Alto University. Her main research interests involve topics that examine how clinical and forensic psychology intertwines. She is also piloting the Spanish translation of the Cultural Assessment of Risk for Suicide (CARS) measure and was involved in several projects focused on the increase in mental health issues (e.g., suicide risk) among Puerto Ricans after Hurricane María.

Cesar Merino-Soto is a psychologist (Lima, Peru); graduated from the Inca Garcilaso de la Vega University (Lima, Peru); Master degree with mention of educational psychology (University of San Martín de Porres, Lima, Peru). Currently, a full-time fellow from the National Council of Science and Technology (CONACYT, Mexico) for doctoral studies at the Center for Transdisciplinary Research in Psychology (CITPSI, Autonomous University of the State of Morelos, Cuernavaca, Mexico). He works as a university professor in several Peruvian universities, mainly at the San Ignacio de Loyola University, and full-time researcher at the University of San Martín dePorres. As a Researcher, he is qualified by the National Council of Science and Technology (CONCYTEC, Peru), received the National Prize of Psychology (Peruvian Psychologist Association, 2017), second prize of research in the Society of Psychology of Uruguay, and second place in the Research contest of the II NANDA International Congress (Peru). He has received recognition in several Peruvian universities for teaching work, outstanding professional work, and academic excellence. He is a member of the scientific board and/or editorial board of national and international journals (between Q4 and Q1). He is a guest reviewer invited in national and international journals (between Q4 and Q1). He is the former President of the National Psychological Test Commission (Peruvian Psychologist Association, 2017). He has been a guest speaker at national and international conferences. He has published more than 180 publications in the form of scientific articles, editorial letters, and chapters in books. He is thematically oriented to research in quantitative and psychometric methodology.

Jillian M. Millar, PhD, has been a practicing psychologist since 2008 and has experience of working in a variety of settings including private practice, public hospitals, and community mental health units in both Queensland and Western Australia. Jillian has worked with clients with a range of presenting problems such as anxiety, depression, adjustment difficulties, family and relationship issues, self-esteem difficulties, anger management and emotion regulation problems, adult traumas, etc. Jillian uses a range of therapeutic interventions to suit the needs of her clients including cognitive-behavior therapy, narrative therapy, transactional analysis, and interpersonal therapies. She received her PhD degree in the year 2012 from Edith Cowan University for her research work on posttraumatic growth carried out under the supervision of Professor Julie Ann Pooley.

Swati Mukherjee, PhD, is a Scientist "E" at the Defence Institute of Psychological Research (DIPR), Delhi. She is involved in many major research projects of the institute, including research on suicide in the Armed Forces. She has to her credit a number of national and international publications in the form of journal articles and book chapters published by reputed publishers. She has coedited a volume on "Recent Developments in Psychology," has coauthored a manual on "Suicide and Fratricide: Dynamics and Management" for the Armed Forces and on "Overcoming Obsolescence and Becoming Creative in R&D Environments." Her areas of interest are social psychology and political behavior. She has a keen research interest in studying social conflicts and identity dynamics. The interface between military psychology and peace psychology also interests her. Her current official assignments are focused on positive mental health practices and developing training modules for inculcating hardiness and flexible thinking among the Armed Forces personnel.

Jhuma Mukhopadhyay, PhD, is an Assistant Professor of Psychology at Department of Psychology, Asutosh College of the University of Calcutta, India. She secured fifth position in her BSc and first position in MSc examination conducted by the University of Calcutta in Psychology. She received her PhD degree from the University of Calcutta in 2006. She has published several articles and research papers in different national and international journals and contributed in book chapters of edited books. She has also presented several research papers in national and international conferences and seminars. She won the second best paper award in International Conference of Neuroscience and Mental Health, 2018, Cambodia.

Anthony Vernon Naidoo, PhD, is a Professor and past Head of the Department of Psychology at Stellenbosch University, Stellenbosch, South Africa. He is an expert in the areas of applied community psychology, career psychology, work-related behavior, and wilderness therapy.

Julijana Nevland, PhD, is a North Dakota Licensed Psychologist, Assistant Professor, and Chair of the Graduate Counseling Program at the University of Mary in Bismarck, North Dakota. She received her doctorate in Counseling Psychology from the University of North Dakota in 2013. Dr. Nevland completed her predoctoral internship with the Federal Bureau of Prisons, where she received specialized training in forensic assessment, suicide prevention, chemical dependence, dual diagnosis, and behavior management. She has worked in various clinical settings, to include state and federal prisons, veteran administration (VA) hospital, community mental health agency, and is currently in private practice with Chambers and Blohm Psychological Services (Bismarck, ND). Dr. Nevland specializes in the treatment of mood disorders (anxiety, depression, PTSD) and suicide prevention. She has presented original research and symposia at regional and national conferences, including the American Psychological Association conference, Society for the Teaching of Psychology conference, National Multicultural Conference and Summit, and the ND Counseling Association conference, among others. Dr. Nevland serves as Chair of the ND Board of Addiction Counseling Examiner, president of the ND Association of Counselor Educators and Supervisors, and member of the University of Mary's Institution Review Board (IRB). Her research interests are in the area of positive psychology, especially related to mindfulness and effective coping strategies.

Meera Padhy received her PhD from Utkal University in 1996. She was a recipient of UGC-JRF fellowship during her PhD She has been working as an Assistant Professor at the Centre for Health Psychology, University of Hyderabad, India since 2007. She has published extensively on Health Psychology, Developmental Psychology, and Occupational Psychology in national and international journals of repute as well as presented in many conferences. Her work has been recognized with awards like the Best Health Psychologist, IPERA in 2013 and Best Paper Presenter Award in 52nd and 21st International Conference of the Indian Academy of Applied Psychology in 2017. Her research interests include behavioral diabetology, occupational health, social support, leisure, and well-being.

Surya Prakash Pati, PhD, is an Assistant Professor in Organisational Behavior and Human Resources at the Indian Institute of Management, Kozhikode, Kerala, India. He has a strong research interest in identifying the strategic role of HR practices in influencing organizational processes and performance. Specifically, he researches on employee engagement, spirit at work, and indigenous management practices. He has presented his research in prestigious platforms such as the Annual Meetings of the Academy of Management. His works have been published in reputed journals such as *Indian Journal of Industrial Relations*, *Journal of Human Values*, *Human Resources Development Review*, and *International Journal of Work, Organization, and Emotion.*

Swati Patra, PhD**,** is currently working as Professor in the discipline of psychology, School of Social Sciences, Indira Gandhi National Open University (IGNOU), New Delhi. She has earlier worked in the National Council for Educational Research and Training (NCERT). She has 16 years of teaching, research, and training experience. Her specializations are counseling psychology, educational psychology, family therapy, distance education, and human development and disadvantage. Her areas of research interest are counseling, mental health, emotional intelligence, positive psychology, parenting, and adolescent development. Her professional activities include among other things course design and development, course writing, course editing, and research guidance for PhD. She has various research publications and has edited a book "Perspectives on Mental Health and Well-being in Indian Context." She has been involved in interactive radio counseling (Gyanvani) and academic counseling through television (Gyandarshan) of IGNOU. She also provides guest lectures and conducts training workshops on behavioral skills for different organizations and institutions.

Julie Ann Pooley, PhD, is currently Professor and the Associate Dean of Teaching and Learning for the Faculty of Engineering, Health, and Science and School of Arts and Humanities at Edith Cowan University. Julie Ann has been involved in teaching in both the undergraduate and postgraduate psychology programs and has been a recipient of a National Teaching Award and Citation by the Australian University Teaching Committee (2003, 2011). Her research focuses on resilience at the individual and community levels. Julie Ann has been involved in and directed many community-based research consultancies, projects, and workshops and has been involved in the generation of many different community-oriented reports for various cities and districts. She holds several awards both at university and national level for her achievements in teaching, research, and community services. She has many outstanding publications in reputed international journals, books, and conference proceedings.

Rabindra Kumar Pradhan, PhD, is currently working as an Associate Professor in the Department of Humanities and Social Sciences, Indian Institute of Technology Kharagpur, India. His area of specialization includes positive psychology, I/O psychology, organizational behavior, and human resource management. His current researches include employee health and well-being, emotional intelligence, emotional labor, employee performance, workplace spirituality, HR flexibility, positive leadership, and HR sustainability. He has more than 20 years of research and teaching experience and has published numerous papers in reputed peer-reviewed journals.

Tomas Prosek is a researcher at the Department of Psychology, Faculty of Arts, Masaryk University, Brno, Czech Republic

Kevin Rodriguez is a second-year PhD student in Clinical Psychology at Palo Alto University. He received his BA in Precounseling and Research from Our Lady of the Lake University in San Antonio, Texas. He has forensic experience working with medical investigators determining the cause and manner of unusual or violent deaths and has conducted research in measuring capital punishment endorsement. Currently, he is working in a project validating the psychometric properties of a suicide scale of the Spanish translation of the Cultural Assessment of Risk for Suicide (CARS), as well as analyzing the efficiency of psychological assessments for immigrants.

Ayesha Saha is a graduate in Psychology from Lady Shri Ram College for Women (LSR), University of Delhi and has pursued her postgraduation in MSc Psychology (Clinical) from Christ University, Bangalore. Passionate in the field of research, Ayesha has been a part of the research project for preparing a teacher's handbook for children with autism at NCERT, New Delhi. With her dynamic interest in the field of research, she has been able to extend her horizons in research related to positive psychology, military psychology, and suicidology. She is currently interning at PsychPress in Melbourne, and is actively involved in various stages of development of new assessments and standardization, along with generating selection and development reports. Ayesha has presented papers on varied topics in the national and international conferences and has original publications in various peer-reviewed journals.

Sanjeev P. Sahni, PhD, is a Professor and Principal Director of the Jindal Institute of Behavioral Sciences at the O.P. Jindal Global University, Sonipat. He is also the Director of "Center for Victimology and Psychological studies" and "Center for Innovative Leadership and Change", Member of the Governing Body and Advisor to the Vice-Chancellor of O.P. Jindal Global University. He has dedicated 14 years as a scientist in Ministry of Human Resource Development (HRD) and Sports Authority of India (SAI).

He served as Head Education of Jindal Steel and Power Limited where he managed the educational institutes of JSPL, which includes O.P. Jindal Global University (OPJGU), Jindal Institute of Technology (OPJIT), Jindal Institute of Power Technology (JIPT), four Community Colleges of Technology and Skills (OPJCCTS), three Industrial Training Institutes (ITIs), Medical College, Physiotherapy courses and ten schools. He is a recipient of many fellowships and has worked in many important projects. Dr. Sahni has 15 years of teaching experience in different subjects of psychology such as criminal psychology, competency mapping, organizational development, and leadership.

Renata Sargon is a third-year student in the forensic track and has been involved in developing a Crisis Intervention Team training program for police officers with Dr. Leonard Beckum. She is also interested in studying help-seeking behaviors in police officers, as well as lone wolf terrorism. Renata is originally from Chicago and is of Assyrian American decent.

Manisha Sawhney, PhD, is an Assistant Professor of Psychology at the University of Mary, Bismarck. She earned her BA in Psychology and Economics from Himachal Pradesh University, India (1999), MA in Clinical Psychology from Himachal Pradesh University, India (2001), and a PhD in Psychology from the University of Delhi, India (2006). She completed her dissertation on the topic of *Depression in University Students: A Neuro-Psychobiological Analysis*. Her primary research interest lies in the understanding the relationship between negative emotions and cardiovascular diseases. She is especially interested in investigating cardio-toxic emotions like depression, heart rate variability, and coping styles. In India, Dr. Sawhney was a research investigator with Defence Institute of Psychological Research (DIPR), India, and her research focused on military psychology, especially designing psychological methods and trainings to combat PTSD and increase the morale of soldiers operating in low intensity conflicts. She has served as a co-investigator on several DRDO projects and has contributed to the development of stress management training modules, which have been extensively used by the DRDO, India.

N.D.S. Naga Seema, PhD, is currently an Assistant Professor at the Centre for Health Psychology, School of Medical Sciences, University of Hyderabad, India. She is involved in teaching students under undergraduate and postgraduate psychology programs for a substantial period of time. She is a recipient of Young Scientist Award 2010 instituted by the Indian Academy of Applied Psychology, one of the oldest professional bodies in India that is dedicated to the field of Psychology. Her research activities are focused primarily on Health Psychology and specifically on Community Health Psychology, Health Communication, and Biopsychosocial health of women, and Yoga. Dr. Seema has been involved in content development and visual aids in e-learning for University Grants Commission-Consortium for Education Communication (UGC-CEC). She has to her credit several national and international publications including edited books.

Surendra Kumar Sia, PhD, is presently working as a Professor and heading the Department of Applied Psychology, Pondicherry University, India. He has done his MA, MPhil and PhD in psychology from the University of Delhi. Prof. Sia has more than 14 years of teaching and research experience and successfully supervised eight PhD students and two MPhil students. To his credit, he has 46 research publications in journals of national and international repute like *South Asian Journal of Human Resource Management*, *Asia-Pacific Journal of Management Research and Innovation*, *Global Business Review*, *Management and Labour Studies*, *Vision: The Journal of Business Perspectives*, *Journal of Indian Academy of Applied Psychology*, *Indian Journal of Industrial Relations*, *Management of Environmental Quality*, *Benchmarking: An International Journal*, etc. His areas of interest include organizational behavior, environmental psychology, and positive psychology.

Tushar Singh, PhD, is currently working as Assistant Professor at the Department of Psychology, Banaras Hindu University. He completed his DPhil in Psychology from the University of Allahabad. He has more than 10 years of teaching experiences in premier institutes like Amity University, Lucknow, Centre for Behavioral and Cognitive Sciences, University of Allahabad, and now in BHU. Besides teaching Dr. Singh is actively involved in quality research and publications. He has worked as a research director in an international research project funded by the Economic and Social Research Council of UK. He has been granted research projects from UGC, NCERT, and UK–India Collaborative Research Project from DBT, India and MRC, UK. He has been guiding four doctoral researches and many masters' dissertations. He has published more than a dozen of research papers and book chapters in national and international journals and has presented papers in national and international conferences. He has also been elected as the treasurer of National Academy of Psychology for the duration 2018–2021 and has served as the member of the Executive Council of National Academy of Psychology for the duration 2014–2017. He is serving in editorial committees of various journals including Indian Journal of Social Science Researches, Psybernews, etc.

Parwinder Singh, PhD, is presently serving as an Assistant Professor in the Department of HSS at IIT Ropar, Punjab, India. With a specialization in counseling psychology, he mainly focuses on devising, executing, and evaluating various psychological interventions targeting different facets of maladaptive behavior. He is working on aggression, depression, and emotion regulation and aspires his work to be focused upon characterizing the cognitive architecture of emotion regulation and on understanding how dysfunctional cognitions lead to an inability to regulate ones' emotions effectively. He has presented more than 40 papers in various seminars/conferences and has 14 publications in research journals to his credit.

Alena Slezackova, PhD, is an Associate Professor of Psychology at Masaryk University, Faculty of Arts, Brno, Czech Republic. She is a founder and director of the Czech Positive Psychology Centre (CPPC) and she conducts her research within the Academic Center for Positive Psychology (ACPP) affiliated to Faculty of Arts, Masaryk University in Brno. Her scientific interests include hope, mental health, and well-being. She is a member of the Advisory Council of the International Positive Psychology Association (IPPA) and the Country Representative for the Czech Republic in the European Network for Positive Psychology (ENPP).

Pieter Snyman, PhD, is a lecturer and program coordinator of the wellness recreation management program at the Department of Marketing Logistics, Supply Chain and Sport, Tshwane University of Technology, Pretoria, South Africa. Pieter is also head of the UCSA/VCSV outdoor education and development center that runs outdoor experiential camps for the community.

G. N. Sumathi is currently working as an Associate Professor at VIT Business School (Human Resource and Organizational Behavior division) in VIT with a total teaching experience of about eight years. She is an Assistant Director at Ranking and Accreditation at VIT, Vellore for the past one year. She also worked at Anna University, Chennai. She earned the alumna status from IIM Ahmedabad through FDP, earned PhD from IIT Madras, MBA from Pondicherry University and BTech from Anna University. She teaches courses on human resource management, organizational behavior, knowledge management, business analytics, business research methods, training and development, organizational change and development, and stress management. Her research interests include human resource practices, emotional intelligence, acculturative stress, knowledge management in service sector, employee behavior in public health system, and mental health among international students. She is a certified HR analytics and metrics professional by Middle Earth CAMI.

Nadezhda V. Tarabrina, PhD, is Professor at the Laboratory of Psychology of Human Development in Normal and Posttraumatic Conditions, Institute of Psychology of the Russian Academy of Sciences (Moscow, Russia). Professor Tarabrina was a Member of Advisory Committee "RUSSIA-NATO on Social and Psychological Consequences of Terrorism" (2002–2009). Her areas of expertise are the psychology of post-traumatic stress, sensitivity to terroristic threat, negative affectivity, psychological well-being, the impact of the media, traumatization, and personality.

Feba Thomas is a Research Scholar, VIT Business School, VIT, Vellore. She specializes in Human Resource Management. She has published many papers in reputed journals and has presented papers in international seminars and conferences.

Catalina Udani is currently an Honors student at the University of Central Florida (UCF). In fall 2017, she served as a Lawrence J. Chastang Global Fellow for the Global Perspectives Office at UCF. She is pursuing a bachelor's degree in Human Communication, with minors in Intelligence and National Security and Terrorism Studies. Her research interests include economic growth in developing countries and the black-market contribution to international trade. Upon graduation, she plans on pursuing a graduate degree in Peace and Conflict Studies with a focus on Ethics. A student of German and Filipino, Udani has traveled throughout Europe, East, and Southeast Asia. Lastly, Catalina completed her Honors thesis titled *A Content Analysis of Jihadist Magazines: Theoretical Perspectives*. Her thesis sampled the textual content and imagery of issues of *Dabiq* and *Rumiyah*, two ISIS publications, using thematic analysis to procedurally code the data by recognizing shared characteristics and concepts. The fundamental goal of Catalina's thesis was to gain a greater understanding of the way ISIS, its members, and the jihadist movement communicates their intentions, with the hope of preventing further recruitment and radicalization.

Kristen J. Vescera, MA, is a third-year graduate student at Palo Alto University. She received a Bachelor of Arts degree at Syracuse University in English & Textual Studies and Geography. After traveling to Latin American countries for a year assisting high school and college students in volunteer projects across Costa Rica, Honduras, and Nicaragua, she developed an interest in working with adolescents. She moved to Denver, Colorado and received a Master of Arts degree from University of Colorado Denver in Counseling Psychology and Counselor Education with an emphasis in School Counseling. Following graduation, she enlisted in the United States Army Reserves as a Psychological Operations Specialist and immediately deployed to Afghanistan in support of a nine-month combat operation. Kristen is currently a reserve Staff Sergeant that teaches psychological operations doctrine to junior enlisted service members. She is interested in working with active duty and veteran populations within the larger context of policy and program development.

Amanda Wallick graduated summa cum laude from the University of Nevada, Las Vegas in 2016, earning her BA in Psychology and a minor in Marriage and Family Therapy. As an undergraduate student, Amanda worked for Volunteers in Medicine of Southern Nevada where she created and administered an assessment measure for patients' primary care physicians to assess for depression, anxiety, suicidal ideation, PTSD, and alcohol dependency. In 2017, she began the Clinical Psychology PhD program at Palo Alto University (PAU), where she is completing the Trauma Area of Emphasis. She is currently serving as secretary for the Association for Traumatic Stress Studies at PAU for the 2018–2019 academic calendar. Her research interests include trauma, PTSD, severe mental illness, and substance abuse. She is a student affiliate of APA Division 56 (Trauma Psychology), the Association for Psychological Science (APS), as well as the International Society for Traumatic Stress Studies (ISTSS).

Rachel N. Ward received her Bachelor of Science in Psychology from Royal Holloway, University of London in 2017. As an undergraduate student, Rachel worked as a lead research assistant at LifeMoves Homeless Shelter Network in Northern California. In 2017, she started the Clinical Psychology PhD program at Palo Alto University, where she will be completing the Trauma Area of Emphasis. Her primary research interests include how various factors moderate risk and resiliency in PTSD, with a particular focus on the effects of social support, social identities, and the nature of the index trauma. Rachel is currently a graduate student affiliate of the APA divisions 56 (Trauma Psychology) and 35 (Society for the Psychology of Women), and the International Society for Traumatic Stress Studies.

Matthew M. Yalch, PhD, HSPP is an Assistant Professor of Psychology and Director of the Trauma and Personality Studies (TAPS) lab at Palo Alto University. Here and elsewhere he has taught courses in statistics/research methods and clinical assessment/diagnosis. Dr. Yalch's research interests include the intersection of trauma and personality, interpersonal theory, and the application of Bayesian statistics. He is also a licensed clinical psychologist specializing in the treatment of comorbid PTSD/SUD, multimethod personality assessment, and motivational interviewing, and has provided training on these topics to health care providers across the country. Prior to his work in clinical psychology, Dr. Yalch served in the US Army as a Military Intelligence Analyst and Officer, completing tours in Korea, Iraq, and Afghanistan.

Conrad S. Zygmont, PhD, is Lecturer and Program Coordinator of the Psychology Department at Helderberg College of Higher Education. He has completed his PhD in psychology focusing on wilderness therapy/adventure programming.

Abbreviations

AARs	after action reviews
ACORN	Association of Community Organizations for Reform Now
ACT	acceptance and commitment therapy
AD	additional decrement
ALTR	altruistic cherishing
APA	American Psychiatric Association
ARS-30	Academic Resilience Scale
AVE	average variance explained
AVO	avoidance
BFI	Big Five Inventory
BIPQ	Brief Illness Perception Questionnaire
BIT	Brief Intervention of Thriving
BPD	borderline personality disorder
CAD	cardiovascular disease
CBSM	cognitive-behavioral stress management
CBT	cognitive behavior therapy
CBT-SP	cognitive-behavioral therapy for suicide prevention
CD-RISC	Connor–Davidson Resilience Scale
CE	constant effect
CFI	Comparative Fit Index
CR	composite reliabilities
CFA	confirmatory factor analysis CFA
COR	conservation of resources COR
COPD	chronic obstructive pulmonary disease
CSI	coping strategy indicator
CVD	cardiovascular disease
DBT	dialectical behavioral therapy
DBT-ST	Dialectical Behavior Therapy group skills training
DERS	Difficulties in Emotion Regulation Scale
DRN	dorsal raphe nuclei
DSM	Diagnostic and Statistical Manual of Mental Disorders
ECP	expected change in parameter
EI	emotional intelligence
ER	emotion regulation
ERQ	Emotion Regulation Questionnaire
EWB	employee well-being
FACT-G	functional assessment of cancer therapy
FAM	cherishing family
FDA	Food and Drug Administration
FRI	cherishing friends
GLHC	God locus of health control
GFI	Goodness of Fit Index
GSE	general self-efficacy
GSR	gross stress reaction
HADS	Hospital Anxiety Depression Scale
HILS	Harmony in Life Scale
HR	human resource
HRM	human resource management
HRQoL	health-related quality of life
ICTs	information and communication technologies
IDV	individualism
IEC	Institute Ethics committee
IES-R	Impact of Events Scale-Revised
IMT	inner mental training
IND	indulgence
IC	interacting cooperatively
IPA	interpretive phenomenological analysis
IPV	intimate partner violence
IRT	item response theory
ISEL	Interpersonal Support Evaluation List

ISIS	Islamic State of Iraq and Sham
LA	living authentically
LOC	locus of control
LTO	long-term orientation
MAS	masculinity
MSV	maximum shared variance MSV
MBSR	mindfulness-based stress reduction
MDD	major depressive disorder
MetS	metabolic syndrome
MHLC-C	multidimensional health locus of control cancer
MI	modification indexes
MISS	Medical Interview Satisfaction Scale
MMI	method of motivational induction
MS	Mississippi scale
MSD	musculoskeletal disorders
MSES	Mastery Self-efficacy Scale
MSPE	Mindful Sport Performance Enhancement
MSPSS	Multidimensional Scale of Perceived Social Support
NCDs	noncommunicable diseases
NFI	Normed Fit Index
NSSI	nonsuicidal self-injury
OHP	Occupational Health Psychology
PCM	partial credit model
PD	power distance
PERS	personal cherishing
PHS	Perceived Hope Scale
PL	prelimbic
PMR	progressive muscular relaxation
POB	positive organizational behavior
PROSO	problem solving
PST	psychological skill training
PsyCap	psychological capital
PTG	post-traumatic growth
PTGI	post-traumatic growth inventory
PTSD	post-traumatic stress disorder
PVN	paraventricular nucleus
PWB	psychological well-being
QoL	quality of life
QPL	question prompt list
QTT	questionnaire of the terrorist threat
QWL	quality of work life
RAW	Resilience at Work
RSA	Resilience Scale for Adults
RIA	reflective inquiry and action plan
RMSEA	Root Mean Squared Error of Approximation
RS	resilience scale
RSDB	reckless and self-destructive behavior
SAMS	Schematic Appraisals Model of Suicide
SCL-90-R	symptom check list-90-revised
SCT	symbolic convergence theory
SD	standard deviation
SES	socioeconomic status
SIT	social identity theory
SMART	Stress Management and Resiliency Training
SOC model	Selective Optimization with Compensation model
SPANE	Scale of Positive and Negative Experience
SPW	scales of psychological well-being
SQ-COREL	squared correlation
SSS	search for social support
SUD	substance use disorder
SWB	subjective well-being
BRS	The Brief Resilience Scale
TS	technostress
UA	uncertainty avoidance
vmPFC	ventral medial prefrontal cortex
WEP	wilderness experiential program
WLSMV	mean-and-variance-adjusted unweighted least squares
WPV	workplace violence
YOD	young onset dementia

About This Book

The sole purpose of a civilization is to ensure human security, well-being, and happiness. The modern civilization, though high on security, has oftentimes tended to jeopardize and undermine human well-being and happiness. The ever-increasing complexities of modern life and the resultant predicaments of modern man have necessitated the discipline of psychology to evolve a focused perspective on human well-being and resilience. The present volume, *Emotion, Well-being, and Resilience: Theoretical Approaches and Practical Applications*, brings together scholars from across six continents, engaged in research and practice of ensuring human well-being and resilience in varied domains of life and society.

The volume consists of 31 chapters organized under four subthemes reaching in impact across domains and various cross-cultural and multicultural environments. From an applications perspective, the first section focuses on the clinical and counseling domain; however, it also highlights theoretical concepts underpinning the constructs of resilience and well-being. Moreover, in doing so it succinctly brings out varied theoretical perspectives existing in the field. Ward, in the first chapter, for example, discusses resilience in the context of stress and coping, while Bongar, in the third chapter, takes a wider view and describes resilience as perseverance through adversity focusing on its role in controlling self-destructive behavior.

The second section adds on the discussion with its focus on the home, school, community, and sports domains. It begins with a systematic analysis of family resilience, resilience among elderly populations in specific cultural contexts, and resilience in school. The chapters in this section emphasize upon viewing resilience not merely as an individual level construct but also as essentially a community-level construct, with the need for creating training and interventions that go beyond the individual and transform the community.

The quintessential theme of resilience in the workplace and organizational contexts is dealt with in the third section. Discussing various theoretical perspectives and empirical observations related to positive emotion, employee well-being, resilient leadership, technostress, meaningful work, and measurement of resilience at work, the section also highlights the importance of fostering resilience at workplace and outlines an overview of the strategies and interventions to be adopted in the for developing personal and team resilience among workers.

The fourth section focuses on emerging concepts and themes in the field of resilience research and places these in the context of resilient response to extreme environments. Incorporating varied contexts such as military resilience, resilience against terrorist threats, or even the resilience of terrorist groups like the Islamic State in Iraq and Syria to continue despite setbacks, this section also includes deliberations about areas beyond resilience, for example, post-traumatic growth.

On the whole, the volume provides an exhaustive and comprehensive collection of readings in research and practice of emotion, resilience, and well-being. It is a valuable resource for researchers and practitioners of resilience in varied domains across the globe.

Foreword

This volume broadly focuses on a very important topic that is concisely described in its title. Its editors, Dr. Rabindra Kumar Pradhan and Dr. Updesh Kumar, have done a stellar job in inviting a distinguished group of international scholars to contribute 31 chapters that comprehensively cover the basic aspects of this topic, making it a "must-read" in the area resilience for theorists, researchers, and practitioners in the field of positive psychology. This is not a new area in psychology but has evolved from the work of Abraham Maslow, Marie Jahoda, and others in the "humanistic movement" of the 1950s, which continued with the work of Martin Seligman and morphed into what is currently referred to as "positive psychology" from the early 1990s.

Many of the contributors to this volume consider resilience to be a complex and multidimensional construct, which is why reliable multifactor measures are needed to study, assess, and guide the way in which it should be strengthened. A consensus of descriptions among the authors suggests that resilience is the ability to cope with, withstand, and recover from challenges and setbacks in life. Others have stressed that this construct also includes the ability to grow and become stronger after one has effectively dealt with these challenging situations, suggesting that resilience is not only the ability to survive but to thrive in the face of adversity. The chapters progress from describing a number of different theories of resilience to the need to empirically study them and then apply the key findings. This desired progression can be described schematically as follows: (1) theory → (2) research → (3) application.

The connection between positive emotions, effective resilience, and overall well-being is found in many of the chapters. Finding meaning in what one does, and in life in general, is thought to generate a range of positive emotions that drive emotional engagement, commitment, and self-motivation, which are significantly important for coping with challenges, resilience, and personal growth as well as a sense of overall well-being. Examples of "negative emotions" observed in attempting to address setbacks are negativism, pessimism, helplessness, and hopelessness, while examples of "positive emotions" are optimism, enthusiasm, passion, engagement, and self-motivation that fuel resilience and well-being. It is also emphasized in many of the chapters that resilience can be described as interpersonalin the family, among friends and colleagues at work as well as in organizations, in addition to being described on the individual level per se. In organizations, for example, resilience is a critically important factor in leadership, which sets the tone for the entire organization. This is observed in "inspirational leaders" who breathe life-giving air into their organizations which helps it flourish, versus "expirational leaders" who are negativistic, pessimistic, and capable of taking the air out of an organization which can lead to organizational death. The former type of leadership facilitates individual and organizational resilience as well as growth, performance, and productivity.

To better understand and eventually enhance resilience, a number of authors emphasize the need to move beyond theory to empirically identifying the key predictors involved. This requires identifying those factors that predict negative reactions to risk and the best predictors of resilience and growth among the following factors and combinations thereof: (a) physical, (b) cognitive, (c) intrapersonal, (d) interpersonal, and/or (e) motivational factors. As such, predictive models will need to be developed by applying sophisticated multivariate statistics such as Multiple Regression Analysis; and separate models will need to be created for various occupational groups, in order to identify the specific resources, attributes, and behaviors that need to be strengthened in the various groups that are examined.

In addition to more accurately understand the nature of resilience; contributors to this volume also stress the need for better assessment, training, and development tools. This, in turn, necessitates creating better multifactor conceptual and psychometric models designed to facilitate the assessment and development of resilience with the potential to enhance growth and well-being in addition to resilience, based on the above-mentioned research procedures. Such tools are thought to more accurately identify those individual, interpersonal, organizational, and occupational resources that help reduce the negative effects of risk factors and increase the positive effects of protective factors. Moreover, a variety of different training programs will need to be created in order to achieve this goal. The effectiveness of these programs will then need to be examined in order to enhance them further, that is, what works best with whom. In addition to assess what needs to be enhanced using these models, they are expected to provide valuable information that can be used in parenting and education to help strengthen these factors in children in order to develop the next generation of more resilient individuals who will be more adept in dealing with the inevitable challenges and setbacks in life. It has also been suggested by a few of the authors that this method should be applied to strengthen individuals working in high-risk occupations, such as law enforcement officers, firefighters and other emergency responders as well as combat soldiers. A few of the contributors have emphasized the importance of creating "resilience-building programs" to be applied at home by parents, at school as part of the required curriculum, and at work in the form of workshops for employees and in leadership development. To reiterate, this would be especially important for individuals working in high-risk occupations. This type of training would also increase their occupational performance as well as productivity. The instructors involved in resilience-building programs would need to be selected among those with the highest levels of resilience themselves.

The above-mentioned training, development, and resilience-building programs will need to begin by factoring in self-awareness, a critically important factor in self-development and self-improvement. As such, it is hoped that the next edition of this volume will also include how best to evaluate and enhance self-awareness.

—Reuven Bar-On, PhD
Author of the Emotional Quotient Inventory™
and the Multifactor Measure of Performance™

Acknowledgments

We wish to acknowledge our gratitude to many people who have made valuable contributions toward the completion of this international handbook. First of all, we are thankful to all the contributors of this international handbook for their time, efforts, and cooperation throughout this academic endeavor. Without their active support and timely help, it would not have been possible.

We are thankful to Apple Academic Press for showing keen interest in publishing this proposed international handbook. We sincerely thank Sandra Jones Sickels, Vice President, Editorial and Marketing, Apple Academic Press Inc., USA for her resilient efforts, constant supervision, and monitoring the progress of the handbook from time to time.

We are grateful to all the reviewers of the book for their intellectual efforts, timely review, and meaningful suggestions for each and every chapter, which have greatly improved the quality of the chapters.

We express our deep sense of gratitude to Professor Reuven Bar-On, Professor Charles R. Figley, Professor Donald Meichenbaum, and Professor Bret A. Moore for their critical analyses and meaningful appreciation about the intellectual contributions of the book as a whole for our readers. We hope their critical evaluations and noteworthy recommendations will attract readers from multiple corners and contexts.

We sincerely thank our colleagues, students, staff, and other office-bearers who have supported and encouraged us to complete this international handbook.

We are grateful to our family members for their continuous moral and emotional support throughout this handbook project.

Last but not least, we invite and welcome suggestions and comments from readers, which will help us to improve in our future endeavors.

Rabindra Kumar Pradhan
Indian Institute of Technology Kharagpur, India

Updesh Kumar
DIPR, DRDO, Delhi, India

Preface

At this juncture of human and technological development, when artificial intelligence and automation are slowly taking over the world, holding on to the study of emotions, well-being, and resilience have become imperative. These qualities influence sustainable performances and growth by individuals as well as organizations. Resilience and well-being can be considered as two sides of the same coin in the process of human survival. These two concepts have been well researched globally in the context of home, school, work, community, and industrial organizations. Rapid increase in fierce competition seems to begin right from the early stages of life. Home, education, profession, interpersonal relationships, age—all have their unique flavors, unique contexts, and hence, unique problems as well. Globalization has posed many work-related challenges for employees at the workplace both for developed and emerging economies in the pursuit of performance excellence thereby causing problems and hampering employees' state of well-being. Particularly, people working in multinational companies are facing tremendous problems in expatriation, and repatriation in terms of socio-cultural adjustments to a new environment and also to the process of readjustment to their native culture. Along with the enhanced exchange of knowledge and business, diversity at international schools, colleges, universities, and business organizations have side by side created a sense of survival threat to the local population leading to a growing state of social apathy, aggression, and violence in many places across the globe. Similarly, frequent natural and man-made disasters have been continuously disturbing and destabilizing human life in society and community. Thriving in an environment where we need to continuously face either natural or man-made challenges requires building up of internal and external resources and also utilizes them efficiently. Keeping this in view, this international handbook aims to discuss the role of emotion, resilience, and well-being in the various context of human life such as home, school, and workplace. It also addresses the role of several other factors such as emotional intelligence, performance, productivity, and employee's health and happiness. The proposed volume endeavors to bring out various theoretical and practical implications of the constructs under various studies. Leading researchers and academicians from various domains such as health, education, information and technology, and manufacturing industries have orchestrated their ideas, views, and empirical findings to make this book more meaningful for its readers. This edition would be useful for research scholars and teachers in psychology, human resource managers, and mental health practitioners working in the field of emotion, resilience, and well-being.

The present handbook, *Emotion, Well-being, and Resilience: Theoretical Approaches and Practical Applications,* is an edited volume and consists of 31 chapters categorized under four sections. These 31 chapters are contributed by more than 70 authors from six continents–Asia, Africa, Australia, Europe, North America, and South America. This handbook provides a common platform for researchers, academicians, and practitioners to learn and practice new concepts, models, and knowledge in their concerned fields. Chapters are written by authors of international repute and are of global standards embodied with emerging concepts, approaches, and findings with practical applications expected to impact a large cross-cultural audience in its totality.

THE CONTENTS

Stress is such an everyday concept that the term is widely used colloquially. Every human being experiences stress—more or less, acute or chronic, internal or external. But it is mostly discussed after the consequences have started their detrimental effects. The first section of this handbook provides a comprehensive understanding of theoretical concepts and models related to emotion, well-being, and resilience in general and its emphasis and applications in clinical and counseling settings in particular. Some of the chapters present empirical findings with practical applications. Beginning with understanding the concept of stress, its relationship with health and mental illness and factors associated with being stress-resilient have all been discussed in the wide array of chapters in this section. Resilience has been discussed as a multidimensional concept. Therefore, understanding its implications in both, the preventive as well as curative settings, is important. In this respect, besides discussing the concept in the light of clinical settings, recent developments in the field of positive psychology have also been included. For example, cherishing as an empirically established construct might be a very interesting and useful subject matter (discussed in Chapter 4). The authors define cherishing as a secure attachment behavior manifesting a strong sense of affect, protectiveness, and care toward certain targets or objects. This chapter reports that cherishing significantly and positively correlates with psychological well-being on a sample of Indian university students. Moving on to the global scenario, this section throws light on resilience and well-being in a culturally and geographically diverse population like the Czech Republic, South Africa, Peru, United States of America, and South Korea, besides India.

Part II of the handbook presents the applications and implications of different studies carried out in the field of resilience and well-being in the contexts of home, school, community, and sports environments. Taking these concepts of psychology outside the clinical setting is also important in order to make their reach more permeating in society. Family being the first and most basic unit of society receives special attention in the form of a discussion of systematic analysis of family resilience which iterates the usefulness of family resources or characteristics in coping with stressful situations (Chapter 13). This section also covers other contexts where promoting resilience is important such as in the old age (Chapter 14), in education (Chapters 15 and 16), the community with special regard to place identity (Chapter 17), and finally in sports (Chapter 18). The varied array of contextual explanations of resilience in this section makes way for the development of skill sets and their implementation in order to reap the benefits of knowledge of the concepts. It also provides thought-provoking ideas for emerging research scholars to study these more extensively and intensively.

Moving on to Part III, it presents a comprehensive understanding of emotion, well-being, and resilience in different work and organizational contexts. As already discussed about the extremely demanding nature of organizational settings due to globalization, understanding how to maintain a productive frame of mind and promote well-being in the organizations is an extremely important requirement. Starting from theoretical underpinnings of positive emotions at work (Chapter 19) to discussing leadership resilience at work (Chapter 20) to organizational health and employee well-being (Chapters 21 and 25), almost all aspects of workplace hazards and well-being has been covered in this section. The readers working in any organization will surely benefit from this section as it covers empirically studied health-promoting aspects in the fields of organizational psychology, organizational behavior, and human resources management, ultimately contributing to work-life balance. The measurement of psychological constructs has always been an area of research and validation of one such measurement tool called the Resilience at Work (RAW) scale, in the Indian scenario (Chapter 23) would prove beneficial and usefully for researchers and organizational practitioners. Along with the daily hassles and other innumerable sources of stress, technostress has become a reality in the era of digital life. This area also receives special attention (Chapter 24) in the context of a rapidly technologically developing country like India.

Finally, coming to the last section of this handbook, Part IV presents very interesting findings and emerging concepts in the field of resilience and well-being studies in the context of extreme environments. A very well-known extreme environment is the military. The importance of building and constructive use of psychological flexibility in the military has been very vividly and distinctively discussed in Chapter 25. Psychological flexibility has been defined as a process of adaptation to fluctuating situational demands, reconfiguration of mental resources, shifting of perspectives, and balancing of competing desires, needs, and life domains (Kashdan & Rottenberg, 2010). Along with the promotion of such qualities in the military, the necessity of coping with the worldwide threat of terrorism has been clearly outlined (Chapters 27 and 28). The next couple of chapters expand our knowledge on post-traumatic growth, drawing attention on post-traumatic growth as an approach to trauma treatment and care. This handbook concludes in the most interesting manner by discussing about the utility of adventure programs for the development of resilience and related constructs in adolescent, corporate, and clinical contexts (Chapter 31). Adventure programming is the purposeful use of adventure activities and/or wilderness for the purpose of cognitive, physical, social, and/or affective change in groups or individuals (Glass & Myers, 2001). This chapter provides evidence of the effectiveness of well designed and implemented adventure programs for the development of resilience and well-being.

We sincerely hope that the efforts of all the authors and editors will meet the enthusiasm of the readers and create a beautiful confluence.

—Rabindra Kumar Pradhan
Updesh Kumar

REFERENCES

Glass, J. S., & Myers, J. E. (2001). Combining the old and the new to help adolescents: Individual psychology and adventure-based counseling. *Journal of Mental Health Counseling*, *23*(2), 104–114.

Kashdan, T. B., &Rottenberg, J. (2010). Psychological flexibility as a fundamental aspect of health. *Clinical Psychology Review*, *30*(7), 865–878.

Winwood, P. C., Colon, R., & McEwen, K. (2013). A Practical Measure of Workplace Resilience. *Journal of Occupational and Environmental Medicine*, *55*(10), 1205–1212. doi:10.1097/jom.0b013e3182a2a60a.

PART I

Emotion, Well-Being, and Resilience in Clinical and Counseling Context

CHAPTER 1

Stress, Resilience, and Coping

RACHEL N. WARD[*], ABBIE J. BRADY, REBEKAH JAZDZEWSKI, and MATTHEW M. YALCH

Department of Clinical Psychology, Palo Alto University, Palo Alto, CA, USA

[*]*Corresponding author. E-mail: rward@paloaltou.edu*

ABSTRACT

Stress can have adverse effects on psychological functioning. This is evident by increases in symptoms of post-traumatic stress disorder, depression, and other psychiatric disorders in those who have experienced high levels of stress. However, there are several factors that make people more at risk or more resilient to developing psychological problems in the aftermath of experiencing high levels of stress. Similarly, there are several methods of coping that are more or less effective in the aftermath of stressful experiences. In this chapter, we will review the nature and impact of stress, risk and resilience factors, and different methods of coping with stress.

1.1 INTRODUCTION

Psychological stress is the physical and/or psychological outcome of a perceived environmental demand on an individual (Cohen et al., 2007). The environmental demands that promote psychological stress (stressors) can come in a number of forms, ranging from chronic stressors such as chronic illness and community violence to time-limited high-severity stressors, such as natural disasters and terrorist attacks. An individual's response to stressors can be mental and/or physical, depending on the individual's psychological makeup, social and cultural resources, and the needs of the situation. In this chapter, we will discuss the nature of psychological stress, the types of factors that make an individual more or less resilient to psychological stress, and methods of coping that may be useful in dealing with psychological stress.

1.2 THE NATURE AND EFFECTS OF STRESSORS

1.2.1 THE NATURE OF STRESSORS

One factor that influences an individual's response to stress is the nature of the stressor(s). Greater perceived severity of the stressor and longer duration of the stressor are associated with more problematic stress responses (Dyb et al., 2014; Gil, 2015; Sayed et al., 2015; Xue et al., 2015). Perceptions of stressor severity are subjective and can be quantified in a number of different ways. For example, research suggests that greater perceived stressfulness and greater interpersonal impact of stressors are associated with a worse stress response (Martin et al., 2013;

Martinez-Torteya et al., 2009). Closer proximity to the stressor is also associated with worse responses (Gil, 2015; Silwal et al., 2018). This is especially evident in the case of vicarious traumas. For example, research suggests that witnessing injury or death first-hand is especially predictive of outcomes such as post-traumatic stress disorder (PTSD; Cheng et al., 2014; Xue et al., 2015). Proximity can also be relational in nature. This is evident in studies suggesting that events such as the death of a family member or colleague as well as exposure to human remains has been found to increase the risk of PTSD in 9/11 first responders (Pietrzak et al., 2014).

Along these lines, stressors can be categorized into three general groups: negative events, chronic strains, and traumatic stressors (Thoits, 2010). Negative events are single aversive demands on a person's life. Meanwhile, chronic strains are characterized by demands that are persistent or repeated. Traumatic stressors occur in situations in which the individual believes that there is a threat to their life, bodily integrity, or sanity (Thoits, 2010). All three types of stressors prompt physiological and psychological responses that impact physical and psychological health.

1.2.2 PHYSIOLOGICAL EFFECTS OF STRESS

Psychological stress cues activation of the fight-or-flight system of the body. This involves the body's production of adrenaline and subsequent release of glucocorticoid hormones such as cortisol, which in turn initiates more overt signals of physiological arousal such as the increased heart rate (McEwen, 1998). Not all stressors result in full activation of the fight-or-flight response (Taylor, 2010). However, an overload of numerous stressors of varied severity may lead to allostasis, the breakdown of physiological processes from the overproduction of cortisol, creating negative effects for brain, immune system, and cardiovascular functioning (McEwen, 1998).

There is a broad literature suggesting an association between psychological stress and physical health problems (for a theoretical review, see van der Kolk, 2014). The increased levels of stress are associated with higher blood pressure (Taylor, 2010), increased likelihood of chronic medical conditions such as cardiovascular, pulmonary, and metabolic diseases (Agorastos et al., 2014; Finch et al., 2001), decreased immune system (Segerstrom & Miller, 2004) and neurological functioning (Qin et al., 2009), and poorer perceived health more generally (Agorastos et al., 2014; Finch et al., 2001). Risks of these adverse health outcomes become more likely as the probability of allostasis increases (Ehlert, 2013). Research further suggests that the pernicious effects of stress may affect health more adversely than tobacco use and excessive drinking (Toussaint et al., 2016).

1.2.3 PSYCHOLOGICAL EFFECTS OF STRESS

Stress also has an impact on psychological health. Higher levels of stress are associated with poorer perceived mental health, as well as increased likelihood of a number of psychiatric disorders (Folkman et al., 1986; Langner & Michael, 1963). With respect to these disorders, the two syndromes most closely associated with stress are PTSD and depression (Campbell et al., 2008; Colodro-Conde et al., 2017).

The PTSD is by definition a disorder that develops following exposure to a severe (traumatic) stressor (American Psychiatric Association, 2013). Research suggests that greater duration and chronicity of exposure to the stressor(s), as well as whether the stressor is man-made (e.g., a trauma perpetrated by a person vs. a natural disaster), are associated with more severe and complex PTSD symptoms (e.g., Kira et al., 2008; for theoretical review, see Herman, 1992). Greater severity of PTSD symptoms is also associated with increases in somatic (e.g., chronic pain), externalizing (e.g., substance use), and internalizing (e.g., anxious, depressive) symptoms (Campbell et al., 2008; Miller et al., 2008).

For its part, depression is a syndrome with a high genetic component (Kendler et al., 2010). However, research suggests that stressful events interact with the genetic vulnerability for depression to produce an increased likelihood of developing depressive symptoms (Colodro-Conde et al., 2017; Kendler et al., 2010). As with PTSD, certain kinds of stressors are particularly associated with depression. Research suggests that single high-intensity stressors, interpersonal stressors, and stressors that have the potential to alter a person's identity, such as a loss of a job or a loved one, are especially likely to lead to depressive symptoms (Brown & Harris, 1986; Soria-Saucedo et al., 2018). Other research suggests that lifetime exposure to stress in general and exposure to childhood trauma in particular are associated with depression (Hammen, 2005).

1.3 RESILIENCE TO STRESS

Not everyone develops physical or psychological problems in the aftermath of stressful experiences, even when the stressors are severe. The degree to which a person can respond adaptively to stressful experiences is referred to as resilience (Southwick et al., 2014). Resilience is an umbrella term encompassing a number of overlapping facets, including social, psychological, and cultural factors (Hobfoll, 1989; Southwick et al., 2014).

1.3.1 SOCIAL ASPECTS OF RESILIENCE

The support a person receives from his or her social network has a strong influence on that person's ability to respond adaptively to stress, with higher levels of social support leading to reduced psychological problems in the aftermath of stressors. A number of studies suggest that social support is negatively associated with symptoms of PTSD in veterans, first responders to the 9/11 terrorist attacks, and survivors of natural and man-made disasters (e.g., Cheng et al., 2014; Dyb et al., 2014; Galea et al., 2008; Pietrzak et al., 2014; Sayed et al., 2015). This may be particularly the case for financial support. Research suggests that a lack of financial support is associated with greater severity of PTSD symptoms following natural disasters (Cheng et al., 2014; Galea et al., 2008).

Findings appear to be similar with respect to the association between social support and depression. Several studies of parenting-related stress reported by mothers indicate that social support is negatively associated with depressive symptoms (Aneshensel & Frerichs, 1982; Cairney et al., 2003). Research further suggests that emotional support is particularly linked to lower levels of depression in expecting mothers (Jones et al., 2005). Thus, the role of social support appears to be beneficial regardless of the stressor, although the specific type of support that is most effective may vary depending on the situation.

1.3.2 PSYCHOLOGICAL ASPECTS OF RESILIENCE

There are a number of psychological factors that also convey resilience to stress. Much of this research has focused on personality traits, defined as stable individual differences between people that influence their responses to the environment (McCrae & Costa, 2008; Widiger & Smith, 2008). One trait of particular relevance with resilience is neuroticism, the general predisposition to experience negative affect, especially in response to stress (Widiger, 2009). Research suggests that not only is neuroticism positively associated with a broad swath of mental and physical health problems (including but not limited to PTSD and depression; see Jakšić al., 2012; Kotov et al., 2010; Lahey, 2009), but at high levels also exacerbates (i.e., reduces resilience to) the effects of stress, resulting in higher levels of PTSD, depression, and related symptoms (Kendler et al., 2004; Lawrence & Fauerbach, 2003; Yalch et al., 2017; Yalch & Levendosky, 2017). Another important trait is dominance, which at high levels buffers the effect of stress on symptoms of PTSD, depression, anxiety, and other psychological problems (Bernard et al., 2019; Yalch et al., 2013; Yalch et al., 2015). High dominance as well as high warmth are also associated with less self-critical appraisals of stress (Yalch & Levendosky, 2016), which may indicate a potential pathway through which the effects of these traits might occur.

These and other personality traits do not act in isolation, but rather interact with each other to provide resilience to stressors (for theoretical review, see Miller, 2003). Specifically, research suggests that low levels of neuroticism combined with high levels of more adaptive traits, such as agreeableness, conscientiousness, and extraversion, is associated with moderate levels of PTSD and other symptoms in individuals exposed to traumatic stressors. In contrast, high neuroticism combined with low extraversion leads to comparatively higher PTSD and comorbid internalizing symptoms, and high neuroticism combined with low agreeableness and conscientiousness leads to higher PTSD and comorbid externalizing symptoms (e.g., Forbes et al., 2010; Miller et al., 2003, 2004; Miller & Resick, 2007). With respect to dominance, results of one study incorporating a clinical sample with a high rate of exposure to traumatic stressors indicated that individuals who were high in dominance had a shorter duration of PTSD symptoms compared to those who were low in both dominance and warmth (Thomas et al., 2014).

Another relevant psychological factor that influences resilience to stress is self-efficacy, which can be defined as an individual's capacity to believe in his or her ability to succeed in specific situations and achieve specific goals (Benight & Bandura, 2004). Social cognitive theorists hypothesize that high self-efficacy increases human agency, thereby improving individuals' ability to respond effectively to traumatic and otherwise stressful experiences (e.g., Benight & Bandura, 2004). Research suggests that high self-efficacy in the face of stress is associated with lower symptoms of PTSD (Ikizer et al., 2015) and depression (Sawatzky et al., 2012; Tsay & Chao, 2002). Similarly, research suggests that an individual's belief in their ability to manage their anxiety also buffers the effects of stress on symptoms of PTSD and depression as well as on the use of drugs (Schroder et al., 2017). There is also a reason to believe that unlike personality traits, which are viewed as largely stable, self-efficacy is more fluid and can be modified to increase resilience to stress. One study found that an intervention to increase self-efficacy in nurses not only increased levels of self-efficacy, but also decreased levels of secondary traumatization (Berger & Gelkopf, 2011).

1.3.3 CULTURAL ASPECTS OF RESILIENCE

In addition to social and psychological influences on resilience, there are also aspects of culture that affect resilience. Boardman (2004) found that the stability of the neighborhood in which an individual lives can mitigate the negative effects of stress on physical health. Living in a stable neighborhood increases social capital, the social resources that can be drawn upon when in need, thus increasing emotional and material resources and creating a sense of meaning through community (Marmot, 2005). Neighborhoods tend to differ in terms demographic factors that affect individuals' responses to stress, including race, immigrant status, and identification as a marginalized sexual minority.

There are also resources related to race and ethnicity associated with increased resilience. Higher levels of socialization with and higher perceived social support from one's racial/ethnic group are related to increased resilience in the face of adversity in African American and Mexican American samples (Berkel et al., 2010; Brown, 2008). Socialization with people who are similar racially/ethnically is postulated to be a protective factor against the negative psychological effects of racial discrimination (Brown & Tylka, 2011; Neblett et al., 2006), which is itself a unique stressor for minority subcultures (Finch et al., 2001; Flores et al., 2008; Williams et al., 1997). Similarly, the research suggests that among immigrant populations, giving and receiving social support, as well as having access to social resources (childcare, language tutors, case managers, etc.), practicing culture consistent with their country of origin, and having a high degree of religiosity also increase resilience (Sajquim de Torres & Lusk, 2018; Goodman et al., 2017). For individuals identifying as sexual minorities, connection to a community with a similar sexual identity also promotes resilience (McConnell et al., 2018; Monin et al., 2017). In sum, this research suggests that although membership of a marginalized identity brings with it unique stressors, affiliation with those who share that marginalized identity can be a form of resilience.

Other factors that affect individuals' response to stress are age and sex. Older age is related to fewer symptoms of anxiety, depression, and general distress (Jorm, 2000). In contrast, younger age has been associated with decreased resilience when undergoing a stressor, which may be related to research suggesting that younger individuals have more pronounced negative affect in response to a stressor than older individuals (Diehl et al., 2012). Other research has supported this finding, suggesting that older adults react in more adaptive ways to stress than younger adults (Schilling & Diehl, 2015). With respect to sex, the research suggests that women are more likely to report exposure to stressful experiences that are interpersonal in nature, including traumatic stressors perpetrated by people close to them, and exhibit more psychological symptoms (e.g., of PTSD and depression) in response to these stressors than men (Goldberg & Freyd, 2006; Martin et al., 2013; Nolen-Hoeksema, 2001; Tang & Freyd, 2012; Tolin & Breslau, 2007). Women are also less resilient to stress compared to men following a loss of social support (Olff, 2017). In contrast, men have been found to be more vulnerable to the cardiovascular effects of stress than women and children (Allen et al., 1997).

1.4 STRESS, RESILIENCE, AND COPING

Coping refers to behaviors to manage stress that crosses a threshold of tolerability (Lazarus & Folkman, 1984). Coping comes in a number of different forms and can be categorized as task-focused versus emotion-focused coping and engaged versus disengaged coping (Booth & Neill, 2017; Feinstein et al., 2017; Chen et al., 2018; Delahaij & van Dam, 2017).

Task- versus emotion-focused coping refers to whether a given coping behavior is targeted toward the stressor itself (e.g., an upcoming exam) or the emotions stemming from recognizing the stressor (e.g., anxiety). To use an upcoming exam as an example of a stressor, task-focused coping might involve studying for the exam, whereas emotion-focused coping might involve taking deep breaths to manage anxiety about how well one might do in the exam. With respect to the efficacy of task- versus emotion-focused coping, research suggests that task-focused coping is more effective for stressors over which an individual might have some element of control, whereas emotion-focused coping is more appropriate in the context of uncontrollable stressors (Delahaij & van Dam, 2017; Lee et al., 2016). However, emotion-focused coping is most beneficial when coupled with task-focused coping (Lee et al., 2016). In the absence of task-focused coping, focusing on negative emotions is associated with symptoms of depression and anxiety, especially among women (Hundt et al., 2013; Kelly et al., 2008). There appears to be an association between age and the use of task- versus emotion-focused coping. Research suggests that as people age they become more likely to use task-focused coping (Chen et al., 2018), which may be related to research suggesting that as people age, they become more effective at coping with stress (Schilling & Diehl, 2015). The preference for task-focused coping in older adults appears to be mediated by a positive affect such that as people age they become more positive, and therefore more likely to focus directly on a stressful task.

Engaged (vs. disengaged) coping refers to how active a coping behavior is. Engaged coping involves actively approaching either the stressful task (e.g., studying for an upcoming exam) or the aversive emotion (actively managing anxiety about an upcoming exam). Disengaged coping involves avoiding the stressor and its sequelae (e.g., by avoiding thinking about an exam and how anxious one is about one's lack of preparation for it by playing video games). Research suggests that engaged coping is generally more adaptive than disengaged coping (Feinstein et al., 2017).

The level of psychopathology may also influence how flexible an individual's use of coping is. The connection between coping styles and psychopathology can be dynamic and multidirectional such that not only can certain styles of coping lead to psychopathology, but psychopathology can predispose individuals to certain coping styles (Heffer & Willoughby, 2017). Research suggests people who are depressed are less likely to seek out and use a broader array of coping strategies (Heffer & Willoughby, 2017). This is problematic as people who are depressed and/or suicidal are more likely to rely on emotion-focused versus task-focused coping. In contrast, someone who is more flexible in their coping (perhaps because they are not overwhelmed by high levels of symptoms), upon realizing that emotion-focused coping (e.g., via using deep breathing) has not reduced their anxiety may change their coping strategy to something more task-focused (e.g., trying to solve the stress-inducing problem; Lee et al., 2016). This suggests that adaptive coping is more related to using a flexible range of coping styles than it is the use of any one specific style (Booth & Neill, 2017; Lee et al., 2016).

1.4.1 COPING AND RESILIENCE

There is a close association between coping and resilience such that a person's resilience affects what coping strategies are used, and the ability to cope effectively with stress affects a person's resilience (Booth & Neill, 2017; Lee et al., 2016). For example, individuals' coping responses are often affected by their appraisals of stress, which as research suggests are influenced by individual differences such as personality traits (Lazarus & Folkman, 1984). There are two general types of stress-related appraisals: primary and secondary appraisals. Primary appraisals concern how severe (e.g., dangerous) the stressor is and secondary appraisals concern whether and how one can manage the stressor (Lazarus & Folkman, 1984). Both kinds of appraisals are influenced by psychological resilience factors, such as personality traits in the case of primary appraisals (e.g., see Yalch & Levendsky, 2016) and self-efficacy in the case of secondary appraisals (e.g., see Benight & Bandura, 2004; Delahaij & van Dam, 2017). Personality traits can also influence the selection of coping strategies directly (Parker & Wood, 2008). For example, high levels of traits such as extraversion may make an individual more likely to select engaged coping strategies, such as utilizing social support, whereas high levels of neuroticism may make an individual more likely to utilize disengaged emotion-focused coping (Pereira-Morales et al., 2018). With respect to the dynamic interaction between resilience and coping, individuals who experience more positive emotions (due in part to personality) utilize more adaptive coping strategies and therefore develop increased resilience (Gloria & Steinhardt, 2014). These individuals with increased resilience in turn tend to utilize multiple coping strategies at once, likely due to an ability to identify the most adequate strategy for each stressor and then employ it, instead of restricting themselves to using a single strategy (Lee et al., 2016).

Resilience and coping may also be influenced by the age at which a person is first called on to be resilient to and cope with stress. There is some research suggesting that some degree of early life stressors exerts a positive influence on resilience (Booth & Neill, 2017). One reason for this is that early life stressors change the perception of stress in later life and influence coping patterns (Turner & Lloyd, 1995). However, research suggests that children who experience highly stressful environments early in their development may be more likely to use avoidant or aggressive coping styles (Repetti et al., 2002), suggesting that in order to be helpful, early life stress should not be too severe.

1.5 CONCLUSION

Stress can have deleterious effects on individuals. This is evident in increases in physical health problems such as chronic medical conditions and psychological problems such as depression and PTSD. The nature and severity of the stressor both influence the degree of stress-related physical and psychological problems. However, individual differences in resilience also influence physical and psychological problems in the aftermath of stressful experiences. Resilience includes social, psychological, and cultural aspects as well as the methods individuals use to cope with stressful events. Resilient individuals have robust social support and cultural networks, high levels of agreeableness, conscientiousness, dominance and extraversion, low levels of neuroticism, and a self-efficacious view of themselves. Resilient individuals use a range of coping methods with a preference for engaged, task-focused coping. Overall, individuals' resilience to stress is idiosyncratic, based upon the interaction of multiple factors and leading to a multidetermined physical and psychological response.

KEYWORDS

- **coping**
- **personality**
- **resilience**
- **self-efficacy**
- **stress**

REFERENCES

Agorastos, A., Pittman, J. O. E., Angkaw, A. C., Nievergelt, C. M., Hansen, C. J., Aversa, L. H., Parisi, S. A., Barkauskas, D. A., Marine Resiliency Study Team, & Baker, D. G. (2014). The cumulative effect of different childhood trauma types on self-reported symptoms of adult male depression and PTSD, substance abuse and health-related quality of life in a large active-duty military cohort. *Journal of Psychiatric Research, 58*, 46–54.

Allen, M. T., Matthews, K. A., & Sherman, F. S. (1997). Cardiovascular reactivity to stress and left ventricular mass in youth. *Hypertension, 30*, 782–787.

American Psychiatric Association. (2013). *Diagnostic and Statistical Manual of Mental Disorders* (5th ed.). Arlington, VA: American Psychiatric Publishing.

Aneshensel, C. S., & Frerichs, R. R. (1982). Stress, support, and depression: A longitudinal causal model. *Journal of Community Psychology, 10*, 363–376.

Benight, C. C., & Bandura, A. (2004). Social cognitive theory of posttraumatic recovery: The role of perceived self-efficacy. *Behavior Research and Therapy, 42*, 1129–1148.

Berkel, C., Knight, G. P., Zeiders, K. H., Tein, J.-Y., Roosa, M. W., Gonzales, N. A., & Saenz, D. (2010). Discrimination and adjustment for Mexican American adolescents: A prospective examination of the benefits of culturally related values. *Journal of Research on Adolescence, 20*, 893–915.

Berger, R., & Gelkopf, M. (2011). An intervention for reducing secondary traumatization and improving professional self-efficacy in well baby clinic nurses following war and terror: A random control group trial. *International Journal of Nursing Studies, 48*, 601–610.

Bernard, N. K., Yalch, M. M., Lannert, B. K., & Levendosky, A. A. (2019). Main and moderating effects of interpersonal style on the association between dating violence and PTSD symptoms. *Violence and Victims, 34*(3), 522–535.

Boardman, J. D. (2004). Stress and physical health: The role of neighborhoods as mediating and moderating mechanisms. *Social Science and Medicine, 58*, 2473–2483.

Booth, J., & Neill, J. (2017). Coping strategies and the development of psychological resilience. *Journal of Outdoor and Environmental Education, 20*, 47–54.

Brown, D. L. (2008). African American resiliency: Examining racial socialization and social support as protective factors. *Journal of Black Psychology, 34*, 32–48.

Brown, G. W., & Harris, T. O. (1986). Establishing causal links: The Bedford College studies of depression. In H. Katsching (Ed.), *Life events and psychiatric disorders: Controversial issues* (pp. 107—187). Cambridge, England: Cambridge University Press.

Brown, D. L., & Tylka, T. L. (2011). Racial discrimination and resilience in African American young adults: Examining racial socialization as a moderator. *Journal of Black Psychology, 37*, 259–285.

Cairney, J., Boyle, M., Offord, D. R., & Racine, Y. (2003). Stress, social support and depression in single and married mothers. *Social Psychiatry and Psychiatric Epidemiology, 38*, 442–449.

Campbell, R., Greeson, M. R., Bybee, D., & Raja, S. (2008). The co-occurrence of childhood sexual abuse, adult sexual assault, intimate partner violence, and sexual harassment: Amediational model of posttraumatic stress disorder and physical health outcomes. *Journal of Consulting and Clinical Psychology, 76*(2), 194–207.

Chen, Y. Y., Peng, Y., Xu, H., & O'Brien, W. H. (2018). Age differences in stress and coping: Problem-focused strategies mediate the relationship between age and positive affect. *International Journal of Aging and Human Development, 86*, 347–363.

Cheng, Y., Wang, F., Wen, J., & Shi, Y. (2014). Risk factors of post-traumatic stress disorder (PTSD) after Wenchuan earthquake: a case control study. *PLoS One, 9,* 1–7.

Cohen, S., Janicki-Deverts, D., & Miller, G. E. (2007). Psychological stress and disease. *JAMA, 298*(14), 1685–1687.

Colodro-Conde, L., Couvy-Duchesne, B., Zhu, G., Coventry, W. L., Byrne, E. M., Gordon, S., Wright, M. J., Montgomery, G. W., Madden, P., Major Depressive Disorder Working Group of the Psychiatric Genomics Consortium, Ripke, S., Eaves, L. J., Heath, A. C., Wray, N. R., Medland, S. E., & Martin, N. G. (2018). A direct test of the diathesis–stress model for depression. *Molecular Psychiatry, 23*(7), 1590–1596.

Delahaij, R., & Van Dam, K. (2017). Coping with acute stress in the military: The influence of coping style, coping self-efficacy and appraisal emotions. *Personality and Individual Differences, 119*, 13–18.

Diehl, M., Hay, E. L., & Chui, H. (2012). Personal risk and resilience factors in the context of daily stress. *Annual Review of Gerontology and Geriatrics, 32*, 251–274.

Dyb, G., Jensen, T. K., Nygaard, E., Ekeberg, O., Diseth, T. H., Wentzel-Larsen, T., & Thoresen, S. (2014). Post-traumatic stress reactions in survivors of the 2011 massacre on Utoya island, Norway. *British Journal of Psychiatry, 204*, 361–367.

Ehlert, U. (2013). Enduring psychobiological effects of childhood adversity. *Psychoneuroendocrinology, 38*, 1850–1857.

Feinstein, B., Davila, J., & Dyar, C. (2017). A weekly diary study of minority stress, coping, and internalizing symptoms among gay men. *Journal of Consulting and Clinical Psychology, 85*, 1144–1157.

Finch, B. K., Hummer, R. A., Kol, B., & Vega, W. A. (2001). The role of discrimination and acculturative stress in the physical health of Mexican-origin adults. *Hispanic Journal of Behavioral Sciences, 23*, 399–429.

Flores, E., Tschann, J. M., Dimas, J. M., Bachen, E. A., Pasch, L. A., & de Groat, C. L. (2008). Perceived discrimination, perceived stress, and mental and physical health among Mexican-origin adults. *Hispanic Journal of Behavioral Sciences, 30*, 401–424.

Folkman, S., Lazarus, R., Gruen, R., & DeLongis, A. (1986). Appraisal, coping, health status, and psychological symptoms. *Journal of Personality and Social Psychology, 50*, 571–579.

Forbes, D., Elhai, J. D., Miller, M. W., & Creamer, M. (2010). Internalizing and externalizing classes in post-traumatic stress disorder: A latent class analysis. *Journal of Traumatic Stress, 23*, 340–349.

Galea, S., Tracy, M., Norris, F., & Coffey, S. F. (2008). Financial and social circumstances and the incidence and course of PTSD in Mississippi during the first two years after Hurricane Katrina. *Journal of Traumatic Stress, 21*, 357–368.

Gil, S. (2015). Personality and trauma-related risk factors for traumatic exposure and for posttraumatic stress symptoms (PTSS): A three-year prospective study. *Personality and Individual Differences, 83*, 1–5.

Gloria, C. T., & Steinhardt, M. A. (2014). Relationships among positive emotions, coping, resilience and mental health. *Stress & Health: Journal of the International Society for the Investigation of Stress, 32*, 145–156.

Goldberg, L. R., & Freyd, J. J. (2006). Self-reports of potentially traumatic experiences in an adult community sample: Gender differences and test-retest stabilities of the items in a Brief Betrayal-Trauma Survey. *Journal of Trauma and Dissociation, 7*, 39–63.

Goodman, R., Vesely, C., Letiecq, B., & Cleaveland, C. (2017). Trauma and resilience among refugee and undocumented immigrant women. *Journal of Counseling and Development 95*, 309–322.

Hammen, C. (2005). Stress and depression. *Annual Review of Clinical Psychology, 1*, 293–319.

Heffer, T., & Willoughby, T. (2017). A count of coping strategies: A longitudinal study investigating an alternative method to understanding coping and adjustment. *PLoS One, 12*, 1–16.

Herman, J. (1992). *Trauma and recovery: The aftermath of violence—from domestic abuse to political terror*. New York: Basic Books.

Hobfoll, S. E. (1989). Conservation of resources. A new attempt at conceptualizing stress. *American Psychologist, 44*, 513–524.

Hundt, N. E., Williams, A. M., Mendelson, J., & Nelson-Gray, R. O. (2013). Coping mediates relationships between reinforcement sensitivity and symptoms of psychopathology. *Personality and Individual Differences, 54*, 726–731.

Ikizer, G., Karanci, N. A., & Doğulu, C. (2015). Exploring factors associated with psychological resilience among earthquake survivors from Turkey. *Journal of Loss and Trauma, 21*, 384–398.

Jakšić, N., Brajković, L., Ivezić, E., Topić, R., & Jakovljević, M. (2012). The role of personality traits in posttraumatic stress disorder (PTSD). *Psychiatria Danubina, 24*(3), 256–266.

Jones, S. M., Bogat, G. A., Davidson, W. S., Eye, A. V., & Levendosky, A. (2005). Family support and mental health in pregnant women experiencing interpersonal partner violence: An analysis of ethnic differences. *American Journal of Community Psychology, 36*, 97–108.

Jorm, A. F. (2000). Does old age reduce the risk of anxiety and depression? A review of epidemiological studies across the adult life span. *Psychological Medicine, 30*, 11–22.

Kelly, M. M., Tyrka, A. R., Price, L. H., & Carpenter, L. L. (2008). Sex differences in the use of coping strategies: Predictors of anxiety and depressive symptoms. *Depression and Anxiety, 25*, 839–846.

Kendler, K. S., Kessler, R. C., Walters, E. E., MacLean, C., Neale, M. C., Heath, A. C., & Eaves, L. J. (2010). Stressful life events, genetic liability, and onset of an episode of major depression in women. *FOCUS, 8*, 459–470.

Kendler, K. S., Kuhn, J., & Prescott, C. A. (2004). The interrelationship of neuroticism, sex, and stressful life events in the prediction of episodes of major depression. *American Journal of Psychiatry, 161*, 631–636.

Kira, I. A., Lewandowski, L., Templin, T., Ramaswamy, V., Ozkan, B., & Mohanesh, J. (2008). Measuring cumulative trauma dose, types, and profiles using a development-based taxonomy of traumas. *Traumatology, 14*, 62–87.

Kotov, R., Gamez, W., Schmidt, F., & Watson, D. (2010). Linking "big" personality traits to anxiety, depressive, and substance use disorders: A meta-analysis. *Psychological Bulletin, 136*, 768–821.

Lahey, B. B. (2009). Public health significance of neuroticism. *American Psychologist, 64*, 241–256.

Langner, T. S., & Michael, S. T. (1963). *Life Stress and Mental Health: II. The Midtown Manhattan Study*. Oxford, England: Free Press Glencoe.

Lawrence, J. W., & Fauerbach, J. A. (2003). Personality, coping, chronic stress, social support and PTSD symptoms among adult burn survivors: A path analysis. *Journal of Burn Care and Rehabilitation, 24*, 63–72.

Lazarus, R. S., & Folkman, S. (1984). *Stress, Appraisal, and Coping*. New York: Springer Publishing Company.

Lee, J. H., Seo, M., Lee, M., Park, S. Y., Lee, J. H., & Lee, S. M. (2016). Profiles of coping strategies in resilient adolescents. *Psychological Reports, 120*, 49–69.

Marmot, M. (2005). Social determinants of health inequalities. *The Lancet, 365*, 1099–1104.

Martin, C. G., Cromer, L. D., DePrince, A. P., & Freyd, J. J. (2013). The role of cumulative trauma, betrayal, and appraisals in understanding trauma symptomatology. *Psychological Trauma: Theory, Research, Practice, and Policy, 5*, 110–118.

Martinez-Torteya, C., Bogat, G. A., von Eye, A., Levendosky, A. A., & Davidson, W. S. (2009). Women's appraisals of intimate partner violence stressfulness and their relationship to depressive and posttraumatic stress disorder symptoms. *Violence and Victims, 24*, 707–722.

McConnell, E. A., Janulis, P., Phillips, G., Truong, R., & Birkett, M. (2018). Multiple minority stress and LGBT community resilience among sexual minority men. *Psychology of Sexual Orientation and Gender Diversity, 5*, 1–12.

McCrae, R. R., & Costa, P. T. (2008). The Five-Factor Theory of personality. In O. P. John, R. W. Robins, & L. A. Pervin (Eds.), *Handbook of Personality: Theory and Research* (3rd ed.). (pp. 159–181). New York: Guilford Press.

McEwen, B. S. (1998). Stress, adaptation, and disease: Allostasis and allostatic load. *Annals of New York Academy of Sciences, 840*, 33–44.

Miller, M. W. (2003). Personality and the etiology and expression of PTSD: A three-factor model perspective. *Clinical Psychology: Science and Practice, 10*, 373–393.

Miller, M. W., Greif, J. L., and Smith, A. A. (2003). Multidimensional Personality Questionnaire profiles of veterans with traumatic combat exposure: Externalizing and internalizing subtypes. *Psychological Assessment, 15*, 205.

Miller, M. W., Kaloupek, D. G., Dillon, A. L., & Keane, T. M. (2004). Externalizing and internalizing subtypes of combat-related PTSD: A replication and extension using the PSY-5 scales. *Journal of Abnormal Psychology, 113*, 636–645.

Miller, M. W., & Resick, P. A. (2007). Internalizing and externalizing subtypes in female sexual assault survivors: Implications for the understanding of complex PTSD. *Behavior Therapy, 38*, 58–71.

Miller, M. W., Fogler, J. M., Wolf, E. J., Kaloupek, D. G., & Keane, T. M. (2008). The internalizing and externalizing structure of psychiatric comorbidity in combat veterans. *Journal of Traumatic Stress, 21*, 58–65.

Monin, J. K., Mota, N., Levy, B., Pachankis, J., & Pietrzak, R. H. (2017). Older age associated with mental health resiliency in sexual minority US veterans. *American Journal of Geriatric Psychiatry, 25*, 81–90.

Neblett, E. W., Philip, C. L., Cogburn, C. D., & Sellers, R. M. (2006). African American adolescents' discrimination experiences and academic achievement: Racial socialization as a cultural compensatory and protective factor. *Journal of Black Psychology, 32*, 199–218.

Nolen-Hoeksema, S. (2001). Gender differences in depression. *Current Directions in Psychological Science, 10*, 173–176.

Olff, M. (2017). Sex and gender differences in post-traumatic stress disorder: an update. *European Journal of Psychotraumatology, 8*(4), 1–2.

Parker, J. D. A., & Wood, L. M. (2008). Personality and the coping process. In G. J. Boyle, G. Matthews & D. H. Saklofske (Eds.), *The Sage Handbook of Personality Theory and Assessment: Volume 1—Personality Theories and Models* (pp. 506–520). London: Sage Publications.

Pereira-Morales, A. J., Forero, D. A., Adan, A., & Lopez-Leon, S. (2018). Personality traits and health-related quality of life: The mediator role of coping strategies and psychological distress. *Annals of General Psychiatry, 17 (1) 25.*

Pietrzak, R. H., Feder, A., Singh, R., Schechter, C. B., Bromet, E. J., Katz, C. L., … Southwick, S. M. (2014). Trajectories of PTSD risk and resilience in world trade center responders: An 8-year prospective cohort study. *Psychological Medicine, 44*, 205–219.

Qin, S., Hermans, E. J., van Marle, H. J., Luo, J., & Fernández, G. (2009). Acute psychological stress reduces working memory-related activity in the dorsolateral prefrontal cortex. *Biological Psychiatry, 66*, 25–32.

Repetti, R. L., Taylor, S. E., & Seeman, T. E. (2002). Risky families: Family social environments and the mental and physical health of offspring. *Psychological Bulletin, 128*, 330–366.

Sajquim de Torres, M., & Lusk, M. (2018). Factors promoting resilience among Mexican immigrant women in the United States: Applying a positive deviance approach. *Estudios Fronterizos, 19*, 1–20.

Sawatzky, R. G., Ratner, P. A., Richardson, C. G., Washburn, C., Sudmant, W., & Mirwaldt, P. (2012). Stress and depression in students: The mediating role of stress management self-efficacy. *Nursing Research, 61(1)*, 13–21.

Sayed, S., Iacoviello, B. M., & Charney, D. S. (2015). Risk factors for the development of psychopathology following trauma. *Current Psychiatry Reports, 17*, 1–7.

Segerstrom, S. C., & Miller, G. E. (2004). Psychological stress and the human immune system: A meta-analytic study of 30 years of inquire. *Psychological Bulletin, 130*(4), 601–630.

Silwal, S., Dybdahl, R., Chudal, R., Sourander, A., & Lien, L. (2018). Psychiatric symptoms experienced by adolescents in Nepal following the 2015 earthquakes. *Journal of Affective Disorders, 234*, 239–246.

Schilling, O. K., & Diehl, M. (2015). Psychological vulnerability to daily stressors in old age. *ZeitschriftFür Gerontologie Und Geriatrie, 48*, 517–523.

Schroder, H. S., Yalch, M. M., Dawood, S., Callahan, C. P., Donnellan, M. B., & Moser, J. S. (2017). Growth mindset of anxiety buffers the link between stressful life events and psychological distress and coping strategies. *Personality and Individual Differences, 110*, 23–26.

Soria-Saucedo, R., Lopez-Ridaura, R., Lajous, M., & Wirtz, V. J. (2018). The prevalence and correlates of severe depression in a cohort of Mexican teachers. *Journal of Affective Disorders, 234*, 109–116.

Southwick, S. M., Bonanno, G. A., Masten, A. S., Panter-Brick, C., & Yehuda, R. (2014). Resilience definitions, theory, and challenges: interdisciplinary perspectives. *European Journal of Psychotraumatology, 5*(1) 25338.

Tang, S. S. S., & Freyd, J. J. (2012). Betrayal trauma and gender differences in posttraumatic stress. *Psychological Trauma: Theory, Research, Practice, and Policy, 4*, 469–478.

Taylor, S. E. (2010). Mechanisms linking early life stress to adult health outcomes. *Proceedings of the National Academy of Sciences of the United States of America, 107*, 8507–8512.

Thoits, P. A. (2010). Stress and health: Major findings and policy implications. *Journal of Health and Social Behavior, 51*, 41–53.

Thomas, K. M., Hopwood, C. J., Donnellan, M. B., Wright, A. G. C., Sanislow, C. A., McDevitt-Murphy, M. E., Ansell, E. B., Grilo, C. M., McGlashan, T. H., Shea, M. T., Markowitz, J. C., Skodol, A. E., Zanarini, M. C., & Morey, L. C. (2014). (2014). Personality heterogeneity in PTSD: Distinct temperament and interpersonal typologies. *Psychological Assessment, 26*, 23–34.

Tolin, D. F., & Breslau, N. (2007). *Sex Differences in Risk of PTSD: (642812007-001)* [Data set]. American Psychological Association.

Toussaint, L., Shields, G. S., Dorn, G., & Slavich, G. M. (2016). Effects of lifetime stress exposure on mental and physical health in young adulthood: How stress degrades and forgiveness protects health. *Journal of Health Psychology, 21*(6), 1004–1014.

Tsay, S. L., & Chao, Y. F. (2002). Effects of perceived self-efficacy and functional status on depression in patients with chronic heart failure. *Journal of Nursing Research, 10*, 271–278.

Turner, R. J., & Lloyd, D. A. (1995). Lifetime traumas and mental health: The significance of cumulative adversity. *Journal of Health and Social Behavior, 36*, 360–376. DOI: 10.2307/2137325.

van der Kolk, B. (2014). *The Body Keeps the Score: Brain, Mind, and Body in the Healing of Trauma*. New York: Penguin Books.

Widiger, T. A. (2009). Neuroticism. In M. R. Leary and R. H. Hoyle (Eds.), *Handbook of Individual Differences in Social Behavior* (pp. 129–146). New York: The Guilford Press.

Widiger, T. A., & Smith, G. T. (2008). Personality and psychopathology. In O. P. John, R. W. Robins, & L. A. Pervin (Eds.), *Handbook of Personality: Theory and Research* (3rd ed.). (pp. 743–769). New York: The Guildford Press.

Williams, D. R., Yu, Y., Jackson, J. S., & Anderson, N. B. (1997). Racial differences in physical and mental health: Socio-economic status, stress and discrimination. *Journal of Health Psychology, 2*, 335–351.

Xue, C., Ge, Y., Tang, B., Liu, Y., Kang, P., Wang, M., & Zhang, L. (2015). A meta-analysis of risk factors for combat-related PTSD among military personnel and veterans. *PLoS One, 10*, 1–21.

Yalch, M. M., Bernard, N. K., & Levendosky, A. A. (2015). An interpersonal theory perspective on intimate partner violence. In E. Bellamy (Ed.), *Partner V : Risk Factors, Therapeutic Interventions and Psychological Impact* (pp. 113–126). Hauppauge, NY: Nova Science Publishers.

Yalch, M. M., Lannert, B. K., Hopwood, C. J., & Levendosky, A. A. (2013). Interpersonal style moderates the effect of dating violence on symptoms of anxiety and depression. *Journal of Interpersonal Violence, 28*, 3171–3185.

Yalch, M. M., & Levendosky, A. A. (2016). The influence of interpersonal style on the appraisal of intimate partner violence. *Journal of Interpersonal Violence, 31*, 2430–2444.

Yalch, M. M., & Levendosky, A. A. (2017). Main and moderating effects of temperament traits on the association between intimate partner violence and hazardous alcohol use in a sample of young adult women. *Psychological Trauma: Theory, Research, Practice, and Policy, 10*(6), 611–618.

Yalch, M. M., Levendosky, A. A., Bernard, N. K., & Bogat, G. A. (2017). Main and moderating influence of temperament traits on the association between intimate partner violence and trauma symptoms. *Journal of Interpersonal Violence, 32*(20), 3131–3148.

CHAPTER 2

Resilience and Reducing Self-Destructive Behaviors

BRUCE BONGAR,* KRISTEN J. VESCERA, PAOLA MENDOZA-RIVERA, and RACHEL BACIGALUPI

Department of Clinical Psychology, Palo Alto University, 1791 Arastradero Road, Palo Alto, CA 94304, USA

Corresponding author. E-mail: bongar@paloaltou.edu

ABSTRACT

Recent developments in positive psychology underscore the potential benefit of focusing on resilience factors rather than psychiatric illness or symptoms. Resilience can be defined as a set of qualities that can be multidimensional and enable an individual to persevere through adversity. Both internal and external stressors are constant in a person's life and one's ability to cope affects their resilience. Sometimes, one's ability to cope is unsuccessful and can result in self-destructive behaviors. Obtaining information on risk factors is important in identifying personal and environmental issues that contribute to self-destructive behaviors. These behaviors have been described as wide-ranging acts of harm to self, including suicidal behaviors, nonsuicidal self-injury, and recklessness. Direct and indirect subtypes of self-destructive behaviors are important to understand and will be explored. Past and childhood trauma is known to be a strong predictor of self-harming behavior into adulthood. Further, post-traumatic stress disorder diagnoses are strongly correlated with prevalence of suicide attempts. The literature indicates that other risk factors include psychiatric illness, socioeconomic disadvantages, early and late adulthood, sexual orientation, and military/veteran status. Additionally, self-harming behavior is a predictor for suicide attempts, with a suicide completion rate of 5% for up to 9 years after hospitalization for self-harm. Research shows that adolescents, military/veteran, and ethnically and sexually diverse populations are most at risk. Several cultural aspects within these groups will be explored to show potential protective factors which can be applied to the individual. It is important to understand the factors that relate to resilience in dealing with negative effects of trauma and stress, such as post-traumatic stress disorder and other mental illness. Successful evidence-based resilience programs within each of these at-risk groups will be discussed along with implications for further research.

2.1 INTRODUCTION

Research to prevent the growing number of suicides in the United States has been increasing significantly. According to the Centers for Disease Control and Prevention (CDC, 2016), suicide completion is the tenth leading cause of death in the United States of America and is rising. The Suicide Prevention Resource Center (SPRC, 2013) stated that prior suicide attempt(s), alcohol and drug abuse, mood and anxiety disorders, and access to lethal means have been identified as major suicide risk factors across all populations. These suicidal behaviors and risk factors are commonly related to suicide completion (Bongar & Sullivan, 2013).

Nock (2010) defined suicidal behavior or self-injury as intentional behaviors that cause some degree of mental or physical injury to the self. He stated that these behaviors can be divided into two categories: direct and indirect. Direct harm refers to intended bodily injury (i.e., cutting) whereas behaviors where bodily injury is an unintentional product are classified as indirectly harmful (i.e., smoking and drinking). While he stated that in participant self-report most people with these self-harmful behaviors state they are not suicidal, he also argued that the levels of severity of these behaviors should not be underestimated in suicide research. Due to the prevalence of these behaviors, more research is warranted to understand why people develop self-harmful behaviors (Nock, 2010) and will be discussed more in depth later in this chapter.

Interest in the concept of resilience in those who exhibit suicidal risk factors is an increasing research area. The study of people that have overcome an adverse trauma in their lives can be beneficial for studying resilience and compared to those at risk for suicide (Windle, 2011; Roy et al., 2007) although low levels of resilience may be linked to risk for suicidal behavior (Roy et al., 2007). This highlights a potential pathway to better understanding of risk factors for suicide: strengths and resilience of an individual. The positive psychology movement has focused on strengths to help individuals and larger institutions identify and understand existing positive qualities in order to promote resilience in one's life (Peterson & Seligman, 2004).

There is wide variation in how resilience has been defined in the literature as evidenced by estimates of the prevalence of resilience that range from 25% to 84% (Windle, 2011). In general, the term resilience is typically used when speaking about the ability to bounce back or recover from illness or misfortunate events (Windle, 2011). The American Psychological Association (2009) refined the definition to incorporate the idea of adapting and moving forward after experiencing difficult situations such as trauma, stress, or other adversity.

The concept of resilience within research has largely been driven by developmental psychologists and they often refer to resilience as one's ability to work toward positive psychological, emotional, physical, and other growth outcomes in light of threat or adversity (Windle, 2011). However, Windle (2011) argued that additionally there may be evidence that resilience throughout the lifespan may be important when looking at the role that protective factors play in one's life. Richardson (2002) defined resilience as strengths or assets that help people survive adverse life events. These characteristics of resilience are also known as protective factors. For the purposes of the present chapter, this definition will be used to define resilience since it is both broad yet focused enough to encompass resilience, risk factors for suicide (i.e., self-destructive behaviors), as well as the differences and similarities in individual cultural factors, including childhood development and development across the lifespan.

McWhirter et al. (2004) conceptualized that there are three contributing factors that develop resilience in childhood. First, and perhaps most importantly, the family system is extremely important to a child's development of healthy resiliency. Next, a child's social network outside the family has been found to help the child overcome obstacles due to ties to the overarching community network. Finally, individual characteristics such as humor and future-oriented thinking can increase overall resilience in the face of adversity. These factors in more resilient children have been found to become coping strategies that engage short- and long-term assessment of their own needs and the needs of others (McWhirter et al., 2004).

Mandleco and Peery (2000) further provided a framework to conceptualize factors that contribute to childhood characteristics and factors that may predict resilience into adulthood. They describe two broad originating categories for resilience to occur which are internal or external to each individual. Internal factors that are created from within the individual are either biological (i.e., "general health, genetic predisposition, temperament, and gender") or psychological factors (i.e., "cognitive capacity, coping ability, and personality characteristics") (Mandleco & Peery, 2000, p. 101).

Factors created externally develop from the environment around the individual and encompass relationships both within and outside of the family system. Mandelco and Peery (2000) noted that the home environment, parenting styles, and family members themselves (i.e., parents, siblings, and grandparents) are considered within the family. Those factors outside of the family include churches, schools, healthcare agencies, daycare programs and youth organizations, as well as other adults and peers. It is argued that while both internal and external factors are important for child development, they are especially important for developing the resilient child as he or she grows into adulthood (Mandelco & Peery, 2000).

2.1.1 COPING SKILLS AND SELF-DESTRUCTIVE BEHAVIORS

In some cases, individuals may engage in self-destructive behaviors in response to difficulties coping with external life stressors or adversity. Some theorists posit that this may be a means of emotional avoidance (Linehan, 1993). Thus, individuals with an inability to cope with distressing emotions may turn to self-destructive behaviors as an adaptive means of coping. Moreover, one's coping skills may serve a vital role in an individual's risk of engaging in self-destructive behaviors. While a common topic in psychological research is the relationship between self-destructive behaviors and personality, additional studies have observed how one's coping skills relate to self-destructive behavior and personality. Researchers found that coping and emotion regulation skills moderated the relationship between self-injury and personality (Hasking et al., 2010). These findings could provide evidence of the potential for coping strategies to serve as a protective factor against self-harming behaviors in individuals, regardless of preexisting personality traits.

2.2 UNDERSTANDING SELF-HARMING BEHAVIORS

It is important to differentiate terminology with regard to self-harming behavior. The term self-harming behavior may encompass nonsuicidal self-injury (NSSI), suicide attempts, and risk-taking or risk-seeking behavior (Claes & Vandereycken, 2007). While suicidal ideation and self-harming behaviors have been shown to be correlated (Haw et al., 2003), it is critical to note that not all self-destructive behaviors are coupled with suicidal thoughts. For example, when researchers interviewed individuals hospitalized for nearly-fatal self-harm, only two-thirds of individuals reported suicidal thoughts (Douglas et al., 2004), although repeated engagement in acts of self-harm has been shown to substantially increase the risk of suicide. In one study, research suggests that in a 9-year follow-up, 5% of individuals hospitalized for self-harm will have completed suicide (Owens et al., 2002).

2.2.1 DIRECT VERSUS INDIRECT SELF-INJURY

There are differential subtypes within NSSI. Overall, NSSI is defined as actions which cause bodily harm without the presence of suicidal intent which encompass a wide range of behaviors. In order to better understand NSSI, researchers have used the terms "direct" and "indirect" self-injury (Germain & Hooley, 2012). Direct self-injury can be conceptualized as any deliberate injuries made to one's body, which are not socially accepted. This may include acts such as cutting, abrading, or burning oneself (Claes & Vandereycken, 2007). In contrast, indirect self-injury is often defined as "behavior that is clearly damaging to the self but does not involve immediate and deliberate damage to body tissue" (Germain & Hooley, 2012, p. 78). Indirect methods may include substance abuse, disordered eating, recklessness, or sexual risk-taking (Claes & Vandereycken, 2007).

Aside from the nature of the action, there are additional differences between indirect and direct self-injury. Research suggests that individuals who engage in direct forms of NSSI may be more internal and self-critical than those who report engaging in more indirect means of self-injury. Those who engage in more direct forms of self-injury displayed higher scores on assessments measuring suicide risk and reported more past suicide attempts while individuals who engage in more indirect methods of self-injury reported more substance abuse and a greater majority were involved in abusive relationships (Germain & Hooley, 2012).

While there are many differences between indirect and direct self-injury, similarities between individuals who engage in these behaviors have been observed. For example, exposure to acts of violence has been found to be a risk factor for both direct and indirect methods of self-injury (Berenson et al., 2001). Additionally, researchers found no difference in scores of impulsivity, depression, and borderline personality disorder (BPD) (Germain & Hooley, 2012). These results further support that self-injury, regardless of the subtype, is often an impulsive act. At times, individuals who engage in self-harming behaviors report that they made the decision to do so only minutes before acting (Simon et al., 2001).

2.2.2 RISK FACTORS

Risk factors for such behaviors are often discussed in research on self-harming behaviors. Some clinicians and theorists suggest a heightened risk for self-injury in response to environmental stressors, such as socioeconomic status (SES). Others propose the importance of examining more personal and cultural risk factors, including age, ethnicity, or sexual orientation. A considerable amount of research has focused on the relationship between different psychiatric disorders and self-harming behaviors. The presence of a psychiatric disorder has been shown to serve as a risk factor for self-harming behaviors. In one study, over 90% individuals hospitalized for self-destructive behaviors were diagnosed with at least one psychiatric disorder (Skegg, 2005). Yet, one could argue that all these factors may contribute to one's risk to engage in self-destructive behaviors. These various risk factors for self-destructive behavior are discussed below.

2.3 CULTURAL CONSIDERATIONS

Research by Downs and Eisenberg (2012) found that a great barrier in assessing risk is that people, particularly men, seldom seek help from a mental health practitioner when contemplating self-harm. Recent research indicates that this barrier is typically influenced by cultural factors (Chu et al., 2010, 2013). Therefore, it is important for psychologists to learn how to identify both risk and protective factors that are generalizable and culture-specific in order to accurately assess and manage an individual's level of risk for self-harm. While it is essential to consider the cultural aspects of risk and protective factors pertaining to self-destructive behavior, it is also necessary to consider how exactly the term *culture* is conceptualized.

Culture is defined differently by various scholars in the literature; however, this chapter's discussion will rely on Hays' (1996) definition of culture: "all of the learned behaviors, beliefs, norms, and values that are held by a group of people and passed on from older members to newer members, at least in part to preserve the group" (p. 333). This definition will be used because it has been widely accepted in the realm of multicultural counseling. As Hays (1996) discussed regarding her ADRESSING model for multicultural counseling, in order to understand an individual's cultural background, one must first identify salient and specific cultural influences or identity factors (i.e., "age," "disability," "religion," "ethnicity/race," "social status," "sexual orientation," "indigenous heritage," "national origin," and "gender").

Even though it could be the case that an individual identifies with one, primary cultural identity (e.g., American), individuals often define themselves as having multiple intersecting identities (Sarno et al., 2015). In some cases, an individual's intersecting identities may even conflict with each other, as can be the case for biracial people or sexual minorities (Sarno et al., 2015). This concept of intersecting and conflicting identities adds further complexity to the study of the cultural aspects associated with risk and protective factors. Therefore, it is of utmost importance to study in detail how diverse cultural identities may impact the resiliency of individuals within diverse groups.

It is important to consider the influence of culture in both resilience and self-destructive behaviors as we will show that culture could indeed account for many of the risk factors described below. It is important to consider the influence of culture when discussing resilience and risk factors of self-destructive behaviors. Factors of resilience and self-destructive behaviors may present differently across varying cultures. Subsequently, levels of risk appear to vary when the US population is broken down by culture (e.g., age, ethnicity, and gender; Chu et al., 2013). These differences in risk level, which take into account all forms of self-destructive behavior, may be due to culturally influenced risk and protective factors that are specific to the individual. In order to understand culture-specific risk and protective factors for self-destructive behavior, it is necessary to know which factors have been shown to generalize across populations in the past. SPRC (2013) identified several protective factors that also generalize across all populations, which include effective mental health care, connectedness (to individuals, family, community, and social institutions), problem-solving skills, and contacts with caregivers. This list of generalizable risk and protective factors could be helpful in assessing someone's risk to commit suicide or engage in additional self-destructive behaviors, but it should by no means be considered exhaustive.

2.3.1 PSYCHIATRIC ILLNESS

Overall, the most wide-ranging risk factor across groups for self-destructive behaviors may be the presence of psychiatric illness. Depression and anxiety have both been found to be correlated with self-destructive behaviors (Hawton et al., 2012). Additionally, substance use and eating disorders have been found to be correlated with indirect self-injury (Germain & Hooley, 2012). However, a considerable amount of research has focused on the relationship between self-destructive behaviors and personality disorders. Studies have shown that personality disorder diagnoses are common among individuals who engage in self-harm (Skegg, 2005). More specifically, BPD is the most commonly seen personality disorder in individuals who engage in self-harm. Persistent self-harm or threats to self-harm is a diagnostic feature of BPD (Haw et al., 2001). Moreover, individuals diagnosed with BPD often suffer from a lack of coping abilities (Linehan, 1993). This may influence why they may be at risk of employing more maladaptive ways of coping, such as self-harm.

2.3.2 SOCIOECONOMIC FACTORS

When considering likely risk factors for self-destructive behaviors, one must consider SES status. While low SES status is a prominent risk factor for wide-ranging medical and mental health problems, this may also be true in regard to behaviors such as self-harming behaviors (Skegg, 2005). The life stressors which accompany low SES may play a part in an individual's likelihood to attempt to cope by means of engaging in self-destructive behaviors. Moreover, research has shown that rates of self-harm are greater in individuals from low SES areas, particularly in adolescent groups (Hawton et al., 2012). Unfortunately, low education levels are generally observed in areas of low SES (Lee & Burkam, 2002). This association can also be seen in the risk for self-destructive behaviors. Thus, researchers have highlighted how a low education level may also serve as a risk factor for self-harming behaviors (Skegg, 2005). Therefore, both SES disadvantages and low education levels should be considered when assessing for risk of self-destructive behaviors.

2.3.3 AGE

An additional risk factor discussed when observing self-destructive behaviors is age. While age cohorts are commonly examined in regard to risk of any disorder or disease, it may also play a prominent role in self-destructive behaviors. Many studies have shown that adolescents and young adults may be at a greater risk for self-harming behavior and elevated rates of self-harm have been seen in these age groups (Fliege et al., 2009). While the rate of self-harm is substantially greater in adolescents, the risk of self-harm also exists in adult populations.

In fact, the risk of completed suicide increases as age increases and self-harming behaviors perpetuate in adult populations (Murphy et al., 2012). Considering all age cohorts, it seems that overall, adolescents and older adults may be at heightened risk for engaging in self-destructive behaviors.

2.3.3.1 AGE AS A CULTURAL FACTOR FOR RISK AND RESILIENCE

The CDC (2016) statistics placed older men (above the age of 65 years) at the highest risk for completed suicides in the US population. However, females appeared to be most at-risk for self-harm in the age group of 45–54 years (CDC, 2016). Also, research shows an increase in self-destructive behaviors among younger individuals (i.e., <18 years of age; CDC, 2016). In the United States, the most common age of onset for self-harm is 16 years (Skegg, 2005). As reported by the National Institute of Mental Health (NIMH, 2018) based on data published by the CDC in 2016, suicide "was the second leading cause of death among individuals between the ages of 10 and 34, and the fourth leading cause of death among individuals between the ages of 35 and 54" ("Suicide is a Leading Cause of Death in the United States," para. 1). Taking these statistics into account, it is important to examine whether different risk and protective factors influence self-destructive behavior within different age groups.

McGirr et al. (2008) found impulsivity and aggression to be risk factors for self-destructive behavior among people under the age of 18, whereas they found depressive symptoms to be risk factors for self-destructive behavior among older adults. It may be possible that this important distinction between younger and older individuals is due to biological components related to child and adolescent development. Literature regarding child and adolescent development states that many neurocognitive and hormonal changes occur during this period (Bridge et al., 2006; Gould et al., 2003; Manceaux et al., 2015). As per previous findings, these changes may lead to chemical imbalances in the brain of younger persons that may predispose them to some depressive symptoms, such as low-self esteem, and impulsive-aggressive behavior (Bridge et al., 2006; Gould et al., 2003; Manceaux et al., 2015). Subsequently, these biological changes that occur during early human development have been linked to self-destructive behaviors, including suicide, in youths (Bridge et al., 2006; Gould et al., 2003; Manceaux et al., 2015). This link between biological development and impulsive-aggressive behavior may explain why the literature has found that self-harming and suicidal behavior in younger individuals show less signs of intent and planning in comparison to their older counterparts.

Research by Pope et al. (2016) indicated that, although suicide attempt rates were lower among the geriatric population, suicide completion rates tend to be higher in comparison to the general population. Similar to McGirr et al. (2008), these researchers also found that a diagnosis of major depressive disorder (MDD) is highly correlated with completed suicide and self-destructive behavior in older adults, which is in part due to the fact that individuals over the age of 65 years "are more likely to seek medical attention for a physical illness than for depression" (Pope et al., 2016, p. 706). In addition, Pope and colleagues (2016) found that having a chronic illness and access to firearms are also specific risk factors among the geriatric population. Research also indicates that chronic illnesses (e.g., chronic pain, diabetes, multiple sclerosis, kidney/heart disease, and cancer) and depression are positively correlated to each other, and they are both positively correlated to self-destructive behavior (Jia et al., 2015). Nevertheless, research by Jia et al. (2015) suggests that depression likely mediates the relationship between the presence of a chronic illness and self-destructive behavior, which confirms McGirr et al.'s (2008) findings that major depression seems to be the most significant risk factor among older adults.

Some age-specific protective factors for older adults against self-destructive behaviors are likely similar to the factors that protect against depression. According to Heisel et al. (2016), good perceived social support, living/volunteering in their communities, and good perceived health-related quality of life are all protective factors against self-destructive behaviors among older adults. Also, de Frias and Whyne (2015) found that mindfulness is another protective factor in older adults. Overall, these findings support the idea that different risk and protective factors may be at play within different age groups. Therefore, different intervention strategies should be implemented to address and manage self-destructive behavior in individuals of different ages.

Research shows that community- and school-based programs focusing on increasing mental health, but not solely suicide awareness work well for decreasing self-destructive behavior in children and adolescents (Bridge et al., 2006; Gould et al., 2003). Additionally, skills training and stepped-care programs (i.e., programs that either decrease or increase the level of care based on the individual's risk) have proven more helpful among children and adolescents than Suicide Awareness Curriculums, which may have the adverse effect of further distressing at-risk youth (Bridge et al., 2006; Gould et al., 2003). Among older adults, programs that include a mindfulness component and focus on increasing the individual's reasons for living (e.g., perceived social support) have received support in the literature with regard to reducing risk for depression, suicide, and self-destructive behaviors (de Frias et al., 2015; Heisel et al., 2016). In addition, it is imperative that physical and mental health professionals that work with geriatric populations receive proper training in how to identify and manage symptoms of depression, risk for self-destructive behavior, and access to firearms (Pope et al., 2016). With regard to young and middle-aged adults, there appears to be a gap in the literature regarding age-specific risk and protective factors for these populations.

2.3.4 SEXUAL AND GENDER IDENTITY

When considering risk factors for self-destructive behaviors, sexual and gender identities certainly need to be considered. Sexual and gender minority groups often face increased societal stressors compared to heterosexual and cisgender individuals (Meyer, 2015). Thus, individuals in sexual and gender minority groups have reported turning to self-harm as a means of coping. Likewise, self-harming behaviors in sexual and gender minority groups are more common in more homophobic or heterosexist areas (Jackman et al., 2016). Additionally, gay, lesbian, and bisexual individuals have been shown to have a higher likelihood to engage in self-harming behaviors than heterosexual individuals (Skegg, 2005). Within sexual minority groups, bisexual or people questioning their sexuality were found to be most at-risk of self-harm and within gender minority groups, transgender and gender nonconforming individuals were found to be most at-risk of engaging in self-harming behaviors (Jackman et al., 2016). These findings emphasize the need to consider sexual and gender minority stress and ways in which one copes with these stressors with regard to self-destructive behaviors.

2.3.4.1 SEXUAL ORIENTATION AND GENDER IDENTITIES AS CULTURAL FACTORS FOR RISK AND RESILIENCE

According to the NIMH (2018), men are nearly four times more likely to commit suicide than women, while women are more likely than men to engage in self-harm. Additionally, women are more likely to commit suicide via more passive methods (e.g., poisoning), while men are more likely to commit suicide via more aggressive methods (e.g., firearm; NIMH, 2018). However, these statistics only look at gender as a binary concept, and they do not take into account sexual orientation. Research on sex and gender minorities indicates that the LGBTQ+ (i.e., lesbian, gay, bisexual, transgender, queer, etc.) population is at a higher risk for suicide, as well as self-destructive behaviors (Aranmolate et al., 2017; Chu et al., 2010, 2013).

Aranmolate et al. (2017) reported that suicide and self-destructive behaviors among individuals identifying as LGBTQ+ has been positively correlated with peer and family rejection (i.e., social discord), discrimination (i.e., sexual minority stress), and school failure. Additionally, researchers reported that (cyber) bullying, hopelessness, cultural norms, religious beliefs, and not wanting to seek mental health services due to stigma are risk factors specific to LGBTQ+ individuals (Aranmolate et al., 2017). Therefore, treatments aiming to reduce LGBTQ+-specific risk factors should focus on strengthening protective factors specific to this population, such as family and community connectedness and school safety (Aranmolate et al., 2017). For such reasons, school-, community-, and family-based interventions (e.g., summer camp programs) may prove most helpful in this population (Gillig et al., 2017). Nonetheless, interventions tailored to sex and gender minorities have received little attention in the literature, and further research should address LGBTQ+-specific interventions to reduce the risk of engaging in self-destructive behaviors among these individuals (Haas et al., 2010).

2.3.5 RACE AND ETHNICITY

Throughout the literature on self-destructive behaviors, some researchers have aimed to observe how race and ethnicity may interact with one's risk of engaging in these behaviors. Similar to individuals from low SES groups or sexual and gender minority groups, ethnic and racial minority groups face increased stressors (Thoits, 2010). Likewise, racial and ethnic minority groups may turn to self-harm as a means of coping with minority stress. This may influence research findings which suggest that racial and ethnic minority groups are at an increased risk for engaging in self-destructive behaviors. For example, one such study found that, in low SES areas of the Southern USA, African–American students were more likely to engage in direct self-harm than Caucasians (Gratz et al., 2012). Similarly, research has shown that ethnic minorities, such as Native Americans, displayed higher rates of self-injurious behavior than non-minority group members (Gholamrezaei et al., 2017). This may further highlight the impact of minority stress on the risk of self-destructive behavior.

2.3.5.1 RACE AND ETHNICITY AS CULTURAL FACTORS FOR RISK AND RESILIENCE

When suicide rates in the US population are examined by race/ethnicity, American Indians/Alaskan Natives (AI/AN) are at the highest risk for suicide (NIMH, 2018). There is also increasing trend toward suicide risk among adolescent Latinas, adolescent African American males, and Asian American females aged 65 and older (Chu et al., 2010, 2013; SPRC, 2013). These findings suggest that racial/ethnic identity, such as age, influences an individual's decision-making process to engage in self-destructive behaviors. A study by Walker et al. (2010) found that patterns of resilience also vary across ethnic groups. Recent research indicates that few risk and resiliency assessments have actually been properly adapted to be culturally sensitive (Chu et al., 2013). This is problematic, particularly because cultural minorities are more likely to be considered *hidden ideators* (Morrison & Downey, 2000). Hidden ideators are individuals who are at-risk for engaging in self-destructive behavior, but they are not likely to disclose any self-harm ideation to their primary care physician or mental health service provider (Morrison & Downey, 2000). Thus, both primary care physicians and mental health service providers need to be thoroughly trained in how to recognize risk and resiliency factors that are specific to an individual's racial/ethnic identity.

Chu et al. (2010, 2013) identified four major categories of culturally specific risk: cultural sanctions, idioms of distress, minority stress, and social discord. Cultural sanctions refer to the acceptability or unacceptability/shame associated with utilizing self-destructive behaviors as a viable option to solve or escape from one's problems (Chu et al., 2010, 2013). For example, African Americans and LatinX individuals are less likely to commit suicide if they view it as amoral or against their religious beliefs (Chu et al., 2010). Idioms of distress refer to how different cultures may express or explain their mental health symptoms (Chu et al., 2010, 2013). This category is particularly salient because hidden ideators are more likely to express their psychological distress via idioms of distress (e.g., somatic symptoms) than by directly self-disclosing self-destructive and suicidal thoughts (Chu et al., 2010; Morrison & Downey, 2000). Minority stress refers to the psychological distress that is uniquely experienced by cultural minorities due to the oppression and prejudice associated with their identity or social position (Chu et al., 2010, 2013). Although minority stress is not a category that is unique to race or ethnicity, it encompasses acculturative stress, which is highly correlated with mental health issues (including suicide) in immigrant populations (Berry, 1997; Chu et al., 2010, 2013; Hovey & King, 1996). Last, social discord (which, once again, is not unique to race or ethnicity) involves family conflict, lack of integration with the community, and alienation, which are culture-specific risk factors predominantly among collectivistic cultures (Chu et al., 2010, 2013).

Although none of these categories is more highly linked to one specific racial/ethnic identity, they are certainly linked to at-risk behavior across racial/ethnic minorities (Chut et al., 2010, 2013). Therefore, it is important to consider all four categories when assessing risk for suicide and self-destructive behaviors in individuals that identify as a racial/ethnic minority. Doing so will not only shed light on any risk factors that may be present, but it will also shed light on how resiliency can be increased or developed during treatment. As an example, "familismo" (family connectedness) is considered a protective factor against acculturative stress and at-risk behavior among LatinX individuals (Hovey & King, 1996). Therefore, assessing for the presence or absence of "familismo" within a LatinX individual may help a mental health professional determine whether social discord and acculturative stress are potential risk factors for that specific individual. Additionally, it could also help the mental health professional determine whether interventions that focus on developing and/or strengthening the individual's sense of "familismo" would be helpful to decrease the level of risk in treatment.

There is a paucity of research that focuses on racial/ethnic-specific interventions that help reduce risk for suicide and self-destructive behaviors (Al-Halabí et al., 2016). This literature gap includes a lack of evidence-based interventions that are specific to race or ethnicity because generalizing these interventions to accommodate all cultural variances would require more time, money, and resources than most researchers are capable of. Therefore, it appears that the best solution for this dilemma is to utilize evidence-based interventions and make pertinent cultural adaptations that are tailored to the individual instead. Individual-specific interventions that address the four major categories of culturally specific risk and focus on the individual's reasons for living appear to work best with racial/ethnic minorities (Chu et al., 2010, 2013; Morrison & Downey, 2000; Walker et al., 2010).

2.3.6 TRAUMA

Another prominent risk factor for self-destructive behavior is the experience of trauma. For example, a diagnosis of post-traumatic stress disorder (PTSD) has been found to be a significant predictor of later self-harming behavior (Dyer et al., 2009). However, it is important to differentiate between types of trauma. For example, military trauma and childhood trauma both have similarities and differences. Military trauma and childhood trauma are frequently characterized as complex trauma, in that they may be prolonged or repeated incidences. Yet, childhood trauma specifically has been largely linked to dissociation (Perry et al., 1995). Moreover, childhood trauma has been found to be a major risk factor for self-destructive behavior and PTSD later in life (Fliege et al., 2009). Additionally, the experience of childhood trauma coupled with dissociation may make an individual more at risk for engaging in self-destructive behaviors (Low et al., 2000). The experience of childhood trauma has proven to be a risk for self-harm not only in adults, but in adolescent populations as well (Hawton et al., 2012). More specifically, childhood sexual trauma has shown to be more strongly related to self-harm than any other form of childhood trauma (Gratz, 2003).

Trauma associated with serving in the military has also shown to put individuals at risk for self-destructive behaviors. For instance, researchers have found that individuals who had been deployed for combat in the military displayed a higher frequency of indirect self-harm, such as risky behaviors, than civilians or personnel who had not deployed (Thomsen et al., 2011). Moreover, NSSI has also been shown to be prevalent in veteran populations both in-treatment and not in-treatment for PTSD (Kimbrel et al., 2016). At times, service members may be asked to commit actions that they may feel are contrary to their religious, moral, or spiritual beliefs. These acts, or learning about these acts, can at times lead to a form of distress called moral injury. Some researchers noted that in veterans, the presence of moral injury may be a major risk factor for engaging in self-harm or suicidal behaviors (Litz & Maguen, 2012). Altogether, these findings emphasize the severity that multiple types of trauma may serve as risk factors for varying forms of self-destructive behaviors. Overall, research findings show that varying forms of trauma may serve as risk factors for engaging in self-destructive behaviors.

2.3.6.1 TRAUMA AS A CULTURAL FACTOR FOR RISK AND RESILIENCE

Exposure to trauma is a known risk factor for self-destructive behavior, such as suicide, in people of all ages (Bridge et al., 2006; Gould et al., 2003; Low et al., 2000). Particularly in early childhood, physical and sexual abuses have been linked to suicidal and other kinds of self-destructive behavior, such as cutting and substance abuse (Bridge et al., 2006; Gould et al., 2003; Low et al., 2000). Additionally, other stressful life events, such as family dysfunction, parental divorce, and bullying at school, also seem to be risk factors for self-harming behavior and suicide in children and adolescents (Bridge et al., 2006; Gould et al., 2003). Nevertheless, research suggests that several age-specific protective factors may exist for children and adolescents as well. For example, studies by Bridge and colleagues (2006) as well as by Gould and colleagues (2003) suggest that family cohesion, healthy relationships with teachers and/or peers at school, and religiosity are protective factors for children and adolescents. However, it is important to note that the studies that have examined religion as a protective factor among children and adolescents do not appear to be well-controlled with regard to confounding variables such as substance use (Gould et al., 2003).

Although being an active duty military service member or a veteran may often be overlooked as a cultural identity, there is a breadth of literature evidencing the importance of understanding military status as an influencing cultural factor (Hall, 2011). Individuals in the military have their own language, structure, and set of values (Hall, 2011). In addition, military jobs may require personnel to be repeatedly exposed to traumatic events and high-stress situations, such as intense training and combat zones (Thomsen et al., 2011). These job requirements have been linked in the literature to a myriad of mental health issues such as chronic stress, PTSD, reckless and self-destructive behavior (RSDB), for example, substance use and suicide (Lusk et al., 2017; Thomsen et al., 2011).

A study by Thomsen et al. (2011) found that deployed military personnel were at a higher risk for self-destructive behavior and suicide when compared to nondeployed military personnel and civilians. In fact, the literature shows that veterans are 41%–61% more likely to commit suicide than their civilian counterparts, and that the risk is elevated when there is a history of deployment (Lusk et al., 2017). These findings depict a serious public health issue with regard to military mental health, and they also emphasize the importance of assessing for risk and protective factors in this population. Moreover, another study by Lusk et al. (2017) noted that PTSD and RSDB are highly correlated with each other, particularly among military personnel. These researchers found that not only can RSDB manifest as a symptom of PTSD, but it can also lead to further re-traumatization and, subsequently, to an increase in PTSD symptoms and risk for suicide (Lusk et al., 2017). Hence, exposure to combat trauma appears to be the most salient risk factor for RSDB and suicide among military personnel and veterans (Lusk et al., 2017). Similarly, having mental health diagnoses such as PTSD and substance use disorder (SUD), particularly if comorbid with each other, as well as having experienced military sexual trauma (i.e., sexual trauma during military service) are also considered military-specific risk factors (Kimerling et al., 2016; MacManus et al., 2015).

Pietrzak et al. (2010) found that family and community support, unit/military support, and spirituality are important protective factors to focus on throughout treatment. Thus, evidence-based practices that have proven useful among the military population involve interventions that are derived from existing training programs, such as the Marine Operational Stress Control and Readiness program, which deliver psychological concepts in a way that is already familiar to this population (Meredith et al., 2011). Also, community-based treatments, family-based interventions, and unit-level programs that focus on building resilience and developing healthy coping skills (e.g., assertive communication, exercise, and meditation) have shown some utility in reducing risk among active military personnel and veterans (Meredith et al., 2011).

2.3.7 CHALLENGES OF INTERSECTING IDENTITIES

This chapter has addressed several major cultural factors that influence risk and resiliency in the individual but do not constitute an exhaustive list. The data on influencing cultural factors have not even considered the challenges that arise when an individual identifies with multiple intersecting identities, some of which may be in conflict. Culture is a rich, complex concept, which is why an individual's cultural background should not be pigeonholed in just one category. When assessing an individual's risk and protective factor to determine the best line of treatment, the individual's cultural background should be examined through a holistic lens. Thus, mental health professionals need to inquire not only about the individual's cultural identities, but also need to inquire about how those identities intersect when evaluating clients.

In order to establish effective treatment plans, mental health professionals need to identify which risk and protective factors are most salient to individual clients and determine which factors should be addressed in treatment. Therefore, mental health professionals must look at their clients with a "bigger picture" perspective. Furthermore, risk and protective factors as well as risk level for suicide and self-destructive behavior may change depending on which cultural identities are most salient to the individual. As an example, Reimann and Mazuchowski (2018) reported that the suicide risk in active military personnel varies by age and gender. These researchers found that women are at a higher risk for suicide between the ages of 17 and 29 years, while men exhibit a lower suicide risk between the ages of 25 and 49 years. Therefore, it is imperative that cultural considerations are factored in when assessing the risk and protective factors for suicide and self-destructive behavior so as to make well-informed, appropriate decisions with regard to treatment.

2.4 CONCLUSION

This chapter focused on the concept of resilience and its impact on self-destructive behaviors. While recent research on resilience has focused on developmental constructs that are internal or external to the individual, clinicians and individuals can also focus on applying strengths-based approaches as a means to build resilience based on the risk and cultural factors of the individual. Focusing on those most at risk for these behaviors will help us understand more clearly a target population in which to detect and screen for self-destructive behaviors. As noted above, identifying these behaviors may also help further identify those at greater risk to complete suicide.

To achieve this goal, further research is necessary to examine the validity of existing culturally competent risk assessments for suicide and self-destructive behavior. Additionally, further research is also necessary to bridge the gap in the literature regarding the usefulness of culturally adapted treatment interventions, which is often overlooked, particularly among certain populations (e.g., gender and sexual minorities). Overall, resilience can provide hope and increase self-efficacy and control (Richardson, 2002), so future research must be geared toward finding better ways to incorporate resilience-building interventions in treatment.

KEYWORDS

- **resilience**
- **self-destructive behaviors**
- **suicide**
- **at-risk**
- **cultural factors**

REFERENCES

Al-Halabí, S., Sáiz, P. A., Burón, P., Garrido, M., Benabarre, A., Jiménez, E., Cervilla, J., Navarrete, M. I., Díaz-Mesa, E. M., García-Álvarez, L., Muñiz, J., Posner, K., Oquendo, M. A., PazGarcía-Portilla, M., & Bobes, J. (2016). Validation of a Spanish version of the Columbia-Suicide Severity Rating Scale (C-SSRS). *Revista de Psiquiatría y Salud Mental (English Edition)*, *9*(3), 134–142.

American Psychological Association. (2009). *The Road to Resilience*. Retrieved from http://www.apa.org/help-center/road-resilience.aspx

Aranmolate, R., Bogan, D. R., Hoard, T., & Mawson, A.R. (2017). Suicide risk factors among LGBTQ youth: Review. *JSM Schizophrenia 2*(2), 1011.

Berenson, A. B., Wiemann, C. M., & McCombs, S. (2001). Exposure to violence and associated health-risk behaviors among adolescent girls. *Archives of Pediatrics & Adolescent Medicine*, *155*(11), 1238–1242.

Berry, J. W. (1997). Immigration, acculturation, and adaptation. *Applied Psychology*, *46*(1), 5–34.

Bongar, B., & Sullivan, G. (2013). *The Suicidal Patient: Clinical and Legal Standards of Care,* 3rd ed. Washington, DC: American Psychological Association.

Bridge, J. A., Goldstein, T. R., & Brent, D. A. (2006). Adolescent suicide and suicidal behavior. *Journal of Child Psychology and Psychiatry*, *47*(34), 372–394.

CDC (Centers for Disease Control and Prevention) (2016). Youth risk behavior surveillance: United States, 2015. *MMWR: Morbidity and Mortality Weekly Report, 65*(6), 73–77.

Chu, J., Floyd, R., Diep, H., Pardo, S., Goldblum, P., & Bongar, B. (2013). A tool for the culturally competent assessment of suicide: The cultural assessment of risk for suicide (CARS) measure. *Psychological Assessment, 25*(2), 424.

Chu, J. P., Goldblum, P., Floyd, R., & Bongar, B. (2010). The cultural theory and model of suicide. *Applied and Preventive Psychology*, *14*(1–4), 25–40.

Claes, L., & Vandereycken, W. (2007). Self-injurious behavior: Differential diagnosis and functional differentiation. *Comprehensive Psychiatry*, *48*(2), 137–144.

deFrias, C. M., & Whyne, E. (2015). Stress on health-related quality of life in older adults: The protective nature of mindfulness. *Aging & Mental Health*, *19*(3), 201–206.

Douglas, J., Cooper, J., Amos, T., Webb, R., Guthrie, E., & Appleby, L. (2004). "Near-fatal" deliberate self-harm: Characteristics, prevention and implications for the prevention of suicide. *Journal of Affective Disorders, 79*(1–3), 263–268.

Downs, M. F., & Eisenberg, D. (2012). Help seeking and treatment use among suicidal college students. *Journal of American College Health, 60*(2), 104–114.

Dyer, K. F., Dorahy, M. J., Hamilton, G., Corry, M., Shannon, M., MacSherry, A., McRobert, G., Elder, R. & McElhill, B. (2009). Anger, aggression, and self-harm in PTSD and complex PTSD. *Journal of Clinical Psychology*, *65*(10), 1099–1114.

Fliege, H., Lee, J. R., Grimm, A., & Klapp, B. F. (2009). Risk factors and correlates of deliberate self-harm behavior: A systematic review. *Journal of Psychosomatic Research*, *66*(6), 477–493.

Germain, S. A. S., & Hooley, J. M. (2012). Direct and indirect forms of non-suicidal self-injury: Evidence for a distinction. *Psychiatry Research*, *197*(1–2), 78–84.

Gholamrezaei, M., De Stefano, J., & Heath, N. L. (2017). Nonsuicidal self-injury across cultures and ethnic and racial minorities: A review. *International Journal of Psychology*, *52*(4), 316–326.

Gillig, T. K., Miller, L. C., & Cox, C. M. (2017). "She finally smiles… for real": Reducing depressive symptoms and bolstering resilience through a camp intervention for LGBTQ youth. *Journal of Homosexuality*, *66*(3), 368–388. https://doi-org.paloaltou.idm.oclc.org/10.1080/00918369.2017.1411693

Gould, M. S., Greenberg, T. E. D., Velting, D. M., & Shaffer, D. (2003). Youth suicide risk and preventive interventions: A review of the past 10 years. *Journal of the American Academy of Child & Adolescent Psychiatry*, *42*(4), 386–405.

Gratz, K. L. (2003). Risk factors for and functions of deliberate self-harm: An empirical and conceptual review. *Clinical Psychology: Science and Practice*, *10*(2), 192–205.

Gratz, K. L., Latzman, R. D., Young, J., Heiden, L. J., Damon, J., Hight, T., & Tull, M. T. (2012). Deliberate self-harm among underserved adolescents: The moderating roles of gender, race, and school-level and association with borderline personality features. *Personality Disorders: Theory, Research, and Treatment*, *3*(1), 39.

Haas, A. P., Eliason, M., Mays, V. M., Mathy, R. M., Cochran, S. D., D'Augelli, A. R., Silverman, M.M., Fisher, P.W., Hughes, T., Rosario, M., & Russell, S. T. (2010). Suicide and suicide risk in lesbian, gay, bisexual, and transgender populations: Review and recommendations. *Journal of Homosexuality*, *58*(1), 10–51.

Hall, L. K. (2011). The importance of understanding military culture. *Social Work in Health Care*, *50*(1), 4–18.

Hasking, P. A., Coric, S. J., Swannell, S., Martin, G., Thompson, H. K., & Frost, A. D. (2010). Brief report: Emotion regulation and coping as moderators in the relationship between personality and self-injury. *Journal of Adolescence*, *33*(5), 767–773.

Haw, C., Hawton, K., Houston, K., & Townsend, E. (2001). Psychiatric and personality disorders in deliberate self-harm patients. *The British Journal of Psychiatry*, *178*(1), 48–54.

Haw, C., Hawton, K., Houston, K., & Townsend, E. (2003). Correlates of relative lethality and suicidal intent among deliberate self-harm patients. *Suicide and Life-Threatening Behavior*, *33*(4), 353–364.

Hawton, K., Saunders, K. E., & O'Connor, R. C. (2012). Self-harm and suicide in adolescents. *The Lancet*, *379*(9834), 2373–2382.

Hays, P. A. (1996). Addressing the complexities of culture and gender in counseling. *Journal of Counseling & Development, 74*(4), 332–338.

Heisel, M. J., Neufeld, E., & Flett, G. L. (2016). Reasons for living, meaning in life, and suicide ideation: Investigating the roles of key positive psychological factors in reducing suicide risk in community-residing older adults. *Aging & Mental Health*, *20*(2), 195–207.

Hovey, J. D., & King, C. A. (1996). Acculturative stress, depression, and suicidal ideation among immigrant and second-generation Latino adolescents. *Journal of the American Academy of Child & Adolescent Psychiatry*, *35*(9), 1183–1192.

Jackman, K., Honig, J., & Bockting, W. (2016). Nonsuicidal self-injury among lesbian, gay, bisexual and transgender populations: an integrative review. *Journal of Clinical Nursing*, *25*(23–24), 3438–3453.

Jia, H., Zack, M. M., Thompson, W. W., Crosby, A. E., & Gottesman, I. I. (2015). Impact of depression on quality-adjusted life expectancy (QALE) directly as well as indirectly through suicide. *Social Psychiatry and Psychiatric Epidemiology*, *50*(6), 939–949.

Kimbrel, N. A., DeBeer, B. B., Meyer, E. C., Gulliver, S. B., & Morissette, S. B. (2016). Nonsuicidal self-injury and suicide attempts in Iraq/Afghanistan war Veterans. *Psychiatry Research*, *243*, 232–237.

Kimerling, R., Makin-Byrd, K., Louzon, S., Ignacio, R. V., & McCarthy, J. F. (2016). Military sexual trauma and suicide mortality. *American Journal of Preventive Medicine*, *50*(6), 684–691.

Lee, V. E., & Burkam, D. T. (2002). *Inequality at the starting gate: Social background differences in achievement as children begin school*. Economic Policy Institute, Washington, DC 20036.

Linehan, M. (1993). *Cognitive-behavioral treatment of borderline personality disorder*. Guilford Press.

Litz, B., & Maguen, S. (2012). Moral injury in Veterans of war. *Research Quarterly*, *23*(1), 1–6.

Low, G., Jones, D., MacLeod, A., Power, M., & Duggan, C. (2000). Childhood trauma, dissociation and self-harming behavior: A pilot study. *British Journal of Medical Psychology, 73*(2), 269–278.

Lusk, J. D., Sadeh, N., Wolf, E. J., & Miller, M. W. (2017). Reckless self-destructive behavior and PTSD in veterans: The mediating role of new adverse events. *Journal of Traumatic Stress, 30*(3), 270–278.

MacManus, D., Rona, R., Dickson, H., Somaini, G., Fear, N., & Wessely, S. (2015). Aggressive and violent behavior among military personnel deployed to Iraq and Afghanistan: Prevalence and link with deployment and combat exposure. *Epidemiologic Reviews, 37*(1), 196–212.

Manceaux, P., Jacques, D., & Zdanowicz, N. (2015). Hormonal and developmental influences on adolescent suicide: A systematic review. *Psychiatria Danubina, 27*(Suppl 1), 300–304.

Mandleco, B. L., & Peery, J. C. (2000). An organizational framework for conceptualizing resilience in children. *Journal of Child and Adolescent Psychiatric Nursing, 13*(3), 99–111.

McGirr, A., Renaud, J., Bureau, A., Seguin, M., Lesage, A., & Turecki, G. (2008).

Impulsive-aggressive behaviors and completed suicide across the life cycle: A predisposition for younger age of suicide. *Psychological Medicine, 38*(3), 407–417.

McWhirter, J. J., McWhirter, B. T., McWhirter, E. H., & McWhirter, R. J. (2004). *At-risk youth: A comprehensive response* (3rd ed.). Belmont, CA: Brooks/Cole–Thomson Learning.

Meredith, L. S., Sherbourne, C. D., Gaillot, S. J., Hansell, L., Ritschard, H. V., Parker, A. M., & Wrenn, G. (2011). Promoting psychological resilience in the US military. *Rand Health Quarterly, 1*(2). Retrieved from https://www.ncbi.nlm.nih.gov/pmc/articles/PMC4945176/

Meyer, I. H. (2015).Resilience in the study of minority stress and health of sexual and gender minorities. *Psychology of Sexual Orientation and Gender Diversity, 2*(3), 209–213.

Morrison, L. L., & Downey, D. L. (2000). Racial differences in self-disclosure of suicidal ideation and reasons for living: Implications for training. *Cultural Diversity and Ethnic Minority Psychology, 6*(4), 374–386.

Murphy, E., Kapur, N., Webb, R., Purandare, N., Hawton, K., Bergen, H., Waters, K. & Cooper, J. (2012). Risk factors for repetition and suicide following self-harm in older adults: multicentre cohort study. *The British Journal of Psychiatry, 200*(5), 399–404.

NIMH (National Institute of Mental Health). (2018, May). *Suicide.* Retrieved from https://www.nimh.nih.gov/health/statistics/suicide.shtml

Nock, M. K. (2010). Self-injury. *Annual Review of Clinical Psychology, 6,* 339–363.

Owens, D., Horrocks, J., & House, A. (2002). Fatal and non-fatal repetition of self-harm: Systematic review. *The British Journal of Psychiatry, 181*(3), 193–199.

Perry, B. D., Pollard, R. A., Blakley, T. L., Baker, W. L., & Vigilante, D. (1995). Childhood trauma, the neurobiology of adaptation, and "use-dependent" development of the brain: How "states" become "traits." *Infant Mental Health Journal, 16*(4), 271–291.

Peterson, C., & Seligman, M. E. P. (2004). *Character strengths and virtues: A handbook and classification.* Oxford: Oxford University Press.

Pietrzak, R. H., Goldstein, M. B., Malley, J. C., Rivers, A. J., Johnson, D. C., & Southwick, S. M. (2010). Risk and protective factors associated with suicidal ideation in veterans of Operations Enduring Freedom and Iraqi Freedom. *Journal of Affective Disorders, 123*(1–3), 102–107.

Pope, N. D., Slovak, K. L., & Giger, J. T. (2016). Evaluating a training intervention to prepare geriatric case managers to assess for suicide and firearm safety. *Educational Gerontology, 42*(10), 706–716.

Reimann, C. A., & Mazuchowski, E. L. (2018). Suicide rates among active duty service members compared with civilian counterparts, 2005–2014. *Military Medicine, 183*(suppl_1), 396–402.

Richardson, G. E. (2002). The metatheory of resilience and resiliency. *Journal of Clinical Psychology, 58*(3), 307–321.

Roy, A., Sarchiapone, M., & Carli, V. (2007). Low resilience in suicide attempters. *Archives of Suicide Research, 11*(3), 265–269. doi:10.1080/13811110701403916

Sarno, E. L., Mohr, J. J., Jackson, S. D., & Fassinger, R. E. (2015). When identities collide: Conflicts in allegiances among LGB people of color. *Cultural Diversity and Ethnic Minority Psychology, 21*(4), 550.

Simon, T. R., Swann, A. C., Powell, K. E., Potter, L. B., Kresnow, M. J., & O'Carroll, P. W. (2001). Characteristics of impulsive suicide attempts and attempters. *Suicide and Life-Threatening Behavior, 32*(supplement to Issue 1), 49–59.

Skegg, K. (2005). Self-harm. *The Lancet, 366*(9495), 1471–1483.

SPRC (Suicide Prevention Resource Center). (2013). *Suicide among racial/ethnic populations in the U.S.:*

Hispanics. Retrieved from http://www.sprc.org/sites/default/files/migrate/library/Hispanics%20Sheet%20Aug%2028%202013%20Final.pdf

Thoits, P. A. (2010). Stress and health: Major findings and policy implications. *Journal of Health and Social Behavior, 51*(1_suppl), S41–S53.

Thomsen, C. J., Stander, V. A., McWhorter, S. K., Rabenhorst, M. M., & Milner, J. S. (2011). Effects of combat deployment on risky and self-destructive behavior among active duty military personnel. *Journal of Psychiatric Research, 45*(10), 1321–1331.

Walker, R. L., Alabi, D., Roberts, J., & Obasi, E. M. (2010). Ethnic group differences in reasons for living and the moderating role of cultural worldview. *Cultural Diversity and Ethnic Minority Psychology, 16*(3), 372–378.

Windle, G. (2011). What is resilience? A review and concept analysis. *Reviews in Clinical Gerontology, 21*(2), 152–169. doi:10.1017/S0959259810000420

CHAPTER 3

Resilience and Mental Health

AMANDA WALLICK*, ANNA FEINMAN, JACIE T. BROWN, and LISA M. BROWN

Department of Clinical Psychology, Palo Alto University, 1791 Arastradero Road, Palo Alto, CA 94304, USA

Corresponding author. E-mail: awallick@paloaltou.edu

ABSTRACT

It is widely acknowledged that most people, at some point in their lives, will experience some type of a traumatic event, such as disaster, epidemic, car accident, or health emergency. Relationship problems, a death of a close friend or loved one, and financial problems, can also be highly distressing and disruptive in day-to-day functioning. Although potentially traumatic events are common occurrences, only a small, but significant number of people go on to develop psychological symptoms that require treatment. In fact, most people will demonstrate considerable resiliency in response to adverse events. Although resilience has been defined in a variety of ways, most definitions include some aspect of bouncing back, adapting, or incorporating new behavioral, cognitive, and psychological ways of being after a stressful or traumatic event. Resilience is a multidimensional construct, influenced by a variety of factors including personality, level of social support, financial resources, and attachment styles. Resilience can occur at an individual or community level. This chapter will describe resilience and explain how it serves as a protective factor against developing some mental health disorders. High levels of resilience attenuate the expression of anxiety and depression, while maladaptive strategies increase both anxiety and depression symptoms. People with low resilience may become "stuck" in a state of nonrecovery making them more susceptible to developing post-traumatic stress disorder. This chapter will also review the importance of resilience building for first responders, humanitarian relief workers, and mental health clinicians. Learning how to confidently deal with and reframe negative experiences can contribute to positive outcomes for those who were exposed to traumatic stressors as well as their health care providers. Finally, this chapter will conclude with a discussion about the potential positive impacts on wellbeing following traumatic events, otherwise known as posttraumatic growth.

3.1 INTRODUCTION

On August 5, 2010, 33 Chilean miners were trapped nearly half a mile underground following a cave-in of the small copper and gold mine near the Chilean city of Copiapo. The miners were trapped for a record 69 days enduring extreme heat, emerging medical problems, and surviving at times on one tablespoon of tuna every 48 hours. Four months after being rescued, a 60-min interview with several of the trapped miners revealed that the miners "lived and slept with thoughts of death." One of the miners told the interviewer that, at one point, an idea had been proposed to move all of the miners into a common area, start one of the work trucks that was trapped underground with them, and have everyone inhale the carbon monoxide until they all passed out from the fumes. The miners rejected the idea and continued to persevere, never giving up hope that they would be rescued. After months of drilling, on October 12, 2010, with a reported one billion persons watching the event live worldwide, the miners were finally rescued. Five years later, a CNN interview revealed that all but one of the miners were reportedly struggling with nightmares, sleeping difficulties, an inability to concentrate, high levels of hopelessness, and a heavy reliance on medication. In some cases, a few of the miners had to be hospitalized because of their overwhelming anxiety and depression.

3.2 RESILIENCY

What factors play into our ability to cope after a traumatic event has occurred? Many researchers propose that it is resiliency that allows us to successfully return to a homeostasis after a traumatic event or major life stressor has occurred. Currently, there are many definitions of resiliency. In some literature, the term resiliency is used interchangeably with post-traumatic growth (PTG), but they are not the same. The PTG, sometimes referred to "meaning making" or "benefit finding," is a positive outcome that occurs after a trauma or major life challenge where the person grows from their adversity and exceeds their pre-event level of functioning. Because research on resiliency is still fairly new, there is much debate about whether resiliency is a process, an outcome, an individual trait, or some mix of all three. Broadly speaking, resiliency can be viewed as positive psychological, cognitive, physiological, affective, and behavioral changes that occur after a traumatic event (Van Voorhees et al., 2018). Resiliency is both an intrinsic and extrinsic process of adaptation that requires endurance, optimism, flexibility, and growth by an individual or community.

Resiliency can serve as a protective factor and has been shown to protect military populations from engaging in suicidal behavior (Hourani et al., 2018) and children against internalizing or externalizing problems (Erikson et al., 2011). It can also be understood as a dynamic process involving a pattern of adaptive functioning in the face of significant adversity, trauma, tragedy, threats, or significant sources of stress (Waldegrave et al., 2016). While many definitions of resiliency exist, most can be broken down to describe, simply, an individual's ability to withstand or bounce back after a traumatic event has occurred.

Traumatic events are most often thought of as occurring on an individual level. Physical or sexual abuse, exposure to military combat, a car accident, the death of a loved one, divorce, or financial problems can all be stressful enough to be considered a traumatic event. Traumatic events can also occur at a community level. In the United States, the major events of the last 20 years (e.g., Hurricane Katrina, the September 11th terrorist attacks, and the Flint water crisis in Michigan) have broadened the definition of trauma and have highlighted the psychological impact that these events can have on day-to-day functioning (Dalenberg et al., 2017). When an individual or a community encounters a potentially traumatic event, it is often their level of resiliency that will predict what mental health challenges they will face once the stressor has subsided or resolved. The most current research tells us that individual levels of resiliency are influenced by genetic and environmental factors, including one's personality (Wagnild, & Young, 1993; Oshio et al., 2018), level of social support (Theron & Van Rensburg, 2018; Abraham et al., 2018; Gagnon & Stewart, 2014), and attachment style (Holmes, 2017).

3.3 PERSONALITY

Resiliency is a multidimensional construct that is influenced by a variety of factors. One factor that influences an individual's level of resiliency is their personality. Every person has characteristic patterns of thoughts, emotions, and behaviors. These patterns, because they are relatively stable over time and consistent over situations, are said to represent our personalities. A common way to measure personality is to use an assessment such as the Big Five Inventory (BFI; Goldberg, 1993). The BFI measures an individual's level of openness (i.e., willingness to try new things), conscientiousness (i.e., an individual's degree of self-control, persistence, and need for achievement), extroversion (i.e., extrovert or introvert), agreeableness (i.e., getting along well with others), and neuroticism (i.e., the tendency to experience negative emotional states; Costa et al., 1991).

Research by Oshio et al. (2018) revealed that resiliency is highly correlated with extraversion, openness, agreeableness, and conscientiousness such that higher levels of any of these personality facets were associated with an increase in a person's level of resiliency. Neuroticism, however, showed a negative correlation with resiliency, indicating that individuals with higher levels of neuroticism, that is poor stress tolerance, have lower levels of resiliency. Wagnild and Young (1993) developed a more direct way of measuring an individual level of resiliency by developing a 25-item Resiliency Scale (RS). The RS measures five different personality characteristics that are said to denote resilience. Scores in the domains of equanimity (i.e., balanced life perspectives and experiences), perseverance (i.e., willingness to continue reconstructing one's life after tragedy), self-reliance (i.e., relying on one's own strengths), meaningfulness (i.e., realization that there is something to live for and that life has a purpose), and existential aloneness (i.e., realization that each person is unique and that some experiences must be faced alone) are all totaled to indicate an individual's specific level of resilience. So, while personality is multifaceted, current research does give us some insight into what personality features may result in higher versus lower levels of resiliency.

3.4 SOCIAL SUPPORT

Another major contributing factor to the development of resiliency is an individual's social support. While many competing definitions of social support exist, most can be thought of as the degree of access to, and use of, strong ties to other individuals who are similar to one's self (Nan et al., 1985). However, social support requires not only that you have relationships with others, but also that these relationships involve the presence of solidarity and trust, intimate communication, and mutual obligation (Pearlin et al., 1981). Many studies show that having strong relationships both within and outside your family is a primary factor for developing resiliency (Theron

& Van Rensburg, 2018; Abraham et al., 2018; Gagnon & Stewart, 2014).

Social support broadly encompasses one's perceptions of their support (perceived support) and their receipt of supportive behaviors (received support; Haber et al., 2007). Of these two factors, perceived social support, which describes one's perceptions of, and satisfaction with, the general availability of support has been shown to affect one's mental health (Haber et al., 2007). Abraham et al. (2018) studied the levels of resiliency in a sample of women who had traumatic immigration journeys from Eritrea to Norway. They found that one of the primary coping strategies was to rely on social support, both during the journey and after arrival in Norway. The Eritrean refugee women created "borrowing networks" once they had arrived in Norway's asylum reception centers. They reported relying on these networks for everything from translation assistance to emotional and material support. Furthermore, the women reported that simply by having contact with others in the same situation allowed them to better accept and cope with their current challenges, which in turn increased their level of resiliency (Abraham et al., 2018).

Social support was also identified as an important component not only in the development, but also in the maintenance of resiliency. Kobiske and Bekhet (2018) studied resiliency levels in caretakers of partners diagnosed with early onset or young onset dementia (YOD), a dementia that is diagnosed before the age of 65 years. They found that social support was a key factor in sustaining resiliency when caring for a loved one diagnosed with YOD. Social support gave caretakers brief periods of relief and rest from their caretaking duties that, in turn, facilitated better adaptation to stress and enhanced their ability to preserve resiliency (Kobiske & Bekhet, 2018). Within adolescents, Theron and Van Rensburg (2018) found that social support moderated levels of resiliency for adolescents. Disadvantaged South African adolescents with the greatest amount of peer support demonstrated the highest levels of resiliency and reported feeling safer within their communities (Theron & Van Rensburg, 2018). While social support does not ensure resilience, what research tells us is that across different ages, cultures, and corners of the world, the quality and quantity of one's social relationships helps to develop and maintain resiliency, mitigating the harmful effects of trauma.

3.5 ATTACHMENT

Attachment is another key contributing factor to the development of resiliency. The unique communication patterns, or attachment styles, we develop with our caregivers from infancy through childhood creates the conditions for a having a sense of security later in life (Turunen et al., 2014). Attachment influences how one regulates their emotional reactions when distressed, how support is sought when threatened, and can be broadly categorized as secure or insecure (Bowlby, 1980; Turunen et al., 2014). Securely attached individuals are confident that when stressed, they will receive protection, comfort, and relief from their parents. It is also through secure attachment that individuals develop self-soothing and problem-solving skills that aid in emotion regulation (Turunen et al., 2014). Holmes (2017) proposed that secure attachment is comprised of four resilience-enhancing elements: mentalizing, activeness, stress inoculation, and the extended self. Mentalizing involves being able to step back from ones and others' thoughts and actions in order to see them objectively. Reviewing and reflecting on ones adversity is thought to be the first step toward overcoming it (Holmes, 2017). Activeness involves fostering a child's autonomy. Through the bonds of secure attachment, parents can encourage children to tackle and overcome minor stressors and setbacks, thereby enabling the development of resiliency from a young age. Less resilient children tend to engage in strategies of compliance or helplessness that are often seen in those with insecure attachment (Holmes, 2017). In securely attached children, stress inoculation involves readily and regularly repaired "ruptures" to the parent–child relationship (Holmes, 2017). For example, ruptures occur with a hungry, scared, hurt, or otherwise unhappy baby who looks to be comforted and calmed by their caretaker. If the baby finds that their caretaker is responsive and soothes and alleviates their distress, then the baby learns from these small micro-doses of stress that the feeling can be tolerable and managed. This primes a child's stress response system to help them better-overcome later-life traumas and tragedies with greater levels of resilience. Homes' (2017) final component of secure attachment relevant to resilience is the idea of the extended self. Our ability to extend ourselves to others (as a mother does for her child) allows better regulation of the affect, both as an infant and as an adult (Holmes, 2017). This co-regulation is often seen in a smiling baby who is able to make her caretaker smile back at them. As the baby's affect is highly influenced by their caretaker, so too is the caretaker's affect influenced by their baby. It is this relationship that begins in infancy as "the extended self" that develops our adult needs for social support during times of tragedy or disaster. When all four of these elements are working in unison, tragedy and trauma may be easier to bear because resiliency has been nurtured from a young age. As the section below describes, individuals with high resiliency are also able to attenuate mental health vulnerabilities that result from traumatic experiences.

3.6 RESILIENCY AND MENTAL DISORDERS

Resiliency has been conceptualized as occurring on a vulnerability continuum, which implies that resilience can help to deter mental illness and psychopathology, but does not provide complete immunity (Ingram & Price, 2001). Although stressful life events do increase the probability of both psychiatric symptoms and mental health disorders (Hjemdal et al., 2006), resilience can help individuals achieve a positive state of mental health and reduce negative symptoms (Hu et al., 2015). In this way, resilience serves as a protective factor for the development of certain psychiatric disorders and mental illnesses.

Traditionally, the complete absence of a mental health disorder was a determining factor in quality of life, physical and psychological health, and social productivity (Linley & Joseph, 2004). More recently, a more nuanced approach of maintaining good mental functioning, by way of resilience, has evolved (Davydov et al., 2010). Research shows that maintaining a stable equilibrium in the face of adversity (i.e., resilience) provided mental health protection and aided in the process of recovery (Davydov et al., 2010). However, the presence of resilience does not guarantee a complete return to normal functioning with no threshold or subthreshold mental health symptomatology (Bonanno, 2004). In the aftermath of a traumatic experience the premise that resilience exists on a continuum helps us to understand how those high in resilience are able to maintain relatively stable levels of both psychological and physical functioning whereas those with low levels of resilience are more vulnerable to mental health disorders (Bonanno, 2004). After a traumatic event a person's mental health is vulnerable to many different threats. One threat is a greater risk for developing a mental health disorder such as post-traumatic stress disorder (PTSD).

The *Diagnostic and Statistical Manual of Mental Disorders* (*DSM*) is the handbook used by healthcare professionals in the United States to diagnose mental disorders and provides a common language for all clinicians to use when communicating about patients (American Psychiatric Association, 2013). From the American Psychiatric Association's (APA) first publication of the DSM in 1952 to the fifth edition published in 2013, the development and diagnosis of PTSD has undergone considerable scrutiny and a flurry of intense research activity. The gross stress reaction (GSR) was a term developed to describe symptomology observed during World War II (WWII). The GSR was included in the DSM-I as a mental health disorder for acute combat trauma and other major disasters (American Psychiatric Association, 1952). GSR assumed that the reactions to trauma would subside relatively quickly, and if symptoms of GSR were still present after 6 months, a different mental disorder had to be considered (Dalenberg et al., 2017). However, WWII veterans who continued to exhibit combat-related trauma reactions beyond the 6-month mark no longer met timeframe criteria for GSR despite still suffering psychological distress as a result of their combat experiences. It was during this time that researchers realized that reactions to a traumatic event could be chronic and endure long after the traumatic event had ceased (DiMauro et al., 2014).

In response to this new understanding of trauma, the first official recognition of PTSD as a diagnosis was its inclusion in the DSM-III (American Psychiatric Association, 1994). Medical professionals soon noted that this disorder affected not only war veterans, but also the medical professionals who treated the wounded soldiers. This realization helped psychologists better conceptualize the diagnosis as applicable to both those directly and indirectly involved in wartime combat (DiMauro et al., 2014).

Since PTSD's introduction in the DSM-III, the diagnostic criteria have been expanded and refined in each subsequent edition, including the current version, DSM-5. All available psychiatric, developmental, neurological, and psychological research was reviewed to inform the development of the current definition for PTSD. In DSM-5, Criterion A now includes a cognitive label for qualifying the traumatic event so that it is no longer necessary to have directly experienced a traumatic event, being threatened is sufficient to meet criteria (American Psychiatric Association, 2013). Lending credit to Anna Freud's interpretation of trauma, that a definition for trauma should be person-specific (Figley et al., 2017), the DSM-5 incorporates a more subjective clinical evaluation as to whether the traumatic event qualifies as meeting criteria for a diagnosis of PTSD.

Recently, researchers have demonstrated that resiliency serves as a protective factor against developing PTSD across multiple contexts and cultures. In one study, researchers examined the development of mental health disorders in women who had been exposed to intimate partner violence and were seeking a temporary restraining order from a Family Justice Center (Jose & Novaco, 2016). These researchers found that the main mitigating factor for the women's perceived stress, mood symptoms, and anger reactions was their individual levels of resiliency. Those who self-reported higher levels of resiliency had decreased impacts to their short- and long-term mental health, including the development of PTSD (Jose & Novaco, 2016). Those who are exposed to traumatic events regularly such as firefighters, paramedics, and police officers would likely be considered most at risk for developing PTSD. However, studies with Korean police officers (Lee et al., 2016), firefighters (Lee et al., 2014), and paramedics from German-speaking Switzerland (Streb et al., 2014) showed that those with higher levels of resilience had significantly lower rates of PTSD symptoms than those with low levels of resilience.

When acute tragedy strikes individuals with high levels of resiliency are better able to ward off mental health disorders such as PTSD by effectively engaging their social support networks, believing in their ability to cope successfully, and perceiving adversity as a challenge rather than as a threat (Tugade & Fredrickson, 2004). Furthermore, resilient individuals are more capable of facing their fears and experiencing positive emotions, which allows them to actively cope when confronted with a stressful experience (Thompson et al., 2018). Those with lower levels of resiliency are more vulnerable to mental health disorders such as PTSD (Thompson et al., 2018). Researchers find that those with lower levels of resiliency tend to internalize the causes of their trauma and socially withdraw, resulting in greater vulnerability to developing PTSD (Thompson et al., 2018). Overall, research continues to indicate that resilience is an important factor for clinicians to consider when assessing for a patient's risk for developing PTSD (Wrenn et al., 2011).

As research has evolved to better understand how resilience serves as a mitigating factor in some mental health diagnoses, many scales and questionnaires have been developed in order to evaluate individual differences in levels of resiliency. One of these scales is the Connor-Davidson Resilience Scale (CD-RISC), a 25-item self-report scale that is said to measure resilience in adults with a psychiatric disorder (Campbell-Sills & Stein, 2007). The CD-RISC has strong psychometric properties as well as high internal consistency and test–retest reliability when tested on clinical samples and in the general population (Connor & Davidson, 2003). A score obtained using the CD-RISC was demonstrated to be a moderator of the relationship between a history of child abuse and adult psychiatric symptomatology (Campbell-Sills & Stein, 2007). Participants who characterized themselves as resilient on the scale reported lower levels of psychiatric symptoms despite a history of child abuse. In fact, their symptom manifestation was reported to be as low as those with little history of child abuse. Lastly, individuals receiving both cognitive behavioral therapy (CBT) and pharmacological treatments to increase resilience showed an increase in scores on the CD-RISC in the post-treatment assessment (Campbell-Sills & Stein, 2007).

Campbell-Sills and Stein (2007) sought to decrease the number of items on the CD-RISC while creating a version that adequately fit the data and minimized redundancy. The final result was a 10-item scale reflecting "the ability to tolerate experiences such as change, personal problems, illness, pressure, failure and painful feelings" (Campbell-Sills & Stein, 2007, p. 1026). The ability to successfully recover from these challenges (i.e. resiliency) is reflected by endorsement of these items on the scale (Campbell-Sills & Stein, 2007). Both the CD-RISC and the CD-RSIC-10 are widely used in the current studies investigating depression, anxiety, PTSD, and many other mental health issues (Min et al., 2013; Scali et al., 2012; Wingo et al., 2010).

Research conducted with a sample of almost 1000, low-income, highly traumatized African–American adults highlighted the contributing effects of childhood abuse and other traumatic incidents (i.e., traffic accidents, natural disasters, or the unexpected death of a loved one) on depressive symptoms (Wingo et al., 2010). The authors reported that traumas experienced across the lifespan contributed to depressive symptomatology while resilience moderated the severity of their depression across all trauma types (Wingo et al., 2010).

In a study focusing on resilience in adult women, Scali et al. (2012) found that higher scores on the CD-RISC-10 were associated negatively with the presentation of anxiety disorders but found no significant association with mood disorders. Moreover, high levels of resilience were found to be positively and independently associated with a previous history of trauma. Lastly, the investigators found no significant association between levels of resilience and past psychiatric diagnoses, which suggests that global resilience scores have the potential to be state dependent (Scali et al., 2012).

Resiliency was also studied to determine if it could act as a possible moderator of suicidal ideation for those with depression and anxiety (Min et al., 2015). Researchers examined childhood trauma history, levels of depression and anxiety, and current suicidal ideation in 436 patients diagnosed with depression or anxiety (Min et al., 2015.) Using a Korean translation of the CD-RISC, Min et al. (2015) found that a higher level of resilience acted as a protective factor against moderate to severe suicidal ideation for patients with higher levels of depression or anxiety symptoms.

In another study investigating characteristics of resilience in patients with anxiety and depression disorders, the authors found that those who endorsed spirituality, positive psychological factors, life-purpose, and physical exercise showed the increased levels of resilience, better coping capabilities, and overall better mental health prognosis (Min et al., 2013). The authors further concluded that not only is resilience a significant coping mechanism, but it is modifiable and can be enhanced with intervention (Min et al., 2013). These studies suggest that not only does resilience serve as a protective factor in a number of mental health disorders by moderating symptom severity (Wingo et al., 2010), aiding in coping and recovery (Min et al., 2013), and generally improving prognosis (Min et al., 2013; Scali et al, 2012), but that it may be possible to cultivate resilience in order to achieve better mental health outcomes. A meta-analysis was conducted that looked at findings related to the association between depression and resilience in a geriatric sample (Ávila, 2016). Studies conducted in the United States, China, and Belgium identified a positive association between greater resilience and less depressive symptomatology in this population (Ávila et al., 2016).

3.7 RESILIENCY IN MENTAL HEALTHCARE WORKERS

Much of the current resilience literature focuses on people commonly seen in and out of treatment such as immigrants, trauma-survivors, and correctional populations. Less research has examined how those who provide treatment such as first responders, crisis workers, nurses, and mental health clinicians fare. At present, the literature on healthcare professionals is primarily comprised of exploratory studies on training programs and nonprofit organizations (Edward et al., 2009; Edward & Hercelinskyj, 2007; McAllister & McKinnon, 2009; Warelow & Edward, 2007). Several studies reported that resilience of mental health professionals can be improved through professional development, peer support, consultation, supervision (Edward, 2005; Edward & Hercelinskyj, 2007), courses, internships and other educational experiences (McAllister & McKinnon, 2009). This is a worthy endeavor as an increase in resilience has the potential to improve clinical outcomes for clinicians (Edward & Hercelinskyj, 2007) and decrease psychological distress (Kinman & Grant, 2010). Research also points to a link between job satisfaction and resilience. For example, one study looking at the exposure of mental health nurses to violence in the line of duty found that life satisfaction was affected more by staff resilience than it was by exposure to verbal and physical violence (Itzhaki et al., 2015). A second study that focused on resilience and job satisfaction among psychiatric nurses working in inpatient units also found a significant positive correlation between resilience and job satisfaction. Moreover, nursing professional status, which refers to attitudes regarding recognition as well as significance, pride, and skill in the profession, was also positively correlated with resilience (Matos et al., 2010).

Edward (2005) used a qualitative research approach to further understand individual levels of resilience in crisis care clinicians. A number of participants expressed the importance of having a sense of self, preserving faith and hope, and championing insight (Edward, 2005). Warelow and Edward (2007) explored the concept of resilience as a practice that is attainable by anyone and achieved through role-modeling and education. In a study of psychiatric nurses, the practice of maintaining resilience helped them to find the positivity in a negative experience, build toward recovery, and take a strength-based approach (Warelow & Edward, 2007).

Several studies have looked at increasing resilience in healthcare workers already employed by organizations through the use of specific interventions (Brennan, 2017; Foureur, et al., 2013; Pipe et al., 2012; Ruotsalainen et al., 2015). Researchers have identified a link between mindfulness and resilience, which eventually led to the improved patient outcomes. Mindfulness-based stress reduction was found to decrease symptoms of stress, anxiety, and depression, which resulted in an increase in resilience scores for the nurse participants (Foureur et al., 2013). A study that looked at building resilience in the healthcare workplace implemented two educational programs, "Transforming Stress" and "5-H." Both programs aimed to teach doctors, nurses, and clinical managers to recognize symptoms of stress and use physiological interventions to decrease and manage the impact of stress. As a result of the implementation of these programs, healthcare workers reported becoming more resilient and effective in their roles (Pipe et al., 2012).

A study intended to increase preventative measures against occupational stress in healthcare workers looked at a variety of resilience-building interventions that were implemented across multiple studies (Ruotsalainen et al., 2015). Researchers found that a variety of interventions have been explored during the last decade, including CBT, massage, aroma-therapy, and a Chinese relaxation technique referred to as Qigong. CBT, both with and without relaxation techniques, was found to reduce stress, although the authors contended that more randomized control studies were necessary to definitively recommend the approach as an evidence-based treatment to enhance resilience (Ruotsalainen et al., 2015).

The overall theme across much of the research seems to be that resilience training and practice has positive implications for mental health and retention of crisis clinicians, psychiatric nurses, and other individuals who work in the helping profession (Warelow & Edward, 2007). An understanding of the concept, encouragement of the practice, and the development of professional programs for resiliency are key to enhancing the effectiveness of healthcare workers as a whole (Edward & Hercelinskyj, 2007). However, it is likely that large-scale, systemic change will be necessary to achieve global results in promoting and improving resilience (McAllister & McKinnon, 2009). Finally, researchers overwhelmingly agree that more research is needed in the field of health-care worker resilience to achieve positive results (Edward, 2005; Foureur et al., 2013, Itzhaki et al., 2015; Matos et al., 2010; Pipe et al., 2012).

3.8 COMMUNITY RESILIENCE

Traumatic events are perceived, and recovered from differently, depending on whether they are experienced by an individual or collectively as a group. For example, trauma can occur on an individual level such as a veteran coming home from a tour of duty or to an entire community after a hurricane. Disasters and a terrorist attack are also examples of traumatic events that are capable of affecting an entire community. When these types of traumas are experienced collectively, resiliency can also develop on a community level. As previously discussed in this chapter, individual characteristics such as personality, social support, and attachment style can all influence a person's ability to "bounce back" after a traumatic experience. Below are characteristics that foster community resilience, or a community's ability to "bounce back" after a traumatic event.

When hurricane Katrina struck Louisiana in August 2005, over 15 million people were impacted. Many of those affected lost their homes, parents, children, friends, and pets. Not only were these losses felt on an individual level, but on a community or macro-level as well. During this time, people were vulnerable to a wide array of mental disorders including PTSD, depression, and anxiety. As researchers and medical practitioners flocked to this area to investigate the psychosocial aftermath of this disaster, they noted that the New Orleans community along with the Association of Community Organizations for Reform Now (ACORN) had banded together to begin the healing process (Glandon et al., 2008). Residents in New Orleans created new social support networks by mobilizing 15,000 volunteers to help preserve their city. Residents provided lawn maintenance for their displaced neighbors to avoid city fines and launched a lead paint remediation program. New Orleans residents and ACORN organized a "human levee" along the Monticello Canal demanding fair flood protection for all city residents and helped to redevelop nearly 150 homes in low- and moderate-income neighborhoods (Glandon et al., 2008). In addition, the community and ACORN advocated for numerous legal and policy measures to protect homeowner rights and increase financial assistance for those who were most affected (Glandon et al., 2008). Researchers examining community resilience found that for people who participated in ACORN the most commonly mentioned sources of strength and support were being able to share resources, commiserate with family, and help others (Glandon et al., 2008).

Another study investigating resiliency looked at how participation in a virtual community such as Facebook affected individual resiliency levels post-hurricane Katrina (Clark, 2014). Virtual communities often mimic physical communities wherein they serve as a place to connect with others who share related experiences. Clark (2014) found that hurricane Katrina survivors who interacted in virtual communities, such as Facebook, frequently reported higher levels of perceived resiliency.

These results highlight the importance of community, whether physical or virtual, in helping foster resiliency. Resilient communities, such as New Orleans, organize their resources quickly and effectively as well as utilize new social support networks (Glandon et al., 2008). Having a community that supports you, or that you are able to support by helping others, helps to develop and maintain resiliency, whereas a lack of community support in the aftermath of trauma can contribute to the development of psychological and social problems.

On September 11, 2001, the United States experienced trauma on a national level. Following the terrorist attacks, the nation found healing through community resiliency. By adapting Walsh's (2007) four tasks in adaptation to loss, we can see how community resilience can be fostered. Their first principle, shared acknowledgment of the reality of the traumatic event and losses, involves clarifying the facts and any ambiguities of the traumatic event. "Project Liberty" was a large scale, government-funded effort in New York City that was created as an immediate response to the psychological and social needs of the community following September 11, 2001 (Henley et al., 2011). It offered psychosocial public education and crisis counseling programs to anyone who needed it, serving over 1.2 million people in a two-and-a-half year period (Henley et al., 2011). This provided people the opportunity to use counseling or be connected to community resources where they could better understand their traumatic event.

The second principle is to share the experience of loss and survivorship (Walsh, 2007). For the survivors of September 11, 2001, one of the ways this was done was through the establishment of The National September 11 Memorial Museum. This memorial serves is a place where anyone can go to mourn those who died that day (Paliewicz, 2017).

The third principle involves reorganization of the community and planning for the survivors' well-being (Walsh, 2007). Reorganization included the clearing of debris from the collapse of the buildings, rebuilding of destroyed homes and buildings, and the creation of a memorial where the Twin Towers once stood. Set within the footprints of the original Twin Towers, a bronze memorial with the names of every person who perished in the terror attacks was erected. Immediately following the attacks, Project Liberty aided in planning for the survivors' well-being by dispatching professionals and laypersons into the community to assess for mental health needs. Simple screeners were used to allow for quick assessments of those who may be showing signs of PTSD. In response to the large need for mental health services, training programs (largely funded through private donations and philanthropic foundations) were initiated that educated approximately 1500 licensed clinicians to treat those who were suffering from PTSD and complicated grief (Henley et al., 2011).

The final principle identified by Walsh (2007) for developing community resilience is the reinvestment in relationships and life pursuits. This includes finding new purpose from the tragic loss and constructing new hopes and dreams.

As many survivors of September 11, 2001, have done, assessing life plans and aspirations as well as finding new meaning in life helped with maintaining resiliency, which in turn contributed to community resiliency.

Recently, Hurricane Maria struck the island of Puerto Rico in September 2017. In the aftermath of Hurricane Maria, the people of Puerto Rico experienced electrical outages throughout the providence, impacting the use of hospitals and other emergency services. Access to clean water was also limited, making it difficult to avoid disease and heal from injuries that were sustained during the hurricane. There was also slow and minimal assistance from the United States government that adversely affected the recovery process.

Researchers who traveled to the island following the devastation of Hurricane Maria noted the remarkable resiliency of the Puerto Rican people during this time (Rodríguez, 2017). Music is often described as "a tool of distraction, of community, of history" for the Puerto Rican people and was used after the hurricane to bring people together to sing familiar songs (Rodríguez, 2017). Singing was used because it required no electrical power yet offered a way for the community to face adversity, unify survivors, and foster resilience. While waiting for supplies to reach the island and for electricity to be restored, the people of Puerto Rico rose up to help one another. Reports of neighbors helping each other clean debris and rebuild streets were common during this time period (Rodríguez, 2017). There were also many examples of altruism including large gatherings of food and water from Puerto Ricans living on the mainland in support of those affected on the island. In general, Puerto Ricans acted together, creating community resilience by way of optimism, hope, adaptation, and positive social networks in order to heal from their traumatic experience. Community resilience underlies the idea that if a community is able to stand together after tragedy, they will be able to heal as a whole.

3.9 POST-TRAUMATIC GROWTH

The traditional assumption has been that negative life events hinder positive adaptation and mental health outcomes (Fletcher & Sarkar, 2013). However, in the past decade, research has shown that individuals with some adversity in their past report better well-being and mental health outcomes than those with no adversity in their background (Neff & Broady, 2011; Seery, 2011; Seery et al., 2010).

Healing after a traumatic event or a major life crisis can sometimes result in positive changes or growth within an individual that exceeds their pre-event level of functioning. The PTG occurs when a traumatic event serves as a catalyst for positive changes within an individual that did not exist before the traumatic event. Although resiliency can be broadly summarized as recovery from stress, PTG includes a sense of thriving after trauma that may include an enhanced appreciation for life, deeper relationships, more compassion for others, increased levels of spirituality, and increased levels of self-reliance (Taku & McLarnon, 2018). The PTG is a somewhat-controversial construct that has only recently begun to be studied in the last two decades. Interest in PTG studies began rising when researchers noticed that after traumatic events, especially in community settings, people would report the increased levels of resilience and positive enhancements to their overall well-being.

Within the last decade, PTG has been studied after mass shootings that have taken place in the United States (Vieselmeyer et al., 2017).

After the Seattle Pacific University shooting, researchers reported that multiple students indicated having higher levels of appreciation for life (Vieselmeyer et al., 2017). This was also found to be true for students in the aftermath of the Virginia Tech school massacre (Day et al., 2017). Virginia Tech counselors who were present on the day of the shooting also reported PTG in the form of a heightened sense of professional and personal self-awareness (Day et al., 2017). In students exposed to a school shooting in Finland, researchers measured PTG along with the students' attachment style (Turunen et al., 2014). The results of their study showed that the students with secure attachments and insecure-preoccupied attachments demonstrated PTG following the school shooting. However, those with insecure-avoidant attachment showed very low levels of growth and the most intrusive and hyperarousal symptoms. These researchers hypothesized that attachment may play a crucial role not only in the development of resiliency (as we discussed at the start of this chapter) but also in the regulation of emotions and use of social support networks. This study highlights that while some people who have experienced trauma are able to convert PTSD to PTG, not all people experience growth as a result of traumatic events. How this process occurs within an individual or community still requires more research to better determine what genetic and environmental influences contribute to PTG.

3.10 CONCLUSION

A traumatic experience can occur at any time, to any individual, or within any community. Returning to the discussion of the trapped Chilean miners, it is clear that the miners were able to establish community resiliency while trapped underground for 69 days. Their underground community enabled the miners to support each other while also being supported by their aboveground community who worked frantically to free them. The miners persevered despite thoughts of suicide, medical injuries, and illnesses, and a severe shortage of food. Once rescued, the miner's individual levels of resiliency—influenced by each of their personalities, levels of social support, and attachment style—most likely influenced the mental health vulnerabilities each miner faced. As all but one of the miners were reportedly still displaying post-traumatic stress symptoms 5 years after being rescued, it is evident that mental health is particularly vulnerable after a traumatic event. Trauma is something that can tear a person or community apart at the seams. However, with the use of individual or community resiliency programs and initiatives, mental health vulnerabilities can be mitigated, and in some cases PTG may occur.

KEYWORDS

- **resilience**
- **mental health**
- **trauma**
- **stress**
- **assessment**
- **treatment**

REFERENCES

Abraham, R., Lien, L., & Hanssen, I. (2018). Coping, resilience and posttraumatic growth among Eritrean female refugees living in Norwegian asylum reception centres: A

qualitative study. *International Journal of Social Psychiatry, 64*(4), 359–366. doi:10.1177/0020764018765237
American Psychiatric Association. (1952). *Diagnostic and statistical manual of mental disorders*. Washington, DC: American Psychiatric Association.
American Psychiatric Association. (1994). *Diagnostic and statistical manual of mental disorders* (4th ed.). Washington, DC: American Psychiatric Association.
American Psychiatric Association (2013). *Diagnostic and statistical manual of mental disorders* (5th ed.). Arlington, VA: American Psychiatric Publishing.
Ávila, M. P. W., Lucchetti, A. L. G., & Lucchetti, G. (2016). Association between depression and resilience in older adults: A systematic review and meta-analysis. *International Journal of Geriatric Psychiatry, 32*(3), 237–246. https://doi.org/10.1002/gps.4619
Bonanno, G. A. (2004). Loss, trauma, and human resilience. *American Psychologist*, 59, 20–28.
Bowlby, J. (1980). Attachment and loss: Loss, sadness, and depression. Vol. 3. New York, NY: Basic Books.
Brennan, E. J. (2017). Towards resilience and wellbeing in nurses. *British Journal of Nursing, 26*(1), 43–47.
Campbell-Sills, L., & Stein, M. B. (2007). Psychometric analysis and refinement of the Connor–Davidson resilience scale (CD-RISC): Validation of a 10-item measure of resilience. *Journal of Traumatic Stress: Official Publication of the International Society for Traumatic Stress Studies, 20*(6), 1019–1028.
Clark, K. (2014). *Hurricane Katrina survivors' virtual community engagement level and its impact on resiliency: A comparison study. Dissertation Abstracts International: Section B: The Sciences and Engineering.* ProQuest Information & Learning. Retrieved from https://search-ebscohost-com.paloaltou.idm.oclc.org/login.aspx?direct=true&db=psyh&AN=2014-99140-268&site=ehost-live
Connor, K. M., & Davidson, J. R. T. (2003). Development of a new resilience scale: The Connor–Davidson Resilience Scale (CDRISC*). Depression and Anxiety*, 18, 76–82.
Costa, P. T., McCrae, R. R., & Dye, D. A. (1991). Facet scales for agreeableness and conscientiousness: A revision of the NEO personality inventory. *Personality and Individual Differences, 12*(9), 887–898.
Dalenberg, C. J., Straus, E., & Carlson, E. B. (2017). Defining trauma. In S. N. Gold, S. N. Gold (Eds.), *APA handbook of trauma psychology: Foundations in knowledge* (pp. 15–33). Washington, DC: American Psychological Association. doi:10.1037/0000019-002
Davydov, D. M., Stewart, R., Ritchie, K., & Chaudieu, I. (2010). Resilience and mental health. *Clinical Psychology Review, 30*(5), 479–495.
Day, K. W., Lawson, G., & Burge, P. (2017). Clinicians' experiences of shared trauma after the shootings at Virginia Tech. *Journal of Counseling & Development, 95*(3), 269–278. https://doi-org.paloaltou.idm.oclc.org/10.1002/jcad.12141
DiMauro, J., Carter, S., Folk, J. B., & Kashdan, T. B. (2014). A historical review of trauma-related diagnoses to reconsider the heterogeneity of PTSD. *Journal of Anxiety Disorders, 28*(8), 774–786.
Edward, K. (2005). The phenomenon of resilience in crisis care mental health clinicians. *International Journal of Mental Health Nursing, 14*(2), 142–148.
Edward, K. L., & Hercelinskyj, G. (2007). Burnout in the caring nurse: Learning resilient behaviors. *British Journal of Nursing, 16*(4), 240–242.
Edward, K. L., Welch, A., & Chater, K. (2009). The phenomenon of resilience as described by adults who have experienced mental illness. *Journal of Advanced Nursing, 65*(3), 587–595.
Erikson I, Cater A, Andershed A-K, Andershed H. (2011). What protects youths from externalizing and internalizing problems? A critical review of research findings and implications for practice. *Australian Journal of Guidance and Counselling (21)*, 113–25.
Figley, C. R., Ellis, A. E., Reuther, B. T., & Gold, S. N. (2017). The study of trauma: A historical overview. In S. N. Gold, S. N. Gold (Eds.), *APA handbook of trauma psychology: Foundations in knowledge* (pp. 1–11). Washington, DC, US: American Psychological Association.
Fletcher, D., & Sarkar, M. (2013). Psychological resilience: A review and critique of definitions, concepts, and theory. *European Psychologist, 18*(1), 12–23.
Foureur, M., Besley, K., Burton, G., Yu, N., & Crisp, J. (2013). Enhancing the resilience of nurses and midwives: Pilot of a mindfulness based program for increased health, sense of coherence and decreased depression, anxiety and stress. *Contemporary Nurse, 45*(1), 114–125.
Gagnon, A. J., & Stewart, D. E. (2014). Resilience in international migrant women following violence associated with pregnancy. *Archives of Women's Mental Health, 17*, 303–310.
Glandon, D. M., Muller, J., &Almedom, A. M. (2008). Resilience in post-Katrina New Orleans, Louisiana: A

preliminary study. *African Health Sciences*, *8*(Suppl 1), S21–S27.

Goldberg, L. R. (1993). The structure of personality traits: Vertical and horizontal aspects. In D. C. Funder, R. D. Parke, C. Tomlinson-Keasey, & K. Widaman (Eds.), *Studying lives through time: Personality and development* (pp. 169–188). Washington, DC: American Psychological Association.

Haber, M. G., Cohen, J. L., Lucas, T., & Baltes, B. B. (2007). The relationship between self-reported received and perceived social support: a meta-analytic review. *American Journal of Community Psychology*, *39*(1–2), 133–144. Retrieved from https://paloaltou.idm.oclc.org/login?url=http://search.ebscohost.com/login.aspx?direct=true&db=mnh&AN=17308966

Henley, R., Marshall, R., & Vetter, S. (2011). Integrating mental health services into humanitarian relief responses to social emergencies, disasters, and conflicts: A case study. *Journal of Behavioral Health Services & Research*, *38*(1), 132–141. https://doi-org.paloaltou.idm.oclc.org/10.1007/s11414-010-9214-y

Hjemdal, O., Friborg, O., Stiles, T. C., Rosenvinge, J. H., & Martinussen, M. (2006). Resilience predicting psychiatric symptoms: A prospective study of protective factors and their role in adjustment to stressful life events. *Clinical Psychology & Psychotherapy, 13*(3), 194–201. https://doi.org/10.1002/cpp.488

Holmes, J. (2017). Roots and routes to resilience and its role in psychotherapy: A selective, attachment-informed review. *Attachment & Human Development*, *19*(4), 364–381. doi:10.1080/14616734.2017.1306087

Hourani, L. L., Williams, J., Lattimore, P. K., Morgan, J. K., Hopkinson, S. G., Jenkins, L., & Cartwright, J. (2018). Workplace victimization risk and protective factors for suicidal behavior among active duty military personnel. *Journal of Affective Disorders*, *23*, 645–651. doi:10.1016/j.jad.2018.04.095

Hu, T., Zhang, D., & Wang, J. (2015). A meta-analysis of the trait resilience and mental health. *Personality and Individual Differences,* 76, 18–27.

Ingram, R. E., & Price, J. M. (2001). The role of vulnerability in understanding psychopathology. In R. E. Ingram, & J. M. Price (Eds.), *Vulnerability to psychopathology: Risk across the lifespan* (pp. 3–19). New York: The Guilford Press.

Itzhaki, M., Peles-Bortz, A., Kostistky, H., Barnoy, D., Filshtinsky, V., & Bluvstein, I. (2015). Exposure of mental health nurses to violence associated with job stress, life satisfaction, staff resilience, and post-traumatic growth. *International Journal of Mental Health Nursing*, *24*(5), 403–412.

Jose, R., & Novaco, R. W. (2016). Intimate partner violence victims seeking a temporary restraining order: Social support and resilience attenuating psychological distress. *Journal of Interpersonal Violence*, *31*(20), 3352–3376.

Kinman, G., & Grant, L. (2010). Exploring stress resilience in trainee social workers: The role of emotional and social competencies. *The British Journal of Social Work*, *41*(2), 261–275.

Kobiske, K. R., & Bekhet, A. K. (2018). Resilience in caregivers of partners with young onset dementia: A concept analysis. *Issues in Mental Health Nursing*, *39*(5), 411–419. doi:10.1080/01612840.2017.1400625

Lee, J., Ahn, Y., Jeong, K., Chae, J., & Choi, K. (2014). Resilience buffers the impact of traumatic events on the development of PTSD symptoms in firefighters. *Journal of Affective Disorders*, *162*, 128–133.

Lee, J., Choi, H., Kim, J., Nam, J., Kang, H., Koh, S., & Oh, S. (2016). Self-resilience as a protective factor against development of post-traumatic stress disorder symptoms in police officers. *Annals of Occupational and Environmental Medicine*, *28*, 58.

Linley, P. A., & Joseph, S. (2004). Positive psychology in practice. Hoboken, NJ, USA: Wiley.

Matos, P. S., Neushotz, L. A., Griffin, M. T. Q., & Fitzpatrick, J. J. (2010). An exploratory study of resilience and job satisfaction among psychiatric nurses working in inpatient units. *International Journal of Mental Health Nursing*, *19*(5), 307–312.

McAllister, M., & McKinnon, J. (2009). The importance of teaching and learning resilience in the health disciplines: A critical review of the literature. *Nurse Education Today*, *29*(4), 371–379.

Min, J.-A., Jung, Y.-E., Kim, D.-J., Yim, H.-W., Kim, J.-J., Kim, T.-S., Lee, C. U., Lee, C., Chae, J.-H. (2013). Characteristics associated with low resilience in patients with depression and/or anxiety disorders. *Quality of Life Research*, 22(2), 231–241. https://doi.org/10.1007/s11136-012-0153-3

Min, J.-A., Lee, C.-U., & Chae, J.-H. (2015). Resilience moderates the risk of depression and anxiety symptoms on suicidal ideation in patients with depression and/or anxiety disorders. *Comprehensive Psychiatry*, *56*, 103–111. https:// zdoi.org/10.1016/j.comppsych.2014.07.022

Nan, L., Woelfel, M. W., & Light, S. C. (1985). The buffering effect of social support subsequent to an important life event. *Journal of Health & Social Behavior*, *26*(3), 247–263.

Neff, L. A., & Broady, E. F. (2011). Stress resilience in early marriage: Can practice make perfect? *Journal of Personality and Social Psychology*, 101, 1050–1067. doi: 10.1037/ a0023809

Oshio, A., Taku, K., Hirano, M., & Saeed, G. (2018). Resilience and big five personality traits: A meta-analysis. *Personality and Individual Differences*, *127*, 54–60. doi:10.1016/j.paid.2018.01.048

Paliewicz, N. S. (2017). Bent but not broken: Remembering vulnerability and resiliency at the National September 11 Memorial Museum. *Southern Communication Journal*, *82*(1), 1–14. https://doi-org.paloaltou.idm.oclc.org/10.1080/1041794X.2016.1252422

Pearlin, L., Lieberman, M. A., Menaghan E. G., & Mullan J. T. (1981). "The stress process." *Journal of Health and Social Behavior*, *22*, 337–356.

Pipe, T. B., Buchda, V. L., Launder, S., Hudak, B., Hulvey, L., Karns, K. E., & Pendergast, D. (2012). Building personal and professional resources of resilience and agility in the healthcare. *Stress and Health*, 28(1), 11–22.

Rodríguez, G. V. M. (2017). Puerto Rico se levanta: Resilience in times of disaster. *Brown University Child & Adolescent Behavior Letter*, *33*(12), 1–5. https://doi-org.paloaltou.idm.oclc.org/10.1002/cbl.30256

Ruotsalainen, J. H., Verbeek, J. H., Mariné, A., & Serra, C. (2015). Preventing occupational stress in healthcare workers. *Cochrane Database Syst Rev*, *4*, 4–152.

Scali, J., Gandubert, C., Ritchie, K., Soulier, M., Ancelin, M.-L., & Chaudieu, I. (2012). Measuring resilience in adult women using the 10-Items Connor-Davidson Resilience Scale (CD-RISC). Role of trauma exposure and anxiety disorders. *PLoS One*, *7*(6), e39879. https://doi.org/10.1371/journal.pone.0039879

Seery, M. D. (2011). Resilience: A silver lining to experiencing adverse life events. *Current Directions in Psychological Science*, *20*, 390–394.

Seery, M. D., Holman, A. E., & Silver, R. S. (2010). Whatever does not kill us: Cumulative lifetime adversity, vulnerability and resilience. *Journal of Personality and Social Psychology*, *99*, 1025–1041. doi: 10.1037/a0021344

Streb, M., Häller, P., & Michael, T. (2014). PTSD in paramedics: Resilience and sense of coherence. *Behavioral and Cognitive Psychotherapy*, *42*(4), 452–463. doi:10.1017/S1352465813000337

Taku, K., & McLarnon, M. J. W. (2018). Posttraumatic growth profiles and their relationships with HEXACO personality traits. *Personality & Individual Differences*, *134*, 33–42. https://doi-org.paloaltou.idm.oclc.org/10.1016/j.paid.2018.05.038

Theron, L., & van Rensburg, A. (2018). Resilience over time: Learning from school-attending adolescents living in conditions of structural inequality. *Journal of Adolescence*, *67*, 167–178. doi:10.1016/j.adolescence.2018.06.012

Thompson, N. J., Fiorillo, D., Rothbaum, B. O., Ressler, K. J., & Michopoulos, V. (2018). Coping strategies as mediators in relation to resilience and posttraumatic stress disorder. *Journal of Affective Disorders*, *225*, 153–159. doi:10.1016/j.jad.2017.08.049

Tugade, M. M., & Fredrickson, B. L. (2004). Resilient individuals use positive emotions to bounce back from negative emotional experiences. *Journal of Personality and Social Psychology*, *86*(2), 320–333. Retrieved from http://dx.doi.org/10.1037/0022-3514.86.2.320

Turunen, T., Haravuori, H., Punamäki, R., Suomalainen, L., & Marttunen, M. (2014). The role of attachment in recovery after a school-shooting trauma. *European Journal of Psychotraumatology*, *5*(1), 22728.

Van Voorhees, E. E., Wagner, H. R., Beckham, J. C., Bradford, D. W., Neal, L. C., Penk, W. E., & Elbogen, E. B. (2018). Effects of social support and resilient coping on violent behavior in military veterans. *Psychological Services*, *15*(2), 181–190. doi:10.1037/ser0000187

Vieselmeyer, J., Holguin, J., & Mezulis, A. (2017). The role of resilience and gratitude in posttraumatic stress and growth following a campus shooting. *Psychological Trauma: Theory, Research, Practice, and Policy*, *9*(1), 62–69. https://doi-org.paloaltou.idm.oclc.org/10.1037/tra0000149

Wagnild, G. M., Young, H. M. (1993). Development and psychometric evaluation of the resilience scale. *Journal of Nursing Measurement 1*, 165–178.

Waldegrave, C., King, P., Maniapoto, M., Tamasese, T. K., Parsons, T. L., & Sullivan, G. (2016). Relational resilience in Māori, Pacific, and European sole parent families: From theory and research to social policy. *Family Process*, *55*(4), 673–688. doi:10.1111/famp.12219

Walsh, F. (2007). Traumatic loss and major disasters: Strengthening family and community resilience. *Family Process*, *46*(2), 207–227. https://doi-org.paloaltou.idm.oclc.org/10.1111/j.1545-5300.2007.00205.x

Warelow, P., & Edward, K. L. (2007). Caring as a resilient practice in mental health nursing. *International Journal of Mental Health Nursing*, *16*(2), 132–135.

Wingo, A. P., Wrenn, G., Pelletier, T., Gutman, A. R., Bradley, B., & Ressler, K. J. (2010). Moderating effects of resilience on depression in individuals with a history of childhood abuse or trauma exposure. *Journal of Affective Disorders*, *126*(3), 411–414. https://doi.org/10.1016/j.jad.2010.04.009

Wrenn, G. L., Wingo, A. P., Moore, R., Pelletier, T., Gutman, A. R., Bradley, B., & Ressler, K. (2011). The effect of resilience on posttraumatic stress disorder in trauma-exposed inner-city primary care patients. *Journal of the National Medical Association*, *103*(7), 560–566. doi:10.1016/S0027-9684(15)30381-3.

CHAPTER 4

Cherishing and Psychological Well-Being: Some Empirical Observations

PRATISHTHA BHATTACHARYYA[1*] and RABINDRA KUMAR PRADHAN[2]

[1]*Xavier School of Human Resource Management, Xavier University Bhubaneswar, Odisha, India*

[2]*Department of Humanities and Social Sciences, Indian Institute of Technology, Kharagpur, West Bengal, India*

[*]*Corresponding author. E-mail: pratishthabhattacharyya@gmail.com*

ABSTRACT

Cherishing refers to a secure attachment behavior manifesting a strong sense of affect, protectiveness, and care toward certain targets or objects. Most of the literatures on cherishing are found to be in the domain of objects of cherishing. However, cherishing as a psycho-social construct has never been investigated thoroughly by any psychologist or any social scientist. A recent study reveals cherishing as a psycho-social construct performs various functions and correlate significantly with psychological well-being. Several other researches also reports that secure attachment style positively associates with several individual outcomes such as coping with stress, social and emotional adjustment and psychological well-being. Based on the review of contemporary literature, the present study explores the association between cherishing and psychological well-being on 105 undergraduate students from a large university in eastern India. The findings of the study show significant positive association between cherishing and the six subcomponents of psychological well-being. This study offers an initial empirical understanding about the possibility of the association between cherishing and psychological well-being. The findings of the study offer several possibilities and opportunities for psychological interventions for nurturing cherishing to strengthen individual capacities for enhancing psychological wellbeing.

4.1 INTRODUCTION

Cherishing is described in terms of a strong sense of effect, protectiveness, and care toward certain targets or objects labeled as "objects of cherishing" (Bhattacharyya & Pradhan, 2019a, 2019b). Despite the frequent use of the word "cherishing" across several behavioral science platforms, cherishing as a concept has hardly received any scientific validation so far. Cherishing as a concept is only frugally touched upon in Broudy's work "enlightened cherishing" (Martin, 1994) and Lee's (1994) work on environmental ethics. These former works explain resonate cherishing as a strong desire mixed with love to protect the object one loves from harm's way. However, cherishing received more empirical clarification with a recent study conducted in the Indian scenario (Bhattacharyya & Pradhan, 2019b; Pradhan & Bhattacharyya, 2019).

4.2 THE PRELIMINARY STUDY

The initial study elucidates the concept of cherishing in terms of a secure attachment behavior manifesting a robust sense of care, protectiveness, and affection toward the object of cherishing (Bhattacharyya, 2018). The study also identified four specific types of objects of cherishing, that is, altruistic cherishing, personal cherishing, cherishing family, and cherishing friends. Cherishing family and cherishing friends refer to the behavioral display of protectiveness, care, and love toward our close circle of family and friends. Personal cherishing conveys a behavioral display of protectiveness, care, love toward our own quality, capacities, strengths, achievements, and other concepts related to our self. In contrast, altruistic cherishing refers to a behavioral display of protectiveness, care, and love for action and intentions to offer unconditional help to others.

The nature of cherishing as a secure attachment behavior is very much visible through the attributes of cherishing displaying a sense of attachment for the objects of cherishing. Moreover, the objects of cherishing reflect secure attachment which is apparent from the behavioral display of love, care, and protectiveness toward close relationships such as family friends, toward the intention of unconditional help for others, and for ourselves.

4.2.1 CHERISHING AND ATTACHMENT THEORY

Attachment theories provide an elaborate and comprehensible platform to explore and understand the basic reasons for differences among adult individuals (Karreman & Vingerhoets, 2012). Interestingly, attachment is also indicated to be of supreme importance when it comes to life parameters such as quality of life and individual well-being (Karreman & Vingerhoets, 2012). Attachment theories centers around attachment security which elaborates about that the sense of assurance individuals possess about whether others would be responsive to their needs (Homan, 2018). This sense of security is what constitutes the base for psychological health (Homan, 2018). Theoretically, attachment security can be both adaptive and maladaptive (Stevenson et al., 2019). The development of a secure attachment can be traced back to an individual's childhood. A responsive and safe environment that allows the child to explore and interact with the environment on their own terms, which is also responsive to the needs of a child, fosters a secure attachment among individuals (Mikulincer & Shaver, 2004). Adaptive or secure attachment is indicative of higher well-being; on the other hand, maladaptive styles such as anxious and avoidant attachment have been found to be associated with diminishing levels of well-being (Stevenson et al., 2019).

4.2.2 CHERISHING AND WELL-BEING

The inner working model in a secure attachment ensures that an individual feels loved, valued, and accepted by their attachment figures (Mikulincer & Shaver, 2005). Consequently, this leads to more positive effect and emotional balance for an individual during stress (Mikulincer & Shaver, 2007). In contrast, individuals with insecure attachment styles develop a negative perception about their relationships and perceive less support from partners, close friends, and romantic partners (Florian et al., 1995), as a consequence they experience more negative affect and diminished life satisfaction (Wei et al., 2011). For instance, a recent study on avoidant attachment style reported it is a significant contributor diminished subjective well-being (SWB) among older adults (Li & Fung, 2014). The concept of well-being is a widely discussed topic in positive psychology. However, the topic of well-being research endorses two overlapping philosophies that are hedonism and eudaimonism (Ryan & Deci, 2001).

4.2.2.1 SUBJECTIVE WELL-BEING

The field of hedonism is concerned more about subjective happiness and experiences of pleasure (Kubovy, 1999). In more measurable terms, hedonism is best explained by Diener et al.'s (1999) conceptualization of SWB, which comprises life satisfaction, the presence of positive effect, and the absence of negative effect (Diener et al., 1999).

Empirical investigations corroborate that secure attachment leads to SWB and healthy social adjustment (Ma & Huebner, 2008). A significant pathway by which attachment security derives SWB is through the process of successful identity formation (Kroger & Marcia, 1980). The feeling of being loved, valued, and accepted by others, which accompanies a secure attachment promotes the development of positive self-regard and consequently SWB (Mikulincer & Shaver, 2007). Therefore, congruent to the contemporary conceptualization of cherishing, the present research considers examining the association between cherishing and SWB by taking into perspective three significant factors that form the integral part of SWB: happiness, life satisfaction, and perceived stress.

4.2.2.2 PSYCHOLOGICAL WELL-BEING (PWB)

However, eudaimonism conveys that well-being comprises of realizing one's true nature or potentials (Waterman, 1993). More objectively eudaimonism is well explained and elaborated with the concept of PWB., educate oneself, pursue a vocation, "leisure interests," and to be able to carry on daily decisions about one's vocation, accommodation, education, and other related factors (WHO, 2013, p. 7). The seminal work of Ryff and Keyes (1995) discusses about six aspects of fulfilling one's true potentials that are self-acceptance, personal growth, purpose in life, positive relations with others, environmental mastery, and autonomy. Several contemporary studies corroborate that secure attachment styles share a significant association with PWB across ages. For instance, one recent study reported that older adults above 70 years of age with secure attachment styles were more self-accepting and this eventually facilitated higher PWB in their lives (Homan, 2018). Therefore, based on the consideration of cherishing as a secure attachment behavior, it can be assumed that cherishing will share a bond with PWB.

Previous studies report that a secure attachment style has a positive association with PWB (Karreman & Vingerhoets, 2012). Therefore, on the basis of the available literature, this study examines whether cherishing as a secure attachment behavior positively relates to PWB. Along with this, the study also verifies whether the four-factor structure of cherishing shows a good fit with the present sample.

4.3 METHOD

4.3.1 PARTICIPANTS

The present study consists of 105 students (male = 69, female = 36) from a subject pool of a social science subject taught at a large university in eastern India. The age of the participants ranged from 18 to 23 years (*M* = 19.93, *SD* = 1.52).

4.3.2 MEASURES

4.3.2.1 CHERISHING

Cherishing was measured by the 14 items self-report measure (Appendix A). Participants were instructed to indicate their agreement on statements based on a 7-point scale (where 1 = almost never true to 7 = almost always true). The internal consistencies for family, friends, altruistic, and personal cherishing sub-scales were 0.80, 0.77, 0.72, and 0.70, respectively. The Cronbach's alpha of the cherishing scale was 0.88.

4.3.2.2 SOCIAL DESIRABILITY

The short form of the Marlowe–Crowne Social Desirability (Strahan & Gerbasi, 1972) scale was used to measure social desirability.

4.3.2.3 PSYCHOLOGICAL WELL-BEING

PWB was measured with the help of a short version of the Ryff (1989) PWB scale. The short versions of the PWB measu similar subscales as the original scale. The PWB scale is comprised of six subscales (three items per scale) measuring autonomy, self-acceptance, purpose in life, personal growth, positive relationship with others, and environmental mastery. The entire scale is measured on a 6-point Likert-type scale (where 1 = strongly disagree to 6 = strongly agree). Moreover, the six subscales balance the right mix of positively and negatively worded items. Previous studies have extensively used these scales across several contexts (Clarke et al., 2001). In the present study, the internal consistency for the three-item sub-scales ranged from 0.53 to 0.70. Although it has not been used in the Indian scenario, a recent validation work by Linley et al., (2009) confirms that the 18-item PWB scale reported to retain the original structure even when used with Indian samples.

4.4 RESULTS

A confirmatory factor analysis was performed to examine whether the four-factor structure of cherishing revealed a good model fit with the present sample. The four factors of cherishing were labeled as ALTR (altruistic cherishing), PERS (personal cherishing), FAM (cherishing family), and FRI (cherishing friends). All the items of the cherishing scale loaded high on their respective factors (i.e., greater than 0.50). The findings reported a good model fit with the present sample: S–B$\chi^2(df)$ = 67.38(73), RMSEA = 0.077, CFI = 1.00, IFI = 1.00, SRMR = 0.060 (Figure 4.1).

Correlation of the cherishing sub-scales (i.e., personal r = 0.09, p>0.05; altruistic r = 0.05, p>0.05; family r = –0.03, p>0.05; friend r = 0.04, p>0.05) revealed a nonsignificant relationship with social desirability.

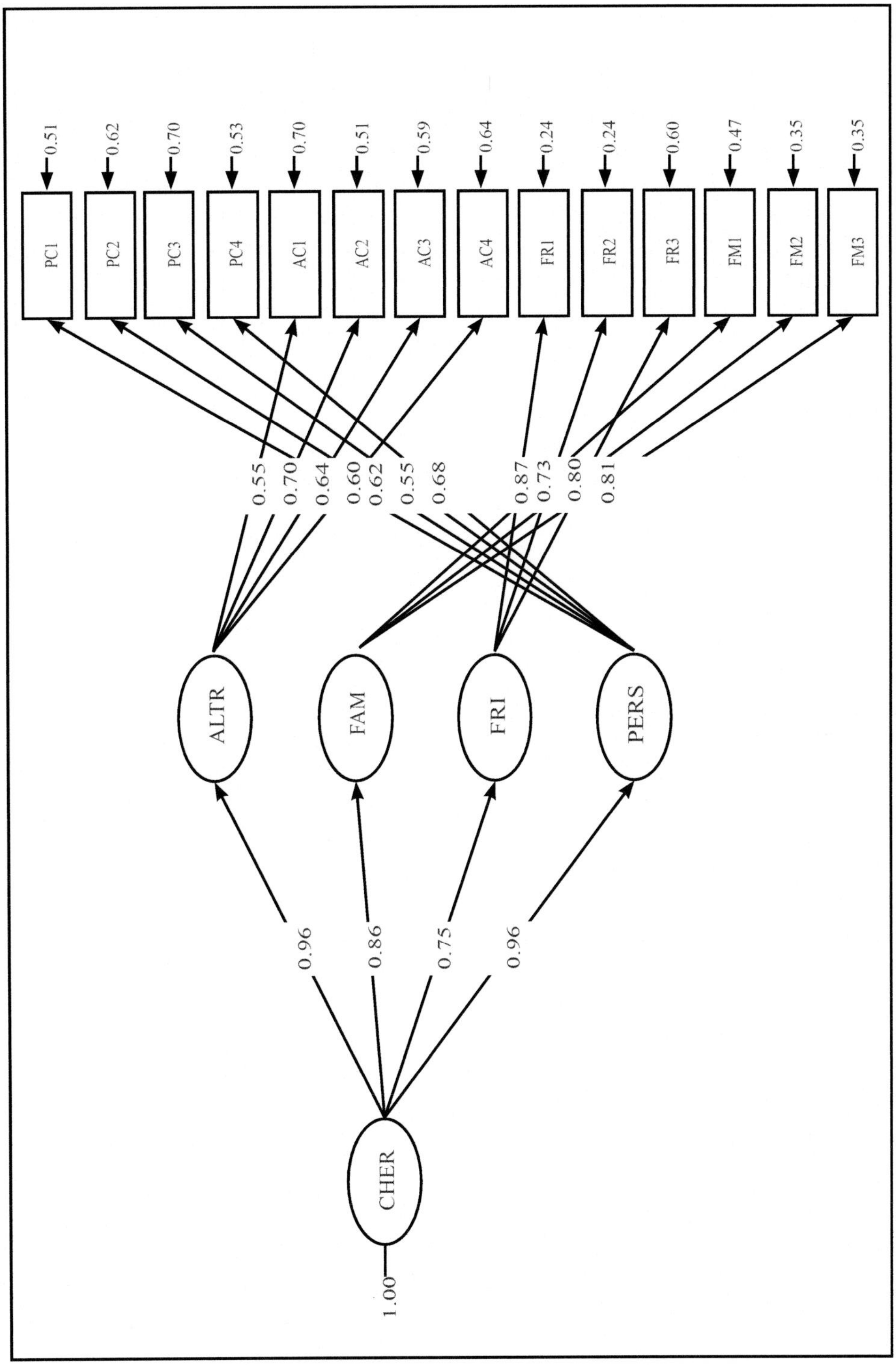

FIGURE 4.1 CFA of the cherishing structure.

4.4.1 CHERISHING AND PWB

A zero-order correlation analysis was performed with PWB and cherishing. This study demonstrated significant positive correlations between each of the cherishing subscales and the PWB subscales (Table 4.1). Based on the results of the correlation analysis, the preset study confirms the hypothesis that cherishing is related to PWB. Although PWB revealed association with cherishing, the correlation coefficients between them were not so high and significant to show any overlap between the two constructs making them redundant, thus, reporting the distinctness of cherishing from PWB.

4.5 DISCUSSION

The purpose of the present study was to explore the relationship between cherishing as a psycho-social construct and its association with PWB. This is an initial attempt at exploring the possibility of an association between the two constructs, and a first attempt at establishing an association between cherishing and PWB. The results of the study establish that cherishing as a psycho-social construct has positive association with PWB. This is congruent to the assumption of cherishing as a secure attachment behavior, as it is corroborated across several popular literature that secure attachment forms strong bond with PWB (Homan, 2018).

The association between cherishing and PWB is indicative of several beneficial implications in the sphere of individual mental health and psychological adjustment. Moreover, when considered from the perspective of secure attachment as a malleable construct, it renders the possibility of fostering cherishing among individuals to enhance emotional and psychological development. Although previous investigations consider attachment styles as a fixed and stable construct, contemporary literature provides several strong corroboration regarding the malleability of attachment styles through investigation on "attachment primes" (Canterberry & Gillath, 2012). In similar lines, cherishing as a secure attachment behavior holds the promise of being malleable and pivotal in fostering positive emotional and social development among individuals.

4.6 CONCLUSION

The present study offers an initial understanding of how cherishing as a psycho-social construct relates to a significant positive psychological construct such as PWB. However, the present study only offers a nuanced understanding of the association between cherishing and PWB as the study was only conducted on a small sample of undergraduate students from urban backgrounds. The study could have benefited immensely if it availed the opportunity to involve perspectives from diverse socioeconomic factors in India. Moreover, the study could have enriched with the inclusion of perspectives from various cross-cultural samples across the world.

KEYWORDS

- **cherishing**
- **psychological well-Being**
- **secure attachment**
- **India**

TABLE 4.1 Result of Zero-Order Correlation.

Variables	M	SD	1	2	3	4	5	6	7	8	9	10
1. AUTO	4.43	0.82	1	0.27^{**}	0.28^{**}	0.26^{**}	0.26^{**}	0.41^{**}	0.27^{**}	0.24^{*}	0.26^{**}	0.23^{**}
2. ENVT	4.30	0.75	–	1	0.30^{**}	0.31^{**}	0.24^{*}	0.20^{**}	0.23^{**}	0.25^{**}	0.23^{*}	0.20^{*}
3. PR_GR	4.79	0.90	–		1	0.23^{**}	0.30^{**}	0.28^{**}	0.23^{*}	0.19^{*}	0.22^{*}	0.24^{*}
4. P_REL	4.75	0.80	–	–	–	1	0.26^{**}	0.27^{**}	0.32^{**}	0.30^{**}	0.26^{**}	0.27^{**}
5. PURP	4.90	0.79	–	–	–	–	1	0.21^{*}	0.25^{**}	0.20^{*}	0.22^{*}	0.20^{*}
6. SLF_ACP	4.50	0.80	–	–	–	–	–	1	0.28^{**}	0.23^{*}	0.20^{*}	0.27^{**}
7. PERS	5.67	0.83	–	–	–	–	–	–	1	0.68^{***}	0.56^{***}	0.55^{***}
8. ALTR	5.57	0.79	–	–	–	–	–	–	–	1	0.52^{***}	0.50^{***}
9. FAM	6.08	0.94	–	–	–	–	–	–	–	–	1	0.59^{***}
10. FRI	5.97	0.88	–	–	–	–	–	–	–	–	–	1

PERS = personal cherishing, ALTR = altruistic cherishing, FRI = cherishing friends, FAM = cherishing family, AUTO = autonomy, ENVT = environmental mastery, PR_GR = personal growth, P_REL = positive relationships with others, PURP = purpose, and SLF_ACP = self acceptance.

REFERENCES

Bhattacharyya, P. (2018). *Cherishing: Exploring the nature, structure, and relationship*. (Unpublished doctoral dissertation), Indian Institute of Technology, Kharagpur, West Bengal, India.

Bhattacharyya, P., & Pradhan, R. K. (2019a). *Exploring the factor structure of cherishing as a psycho-social construct.* Paper Presented in the 54th National & 23rd International Conference of Indian Academy of Applied Psychology (IAAP) Conference, February 14–16, 2019.

Bhattacharyya, P., & Pradhan, R. K. (2019b). Exploring cherishing: A qualitative approach. *The Qualitative Report, 24*(7), accepted on 25th March 2019 for publication.

Canterberry, M & Gillath, O. (2012). Attachment and caregiving: Functions, interactions, and implications. In P. Noller & G. C. Karantzas (Eds.) *The Wiley-Blackwell handbook of couples and family relationships* (1st ed). UK: Blackwell Publishing Ltd.

Clarke, P. J., Marshall, V. M., Ryff, C. D., & Wheaton, B. (2001). Measuring psychological well-being in the Canadian study of health and aging. *International Psychogeriatrics, 13*(1), 79–90.

Diener, E., Suh, E. M., Lucas, R. E., & Smith, H. L. (1999). Subjective well-being: Three decades of progress. *Psychological Bulletin, 125*(2), 276–302.

Florian, V., Mikulincer, M., & Bucholtz, I. (1995). Effects of adult attachment style on the perception and search for social support. *Journal of Psychology, 129*(6), 665– 676.

Homan, K. J. (2018). Secure attachment and eudaimonic well-being in late adulthood: The mediating role of self-compassion. *Aging and Mental Health, 22*(3), 363–370.

Kubovy, M. (1999). On the pleasures of the mind. In D. Kahneman, E. Diener, and N. Schwarz (Eds.), *Well-being: The foundations of hedonic psychology* (pp. 134–154). New York: Russell Sage Foundation.

Lee, K. (1994). Awe and humility: Intrinsic value in nature. Beyond an earthbound environmental ethics. *Royal Institute of Philosophy Supplements,* 36, 89–102.

Li, T., & Fung, H. H. (2014). How avoidant attachment influences subjective well-being: An investigation about the age and gender differences. *Aging and Mental Health, 18*(1), 4–10.

Kroger, J., & Marcia, J. E. (1980). The identity statuses: Origins, meanings, andinterpretations. In S. J. Schwartz., K. Luyckx, and V. L. Vignoles (eds.), *Handbook of identity theory and research* (pp. 31–53, Vol. 1). New York, USA: Springer.

Linley, P. A., Maltby, J., Wood, A. M., Osborne, G., & Hurling, R. (2009). Measuring happiness: The higher order factor structure of subjective and psychological well-being measures. *Personality and Individual Differences, 47*(8), 878–884.

Ma, C. Q., & Huebner, E. S. (2008). Attachment relationships and adolescents' life satisfaction: Some relationships matter more to girls than boys. *Psychology in the Schools, 45*(2), 177–190.

Martin, M. W. (1994). *Virtuous giving: Philanthropy, voluntary service, and caring. Bloomington,* Indianapolis: Indiana University Press.

Mikulincer, M., Dolev, T., & Shaver, P. R. (2004). Attachment-related strategies during thought suppression: ironic rebounds and vulnerable self-representations. *Journal of Personality and Social Psychology, 87*(6), 940–956.

Mikulincer, M., & Shaver, P. R. (2005). Attachment security, compassion, and altruism. *Current Directions in Psychological Science, 14*(1), 34–38.

Mikulincer, M., & Shaver, P. R. (2007). *Attachment in adulthood: Structure, dynamics, and change.* New York: Guilford Press.

Pradhan, R. K. & Bhattacharyya, P. (2019). *Cherishing as a mediator between emotional stability and subjective happiness.* Paper presented in the 54th National & 23rd International Conference of Indian Academy of Applied Psychology (IAAP) conference, February, 14–16, 2019.

Ryan, R., & Deci, E. (2001). On happiness and human potentials: A review of research on hedonic and eudaimonic well-being. *Annual Review of Psychology, 52*(1), 141–166.

Ryff, C. D. (1989). Happiness is everything, or is it? Explorations on the meaning of psychological well-being. *Journal of Personality and Social Psychology, 57*, 1069–1081.

Ryff, C. D., & Keyes, C. L. M. (1995). The structure of psychological well-being revisited. *Journal of Personality and Social Psychology, 69*(4), 719–727.

, J. C., Millings, A., & Emerson, L. M. (2019). Psychological well-being and coping: The predictive value of adult attachment, dispositional mindfulness, and emotion regulation. *Mindfulness, 10*(2), 256–271.

Strahan, R., & Gerbasi, K. C. (1972). Short, homogenous versions of the Marlowe–Crowne social desirability scale. *Journal of Clinical Psychology, 28*(2), 191–193.

Waterman, A. S. (1993). Two conceptions of happiness: Contrasts of personal expressiveness (eudaimonia) and hedonic enjoyment. *Journal of Personality and Social Psychology, 64*(4), 678–691.

Wei, M., Liao, K. Y., Ku, T., & Shaffer, P. A. (2011). Attachment, self-compassion, empathy, and subjective well-being among college students and community adults. *Journal of Personality*, *79*(1), 191–221.
WHO (2013). *Investing in mental health: Evidence for action* (NLM classification: WM 30). Geneva, Switzerland.

APPENDIX A

Cherishing Item Pool

What is This Questionnaire About?

We may have experienced or still experience the feeling to take care and protect from harm or damage those that we love (such as any person(s), ourselves, event(s), object(s), action(s), and intention(s)]. On the next page, there are a collection of statements that asks you to think about yourself, any person(s), event(s), object(s), action(s), and intention(s) that you love dearly and feel like taking care and protecting them from harm or damage.

Please indicate the extent to which you agree or disagree with each statement by using a circling a number between 7 "almost always true" and 1 "almost never true" (almost never true = 1, usually not true = 2, rarely true = 3, occasionally true = 4, often true = 5, very often true = 6, almost always true = 7). There are no rights or wrong answers, and your own personal views are important.

Cherishing family

1. I treasure the memory of my loved ones.
2. I treasure time spent with my family.
3. I reciprocate the love and affection of my loved ones.

Cherishing friends

4. It is difficult for me to disconnect from my friends.
5. I value my connections with my friends.
6. I value the moments I share with my friends.

Altruistic cherishing

7. I feel happy to nurture others.
8. Caring for others gives my life a meaning.
9. I collaborate in activities for a better world.
10. I love to work for people who are less privileged than us.

Personal Cherishing

11. I appreciate my personal strengths.
12. I feel I am worthy of being loved by others.
13. I value myself as a person.
14. I see the best in me even in hardship.

Source: Developed by Authors.

CHAPTER 5

Social Support and Emotion Regulation as Predictors of Well-Being

MEERA PADHY[1*] and PADIRI RUTH ANGIEL[2]

[1]*Centre for Health Psychology, University of Hyderabad, Central University, Hyderabad, Telangana*

[2]*Department of Applied Psychology, School of Behavioral Sciences, Central University of Tamil Nadu, Thiruvarur, Tamil Nadu*

[*]*Corresponding author. E-mail: meerapadhy@yahoo.com*

ABSTRACT

The study examined facets of social support (tangible, appraisal, and belonging) and two dimensions of emotion regulation strategy (cognitive reappraisal and expressive suppression) as predictors of well-being among 375 participants in India. Although there is overlap across these phenomena, previous research has not examined them together among Indian population. The participants completed emotion regulation questionnaire, interpersonal support evaluation list, and brief intervention of thriving scale. The results of the present study showed that only belongingness support and cognitive reappraisal were found to be significant predictors of well-being. This indicates that belongingness support and cognitive reappraisal play significant role in an individual's well-being. The study has notably highlighted the importance of emotional regulation along with social support in the well-being of an individual.

5.1 INTRODUCTION

5.1.1 WELL-BEING

Over the years, researchers have shifted their focus of study from negative aspects of life to the positive, devoting their attention to positive psychology (Seligman & Csikszentmihalyi, 2000). Thus, in the present century, well-being research has emerged as an important area in the field of psychology. "Well-being is the state of successful performance throughout the life course integrating physical, cognitive, and social-emotional functions that result in productive activities deemed significant by one's cultural community, fulfilling social relationships, and the ability to transcend moderate psychosocial and environmental problems" (Pollard & Davidson, 2001). Well-being has a subjective dimension, the sense of satisfaction associated with fulfilling one's potential (Bornstein, et al., 2003). It is a complex construct that concerns optimal experience and functioning

(Ryan & Deci, 2001) and the realization of one's physical, mental, social, and spiritual potential (Awartani et al., 2008). According to Kesebir and Diener (2008, p. 118), the components of subjective well-being "include life satisfaction (global judgments of one's life), satisfaction with important life domains (satisfaction with one's work, health, marriage, etc.), positive affect (prevalence of positive emotions and moods), and low levels of negative affect (prevalence of unpleasant emotions and moods)."

The eudemonic theory of Aristotle states that to live a good life, a person must possess certain virtues. The model of psychological well-being (Ryff, 1989, 1995; Ryff & Singer, 1996) embodies the eudemonic concept by substituting virtues with six dimensions of positive mental health that a person must possess to experience the highest levels of well-being. Ryff named these dimensions self-acceptance, positive relations with others, autonomy, environmental mastery, purpose in life, and personal growth. "Well-being is a positive physical, social, and mental state; it is not just the absence of pain, discomfort and incapacity. It arises not only from the action of individuals, but from a host of collective goods and relationships with other people. It requires that basic needs of the individuals are met, that they have a sense of purpose, and feel capable to achieve important personal goals and participate in society. It is enhanced by conditions that include supportive personal relationships, involvement in empowered communities, good health, financial security, rewarding employment, and a healthy and attractive environment" (Steuer & Marks, 2008). Seligman (2011) states that, "if we want to flourish and if we want to have well-being, we must indeed minimize our misery; but in addition, we must have positive emotion, meaning, accomplishment, and positive relationships" (p. 57).

5.1.2 SOCIAL SUPPORT AND WELL-BEING

Social support is a vital source of sustenance that protects our health and nourishes our well-being. A thorough review conducted by Steger (2009) on research work done on the meaning of life shows that "most people have indicated that relationships with others are the most important source of meaning in their lives" (p. 683). Social support refers to the interpersonal processes that help in maintaining and promoting health and well-being of the individual. The source of social support involves individual's social network, composed of intimate friends and family members. The role of one's social network is to provide a sense of belonging, assurance of being loved, and cared for (Cohen & Wills, 1985). Caplan (1974) defines support systems as "social interactions with another individual, a network, a group, or an organization that provide individuals with opportunities for feedback about themselves and validation from others."

The construct of social support is multi-dimensional and is categorized into several subtypes. The commonly used framework proposed by House (1981) delineates four distinct subtypes—tangible support, informational or appraisal support, emotional support, and belongingness support. Tangible support refers to providing tangible or concrete assistance in terms of financial aid, material resources, or needed/instrumental services (Sanderson, 2013). Informational or appraisal support refers to the process of receiving information, advice and guidance from others. This includes providing information about the root cause of the problem, providing suggestions about one's decision or information about organizations that provide help. It helps the individual in understanding or appraising the problem with a different perspective and find out which resources and coping strategies they need to use to deal with it (Friedman & Silver, 2007). Emotional support refers to the expression of concern, care, warmth, and empathy and providing love, reassurance, comfort, nurturance, and encouragement to the person in a stressful state (Friedman & Silver, 2007). Individuals facing stress need someone to provide a listening ear, as ventilating their feelings helps them calm down and feel accepted. Belongingness support provides social companionship or belongingness where the individuals enjoy having the company of others to engage in social activities such as going out for a dinner/party, movie/concert, or camping/sports activities. This contributes to positive mood of the person (Maestas et al., 2007).

Review of research done on the impact of social support emphasizes that everyone requires a social network to satisfy the human need of being cared for, accepted, and emotionally supported, particularly in times of stress. Physical and psychological comfort provided by friends and relatives to the people who are facing stress plays a significant role. Efforts were made by researchers to explore the effects of social support on psychological well-being. Social support was categorized under two broad categories—aid-related and appreciation-related social support. Aid-related support was more related to potential for receiving help from others when needed, and being cared for by others. Appreciation-related support was more related to being recognized by others as an efficient source of help and reassurance of worth (Gencoz & Ozlale, 2004). Support enables individuals to enjoy and cherish life and helps them cope with strenuous and stressful situations acting as a buffer against traumatic life events (Diener & Seligman, 2004). Individuals, who receive psychological and material support from family and friends when in need, are found to be happier and in better health than those who are socially less connected (Lyuobomirsky et al., 2005).

5.1.3 EMOTION REGULATION AND WELL-BEING

Emotion regulation (ER), which refers to the ability to identify, experience, modulates, and express emotions, is critical to healthy psychological functioning. Emotion regulation is defined by Gross (1998) as regulation of affective states which includes overt to covert, explicit to implicit, and voluntary to automatic dimensions. Emotion regulation requires one's capacity to tactically adjust each aspect of emotions in agreement with situational demands (Bonanno et al., 2004), immediately as well as in anticipation or retrospectively (Gross, 2001; Magai et al., 2006). It refers to "the process by which individuals infer which emotions they have, when they have them and how they experience and express these emotions" (Gross, 1998, p. 275). It involves both purposeful expression of emotions as well as control and inhibition of emotions. Emotion regulation strategies can be conceptualized along a continuum from effortful, consciously executing and altering one's emotional response (Gyurak et al., 2011), to automatic or implicit, unconsciously managing emotional responses (Gyurak et al., 2011; Mauss et al., 2007). The effortful or deliberate emotion regulation strategies are the antecedent and response-focused strategies which refer to cognitive reappraisal and suppression, respectively. Cognitive reappraisal refers to changing the way one thinks about a potential emotion-eliciting event where as suppression refers changing the way one responds behaviorally to an emotion-eliciting event. In emotional regulation strategy of suppression, there is a deliberate inhibition of emotional expression when emotion is aroused (Gross, 1998). Though inhibition of emotion expression strategy is sometimes useful not to disrupt social interactions, yet research studies have shown the decreased positive emotions (Gross & Levenson, 1997), poor interpersonal functioning (Butler et al., 2003), lowered well-being (Gross & John, 2003), and increased rumination regarding negative mood (Gross & John, 2003) associated with this strategy. Experimental studies indicate that cognitive reappraisal is more effective at decreasing emotional experience and behavioral expression (Gross, 2002), associated with healthier profiles of affective, cognitive, and social consequences (Gross & John, 2003) and enhanced well-being (Haga et al., 2009). In contrast, emotional suppression, which refers to attempts to control or conceal the experience and expression of emotion, has been associated with poor memory for experiences that involve suppression (Richards & Gross, 2000) and less positive interpersonal interactions (Butler et al., 2002). A significant body of experimental research studying the use of suppression in response to stress-provoking stimuli has linked this strategy to a host of negative outcomes, including restricted positive affect, increased negative affect, poorer social adjustment, increased physiological arousal, and decreased well-being (Dalgleish et al., 2009; Campbell-Sills et al., 2006; Gross & John, 2003; Roberts et al., 2008; Tull et al., 2010). In comparison with cognitive reappraisal, suppression has been associated with depression and anxiety (Geraerts et al., 2006).

The effect of emotion regulation on well-being depends on several factors such as physiological consequences of the expression of emotions, repercussion of emotional experience and emotional expression of one's self-image, the nature of the relationship with the person one expresses to, and one's overt reaction. Individual difference is noted in the use of these emotion regulation strategies and these individual differences influence one's emotion, relationships with others, and well-being (Gross & John, 2003).

Emotion regulation is considered as a set of skills that are acquired throughout the lifespan (Gross & Levenson, 1997; Labouvie-Vief et al., 2003). Research findings suggest that elders are more conscious and reflective of emotions as well as themselves. They have considerable knowledge about emotions, the relations between emotions and situational exposures, and the influence of emotions on others (Schieman, 1999). It is supported that self-reflective and knowledge-based capacities relevant to emotion regulation may increase with age, though recognition ability of emotion seems to decline (MacPherson et al., 2006). Fewer adaptive resources are generally seen among older adults (Panksepp & Miller, 1995). Therefore, it seems likely that such emotional regulation may become more difficult in later life.

5.1.4 SOCIAL SUPPORT AND EMOTION REGULATION

Social interactions generate emotions affecting the internal and interpersonal experiences of the individuals. These interactions generate numerous emotional responses either positive or negative such as anger at being treated unfairly, shame at being exposed, fear at being threatened, or sadness when rejected. Thus, regulating emotions and emotionally driven behaviors are essential for successfully creating healthy relationships. Individuals turn to others for help in managing their distressing emotions (Zaki & Williams, 2013) and reducing stress (Lazarus & Folkman, 1984; Uchino et al., 1996). Interpersonal situations foster emotion regulation and people respond to others' requests with supportive and empathic behaviors (Batson, 2011). The results of different studies showed that simply holding the hand of an intimate person helps regulate incoming aversive affects (Coan et al., 2006; Coan, 2011). This is also confirmed by the observations that when people experience stressful events they usually seek the help of others (Taylor et al., 2000). According to Zaki and Williams (2013), sharing of emotions verbally with others helps in regulating emotions regardless of the other person's regulatory goal. Disclosing and labeling emotions, therefore, help a person to emotionally regulate at a neural and psychological level (Lieberman et al., 2011; Kircanski et al., 2012).

5.1.5 THE CURRENT STUDY

As summarized above, extensive research studies on social support and well-being have been done, but their possible relationships with emotion regulation have not received much focus, especially among Indian population. Regulation of emotions is essential for adaptive functioning whereas dysfunctional emotional regulation has adverse consequences, leading to poor well-being. In the past, research on emotions has focused mainly on intraindividual processes. However, several emotions are highly social in nature. Various emotions and emotional processes are elicited as well as experienced in different social contexts. Though previous studies have highlighted the association between social support and well-being as well as between emotion regulation and well-being yet few studies have investigated the causal pathways produced by these associations. Understanding the triadic reciprocity among these three psychological constructs is important in the evaluation of strategies for promoting well-being among individuals. Taking all the above points into account, the objectives of this study were to examine the relationship between facets of social support and dimensions of emotional regulation strategy and well-being, and to investigate the effects of social support and emotion regulation on well-being. The following hypotheses were formulated based on the objectives of this study.

Hypothesis 1: Social support, emotion regulation strategies, and well-being would be related to each other.

Hypothesis 2: Social support and emotion regulation strategies would predict well-being.

5.2 METHOD

5.2.1 STUDY SAMPLE AND PROCEDURE

The study consisted of 375 subjects (174 male and 201 female), with an age between 18 and 60 years ($M = 35.05$, $SD = 12.82$). In terms of occupation, 53% were employed, whereas 47% were students. All the participants agreed to participate in the study voluntarily. After agreeing to participate in this study, the objectives of the study were explained to the participants. The Emotion Regulation Questionnaire (ERQ), Interpersonal Support Evaluation List (ISEL), and Brief Intervention of Thriving (BIT) were administered individually and in groups as well. Appropriate statistics were used to analyze the data gathered from the participants.

5.2.2 MEASURES

5.2.2.1 INTERPERSONAL SUPPORT EVALUATION LIST

The ISEL-15 is derived from the long form of the ISEL and contains 15 items which assess the perceived availability of tangible, appraisal, and belonging social support on a two-point scale of "True" and "False" (Cohen et al., 1985). Some of the items measuring each dimension of social support are "If I need help moving, I would be able to find someone to help me" (tangible support), "There is at least one person I know whose advice I really trust" (appraisal support), and "When I feel lonely, there are several people I could talk to" (belonging). The internal reliability of the total general population ISEL as given by the developers ranges from 0.88 to 0.90 and for subscales it ranges from 0.70 to 0.82. For the present sample in this study, Cronbach's alpha value was 0.72, and for the dimensions of tangible support, appraisal support and belonging support, it was found to be 0.45, 0.60, and 0.59, respectively.

5.2.2.2 EMOTION REGULATION QUESTIONNAIRE

The 10-item Emotion Regulation Questionnaire was administered to measure an individual's ability in utilizing suppression and reappraisal strategies for emotion regulation (Gross & John, 2003). The cognitive reappraisal subscale consisted of six items, Cronbach's alpha value was 0.79, and a sample item is "I control my emotions by changing the way I think about the situation I'm in". In this study, 0.64 was Cronbach's alpha value for the reappraisal subscale. The suppression subscale has four items, Cronbach's alpha value was 0.73, and a sample item is "I keep my emotions to myself." For the present study, Cronbach's alpha value for this subscale was 0.59. Each item on the scale has a 7-point Likert scale, ranging from "1" (strongly disagree) to "7" (strongly agree).

5.2.2.3 BRIEF INVENTORY OF THRIVING

The BIT was used to measure psychological well-being of the participants (Su et al., 2014). The BIT has 10 items, measuring 6 of the 7 core psychological well-being dimensions (positive relationships, engagement, mastery, meaning, optimism, and subjective well-being). An example of the items in the scale—"My life is going well." The responses for each item were in the form of a 5-point Likert scale ranging from 1 (strongly disagree) to 5 (strongly agree). The tool has high internal consistency with reliability coefficients above 0.90, and test–retest reliability of 0.83. In the current study, Cronbach's alpha for this subscale was 0.85.

5.2.3 RESULTS

The obtained quantitative data were analyzed using descriptive statistics, Pearson's product–moment correlation coefficient (Pearson's r), and multiple hierarchical regression analysis using IBM SPSS Statistics 20. In the analysis, the predictors were demographic variables (age and gender), dimensions of social support (tangible, appraisal, and belonging), and emotion regulation strategies (cognitive reappraisal and expressive suppression). The criterion was well-being. Prior to the hierarchical regression, Pearson's r was computed to find a linear relationship between the predictors and criterions, and to identify the suitable predictors to be entered into the model.

5.2.4 RELATIONSHIP AMONG THE MEASURES

The values of Pearson's "*r*" along with descriptive statistics (*M* and *SD*) are presented in Table 5.1. From Table 5.1, it was found that all predictors except age and gender had significant positive correlations with well-being. Age was found to be significantly positively related to appraisal support and expressive suppression strategy and negatively related to social support as a whole. Gender was negatively related to expressive suppression strategy. This indicates that women tend to use the expressive suppression strategy lesser compared to men. All the dimensions of social support were positively related to cognitive reappraisal emotion regulation strategy and well-being. On the other hand, all the dimensions of social support except belonging support were negatively related to expressive suppression. Both cognitive reappraisal and expressive suppression strategies were positively related to well-being. The significant correlation coefficient between predictors and criterion varied between 0.13 and 0.48. Therefore, dimensions of social support and dimensions of emotion regulation were selected to be entered into the hierarchical model as predictors.

5.2.5 PREDICTORS OF WELL-BEING

Based on the findings of the above analysis, hierarchical regression analysis was performed. As the linearity assumption was tested, other essential assumptions such as normality, homoscedasticity, and the absence of multicollinearity were also verified. The results of the model are presented in Table 5.2.

TABLE 5.1 Correlation Between Demographic Variables, Dimensions of Social Support, Dimensions of Emotion Regulation, and Well-Being

	1	2	3	4	5	6	7	8	9	M	SD
Demographic											
1. Age	1	0.19**	−0.06	0.20*	−0.06	−0.14**	−0.03	0.11*	0.02	35.08	12.88
2. Gender		1	0.007	0.08	0.002	0.04	0.04	−0.10*	0.02		
Social support											
3. Tangible			1	0.40**	0.38**	0.76**	0.21**	−0.12*	0.25**	3.51	1.26
4. Appraisal				1	0.36**	0.78**	0.21**	−0.14**	0.23**	3.51	1.39
5. Belonging					1	0.76**	0.32**	0.01	0.48**	3.63	1.33
6. Total						1	0.32**	−0.11*	0.41**	10.65	3.05
Emotion Regulation											
7. Cognitive reappraisal							1	0.18**	0.42**	28.15	5.74
8. Expressive suppression								1	0.13*	16.43	4.63
9. Well-being									1	37.04	7.08

$^{**}p < 0.01$; $^{*}p < 0.05$. Male = 1, female =2.

TABLE 5.2 Summary Table of Hierarchical Regression Analysis for Dimensions of Social Support and Dimensions of Erq Predicting Well-Being ($N = 375$)

Model	B	SE	B	T	Adjusted R^2	R^2	ΔR^2
Model 1 ($C = 26.53$, $F = 37.52^{***}$)					0.23	0.23	
Tangible	0.36	0.29	0.06	1.23			
Appraisal	0.23	0.26	0.05	0.90			
Belonging	2.32	0.27	0.44	8.60^{***}			
Model 2 ($C = 16.92$, $F = 34.07^{***}$)					0.31	0.32	0.09^{***}
Tangible	0.28	0.28	0.05	1.02			
Appraisal	0.16	0.25	0.03	0.65			
Belonging	1.9	0.26	0.36	7.21^{***}			
Cognitive reappraisal	0.34	0.06	0.28	5.90^{***}			
Expressive suppression	0.13	0.07	0.08	1.83			

C = constant, B = unstandardized beta coefficient, SE = standard error, β = standardized beta coefficient, t = t values of beta, $\Delta R^2 = R^2$ change.

$^{*}p < 0.05$, $^{**}p < 0.01$, $^{***}p < 0.001$

As seen in Table 5.2, the significantly correlated predictors were entered hierarchically in two blocks—Block 1 (dimensions of social support), Block 2 (dimensions of emotion regulation) —in respect of well-being. Model 1, consisting of dimensions of social support, was found to be significant, $F(3, 371) = 37.52$, $p < 0.001$, which explained 23% significant proportion of variance (*Adjusted* $R^2 = 0.23$) of well-being. From the analysis, only belonging social support was found to be significant predictor for well-being ($\beta = 0.44$, $p < 0.001$) in the model. In addition to dimensions of social support, when cognitive reappraisal and expressive suppression emotion regulation dimensions were entered, Model 2 was found to be significant, $F(5, 369) = 34.07$, $p < 0.001$, and the model explained 9% more significant proportion of variance ($\Delta R^2 = 0.09$, $p < 0.001$) amounting to total 32% significant proportion of variance of well-being (*Adjusted* $R^2 = 0.31$). The results revealed that in Model 2, cognitive reappraisal was found to be a significant predictor ($\beta = 0.28$, $p < 0.001$) for well-being in addition to belonging support.

5.3 DISCUSSION

The first objective of this study was to investigate the relationship between facets of social support, dimensions of emotional regulation strategy, and well-being. Based on this, the first hypothesis stated that social support, emotion regulation strategies, and well-being would be related to each other. Pearson's product moment correlation was calculated, and the hypothesis was accepted as the results showed that there exists a relationship among these variables. The demographic variables of age and gender were also included in the analysis. Age was found to have a significant positive relationship with appraisal support and expressive suppression which implies that as age increases, appraisal support and expressive suppression increases. The results indicate that as age increases, individuals receive greater appraisal support. In the context of a stressful event, an older individual is better at judging and appraising a situation when compared to a younger individual. On the contrary, when older individuals are unable to appraise the problem appropriately, they seek advice and guidance from others. Therefore, older individuals receive greater appraisal support in understanding a problem from various perspectives and are able to cope well with the problem. As people advance in age, though their social networks decrease, the satisfaction with their social support increases which has a buffering effect on their mental health (Luong et al., 2011). This helps them to cope better and appraise events as less threatening (Cohen, 2004). In a study conducted by Kafetsios and Sideridis (2006), middle-aged adults (35–66 years) when compared to younger adults (18–34 years) found to have more satisfaction with social support which increased their well-being.

The demographic variable of age was also found to have a significant positive relationship with the emotional regulation strategy of expressive suppression. This indicates that as age increases, older adults try to control or cover up the experience or expression of emotion. Research suggests that using suppression strategy in the face of stress-provoking situation may lead to reduced positive affect and increased negative affect, poor social adjustment, and decreased well-being (Dalgleish et al., 2009). However, if we analyze this result which states that as age increases, the use of expressive suppression also increases, in the light of the previously discussed result stating that as age increases appraisal support increases, we can justify that as older adults are able to appraise a situation differently because of the support they have, they engage in suppression of their emotions rather than expressing them. In a study conducted by Charles and Piazza (2007), when compared to younger adults, older adults recalled more positive emotions and fewer negative emotions with their close social partners. Older adults report fewer problematic relationships (Fingerman et al., 2004) and interpersonal stressors than younger adults (Almeida & Horn, 2004). This viewpoint is supported by the socio-emotional selectivity theory, which states that as individuals' age, they become aware of the limitations in life and focus more on maintaining close relationships and reducing negative emotions (Carstensen & Charles, 1998). Even in the case of conflicts, older adults tend to view their partner's behavior favorably and report lesser negative emotions such as anger regarding the conflict (Blanchard-Fields & Coats, 2008). Therefore, older adults use more of expressive suppression strategy to successfully overlook and circumvent interpersonal problems (Blanchard-Fields, 2007).

On the contrary, the results in the current study also showed a significant negative relationship between age and social support as a whole. This indicates that as age increases, overall social support decreases. Studies show that the effectiveness of social support received from family and friends varies across the life span. Older adults tend to minimize their social networks and maintain only few rewarding social relationships (Lang, 2004). According to a study conducted by Segrin (2003), older adults are not as dependent as younger adults on receiving social support from different sources in order to maintain their sense of well-being.

Further analysis of the relationship among the variables shows that there exists a significant negative relationship between gender and expressive suppression strategy. This indicates that women tend to use the expressive suppression strategy lesser compared to men. Expressive suppression is a behavioral strategy directed toward inhibiting emotional responses, which may include facial expressions, gestures, or engaging in verbal dialogue (Gross, 2002). Women by nature are expressive and vocal about their feelings and emotions. It is a popular belief that women are more emotional than men especially in behaviorally expressing their emotions (Hess et al, 2000). These findings are supported by another study conducted by Williams and Barry (2003), which concluded that women experience emotions more intensely and express them more often and exercising control or suppression led to more psychological distress and physical symptoms. Research shows that men showed greater ability to suppress when compared to women (Gross & John, 2003). In a neuroscience research conducted by McRae et al., (2008), it was found that men regulate their emotions with greater efficiency than women. Men tend to use the emotion regulation strategy of cognitive reappraisal with ease when compared to women. This leads to minimal experience of negative affect and the need to express it.

The correlation analysis also revealed that there exists a significant positive relationship between social support and well-being, implying that as social support increases, well-being of an individual will also increase. This finding is consistent with previous research work. As mentioned earlier in this paper, social support is vital for an individual's health and well-being. It provides nurturance both psychologically and socially through healthy and meaningful relationships. The four aspects of social support namely tangible support, appraisal support, emotional support and belongingness support cater to all needs and aspects of an individual, that is, the physical, cognitive, emotional, and social needs. Having social support is beneficial both in the presence and absence of stress. According to the direct effect hypothesis, social support has a significant impact on the health and well-being of an individual regardless of the amount of stress they face (Cassel, 1976). Also, according to the buffering hypothesis, social support acts as a protective shield against the impact of high stress (Cohen & Wills, 1985). Studies have shown that one of the most significant contributors to well-being is social support (Diener, & Seligman, 2002). Individuals who have fulfilling relationships report being happy and satisfied when compared to those who do not have fulfilling relationships. A meta-analysis on variables related to subjective well-being showed that those who had social contact quality had better life satisfaction, self-esteem, and happiness, when compared to those who had social contact quantity (Pinquart & Sorensen, 2000).

The emotional regulation strategies of cognitive reappraisal and expressive suppression both were found to have a significant positive relationship with well-being. Emotions are the important aspect of an individual in addition to cognition and behavior. These three aspects are interrelated and interact with each other. Any change positive or negative in one aspect influences the other two aspects. Emotions color our every part of life, down to every minute. Emotions arise out of both cognitions and behaviors, and also from social interactions and events in our life. Every individual experiences both good and bad relationships and faces positive and negative events. Regulating emotions is vital in maintaining equilibrium and well-being. Emotional regulation helps us to be flexible and adjust with the situational demands by modulating emotions and expressing them purposefully. The strategy of cognitive reappraisal helps an individual to appraise the situation as less threatening and in a positive light which helps them to deal with the situation in a more effective manner. Reappraisal was found to decrease both the experience and behavioral expression of the negative emotion (Gross & Levenson, 1997). Using the reappraisal strategy leads to coping better with stressful event and more positive mood and enhanced psychological well-being (Gross & John, 2003). The use of expressive suppression leads to more negative effect and unresolved issues. However, there are some studies which show that using expressive suppression may be helpful. As mentioned earlier, even in the case of conflicts, older adults may view their partner's behavior favorably and report lesser negative emotions such as anger regarding the conflict (Blanchard-Fields & Coats, 2008). Therefore, older adults use more of expressive suppression strategies to successfully overlook and circumvent interpersonal problems (Blanchard-Fields, 2007) and experience greater satisfaction and well-being.

The second objective of this study was to assess the role of social support and emotional regulation in the well-being of individuals. Likewise, it was hypothesized that social support and emotional regulation strategies would predict well-being. The results show that among the three dimensions of social support, namely tangible support, appraisal support, and belongingness support, only belongingness support was found to be a significant predictor in well-being. Also, among the two strategies of emotional regulation namely cognitive reappraisal and expressive suppression, only cognitive reappraisal was found to be a significant predictor of well-being. This indicates that belongingness support and cognitive reappraisal play a significant role in an individual's well-being.

Belongingness support provides social companionship where individuals enjoy the company of others and engage in social activities such as shopping, watching a movie, attending a concert or a show or sports. Engaging in activities helps individuals spend quality time together and thereby gives them more access to other kinds of social support, that is, tangible and emotional support. There is an exchange of mutual support and reciprocity, which is a major contributor to positive mood (Maestas et al., 2007). Social cohesion, social participation, and belongingness support were found to enhance well-being among the elderly (Sampson, 1988). A study was conducted on older adults to examine if social cohesion and belonging would predict well-being among older adults (Cramm & Nieboer, 2015). The researchers gathered data after controlling for baseline well-being level and other demographic variables. Results showed that the belongingness support that they received from the social groups predicted well-being among them. The study also indicated that since they belong to these groups, they have a greater level of instrumental support from the community which further enhances their well-being. Another study was conducted on women suffering from ovarian cancer to measure the benefits of using online social networking sites and their role in enhancing their psychological well-being (Erfani et al., 2016). These women were able to exchange information and received support on these networking sites. The results showed that the women felt a sense of belongingness and received social support which enhanced their psychological well-being.

Finally, the results also showed that using cognitive reappraisal predicted well-being among individuals. Cognitive reappraisal is one of the ways to cope and regulate one's emotions while facing a stressful situation. It is to change one's thinking and reappraise an emotion-eliciting situation, event or perception of a person in a manner that it reduces its emotional impact (Lazarus & Alfret, 1964). Since it is an antecedent-focused strategy, it intervenes and modifies one's tendency to emotionally react to something or someone. Extensive research conducted by Gross and John (2003) indicates that reappraisal reduces the experience of negative affect and the impact of adversity. Those who engage in cognitive reappraisal tend to experience greater positive mood and response. As a result, individuals are able to adapt better to changing situations. Reappraising acts as an antidote for depressive symptoms and enhances optimism, greater life satisfaction, high self-esteem, and high self-efficacy. Gross and his colleague's research measured Ryff's (1989) domains of well-being among individuals who frequently used the strategy of cognitive reappraisal. They found that those who used the reappraisal strategy had high levels of well-being, that is, environmental mastery, personal growth, self-acceptance, and clearer purpose in life. They took charge of their environment, had enhanced sense of autonomy, positive relationship with others, and healthy social functioning. Additionally, those individuals who used suppression had low well-being and more depressive symptoms. These findings were ascertained by the study conducted on university students to find out the relationship between emotional regulation strategies and well-being (Haga et al., 2009). Results showed that cognitive reappraisal predicted well-being.

5.4 CONCLUSION

The purpose of this study was to examine the relationship between facets of social support, dimensions of emotional regulation strategy, and well-being, as well as to investigate the effects of social support and emotion regulation strategies on well-being. Based on the results of this study, we can draw the following conclusions. First, a significant relationship exists between social support and well-being, social support and emotion regulation strategies, as well as emotional regulation strategies and well-being in this sample. Second, social support and emotion regulation together explain a significant proportion of variance in well-being (32%). Social support and emotion regulation independently contribute to well-being. In this analysis, belonging support and cognitive reappraisal emerge as the most significant predictors of well-being among individuals.

5.5 IMPLICATIONS

The findings of the study provide us an insight about the benefit of having social support and being able to regulate one's emotions in a healthy way and how these factors have an impact on well-being. The findings suggest that appraisal support and expressive suppression increase with age. This finding has a noteworthy implication for daily living and for life at large. Young or old, all individuals can be given psycho-education to inculcate the practice of seeking appraisal support when faced with any kind of problem. In contemporary times, where people are extremely busy with their work and routines, very few people have the inclination of providing support to those in need. Therefore, individuals facing stress cannot wait for others to come and support them but have to actively seek support in order to face challenging situations. Under stress, individuals get entangled in their own perspective of the situation. Seeking support from others (family, friends, or mentors) helps them to appraise the situation in a different manner and gain more information about the challenges. This in turn helps in not being impulsive at making a decision or taking a particular course of action or an unhealthy emotional reaction. Individuals will be able to engage in emotional suppression and tackle the situation or conflict in an objective manner. Though many studies have shown that expressive suppression is harmful, it depends on the context of using this emotion regulation strategy. Younger generation must cultivate the habit of seeking for appraisal support. One can receive appraisal support not just from dear ones, but can also receive it from counselors or psychologists. Currently, there is increasing awareness and lesser stigma among the general public about the benefits of seeking professional help from counselors or psychologists. Therefore, receiving appraisal not only helps in gaining better understanding about a challenge but helps in not reacting to the situation without giving it a careful thought.

In this study, expressive suppression has been found to have a positive relationship with well-being but women seem to be using it lesser than men. Women can be empowered to regulate their emotions and to use cognitive reappraisal strategy, seek for appraisal support when needed and respond to the situation in a healthy manner.

The findings also suggest that as age increases social support decreases, this has been justified by stating as people age they maintain few but meaningful relationships. With the ever increasing usage of social media and having a large network of friends, people lack meaningful and intimate relationships. Workshops or awareness sessions can be organized in institutions and workplaces, where individuals can be exposed to the scientific information regarding the benefits of maintaining close relationships.

The study has highlighted the importance of social relationships, social support, and emotional regulation strategies and the positive impact it has on one's well-being. Specifically belongingness support and cognitive reappraisal strategy have shown to significantly predict well-being. These findings can be incorporated in intervention modules pertaining to life-skills training programs or social skills development. The study also provides an important insight that along with cognitive reappraisal strategy, expressive suppression is also important in some contexts. This can be used and considered as an important variable while studying conflict management. The use of belongingness support and cognitive reappraisal strategy can be highlighted by counselors and psychologists in clinical settings.

The study has notably highlighted the importance of emotional regulation along with social support in the well-being of an individual.

5.6 LIMITATIONS

Data were collected from only one city in India; collecting it from different major cities of the country could have made it more representative of the culture. The sample size could have been larger and well-defined. This would make it possible to perform statistical analyses to assess the impact of the predictor variables on the criterion for a particular group, rather than assess the impact of these variables on the total sample. This would clearly enable one to see the major indictors that would predict well-being for each group.

The study used a quantitative approach; it could have used a mixed method designto obtain qualitative data which would provide better insight and in-depth information.

KEYWORDS

- **social support**
- **emotion regulation**
- **cognitive reappraisal**
- **expressive suppression**
- **well-being**

REFERENCES

Almeida, D. M., & Horn, M. C. (2004). Is daily life more stressful during middle adulthood? In O. G. Brim & C. D. Ryff (Eds.), *How healthy are we? A national study of well-being at midlife* (pp. 425–451). Chicago: University of Chicago Press.

Awartani, M., Whitman, C., & Gordon, J. (2008). Developing instruments to capture young people's perceptions of how school as a learning environment affects their well-being. *European Journal of Education,* 43(1), 51–70.

Batson, C. D. (2011). *Altruism in humans.* Oxford, UK: Oxford University Press.

Blanchard-Fields, F. (2007). Everyday problem solving and emotion: An adult developmental perspective. *Current Directions in Psychological Science, 16*(1), 26–31.

Blanchard-Fields, F. & Coats, A. H. (2008). The experience of anger and sadness in everyday problems impacts age differences in emotion regulation. *Developmental Psychology, 44*(6), 1547–1556.

Bonanno, G. A., Papa, A., Lalande, K., Westphal, M., & Coifman, K. (2004). The importance of being flexible: The ability to both enhance and suppress emotional expression predicts long term adjustment. *Psychological Science, 15*(7), 482–487.

Bornstein, M. H., Davidson, L., Keyes, C. L. M., & Moore, K. A. (2003). *Well-being positive development across the life course.* Mahwah: Lawrence, Erlbaum Associates.

Butler, E. A., Egloff, B., Wlhelm, F. H., Smith, N. C., Erickson, E. A., & Gross, J. J. (2003). The social consequences of expressive suppression. *Emotion, 3*(1), 48.

Campbell-Sills, L., Barlow, D. H., Brown, T. A., & Hofmann, S. G. (2006). Effects of suppression and acceptance on emotional responses of individuals with anxiety and mood disorders. *Behavior Research and Therapy, 44*(9), 1251–1263.

Caplan, G. (1974). *Social systems and community mental health: Lectures on concept development.* New York: Behavioral Publications.

Carstensen, L. L. & Charles, S. (1998). Emotion in the second half of life. *Current Directions in Psychological Science, 7*(5), 144–149.

Cassel, J. C. (1976). The contribution of the social environment to host resistance. *American Journal of Epidemiology, 104*(2), 107–123.

Charles, S. T & Piazza, J. R. (2007). Memories of social interactions: Age differences in emotional intensity. *Psychology and Aging, 22*(2), 300–309.

Coan, J. (2011). The social regulation of emotion. In J. Decety & J. T. Cacioppo (Eds.), *Handbook of social neuroscience* (pp. 614–623). New York: Oxford University Press.

Coan, J. A., Schaefer, H. & Davidson, R. (2006). Lending a hand: Social regulation of the neural response to threat. *Psychological Science, 17*(12), 1032–1039.

Cohen, S. (2004). Social relationships and health. *American Psychologist, 59*(8), 676–684.

Cohen, S., & Wills, T. A. (1985). Stress, social support, and the buffering hypothesis. *Psychological Bulletin, 98*(2), 310–357.

Cohen, S., Mermelstein, R., Karmarck, T., & Hoberman, H. M. (1985). Measuring the functional components of social support. In I. G. Sarason & B. R. Sarason (Eds.), *Social support: Theory, research, and applications* (pp. 73–94). The Hague, Netherlands: Martinus Nijhoff.

Cramm, J. M., & Nieboer, A. P. (2015). Social cohesion and belonging predict the well-being of community-dwelling older people. *Bio Medical Central Geriatrics, 15*(30), 1–10.

Dalgleish, T., Yiend, J., Schweizer, S., & Dunn., B. D. (2009). Ironic effects of emotion suppression when recounting distressing memories. *Emotion, 9*(5), 744–749.

Diener, E., & Seligman, M. E. P. (2002). Very happy people. *Psychological Science, 13*(1), 81–84.

Diener, E., & Seligman, M. E. P. (2004). Beyond money: Toward an economy of well-being. *Psychological Science in the Public Interest, 5*(1), 1–31.

Erfani, S. S., Abedin, B., & Blount, Y. (2016). Social support, social belongingness, and psychological well- being: Benefits of online healthcare community membership.

Pacific Asia Conference on Information System, Pacific Asia Conference on Information Systems 2016, pp. 1–9.

Fingerman, K. L., Hay, E. L., & Birditt, K. S. (2004). The best of ties, the worst of ties: Close, problematic, and ambivalent relationships across the lifespan. *Journal of Marriage and Family, 66*(3), 792–808.

Friedman, H. S., & Silver, R. C. (2007). *Foundations of health psychology.* New York, USA: Oxford University Press.

Gencoz, T., & Ozlale, Y. (2004). Direct and indirect effects of social support on psychological well-being. *Social Behavior and Personality: An international Journal, 32*(5), 449–458.

Geraerts, E., Merckelbach, H., Jelicic, M., & Smeets, E. (2006). Long term consequences of suppression of intrusive anxious thoughts and repressive coping. *Behavior Research and Therapy, 44*(10), 1451–1460.

Gilhooly, K. J., Phillips, L. H., Wynn, V., Logie, R. H., & Della Sala, S. (1999). Planning processes and age in the five-disc Tower of London task. *Thinking and Reasoning, 5*(4), 339–361.

Gross, J. J. (2002). Emotion regulation: Affective, cognitive, and social consequences. *Psychophysiology, 39*(3), 281–291.

Gross, J. J. (1998). The emerging field of emotion regulation: An integrative review. *Review of General Psychology, 2*(3), 217–299.

Gross, J. J. (2001). Emotion regulation in adulthood: Timing is everything. *Current Directions in Psychological Science,10*(6), 214–219.

Gross, J. J., & John, O. P. (2003). Individual differences in two emotional regulation processes: Implications for affect, relationships, and well-being. *Journal of Personality and Social Psychology, 85*(2): 348–362.

Gross, J. J., & Levenson, R. W. (1997). Hiding feelings: The acute effects of inhibiting negative and positive emotion. *Journal of Abnormal Psychology, 106*(1), 95.

Gyurak, A., Gross, J. J., & Etkin, A. (2011). Explicit and implicit emotion regulation: A dual- process framework. *Cognition and Emotion, 25*(3), 400–412.

Haga, S. M., Kraft, P., & Corby, E. K. (2009). Emotion regulation: Antecedents and well- being outcomes of cognitive reappraisal and expressive suppression in cross-cultural samples. *Journal of Happiness Studies, 10* (3), 271–291.

Hess, U., Senécal, S., Kirouac, G., Herrera, P., Philippot, P. & Kleck, R. E. (2000). Emotional expressivity in men and women: Stereotypes and self-perceptions. *Cognition and Emotion, 14*(5), 609–642.

House, J. S. (1981). *Work stress and social support.* Reading, MA: Addison-Wesley.

Kafetsios, K., & Sideridis, G. D. (2006). Attachment, social support, and well-being in youth and older adults. *Journal of Health Psychology, 11*(6), 836–875.

Kesebir, P., & Diener, E. (2008). In pursuit of happiness: Empirical answers to philosophical questions. *Perspectives on Psychological Science, 3*(2), 117–125.

Kircanski, K., Lieberman, M. D., & Craske, M. G. (2012). Feelings into words: Contributions of language to exposure therapy. *Psychological Science, 23*(10), 1086–1091.

Kray, J., & Lindenberger, U. (2000). Adult age differences in task switching. *Psychology and Aging, 15*(1), 126–147.

Labouvie-Vief, G., Lumley, M. A., Jain, E., & Heinze, H. (2003). Age and gender differences in cardiac reactivity and subjective emotion responses to emotional autobiographical memories. *Emotion, 3*(2), 115–126.

Lang, F. R. (2004). Social motivation across the life span. In F. R. Lang & K. L. Fingerman (Eds.), *Growing together: Personal relationships across the life span* (pp. 341–367). New York: Cambridge University Press.

Lazarus, R. S., & Alfert, E. (1964). The short-circuiting of threat. *Journal of Abnormal Social Psychology, 69,* 195–205.

Lazarus, R. S., & Folkman, S. (1984). *Stress, appraisal, & coping.* New York, NY: Springer Publishing.

Lieberman, M. D., Inagaki, T. K., Tabibnia, G., & Crockett, M. J. (2011). Subjective responses to emotional stimuli during labeling, reappraisal, and distraction. *Emotion, 11*(3), 468–480.

Luong, G., Charles, S. T., & Fingerman, K. L. (2011). Better with age: Social relationships across adulthood. *Journal of Social and Personal Relationships, 28*(1), 9–23.

Lyuobomirsky, S., King, L., & Diener, E. (2005). The benefits of frequent positive affect: Does happiness lead to success? *Psychological Bulletin, 131*(6), 803–855.

MacPherson, S. E., Phillips, L. H., & Della Sala, S. (2006). Age related differences in the ability to perceive sad facial expressions. *Aging Clinical and Experimental Research, 18*(5), 418–424.

Maestas, R., Vaquera, G. S., & Munoz Zehr, L. (2007). Factors impacting sense of belonging at a Hispanic-serving institution. *Journal of Hispanic Higher Education, 6*(3), 237–256.

Magai, C., Consedine, N. S., Krivoshekova, Y. S., McPherson, R., & Kudadjie-Gyamfi, E. (2006). Emotion experience and expression across the adult lifespan: Insights from a multi-modal assessment study. *Psychology & Aging, 21*(2), 303–317.

Mauss, I. B., Cook, C. L., & Gross, J. J. (2007). Automatic emotion regulation during anger provocation. *Journal of Experimental Social Psychology*, *43*(5), 698–711.

Maylor, E. A., & Lavie, N. (1998). The influence of perceptual load on age differences in selective attention. *Psychology & Aging*, *13*(4), 563–573.

McRae, K., Ochsner, K. N., Mauss, I. B., Gabrieli, J. J. D., & Gross, J. J. (2008). Gender differences in emotion regulation: An fMRI study of cognitive reappraisal. *Group Processes & Intergroup Relations*, *11*(2), 143–162.

Panksepp, J., & Miller, A. M. (1995) Emotions and the aging brain: Regrets and remedies. In C. Magai & S. H. McFadden (Eds.), *Handbook of emotion, adult development and aging*. San Diego, CA: Academic.

Pinquart, M., & Sörensen, S. (2000). Influences of socioeconomic status, social network, and competence on subjective well-being in later life: A meta-analysis. *Psychology and Aging, 15*(2), 187–224.

Pollard, E. L. & Davidson, L. (2001). *Foundations of child well-being.* (UNESCO Education Sector Monograph No. 18). USA: Centre for Child Well-Being.

Richards, J. M., & Gross, J. J. (2000). Emotion regulation and memory: The cognitive costs of keeping one's cool. *Personality Processes and Individual Differences*, *79* (3), 410–424.

Roberts, N. A., Levenson, R. W., & Gross, J. J. (2008). Cardiovascular costs of emotion suppression cross ethnic lines. *International Journal of Psychophysiology, 70*(1), 82–87.

Ryan, R. M., & Deci, E. L. (2001). On happiness and human potentials: A review of research on hedonic and eudemonic well-being. *Annual Review of Psychology,* 52, 141–166. Mahwah, NJ: Erlbaum.

Ryff, C. D. (1989). Happiness is everything or is it? Explorations on the meaning of psychological well-being. *Journal of Personality and Social Psychology, 57*(6), 1069–1081.

Ryff, C. D. (1995). Psychological well-being in adult life. *Current Directions in Psychological Science, 4*(4), 99–104.

Ryff, C. D., & Singer, B. (1996). Psychological well-being: Meaning, measurement, and implications for psychotherapy research. *Psychotherapy and Psychosomatics, 65*(1), 14–23.

Sampson, R. J. (1988). Local friendship ties and community attachment in mass society: A multilevel systemic model. *American Sociological Review, 53*(5), 766–779.

Sanderson, C. A. (2013). *Health psychology* (2nd ed.). New Jersey, USA: John Wiley & Sons. Inc.

Schieman, S. (1999) Age and anger. *Journal of Health and Social Behavior*, *40*(3), 273–289.

Segrin, C. (2003). Age moderates the relationship between social support and psychosocial problems. *Human Communication Research, 29*(3), 317–342.

Seligman, M. E. P. (2011). *Flourish: A visionary new understanding of happiness and well-being*. New York: Simon & Schuster.

Seligman, M. E. P., & Csikszentmihalyi, M. (2000). Positive psychology: An introduction. *American Psychologist, 55*(1), 5–14.

Steger, M. F. (2009). Meaning in life. In S. J. Lopez & C. R. Snyder (eds.), *Handbook of positive psychology* (2nd ed., pp. 679–687). New York: Oxford University Press.

Steuer, N. & Marks, N. (2008). Local well-being: Can we measure it? *The Young Foundation and the New Economics Foundation,* London.

Su, R., Tay, L., & Diener, E. (2014). The development and validation of the Comprehensive Inventory of Thriving (CIT) and the Brief Inventory of Thriving (BIT). *Applied Psychology: Health and Well-Being*, *6*(3), 251–279.

Taylor, S. E., Klein, L. C., Lewis, B. P., Gruenewald, T. L., Gurung, R. A. R., & Updegraff, J. A. (2000). Biobehavioral responses to stress in females: Tend-and-befriend, not fight-or-flight. *Psychological Review, 107*(3), 441–429.

Tull, M. T., Jakupcak, M., & Roemer, L. (2010). Emotion suppression: A preliminary experimental investigation of its immediate effects and role in subsequent reactivity to novel stimuli. *Cognitive Behavior Therapy, 39*(2), 114–125.

Uchino, B. N., Cacioppo, J. T., & Kiecolt-Glaser, J. K. (1996). The relationship between social support and physiological processes: A review with emphasis on underlying mechanisms and implications for health. *Psychological Bulletin, 119*(3), 488–531.

Williams, L. M., & Barry, J. (2003). Do sex differences in emotionality mediate sex differences in traits of psychosis-proneness? *Cognition and Emotion, 17*(5), 747–758.

Zaki, J., & Williams, C. (2013). Interpersonal emotion regulation. *Emotion, 13*(5), 803–810.

CHAPTER 6

The Role of Perceived Hope in Harmony in Life Among Czech and South African Adults

ALENA SLEZACKOVA[1*], THARINA GUSE[2], and TOMAS PROSEK[1]

[1]*Department of Psychology, Faculty of Arts, Masaryk University, Brno, Czech Republic*

[2]*Department of Psychology, University of Pretoria, Pretoria, South Africa*

[*]*Corresponding author. E-mail: slezackova@phil.muni.cz*

ABSTRACT

In view of the globalized nature of today's world, we aimed to explore whether and to what extent there were differences in levels of perceived hope, positive and negative emotions, as well as harmony in life between two national samples, Czech and South African. Although the Czech Republic (CZ) and South Africa (SA) significantly differ in terms of history, culture, language, and socioeconomic background, they both experienced recent political and macrosocial changes. It was assumed that remaining hopeful might play an important role in attaining subjective well-being despite the perceived socio-political issues in both countries. As the concepts of perceived hope and harmony in life are relatively new, the main aim of this exploratory study was to examine the role of perceived hope and positive and negative affect in the level of harmony in life among CZ and SA. The specific aims were to (1) Compare differences in perceived hope, emotional experiences, and harmony in life between the CZ and SA samples; (2) examine the predictive power of perceived hope in harmony in life and (3) investigate the possible mediating role of perceived hope in the relationship between positive affect and harmony in life. Data from the international Hope Barometer survey, collected through an online survey between 2016 and 2017, were employed and processed using IBM SPSS Statistics. The CZ sample consisted of 310 adults (63.5% females, mean age = 29.6) while the SA sample consisted of 414 adults (69.6% females, mean age = 38.9).

Perceived hope proved to be an independent predictor of harmony in life in both samples. Further analyses revealed interesting culture-specific differences in the measured variables. Better understanding of cross-cultural similarities and differences in hope and subjective well-being might be useful in understanding and promoting harmony, resilience, and respect for diversity in today's multicultural society.

6.1 INTRODUCTION

Researchers' interest in positive emotions and subjective well-being (SWB) has increased substantially over the last couple of decades (DelleFave et al., 2011; Diener et al., 2017; Fredrickson & Joiner, 2018; Hudson et al., 2017). While this research yielded several findings that increased our understanding of positive human functioning, it also produced revised perspectives and novel constructs that need to be further explored. This is especially true of the "transcendental" conceptualizations of hope (i.e., perceived hope; Krafft et al., 2017) and well-being (i.e., harmony in life; Kjell et al., 2016), which are expected to differ from their much more thoroughly explored dispositional or cognitive counterparts (Snyder, 2000; Diener et al., 1985).

This chapter discusses the new concept of perceived hope in relation to harmony in life and positive and negative emotions between two culturally different samples, Czech and South African. Although both the Czech Republic (CZ) and South Africa (SA) significantly differ in terms of history, culture, language, and socio-economic background, they have undergone major socio-political changes in last decades, which may have affected the quality of life of their populations. Thus, it can be assumed that remaining hopeful, as evident in levels of perceived hope, may play an important role in maintaining SWB despite socio-political issues.

As the concept of perceived hope is relatively new, this study aimed to investigate its relationship with positive and negative affect and its predictive power in harmony in life. A further aim was to examine whether levels of hope and other measured variables vary across countries.

6.2 HARMONY IN LIFE

One of the key subjects of research in positive psychology is SWB, which is defined as one's overall evaluation of own life and emotional experiences (Diener, 1984). SWB contains a cognitive element, that is, a rational assessment of a person's life satisfaction based on a comparison of one's life circumstances to one's expectations, and an affective element, referring to the presence of positive and negative affect (Diener et al., 1985). So far, the concept of life satisfaction has probably been the most thoroughly researched conceptualization of well-being (Diener et al., 2017; Pavot & Diener, 2008).

Recently, however, Kjell et al. (2016) introduced a new, more holistic and flexible perspective on well-being referred to as *harmony in life*, which differs from SWB by emphasizing peaceful acceptance over satisfaction. Harmony in life is a global assessment of experienced balance, mindful, nonjudgmental acceptance, fitting in, and a sense of being attuned with one's life (Delle Fave, 2014; Garcia et al., 2014).

Kjell and colleagues (2016) found that higher levels of harmony in life were related to higher subjective happiness and psychological well-being and lower depression, stress, and anxiety. In comparison to life satisfaction, harmony in life showed stronger correlations with all the above variables and explained unique variance in these various types of well-being. Kjell et al. (2016) also reported that, although the constructs of life satisfaction and harmony overlap substantially, people tend to associate them with different concepts and meanings. While life satisfaction is closely linked with the ideas of personal goal attainment and social success, harmony tends to evoke ideas of peace, belongingness, beauty, and solidarity. Harmony in life is conceived as a holistic rather than individualistic experience, characterized by the acceptance of one's limits and the fact of "being one part of a bigger picture" rather than as a simple outcome of personal successes and failures.

Garcia et al. (2014) also examined harmony in life in relation to positive and negative affect and psychological well-being. They found that individuals with an affective profile of high positive and low negative affect showed higher levels of harmony, whereas those reporting low positive and high negative affect reported lower harmony in life.

Another study which highlighted the importance of harmony implemented a mixed-method approach to explore qualitative aspects of happiness as well as a quantitative evaluation of the degree of happiness experienced in various life domains (DelleFave et al., 2011, 2016). The authors examined psychological definitions and experiences of happiness across six countries using open-ended questions. Results of the coding procedure and categorization revealed that happiness was primarily defined as a condition of harmony and psychological balance, which comprised items such as harmony, feeling of balance, inner peace, positive relationship with oneself, contentment, and serenity.

6.3 PERCEIVED HOPE

Hope has always been the center of attention for philosophers, theologists, sociologists and, recently psychologists (Krafft & Walker, 2018a; Scioli & Biller, 2009). Most broadly, hope can be described as a tendency to have positive expectations about the future (Snyder, 2000). In psychology, numerous perspectives on hope have been employed: as hopeful thinking (Snyder, 2000), a positive emotional experience (Fredrickson, 2009), a character strength (Peterson & Seligman, 2004), or a transcendental phenomenon (Emmons, 2005).

One of the most prominent theories of hope, Snyder's (2000) cognitive conceptualization of hope, construed *dispositional hope* as a "positive motivational state that is based on an interactively derived sense of (a) agency (goal-oriented energy) and (b) pathways (planning to meet the goals)" (Snyder et al., 1991, p. 287). Hope is, therefore, directly linked to personal goals and a sense of control necessary to meet these goals through one's own efforts.

In contrast, Krafft et al., (2017, 2018) recently developed a much broader concept of *perceived hope*, which especially accounts for phenomena that are beyond human control. Perceived hope is the sense of deep trust that things will turn out well regardless of the outcome of one's personal strivings. Since the concept is most prominent in difficult life situations that one cannot control, perceived hope is closely related to experiencing meaning in life, helping other people, enjoying close and trusted relationships, and creating spiritual or religious experience.

Since perceived hope is a transcendental strength characterized by a general sense of acceptance and prosocial orientation, one might expect it to be positively related to harmony in life. Although published research regarding the construct's predictors and correlates is limited, this association has recently been confirmed in German, Swiss, and Indian samples (Krafft & Choubisa, 2018; Krafft & Walker, 2018b).

Apart from these direct associations, perceived hope displayed significant relationships with other positive variables. Slezackova and Krafft (2016) reported significant positive interrelationships among perceived hope, dispositional hope, life satisfaction, and meaning in life among Czech adults. Both types of hope served as independent predictors of life satisfaction, with perceived hope being the strongest predictor. Similarly, in a study among Swiss and German samples, perceived hope had stronger predictive power with respect to life satisfaction and happiness than dispositional hope (Krafft et al., 2017).

Recently, Slezackova et al. (2018) explored psychosocial correlates and predictors of perceived hope among Czech and Maltese adults. In both samples, the most important predictor of perceived hope was dispositional optimism, a characteristic responsible for one's positive outlook on life. However, the predictive capability of the varying predictors in relation to perceived hope was found to be culturally dependent. While in the Czech sample, higher perceived hope was predicted by higher generativity and lower loneliness, in the Maltese sample an important role was played by spirituality, which was found to be the second independent predictor of perceived hope. Finally, in a study among a sample of South African university students, perceived hope had a significant indirect effect on well-being through the relationship with meaning in life (Guse & Shaw, 2018). Thus, while research on perceived hope has been expanding, there is still a need for studies on the dynamics of perceived hope in the experience of well-being in diverse contexts.

6.4 POSITIVE AND NEGATIVE FEELINGS

Both positive and negative emotions are important to well-being and reflect the affective component of SWB (Lucas & Diener, 2015). In particular, experiencing relatively more positive than negative feelings is associated with higher levels of well-being (Pavot & Diener, 2013).

Positive feelings such as happiness, joy, and excitement indicate an enjoyable engagement with the environment (Cohen & Pressman, 2006). These positive emotions can also be viewed as multisystem responses related to the way in which individuals interpret changes in their current circumstances (Fredrickson, 2013). If this change is interpreted as good, positive feelings occur. Positive emotions seem to both characterize and enhance optimal functioning (Fredrickson & Joiner, 2018). According to Fredrickson's (2001, 2013) broaden-and-build theory, positive emotions strengthen social, cognitive, and behavioral resources, which ultimately facilitate an upward spiral toward greater well-being. A large body of research has since supported this theory (Fredrickson & Joiner, 2018).

Negative affect consists of feelings of high unpleasantness and high arousal, including being nervous, afraid, and distressed (Lucas et al., 2009) and is associated with lower levels of well-being. In particular, negative emotions have been linked to impaired mental health and the presence of psychiatric disorders (Watson et al., 1988). Recent research further found that experiencing negative emotions in response to daily stressors could have a long-term negative effect on physical health (Leger et al., 2018).

Research suggests that there may be cultural differences in the psychological meanings ascribed to positive and negative emotions (Wirtz et al., 2009) as well as the value placed on specific feelings (Diener et al., 2017). However, most studies have investigated differences between Western (mainly USA) and Eastern countries such as Japan (e.g., Tsai et al., 2006). Therefore, it is important to also explore possible differences in the experience of emotions in our study.

6.5 CULTURAL BACKGROUND: CZECH REPUBLIC AND SOUTH AFRICA

Over the past few years, researchers have increasingly focused on interplay between SWB and culture (e.g., Diener et al., 2003; Oishi & Gilbert, 2016) and it is generally accepted that well-being needs to be understood in cultural context. Accordingly, we set out to explore well-being, conceptualized as harmony in life, as well as positive and negative affect, among Czech and South African participants. Both countries were part of the International Hope Barometer Survey. Despite many profound differences in culture, language, and history, both countries have undergone significant socio-political changes in last decades, which may have affected the quality of life and well-being of their citizens.

The Czech Republic (CZ) is a developed Central European state with a parliamentary democracy which has over 10.6 milion inhabitants. The majority of the inhabitants are Czechs (64.3%), followed by Moravians (5%), and Slovaks (1.4%), 27.5% are unspecified ethnic groups[1] (IndexMundi, 2018). Most people (88.5%) identify themselves as nonreligious (also undeclared religion or spiritual but not religious); over 10% are Roman Catholics.

The CZ is one of the European countries that have undergone several macro-social changes in the past decades related to political, economic, or cultural transitions (Klingemann et al., 2006). The Czechoslovak Republic was established in 1918 and until 1938 was one of the 10 most developed countries of the world. After World War II democracy ended with the communist takeover in 1948, which resulted in human and political rights being suppressed and private property was expropriated. An attempt to change the totalitarian communist regime (known as Prague Spring) failed when the Soviet Army invaded the country in 1968. The gradual decay of the communist regimes in Eastern Europe in the late 1980s, and the mass protests of the Czechoslovak people, resulted in the so-called Velvet Revolution and the overthrow of the communist government in 1989. In 1993, two independent states were peacefully created out of the former Czechoslovakia: the Czech and the Slovak Republic. Alhough the transition in 1989 brought a number of positive changes (e.g., freedom of faith and religion, higher level of education, and free travel abroad), some negative changes also occurred, such as higher unemployment, increased divorce rates, and loss of social security, especially in the early 1990s (Klicperova et al., 1997). In the early years of transition, the CZ went through deep recession and life satisfaction dropped and recovered somewhat few years later (Easterlin, 2009). Nowadays, in the time of the economic growth and almost zero unemployment, many Czechs are rather pessimistic as concerns national politics and burning social issues; however, they remain optimistic about their personal life (Slezackova et al., 2018).

South Africa (SA) is a country at the southern tip of the African continent with a relatively young parliamentary democracy. It has around 60 milion inhabitants. The population consists of four main ethnic groups: Black African 80.2%, White 8.4%, Colored[2] 8.8%, and Indian/Asian 2.5% (IndexMundi, 2018). As per religious beliefs, 36.6% of inhabitants belong to a Protestant church, 36% to other Christian churches, 7.1% to Catholic religion, 1.5% are Muslims, and 15.1% are nonreligious (IndexMundi, 2018).

Until 1994, South Africa was governed by a political system referred to as apartheid. This sociopolitical model entailed oppression of all "non-white" ethnic groups (Adams et al., 2012).

[1] Due to a high degree of nonresponse to the question of nationality

[2] A term used in South Africa for persons of mixed race ancestry

Democracy was established in 1994, with the first free elections that included all the ethnic groups. Since then, the Black African group gained political dominance, but the majority of Blacks remain less affluent (Adams et al., 2016). In contrast, the minority White group still seems to remain economically dominant. Despite expectations of a bright future with more opportunities for all, as well as the establishment of a "rainbow nation," consisting of all the ethnic groups, South Africans are concerned about the economy and the political situation (Guse, 2018). Corruption, economic disparity, and a low skill base due to an ineffective educational system are some of the key features of society (Kagee, 2014). With an unemployment rate of 26.6% (Statistics South Africa, 2018) of which many are young, unskilled people, South Africa also remains a very unequal society.

To understand possible cultural differences between the two countries, Hofstede's (2001) six-dimensional model of national culture was used as the point of departure. The six dimensions include power distance (PD), individualism (IDV) versus collectivism, masculinity (MAS) versus femininity, uncertainty avoidance (UA), long-term orientation (LTO) versus short-term normative orientation, and indulgence (IND) versus restraint. Hofstede (2001) evaluated several countries on these dimensions. CZ and SA seemed similar on some dimensions, but differed on others.

According to Hofstede (2001), the CZ and SA both scored relatively high on individualism (CZ scored 58 out of 100; SA: 65) and PD (CZ: 57; SA: 49). This implies that both countries are individualist societies whose members are expected to take care primarily of themselves and their immediate families. The PD scores reflect the degree to which the less powerful members of the society accept unequal distribution of power. Cultures which score higher on PD support the notion that there should be clear divides between superiors and subordinates, and that those in power are entitled to considerable privileges and influence.

The CZ and SA also scored similarly on MAS (CZ: 57; SA: 63) which indicates that achievement, material rewards for success, and assertiveness are valued highly in both societies. However, higher UA (CZ: 74; SA: 49) among Czechs suggests that Czech people are more uncomfortable with ambiguity and uncertainty then South Africans. Czech society also tends to manifest greater intolerance of unorthodox ideas and behavior and maintain strict codes of belief and behavior.

The biggest differences between the Czech and South African cultures are on IND (CZ: 29; SA: 63) and LTO (CZ: 70, SA: 34). The higher IND score obtained by the South African sample in Hofstede's (2001) study suggests a greater tendency toward optimism and willingness to enjoy life. Czechs, on the other hand, tended to be less indulgent and to have a tendency toward cynicism and pessimism. The LTO dimension is a "pragmatic" one, with the high score implying that Czech society encourages thrift and other means of preparation for the future. Conversely, the low score on the LTO dimension in the South African sample suggests that they exhibit great respect for traditions, a relatively small propensity to save for the future, and a focus on achieving quick results.

Although the above dimensions pointed to several cultural differences between the two countries, one has to keep in mind that the South African results were computed for the White population of SA, and Hofstede (2001) warns that, as the majority of the South African population is Black African, their scores may be different.

6.6 OBJECTIVES

Against the contextual and cultural backgrounds described above, we find it pertinent to consider macrosocial influences on adults'optimal functioning and SWB, specifically as it relates to their expectations about the future. Accordingly, the aim of the study was to investigate the role of perceived hope and emotional experiences in the level of harmony in life among Czech and South African samples.

Although the beneficial effect of positive emotions has been widely acknowledged in research (Fredrickson, 2009, 2013), little is known about its relationships with perceived hope and harmony in life. As the concepts of perceived hope (Krafft et al., 2017, 2018) and harmony in life (Kjell et al., 2016) are relatively new, no research has yet explored differences in these variables between the culturally different Czech and South African populations.

Considering these research gaps and taking into account findings from recent research studies on perceived hope (Krafft et al., 2018), the primary aim of the study is to shed light on the role of perceived hope in the relationship between positive feelings and harmony in life between these two groups.

The following objectives were set for the study:

- First, to establish levels of perceived hope, harmony in life, and positive and negative affect among Czech and South African research samples, and to investigate whether there are significant differences between the two groups on these variables.
- Second, to explore the relationship among these variables in both samples. Based on previous research, we expected significant intercorrelations among perceived hope, SWB (operationalized as harmony in life), and positive feelings.
- Third, to examine the mediating effect of perceived hope on the relationship between positive affect and harmony in life.

6.7 METHOD

6.7.1 *SAMPLE*

The total sample consisted of 724 adults. The Czech sample consisted of 310 respondents (63.5% females, 36.5% males, aged 18–72 years, mean age = 29.6). The South African sample comprised 414 adults (69.6% females, 30.4% males, aged 18–79 years, mean age = 38.9). To facilitate further analyses, the participants were divided into four age groups according to developmental stages (Arnett, 2004; Colarusso, 1992): 18–25 years (youth/emerging adulthood), 26–39 years (young adulthood), and 40–55 years (middle age/middle adulthood). Due to the low number of respondents above 55 years, these were merged together into a single "55+"group (later adulthood) for each of the two samples (see Table 6.1).

TABLE 6.1 Distribution of Age Groups in the Whole Sample ($N = 724$)

Age	Czech Republic		South Africa	
	N	%	N	%
18–25	179	57.7	90	21.7
26–39	70	22.6	135	32.6
40–55	39	12.6	125	30.2
56+	22	7.1	64	15.5
Total	310	100	414	100

Most participants had a university degree (CZ: 44.5%; SA: 67.1%) or had completed standard secondary education (CZ: 49.4%; SA: 30.6%); the rest had primary education only (CZ: 1.6%; SA:1.7%).

The distribution of the marital status in the sample was as follows: a large part of the participants in the Czech sample were single (CZ: 43.2%; SA: 37%) or living in a partnership (CZ: 33.5%; SA: 14%), while the majority in the SA sample was married (CZ: 18.4%; SA:41.5%). A small percentage of the sample was divorced or widowed (CZ: 4.8%; SA: 6.5%).

As for religious beliefs, most of the SA participants were Christian (CZ: 19.7%; SA: 62.5%); others were members of a different religion (such as Buddhism, Islam, or Judaism; CZ: 5.1%; SA: 11.4%). In the Czech sample, a significant part was spiritual but not religious (CZ: 32.6%; SA: 19.1%), or atheists (CZ: 42.6%; SA: 7%).

6.7.2 MEASURES

The Harmony in Life Scale (HLS; Kjell et al., 2016) is a five-item measure of psychological experiences of inner balance, peace of mind, calm, and unity. The items are scored on a seven-point Likert-type response scale (1—strongly disagree, 7—strongly agree). The higher the total score, the higher the overall degree of the respondent's experience of harmony in life.

The Perceived Hope Scale (PHS; Krafft et al., 2017) is a six-item tool covering aspects such as level of hope, effect of hope, fulfillment of hope, and hope/anxiety duality. The scale employs a five-point Likert-type response scale (0—strongly disagree, 5—strongly agree). Higher total scores indicate higher levels of perceived hope. The PHS showed good psychometric properties (Krafft et al., 2017; Slezackova et al., 2020).

The Scale of Positive and Negative Experience (SPANE; Diener et al., 2010) is a 12-item scale with six measuring positive emotional experiences and feelings (e.g.,"good," "pleasant," and "happy") and six items to assess a level of negative emotional experiences (e.g.,"sad," "angry," and "afraid"). The participants score the items on a 5-point scale (1—very rarely or never, and 5—very often or always). Because of the partial independence of the two types of feelings, the positive and negative scales are scored separately. The SPANE previously demonstrated favorable psychometric properties in both Czech (Singh et al., 2016) and South African samples (Du Plessis & Guse, 2017).

All measures used showed very good reliability (see Table 6.2).

6.7.3 PROCEDURE AND DATA ANALYSIS

Data from the *Hope Barometer* international research program which aims to explore the relationships of hope to various psychosocial variables in several countries with different cultural backgrounds (Krafft et al., 2018) were employed. It was collected using an anonymous online survey advertized among the general adult populations of CZ and SA (voluntary sampling) in November 2016 and 2017. A link to the questionnaire was distributed through e-mail, websites, and social networks. Principles of research ethics were adhered to during collecting and processing of data. Participation in the research was voluntary and participants were informed of the subject of research and of the estimated amount of time needed to complete the questionnaire. Research participants could withdraw from the research at any time by not completing or submitting a questionnaire.

All analyses have been conducted using IBM SPSS Statistics version 18, SPSS version

25, and JASP (JASP Team, 2018). We used descriptive statistics to describe the characteristics of the sample as well as the mean values of variables, and a histogram to estimate the probability distribution of the variables. With respect to recent scientific discussion on advanced methods in the assessment of scale quality (Dunnet et al., 2014; Peters, 2014), we used both Cronbach's alpha and McDonald's omega to assess the reliability of the instruments.

We examined the association between variables using Pearson's correlation coefficient and multiple linear regression analysis. A Plugin Process (Hayes, 2012) was used to run the mediation model.

6.8 RESULTS

6.8.1 DESCRIPTIVE RESULTS AND DIFFERENCES BETWEEN THE CZECH AND SOUTH AFRICAN SAMPLES

First, we examined the levels of perceived hope, harmony in life, and positive and negative feelings among the Czech and South African samples, and compared the two samples on these variables. Means and standard deviations for the variables of interest are shown in Table 6.2.

The results of the comparative analysis revealed statistically significant differences between Czech and South African samples in levels of perceived hope ($t = -4.82$; $df = 585.99$[3]; $p < 0.001$; *Cohen* $d = -0.37$), harmony in life ($t(722) = -4.39$; $p < 0.001$; $d = -0.33$), and positive feelings ($t(625.45) = -3.22$; $p < 0.001$; $d = -0.24$). No significant difference was found in the level of negative feelings ($t(722) = -0.31$; $p = 0.760$; $d = -0.02$).

Czech participants showed lower perceived hope (M = 3.40, SD = 1.07) compared to South Africans (M = 3.76, SD = 0.88). They also reported lower levels of positive feelings (M = 3.62, SD = 0.79) than South African respondents (M = 3.80, SD = 0.71) and showed lower harmony in life (M = 4.84, SD = 1.22) in comparison with the South African group (M = 5.24, SD = 1.19).

TABLE 6.2 Descriptive Statistics of Methods Used and Their Reliability in Czech (N = 310) and South African (N = 414) Samples

Measure	Country	M	SD	MIN	MAX	Cronbach's Alpha	McDonald's Omega
PHS	CZ	3.40	1.07	0	30	0.90	0.90
(6 items)	SA	3.76	0.88	0	30	0.89	0.89
SPANE P	CZ	3.62	0.79	6	30	0.91	0.91
(6 items)	SA	3.80	0.71	9	30	0.90	0.90
SPANE N	CZ	2.54	0.78	6	29	0.83	0.84
(6 items)	SA	2.56	0.79	6	30	0.87	0.86
HILS	CZ	4.84	1.22	5	35	0.88	0.89
(5 items)	SA	5.24	1.19	6	35	0.89	0.89

Note: CZ = Czech sample, SA = South African sample, PHS =Perceived Hope Scale, SPANE P = positive feelings, SPANE N = negative feelings, HILS = Harmony in Life Scale.

[3] Unfulfilled assumption of homogeneity of variances

6.8.2 CORRELATES AND PREDICTORS OF HARMONY IN LIFE

Table 6.3 shows the relationships among all measured variables in the two samples. In addition to the expected significant correlations among perceived hope, harmony in life, and positive feelings, we also found significant negative relationships among these variables and negative feelings. Interestingly, all correlations were slightly stronger in Czech sample than in the South African sample, except for the relationships between positive and negative feelings. In both samples, the strongest relationship with harmony in life was the experience of positive feelings. The strongest correlate of perceived hope in the Czech sample was harmony in life, whereas in the South African sample, it was positive feelings.

Multicollinearity diagnostics for both samples were implemented before performing regression analyses. In both samples, there was no problem with multicollinearity, as evident from the variance inflation factor values.

Two separate regression analyses were performed to determine which of the examined variables were independent predictors of harmony in life and to examine possible significant differences between the two samples. The results are presented in Table 6.4.

The first regression analysis showed that, in the Czech sample, harmony in life was primarily predicted by positive feelings (β = 0.394, $p < 0.001$) and perceived hope (β = 0.328, $p < 0.001$). Negative feelings also served as a significant albeit weaker independent predictor of harmony in life (β = −0.201, $p < 0.001$). The model was significant (F (3, 306) = 170.33; $p < 0.001$) and explained 62.5% of variance of harmony in life (R = 0.79; R^2 = 0.625; R^2 adjusted = 0.622).

The second regression analysis identified positive feelings (β = 0.370, $p < 0.001$) and perceived hope (β = 0.275, $p < 0.001$) as the independent predictors of harmony in life in the South African sample. Negative feelings seemed to be a significant, albeit a weaker

TABLE 6.3 Pearson's Correlation Coefficients for All Variables in the Czech (N = 310) and South African Samples (N = 414)

	Country	PHS	SPANE P	SPANE N
SPANE P	CZ	0.66*** [0.59, 0.72]	1	
	SA	0.58*** [0.51, 0.64]	1	
SPANE N	CZ	−0.47*** [−0.55, −0.37]	−0.58*** [−0.65, −0.50]	1
	SA	−0.41*** [−0.49, −0.32]	−0.62*** [−0.67, −0.55]	1
HILS	CZ	0.68*** [0.62, 0.74]	0.73*** [0.67, 0.78]	−0.58*** [−0.65, −0.50]
	SA	0.56*** [0.48, 0.62]	0.63*** [0.57, 0.68]	−0.50*** [−0.57, −0.43]

Note: ***$p < 0.001$

95% confidence intervals are stated in brackets.

PHS = Perceived Hope Scale, SPANE P = positive feelings, SPANE N = negative feelings, HILS = Harmony in Life Scale

TABLE 6.4 Regression Model Predicting Harmony in Life (HILS) for the Czech and South African Samples

Predictors of harmony in Life	Czech Republic				South Africa			
	B	SE	β	p	B	SE	β	P
PHS	0.312 [−0.224, −0.399]	0.045	0.328	<0.001	0.311 [0.213, .410]	0.050	0.275	<0.001
SPANE P	0.510 [0.380, 0.639]	0.066	0.394	<0.001	0.518 [0.377, 0.659]	0.072	0.370	<0.001
SPANE N	−0.262 [−0.374, −0.151]	0.057	−0.201	<0.001	−0.204 [−0.317, −0.092]	0.057	−0.163	<0.001

Note: 95% confidence intervals are stated in brackets.

PHS = Perceived Hope Scale, SPANE P = positive feelings, SPANE N = negative feelings.

independent predictor of harmony ($\beta = -0.16$, $p < 0.001$). The model was significant ($F(3, 410) = 120.03$; $p < 0.001$) and explained 46.8% of variance of harmony in life ($R = 0.68$; $R^2 = 0.468$; R^2 adjusted = 0.464).

6.8.3 *MEDIATION MODEL*

Following the results presented thus far, the relationship between hope, positive and negative feelings, and harmony in life by partial mediation modelling was analyzed, yielding the following results (see Figures 6.1 and 6.2).

The model in Figure 6.1 demonstrates the role of perceived hope as partial mediator ($b = 0.302$, $p < 0.001$) between positive feelings and harmony in life in the Czech sample.

Figure 6.2 exhibits the partial mediation model for the South African sample in which positive feelings predict harmony in life via the partial effect of perceived hope ($b = 0.232$; $p < 0.001$).

6.9 DISCUSSION

6.9.1 *COMPARISON BETWEEN THE CZECH AND SOUTH AFRICAN SAMPLES*

The first objective of this study was to assess and compare the levels of perceived hope, harmony in life, and positive and negative feelings among Czech and South African participants.

It turned out that the South African group ($M = 3.76$) showed significantly higher levels of perceived hope than Czech participants ($M = 3.40$). Comparison with the results of previous studies showed that other samples from other nationalities such as Indian ($M = 3.36$) or German ($M = 3.29$) also demonstrated slightly lower levels of perceived hope (Krafft & Choubisa, 2018). This result was unexpected, and, with limited research available on perceived hope, also difficult to explain. One possible explanation may lie in the fact that the majority of the South African sample showed an affiliation with a particular religion (Christianity). As perceived, hope is associated with the transcendental and spiritual domains of hope (Krafft et al., 2017), religious affiliation, through its association with spirituality, could have strengthened perceived hope in the South African sample. Another possibility to be considered is that South Africans seem to be more tolerant of ambiguity than Czechs (Hofstede, 2001). Since perceived hope is particularly important in situations that could be seen as being out of the individual's control (Krafft et al., 2018a), the abililty to tolerate uncertainty may have increased South Africans' levels of perceived hope.

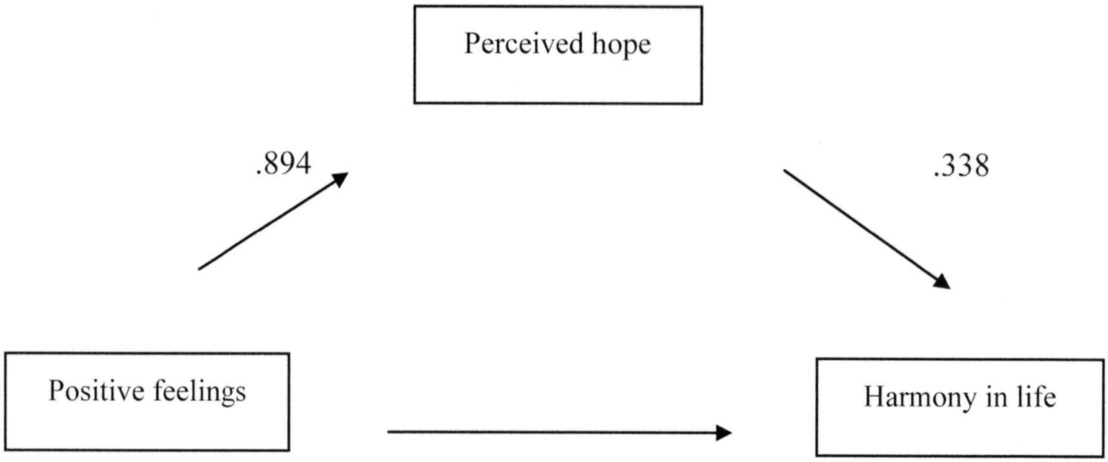

Direct effect: b=.634, 95% CIs[0.513, 0.755]
Indirect effect: b=.302, 95% CIs[0.191, 0.421], Z=6.69

FIGURE 6.1 Perceived hope as a partial mediator between positive feelings and harmony in life in the Czech sample (all standardized coefficients sig. at $p < 0.001$).

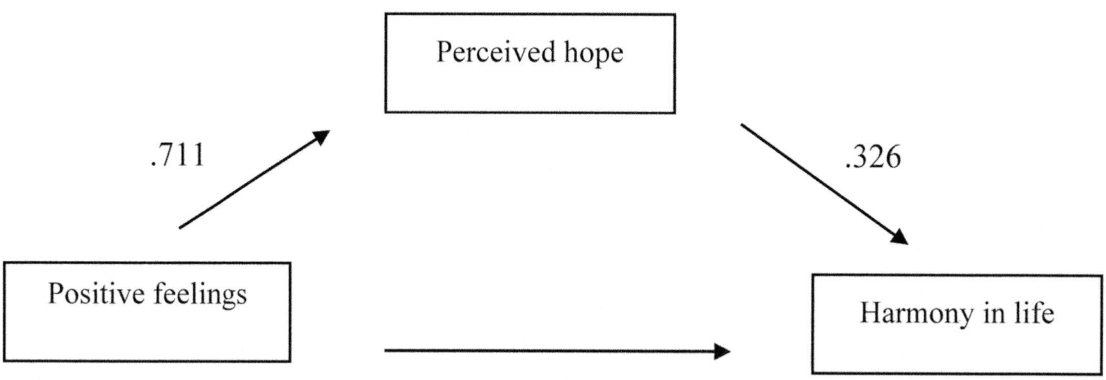

Direct effect: b=.648, 95% CIs [.525, .771]
Indirect effect: b=.232, 95% CIs [.191, .421], Z=5.87

FIGURE 6.2 Perceived hope as a partial mediator between positive feelings and harmony in life in the South African sample (all standardized coefficients sig. at $p < 0.001$).

South Africans (M = 5.24) scored significantly higher on harmony in life as compared to Czech participants (M = 4.84), and also higher than Indian (M = 4.72) and German (M = 4.63) participants in the study by Krafft and Choubisa (2018). This result was again, unexpected, as Czechs (6.6) and South Africans (6.3) reported relatively similar levels of happiness, conceptualized as life satisfaction, according to The World Database of Happiness (Veenhoven, 2018a). Kjell et al. (2016) found that harmony in life was positively correlated with life satisfaction, thus similar levels of harmony in life in the two groups were expected. The current finding could possibly be sample specific and more research is needed to fully understand levels of perceived hope among South Africans.

Moreover, South African respondents reported significantly higher levels of positive feelings (M = 3.80) than Czechs (M = 3.62), who, in turn, reported scores similar to German participants (M = 3.66) and slightly higher than Indian respondents (M = 3.52) (Krafft & Choubisa, 2018). Similar mean scores of positive feelings (M = 3.68) were also obtained in an American sample (Diener et al., 2010). One possible explanation lies in the fact that South Africans scored higher than Czechs on Hofstede's (2001) dimension of IND, which suggests that they have a greater tendency toward optimism and willingness to enjoy life. In contrast, Czechs tend to be much less indulgent and more pessimistic and cynical, which may be reflected on the lower level of positive feelings. Several studies reported a positive correlation between optimism and positive affect (e.g., Krafft & Walker, 2018b; Martínez-Marti & Ruch, 2017), which may partially account for the results.

No statistically significant difference was found in the level of negative feelings between Czech (M = 2.54) and South African (M = 2.56) samples, which were comparable with the mean value obtained on the US sample (M = 2.60) and reported by Dieneret al. (2010).

The current study's results are in some respects in contrast with the outcomes of large international surveys on well-being (Helliwell et al., 2018; Jeffrey et al., 2016; Veenhoven, 2018b, 2018c), which indicated lower levels of well-being in South Africa compared to the CZ. However, one has to keep in mind that most of these surveys contained, apart from the elements of psychological well-being, several socio-economic and environmental indicators of quality of life and overall well-being that could affect the outcomes. The South African sample in the current study consisted mostly of highly educated South Africans, which could have influenced the results.

6.9.2 CORRELATES AND PREDICTORS OF HARMONY IN LIFE

As expected, there were statistically significant positive relationships between perceived hope, harmony in life, and positive feelings, congruent with the results of recent studies conducted among Swiss and German participants (Krafft & Walker, 2018b) as well as Indian and German-speaking respondents (Krafft & Choubisa, 2018).

Harmony in life displayed a higher correlation coefficient for perceived hope in the Czech (r = 0.68) group, compared to the South African group (r = 0.56). Earlier Krafft and Choubisa (2018) reported correlation values significantly weaker for the Indian group (r = 0.33) than for the German group (r = 0.62). The variable assessing positive feelings has shown moderate to strong correlations with perceived hope, especially for the Czech sample (r = 0.66). The relationship between positive feelings and hope was slightly weaker for the South African sample ($r = 0.58$), comparable to the German sample ($r = 0.58$), but higher than the correlation coefficient obtained in the Indian ($r = 0.28$) group (Krafft & Choubisa, 2018). Harmony in life was associated with positive and negative feelings in both samples. This is in line with earlier studies exploring relationships between harmony and other measures of well-being (Garcia et al., 2014; Kjell et al., 2016). Generally, the current results are in accordance with several studies which reported a significant relationship between hope and SWB (Demirli et al., 2015; Kato & Snyder, 2005, Krafft & Walker, 2018b; Slezackova & Krafft, 2016).

Results of the regression analyses confirmed that perceived hope explained unique variance of harmony in life among both Czech and South African respondents. Perceived hope was a stronger independent predictor of harmony in life in the Czech sample than in the South African group. Further analyses revealed the role of perceived hope as a partial mediator between positive feelings and harmony in life in the Czech and the South African samples separately. This means that feeling good and having positive emotional experiences may increase perceived hope, which significantly contributes to higher levels of harmony in life. These findings shed further light on mechanisms that could facilitate harmony in life. In particular, Garcia et al. (2014) found that individuals who have high levels of positive affect and low levels of negative affect experienced more psychological well-being and harmony in life. The current results suggest that a possible mechanism, which could explain this relationship, is the fact that positive feelings could enhance perceived hope, and, in turn, harmony in life. Thus, in accordance with Fredrickson's (2001, 2013) broaden-and-build theory, positive feelings may facilitate the experience of hope, especially in contexts where the individual has limited control over what happens in future (see Krafft & Walker, 2018a).

6.10 CONCLUSION

The broad aim of our study was to examine perceived hope and well-being (conceptualized as positive feelings, negative feelings, and harmony in life) between Czech and South African samples. Both countries have undergone major socio-political change in the past decades, which Lawson (2017) described as similar negotiated transitions. Since these changes may have affected individuals' well-being and future expectations, we wanted to investigate these variables between the two groups.

Our study found that the South African sample showed higher levels of perceived hope, positive feelings, and harmony in life than the Czech group. This may partially be due to cultural differences, as explained by Hofstede's (2001) dimensions of culture. In particular, we argued that South Africans may experience more perceived hope and positive feelings due to being higher than Czechs on the dimension of Indulgence, which reflects optimism and a tendency to enjoy life. However, we are aware that other factors, which could be sample-specific, may have influenced our results. For example, the South African sample was, on average, 10 years older than the Czech sample, and some research indicate that perceived hope (Krafft & Walker, 2018b), positive affect (Charles et al., 2001) and experiences of harmony in life (Carlquist et al., 2017) could increase with age. Nonetheless, the study extended existing knowledge on cross-cultural differences in facets of well-being in a European and African context, which has previously received less research attention.

It was also found that perceived hope predicted most of the variance in harmony in life for both samples. Further, perceived hope mediated the relationship between positive feelings and harmony in life. Therefore, in accordance with broaden-and-build theory (Fredrickson, 2001, 2013), positive emotions seem to facilitate perceived hope (which could be seen as a resource) and thus increase harmony in life. This study shed further light on the role of positive feelings in enhancing specific facets of well-being, which were not previously explored together (perceived hope and harmony in life).

Despite yielding potentially interesting results, there are some limitations to the study. The main limitation lies in the fact that the South African sample mainly consisted of White, female participants with a high level of education. More research is needed, including obtaining more participants from other South African population groups as well as those with lower levels of education and from lower socio-economic groups. In conclusion, this study provided some insight on perceived hope and harmony in life in cross-cultural context, thereby contributing to the extant literature on these relatively new constructs in the field of positive psychology.

KEYWORDS

- **emotions**
- **cross-cultural differences**
- **harmony in life**
- **perceived hope**

REFERENCES

Adams, B. G., Abubakar, A., Van de Vijver, F. J., De Bruin, G. P., Arasa, J., Fomba, E., Omri, G., Hapunda, G., Looh, J., Mazrui, L., & Murugami, M. (2016). Ethnic identity in emerging adults in sub-Saharan Africa and the USA, and its associations with psychological well-being. *Journal of Community and Applied Social Psychology, 26*(3), 236–252.

Adams, B. G., F. J., & De Bruin, G. P. (2012). Identity in South Africa: Examining self-descriptions across ethnic groups. *International Journal of Intercultural Relations, 36*(3), 377–388.

Arnett, J. J. (2004). *Emerging adulthood: The winding road from the late teens through the twenties.* New York: Oxford University Press.

Carlquist, E., Ulleberg, P., DelleFave, A., Nafstad, H. E., & Blakar, R. M. (2017). Everyday understandings of happiness, good life, and satisfaction: Three different facets of well-being. *Applied Research in Quality of Life, 12*(2), 481–505.

Charles, S. T., Reynolds, C. A., & Gatz, M. (2001). Age-related differences and change in positive and negative affect over 23 years. *Journal of Personality and Social Psychology, 80*(1), 136–151.

Cohen, S., & Pressman, S. D. (2006). Positive affect and health. *Current Directions in Psychological Science, 15*, 122–125.

Colarusso, C. A. (1992). Middle Adulthood (ages 40–60). In *Child and Adult Development. Critical Issues in Psychiatry* (An Educational Series for Residents and Clinicians). Springer: Boston, MA.

DelleFave, A. (2014). Harmony. In A. C. Michalos (Ed.), *Encyclopedia of quality of life and well-being research* (pp. 2695–2697), Dordrecht: Springer.

DelleFave, A., Brdar, I., Freire, T., Vella-Brodrick, D., & Wissing, M. P. (2011). The eudaimonic and hedonic components of happiness: Qualitative and quantitative findings. *Social Indicators Research, 100*(2), 185–207.

DelleFave, A., Brdar, I., Wissing, M. P., Araujo, U., Castro Solano, A., Freire, T., Hernández-Pozo, M. D. R., Jose, P., Martos, T., Nafstad, H. E., Nakamura, J., Singh, K., & Soosai-Nathan, L. (2016). Lay definitions of happiness across nations: The primacy of inner harmony and relational connectedness. *Frontiers of Psychology, 7*, 30.

Demirli, A., Türkmen, M., & Arık, R. (2015). Investigation of dispositional and state hope levels' relations with student subjective well-being. *Social Indicators Research, 120*(2), 601–613.

Diener, E. (1984). Subjective well-being. *Psychological Bulletin, 95*(3), 542–575.

Diener, E., Gohm, C. L., Suh, E., & Oishi, S. (2000). Similarity of the relations between marital status and subjective well-being across cultures. *Journal of Cross-Cultural Psychology, 31*, 419–436.

Diener, E., Heintzelman, S. J., Kushlev, K., Tay, L., Wirtz, D., Lutes, L., & Oishi, S. (2017). Findings all psychologists should know from the new science on subjective well-being. *Canadian Psychology/Psychologie Canadienne, 58*(2), 87–104.

Diener, E., Oishi, S., & Lucas, R. E. (2003). Personality, culture, and subjective well-being: Emotional and cognitive evaluation of life. *Annual Review of Psychology, 54*(1), 403–425.

Diener, E., Wirtz, D., Tov, W., Kim-Prieto, C., Choi, D. W., Oishi, S., & Biswas-Diener, R. (2010). New well-being measures: Short scales to assess flourishing and positive and negative feelings. *Social Indicators Research, 97*(2), 143–156.

Dunn, T. J., Baguley, T., & Brunsden, V. (2014). From alpha to omega: A practical solution to the pervasive problem of internal consistency estimation. *The British Journal of Psychology, 105*(3), 399–412.

Du Plessis, G. A., & Guse, T. (2017). Validation of the scale of positive and negative experience in a South African student sample. *South African Journal of Psychology 47*(2), 184–197.

Easterlin, R. A. (2009). Lost in transition: Life satisfaction on the road to capitalism. *Journal of Economic Behavior and Organization, 71*(2), 130–145.

Emmons, R. A. (2005). Emotion and religion. In R. F. Paloutzian & C. L. Park (Eds.), *Handbook of the psychology of religion and spirituality* (pp. 235–252). New York: Guilford Press.

Fredrickson, B. L. (2001). The role of positive emotions in positive psychology. The broaden-and-build theory of positive emotions. *American Psychologist, 56*, 218–226.

Fredrickson, B. L. (2009). *Positivity: Top-notch research reveals the 3 to 1 ratio that will change your life*. New York: Three Rivers Press.

Fredrickson, B. L. (2013). Positive emotions broaden and build. In P. Devine & A. Plant (Eds.) *Advances in experimental social psychology* (Vol. 47, pp. 1–53). Burlington, VA: Academic Press.

Fredrickson, B. L., & Joiner, T. (2018). Reflections on positive emotions and upward spirals. *Perspectives on Psychological Science, 13*(2), 194–199.

Garcia, D., Al Nima, A., & Kjell, O. N. E. (2014). The affective profiles, psychological well-being, and harmony: Environmental mastery and self-acceptance predict the sense of a harmonious life. *Peer J, 2*, e259–e259. DOI 10.7717/peerj.259

Guse, T. (2018). *Hope among South African adults: Preliminary findings from the Hope-Barometer Survey 2017*. Paper presented at the 9th European Conference on Positive Psychology, Budapest, Hungary, June.

Guse, T., & Shaw, M. (2018). Hope, meaning in life and well-being among a group of young adults. In A. Krafft, P. Perrig-Chiello, & A. Walker (Eds.). *Hope for a good life. Results of the Hope-Barometer international research program* (pp. 63–77). Cham: Springer.

Hayes, A. F. (2012). *PROCESS: A versatile computational tool for observed variable mediation, moderation, and conditional process model*ing [White paper]. Retrieved from http://www.afhayes.com/public/process2012.pdf

Helliwell, J., Layard, R. & Sachs, J. (Eds.) (2018). *World happiness report*. The Earth Institute, New York: Columbia University. Retrieved on July 20, 2018.

Hofstede, G. (2001). *Culture's consequences: Comparing values, behaviors, institutions, and organizations across nations*. Thousand Oaks CA: Sage Publications.

Hudson, N. W., Lucas, R. E., & Donnellan, M. B. (2017). Day-to-day affect is surprisingly stable: A 2-year longitudinal study of well-being. *Social Psychological and Personality Science, 8*(1), 45–54.

IndexMundi (2018). Retrieved from http://www.indexmundi.com/, July 20, 2018.

JASP Team (2018). *JASP* (Version 0.86) [Computer software].

Jeffrey, K., Wheatley, H., & Abdallah, S. (2016). *Happy Planet Index: 2016 Report. A global index of sustainable well-being*. London, UK: New Economic Foundation. http://www.happyplanetindex.org/. Retrieved on July 20, 2018.

Kagee, A. (2014). South African psychology after 20 years of democracy: Criticality, social development, and relevance. *South African Journal of Psychology, 44*(3), 350–363.

Kato, T., & Snyder, C. (2005). Relationship between hope and subjective well-being: Reliability and validity of the Dispositional Hope Scale, Japanese version. *Shinrigaku Kenkyu: The Japanese Journal of Psychology, 76*(3), 227–234.

Kjell, O. N. E., Daukantaitė, D., Hefferon, K., & Sikström, S. (2016). The harmony in life scale complements the satisfaction with life scale: Expanding the conceptualization of the cognitive component of subjective well-being. *Social Indicators Research, 126*(2), 893–919.

Klicperova, M., Feierabend, I. K., & Hofstetter, C. R. (1997). In the search for a post-communist syndrome: A theoretical framework and empirical assessment. *Journal of Community and Applied Social Psychology, 7*, 39–52.

Klingemann, H. D., Fuchs, D., & Zielonka, J. (Eds.) (2006). *Democracy and political culture in Eastern Europe*. New York: Routledge.

Krafft, A. M., & Choubisa, R. (2018). Hope in the Indian psychology context: Philosophical foundations and empirical findings. In A. Krafft, P. Perrig-Chiello, & A. Walker (Eds.), *Hope for a good life. Results of the Hope-Barometer international research program* (pp. 131–163). Cham: Springer.

Krafft, A. M., Martin-Krumm, C. & Fenouillet, F. (2017). Adaptation, further elaboration, and validation of a scale to measure hope as perceived by people: Discriminant value and predictive utility vis-à-vis dispositional hope. *Assessment,* 24, 1–16.

A. M. & Walker, A. (2018a). *Positive Psychologie der Hoffnung: GrundlagenausPsychologie, Philosophie, Theologie und ErgebnisseaktuellerForschung [Positive Psychology of Hope: Psychological, Philosophical, Theological Foundations and Results of Current Research].* Berlin: Springer.

Krafft, A. M., & Walker, A. (2018b). Exploring the concept and experience of hope— Empirical findings and the virtuous circle of hope. In A. Krafft, P. Perrig-Chiello, & A. Walker (Eds.), *Hope for a good life. Results of the Hope-Barometer international research program* (pp. 21–61). Cham: Springer.

Krafft, A. M., Perrig-Chiello, P., & Walker, A. (2018). *Hope for a good life. Results of the Hope-Barometer international research program.* Social Indicators Research Series, Vol. 72. Cham: Springer.

Lawson, G. (2017). *Negotiated revolutions: The Czech Republic, South Africa and Chile*. London, UK: Routledge.

Leger, K. A., Charles, S. T., & Almeida, D. M. (2018). Let it go: Lingering negative affect in response to daily stressors is associated with physical health years later. *Psychological Science*, *29*(8), 1–8.

Lucas, R. E., Clark, A. E., Georgellis, Y., & Diener, E. (2003). Reexamining adaptation and the set point model of happiness: Reactions to changes in marital status. *Journal of Personality and Social Psychology, 84*(3), 527–39.

Lucas, R. E., & Diener, E. (2015). Personality and subjective well-being: Current issues and controversies. In M. Mikulincer, P. R. Shaver, M. L. Cooper, & R. J. Larsen (Eds.), *APA handbook of personality and social psychology*, Vol. 4 (pp. 577–599). Washington, DC: APA.

Lucas, R. E., Diener, E., & Larsen, R. J. (2009). Measuring positive emotions. In E. Diener (Ed.) *Assessing well-being: The collected works of Ed Diener* (pp. 139–155). Dordrecht, The Netherlands: Springer.

Martínez-Martí, M. L., & Ruch, W. (2017). Character strengths predict resilience over and above positive affect, self-efficacy, optimism, social support, self-esteem, and life satisfaction. *The Journal of Positive Psychology, 12*(2), 110–119.

Meeks, S., & Murrell, S. A. (2001). Contribution of education to health and life satisfaction in older adults mediated by negative affect. *Journal of Aging and Health, 13*(1), 92–119.

Mikels, J. A., Reed, A. E., Hardy, L. N., & Lockenhoff, C. E. (2014). Positive emotions across the life span. In M. M. Tugade, M. N. Shiota, & L. D. Kirby (Eds). *Handbook of Positive Emotions* (pp. 256–272). New York. The Guilford Press.

Oishi, S., & Gilbert, E. A. (2016). Current and future directions of culture and happiness research. *Current Opinion in Psychology*, *8*, 54–58.

Pavot, W., & Diener, E. (2008). The Satisfaction with life scale and the emerging construct of life satisfaction. *The Journal of Positive Psychology, 3*(2), 137–152.

Pavot, W., & Diener, E. (2013). Happiness experienced: The science of subjective well-being. In I. Boniwell & S. David (Eds.) *The Oxford handbook of happiness* (pp. 134–151). Oxford, UK: Oxford UniversityPress.

Perrig-Chiello, P., Spahni, S., & Margelisch, K. (2018). The older, the better? The role of hope for the regulation of subjective well-being over life-span. In A. Krafft, P. Perrig-Chiello, & A. Walker (Eds.), *Hope for a good life. Results of the Hope-Barometer international research program* (pp. 79–94). Cham: Springer.

Peters, G.-J. Y. (2014). The alpha and the omega of scale reliability and validity. *The European Health Psychologist, 16*(2), 56–69.

Peterson, C., & Seligman, M. E. P. (2004). *Character strengths and virtues. A handbook and classification.* New York: Oxford University Press.

Scioli, A., & Biller, H. B. (2009). *Hope in the age of anxiety: A guide to understanding and strengthening our most important virtue*. New York, NY: Oxford University Press.

Singh, K., Jain, A., Kaur, J., Junnarkar, M., & Slezackova, A. (2016). Cross-cultural differences on Gunas and other well-being dimensions. *Asian Journal of Psychiatry*, *24*, 139–146.

Slezackova, A. (2017). *Hope and well-being. Psychosocial correlates and benefits*. University of Malta: Centre for Resilience and Socio-Emotional Health.

Slezackova, A., Cefai, C., & Prosek, T. (2018). Psychosocial correlates and predictors of perceived hope across cultures: A study of Czech and Maltese contexts. In A. M. Krafft, P. Perrig-Chiello & A. M. Walker (Eds.). *Hope for a good life. Results of the Hope-Barometer international research program* (pp. 165–197). Cham: Springer.

Slezackova, A., & Krafft, A. (2016). Hope: A driving force of optimal human development. In J. Mohan & M. Sehgal (Eds.), *Idea of excellence: Multiple perspectives* (pp. 1–12). Chandigarh: Panjab University.

Slezackova, A., Prosek, T., Malatincova, T., & Krafft, A. M. (2020). Psychometricke vlastnosti ceske verze skaly prozivane nadeje: faktorova struktura a vnitrni konzistence [Psychometric characteristics of the Czech version of The Perceived Hope Scale: factor structure and internal consistency]. Ceskoslovenska psychologie, 64(3), 288–305.

Snyder, C. R. (2000). *Handbook of hope: Theory, measures and applications*. San Diego, CA: Academic Press.

Snyder, C. R., Irving, L. M., & Anderson, J. R. (1991). Hope and health. In C. R. Snyder & D. R. Forsyth (Eds.), *Handbook of social and clinical psychology: The health perspective* (pp. 285–305). New York, USA: Pergamon Press.

Statistics South Africa (2018, May). *Youth unemployment still high*. Retrieved from http://www.statssa.gov.za/?p=11129.

Tsai, J. L., Knutson, B., & Fung, H. H. (2006). Cultural variation in affectvaluation. *Journal of Personality and Social Psychology, 90*, 288–307.

Veenhoven, R. (2018a). *World database of happiness*, Erasmus University Rotterdam, The Netherlands. Retrieved September 31, 2018, from http://worlddatabaseofhappiness.eur.nl

Veenhoven, R. (2018b). *Happiness in Czech Republic (CZ)*, World Database of Happiness, Erasmus University Rotterdam, The Netherlands. Retrieved September 31, 2018, from http://worlddatabaseofhappiness.eur.nl.

Veenhoven, R. (2018c). *Happiness in South Africa (ZA)*, World Database of Happiness, Erasmus University Rotterdam, The Netherlands. Retrieved September 31, 2018, from http://worlddatabaseofhappiness.eur.nl.

Watson, D., Clark, L. A., & Carey, G. (1988). Positive and negative affectivity and their relation to anxiety and depressive disorders. *Journal of Abnormal Psychology, 97*, 346–353.

Wirtz, D., Chiu, C. Y., Diener, E., & Oishi, S. (2009). What constitutes a good life? Cultural differences in the role of positive and negative affect in subjective well-being. *Journal of Personality, 77*(4), 1167–1196.

CHAPTER 7

Difficulties in Emotion Regulation as a Barrier to Resilience: Therapeutic Implications

PARWINDER SINGH

Indian Institute of Technology, Ropar, Punjab, India. E-mail: singh_parwinder@iitrpr.ac.in

ABSTRACT

Enhancing psychological well-being and human functioning is not just a matter of eradicating weaknesses but instilling positive attributes among individuals. Researchers have been exploring factors associated with psychological well-being and findings ways to incorporate those positive ones into the organism's behavioral system. Among other factors, resilience emerged as a significant correlate of psychological well-being, and advantageously research demonstrates that resilience can be enhanced by Connor and Davidson. Resilience is considered as a dynamic process referring to the maintenance of positive adjustment under threatening life circumstances. It is the ability to bounce back after disappointments or setbacks, to be adaptable and flexible and to renew a sense of vitality. Psychological resilience promotes personal assets and protects an individual from the potential negative effect of stressors. Thus, enhancing resilience can be an outcome goal of the interventions applied by psychologists to improve well-being of the individuals in different work settings. To achieve the goal, factors associated with resilience must be understood and manipulated. Among others, one such potential factor may be the ability to regulate one's emotion. In the present study, difficulties in emotion regulation were studied in relation to resilience among adults. The primary objective of this study was to explore the relationship among difficulties in emotional regulation and resilience. It was hypothesized that resilience would be negatively associated with different types of difficulties pertaining to emotional regulation. To test the hypotheses, 135 individual working in different sectors within the age range of 24–42 years ($M = 30.04$, SD = 2.24) were administered two scales namely Difficulties in Emotion Regulation Scale (DERS) and Brief Resilience Scale (BRS). The correlational and regression analysis was applied to test the hypotheses. Findings indicate that difficulties in emotion regulation are significantly associated with resilience among Indian adults. Individuals facing more difficulties in regulating their emotions scored less on resilience scales. The result implies that if effective emotion regulation skills can be instilled through some training, their resilience would also be affected. Future studies targeting those difficulties can be designed and tested for the efficacy of certain intervention programmes to enhance emotion regulation skills and its subsequent effects on the resilience. Some of the key ingredients of an effective intervention programme would be discussed in the paper.

7.1 INTRODUCTION

Psychological resilience capacitates an individual to deal with the potential adverse effect of stress (Fletcher & Sarkar, 2012). Resilience is a dynamic characteristic that facilitates adaptation to stressful situations and building of adequate resistance to the adverse impact of stress. This successful adaptation is contingent upon the use of adaptive cognitive processes that guide behavior while confronting a challenge. It can also be influenced by the nature and outcome of previous stressful experiences. The significant aspects of the process of overcoming the adverse impact of stress that can be inferred from the available literature are that cognitive processes and past experiences interact during a resilient response. Additionally, the outcome of this interaction is the effective management of potential adversity. These features affect an individual's capacity to build and utilize a shield against stress and facilitate the adaptive use of coping techniques. Fortunately, this resilience can be built over time, subject to the availability of the right circumstances (Richardson, 2002).

Psychological resilience is also usually conceptualized as an ability to adapt in highly adverse conditions effectively and bounce back with adequate functioning. Adverse conditions can come in the form of family or relationship problems, failures, poor health, financial worries, and so on. It is pertinent to mention here that highly resilient people are not free from negative emotions or thoughts, but with experience, they have developed appropriate coping mechanisms that facilitate the effective transition from the crisis. They tend to have an optimistic attitude and, with practice, able to effectively balance negative emotions with positive ones. In other words, resilience is characterized by adaptive behaviors, thoughts, and actions that can be learned by and developed in anyone.

7.1.1 THEORETICAL PERSPECTIVES ON RESILIENCE

Resilience theories describe the mechanisms for the impact of resilience on quality adaptation in stressful situations. Different terminology has been used for these resilience models. The most common models described in the literature are the "compensatory model," "the challenge model," and "the protective factor of immunity versus vulnerability model" (O'Leary, 1998). Masten and Reed (2002) also outlined a variety of resilience models supported in the literature. The applicability of any given model would depend on individual circumstances, but all the models incorporate risk and protective factors that are processed in some way to result in a successful outcome.

While theorizing resilience, Richardson (2002) proposed a *meta-theory of resilience and resiliency*, which, as he mentioned, developed through three different phases of inquiry. The first phase tried to identify the attributes of people that help them to adapt successfully in adverse conditions. The next phase enquired into the processes through which people internalize these attributes. The third phase was related to the recognition of our innate capacity to develop resilience and grow. This extensive exploration led researchers to conceptualize resilience as a strength within everyone that drives them to seek wisdom, altruism, harmony with a spiritual source of strength, and ultimately, self-actualization (Richardson, 2002).

In one mega research project with children of schizophrenic parents, Garmezy et al. (1984) found that the majority of the children did not develop the maladaptive behavioral pattern despite living in adverse conditions. Instead, they grew up to be warm and competent people. Researchers were keen to understand the dynamics behind such a peculiar development. Those children were then assessed and found to be high on effectiveness and expectancies. They had excellent problem-solving skills, an internal locus of control, positive outlook, high self-esteem, self-discipline, critical thinking skills, and a good sense of humor. Researchers summarized their findings stating that an external support system, personality disposition, and a supportive family environment affects resiliency development. Further, the investigators, in order to understand the relationship between stress, resilience, and adaptation, tried to model resilience and its development by classifying a particular group of factors. In one model, known as the *compensatory model*, researchers postulated that some compensatory factors counterbalance exposure to risk. A compensatory factor directly and independently influences the outcome of interest without interacting with a risk factor. An additive effect of both the risk and compensatory factors can contribute to the outcome. These compensatory factors may contribute as an active approach to solving life's problems, or it can be a tendency to perceive or construct experiences positively. These factors either decrease the impact of stress and subsequent risk initially or restructure risk throughout development.

In their other model, that is, "the challenge model, "Garmezy et al. (1984) asserted that individuals high on resilience tend to perceive a risk factor as a potential challenge that would help in successful long-term adaptation if it is not excessive. Less than optimal stress is not motivating enough, and more than optimal stress deteriorates performance. Optimal or moderate level of stress presents a challenge that, when overcome, increases competence. Resiliency develops by not avoiding risks, but thorough successfully engaging in it. The third model proposed by the same investigators was the *protective factor model*. As per this model, the interaction of protective and risk factor reduces the likelihood of adverse consequences. Protective factor moderates the effect of exposure to risk.

Later, Rutter (1987, 1990) synthesized existing models and understanding of various dynamics associated with resilience, considering developmental literature and systems theory, and asserted that any protective mechanism is essentially an interactive process in which one variable moderate the effect of another. It indicates that protective factors such as skilfulness in managing emotional, intrapersonal, academic and job-related issues, the ability to restore self-esteem, and so on, foster desirable consequences and healthy personality characteristics (Bonanno, 2004; Ungar, 2004). This theory triggered various studies in different domains and received empirical support (Masten et al., 1988; Fergusson et al., 1994; Fergusson & Lynskey, 1996; Cowen et al., 1997; Luthar, 1999; Werner & Smith, 2001; Ungar, 2004; Bonanno, 2004).

According to *Interactive Models of Resilience* (Rutter, 1990; Freitas & Downey, 1998), it is the interaction of many potential risk and protective factors that make one resilient. These models maintained that variables such as temperament, gender, parental relationships, significant life events, and so on, interact differentially with risk factors. The effect of this interaction depends on how these factors are considered,, that is, protective or vulnerability factors. Thus, resilience is determined by multiple factors and affected by the direct and indirect relationship between risk and protective factors.

Similar to various other constructs, resilience also attracts nature versus nurture debate on its development. In order to manipulate its development, it is essential to understand whether resilient characteristics are learned, inherited, or affected by both genes and environment. A common notion of resilience in psychology is that it is dispositional and has trait-like qualities such as enduring and consistent across time and space (Robins et al., 1996). However, some researchers assert that it is the state-like quality and can be learned (Banyard, 2004; Luthans et al., 2006). Researchers, as well as applied psychologists, can benefit from the understanding that what is unique in the lives of resilient individuals that make them so (Masten & Reed, 2002; Greene & Conrad, 2002). This knowledge may help in developing effective preventive strategies aimed at promoting specific processes associated with resilience (Lambie et al., 2002; Masten & Reed, 2002).

Schetter and Dolbier (2011) classified resilience resources into six categories.

1. Dispositional resources, for example, emotional stability, positive affectivity, hardiness, and so on.
2. Reinforcing resources, for example, high self-esteem, self-concept flexibility, autonomy, and so on.
3. Social and interpersonal, for example, perceived social support and healthy relationships within and outside the family.
4. Adaptive beliefs and values, for example, assumptions about the world, spirituality, personal purpose in life, and so on.
5. Cognitive and behavioral skills, for example, cognitive flexibility, cognitive reappraisal, emotion regulation, social skills, and so on.
6. Other resources, for example, the genetic predisposition to good health, intelligence, social status, and so on.

Researchers further maintained that these resources can be the internal quality or the characteristics of ones' surrounding environment. The motivation to use these resources can also be intrinsic or extrinsic. All these dynamics involved and mentioned above can be stable over a long time or may change due to experience (Segerstrom, 2007; Bonanno et al., 2011).

According to Connor and Davidson (2003), resilience can be learned and enhanced by interventions. Resilience is developed and affected by the presence of a range of factors. For instance, perceived social support and good analytical ability may facilitate the development of resilience. Several other additional factors are also associated with resilience; one such potential factor may be the capacity to regulate ones' emotions. Considering the potential role of emotions and its regulation in the resilience process and lack of sufficient substantial support for the same, the present study has been designed to explore it empirically. Before addressing the method used in this study, a brief introduction to emotions and its regulation is required and therefore given below.

7.1.2 EMOTIONS AND ITS REGULATION

Understanding of the dynamics of emotion regulation has always been a quintessential concern for intellects since the time immemorial. Even some asked whether we should attend them or disregard, respect them or condemn them, express them or suppress them. Different opinions are provided to these queries, but there is a common consent that we need to have some control to regulate our emotions. The understanding of emotion regulation requires a thorough understanding of what emotions are all about. In the present context, we will follow William James' and Buck's viewpoint about emotions as they assert that modulation of emotions is possible as per the requirement of the situation which further facilitates in the resilience process.

James (1884, 1894) delineated emotions in terms of adaptive response tendencies in revolutionarily significant situations. It maintains that individuals may modulate their emotional response tendencies. Inconsistencies between emotional response tendencies and actual behavior arouse interest in knowing about how emotions are regulated. In relatively recent times, Buck (1994) also maintained that emotions are flexible response sequences, triggered by the valuation of a situation as challenging. The duration of these reactions may be short but involve changes in the various systems such as behavioral, experiential, autonomic, and neuroendocrine (Lang, 1995). Gross (1998b) maintained that these reaction tendencies could be altered and it is this moderation that shapes the emotional reaction. With these views, comes the idea of regulating emotional responses. If we possess the ability to modulate emotional responses, we can also learn to regulate and control it adaptively despite maladaptive emotional response tendencies originated from negative encounters and ultimately can be resilient. On the other hand, lack of ability to effectively regulate ones' emotion may prevent the development of healthy traits such as resilience, hardiness, and spontaneous use of effective coping strategies in adverse situations. Thus, understanding of the dynamics of emotion regulation becomes an essential prerequisite for developing any intervention program targeting resilience among individuals. A brief description of emotion regulation and its dynamics is as follows.

Emotion regulation involves influencing emotion dynamics (Thompson, 1990),, that is, influencing, which emotions one experience, changing duration, latency, magnitude, and offset of behavioral, experiential, or physiological responses. According to Gross (1998a), these processes may be automatic or controlled and may have their effects at one or more points in the emotion generative process. In his *process model of emotion regulation*, Gross (1999) distinguished five sets of emotion-regulatory processes: situation selection, situation modification, attention deployment, cognitive change, and response modulation. These processes were once classified as antecedent-focused emotion-regulation and response-focused emotion-regulation (Gross & Munoz, 1995). Antecedent-focused emotion-regulation involves processes that occur before the onset of emotional experience. On the other hand, response-focused emotion-regulation processes mainly occur after emotions are generated (Gross, 1998b). In the process, the first step is the selection of the situation among possible available alternatives. It involves takings specific actions that affect the possibility of being in a particular situation. This selection may initiate a desirable or undesirable emotional experience. After selection, in the second step, a situation may be modified for having the desired emotional impact. In the process model, this step was referred to as situation modification. It is important to mention here that situations vary regarding the extent to which they can be modified. Situations also vary in complexity. Sometimes individuals are unable to select or modify a situation; when situations are inevitable, individuals employ internal regulation strategies. These internal strategies try to direct an individual's attention to some other stimuli in order to prevent emotion from fully emerging. One such strategy and the third step as per the given model is "attention deployment." This strategy may assist an individual to select which aspect of a situation a person attends to in order to avoid the undesirable emotional experience. After deploying attention to selected domains, the individual can select an appropriate meaning out of possible meanings and attach to a given situation. Gross (1999) refers to this process of assigning different meaning as "cognitive change." In the last step,, that is, in "response modulation," individuals can influence the response tendencies once they have been elicited.

Above-mentioned processes may be organized as three categories of emotion regulation, with each category targeting the different domains of the emotional experience. One category may be related to the external situation, in which changing the environment itself is the main target. Another category of the emotion regulation process may focus on the internal situation and tries to change the way one perceives, evaluates or thinks about a situation. Some processes can be related to managing the outward response of emotional expression. Deciding among various possible reactions and enacting effectively may come under the management of overt behavior.

The review of relevant literature, as mentioned, suggests that somehow emotion regulation and resilience are interrelated processes; for instance, Troy and Mauss (2011) tried to explicitly connect specific emotion regulation strategies, taken from Gross's framework, with resilience. However, there are very few such studies conducted so far in which this phenomenon has been explored. Even though the relationship between emotion regulation and resilience seems obvious, no systematic study has been conducted in the Indian context so far as per our best knowledge. Therefore, we do not have sufficient empirical evidence about the nature of the relationship between resilience and emotion regulation among the Indian population. To fill the gap in knowledge, the present study has been planned. In the present study, emotional regulation has been operationalized as the minimum difficulties in regulating one's emotions. The primary aim of the present study is to explore the nature of the relationship between difficulties in emotion regulation and resilience. Specifically, six types of difficulties, as suggested by Gratz and Roemer (2004) pertaining to emotion regulation, were explored in relation to resilience scores of individuals. It has been hypothesized that these difficulties would be negatively related to the resilience of the respondents. In other words, effective emotion regulation would be positively associated with resilience. To test the stated hypotheses, the following method has been adopted.

7.2 METHOD

7.2.1 *SAMPLE*

The sample for the present research work comprises 135 respondents (62 males) with the mean age of 30.04 years (SD = 2.24 years), selected randomly from the three service sectors, that is, education, banking, and medical sector functioning in Punjab and Haryana. Though stress is part and parcel of everybody's life but employed individuals serving in these professions are more likely to experience stress, owing to the growing uncertainty in jobs, role conflicts, overload, time pressure, poor work–family balance, anxiety in personal lives, for example, relationship issues, pressure to match up particular lifestyle, and so on, therefore require immediate attention and some way out. Almost equal representation of the urban (49.6%) and rural population (50.4%) was there in the sample. Demographic characteristics of the sample are presented in Table 7.1.

7.2.2 INSTRUMENTS USED

7.2.2.1 DIFFICULTIES IN EMOTION REGULATION SCALE

Difficulties in regulating emotions were measured using Difficulties in Emotion Regulation (DERS) (Gratz & Roemer, 2004). Six types of difficulties are measured with this 36-item self-report measure in which individuals are asked to indicate how often the items apply to themselves. Respondents are supposed to give their responses on a scale ranging from 1 to 5, where 1 is "almost never (0%–10%)," 2 is "sometimes (11%–35%)," 3 is "about half the time (36%–65%)," 4 is "most of the time (66%–90%)," and 5 is "almost always (91%–100%)." It provides six subscale scores as well as a total score. The subscales of this measure are nonacceptance of emotional responses (six items); difficulties engaging in goal-directed behaviors when distressed (five items); difficulties controlling impulsive behaviors when distressed (six items); lack of emotional awareness (six items); limited access to emotion regulation strategies perceived as useful (eight items); and lack of emotional clarity (five items). Scoring is done in a manner that higher scores reflect more significant difficulties in emotion regulation. Total as well as subscale scores have been found to have high internal consistency (e.g., Gratz et al., 2008; Fox et al., 2007). In addition, the scale has demonstrated good test–retest reliability also (ρI = 0.88; Gratz & Roemer, 2004).

7.2.2.2 THE BRIEF RESILIENCE SCALE

The Brief Resilience Scale is used to assess an individual's ability to recover from stress or bounce back in the usual routine (Smith et al., 2008, 2010). It consists of six items with responses on a five-point scale ranging from "strongly disagree" to "strongly agree." In the scale, three items are negatively worded and therefore scored in reverse (item nos. 2, 4, and 6). Internal consistency was good in all four samples, with Cronbach's alpha ranging from 0.80 to 0.91. The test–retest reliability was also found to be satisfactory (Smith et al., 2008). The authors calculated the test–retest reliability with Intraclass correlation and found it to be 0.69 for 1 month and 0.62 for 3 months.

7.2.3 PROCEDURE

Two districts each from Punjab and Haryana were randomly selected, and in each district, eight establishments,, that is, schools, colleges, banks, and hospitals were chosen for data collection. Nearly 12 participants in each establishment were contacted and asked for participating in the study. After taking consent from all the concerned, questionnaires were then sent through e-mail with proper instructions to complete the questionnaires and with information regarding conditions of confidentiality.

Participants were free to contact in case of any confusion about the questionnaires. On successful submission of the questionnaires, all participants were debriefed and were appreciated for their cooperation. Data were then subject to correlation and regression analysis with resilience being the criterion variable and difficulties in emotion regulation as predictors. Findings are presented in Tables 7.1–7.3. Results and implications were discussed.

7.3 RESULTS

The findings of the study are summarized in Tables 7.1–7.3. Table 7.2 presents correlation coefficients among six types of emotion regulation difficulties and resilience. As depicted, the significant positive correlation has been observed between almost all types of difficulties except one, that is, limited access to emotion regulation strategies. Individuals having more difficulties in regulating their emotions scored poorly on the resilience scale. Resilience has been found to be negatively correlated with "nonacceptance of emotional responses" ($r = -0.52$, $p < 0.01$), "difficulty engaging in goal-directed behavior" ($r = -0.59$, $p < 0.01$), impulse control difficulties ($r = -0.45$, $p < 0.01$), lack of emotional awareness ($r = -0.11$, $p > 0.05$), limited access to emotion regulation strategies ($r = -0.52$, $p < 0.01$), lack of emotional clarity ($r = -0.40$, $p < 0.01$), and total score on Difficulties in Emotion Regulation Scale ($r = -0.40$, $p < 0.01$). Thus, the studied difficulties in regulation ones' emotion are significantly related to resilience.

Data have been further analyzed using regression analysis to explore the significant predictors of resilience, and the findings of regression analysis are presented in Table 7.3. Table 7.3 depicts that among all tested variables, "difficulty engaging in goal-directed behavior" came out to be the most significant predictor of resilience, causing 34.5% {F-change (1, 133) = 70.07, $p<0.01$} variation alone in resilience scores of the sample. "Non-acceptance of emotional responses" has been found as the second most crucial predictor causing 39.9% of variation along with "difficulty engaging in goal-directed behavior" and 5.4% {F-change (1, 133) = 11.86, $p<0.01$} of variation alone in resilience. Regression analysis of the present data has come up with two models explaining 39.9% variation in resilience, which indicates the significance of these two types of difficulties in predicting among individuals.

7.4 DISCUSSION

The present study was intended to explore whether emotion regulation processes are somehow related or not to resilience characteristics among Indian adults. It was hypothesized that difficulties in regulating one's emotion would be negatively associated with resilience characteristics of individuals. Findings of the present study support the hypotheses. Resilience is negatively associated with difficulties in emotion regulation. Though almost all types of difficulties are found to be negatively associated with resilience, two specific difficulties have come out to be significant predictors of resilience characteristics: one in "difficulties in engaging in goal-directed behavior" and another one is "nonacceptance of emotional responses."

The findings of the study imply that individuals having more difficulties in regulating one's emotions tend to be less resilient. Specifically, individuals, who are not able to engage themselves into goal-directed activities during the

TABLE 7.1 Demographic Characteristics of Sample and Descriptive Statistics of All Measures

Demographic Characteristics	N	%age	Minimum	Maximum	Mean	SD
Age	135		24	42	30.04	2.24
Gender						
Male	62	45.9				
Female	73	54.1				
Residence						
Rural	67	49.6				
Urban	68	50.4				
A. Nonacceptance of emotional responses			6	27	13.02	5.01
B. Difficulty engaging in Goal-directed behavior			5	24	13.89	4.52
C. Impulse control difficulties			5	23	11.39	4.12
D. Lack of emotional awareness			6	28	15.56	4.02
E. Limited access to emotion regulation strategies			9	36	18.81	6.08
F. Lack of emotional clarity			4	19	8.29	3.33
G. DERS			47	143	80.96	20.51
H. Resilience			9	28	20.51	4.66

TABLE 7.2 Correlation Coefficients among Various Difficulties in Emotion Regulation and Resilience

Variables	A	B	C	D	E	F	G	H
A. Nonacceptance of emotional responses	1.00							
B. Difficulty engaging in goal-directed behavior	0.56**	1.00						
C. Impulse control difficulties	0.66**	0.60**	1.00					
D. Lack of emotional awareness	0.18*	0.01	0.06	1.00				
E. Limited access to emotion regulation strategies	0.77**	0.74**	0.64**	0.12	1.00			
F. Lack of emotional clarity	0.62**	0.44**	0.59**	0.35**	0.61**	1.00		
G. Total DER	0.86**	0.77**	0.79**	0.35**	0.90**	0.78**	1.00	
H. Resilience	−0.52**	−0.59**	−0.45**	−0.11	−0.52**	−0.40**	−0.59**	1.00

DERS, Difficulties in Emotion Regulation.

*$p < 0.05$

**$p < 0.01$

TABLE 7.3 Stepwise Regression Analysis Predicting Resilience from Six Types of Difficulties in Emotion Regulation

Variables	Model 1			Model 2		
	Unstandardized Coefficients		Standardized Coefficients	Unstandardized Coefficients		Standardized Coefficients
	B	SE(B)	β	B	SE(B)	β
Difficulty engaging in goal-directed behavior	−0.600	0.072	−0.587	−0.441	0.083	−0.432
Nonacceptance of emotional responses				−0.257	0.075	−0.279
R	0.587			0.632		
R^2	0.345			0.399		
F	70.07*			43.82*		
R^2 change				0.54		
F R^2 change				11.86*		

*$p < 0.01$

emotional disturbance phase or do not endure in the same persistently, possess less resilient characteristics as compared to those who can keep themselves motivated in achieving their aspirations in similar situations. Individuals facing these difficulties are not able to resist the demands of the stressful situations, succumb to the pressure involved, give up easily, and experience distress. Additionally, when in emotion arousing situations, such individuals are not able to focus or concentrate on their assignments, got it delayed, which further mounts up the pressure for them. This tendency to let their arousal state, hinder goal-directed activities that may make them less resilient.

Another factor pertaining to resilience that came out to be a significant predictor is the nonacceptance of one's emotional responses. After appropriate awareness and clarity of emotions, emotion regulation requires acceptance of emotions which would further help in devising and using strategies for responding to those emotions. In this study, individuals who perceived themselves as nonaccepting their emotional responses also found to be less resilient as compared to those who perceived in the other way. Individuals, who do not accept their ongoing affective states as normal reactions, become angry, embarrassed, and irritable and consider themselves as weak. They are not able to accept their upsetting emotions as normal reactions in a (perceived) threatening situation. Acceptance of these emotional states may reduce their harmful after-effects. One severe effect of this nonacceptance may be the poor development of resilient characteristics.

The findings of the present studies are in line with previous similar studies (though very few for example, Troy & Mauss, 2011) conducted in western countries. Troy and Mauss (2011) found in their study that individuals high on internal emotion regulation capacity display greater resilience after the adverse experience, as compared to their counterparts. They argued that cognitive reappraisal strategies and attention control affect the onset of adaptive emotions and therefore contribute to resilience. More recent studies, as presented in a meta-analysis, also support this notion that cognitive appraisal and problem-focused coping strategies are psychological protective factors and are essential components of resilience construct (Vanhove et al., 2016).

As the previous studies and the present findings support the idea that emotion regulation strategies and resilience are interrelated, researches are now needed in India to explore the efficacy of different interventions (e.g., Emotion Regulation Therapy) to modify emotion regulation strategies of individuals in order to make them more resilient and ultimately more productive members of the society. Researches also support the notion that both resilience (Connor & Davidson, 2003) and emotion regulation processes can be altered with specific interventions because people can control the way they process their emotional dynamics. For instance, Rothermund et al., (2008) concluded in their study that by directing ones' attention away from the adverse stimuli may help individuals in preventing undesirable emotion experience. Similarly, Gross (1998a) emphasized the role of cognitive appraisals in shaping emotional experience. Porges (2007) studied another dimension of emotion dynamic and found that physiological consequences of emotion can be managed under the right circumstances.

It can be assumed that modification in emotion regulation strategies of individuals would lead to some alterations in their resilience processes, too. The process model of emotion regulation (Gross, 1998b) may also be influential in guiding interventions plans to amend emotion regulation strategies. As mentioned earlier, situation selection, situation modification, attentional deployment, cognitive change, and response modulation, the five main steps in emotion regulation dynamics, can be grouped into three categories. One category may be related to the external situation, in which changing the environment itself is the main target. Another category of the emotion regulation processes may focus on the internal situation and tries to change the way one perceives, evaluates, or thinks about a situation. Some processes can be related to managing the outward response of emotional expression. Deciding among various possible reactions and enacting effectively may come under the management of overt behavior.

7.5 THERAPEUTIC IMPLICATIONS

The findings of the present study suggest that any intervention targeting emotion regulation strategies should have the following components:

- Training to identify, differentiate, and describe emotions, especially in emotion-arousing situations.
- As nonacceptance of emotional responses has been found to be negatively related to resilience, reduction in nonacceptance or augmentation of acceptance can have a positive impact on the resilience of individuals. Therefore, intervention should focus on enhancing the acceptance of one's emotions as a normal reaction to a particular situation. Emotion relabeling, that is, to make it more neutral/normal/accepting, can be instrumental in achieving this goal.
- It is observed in the study that individuals, who are not able to resist the demands of the stressful situations, succumb to the pressure involved, give up easily and let their arousal state hinder their goal-directed activities, are less resilient and vice-versa. Therefore, training through some intervention to uphold goal-directed activities even in the aversive situation and to promote persistence in efforts to manage emotions can make individuals more resilient.

- Training to withstand emotional hijacking over rational issues and think about the consequences of their impulsive behavior. It would facilitate self-control in provoking situations.
- Exercises to strengthen the ability to decide which social situation to seek out and which to avoid; it requires self-knowledge and situational awareness.
- Learning of strategies that can equip individuals with the skills to modify the situation to alter its emotional impact.
- Training to help individuals in refocusing their attention on nonemotional aspects available in aversive social situations.
- Restructuring cognitive styles to interpret social situations, especially negative ambiguous ones more accurately.
- Exercises to keep individuals relaxed in an emotional situation, for example, bio-feedback, meditation, music therapy, progressive muscular relaxation therapy, and so on.
- Learning to minimally use emotional avoidance strategies, such as worry, self-criticism, and so on.

As discussed, emotion regulation involves moderating the experience of both positive and negative emotions as well as changes in emotional responding (Gross, 1999). Any intervention, targeting some specific emotion regulatory process, to make it more adaptive, may result in changes in emotional experience as well as in emotional responding. Learning of effective emotion regulation strategies may further affect the development of resilience, both as an ability and as a dynamic process.

The findings and implications must be generalized with caution as the present study, such as many other studies, has certain limitations. The measures used in this study are self-report measures, so the chances of socially desirable responses cannot be ruled out. However, many researchers are of the view that self-report ratings may provide better information to understand emotional tendencies, as individuals are better equipped to describe their own emotional tendencies than external observers (Littleet al., 2003; Raine et al., 2006). As the current study was conducted on a modest sample of young adults only; therefore, the findings need more replications with a larger sample of all age groups such as adolescents and late adulthood to substantiate conclusive remarks.

7.6 CONCLUSION

The study shows that difficulties in emotion regulation are significantly associated with resilience among Indian adults. Individuals facing more difficulties in regulating their emotions were found to be less resilient and vice-versa. Resilience has been found to be negatively associated with difficulties in emotion regulation. Though almost all types of difficulties were found to be negatively associated with resilience, two specific difficulties have come out to be significant predictors of resilience characteristics: one in "difficulties in engaging in goal-directed behavior" and another one is "nonacceptance of emotional responses." The findings imply that if practical emotion regulation skills can be instilled through some specific training, their resilience would also be affected. Specific intervention for each emotion regulatory process, to make it more adaptive, may result in changes in emotional experience as well as in emotional responding. Learning of effective emotion regulation strategies may further affect the development of resilience, both as an ability and as a dynamic process.

KEYWORDS

- **resilience**
- **emotion regulation**
- **psychological well-being**

REFERENCES

Banyard, V. (2004). Adjustment to college among trauma survivors: An exploratory study of resilience. *Journal of College Student Development, 3*, 185–197.

Bonanno, G. A. (2004). Loss, trauma, and human resilience. *American Psychologist, 59*, 20–28.

Bonanno, G. A., Westphal, M., & Mancini, A. D. (2011). Resilience to loss and potential trauma. *Annual Review of Clinical Psychology, 7*, 511–535. Doi: 10.1146/annurev-clinpsy-032210-104526

Buck, R. (1994). Social and emotional functions in facial expression and communication: The readout hypothesis. *Biological Psychology, 38*, 95–115.

Connor, K. M. & Davidson, J. R. (2003). Development of a new resilience scale: the Connor-Davidson Resilience Scale (CD-RISC). *Depression and Anxiety, 18*, 76–82.

Cowen, E. L., Work, W. C., & Wyman, P. A. (1997). The Rochester Child Resilience Project (RCRP): Facts found, lessons learned and future directions divined. In S. Luthar, J. Burack, J. Weisz, & D. Cicchetti (Eds.), *Developmental psychopathology: Perspectives on risk and disorder* (pp. 527–547). New York: Cambridge University Press.

Fergusson, D. M., Horwood, L. J., & Lynskey, M. T. (1994). The childhoods of multiple problem adolescents: A 15-year longitudinal study. *Journal of Child Psychology and Psychiatry, 35*, 1123–1140.

Fergusson, D. M., & Lynskey, M. T. (1996). Adolescent resiliency to family adversity. *Journal of Child Psychology and Psychiatry, 37*, 281–292.

Fletcher, D., & Sarkar, M. (2012). A grounded theory of psychological resilience in Olympic champions. *Psychology of Sport and Exercise, 13*, 669–678. doi: 10.1016/j.psychsport.2012.04.007

Fox, H. C., Axelrod, S. R., Paliwal, P., Sleeper, J., & Sinha, R. (2007). Difficulties in emotion regulation and impulse control during cocaine abstinence. *Drug and Alcohol Dependence, 89*, 298–301.

Freitas, A. L., & Downey, G. (1998). Resilience: A dynamic perspective. International *Journal of Behavioral Development, 22*, 263–285.

Garmezy, N., Masten, A. S., & Tellegen, A. (1984). The study of stress and competence in children: A building block for developmental psychopathology. *Child Development, 55*, 97–111.

Gratz, K. L., & Roemer, L. (2004). Multidimensional assessment of emotion regulation and dysregulation: Development, factor structure, and initial validation of the Difficulties in Emotion Regulation Scale. *Journal of Psychopathology and Behavioral Assessment, 26*, 41–54.

Gratz, K. L., Tull, M. T., Baruch, D. E., Bornovalova, M. A., & Lejuez, C. W. (2008). Factors associated with co-occurring borderline personality disorder among inner-city substance users: The roles of childhood maltreatment, negative affect intensity/reactivity, and emotion dysregulation. *Comprehensive Psychiatry, 49*, 603–615.

Greene, R. R. & Conrad, A. P. (2002) Basic assumptions and term. In Greene, R. R., (Ed.), *Resiliency: An Integrated Approach to Practice, Policy, and Research* (pp. 29–62). Washington DC: NASW Press.

Gross, J. J., & Munoz, R. E (1995). Emotion regulation and mental health. *Clinical Psychology: Science and Practice, 2*, 151–164.

Gross, J. J. (1998a). Antecedent- and response-focused emotion regulation: Divergent consequences for experience, expression, and physiology. *Journal of Personality and Social Psychology, 74*, 224–237.

Gross, J. J. (1998b). The emerging field of emotion regulation: An integrative review. *Review of General Psychology, 2*, 271–299.

Gross, J. J. (1999). Emotion and emotion regulation. In L. A. Pervin & O. P. John (Eds.), *Handbook of personality: Theory and research* (2nd ed., pp. 525–552). New York: Guilford Press.

James, W. (1884). What is an emotion? *Mind, 9*, 188–205.

James, W. (1894). The physical basis of emotion. *Psychological Review, 101*, 205–210.

Lambie, R. A., Leone, S. D., & Martin, C. K. (2002). Fostering resilience in children and youth. In J. Carlson & J. Lewis (Eds.), *Counseling the adolescent*, 4th ed., (pp. 87–119). Denver, CO: Love Publishing.

Lang, P. J. (1995). The emotion probe: Studies of motivation and attention. *American Psychologist, 50*, 372–385.

Little, T. D., Jones, S. M., Henrich, C. C., & Hawley, P. H. (2003). Disentangling the "whys" from the "whats" of aggressive behavior. *International Journal of Behavioral Development, 27,* 122–133. https://doi.org/10.1080/01650250244000128

Luthans, F., Vogelgesang, G., & Lester, P. (2006). Developing the psychological capital of resiliency. *Human Resource Development Review, 5*(1), 25–44.

Luthar, S. S. (1999). Measurement issues in the empirical study of resilience: An overview. In M. Glanz, & J. L. Johnston (Eds.), *Resiliency and development: Positive life adaptations* (pp. 129–160). New York: Plenum.

Masten, A. S., Garmezy, N., Tellegen, A., Pellegrini, D. S., Larkin, K., & Larsen, A. (1988). Competence and stress in school children: The moderating effects of individual and family qualities. *Journal of Child Psychology and Psychiatry, 29,* 745–764.

Masten, A. S., & Reed, M. J. (2002). Resilience in development. In C. R. Snyder & S. J. Lopez (Eds.), *The handbook of positive psychology* (pp. 74–88). London, England: Oxford University Press.

O'Leary, V. E. (1998). Strength in the face of adversity: Individual and social thriving. *Journal of Social Issues, 54,* 425–446.

Porges, S. W. (2007). The polyvagal perspective. *Biological Psychology, 74,* 116_143.

Raine, A., Dodge, K., Loeber, R., Gatzke-Kopp, L., Lynam, D., Reynolds, C., Stouthamer-Loeber, M., Liu, L. (2006). The reactive-proactive aggression questionnaire: Differential correlates of reactive and proactive aggression in adolescent boys. *Aggressive Behavior, 32,* 159–171. https://doi.org/10.1002/ab.20115

Richardson, G. E. (2002). The metatheory of resilience and resiliency. *Journal of Clinical Psychology, 58* (3), 307–321.

Robins, R., John, O., Caspi, A., Moffitt, T., & Stouthamer-Loeber, M. (1996). Resilient, over-controlled, and under-controlled boys: Three replicable personality types. *Journal of Personality and Social Psychology, 70,* 157–171.

Rothermund, K., Voss, A., & Wentura, D. (2008). Counter-regulation in affective attentional bias: A basic mechanism that warrants flexibility in motivation and emotion. *Emotion, 8,* 34–46.

Rutter, M. (1987). Psychosocial resilience and protective mechanisms. *American Journal of Orthopsychiatry, 57,* 316–331.

Rutter, M. (1990). Psychosocial resilience and protective mechanisms. In J. Rolf, A. S. Masten, D. Cicchetti, K. H. Nuechterlein, & S. Weintraub. (Eds), *Risk and protective factors in the development of psychopathology,* (pp. 181–214). New York: Cambridge University Press.

Schetter, C. D., & Dolbier, C. (2011). Resilience in the context of chronic stress and health in adults. *Social and Personality Psychology Compass, 5*(9), 634–652. doi:10.1111/j.1751-9004.2011.00379.x

Segerstrom, S. C. (2007). Optimism and resources: Effects on each other and onhealth over ten years. *Journal of Research in Personality, 41*(4), 772–786.

Smith, B. W., Dalen, J., Wiggins, K., Tooley, E., Christopher, P., & Bernard, J. (2008). The brief resilience scale: assessing the ability to bounce back. *International Journal of Behavioral Medicine, 15*(3), 194–200.

Smith, B. W., Tooley, E. M., Christopher, P., & Kay, V. S. (2010). Resilience as the ability to bounce back: A neglected personal resource? *Journal of Positive Psychology, 5,* 166–176.

Thompson, R. A. (1990). Emotion and self-regulation. In R. A. Thompson (Ed.), *Socio-emotional development. Nebraska symposium on motivation* (pp. 367–467). Lincoln: University of Nebraska Press.

Troy, A. S., & Mauss, I. B. (2011). Resilience in the face of stress: Emotion regulation as a protective factor. In S. M. Southwick, B. T. Litz, D. Charney, & M. J. Friedman (Eds.), *Resilience and mental health: Challenges across the lifespan* (pp. 30–44). Cambridge, UK: Cambridge University Press.

Ungar, M. (2004). A constructionist discourse on resilience. *Youth & Society, 35,* 341–365.

Vanhove, A. J., Herian, M. N., Perez, A. L. U., Harms, P. D., & Lester, P. B. (2016). Can resilience be developed at work? A meta-analytic review of resilience-building programme effectiveness. *Journal of Occupational and Organizational Psychology, 89*(2), 278–307. http://doi.org/10.1111/joop.12123

Werner, E. E., & Smith, R. S. (2001). *Journeys from childhood to midlife: Risk resilience and recovery.* New York: Cornell University Press.

CHAPTER 8

Assessment of Coping with Stress: A Short Version of a Coping Strategy Indicator (CSI) in Peruvian Adults

CÉSAR MERINO-SOTO[1,2*] and ARTURO JUÁREZ-GARCÍA[2]

[1]*Universidad de San Martín de Porres, Lima, Peru*

[2]*Universidad Autónoma del Estado de Morelos, Cuernavaca, México*

[*]*Corresponding author. E-mail: sikayax@yahoo.com.ar*

ABSTRACT

There is a need for valid but parsimonious measures for the identification of individual resources that reduce the effects of stress in multiple contexts. The objective of this study was to obtain an abbreviated version of the Coping Stress Indicator (CSI), a self-report measure for international use, which evaluates three global strategies: Problem Solving, Support Search, and Avoidance. A total of 607 young adults participated, from Metropolitan Lima City (Perú). The 33 item CSI was used. The CSI was administered in groups, and the analysis was done using a substantive (theoretical analysis of the selected items to reduce the dependence on the results on the sample); and empirical approach (modeling of structural equations for categorical variables, as well as Rasch model analysis for ordinal items). The measurement invariance and the long–short equivalence of the CSI versions were also evaluated. An acceptable model fit for the full version was found; but the derived short version showed a superior fit, with good reliability (but moderate for the avoidance dimension). The equivalence between the long and the short version was acceptable to classify the sample in tertiles and quartiles. Measurement invariance, comparing males and females, reached strict invariance; and the scores of the scales show acceptable properties from the Rasch analysis. The theoretical and practical implications, as well as the potential international expectations of the obtained version are discussed.

8.1 INTRODUCTION

The importance of cognitive processes in the phenomenon of stress became very important from the initial contributions of Lazarus, who revolutionized the concept of stress under an emerging common denominator in its dynamics: the coping process (Lazarus, 1966). Basically, coping is understood as a reactive-adaptive response to a situation evaluated as stressful (Folkman, 2011).

The characteristics of how a person uses coping strategies to stress seem to have an incremental effect on the understanding of the effects of stress itself, and therefore, the evaluation of stress seems incomplete if some measure of coping is not included. The predominant strategy used in the measurement of coping strategies is the self-report, and it is common to find a wide variety of questionnaires in the scientific literature for children and adolescents (Garcia, 2010; Skinner et al., 2003) and adults (Greenaway et al., 2014). The present study focuses on the Coping Strategy Indicator (CSI; Amirkhan, 1990), a measure that quantifies three general coping strategies: (a) problem solving (PROSO), referring to the instrumental approach to the problem in which the person evaluates and manipulates the situations to deal with the problem; (b) search for social support (SSS), in which human contact is actively sought either as instrumental support to face the problem or as a palliative of the effects of stressors, and (c) avoidance (AVO), which reflects the person's emotion-focused response, but involving some form of physical or psychological withdrawal from the problem. The CSI is one of the available coping measures frequently used in research (Greenaway et al., 2014), and was constructed as a measure of the coping state with a dynamic approach where coping behavior emerges in relation to a specific situation. This is relevant from the theoretical–methodological point of view, since it has been found that there is a lack of correspondence between the remembered report of specific coping modes and daily or more general coping (Ptacek et al., 1994), which is explained by the type of predominant or specific stressor that a person faces and the orientation of the report toward the coping-state, implying all this a common bias in many coping assessment scales. In this way, the CSI is a measure that adheres to the original theoretical conceptions of the coping process and positions it as one of the best scales for assessing coping with specific stressors (Amirkhan, 1990; Lazarus, 1966; Greenaway et al., 2014; Terry, 2014).

Based on a heterogeneous sample, the CSI was developed iteratively through mainly empirical decisions, with the application of factorial analysis to successive heterogeneous samples of North Americans. These results allowed reducing the initial number of items (161), until successively selecting 63, 36, and finally 33 items (Amirkhan, 1990). Subsequently, the three-dimensional CSI model was successfully adjusted in several studies with samples of different cultural characteristics, such as Arabs (Aitcheson et al., 2017), Belgians (Bijttebier & Vertommen, 1997), Africans from various racial and ethnic groups (Ager & Maclachlan, 1998; Wissing & Van Eeden, 2002), Spaniards (Soriano & Zorroza, 1999), Chinese (Lo, 2017), Koreans (Shin & Kim, 2002), Peruvians (Merino et al., 2007), Serbs (Žuljević et al., 2015), Turks (Aysan, 1994), South Africans (Spangenberg & Orpen-Lyall, 2000), and an independent replication in North Americans (Clark et al., 1995). Other measurement models also produced a better comparative fit compared to the original model (Ager & Maclachlan, 1998, Desmond et al., 2006), but with low reliability in their scores.

A recent study strengthened the distribution of items in the three constructs; through bi-factor modeling (Žuljević et al., 2015). In this study, a corresponding general latent factor loaded on by the AVO items had greater reliable variance compared with its specific factors (AVO and distraction strategies). This is particularly important to clearly identify the underlying commonality of AVO behaviors, where there is usually instability of factorial outcomes (e.g., Ager & Maclachlan, 1998; Desmond et al., 2006; Žuljević et al., 2015). Although these studies adjusted their data to a 4-factor model, the predominant use of CSI has been with the three original factors even in those studies where there was dimensional discrepancy (e.g., Desmond & MacLachlan, 2006).

Some criticisms of the development of the CSI have indicated that its construction was limited to the emergence of high-order factors and those concurrent correlations with other coping measures did not appear to be sufficiently discriminative of the intended constructs. Specifically, SSS correlated more with the PROSO scale than with the scale of the same name in another instrument (Schwarzer & Schwarzer, 1996). However, this seems reasonable since the contents of the CSI SSS scale sample the instrumental and palliative function of this approach, that is, to seek human contact to reorient oneself toward PROSO and emotional protection (Amirkhan, 1990). Some studies have empirically separated these two expressions of avoidant coping, with moderate covariation among them, but with comparatively low reliability with respect to the other CSI scales such as PROSO and AVO (Ager & Maclachlan, 1998; Desmond et al., 2006). Although it is theoretically and empirically justified, the practical use of this is questionable and limited, mainly due to the high amount of measurement error, and lack of studies that directly replicate these discrepant results.

Psychometric evidence of the relationships between the approach of seeking support and facing the problem have also reported that expressions of support would cooperate more with emotional expression than with some facets of PROSO, such as cognitive restructuring and direct PROSO (Cano et al., 2007; Nava et al., 2010; Rodríguez-Díaz et al., 2014; Tobin et al., 1989). However, the form of this covariation is less clear and has not been discussed by its authors. According to other results, the strategies of logical analysis and positive reevaluation of problems are positively associated with the strategy of solving problems, and negatively with a resignation orientation (Endler & Parker, 1994; Kirchner et al., 2008); while the active search for others, and emotional discharge and AVO are associated with general strategies of seeking support and avoiding the stressful problem (Kirchner et al., 2008). Therefore, perhaps a dimension of a higher order, encompassing the specific coping behaviors, not only allows the instrument to be more generalizable, but also more robust. The replicability of the basic dimensions of coping has also been observed in other instruments adapted to Spanish language (e.g., *Coping Strategies Inventory*, CSI-T; Tobin et al., 1989), specifically in the Spaniard (Cano et al., 2007) and Mexican contexts (Nava et al., 2010; Rodríguez-Díaz et al, 2014).

These studies suggest that the differentiation of coping strategies to stress into three large dimensions (PROSO, SSS, and AVO) is not only theoretically consistent, but predominantly inter-culturally generalizable. Although there is evidence that the structure of this model may not be the best construct representation in every cultural context, it is also not the worst adjustment, and conceptually, it is stable. This representation of coping may be sufficient for intervention purposes, group description, and scientific research. The reliability coefficients obtained in the studies that validated the three dimensions also tend to be large enough to commend their use. In all the studies reviewed the PROSO and SSS scales tended to have high internal consistency (around or greater than 0.80) whereas the AVO scale had moderate to acceptable reliability coefficients (around 0.70).

The replicability of the three dimensions of the CSI has not been investigated in a smaller or shorter version to the latest version of 33 items (Amirkhan, 1990), which can help to define the three-dimensional theoretical relevance of the instrument, or even more, it would have a mainly practical value because of its obvious advantages in scientific research and professional practice with respect to achieving its evaluative efficiency with fewer indicators (Stanton et al., 2002).

In scientific research, it has been recognized that the modeling of relationships requires that the variables be parsimonious in extension of their content, so as not to add more degrees of freedom, capitalize the mismatch of some item, among others (Yang et al., 2010). In professional practice, the main reasons for creating abbreviated versions of existing scales have included: (a) greater parsimony of measurement, (b) reduction of time and logistical problems that accompanying test administration as well as (c) evaluative efficiency; whereby a reduced version reaches maximum validity and precision in scientific research and in clinical practice (Stanton et al., 2002; Yang et al., 2010). In the area of coping assessment, there are a number of examples of shortened tests that have demonstrated acceptable psychometric properties (e.g., Addison et al., 2007; Speyer et al., 2016). However, such adaptation of the CSI is practically nonexistent in the Hispanic and in particular, Latin American context. Some examples of the reduction of the CSI have appeared in the Anglo-Saxon literature (Marsh et al., 2011, Patrick et al., 2009), but as secondary objectives, and therefore, with only a partial report of the item reduction procedure. Additionally, it is worth noting that CSI research in the Latin American context is relevant not only for the contribution to the assessment of coping, where its study is scarce, but because its analysis implies a transcultural validation of the dimensions that constitute it and the universality of this theory, which is a high priority in the case of standardized instruments (Choi & Juárez-García, 2017). Considering that a short version of the CSI can be an important contribution to research and provide a more efficient instrument for measuring coping with stress in Latin America, the objective of this study was to develop an abridged version of the CSI in a sample of Peruvian adults.

8.2 METHOD

8.2.1 PARTICIPANTS

The sample was 607 participants (university A: 389, 64.1%, university B: 218, 35.9%), of which 401 were women (66.1%, males: 206, 33.9%). The average age was 19.6 (SD = 5.4) and ranged from 15 to 53 years old. The age difference according to sex was trivial (Mann–Whitney-Z = 0.833, p = 0.405, r = 0.03), and differences between university were small (Mann–Whitney-Z = 8.705, $p < 0.001$, r = 0.354). The distribution of the participants according to study program was as follows: administration of international business (37, 6.1%) and tourism (14, 2.3%), architecture (19, 3.1%), biology (28, 4.6%), communication (23, 3.8%), law (11, 1.8%), stomatology (20, 3.3%), engineering (47, 7.7%), marketing (28, 4.6%), human medicine (149, 24.5%), psychology (13, 2.1%), and nursing (218, 35.9%). The distributional differences (quantified with h; Cohen, 1992) between males and females that occurred in nursing care was predominant ($p_{mujeres} > p_{varones}$, $h = 0.743$), and in the rest of the groupings, they were trivial or small.

8.2.2 INSTRUMENT

8.2.2.1 COPING STRATEGY INDICATOR (CSI)

It is an instrument that uses a self-report format to evaluate adult coping strategies in the face of identified stressful events (Amirkhan, 1990). It is composed of three theoretical and factorially developed subscales: PROSO, SSS, and AVO. There are 33 items distributed in 11 items in each subscale. The score is obtained from the simple sum of the items in each subscale. The instructions first request that the participant choose a stressful event that has occurred in his or her life and briefly describe it; then, based on that stressor, the subject responds to the items. The criterion for choosing the stressful event is that it is a relatively recent event (within the previous 6 months) and an important one ("a problem that caused you concern"). The items use a three-point response format (a lot, a little, or not at all). In the original instrument, the internal consistency ($\alpha > 0.80$) and stability ($r > 0.80$) reliability coefficients were high, and have been replicated in several studies, including in Peru (Merino, 2010; Merino et al., 2007). The reports on criterion validity and relations with other attributes were satisfactory (Amirkhan, 1994a, b, 1998). The derivative or abbreviated version was obtained using certain criteria described below, where the short version obtained is also specified.

8.2.3 PROCEDURE

The test administration took place in the students' classrooms and consisted of the voluntary participation form, a demographic sheet, and the CSI, presented in this order. Oral instruction for participation and response to the material was maintained in a standardized manner; in the same way, anonymity was guaranteed, and participants were encouraged to respond honestly to all the items.

The analysis procedure began with an evaluation of the replication of the complete CSI model (33 items), by means of structural equation modeling (SEM). For this, the robust *mean-and-variance-adjusted unweighted least squares* (WLSMV; Asparouhov & Muthén, 2010) estimator for categorical variables was applied, based on adjusting the first and second moments of the χ^2 statistic (Muthén, 1984). This method uses a probit link to define the functional relationship between the items and the construct, as well as polychoric correlations between the items, and the estimation of thresholds between response categories. This method provides more precise parameters estimates (e.g., factor loads) when the number of response options is less than five and the distribution asymmetry is strong (Li, 2016a, 2016b; Sass et al., 2014). All the modeling was done with theta parameterization and through the MPLUS 7 program (Muthén & Muthén, 1998–2012). The initial specification was that, for each group, the loads and thresholds (two for each item) of all the items were freely estimated, the latent variance of the factors was set to 1, the latent means set to zero, and the residuals fixed at 1. The short version was obtained by applying several selection criteria for identifying preferred items: (a) items with the highest loads in their corresponding dimension, which would maximize the internal validity of the new CSI-S; (b) the theoretical rationality of the chosen item, to maintain consistency with the definition of the construct and reduce the variability from sample to sample; and (c) the modification indexes (MI) should be below 3.840, to identify the lowest possible potential changes of parameters (e.g., correlated errors), and to reduce and over-parameterize the model fit of the new CSI-S items. The item number in each subscale was set to 5, given that this is an optimal number of items to include in parcels used in SEM modeling (Yang et al., 2010).

For a deeper analysis of the discriminative capacity of the short version obtained, an item response theory (IRT) approach using the partial credit model (PCM; Masters, 1982, 1988), which comes from the Rasch family of models (Luo, 2005; Smith et al., 2008), was applied. There are several models also suitable for polytomous items; however, PCM tends to be useful, efficient, and parsimonious for measures based on questionnaires or surveys whose item structure is ordered (Leis et al., 2015; Luo, 2005; Masters, 1988). Additionally, conceptual equivalence has been reported against another polytomous IRT models (Luo, 2005). The *eRm* package (Mair & Hatzinger, 2007), operated within the *iana* graphic interface (Hock, 2015) in the *R* statistical environment (R Core Team, 2018), was used to fit the PCM model. At the item level, the model fit was determined with the goodness of fit test χ^2, the OUTFIT and INFIT coefficients and their corresponding statistical tests based on the *t* distribution (Linacre, 2002). However, due to the sensitivity of *t*-tests to the sample size (Smith et al., 2008), the OUTFIT and INFIT coefficients were prioritized to decide on the adjustment of the items. The sample size for this analysis was sufficient, since modeling in the Rasch family estimates fewer parameters (Linacre, 1994).

The second part of the analysis consisted of evaluating equivalence of the shortened and full version-through (a) the correlation between the direct scores; (b) the estimation of the agreement of the classification of the level of coping based on quartiles, derived from the direct scores; and (c) the agreement through the analysis of the differences for continuous variables (Giavarina, 2015). The correlation was estimated using the Pearson correlation coefficient with correction for spurious correlation (Merino & Angulo, 2013). The agreement of quantiles was quantified by the *AC* coefficient (Gwet, 2014), and Kappa (κ; Fleiss, 1971). Whereas equivalence based on differences in continuous variables was implemented by constructing the limits of the agreement between two continuous quantitative measures (Altman & Bland, 1983), but due to the linear trend between both measures, the limits of the agreement were adjusted by a regression method (Mander, 2012).

The final analysis consisted of evaluating the measurement invariance between men and women on the short version. Because items were measured on an ordinal scale the parameter of item thresholds was also included, along with their factorial and residual loadings (Vandenberg & Lance, 2000). A procedure of restrictions applied successively to the parameters of the items was followed in a general-specific approach (Millsap & Tein, 2004). The configurational invariance was first tested, with the only restriction being on the number of equal dimensions in both groups. Once this was achieved, we continued with testing the metric invariance, by restricting the equality of factor loading between the groups. In the next level of invariance, the equality of the thresholds was tested (scalar invariance) but keeping the factor loadings between the groups constant. Finally, the residual invariance was tested, specifying the equality of the residuals of the items between both groups.

This sequence of four steps allows a better evaluation of the potential source of noninvariance (Bowen & Masa, 2015), since the metric invariance (factor loadings) and scalar (thresholds) are independently tested. At each step of the described sequence, the least restricted model was compared with the consecutive restricted model. A significant change in model fit indices would indicate that some of the parameters evaluated do not remain invariant between the groups. The statistical test of invariance was evaluated by comparing nested models with the DIFFTEST procedure (Asparouhov & Muthen, 2006; Muthén & Muthén, 1998–2012). Due to the sensitivity of ΔWLSMV $- \chi^2$ regarding the sample size and the degree of incorrect specification of the models (Meade & Lautenschlager, 2004), the evaluation of the adjustment was complemented by the practical difference between the CFI and RMSEA.

8.2.4 RESULTS

8.2.4.1 CSI FULL VERSION

8.2.4.1.1 Internal Structure

The model fit of the full version was moderately satisfactory, considering the discrepancy between the model fit indices obtained: WLSMV $- \chi^2 =$ 1557.620 (df = 492, $p < 0.001$), RMSEA = .060 (IC 90% = 0.056, 0.063), CFI = 0.908, WRMR = 1.813. The essential source of this discrepancy seemed to be associated with the unmodeled correlations errors, given that 9 possible respecifications were detected in correlated errors (IM > 20.0), and two IM >-50.0 with respect to potential changes in the beta parameters between latent variables and noncorresponding items, even with IM > 150.0; their corresponding standardized changes were around 0.300.

8.2.4.1.2 Reliability

The coefficient ω was high for SSS and PROSO, and moderate for AVO; the coefficient α was comparatively low with respect to its corresponding ω in each score. The difference between both ($\Delta_{\alpha-\omega}$) coefficients can be considered significant, ($\Delta_{\alpha-\omega} > |0.060|$; Gignac et al., 2007) only in AVO. Overall, the obtained coefficients tend to be acceptable. The parameters obtained are shown in Table 8.1.

8.2.5 CSI SHORT VERSION

8.2.5.1 INTERNAL STRUCTURE

According to the results shown in Table 8.1, for the SSS construct, the five items with the highest factor loadings were chosen: csi7, csi23, csi24, csi31, and csi32. However, although item 14 retained greater variance compared to the immediately lower item (this is, item 7, selected for the final short version), it was not chosen because its content describes the interaction with other people to obtain information or to orientate toward how to resolve the stressor. Although the SSS may have this purpose, its content was associated predominantly with an active behavioral orientation described by the PROSO construct, and therefore it would add factorial complexity or irrelevant variance. Instead, item 7 described the verbal interaction behavior to attenuate the emotional impact of the stressor, which conceptually is different than what is measured by PROSO. With respect to the PROSO construct, the following items were retained with strong factor loadings: csi8 csi9 csi15 csi16, and csi29. Finally, for AVO, the selected items were csi13, csi18, csi21, csi26, and csi30. The correlations between the constructs were PROSO–SSS = 0.391 ($z = 8.631$, $p < 0.01$), PROSO–AVO = −0.128 ($z = -2.261$, $p < 0.05$), and SSS–AVO = −0.070 ($z = -1.349$, $p > 0.05$). This correlational pattern essentially retains the correlations between the constructs evaluated in the full version of the CSI (see bottom of Table 8.2).

TABLE 8.1 Dimensional Analysis and Descriptive Statistics of the Full CSI (33 Items)

	M	SD	r_{itc}	SSS	PROSO	AVO	h^2	Ratio STN
CSI1	2.34	0.747	0.595	0.699			0.489	0.955
CSI5	2.45	0.713	0.589	0.733			0.537	1.161
CSI7	2.31	0.739	0.672	0.789			0.622	1.649
CSI12	2.41	0.719	0.672	0.784			0.614	1.595
CSI14	2.31	0.715	0.692	0.810			0.657	1.908
CSI19	2.11	0.843	0.637	0.758			0.574	1.351
CSI23	2.14	0.811	0.691	0.840			0.705	2.397
CSI24	2.07	0.806	0.695	0.843			0.711	2.456
CSI25	2.27	0.776	0.623	0.740			0.548	1.210
CSI31	2.37	0.744	0.729	0.861			0.742	2.866
CSI32	2.40	0.738	0.667	0.813			0.660	1.950
CSI2	2.35	0.630	0.456		0.570		0.325	0.481
CSI3	2.63	0.610	0.497		0.667		0.445	0.801
CSI8	2.62	0.622	0.623		0.759		0.577	1.359
CSI9	2.57	0.608	0.610		0.734		0.538	1.168
CSI11	2.49	0.662	0.536		0.696		0.484	0.940
CSI15	2.61	0.589	0.623		0.815		0.665	1.978
CSI16	2.59	0.603	0.624		0.769		0.592	1.447
CSI17	2.47	0.672	0.584		0.703		0.494	0.977
CSI20	2.58	0.614	0.468		0.594		0.353	0.545
CSI29	2.66	0.590	0.601		0.776		0.603	1.514
CSI33	2.31	0.702	0.535		0.684		0.468	0.879
CSI4	2.20	0.720	0.371			0.484	0.234	0.306
CSI6	2.17	0.755	0.266			0.376	0.142	0.165
CSI10	2.28	0.705	0.196			0.275	0.076	0.082
CSI13	1.94	0.708	0.357			0.529	0.280	0.389
CSI18	1.56	0.684	0.350			0.517	0.267	0.365
CSI21	1.59	0.671	0.348			0.582	0.339	0.512
CSI22	1.92	0.836	0.231			0.318	0.101	0.113
CSI26	1.73	0.750	0.394			0.567	0.322	0.474
CSI27	2.51	0.652	0.201			0.308	0.095	0.105
CSI28	1.55	0.745	0.327			0.482	0.232	0.303
CSI30	1.63	0.715	0.382			0.639	0.408	0.690
Correlation								
SSS				1				
PROSO				0.445	1			
AVO				0.087	−0.046	1		
ω				0.948	0.917	0.752		
α				0.911	0.863	0.660		
$\Delta_{\alpha\text{-}\omega}$				−0.037	−0.054	−0.092		
M				25.173	27.891	21.09		
SD				6.077	4.492	3.792		
Sk				−0.533	−1.162	0.056		
Ku				−0.609	1.205	−0.189		

Note: SSS, social support seeking; PROSO, problem solving; AVO, avoidance; α and ω, reliability coefficients; $\Delta_{\alpha\text{-}\omega}$, difference α and ω reliability coefficients; r_{itc}, corrected item-test correlation; SK, skew; Ku, Kurtosis.

TABLE 8.2 Confirmatory Factor Analysis of Final Short Version of the CSI

	r_{itc}	SSS	PROSO	AVO	h^2	Ratio STN
CSI7	0.571	0.747			0.558	1.262
CSI23	0.612	0.684			0.469	0.879
CSI5	0.562	0.759			0.577	1.359
CSI31	0.731	0.916			0.838	5.213
CSI32	0.655	0.836			0.699	2.321
CSI8	0.610		0.771		0.595	1.466
CSI9	0.543		0.710		0.504	1.017
CSI15	0.615		0.840		0.705	2.397
CSI16	0.591		0.765		0.585	1.411
CSI29	0.590		0.801		0.641	1.790
CSI13	0.353			0.505	0.255	0.342
CSI18	0.264			0.370	0.137	0.159
CSI21	0.469			0.726	0.527	1.115
CSI26	0.390			0.491	0.241	0.318
CSI30	0.493			0.836	0.699	2.321
ω		0.893	0.885	0.732		
α		0.828	0.805	0.639		
$\Delta_{\alpha-\omega}$		0.065	0.080	0.093		
M		11.667	13.06	8.458		
SD		2.884	2.258	2.258		
Sk		−0.661	−1.414	0.470		
Ku		−0.514	1.603	−0.220		

Note: SSS, social support seeking; PROSO, problem solving; AVO, avoidance; α and ω, reliability coefficients; $\Delta_{\alpha-\omega}$, difference α and ω reliability coefficients; r_{itc}, corrected item-test correlation; SK, skew; Ku, Kurtosis. Items were choose based in higher factor loadings (see text).

The resulting model fit was satisfactory, WLSMV-χ^2 = 330.365 (df = 87), CFI = 0.955, RMSEA = .068 (IC 90% = .060, .076), WRMR = 1.489; however, a very strong residual covariance was detected between items 24 and 23 of SSS: MI − χ^2 = 94.80, expected change in parameter (ECP) = 1.503. Similar to item 17, the content of item 24 (went to a friend for advice on how to change the situation) describes the instrumental support obtained from others, but for the purpose of changing the situation (as is consistent with PROSO), and so was replaced with item 5 (accepted sympathy and understanding from someone), which had a high factor loading and was judged to have better correspondence with the SSS construct. The new respecified scales showed incremental model fit over the previous version: WLSMV − χ^2 = 253.096 (df = 87), CFI = 0.961, RMSEA = .056 (IC 90% = .048, .064), WRMR = 1.267 (the final version of the SSS scale included items 5, 7, 23, 31 and 32). The intersubscale correlations of the final version were PROSO–SSS = 0.457 (z = 10.140, $p < 0.01$), PROSO–AVO = −0.129 (z = −2.261, $p < 0.05$), and SSS–AVO = −0.157 (z = −2.983, $p < 0.01$).

8.2.5.2 RELIABILITY

The difference between the estimated coefficients ($\Delta_{\alpha\text{-}\omega}$) suggests that these can be considered significant, ($\Delta_{\alpha\text{-}\omega} > |0.060|$; Gignac et al., 2007). But as a whole, and except for the AVO scores, the coefficients are acceptable (<0.80). These parameters are displayed in Table 8.2.

8.2.5.3 INVARIANCE OF MEASUREMENT PARAMETERS

In Table 8.3, it is shown that the consecutive restrictions applied to the parameters of the items did not produce substantial decrements to the model fit; even the invariance of strict measurement was achieved, an unusual result in measures applied in the behavioral sciences (Vandenberg & Lance, 2000). Considering that the statistical difference ($\Delta_{WLSMV-\chi^2}$) between the models was sufficiently small and the sample size moderately large in each group, the probability of a type II error was judged to be sufficiently low. No modifications were added so as not to capitalize the sampling error, especially when they were of small magnitude.

8.2.5.4 RASCH FIT

In the SSS scale, the values of OUTFIT and INFIT (see Table 8.4) are in a range that suggests that the performance of the score can be identified as a productive measure, and appropriate for measurements using ordinal items (range |0.05|–|1.5|; Wright & Linacre, 1994). Item 31 seems to show slight misfit from the perspective of optimal item selection (Wright & Linacre, 1994) and the revised range (|0.70 |–|1.3|) proposed by Smith et al. (2008), but overall, this scale is a useful measure with good discriminative value of the attribute. The results of the adjustment in the PROSO and AVO scales also showed satisfactory fit, because their OUTFIT and INFIT values were within the range of both criteria (Smith et al., 2008; Wright & Linacre, 1994).

On the other hand, all thresholds maintained their expected order, and were clearly separated (Table 8.4). The location of the items in the latent attribute was usually concentrated within ±1 standard deviation of the latent attribute, suggesting a narrow range of measurements. With respect to how well the test measures the latent trait at various levels of the attribute, the information curve of the test (Figure 8.1) shows that the greater informative capacity of the subscales scores is found mainly at the midpoint of the latent attribute, and tends to be symmetric on both sides of the continuum.

To assess the stability of the parameters, the ML test (Martin-Löf, 1973) was applied, using the median criterion to divide the sample into two subgroups, high and low, and compare the invariance between them and a unidimensional specification of the subtest (Christensen et al., 2002). Thus, for SSS the ML test was not statistically significant ($\chi^2 = 29.75$, df = 23, $p = 0.157$); for PROSO, the result was similar; ($\chi^2 = 22.031$,

TABLE 8.3 Results of the Measurement Invariance Analysis According to Sex, for the Short CSI (CSI-S)

	WLSMV – χ^2 (df)	Δ_{WLSMV} (df)	CFI	Δ_{CFI}	RMSEA (95% CI)	Δ_{RMSEA}
Configurational	345.342 (174)	–	0.956	–	0.057 (.048, .066)	–
Metrics (factor loadings)	353.241 (186)	17.122 (12)	0.960	0.004	0.054 (.046, .063)	0.003
Scalar (thresholds)	387.584 (213)	35.763 (27)	0.958	0.002	0.052 (.044, .060)	0.002
Residual	387.584 (2013)	19.453 (15)	0.958	0.000	0.052 (0.044, .060)	0.000

TABLE 8.4 Fit of CSI-S to PCM

	$\chi^2 a$	Outfit		Infit		Parameters for Items (PCM)				
		MSQ	T	MSQ	T	Location	Threshold 1	Threshold 2	B1	B2
SSS										
csi5	471.242	1.031	0.440	0.991	−0.130	0.243	−1.233	0.746	1.233	0.487
csi7	429.910	0.941	−0.940	0.922	−1.33	0.424	−0.695	1.544	0.695	−0.849
csi23	476.278	1.042	0.620	1.007	0.140	1.108	0.221	1.995	−0.221	−2.217
csi31	296.838	0.650	−5.77	0.648	−6.60	0.241	−0.820	1.304	0.821	−0.483
csi32	378.523	0.828	−2.50	0.807	−3.32	0.127	−0.789	1.044	0.790	−0.255
PROSO										
CSI8	336.578	0.859	−1.720	0.836	−2.38	0.470	−0.374	1.315	0.374	−0.942
CSI9	386.545	0.986	−0.190	0.985	−0.200	0.477	−0.839	1.795	0.840	−0.955
CSI15	340.656	0.869	−1.92	0.841	−2.47	0.290	−1.019	1.600	1.019	−0.582
CSI16	359.708	0.918	−1.20	0.892	−1.65	0.414	−0.854	1.682	0.854	−0.828
CSI29	341.529	0.871	−1.52	0.884	−1.61	0.244	−0.709	1.198	0.709	−0.489
AVO										
CSI13	516.270	0.946	−1.02	0.930	−1.35	−0.385	−1.439	0.669	1.439	0.770
CSI18	575.239	1.054	0.790	1.066	1.17	0.586	−0.074	1.247	0.075	−1.173
CSI21	430.655	0.789	−3.65	0.772	−4.50	0.561	−0.302	1.425	0.303	−1.122
CSI26	490.575	0.898	−1.69	0.887	−2.25	0.111	−0.459	0.681	0.459	−0.222
CSI30	399.091	0.731	−4.56	0.736	−5.37	0.382	−0.235	0.999	0.236	−0.764

Note: [a]The degree of freedom for SSS, PROSO, and AVO, was: 267, 391 and 545, respectively. PCM, partial credit model.

df = 23, $p = 0.518$).-In regards to the AVO subtest, ML was statistically significant ($\chi^2 = 38.48$, df = 23, $p = 0.023$), but the ratio χ^2/df was small, so this result was dismissed. Finally, the information function of the scores of each subscale (Figure 8.1) was predominantly concentrated slightly above the midpoint of the latent attribute, between 0.0 and 1.0. For the items, the information curves are essentially similar in location and distributional form.

Although the PCM requires equality of discrimination parameters for a valid fit, and inequality of the factor loadings was found, the results of the fit indices were satisfactory. This seems to be consistent with insubstantial inequality, even though the difference in reliability coefficients was significant. Therefore, the obtained model fit and the presentation of the parameters from this model can be considered satisfactory.

8.2.6 FULL-CSI VS. SHORT CSI EQUIVALENCE

8.2.6.1 INTER-VERSION CORRELATION

The correlations between the corresponding scales between the CSI and CSI-S (e.g., AVO in full CSI and short CSI) were higher than 0.80, indicating that the shared variance is equal to or greater than 64%. After applying the correction for spurious

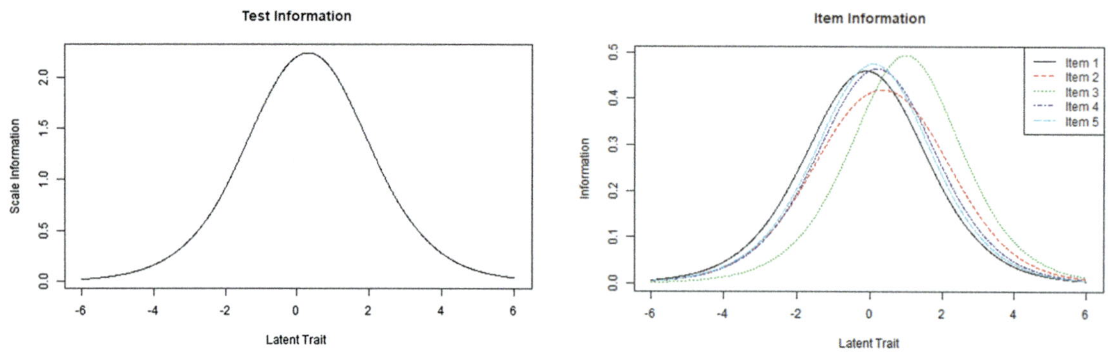

(a) Social Support Seeking

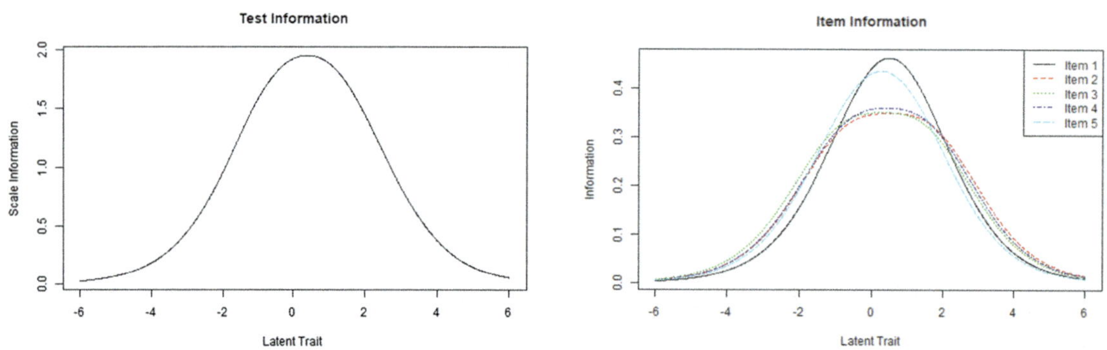

(b) Solving problems

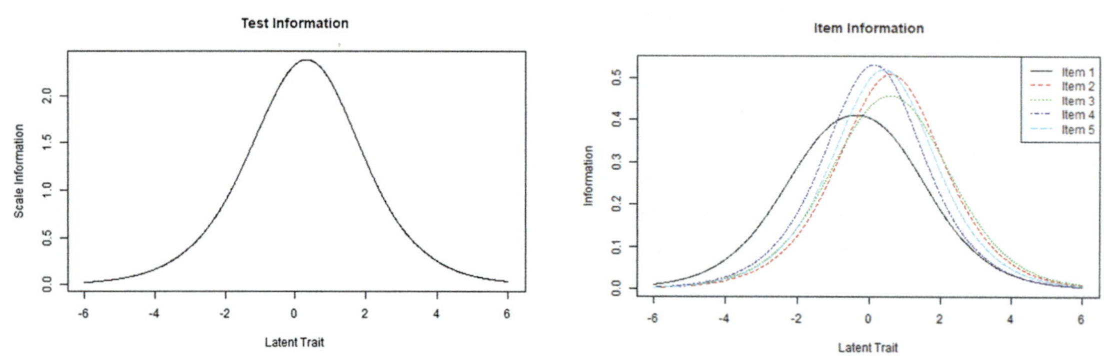

(c) **Avoidance**

FIGURE 8.1 Information curve for test and items of CSI-S. (a) SSS. (b) PROSO. (c) AVO.

variance (Merino & Angulo, 2013), the correlations between the versions of SSS, PROSO, and AVO were, respectively, 0.904 (95% CI [0.788, 0.958]), 0.874 (95% CI [0.727, 0.944]) and 0.747 (95% CI [0.492, 0.884]). On the other hand, the magnitude of the differences between the correlations (Cohen, 1992) of the paired subscales CSI–CSI-S were small ($q < 0.30$) or trivial ($q \leq 0.10$), and correspondingly indicate an essentially similar pattern between the correlations matrices of each version–(SSS_{full}–$PROSO_{full}$ vs. SSS_{short}–$PROSO_{short}$, $q = 0.042$; SSS_{full}–AVO_{full} vs. SSS_{short}–AVO_{short}, $q = 0.195$, and AVO_{full}–$PROSO_{full}$ vs. AVO_{short}–$PROSO_{short}$, $q = 0.103$).

8.2.6.2 CLASSIFICATION AGREEMENT

Table 8.6 shows the results of the classification agreement. The first general trend is that, significance tests were high ($t > 21.0$, $p < 0.001$), and raw agreement was high (>0.85). Taking into account the random variation, the coefficients remained high, except for the moderate tendency of the agreement in AVO. Also, the general trend of the agreement was that the classification based on the SSS score tends to be higher, regardless of the coefficient used. The second highest was PROSO, and AVO shows the least strong agreement, but it can still be considered acceptable, given that the lower limit is higher than 0.60 in each coefficient. In an additional exploratory analysis, the analysis was repeated using three divisions (tertiles), and the results tended to be slightly smaller, but essentially similar to the analysis of the quartile-based classifications.

8.2.6.3 LIMITS OF THE AGREEMENT

The quantification of the agreement between both versions was visualized by means of the

TABLE 8.5 Scores Correlations Between Full CSI and Short CSI with Confidence Intervals (95%)

	CSI			CSI-S		
	SSS	**PROSO**	**AVO**	**SSS**	**PROSO**	**AVO**
CSI						
SSS	1.00					
PROSO	0.386** [0.316, 0.451]	1.00				
AVO	0.116** [0.036, 0.193]	0.012 [−0.067, .091]	1.00			
CSI-S						
SSS	0.955** [0.947, 0.961]	0.386** [.316, .451]	0.084* [0.004, .162]	1.00		
PROSO	0.343** [0.270, 0.411]	0.928** [0.916, 0.938]	0.019 [−0.060, 0.098]	0.349** [0.277, 0.416]	1.00	
AVO	−0.047 [−0.126, 0.032]	−0.106** [−0.184, −0.026]	0.816** [0.787, 0.840]	−0.079 [−0.157, 0.000]	−0.091* [−0.169, −0.011]	1.00

Note: $^{**}p < 0.01$. $^{*}p < 0.05$. CSI, complete version (33 ítems); CSI-S, shortened version (15 ítems).

TABLE 8.6 Classification Agreement Coefficients Between Full CSI and Short CSI

	Segmentation in Tertiles				Segmentation in Quartiles			
	Coef.	s.e.	CI 95%		Coef.	se	CI 95%	
			Low	Upper.			Low	Upper.
SSS								
% agreement	0.946	0.005	0.937	0.956	0.960	0.003	0.954	0.966
Gwet's AC	0.856	0.013	0.830	0.882	0.872	0.009	0.854	0.891
κ	0.850	0.014	0.821	0.879	0.868	0.011	0.847	0.890
PROSO								
% agreement	0.917	0.005	0.905	0.928	0.935	0.004	0.927	0.943
Gwet-AC	0.776	0.015	0.746	0.807	0.796	0.012	0.770	0.821
Kappa de Cohen	0.773	0.017	0.739	0.808	0.798	0.014	0.770	0.827
AVO								
% agreement	0.869	0.008	0.853	0.885	0.908	0.005	0.897	0.918
Gwet-AC	0.648	0.022	0.605	0.692	0.717	0.016	0.683	0.750
κ	0.661	0.002	0.616	0.705	0.714	0.018	0.676	0.751

Note: Coeff., coefficient; se, standard error of the coefficient; CI 95%, confidence interval in 95% for the specific coefficient; % agreement, direct agreement percentage; Gwet-AC, Gwet coefficient; κ, Cohen kappa coefficient.

construction of the limits of the agreement between two continuous quantitative measures (Altman & Bland, 1983). The agreement criterion, in this method, is that the paired data should be within the range of ±2 standard deviations of the average difference between the two. In Figure 8.2(a–c), the tendency of the agreement is shown. The percentage of comparisons outside the limits of the agreement, that is, significant disagreements in SSS, PROSO, and AVO, were respectively 5.11%, 5.27%, and 3.29%. These percentages are small, and they are located around the first quartile in SSS and PROSO, and in the last quartile of AVO.

8.3 DISCUSSION

The objective of the present study was to propose an abbreviated or derivative version of the CSI in a sample of Peruvian adults, which involved a thorough psychometric analysis of both versions and their potential equivalences. In the full version, the factor loadings showed an important variability, indicating that the validity of each item is not homogeneous and should be treated differently. This also impacted on the estimation of reliability, given its relationship with the differences between the alpha and omega coefficients (Fernández & Merino, 2014). This pattern was found in both the long and the short version; although this was greater for the case of the AVO dimension. Notwithstanding the above, the dimensionality was moderate and was associated with other parameters that were not modeled but that contribute to better reflect the covariation between the items and the latent dimensions. The invariance analyses of the scale showed equivalences between men and women, suggesting strict measurement invariance.

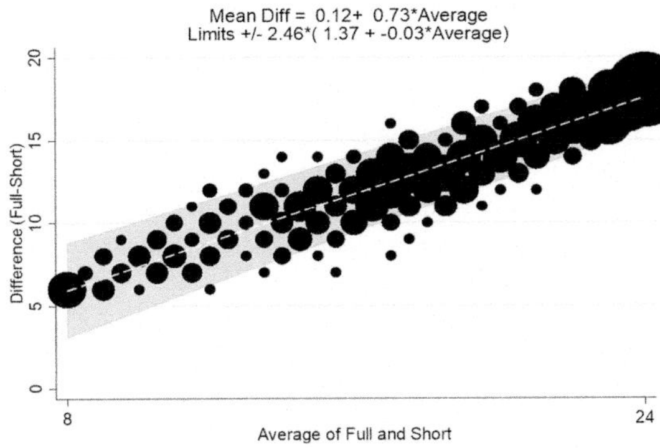

(a) Social Support Seeking

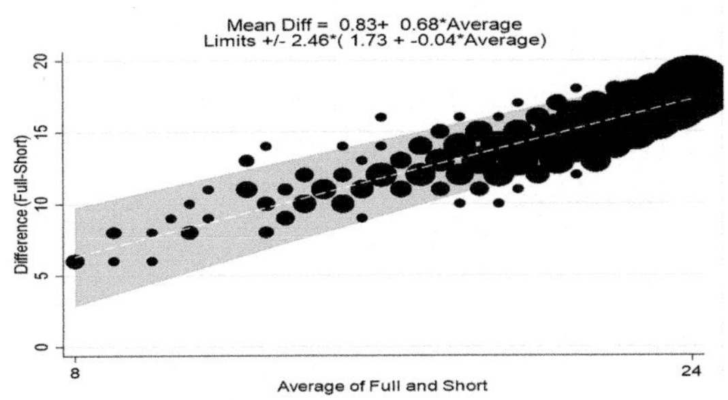

(b) Problem Solving

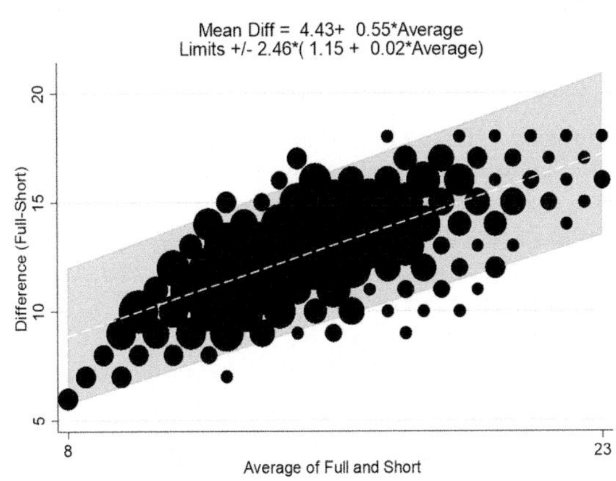

(c) Avoidance

FIGURE 8.2 Bland–Altman graph with adjustment for regression in the agreement limits.

Likewise, the Rasch analyses allowed us to identify that the CSI-S scale is a useful measure with a good discriminative value of the measured attribute, and with floor and ceiling indicators, according to what is theoretically expected. Along with the satisfying scalability properties, the CSI-S can provide interpretable scores to quantify the intensity of the different coping strategies used by the test-takers. However, the greatest informative and reliable capacity is located in the central part of the distribution of the latent attribute. The implication for practice is that additional measures may be required to corroborate the effectiveness of the score in subjects located at the ends of the distribution, that is, who scores very low or high.

With respect to the equivalence of the complete and short CSI, the correlations were higher than 0.80, with small differences in the magnitudes of the correlations of their subscales, which indicates the representation of a fairly homogeneous three-dimensional construct. Although correlation coefficients closer to 1.0 are ideal for clinical applications and would achieve fully interchangeable results between the CSI-S and its full version, the agreement reached is sufficient to describe groups in situations of applied research and professional activity. If it is required to classify the subjects in ordinal categories, it is recommended to create groups based on quartiles, given that the agreement was comparatively better at this level of categorization. If continuous score is required, then acceptable results can also be obtained using the CSI-S compared to the full CSI.

These findings allow us to suggest that, in the evaluation of stress coping styles, particularly in a Peruvian population, the short CSI version is adequate to represent the measured constructs, and has potential for its use in research, and plausibly as a screening assessment in professional practice. However, future studies are necessary to replicate our results in a non-Peruvian context, as well as to provide further evidence of construct validity. Although the present study did not make any direct intercultural comparison with regard to the psychometric parameters obtained, optimistically it can be estimated that, indirectly, the intercultural representation of the constructs measured by the CSI can also be stable, since studies with the full version tend to be stable too. It is clear that this must be corroborated.

The interpretation of our results should consider the limitations of the study. Sample representativeness is not guaranteed in relation to its reference populations, particularly because it is a sample of students (Hanel & Vione, 2016), and considering that Peru is regionally heterogeneous, our results can be valid only for reasonably similar regions. However, this must be balanced with the sample diversity achieved (Allmark, 2004), specifically in relation to the careers and sex of the participants; and with the aim of the research, which was scientific validation of a construct and its measurement (Pernice et al., 2008), in which representativeness may not be a necessary research aim (Rothman et al., 2013).

The results are useful references to generate hypotheses in representative samples and with adequate resources for a deliberately representative sampling (Martínez-Mesa et al., 2016). Overall, the results obtained can be considered satisfactory and stimulating for replication studies in Hispanic and non-Hispanic participants.

KEYWORDS

- **coping stress**
- **stress**
- **assessment**
- **validity**

REFERENCES

Addison, C. C., Campbell-Jenkins, B. W., Sarpong, D. F., Kibler, J., Singh, M., Dubbert, P., Taylor, H. (2007). Psychometric evaluation of a Coping Strategies Inventory Short-Form (*CSI-SF*) in the Jackson Heart Study Cohort. *International Journal of Environmental Research and Public Health, 4*(4), 289–295.

Ager, A., & Maclachlan, C. (1998). Psychometric properties of the Coping Strategy Indicator (CSI) in a study of coping behavior amongst Malawian students. *Psychology & Health, 13*(3), 399–409.

Aitcheson, R. J., Abu-Bader, S. H., Howell, M. K., Khalil, D., & Elbedour, S. (2017). Resilience in Palestinian adolescents living in Gaza. *Psychological Trauma: Theory, Research, Practice, and Policy, 9*(1), 36–43.

Allmark, P. (2004). Should research samples reflect the diversity of the population? *Journal of Medical Ethics, 30*, 185–189.

Altman, D. G., & Bland, J. M. (1983). Measurement in medicine: the analysis of method comparison studies. *Statistician, 32*, 307–317.

Amirkhan, J. H. (1990). A factor analytically derived measure of coping: The Coping Strategy Indicator. *Journal of Personality and Social Psychology, 59*(5), 1066–1074.

Amirkhan, J. H. (1994a). Criterion validity of a coping measure. *Journal of Personality Assessment, 62*(2), 242–261.

Amirkhan, J. H. (1994b) Seeking person-related predictors of coping: Exploratory analysis. *European Journal of Personality, 8*, 13–30.

Amirkhan, J. H. (1998) Attributions as predictors of coping and distress. *Personality and Social Psychology Bulletin, 24*(9), 1006–10118.

Asparouhov, T. & Muthen, B. O. (2006). Robust chi-square difference testing with mean and variance adjusted test statistics. Mplus Web Notes: No. 10. Available in: http://www.statmodel.com/download/webnotes/webnote10.pdf

Asparouhov, T., & Muthén, B. (2010). Simple second order chi-square correction. Retrieved from http://www.statmodel.com/download/WLSMV new chi21.pdf

Aysan F. (1994). Creating strategies to go out to deal with Turkish version of. *Journal of Educational Science Ege University, 3*(1), 113–12.

Bijttebier, P., & Vertommen, H. (1997). Psychometric properties of the coping strategy indicator in a flemish sample. *Personality and Individual Differences, 23*(1), 157–160.

Bowen, N. K. & Masa, R. D. (2015). Conducting measurement invariance tests with ordinal data: a guide for social work researchers. *Journal of the Society for Social Work and Research, 6*(2), 229–249.

Cano, F. J., Rodríguez-Franco, L., & García-Martínez, J. (2007). Adaptación española del Inventario de Estrategias de Afrontamiento. *Actas Españolas de Psiquiatría, 35*, 29–39.

Carver, C. S., Scheier, M. F., & Weintraub, J. K. (1989). Assessing coping strategies: a theoretically based approach. *Journal of Personality and Social Psychology, 56*, 267–283.

Choi, B., & Juárez-García, A. (2017). Language issues in standard questionnaires for assessing psychosocial working conditions: The case of the JCQ and the ERIQ. In *Psychosocial health, work and language* (pp. 3–18). Springer, Cham.

Christensen, K. B., Bjorner, J. B., Kreiner, S. & Petersen, J. H. (2002). Testing unidimensionality in polytomous Rasch models. *Psychometrika, 67*(4), 563–574.

Clark, K. K., Bormann, C. A., & Cropanzano, R. S., &, James, K. (1995). Validation evidence for three coping measures. *Journal of Personality Assessment, 65*(3), 434–455.

Cohen, J. (1992). A power primer. *Psychological Bulletin, 112*, 155–159.

Desmond, D. M., & MacLachlan, M. (2006). Coping strategies as predictors of psychosocial adaptation in a sample of elderly veterans with acquired lower limb amputations. *Social Science & Medicine, 62*, 208–216.

Desmond, D., Shevlin, M., & MacLachlan, M. (2006). Dimensional analysis of the coping strategy indicator in a sample of elderly veterans with acquired limb amputations. *Personality and Individual Differences, 40*, 249–259.

Endler, N., & Parker, J. (1994). Assessment of multidimensional coping: Task, emotion, and avoidance strategies. *Psychological Assessment, 6*, 50–60.

Fernández, M., & Merino, C. (2014). Error de medición alrededor de los puntos de corte en el MBI-GS. *Liberabit, 20*, 209–218.

Fleiss, J. L. (1971). Measuring nominal scale agreement among many raters. *Psychological Bulletin, 76*, 378–382.

Folkman, S. (2011). Stress, health, and coping: Synthesis, commentary, and future directions. In S. Folkman (Ed.), *The Oxford handbook of stress, health, and coping* (pp. 453–462). New York: Oxford University Press Inc. https://doi.org/10.1093/oxfordhb/9780195375343.013.0022

Garcia, C. (2010). Conceptualization and measurement of coping during adolescence: a review of the literature. *Journal of Nursing Scholarship, 42*(2), 166–185.

Giavarina, D. (2015). Understanding Bland Altman analysis. *Biochemia Medica, 25*(2), 141–151.

Gignac, G. E., Bates, T. C., & Jang, K. (2007). Implications relevant to CFA model misfit, reliability, and the Five Factor Model as measured by the NEO-FFI. *Personality and Individual Differences, 43*, 1051–1062.

Greenaway, K. H., Louis, W. R., Parker, S. L., Kalokerinos, E. K., Smith, J. R., & Terry, D. J. (2014). Measures of coping for psychological well-being. In G. J. Boyle, D. H. Saklofske, & G. Matthews (Eds.), *Measures of Personality and Social Psychological Constructs* (pp. 322–351). USA: Academic Press.

Gwet, K. L. (2014). *Handbook of inter-rater reliability*. Gaithersburg, MD: Advanced Analytics, LLC.

Hanel, P. H. P., & Vione, K. C. (2016). Do student samples provide an accurate estimate of the general public? *PLoS One, 11*(12), e0168354.

Hock, M. (2017). iana: GUI for Item Analysis. R package version 0.1: https://github.com/mihock/iana

Kirchner, T., Forns, M., Muñoz, D., & Pereda, N. (2008). Psychometric properties and dimensional structure of the Spanish version of the Coping Responses Inventory—Adult Form. *Psicothema, 20*(4), 902–909.

Lazarus, R. S. (1966). *Psychological stress and the coping process*. New York, NY, USA: McGraw-Hill.

Leis, M., Schmidt, K. M., & Rimm-Kaufman, S. E. (2015). Using the partial credit model to evaluate the student engagement in mathematics scale. *Journal of Applied Measurement, 16*(3), 251–267.

Li, Ch. (2016a). Confirmatory factor analysis with ordinal data: Comparing robust maximum likelihood and diagonally weighted least squares. *Behavior Research Methods, 48*(3), 936–949.

Li, Ch. (2016b). The performance of ML, DWLS, and ULS estimation with robust corrections in structural equation models with ordinal variables. *Psychological Methods, 21*(3), 369–387.

Linacre, J. M. (1994). Sample size and item calibration stability. *Rasch Measurement* Transactions, *7*(4), 328. Retrieved from https://www.rasch.org/rmt/rmt74m.htm

Linacre, J. M. (2002). What do infit and outfit, mean-square and standardized mean? *Rasch Measurement Transactions, 16*(2), 878. Retrieved from: https://www.rasch.org/rmt/rmt162f.htm

Linacre, J. M. (2003). Rasch power analysis: size vs. significance: standardized chi-square fit statistic. *Rasch Measurement Transactions, 17*(1), 918. Retrieved from https://www.rasch.org/rmt/rmt171n.htm

Lo, C.-F. (2017). Stress and coping strategies among university freshmen in Hong Kong: Validation of the Coping Strategy Indicator. *Psychology, 8*, 1254–1266.

Luo, G. (2005). The relationship between the Rating Scale and Partial Credit Models and the implication of disordered thresholds of the Rasch models for polytomous responses. *Journal of Applied Measurement, 6*(4), 443–455.

Mair, P., & Hatzinger, R. (2007). Extended Rasch modeling: The eRm package for the application of IRT models in R. *Journal of Statistical Software, 20*(9), 1–20.

Mander, A. (2012). BATPLOT: Stata module to produce Bland-Altman plots accounting for trend. Retrieved from https://econpapers.repec.org/RePEc:boc:bocode:s448703

Marsh, H. W., Nagengast, B., Morin, A. J. S., Parada, R. H., Craven, R. G., & Hamilton, L. R. (2011). Construct validity of the multidimensional structure of bullying and victimization: An application of exploratory structural equation modelling. *Journal of Educational Psychology, 103*(3), 701–732.

Martínez-Mesa, J., González-Chica, D. A., Duquia, R. P., Bonamigo, R. R., & Bastos, J. L. (2016). Sampling: How to select participants in my research study? *Anais Brasileiros de Dermatologia, 91*(3), 326–330.

Martin-Löf, P. (1973). Statistiska modeller [Statistical models.] Anteckningar från seminarier lasåret 1969–1970, utarbetade av Rolf Sundberg. Obetydligt ändrat nytryck, October 1973. Stockholm: Institütet för Försäkringsmatemetik och Matematisk Statistisk vid Stockholms Universitet.

Masters, G. N. (1988). The analysis of partial credit scoring. *Applied Measurement in Education, 1*(4), 279–297.

Masters, G. N. 1982. A Rasch model for partial credit scoring. *Psychometrika, 47*, 149–174.

Meade, A. W., & Lautenschlager, G. J. (2004). A comparison of item response theory and confirmatory factor analytic methodologies for establishing measurement equivalence/invariance. *Organizational Research Methods, 7*(4), 361–388.

Merino, C. (2010). Consistencia interna de una escala de evaluación de las estrategias de afrontamiento del estrés. *Revista Científica, 4*(1), 260–271.

Merino, C., & Angulo, M. (2013). Validación en Chile de la Escala de Sobrecarga del Cuidador de Zarit en sus versiones original y abreviada: Corrección [carta al Editor]. *Revista Médica de Chile, 141*, 1083–1084.

Merino, C., Manrique, G., Angulo, M. & Isla, N, (2007). Indicador de Estrategias de Afrontamiento al Estrés: Exploración normativa y de su estructura factorial. *Ansiedad y Estrés, 13*(1), 25–40.

Millsap, R. E., & Tein, J.-Y. (2004). Assessing factorial invariance in ordered-categorical measures. *Multivariate Behavioral Research, 39*, 479–515.

Muthén, B. (1984). A general structural equation model with dichotomous, ordered categorical, and continuous latent variable indicators. *Psychometrika, 49*, 115–132.

Muthén, L. K. and Muthén, B. O. (1998–2012). *Mplus user's guide* (7th ed.). Los Angeles, CA: Muthén & Muthén.

Nava, C., Ollua, P., Vega, C. Z., & Soria, R. (2010). Inventario de Estrategias de Afrontamiento: una replicación. *Psicología y Salud, 20*(2), 213–220.

Patrick, V. M., Lancellotti, M. P., & Demello, G. (2009). Coping with non-purchase: Managing the stress of inaction regret. *Journal of Consumer Psychology, 19*, 463–472.

Pernice, R. E., van der Veer, K., Ommundsen, R., & Larsen, K. (2008). On use of student samples for scale construction. *Psychological Reports, 102*(2), 459–464.

Ptacek, J. T., Smith, R. E., Espe, K., & Raffety, B. (1994). Limited correspondence between daily coping reports and retrospective coping recall. *Psychological Assessment, 6*, 41–44.

R Core Team (2018). R: A language and environment for statistical computing. R Foundation for Statistical Computing, Vienna, Austria. Retrieved from https://www.R-project.org/.

Rodríguez-Díaz, F. J., Estrada-Pineda, C., Rodríguez-Franco, L., & Bringas-Molleda, C. (2014). Adaptación del Inventario de Estrategias de Afrontamiento (CSI) a la población penitenciaria de Mexico. *Psicologia: Reflexão e Crítica, 27*(3), 415–423.

Rothman, K. J., Gallacher, J. E., & Hatch, E. E. (2013). Why representativeness should be avoided. *International Journal of Epidemiology, 42*(4), 1012–1014.

Sass, D. A., Schmitt, T. A., & Marsh, H. W. (2014). Evaluating model fit with ordered categorical data within a measurement invariance framework: A comparison of estimators. *Structural Equation Modeling: A Multidisciplinary Journal, 21*(2), 167–180.

Schwarzer, R., & Schwarzer, C. (1996). A critical survey of coping instruments. In M. Zeidner, & N. S. Endler (Eds.), *Handbook of coping: Theory, research, applications* (pp. 107–132). New York: John Wiley and Sons.

Shin, H. J., & Kim, C. D. (2002). A validation study of coping strategy indicator (CSI). *Korean Journal of Counseling and Psychology, 14*(4), 919–935.

Skinner, E. A, Edge, K., Altman, J., & Sherwood, H. (2003). Searching for the structure of coping: a review and critique of category systems for classifying ways of coping. *Psychological Bulletin, 129*(2), 216–269.

Smith, A. B., Rush, R., Fallowfield, L. J., Velikova, G., & Sharpe, M. (2008). Rasch fit statistics and sample size considerations for polytomous data. *BMC Medical Research Methodology*, *8*, 33.

Soriano, J., & Zorroza, J. (1999). Análisis comparativo de tres cuestionarios de afrontamiento al estrés: CSI, CM y WOC. *Boletín de Psicología, 62*, 43–64.

Spangenberg, J. J., & Orpen-Lyall, M. R. (2000). Stress and Coping Strategies in a Sample of South African Managers Involved in Post-Graduate Managerial Studies. *Journal of Industrial Psychology*, *26*, 6–10.

Speyer, E., Morgenstern, H., Hayashino, T. Kerr, P. G., Rayner, H., Robinson, B. M., et al. (2016). Reliability and validity of the coping strategy inventory-short form applied to hemodialysis patients in 13 countries: Results from the Dialysis Outcomes and Practice Patterns Study (DOPPS). *Journal of Psychosomatic Research, 91*, 12–19.

Stanton, J. M., Sinar, E. F., Balzer, W. K., & Smith, P. C. (2002). Issues and strategies for reducing the length of self-report scales. *Personnel Psychology, 55*, 167–194.

Tobin, D. L, Holroyd, K. A, Reynolds, R. V., & Wigul, J. K. (1989). The hierarchical factor structure of the Coping Strategies Inventory. *Cognitive Therapy and Research, 13*, 343–361.

Vandenberg, R. J., & Lance, C. E. (2000). A review and synthesis of the measurement invariance literature: Suggestions, practices, and recommendations for organizational research. *Organizational Research Methods, 3*, 4–69.

Wissing, M. P., & Van Eeden, C. (2002). Empirical clarification of the nature of psychological well-being. *South African Journal of Psychology, 32*, 32–44.

Wright, B. D., & Linacre, J. M. (1994). Reasonable mean-square fit values. Rasch *Measurement Transactions, 8*(3), 370. Retrieved from https://www.rasch.org/rmt/rmt83b.htm

Yang, C., Nay, S., & Hoyle, R. H. (2010). Three approaches to using lengthy ordinal scales in structural equation models: Parceling, latent scoring, and shortening scales. *Applied Psychological Measurement, 34*(2), 122–142.

Žuljević, D., Jovanović, V., & Jerković, V. G. (2015). Latentna struktura srpske adaptacije indikatora strategija prevladavanja. *Primenjena Psihologija, 8*(3), 265–288.

CHAPTER 9

Are You Satisfied with Your Care? A Study to Examine the Predictors of Indian Cancer Patients' Satisfaction with Their Medical Consultation

SHWETA CHAWAK* and MAHATI CHITTEM

Indian Institute of Technology Hyderabad, Telangana, 502285, India

Corresponding author: E-mail: la15resch11004@iith.ac.in /chawakshweta@gmail.com

ABSTRACT

The aim of this study was to examine the factors (such as quality of life [QoL], anxiety, depression, illness perceptions, and locus of control [LOC]) that predict patient satisfaction with the doctor consultation. Around 196 Indian patients undergoing cancer treatment participated in the study. The participants answered questionnaires on QOL, psychological distress, illness perception, LOC, and satisfaction with communication. A series of multiple regressions were conducted to understand the factors (such as QoL, anxiety, depression, illness perception, and LOC) that predicted patient satisfaction with consultation. The study revealed that: (1) relationship with doctor, functional well-being, anxiety, and external LOC predicted relief from distress; (2) relationship with doctor, functional well-being, and doctor's LOC predicted comfort in communication; (3) relationship with doctor, internal, and chance LOC predicted doctor–patient rapport; (4) doctor LOC, communication comfort, and rapport predicted compliance intent. This highlights the importance of the relationship with the doctor on various aspects of patient's satisfaction with consultation such as distress relief, communication comfort, and rapport. The study implies the need for introducing communication skills training programs for doctors in the Indian cancer setting.

9.1 INTRODUCTION

Patient satisfaction with medical care is an important indicator of health and behavioral outcomes, psychological well-being, and quality of care (Butow et al., 1996; Cleary & McNeal, 1988; Fallowfield et al., 1990). Patient's dissatisfaction with their medical visit can lead to poorer understanding and recall of information, poor compliance, and delay in recovery (Fallowfield, 1992). Similarly, patient's satisfaction with their doctor and treatment was observed to influence patient's compliance to treatment (Blanchard et al., 1990; Griffin et al., 2004), highlighting the importance of the doctor's interaction and communication skills. Therefore, identifying indicators of patients' satisfaction and dissatisfaction with care provision is crucial in understanding areas of prioritization in healthcare as well as those which are in need of improvement.

The main assessment tools for patient satisfaction include the Patient Satisfaction Questionnaire (Marshall & Hays, 1994), Patient Satisfaction with Cancer care (Jean-Pierre et al., 2011), Picker Patient Experience Questionnaire (Jenkinson et al., 2002), and Medical Interview Satisfaction Scale (MISS; Meakin & Weinman, 2002). Using these questionnaires, a range of predictors of patient (dis)satisfaction were identified in medical psychology research. For example, older age was associated with increased satisfaction among patients while increased levels of depression, poor quality of life (QoL), and negative illness perceptions predicted patients' dissatisfaction (Frostholm et al., 2005; Homma et al., 2018).

9.1.1 PREDICTORS OF PATIENT'S SATISFACTION WITH DOCTOR COMMUNICATION

A theoretical framework by Ben-Sira (1976) called "social interaction model" suggests that patients enter the medical communication process with anxiety and less knowledge regarding the illness and the treatment as compared to the doctor. The affective component in the doctor's style of communication is considered to be an important factor in patient's evaluations of their medical care. Elaborating on the affective aspects of the doctor's communication, Ben-Sira (1976) described this as acknowledging patient as an individual rather than a case by the doctor.

Research examining the link between doctor–patient communication and patient satisfaction indicated that doctors tend to show two kinds of communication styles: controlled and affiliation (Buller & Buller, 1987). Patients reported more satisfaction with their medical visit when the doctor adopted affiliating communication style such as asking open-ended questions, communicating in a patient-centric manner, spent more time with patients, and provided a combination of medical information and psycho-emotional support to patients (Ishikawa et al., 2002; Fujimori et al., 2007; Lim et al., 2011; Tang & Lee; 2004; Husson & Mols, 2011, Zachariae et al., 2003). On the other hand, the doctor's controlled communicative behavior such as maintaining authority and the use of medical jargons was associated with lower patient satisfaction (Mehra, 2018; Ishikawa et al., 2002).

Research on communication and patient satisfaction suggested that doctors need to reflect on the motivations for patients' question-asking and enquire after patients' preferences and opinions. Consequently, interventions to improve patients' question-asking, such as the question prompt list, were developed which focused on providing patients with a range of questions they could ask their treating doctor (Lim et al., 2011; Shirai et al., 2012). This question-aid helped patients to acquire information regarding their illness and make illness- and treatment-related decisions, which was a reflection of an improved doctor–patient communication (Glynne-Jones et al., 2006).

Another predictor for patient satisfaction was their illness perception. In a study by Frostholm et al. (2005) following dimensions of illness perception were considered: emotional representations, consequence of the illness, perception of longer timeline, personal control, perception of cyclical timeline, and certainty. The results indicated that patients with negative illness perceptions before the consultation especially related to the uncertainty of the illness and treatment were likely to experience patient's dissatisfaction (Frostholm et al., 2005). This sense of illness-related uncertainty created feelings of helplessness and depression among patients. These findings suggest the need to challenge and change negative illness beliefs patients may have in order to improve patients' satisfaction with the doctor consultation. Further, the patient's levels of psychological distress were observed to be related to their levels of satisfaction. Patients with higher levels of distress (Greenley et al., 1982) and depression (Linn & Greenfield, 1982) were found to be less satisfied with their medical visit.

Strikingly, there is a lack of research on examining the role of QoL in patient satisfaction. However, Ong et al. (2000) reported that QoL was associated with patients' psychological well-being (anxiety and anger) which, in turn, could be related to patients' satisfaction with the doctor consultation. In an intervention study by Detmar et al. (2002) patients and caregivers were introduced to Health-related Quality of Life (HRQoL) questionnaire and taught to interpret the results. The HRQoL scale was administered on the patients before the consultation and the results were discussed by the oncologist during the consultation. The results indicated that incorporating HRQoL in consultation led to an increase in patient satisfaction with communication. Furthermore, the author found that patients who reported perceiving doctors to be providing emotional support were more likely to report higher satisfaction (Detmar et al., 2002), indicating that doctors played a key role in alleviating illness-related distress.

Although there is a dearth of research in the understanding locus of control (LOC) as a predictor of patient satisfaction, patient's sense of control has found to determine patient's preference in doctor–patient communication (Murray & McMillan, 1993; Tennstedt, 2000; Zachariae et al., 2003). Indeed, Tennstedt (2000) reported that patients were more likely to have higher doctor LOC (i.e., the illness was in the hands of the doctor) than internal LOC (i.e., the patient had personal control over the illness) and lower self-efficacy which was associated with the patient's involvement in communication with the doctor.

9.1.2 INDIAN PSYCHO-ONCOLOGY RESEARCH IN DOCTOR–PATIENT COMMUNICATION

Cancer incidence in India has been on the rise with the most recent registry in 2010 showing a total of 979,786 reported cases of cancer. From ICMR data (1999–2001) cancer prevalence among five metropolitan cities reported that, Mumbai was found to have the second-highest number of cancer patients with 8505 total number of reported cases (Marimuthu, 2008). The most common cancers in India among males are lung, esophagus, stomach, oral, and pharyngeal cancer, while among females, cervix, and breast are found to be a common type of cancer followed by stomach and esophagus (Murthy & Mathew, 2004). Health-related research in India suggests that patient satisfaction is related to the quality of healthcare such as availability of medicines, accessibility of facilities especially in terms of time invested in availing medical procedures, effective communication of medical information, doctors' and healthcare staff behavior toward care seekers, and the hospital infrastructure (Kumari et al., 2009; Rao et al., 2006; Sodani et al., 2010).

At the national level, statistics revealed that there is one government doctor for every 11,528 people and one nurse for every 483 people in India (Bagcchi, 2015). In India, doctors spend about 10 s–1 min communicating with each patient (Agarwal et al., 2011). Time was a contributing factor in determining patient's satisfaction with consultation. A study by Mehra

(2018) reported patients who perceived longer length of consultation were more satisfied with the doctor's communication and were likely to refer the doctor through word-of-mouth. However, due to the scarcity of availability of doctors in India, the patients may prefer to ignore the problem related to doctor's time spent and communication style. This is reflected in a study by Khadir and Swamynathan (2014) which reported that majority (78.6%) of dissatisfied patients preferred to not complain regarding healthcare services and the two major reasons for noncomplaining: "perceived relative inability" and decision to change hospital.

These findings suggest that hospitals seeking patient satisfaction need to motivate patients to express their dissatisfaction with the healthcare service. It also indicates that empathic behavior (e.g., honest communication) on the part of medical care providers can play a role in the patient's satisfaction. In line with these findings, recent psycho-oncology research in India found that oncologists viewed communication as a complex and delicate negotiation between their ethics of ensuring patient autonomy and family caregivers' desire to protect their patient from harm (Datta et al., 2017).

However, to the best of our knowledge, no research in India sought to understand psychological factors contributing to patients' satisfaction with their oncologist. This is an especially important area of research since Indian patients consider doctors to be God-like figures (Quill, 1989) and most cancer patients in India are informed about their cancer diagnosis by their doctors (Purrakkal et al., 2004), thereby highlighting the important role a doctor plays in a patient's life. Therefore, the current study aims to examine the factors (such as QoL, anxiety, depression, illness perceptions, and LOC) that predict patient satisfaction with the doctor consultation.

9.2 METHOD

9.2.1 PARTICIPANTS

One hundred and ninety-six patients received treatment at a hospital in Mumbai, India were recruited into the study. Patients who were unable to speak English, Marathi, or Hindi, or were diagnosed with a psychiatric condition (as indicated by the hospital nurse) were excluded from the study. Table 9.1 outlines patient demographic and medical details.

9.2.2 MEASURES

9.2.2.1 DEMOGRAPHIC AND MEDICAL DETAILS

Participants were asked about their age, gender, marital status, number of children, level of education, and household income. Details about the cancer site, stage, types of treatment, time since diagnosis, and time since treatment were obtained from patients' medical records.

9.2.2.2 MISS-21

The MISS-21 was used to assess medical satisfaction with the doctor consultation among patients (Meakin & Weinman, 2002). The questionnaire consists of 21 items with four subscales, scored on a 5-point Likert scale: distress relief (6 items), communication comfort (4 items), rapport (8 items), and compliance intent (3 items). A high score in a subscale indicates increased levels of satisfaction in that aspect of the doctor consultation. Scores can be interpreted as follows: (i) distress relief: 0–15 = low, 16–36 = moderate and 37–42 = high relief from distress; (ii) communication comfort: 0–10 = lower, 11–22 = moderate and 23–28 = high comfort in communication; (iii) rapport: 0–20 = low, 21–44 = moderate and 45–56 = high rapport (iv) compliance intent: 0–7 = low, 8–16 = moderate and 17–21 = high compliance intent. Inter-consistency between subscales of MISS-21 ranged from $\alpha = 0.46$ to $\alpha = 0.65$ (Meakin & Weinman, 2002).

9.2.2.3 BRIEF ILLNESS PERCEPTION QUESTIONNAIRE (BIPQ)

The BIPQ was used to measure patients' emotional and cognitive representations of a health threat (Broadbent et al., 2006). Using a 10-point response scale, the questionnaire consists of eight items assessing the dimensions of illness identity, perceived consequences, timeline, personal control, treatment control, worry, illness comprehension, and emotional response. High overall or individual scores indicated negative illness perception(s). Scores between 0 and 21 is perceived as lower illness threat, 22–59 average illness threat, and 60 and above perceived high illness threat. The BIPQ has good test–retest reliability and construct, discriminant and predictive validity (Broadbent et al., 2006). The BIPQ was used in a previous study with Indian cancer patients which showed good internal reliability ($\alpha = 0.79$) (Chittem et al., 2012).

9.2.2.4 HOSPITAL ANXIETY DEPRESSION SCALE (HADS)

The HADS assesses an individual's levels of anxiety and depression (Zigmond & Snaith, 1983). The HADS consists of 14 items with two subscales of anxiety (7 items) and depression (7 items). The items are scored on a 4-point Likert scale (0–3), with a high score indicating increased levels of anxiety and depression. Anxiety and depression scores can be interpreted as follows: 0–7 = normal, 8–10 = borderline case, 11–21 = abnormal. The HADS has been used extensively in psycho-oncology research in India showing good psychometric properties (Chittem et al., 2012; Thomas et al., 2005).

9.2.2.5 FUNCTIONAL ASSESSMENT OF CANCER THERAPY (FACT-G)

The FACT-G measures HRQoL among cancer patients. Using a 5-point Likert scale, the questionnaire comprises 5 subscales of physical well-being (7 items), social well-being (7 items), relationship with the doctor (2 items), emotional well-being (5 items), and functional well-being (7 items) (Cella et al., 1993). The scores for physical, social, and emotional well-being can be interpreted as: 0–10 = low, 11–27 = moderate and 28–38 = high well-being. Relationship with doctor: 0–5 = low, 6–12 = moderate, 13 above = high well-being. Emotional well-being: 0–8 = low, 9–21 = moderate, 22 above = high well-being. The FACT-G exhibited good internal reliability of the total scale ($\alpha = 0.89$) and the correlation between the subscales ranged from $\alpha = 0.78$ to $\alpha = 0.90$ (Cella et al., 1993).

9.2.2.6 MULTIDIMENSIONAL HEALTH LOCUS OF CONTROL CANCER (MHLC-C)

The MHLC-C assesses self-reported health LOC among cancer patients (Wallston et al., 1994). The questionnaire consists of four subscales each measuring unique dimensions within LOC, scored on a 6-point Likert scale: internal (6 items), chance (6 items), doctor (3 items), and other people (3 items). The scores for the subscales (internal and chance) can be interpreted as follows: 6–12 = low LOC, 13–27 = average LOC, and 28–36 = high LOC; scores for subscales (doctor and other people) can be interpreted as follows: 3–6 = low LOC, 7–13 = average LOC, and 15–18 = high LOC. The MHLC-C reported good internal consistency reliability (α = 0.70) (Wallston et al., 1994).

9.2.2.7 GOD LOCUS OF HEALTH CONTROL (GLHC)

The GLHC assesses patients' beliefs about the extent to which God controls their health. The GLHC can be used as an addition to the MHLC scales (Wallston et al., 1999), and consists of 6 items. The score for GLHC indicates that 6–12 = low God LOC, 13–27 = average God LOC, and 28–36 = high God LOC. The scale exhibits excellent reliability (α = 0.91) (Wallston et al., 1999).

9.2.3 PROCEDURE

Participants who fit the inclusion criteria were approached by the lead researcher (SC) during their hospital appointment. Patients who indicated an interest in participating were explained the study in further detail and any questions they had were answered by SC. Following this, informed consent was obtained from participants and SC administered the questionnaires in the patients' preferred language. All questionnaires were forward- and back-translated into Hindi and Marathi.

Of the 226 patients who were approached, 30 patients were unwilling to disclose information about their illness. Therefore, a total of 196 patients participated in the current study. Of the 196 participants, SC orally administered the questionnaires to 193 patients as they were either physically weak or illiterate, using response cards as necessary. Ethical approval for the study was obtained from the Institute Ethics committee (IEC), Indian Institute of Technology Hyderabad (IITH), Hyderabad, India, and the ethics committee of Dr. Balabhai Nanavati Hospital, Mumbai, India.

9.3 RESULTS

9.3.1 PATIENT CHARACTERISTICS

The sample comprised 94 male and 102 female patients with an average age of 55.9 years. The three most frequent cancers in the sample were head and neck, breast, and uterine. The mean time since diagnosis was 2.19 months and the time since receiving treatment was 1.50 months. Table 9.1 describes the demographic and medical details of the patients.

Overall mean scores of the psychological variables revealed that patients reported to be moderately satisfied with the consultation; they experienced moderate relief from distress (*mean* = 29.62; SD = 4.26), comfort in communication with the doctor (*mean* = 13.38; SD = 3.88), and shared moderate rapport with the doctor (*mean* = 37.79; SD = 6.26) and were also moderate in compliance intent (*mean* = 12.26; SD = 2.72).

Patients were also found to report average levels of HRQoL; moderate physical (*mean* = 16.00; SD = 2.74) and social well-being (*mean* = 16.25; SD = 2.78), average relationship with doctor (*mean* = 12.51; SD = 2.29), moderate levels of emotional (*mean* = 13.38; SD = 2.90), and functional well-being (*mean* = 18.77; SD = 4.84). Patients reported average perceived threat toward their illness (*mean* = 40.69; SD = 12.53), borderline anxiety (*mean* = 10.11; SD = 3.07), and depression (*mean* = 9.72;

SD = 2.67). Patients reported to have moderate internal LOC (*mean* = 22.55; SD = 3.43), high on chance LOC (*mean* = 27.39; SD = 3.10), high on doctor LOC (*mean* = 15.82; SD = 1.96), average on other people LOC (*mean* = 8.15; SD = 1.72), and God LOC (*mean* = 24.56; SD = 6.17).

TABLE 9.1 Demographic and Medical Details

Characteristic		Number
Age (years)	Mean	55.91
	Range	84–20
Gender	Male	94
	Female	102
Average monthly income		₹9204
Time since diagnosis	Below 3 months	29
	Above 3 months	152
	Reoccurrence	15
Treatment	Surgery	2
	Chemotherapy	11
	Radiation therapy	37
	Surgery + chemotherapy	3
	Surgery + radiation therapy	81
	Chemotherapy + radiation therapy	20
	Surgery + chemotherapy + radiation therapy	42
Types of cancer	Head and neck	60
	Breast	56
	Uterus	34
	Prostate	11
	Lung	9
	Stomach	7
	Metastatic	6
	Brain	3
	Rectal	5
	Kidney	5

9.3.2 ASSOCIATIONS WITH PATIENT SATISFACTION WITH THE MEDICAL CONSULTATION

A Pearson product–moment correlation coefficient was calculated to understand the association of the demographic and psychological variables with patients' satisfaction with the medical consultation. Table 9.2 depicts the correlation matrix for the demographic and the psychological variables.

The significant correlates for subscales of MISS-21 (distress relief, communication comfort, rapport, and compliance intent) are discussed in turn. There was positive correlation between distress relief subscale of MISS-21 and psychological variables of relationship with doctor ($r = 0.40$, $n = 196$, $p < 0.01$), functional well-being ($r = 0.20$, $n = 196$, $p < 0.01$), anxiety ($r = 0.\ 36$, $n = 196$, $p < 0.01$), chance ($r = 0.16$, $n = 196$, $p < 0.05$), and doctor LOC ($r = 0.17$, $n = 196$, $p < 0.05$).

Communication comfort was positively correlated with psychological variables of relationship with doctor ($r = 0.29$, $n = 196$, $p < 0.01$), emotional well-being ($r = 0.17$, $n = 196$, $p < 0.05$), anxiety ($r = 0.16$, $n = 196$, $p < 0.05$), depression ($r = 0.15$, $n = 196$, $p < 0.05$), and MISS-21 subscale of compliance intent ($r = 16$, $n = 196$, $p < 0.05$). Communication comfort was negatively correlated with MISS-21 subscale of rapport ($r = -0.21$, $n = 196$, $p < 0.01$), and doctor LOC ($r = -0.18$, $n = 196$, $p < 0.05$).

Rapport subscale of MISS-21 was positively correlated with psychological variables of relationship with doctor ($r = 0.28$, $n = 196$, $p < 0.01$), internal ($r = 0.21$, $n = 196$, $p < 0.01$), chance ($r = 0.16$, $n = 196$, $p < 0.05$), and negatively correlated with compliance intent ($r = -0.23$, $n = 196$, $p < 0.01$) and other people LOC ($r = -0.18$, $n = 196$, $p < 0.05$). Compliance

TABLE 9.2 Correlation Matrix of the Variables Included in the Study

	Correlation Matrix																					
	BP	FP	FS	FR	FE	FF	A	D	DR	CC	R	CI	I	C	D	OP	G	AGE	GE	TY	DIG	TRE
BP	1																					
FP	0.48**	1																				
FS	0.21**	0.08	1																			
FR	0.05	0.07	0.21**	1																		
FE	0.13	0.18**	0.01	0.05	1																	
FF	−0.17*	−0.12	0.01	0.06	−0.23**	1																
A	0.27**	0.03	−0.60	0.04	0.26**	−0.30**	1															
D	0.05	0.15*	−0.11	−0.04	0.26**	−0.37**	0.32**	1														
DR	−0.07	0.12	−0.01	0.40**	−0.09	0.20**	0.36**	−0.13	1													
CC	0.06	0.02	−0.06	0.29**	0.17*	−0.29	0.16*	0.15*	−0.10	1												
R	−0.04	0.00	−0.09	0.28**	−0.03	0.08	−0.05	0.01	0.13	−0.21**	1											
CI	0.05	0.10	−0.04	−0.16**	0.10	−0.10	0.07	0.13	−0.02	0.16*	−0.23**	1										
I	0.01	0.00	0.02	−0.13	−0.09	0.00	−0.00	−0.00	0.00	−0.01	0.21**	−0.14	1									
C	−0.00	−0.04	−0.05	0.11	0.26**	0.14*	0.05	−0.04	0.16*	0.04	0.16*	−0.16	−0.10	1								
D	−0.09	−0.02	−0.05	0.16*	−0.05	0.12	−0.15*	−0.05	0.17*	−0.18*	0.01	0.30*	−0.16*	−0.15*	1							
OP	0.00	0.02	0.07	0.03	0.01	−0.09	0.16*	0.07	0.09	0.13	−0.18*	0.27*	−0.17*	−0.11	0.19**	1						
G	0.12	0.11	0.04	0.12	0.12	0.06	0.12	−0.06	0.02	0.05	0.12	−0.12	0.03	0.22**	0.01	−0.14*	1					
AGE	0.09	0.06	0.06	−0.04	0.11	−0.09	0.07	0.01	−0.08	−0.03	0.04	0.01	0.02	0.05	−0.07	0.00	0.09	1				
GE	−0.03	−0.02	−0.12	−0.10	−0.13	0.00	0.09	−0.02	−0.06	−0.08	0.09	−0.00	0.15*	0.04	−0.07	−0.12	0.08	0.09	1			
TY	0.13	0.12	0.03	0.01	0.11	0.03	0.02	0.04	0.04	0.03	0.02	0.07	0.03	0.06	−0.12	−0.20**	0.17*	0.13	0.07	1		
DIG	−0.04	0.05	0.14*	0.11	0.06	−0.05	−0.02	−0.11	0.06	−0.01	−0.12	0.01	−0.13	0.08	0.10	0.06	0.05	−0.14*	−0.16*	0.03	1	
TRE	0.03	0.02	0.06	0.12	0.07	0.02	−0.00	−0.10	0.09	0.03	0.04	0.03	−0.07	0.10	0.13	0.11	0.05	−0.11	−0.22**	−0.02	0.76**	1

**Correlation is significant at the 0.01 level (2 tailed)
*Correlation is significant at the 0.05 level (2 tailed)
Variables: Illness perception (BP), Physical Well-being (FP), social well-being (FS), Relationship with doctor (FR), Functional well-being (FF), Emotional well-being (FE), Anxiety(A), Depression (D), Distress relief (DR), Communication comfort (CC), Rapport (R), Compliance Intent (CI), Internal (I), Chance (C), Doctor (D), Other people (OP), God LOC (G), Age (A), Gender (GE), Type of cancer (TY), Time since diagnosis (DIG), Time since treatment (TRE)

intent was positively associated with doctor ($r = 0.30$, $n = 196$, $p < 0.05$) and other people LOC ($r = 0.27$, $n = 196$, $p < 0.05$), and negatively associated with relationship with doctor ($r = -0.16$, $n = 196$, $p < 0.01$).

There was no significant association found between demographic variables (age, gender, type of cancer, time since diagnosis, and time since treatment) and MISS-21 subscales of distress relief, communication comfort, rapport, and compliance intent.

9.3.3 MULTIPLE PREDICTORS OF PATIENT SATISFACTION WITH CONSULTATION

Multiple regression analysis was carried out to understand the influence of each predictor variable in relation to patients' satisfaction with the doctor consultation (i.e., distress relief, communication comfort, rapport, and compliance intent). Variables which significantly correlated with the MISS-21 subscales were entered into each multiple regression model.

Relief from distress was regressed on the relationship with doctor, functional well-being, anxiety, doctor LOC, and chance LOC. The variables explained 32.2% of variance for relief from distress (see Table 9.3). Relationship with doctor ($b = 0.39$, $p < 0.001$), functional well-being ($b = 0.05$, $p < 0.05$), anxiety ($b = -0.35$, $p < 0.001$), and chance LOC ($= 0.14$, $p < 0.05$) were found to be significant predictors of relief from distress. Individual's relationship with doctor was found to be the greatest predictor for distress relief.

Comfort with communication was regressed on the relationship with doctor, emotional well-being, anxiety, depression, rapport, compliance intent, and doctor LOC. The variables explained 20.6% of the variance in comfort in communication (see Table 9.3). Relationship with doctor ($b = -0.21$, $p < 0.01$), functional well-being ($b = -0.19$, $p < 0.01$), and doctor LOC ($b = -0.14$,

TABLE 9.3 Summary Table With *R*, *R* Squares and Change Statistics of Outcome Variables (Distress Relief, Communication Comfort, Rapport, and Compliance Intent)

Summary Table										
Outcome Variables	**R**	**R Square**	**Adjusted R Square**	**Standard Error of Estimate**	**Change Statistics**					**Durbin Watson**
					R Square Change	**F Change**	**df1**	**df2**	**Sig F Change**	
Distress relief	0.57[a]	0.32	0.30	3.56	0.32	18.03	5	190	0.000	2.12
Communication comfort	0.45[b]	0.21	0.17	3.53	0.21	6.06	8	187	0.000	1.76
Rapport	0.45[c]	0.21	0.18	5.67	0.21	8.16	6	189	0.000	1.95
Compliance intent	0.47[d]	0.22	0.20	2.44	0.22	8.89	6	189	0.000	1.91

[a]Predictors: relationship with doctor, functional well-being, anxiety, doctor LOC, and chance LOC.
[b]Predictors: relationship with doctor, emotional wellbeing, anxiety, depression, rapport, compliance intent, and doctor LOC.
[c]Predictors: relationship with doctor, communication comfort, compliance intent, internal loc, chance loc, and other people LOC.
[d]Predictors: relationship with doctor, communication comfort, rapport, doctor loc, and other people LOC.

$p < 0.05$) were found to be the significant predictors of comfort with communication. Relationship with the doctor was found to be the greatest predictor of comfort with communication.

Rapport was regressed on relationship with doctor, communication comfort, compliance intent, internal LOC, chance LOC, and other people LOC. The variables explained 20.6% of variance in rapport (see Table 9.3). Relationship with doctor ($b = 0.25$, $p < 0.001$), internal LOC ($b = 0.23$, $p < 0.001$), and chance LOC ($b = 0.14$, $p < 0.05$) were found to be significant predictor for rapport. Relationship with doctor was found be the greatest predictor of rapport.

Compliance intent was regressed on relationship with doctor, communication comfort, rapport, doctor LOC, and other people LOC. The variables were found to contribute 22.0% of variance on compliance intent (see Table 9.3). Doctor LOC ($b = 0.31$, $p < 0.01$), communication comfort ($b = 0.12$, $p < 0.05$), and rapport ($b = -0.13$, $p < 0.05$) were found to be predictors of compliance intent. Doctor LOC was found to be the greatest predictor of compliance intent.

9.4 DISCUSSION AND IMPLICATIONS

The current study examined the factors (such as QOL, anxiety, depression, illness perceptions, and LOC) that predicted patient satisfaction with the doctor consultation. The findings revealed that the relationship with the doctor, functional well-being, anxiety, and chance LOC predicted patients' relief from distress. The relationship with the doctor, functional well-being, and doctor LOC predicted patients' comfort with the doctor's communication. The relationship with the doctor, internal LOC, and chance LOC predicted satisfaction with the rapport shared within the encounter. Finally, doctor LOC, communication comfort, and rapport shared with the doctor predicted patients' intent to comply with the doctor's recommendations.

An overarching finding of the current study was that the relationship with the doctor was found to predict patient satisfaction indices of decreased distress and increased comfort in communicating with the doctor and rapport. This finding adds to existing literature which suggests that effective communication between the doctor and patient contributes to the patient's satisfaction with medical consultation (Ong et al., 2000). The doctor–patient relationship is crucial among vulnerable patients as they may heavily rely on the doctor's skills and communication behavior. This is in line with Ben-Sira's (1976) theoretical framework, which emphasizes the role of doctor. However, in India, there is little research and practice around the area of communication skills training for doctors. Future research will benefit from developing and testing the effectiveness of communication skills training for doctors with the aim of improving patients' satisfaction and well-being.

The current study also reported LOC dimensions of external (i.e., chance and doctor) and internal LOC predicted components of satisfaction scale (distress relief, communication comfort, rapport, and compliance intent). The patient's control attribution to chance predicted the patient's relief from distress and contributed in building rapport with the doctor. Further, believing the disease was controlled by the doctor predicted patients' comfort with the doctor's communication, rapport, and the willingness to comply with the doctor's suggestions. The majority of the participants in this study were found to be high on chance and doctor LOC. These results were in line with a study by Tennstedt (2000) that indicated patients tend to have external LOC as compared to internal LOC. One of the reasons for this finding could be related to the patient's perception regarding their doctor. They may see their doctor as a medical provider and expert in the subject matter. This may have helped reduce their distress and help build rapport with the doctor. Secondly, external LOC is often linked with lower levels of self-efficacy (Ashagi & Beheshtifar, 2015); therefore, patients may have relied extensively on the doctor which, in turn, led them to derive satisfaction with their interaction with the doctor. Past studies have also reported that patients with high doctor LOC were likely to report higher familiarity with the doctor. This could determine patients' comfort in communication with the doctor (Arntson et al., 2009).

Interestingly, internal LOC predicted only one satisfaction dimension, namely the rapport shared with the doctor. This may be because patients with internal LOC felt responsible to ask questions and maintain a positive doctor–patient relationship. Reflecting this, Arntson et al. (2009) found that patients with higher internal LOC were likely to report that their doctor encouraged patient involvement, provided information, and shared responsibility. Patients with high internal LOC were also less likely to report concerns before and after the consultation with their doctor. This could contribute in building rapport with the doctors.

Among the subscales of QoL only functional well-being was found to predict with two patient satisfaction indices, that is, distress relief, and communication comfort. Research evidences show that cancer treatment can have a negative impact on patients' abilities to perform daily activities (Fangel et al., 2013; Hofman et al., 2007). However, patients whose functional well-being is not affected may have more positive approach toward their treatment and have accepted their illness; this in turn, may reduce their distress. This finding implies that in order to reduce patient's distress and increase comfort in communication, doctors may provide patients with information related to physical limitations ensuing the treatment and providing timely intervention (e.g., information-sharing about physiotherapy).

The current study also reported that patient's hospital-related anxiety predicted decreased relief from distress during the consultation. This finding is in line with the previous research (Roter et al., 1995; Zachariae et al., 2003), suggesting that decreased levels of anxiety may have permeating effects in medical care such as positive perceptions of the doctor interaction. This may contribute in increasing patient satisfaction with the medical consultation. Hence, it can be beneficial to further explore effectiveness of using positive health psychology interventions to help reduce levels of anxiety among cancer patients. These interventions can focus on a variety of techniques such as using a strength-based approach in identifying and engaging in coping strategies, establish support groups which help harness patients' resilience and positive emotions to manage the illness, and providing psychological support to patients who are highly motivated to improve their well-being.

Communication comfort and rapport shared with the doctor were found to predict patients' intent to comply with the doctor's recommendations in the current study. This finding supports existing research which indicates that the doctor's behavior increases patients' adherence to their advice (Zolnierek & Dimatteo, 2009). This could be due to several reasons. First, by virtue of the focus of the study, the patients may have conceptualized satisfaction with communication as doctor-driven rather than patient-driven. Secondly, the Indian healthcare environment is doctor-centric with patients often relying on the doctor to steer their relationship and, consequently, communication and care plans. Furthermore, the study findings showed that closer rapport was associated with both internal LOC and external LOC of chance and other people. As discussed above, this could be because of the nature of the items in the rapport subscale that call on a combination of patient-related ability to form a personal relationship with the doctor and doctor-centric affective behavior.

The current study results further strengthen the importance of introducing communication skills training to doctors in India. This training can focus on key skills such as exhibiting empathy, encouraging question-asking on the part of the patient, not interrupting patients, and using simple language so that patients are able to understand the doctor's recommendations. It contributes to the existing research on doctor–patient communication, patient satisfaction, and psychological well-being; and it provides direction for future research and clinical care.

9.4.1 LIMITATIONS

A major limitation of this study is that owing to the cross-sectional nature of the data, causal interpretations cannot be drawn. Thus, limiting our understanding of how patients' satisfaction with doctor consultation is influenced or impacted by patient characteristics. Second, the study did not include doctors' perspectives on patient satisfaction which could provide for a more comprehensive understanding of doctor–patient interaction and expectations.

9.4.2 CONCLUSIONS

The current study found that the patient's relationship with the doctor, functional well-being, anxiety, internal LOC and external LOC (doctor and chance), comfort in communication, and rapport shared with the doctor predicted patients' satisfaction with the doctor consultation in India. These findings highlight the integral role of the doctor in patient satisfaction, suggesting a need to closely examine this relationship and its impact on patients' satisfaction. Importantly, this study emphasizes the need to introduce communication skills to doctors as well as using positive health psychology principles within the Indian cancer setting.

ACKNOWLEDGMENTS

We would like to thank the participants who gave their precious time and willingly shared their experiences. We would like to extend our gratitude to Dr. Nagraj Huligol, chief radiation oncologist and hospital staff at Dr. Balabhai Nanavati Hospital for their support during the data collection.

Funding source: This research was funded by MHRD as part of the first author's MPhil program.

KEYWORDS

- **psycho-oncology**
- **cancer communication**
- **doctor–patient relationship**
- **cancer experience**

REFERENCES

Agarwal, A., Agarwal, A., Nag, K., Chakraborty, S., & Ali, K. (2011). Doctor patient communication—a vital yet neglected entity in Indian medical education system. *Indian Journal of Surgery*, *73*(3), 184–186.

Arntson, P., Makoul, G., Pendleton, D., & Schofield, T. (1989). Patients' perceptions of medical encounters in Great Britain: variations with health loci of control and sociodemographic factors. *Health Communication*, *1*(2), 75–95.

Ashagi, M. M., & Beheshtifar, M. (2015). The relationship between Locus of Control (Internal–External) and self-efficacy beliefs of Yazd University of Medical Sciences. *International Journal of Engineering and Applied Sciences*, *2*(8), 2394–3661.

Bagcchi, S. (2015). India has low doctor to patient ratio, study finds. *British Medical Journal*, *351,* h5195.

Ben-Sira, Z. (1976). The function of the professional's affective behavior in client satisfaction: A revised approach to social interaction theory. *Journal of Health and Social Behavior*, *17*(1), 3–11.

Blanchard, C. G., Labrecque, M. S., Ruckdeschel, J. C., & Blanchard, E. B. (1990). Physician behaviors, patient perceptions, and patient characteristics as predictors of satisfaction of hospitalized adult cancer patients. *Cancer*, *65*(1), 186–192. https://doi.org/10.1002/1097-0142(19900101)65:1<186::AID-CNCR2820650136>3.0. CO;2-4

Buller, M K., & Buller, D B. (1987). Physicians' communication style and patient satisfaction. *Journal of Health and Social Behavior*, *28*(4), 375–388.

Butow, P. N., Kazemi, J. N., Beeney, L. J., Griffin, A.-M., Dunn, S. M., & Tattersall, M. H. N. (1996). When the diagnosis is cancer: Patient communication experiences and preferences. *Cancer*, *77*(12), 2630–2637. https://doi.org/10.1002/ (SICI)1097-0142 (19960615) 77:12< 2630::AID-CNCR29>3.0.CO;2-S

Broadbent, E., Petrie, K., Main, J., & Weinman, J. (2006). The brief illness perception questionnaire. *Journal of Psychosomatic Research*, *60*, 631–637.

Cella, F. D., Tulsky, D. S., Gray, G., Sarafian, B., Linn, E., & Bonomi, A. (1993). The functional assessment of cancer therapy scale: Development and validation of the general measure. *Journal of Clinical Oncology*, *11*(3), 570–579. doi: 0732-183X/93/1 103-0027$3.00/0

Chittem, M., Norman, P., & Harris, R. (2012). Relationship between perceived diagnostic disclosure, patient characteristics, psychological distress and illness perception in Indian cancer patients. *Psycho-Oncology*, *22*(6), 1375–1380. doi: 10.1002/pon.3149

Cleary, P. D., & McNeil, B. J. (1988). Patient satisfaction as an indicator of quality care. *Inquiry : A Journal of Medical Care Organization, Provision and Financing*, *25*(1), 25–36. Retrieved from http://www.ncbi.nlm.nih.gov/pubmed/2966123

Datta, S. S., Tripathi, L., Varghese, R., Logan, J., Gessler, S., Chatterjee, S., & Menon, U. (2017). Pivotal role of families in doctor–patient communication in oncology: A qualitative study of patients, their relatives and cancer clinicians. *European Journal of Cancer Care*, *26*(5), e12543.

Detmar, S B., Muller M J., Schornagel, J H., Lidwina D V., & Aaronson N K. (2002) Health-related quality-of-life assessment and patient-physician communication: A randomised controlled trial. *The Journal of American Medical Association.* Doi:10.1001/jama.288.23.3027.

Fallowfield, L., Hall, A., Maguire, G. P., & Baum, M. (1990). Psychological outcomes in women with early breast cancer. *BMJ (Clinical Research Ed.)*, *301*(6765), 1394. Retrieved from http://www.ncbi.nlm.nih.gov/pubmed/2271895

Fallowfield, L. (1992). The ideal consultation. *British Journal of Hospital Medicine*, *47*(5), 364–367.

Fangel, L. M. V., Panobianco, M. S., Kebbe, L. M., Almeida, A. M. de, & Gozzo, T. de O. (2013). Qualidade de vida e desempenho de atividadescotidianasapóstratamento das neoplasiasmamárias. *ActaPaulista de Enfermagem*, *26*(1), 93–100. https://doi.org/10.1590/S0103-21002013000100015

Fujimori, M., Akechi, T., Morita, T., Inagaki, M., Akizuki, N., Sakano, Y., & Uchitomi, Y. (2007). Preference of cancer patients regarding the disclosure of bad news, *Psycho-oncology*, *16* (6), 573–581.

Frostholm, L., Fink, P., Oernboel, E., Christensen, K. S., Toft, T., Olesen, F., &Weinman, J. (2005). The uncertain consultation and patient satisfaction: The impact of patients' illness perceptions and a randomized controlled trial on the training of physicians' communication skills. *Psychosomatic Medicine*, *67*(6), 897–905. https://doi.org/10.1097/01.psy.0000188403.94327.5b

Glynne-Jones, R., Ostler, P., Lumley-Graybow, S., Chait, I., Hughes, R., Grainger, J., & Leverton, T. J. (2006). Can I look at my list? An evaluation of a 'prompt sheet' within an oncology outpatient clinic. *Clinical Oncology*, *18*(5), 395–400.

Greenley, J. R., Young, T. B., & Schoenherr, R. A. (1982). Psychological distress and patient satisfaction.

Medical Care, *20*(4), 373–385. https://doi.org/10.1097/00005650-198204000-00003

Griffin, S. J., Kinmonth, A. L., Veltman, M. W., Gillard, S., Grant, J., & Stewart, M. (2004). Effect on health-related outcomes of interventions to alter the interaction between patients and practitioners: a systematic review of trials. *The Annals of Family Medicine*, *2*(6), 595–608.

Hofman, M., Ryan, J. L., Figueroa-Moseley, C. D., Jean-Pierre, P., & Morrow, G. R. (2007). Cancer-related fatigue: the scale of the problem. *The Oncologist*, *12 Suppl 1*(Supplement 1), 4–10. https://doi.org/10.1634/theoncologist.12-S1-4

Homma, M., Ishikawa, H., & Kiuchi, T. (2018). Illness perceptions and negative responses from medical professionals in patients with fibromyalgia: Association with patient satisfaction and number of hospital visits. *Patient Education and Counseling*, *101*(3), 532–540. https://doi.org/10.1016/j.pec.2017.08.014

Husson, O., Mols, F., & Van de Poll-Franse, L. V. (2010). The relation between information provision and health-related quality of life, anxiety and depression among cancer survivors: A systematic review. *Annals of Oncology*, *22*(4), 761–772.

Ishikawa, H., Takayama, T., Yamazaki, Y., Seki, Y., & Katsumata, N. (2002). Physician-patient communication and patient satisfaction in Japanese cancer consultations. *Social Science & Medicine, 55*, 301–311.

Jean Pierre, P., Fiscella, K., Freund, K. M., Clark, J., Darnell, J., Holden, A., & Patient Navigation Research Program Group. (2011). Structural and reliability analysis of a patient satisfaction with cancer-related care measure: a multisite patient navigation research program study. Cancer, *117*(4), 854–861.

Jenkinson, C., Coulter, A., & Bruster, S. (2002). The Picker Patient Experience Questionnaire: Development and validation using data from in-patient surveys in five countries. *International Journal for Quality in Health Care: Journal of the International Society for Quality in Health Care*, *14*(5), 353–358. Retrieved from http://www.ncbi.nlm.nih.gov/pubmed/12389801

Khadir, F., & Swamynathan, R. (2014). Deterrents of complaining: An empirical study of in patients. *Studies on Ethno-Medicine*, *8*(3), 259–267. https://doi.org/10.1080/09735070.2014.11917642

Kumari, R., Idris, M. Z., Bhushan, V., Khanna, A., Agarwal, M., & Singh, S. K. (2009). Study on patient satisfaction in the government allopathic health facilities of Lucknow district, India. *Indian Journal of Community Medicine*, *34*(1), 35.

Lim, L., Chow, P., Wong, C. Y., Chung, A., Chan, Y. H., Wong, W. K., & Soo, K. C. (2011). Doctor–patient communication, knowledge, and question prompt lists in reducing preoperative anxiety—A randomized control study. *Asian Journal of Surgery*, *34*(4), 175–180.

Linn, L. S., & Greenfield, S. (1982). Patient suffering and patient satisfaction among the chronically ill. *Medical Care*, 425–431.

Marimuthu, P. (2008). Projection of cancer incidence in five cities and cancer mortality in India. *Indian Journal of Cancer*, *45*(1), 4–7.

Marshall, G. N., & Hays, R. D. (1994). The Patient Satisfaction Questionnaire Short Form (PSQ-18). Retrieved from https://www.rand.org/pubs/papers/P7865.htm

Meakin, R., & Weinman, J. (2002). The "Medical Interview Satisfaction Scale" (MISS-21) adapted for British general practice. *Family Practice*, *19*(3), 257–263. Retrieved from http://www.ncbi.nlm.nih.gov/pubmed/11978716

Mehra, P. (2018). Does consultation time influence patient perception of provider communication style, patient satisfaction and word of mouth recommendation in India?: A mediation analysis. *Asia Pacific Journal of Health Management*, *13*(1), 32.

Murray, M., & McMillan, C. (1993). Health beliefs, locus of control, emotional control and women's cancer screening behavior. *British Journal of Clinical Psychology*, *32*(1), 87–100.

Murthy, N. S., & Mathew, A. (2004). Cancer epidemiology, prevention and control. *Current Science*, *86*(4), 518–527.

Ong, L. M., Visser, M. R., Lammes, F. B., & de Haes, J. C. (2000). Doctor-patient communication and cancer patients' quality of life and satisfaction. *Patient Education and Counseling*, *41*(2), 145–156. Retrieved from http://www.ncbi.nlm.nih.gov/pubmed/12024540

Quill, T. E. (1989). Recognizing and adjusting to barriers in doctor-patient communication. *Annals of Internal Medicine*, *111*(1), 51–57.

Purrakkal, D., Pulasseri, D., & Ravindran, S. (2004). Are oncology patients aware of their diagnosis? A survey from Calicut. *Indian Journal of Palliative Care*, *10*(1), 39.

Rao, K. D., Peters, D. H., & Bandeen-Roche, K. (2006). Towards patient-centered health services in India—a scale to measure patient perceptions of quality. *International Journal for Quality in Health Care*, *18*(6), 414–421.

Roter, D. L., Hall, J. A., Kern, D. E., Barker, L. R., Cole, K. A., & Roca, R. P. (1995). Improving physicians' interviewing skills and reducing patients' emotional distress. *Archives of Internal Medicine*, *155*(17), 1877. https://doi.org/10.1001/archinte.1995.00430170071009

Shirai, Y., Fujimori, M., Ogawa, A., Yamada, Y., Nishiwaki, Y., Ohtsu, A., & Uchitomi, Y. (2012). Patients' perception of the usefulness of a question prompt sheet for advanced cancer patients when deciding the initial treatment: A randomized, controlled trial. *Psycho-Oncology*, *21*(7), 706–713.

Sodani, P R., Kumar, R K., Srivastava, J., & Sharma, L. (2010). Measuring patient satisfaction: A case study to improve quality of care at public health facilities. *Indian Journal of Community Medicine, 35*(1), 52–56. DOI- 10.4103/0970-0218.62554

Tang, S. T., & Lee, S. Y. C. (2004). Cancer diagnosis and prognosis in Taiwan: Patient preferences versus experiences. *Psycho-Oncology*, *13*(1), 1–13.

Tennstedt, S. L. (2000). Empowering older patients to communicate more effectively in the medical encounter. *Clinics in Geriatric Medicine*, *16*(1), 61–70.

Thomas, B. C., Devi, N., Sarita, G. P., Rita, K., Ramdas, K., Hussain, B. M., & Pandey, M. (2005). Reliability & validity of the Malayalam hospital anxiety & depression scale (HADS) in cancer patients. *Indian Journal of Medical Research*, *122*(5), 395.

Wallston, K. A., Malcarne, V. L., Flores, L., Hansdottir, I., Smith, C. A., Stein, M. J., & Weisman, M. H. (1999) Clements PJ. Does God determine your health? The God locus of health control scale. *Cognitive Therapy and Research*, *23*(2), 131–142.

Wallston, K. A., Stein, M. J., & Smith, C. A. (1994). Form C of the MHLC Scales: A condition- specific meausre of locus of control. *Journal of Personal Assessment*, *63*(3), 534–553. doi: 10.1207/s15327752jpa6303_1

Zachariae, R., Pedersen, C. G., Jensen, A. B., Ehrnrooth, E., Rossen, P. B., & Von Der Maase, H. (2003). Association of perceived physician communication style with patient satisfaction, distress, cancer-related self-efficacy, and perceived control over the disease. *British Journal of Cancer*, *88*(5), 658–665. https://doi.org/10.1038/sj.bjc.6600798

Zigmond, A. S., & Snaith, R. P. (1983). The hospital anxiety and depression scale. *Acta Psychiatrica Scandinavica*, *67*, 361–70.

Zolnierek, K. B. H., & Dimatteo, M. R. (2009). Physician communication and patient adherence to treatment: a meta-analysis. *Medical Care*, *47*(8), 826–834. https://doi.org/10.1097/MLR.0b013e31819a5acc

CHAPTER 10

Self-Efficacy and Resilience: Correlates of Happiness and Well-Being Among Adolescents with Disability

SWATI PATRA

Indira Gandhi National Open University (IGNOU), New Delhi, India

E-mail: swatipatra@ignou.ac.in

ABSTRACT

Disability and happiness—can they go together? Disability is viewed as a tragedy. It is usually believed that disability leads to suffering and it makes the person with disability less happy and satisfied with life. However, adolescents with disability are individuals in their own right and their belief in their capability can lead to their resilience and happiness. Self-efficacy and resilience can influence one's happiness. The present study attempts to explore this question of whether general self-efficacy and resilience in adolescents with disability can lead to their happiness and well-being? Resilience and self-efficacy are psychological resources having significant implications for one's happiness and well-being. Self-efficacy determines an individual's resiliency to adversity and his/her vulnerability to stress and depression. General self-efficacy aims at a broad and stable sense of personal competence to deal effectively with a variety of stressful situations and promote psychological well-being. Resilience is viewed as a measure of successful stress-coping ability or the ability to thrive in the face of adversity. Happiness and well-being is viewed in terms of positive human functioning. The sample in the present study consisted of 48 adolescents with hearing disability selected purposively. Following tools were given: General self-efficacy, Brief Resilience Scale, and Mental Health Continuum Short Form. The participants were also given a single-item scale on happiness and interviewed with an open-ended questionnaire regarding happiness. Findings revealed that general self-efficacy significantly correlates and predicts mental health and happiness. Qualitative analysis of the questionnaire on happiness revealed a central theme of support by family and friends being a major factor in contributing to the happiness of the adolescents with hearing disability. The study has implications for the adolescents with hearing disability in specific and people with disability in general for enhancing their well-being.

10.1 INTRODUCTION

Resilience and self-efficacy are psychological resources having significant implications for one's happiness and well-being. Resilience is viewed as a measure of successful stress coping ability or the ability to thrive in the face of adversity (Connor and Davidson, 2003). Self-efficacy determines an individual's resiliency to adversity and his/her vulnerability to stress and depression (Bandura et al., 2003). General self-efficacy (GSE) aims at a broad and stable sense of personal competence to deal effectively with a variety of stressful situations and promote psychological well-being (PWB) (Adeyemo & Adeleye, 2008; Schwarzer, 1994). Thus self-efficacy is related to the development of resilience and together these two aspects of one's personality can influence one's happiness and well-being.

The present study has attempted to study GSE and resilience as correlates of happiness and well-being in adolescents with hearing disability. The focus in the field of disability has shifted from an earlier medical model, where the focus was on the biological impairments of an individual and their functional limitations, to a social model. The latter approach views disability in the social context, for example, the social disadvantage experienced by the person due to inaccessibility issues. This study advocates that along with the social model, there needs to be an individual model also, putting emphasis on the positive individual resources and strengths.

Research has started to focus on positive human functioning and well-being with the advent of positive psychology toward the end of 20th century. Earlier, the emphasis was on human suffering and mental illness, focusing on the deficits of individuals. Positive psychology has put the emphasis on human strengths and capacities and highlights flow, meaning in life, optimism, values, hope, resilience, happiness, and well-being. Thus, the preoccupation has changed from treating illness to promoting well-being. However, it may be noted that positive psychology is not a new field. Humanistic psychology in the 1960s and 1970s has studied concepts related to positive psychology. It was Maslow who used the term positive psychology in 1954 in his book *Motivation and Personality* (p. 354). Other researchers have also talked about positive human characteristics (Allport, 1958), meaning (May, 1953; Yalom, 1980), self-actualization (Maslow, 1971), and healthy mindedness (William James, 1902). However, it was Martin Seligman (1998) who popularized the term positive psychology and brought it into the limelight. The goal of positive psychology which is to flourish, thrive, and achieve meaning and well-being in life is the ultimate aim of human life. Positive psychology mainly focuses on three areas: positive subjective states or positive emotions (e.g., happiness, love, and contentment); positive individual traits (e.g., courage, wisdom, and determination); and positive institutions, which are studied at a society level and include healthy family and work environments (Seligman & Csikszentmihalyi, 2000).

In the changing complexities of the present-day world, it is very crucial that we emphasize on the happiness and well-being of people. This becomes all the more important in the case of people with disability. Disability is seen as a tragedy and weakness that colors all our assumptions, expectations, perception, and interactions with people with disabilities. It is usually believed that disability leads to suffering and it makes the person with disability less happy and satisfied with life. Thus, generally our efforts are directed toward making them less sad and reducing their worries and suffering. However, with a positive psychology approach, our focus needs to be on their happiness, satisfaction, joy, hopes, and well-being. The belief is that those who suffer want to build the strengths and not just correct their weaknesses (Duckworth et al., 2005). Hence, studies need to be conducted highlighting the positive attributes of people with disability and emphasizing their happiness and well-being; after all, they are also individuals in their own right and an equal part of our society.

10.2 HAPPINESS AND WELL-BEING

Happiness is a positive emotion experienced by an individual. It is being in a state of joy, satisfied with life, and in general, refers to subjective well-being (SWB). As defined by Diener (1984), happiness refers to the preponderance of positive affect over negative affect with a distinct focus on the affective evaluation of one's life situation. Diener and his colleagues define happiness as a combination of life satisfaction (a cognitive judgment) and the balance of the frequency of positive and negative affect (i.e., hedonic tone) (Diener et al., 1991; Larsen et al., 1985).

Happiness refers to an individual's overall judgment about life as a whole. It takes a global view of life. Veenhoven (1989) has defined happiness as the "overall appreciation of one's life, as a whole" (p. 5). Thus it is a subjective evaluation by the individual only in the context of his/her life. Further, it is not about any specific aspect of the individual's life such as work or family, etc., but an overall assessment or evaluation of the life, including the past, present, and anticipated future. The word "happiness" is not used to denote an optimal appreciation of life; rather, it refers to a degree, like the concepts of "length" or "weight" which denotes more or less of something (Joshi, 2015). The level of happiness of a person is influenced by three factors: the person's set point of happiness determined by genetics; circumstantial factors that include personal qualities (e.g., age, gender, and life history), events (e.g., marriage, unemployment, and retirement), and situations (e.g., nation, location, and culture); and intentional actions, such as what people do and how they think in everyday life (Lyubomirsky et al., 2005). They term these three factors as constituting the "architecture of sustainable happiness."

Happiness has been mainly viewed from two approaches: hedonic and eudaimonic approach. The hedonic approach focuses on experiencing positive emotions and being satisfied with life. It is concerned with the pleasures of life. On the other hand, eudaimonic goes beyond the hedonistic pleasures and is concerned with being in a flow and achieving meaning in life. In terms of well-being, it can be said that hedonic well-being refers to emotional well-being or SWB, whereas eudaimonic well-being involves social well-being and PWB. Seligman has written about the pleasant life, the good life, and the meaningful life. As per Seligman's views on happiness, the eudaimonic approach refers to the good life and the meaningful life. Good life is being engaged in any work, being in a flow that relates to one's personal growth and development. Meaningful life goes beyond the individual and looks for a greater cause and contributes to the community and society. The hedonic approach relates to the pleasant life as viewed by Seligman.

Happiness, life satisfaction, and SWB have been used interchangeably in the literature. However, to differentiate, happiness is related more to SWB rather than PWB. SWB denotes the limited aspect of subjective enjoyment of life, happiness, and life satisfaction whereas PWB includes a broader perspective. According to Diener et al. (2002, p.63), SWB consists of both cognitive and affective evaluation of one's life. "The cognitive component usually corresponds to the individual's evaluation of life satisfaction, according to subjectively determined standards, that may be formulated at a general level when they refer to life as a whole, or at more specific levels when they refer to particular life domains"

(Caprara et al., 2006). The affective element indicates one's emotions, moods, and feelings. Thus, SWB refers to how people think and feel about their lives. It can be life satisfaction experienced by an individual in global terms as well as in domain terms, that is, specific areas of life.

SWB reflects the hedonic perspective (Watson et al., 1988; Diener et al., 1999) and focuses on positive and negative affect, and life satisfaction; whereas PWB is consistent with the eudaimonic perspective (Ryff, 1989; Ryff & Singer, 1996; Kashdan et al., 2008) and includes the dimensions of self-actualization, personal development, and relations with environment. Ryff (1989) has conceptualized PWB as being multidimensional and has incorporated six dimensions in her PWB scales. These are: independence and self-determination (autonomy); having satisfying, high-quality relationships (positive relations with others); the ability to manage one's life (environmental mastery); being open to new experiences (personal growth); believing that one's life is meaningful (purpose in life); and a positive attitude toward oneself and one's past life (self-acceptance) (Ryff & Keyes, 1995; Keyes et al., 2002).

Well-being has been indicated in the present study in terms of mental health. Mental health can be considered as an important aspect of well-being. According to WHO (2007), mental health is a state of well-being in which the individual realizes his/her abilities, copes with the normal stresses of life, works productively and fruitfully, and contributes to society. In a similar line, Bornstein et al. (2003) defined well-being as "the state of successful performance throughout the life course integrating psychical, cognitive, and socio-emotional functions that result in productive activities deemed significant by one's cultural community, fulfilling social relationships, and the ability to transcend moderate psychosocial and environmental problems."

Studies have pointed out that mental health difficulties in students with disabilities are a significant barrier to their learning (Coughlan & Carpenter, 2013), and proactive strategies need to be initiated in this regard in the form of developing emotional resilience and the development of well-being teams in schools. Mpofu et al. (2017) found in their study that participants who were adolescents with physical disabilities living in inclusive community settings of Makonde Urban in Zimbabwe were having high levels of PWB (autonomy and choice, purpose in life, positive relations with others and good personal growth and self-acceptance). Canha et al. (2016) found that students with disabilities reporting a better health self-rating were happier and more satisfied with their lives. Further, the comparison between groups showed that students with disabilities presented more symptoms and lower health perception than their nondisabled peers. Hence studies need to explore the use of positive interventions in the context of disability as they have been found to be effective in reducing depressive symptoms and increasing happiness in nondisabled people (Seligman et al., 2005).

10.3 GSE, HAPPINESS, AND WELL-BEING

Bandura (2001) defines self-efficacy as people's beliefs in their capacity to exercise some measure of control over environmental events. It refers to a person's belief in his/her ability to organize and execute a required course of action to achieve desired goals (Bandura, 1997). It affects one's thought process, attitude, motivation, and behavior. Self-efficacy can be of two types: global or GSE, and self-efficacy in specific domains such as academic, social, and emotional self-efficacy (Bandura et al., 2001). GSE refers to one's belief in one's competence to use various strategies to deal with problems of life in general and cope with various situations; whereas specific self-efficacy refers to the use of strategies to handle problems in specific life domains such as education, career or personal, etc.

Studies have found that students with high levels of self-efficacy and well-being performed well in academic achievement and were more successful at school than students with low self-efficacy (Lyubomirsky, 2001; Khramtsova et al., 2007; Zimmerman et al., 1992). In a study by Caroli and Sagone (2014), it was found that boys expressed greater GSE and PWB than girls—specifically, mastery and self-acceptance.

Yazici et al. (2011) investigated the influence of self-efficacy beliefs and emotional intelligence on the academic achievement of high school students. In another study on 354 university students, Capri et al. (2012) examined the relation between GSE beliefs, life satisfaction, and burnout. Yet another study by Santos et al. (2014) on 969 students found higher levels of SWB for students with higher levels of GSE.

Studies thus point out the importance of self-efficacy. Children with disabilities face a challenging situation in living up to their potential. Hence, it is important that they have an increased self-efficacy so that they have a sense of control. Low self-efficacy has been found to be correlated with higher anxiety and depression (Faure & Loxton, 2003; Kashdan & Roberts, 2004). Since emotional states are believed to be one of the substrates of self-efficacy (Bandura, 2001), promoting self-efficacy beliefs will help protect children and adolescents in counteracting depressive states (Bandura et al., 1999). In another study, Strobel et al. (2011) report that GSE functionally connects personality factors and two components of SWB (life satisfaction and subjective happiness). The findings of their study demonstrate the mediating role of self-efficacy in linking personality factors and SWB. Personality traits predispose individuals to experience varying levels of SWB. The study pointed out that self-efficacy mediated the influence of neuroticism, extraversion, openness, and conscientiousness on life satisfaction. Furthermore, self-efficacy mediated the influence of openness and conscientiousness, but not that of neuroticism and extraversion, on subjective happiness. Results highlight the importance of cognitive beliefs in functionally linking personality traits and SWB.

10.4 RESILIENCE, HAPPINESS, AND WELL-BEING

Resilience asks the basic question, "why some people are able to stay put in the face of adversity, show persistence, and overcome any challenge or hardship whereas others in such circumstance succumb to it?" Resilience is the key to overcome adversity and cope with stressful situations. It is the capacity to bounce back. It involves not only adapting or coping with the negative event but also learning from and getting transformed by their experiences (Grotberg, 2003). According to Zautra et al. (2010), resilience is best defined as a successful adaptation to adverse circumstances. Resilience can be thought of as a person's ability to cope effectively with negative life experiences, and it can develop as a person experiences positive personal growth through negative events (Dunn et al., 2009).

Studies (Pajares & Schunk, 2001; Salami, 2010) have found that students with high PWB and life satisfaction were more resilient, were more committed to their academic goals, and pursued success rather than avoiding failures. Several other studies (Carver et al., 2010; Yu

& Zhang, 2007; Yee Ho et al., 2010) have also found resilience to be related to PWB. Another study by Souri and Hasanirad (2011) examined the relationships between resilience, optimism, and PWB in 414 students of medicine using the Ryff Scale of Psychological Well-Being, Connor–Davidson Resilience Scale, and Scheier and Carver's Life Orientation Test. The results revealed that resilience is able to predict 27% of PWB, and resilience and optimism together explain and predict 35% of PWB variance.

Further, Ganguly et al. (2015) in their study on University students having both "visible" disabilities (sensory and physical disabilities) and "hidden" disabilities (psychological disabilities) reported resilience to be directly and significantly related to academic satisfaction, well-being, and career optimism. Findings of the study by Faisal and Mathai (2017) showed a significant positive relationship between resilience and spirituality, and resilience and PWB. It also showed a significant gender difference in resilience. The results implied enhancing of resilience at a young age, so that people will experience a better PWB. A study by Sagone and Indiana (2017) indicated that adolescents with high positive affect reported higher levels of perceived self-efficacy in life skills than those with low positive affect; in addition, adolescents with high positive affect reached higher levels of resilience (that is, adaptability and engagement) than those with low positive affect. Relationship of positive affect with both resilience and perceived self-efficacy in life skills highlights the importance of these variables for the adolescents' well-being.

The present study was undertaken to focus on the happiness and well-being of people with disabilities, specifically children with hearing disability. Usually, it is difficult for the "outsiders" (those not having a disability) to accept that people with disability (the "insiders") can adjust to their situation relatively well, and experience happiness in their life (Dunn et al., 2012). Delle (2006) and Albrecht (1996) have found that perceived quality of life can be high in spite of physical disabilities. Disability has always been viewed negatively, highlighting only the difficulties, distress, and depression. It portrays the role of a victim for a person with disability. It not only impacts the individual by lowering the self-esteem, but also leads to pity and prejudice in people around and the society in general. The present study derives from the theoretical approach of positive psychology. It aims at approaching the field of disability from a positive psychology perspective by focusing on the subjective states and individual character strengths.

The challenges faced by adolescents and youth in the present times are increasing manifold. As Achenbach and Howell (1993) point out, research evidence indicates a rise in emotional and behavioral problems of children and adolescents. The increasing challenges of adolescence stage affect the well-being of the adolescents. However, the risk and protective model advocates analyzing the risk and protective factors during adolescence so that one can take steps for mitigating the negative effect of the risk factors. Disability as a risk factor can lead to the maladaptive developmental pathway in the child with a disability. The objective here needs to have an adaptive developmental pathway for the child so that the focus shifts from the adverse life event to adaptive coping. The protective factors play an important role here. These include psychological processes and the psychosocial environment of the individual. However, it may not always be possible to control or alter the social environment. In such cases, it is the personal psychological resources that the child is having which have a significant influence on the PWB of the individual. Hence in the present study, self-efficacy and resilience were taken as two key psychological resources that adolescents with disabilities can have as protective factors impacting the happiness and well-being of the individual.

10.5 OBJECTIVES

Thus the specific objectives of the present research are as follows:

- To study GSE, resilience, happiness, and well-being of adolescents having a hearing disability.
- To study the relationship between GSE, happiness, and well-being of adolescents having a hearing disability.
- To study the relationship between resilience, happiness, and well-being of adolescents having a hearing disability.
- To study gender difference in GSE, resilience, happiness, and well-being of adolescents having a hearing disability.

10.6 METHODOLOGY

The sample consisted of 48 adolescents (24 girls and 24 boys) with hearing disability studying in class 10th–12th in a CBSE school meant for the deaf children in New Delhi. All of them had 100% hearing impairment since birth. They belonged to lower socioeconomic status. They were not having any other disability. The following tools were administered to them: GSE Scale (Schwarzer & Jerusalem, 1995), Brief Resilience Scale (Smith et. al., 2008), and Mental Health Continuum Short Form (Keyes, 2002) to assess GSE, resilience, and well-being, respectively. Mental health scale consisted of three dimensions such as emotional well-being which relates to the hedonic aspect, and social well-being and PWB that relate to the eudaimonic well-being. Thus the tool provides a comprehensive view on mental health. The participants were also given a single-item scale on happiness (Abdel-Khalek, 2006) and interviewed with an open-ended questionnaire to assess their happiness. All these were given to adolescents with the help of a sign language expert in the presence of the researcher.

10.7 RESULTS AND DISCUSSION

Analysis of the data revealed the results as shown in Table 10.1. The mean and standard deviation (SD) on the various measures are presented in Table 10.1. The table also shows the "*t*" values obtained on different measures.

The results did not show any significant gender difference on any of the measures taken. However, if we observe the mean scores, it can be seen that male students scored a little higher on GSE, emotional well-being and social well-being; whereas females scored more on happiness and PWB. The mean scores are more or less same on resilience and mental health (total).

Intercorrelation between general self-efficacy, resilience, happiness, and mental health variables are given in Table 10.2.

It can be observed from Table 10.2 that GSE has significant correlation with mental health total ($r = 0.51, p < 0.01$) as well as the subcomponents of mental health, that is, social well-being ($r = 0.38, p < 0.01$) and PWB ($r = 0.54, p < 0.01$). These findings become important as indicated by the Bamford Review (2006) which reports that the incidence of poor mental health among vulnerable children, such as those with a disability or experiencing poverty

TABLE 10.1 Mean, SD, and "t" Value on GSE, Resilience, Happiness, and Mental Health Among Adolescents with Hearing Disability ($N = 48$)

Variables	Mean			SD	"t" value (df = 46)
	Female (N = 24)	Male (N = 24)	Total (N = 48)		
GSE	30.46	31.29	30.88	5.31	−0.54
Resilience	3.17	3.00	3.08	0.58	1.00
Happiness	7.17	6.83	7.00	2.10	0.55
Emotional well-being (Hedonic)	8.92	10.13	9.52	3.30	−1.28
Social well-being (Eudaimonic)	13.25	14.54	13.90	4.84	−0.92
Psychological well-being (Eudaimonic)	19.88	18.08	18.98	5.40	1.14
Mental health (Total)	42.04	42.75	42.40	9.93	−0.25

TABLE 10.2 Correlation Values Between GSE, Resilience, Happiness, and Mental Health among Adolescents with Hearing Disability ($N = 48$)

	GSE	Resilience	Happiness	Emotional Well-being (Hedonic)	Social Well-being (Eudaimonic)	PWB (Eudaimonic)	Mental Health (Total)
GSE	1	0.17	0.46**	0.11	0.38**	0.54**	0.51**
Resilience		1	−0.08	0.08	0.04	0.20	0.15
Happiness			1	0.10	0.19	0.18	0.22
Emotional well-being (Hedonic)				1	0.49**	0.22	0.69**
Social well-being (Eudaimonic)					1	0.21	0.77**
Psychological well-being (Eudaimonic)						1	0.73**
Mental health (Total)							1

**$p < 0.01$.

is disproportionately high. Since the findings of the present study indicate a positive relation between self-efficacy and mental health, it has implications for developing self-efficacy in children with disabilities for promoting their mental health.

However, the correlation of GSE with emotional well-being, the dimension of mental health was not found to be significant ($r = 0.11$). It may be noted here that emotional well-being refers to the hedonic aspect whereas the other two dimensions of social well-being and PWB represent the eudaimonic approach to well-being. So, GSE significantly correlates with the eudaimonic dimension of mental health rather than the hedonic aspect. As discussed earlier in the chapter, hedonic well-being relates to SWB whereas eudaimonic approach relates to the PWB.

Self-efficacy can be said to play a role in well-being and happiness in life as it helps in believing in one's competence to address various problems in life. Addressing life's problems will lead to happiness in life. As pointed out by Magaletta and Oliver (1999), one's level of self-efficacy determines PWB.

Though the findings indicate no significant correlation of GSE with emotional well-being (the hedonic aspect or SWB), they point out a significant correlation of GSE with happiness ($r = 0.46, p < 0.01$). Happiness is viewed as more of SWB than PWB. Studies (Cakar, 2012) indicate that high self-efficacy or an increased belief in one's capacity to control events in one's environment may tend to increase SWB. However, the present findings point out lack of significant correlation between GSE and emotional well-being (or SWB); and significant correlation between GSE and happiness. The findings may be explained that happiness was assessed here in terms of an overall evaluation of happiness in life through a single-item scale of happiness.

Happiness or SWB is frequently defined as having three interrelated qualities: regular positive affect, elevated life satisfaction, and sporadic negative affect (Diener, 1984). The study did not assess happiness along with these three aspects, but employed a global evaluation of happiness. Hence, future studies may explore it further by using different tools on happiness.

The overall findings thus highlighted the role of GSE in mental health as well as happiness in life. On the other hand, resilience did not have a significant correlation with happiness or mental health.

10.8 FINDINGS OF REGRESSION ANALYSIS

The data were further analyzed to find out to what extent GSE is predicting mental health and happiness. Results of regression analysis are given below:

It can be seen from Tables 10.3–10.6 that GSE explains and predicts 25% of mental health (total), 12% of social well-being, 27% of PWB, and 20% of happiness. Thus, enhancing self-efficacy as an individual asset or strength can contribute to the happiness and well-being of children with disabilities.

TABLE 10.3 GSE Predicting Mental Health (Total) among Adolescents with Hearing Disability

Model	Variable	Beta	F (df)	Adjusted R Square
1	GSE	0.51**	16.52 ** (1,46)	0.25

**$p < 0.01$.

TABLE 10.4 GSE Predicting Social Well-being among Adolescents with Hearing Disability

Model	Variable	Beta	F (df)	Adjusted R Square
1	GSE	0.38**	7.64** (1,46)	0.12

**$p < 0.01$.

TABLE 10.5 GSE Predicting PWB among Adolescents with Hearing Disability

Model	Variable	Beta	F (df)	Adjusted R Square
1	GSE	0.54**	18.55** (1,46)	0.27

**$p < 0.01$.

TABLE 10.6 GSE Predicting Happiness among Adolescents with Hearing Disability

Model	Variable	Beta	F (df)	Adjusted R Square
1	GSE	0.46^{**}	12.43^{**} (1, 46)	0.20

$^{**}p < 0.01$.

10.9 QUALITATIVE ANALYSIS OF RESPONSES TO HAPPINESS QUESTIONNAIRE

The study also interviewed 13 adolescents out of the total of 48 using an open-ended questionnaire on happiness. It consisted of eight questions and their responses can be grouped under four major aspects as given below:

10.9.1 PROBLEMS/ISSUES FACED

The adolescents reported less clear or no communication from parents and family. They also observed that they face lots of problems in life and are less happy in life. Their parents get angry and scold them. Almost all of them said their parents and family are not supportive and understanding of them.

e.g., "my mother beats me, I get sad"
"parents angry on me frequently"
"very less time I get clear communication"
"less communication with parents"
"less communication in family and outside"

10.9.2 MEANING OF HAPPINESS AND WHAT MAKES US HAPPY

Happiness according to some of the adolescents was getting good food and gifts. For a majority of them, however, it was talking to friends and good communication with family and friends that make them happy. Family support and love of parents also make them happy.

Some of them also reported that giving thanks to others and providing service to others made them happy.

e.g., "communication and talk make me clear inside and happy"
"I want to express my inside feeling to my family fully that make me happy"
"friend talk make me happy. I can express myself well to them"
"talking with my deaf friends"
"helping poor and needy people and kids"

10.9.3 CHARACTERISTICS OF A HAPPY PERSON

Happy person gives food, clothes to the poor and needy; feeds birds and animals; and happy from inside.

10.9.4 INCREASING HAPPINESS

All adolescents agreed that happiness can be increased. Further, most of them responded that happiness can be increased by the help and support of friends and family. Only one student reported that money and a good job can increase happiness.

Analyzing the responses of the interview participants, it can be seen that the overarching theme coming out of this qualitative analysis is the importance of family support for their happiness. Most of them reported having friends as their support and talking to them gives them happiness. At the same time, they expressed their strong desire for love and support from their family that will make them happier.

Findings point to the need for more in-depth interview using a full-fledged qualitative approach to study the children with hearing disability along with their parents and siblings. This will give a comprehensive picture of the situation and suggest suitable intervention strategies for mental health and well-being of such children.

10.10 SUMMARY AND IMPLICATION OF THE FINDINGS

As Seligman (2002) points out, there is a need to create a science of human strength. This will enable to inculcate good values, focus on what is positive, and can be possible with the individual, having a significant implication for the happiness and well-being of the individual. Adolescents in general, and specifically with disabilities, face greater challenges in their life. Beliefs of self-efficacy will make them more confident and thrive in the face of adversity, and will positively influence their happiness and well-being.

The finding of the present study has singled out the role of GSE in happiness and mental health. However, resilience was not found to have a significant correlation with happiness or mental health. Existing literature, however, points out the contributing role of resilience to mental health. For instance, Abbe et al. (2003) report that happy people engage in less negative self-reflection than unhappy individuals, dwell less on adversity and rebound more quickly from it, and react to distressing life events positively. Hence, resilience can be explored further in future studies.

The study employed self-report inventories that have their own limitation of social desirability factor. However, it has also the advantage of providing the 'insider' perspective. Though the study also used a qualitative method by interviewing a sub-sample with an open-ended questionnaire on happiness, it did not use a full-fledged qualitative approach. Hence, studies can be conducted in the future employing a qualitative research design.

The findings highlight the importance of positive psychology and its implications for the field of disability. The focus needs to shift from disabled people to people with disabilities. This person-first focus will help separate the individual and the condition or disability instead of identifying with the condition. Recognizing that disability is only one aspect of a person's life will enable to focus more on their well-being and happiness. No matter how severe or disruptive a disability may be, no physical condition eliminates people's other "assets" (Dunn & Dougherty, 2005). Assets here refer to a broad collection of existing or potential resources including resources related to the individual. In line with the emphasis of positive psychology on individual strengths, the present study has pointed out the importance of GSE as an asset to enhance the well-being of children with hearing disabilities.

Any disability can be approached from two perspectives: that it is insurmountable and one succumbs to it, feels helpless and hopeless; and the other way is to view it as a surmountable challenge (Wright, 1983) that can be coped with. It may be noted that the sample of the present study had a hearing disability since birth, it was not acquired disability. This might have had an implication for their mental health and self-efficacy. Since it is by birth, they might have adjusted and coped well with their disability due to high-level self-efficacy. However, further studies are required to explore it by taking samples having hearing disability by birth and acquired disability.

10.11 CONCLUSION

Mental health issues and well-being concerns of students with disability are usually marginalized in the context of a focus on the population of general children who are not affected with any type of "disability." However, it is all the more important to highlight the well-being of children with a disability because of the likelihood of their neglect in integrating them. True inclusiveness and equality will happen when we focus on their well-being and happiness as any other human being. Mental health vis-à-vis disability needs to be included in designing of any school-based mental health program.

Review by McMillan and Jarvis (2013) highlights the need for mental health promotion in schools to incorporate targeted approaches for at-risk students within the context of universal, whole-school approaches, and in particular to consider the mental health needs of students with disabilities. Evidence-based findings focusing on the contributing role of various variables toward happiness and well-being of children with disability will enable to undertake appropriate measures. The present study was a step in that direction and highlighted the significant role of GSE in the well-being of children with hearing disability.

Many factors influence the goal of achieving well-being and a better life. Many times, especially in the case of individuals with disabilities, situational factors may not be under one's control. It is crucial that family and the community in general should emphasize a coping rather than succumbing orientation (Elliott et al., 2001; Olkin, 1999). Attempts need to be made to strengthen individual resources to enhance the well-being of the individual. Here, self-efficacy as an individual resource that was found to be significantly correlated with happiness, mental health, and PWB in the present study becomes important. This highlights the two important focus of the positive psychology field, that is, positive emotional experiences relating to happiness, and positive personality or character strengths.

ACKNOWLEDGMENT

The author acknowledges the contribution of Ms. Sonu Pandey in the data collection through sign language for the study.

KEYWORDS

- **resilience**
- **hearing disability**
- **self-efficacy**
- **well-being**
- **happiness**
- **adolescents**

REFERENCES

Abbe A., Tkach C., & Lyubomirsky (2003). The art of living by dispositionally happy people. *Journal of Happiness Studies*, *4*(4), 385–404.

Abdel-Khalek, A. (2006). Measuring happiness with a single-item scale. *Social Behavior and Personality: An International Journal*, *34*, 139–150.

Achenbach T. M. & Howell C. T. (1993). Are American children's problems getting worse? A 13-year comparison. *Journal of the American Academy of Child and Adolescent Psychiatry*, *32*, 1145–1154

Adeyemo D. A. & Adeleye A. T. (2008). Emotional intelligence, religiosity and self-efficacy as predictors of psychological well-being among secondary school adolescents in Ogbomoso, Nigeria. *Europe's Journal of Psychology*, *4*(1), 22–31.

Albrecht, G. L. (1996). Using subjective health assessments in practice and policy making. *Health Care Analysis, 4,* 284–292.

Allport, G. W. (1958). Personality: normal and abnormal. *The Sociological Review*, *6*(2), 167–180.

Bamford (2006). The Bamford review of mental health and learning disability (Northern Ireland): A vision of a Comprehensive Child and Adolescent Mental Health Service. http://www.dhsspsni.gov.uk/index/bamford/published-reports.htm

Bandura, A. (1997). *Self-efficacy: The exercise of control.* New York: W. H. Freeman & Co.

Bandura, A. (2001). Social cognitive theory: An agentic perspective. *Annual Review of Psychology, 52,* 1–26.

Bandura, A., Barbaranelli, C., Caprara, G. V. & Pastorelli, C. (2001). Self-efficacy beliefs as shapers of children's aspirations and career trajectories. *Child Development, 72*(1), 187–206.

Bandura, A., Caprara G. V., Barbaranelli C., Gerbino M., & Pastorelli C. (2003). Role of affective self-regulatory efficacy in diverse spheres of psychosocial functioning. *Child Development*, *74*(3), 769–782.

Bandura, A., Pastorelli C., Barbaranelli C. & Caprara G. V. (1999). Self-efficacy pathways to childhood depression. *Journal of Personality and Social Psychology, 76,* 258–269.

Bornstein, M. H., Davidson, L., Keyes, C. L. M., & Moore, K. A. (2003). *Well-being: Positive development across the life course*. Mahwah, NJ: Lawrence Erlbaum Associates.

Canha L., SimoesC., Matos M. G., & Owens L. (2016). Well-being and health in adolescents with disabilities. *Psicologia Reflexão e Crítica, 29,* 1–8.

Cakar, F. S. (2012). The relationship between the self-efficacy and life satisfaction of young adults. *International Education Studies, 5*(6), 123–130.

Caprara, G. V., Steca, P., Gerbino, M., Paciello, M. & Vecchio, G. M. (2006). Looking for adolescents' well-being: Self-efficacy beliefs as determinants of positive thinking and happiness. *Epidemiologia e Psichiatriasociale, 15*(1), 30–43.

Capri B.; Ozkendir O. M.; Ozkurt B.; & Karakus F. (2012). General self-efficacy beliefs, life satisfaction and burnout of university students. *Procedia—Social and Behavioral Sciences*, *47*, 968–973.

Caroli M. E. De & Sagone E. (2014). Generalised self-efficacy and well-being in adolescents with high vs. low scholastic self efficacy. *Procedia—Social and Behavioral Sciences*, *141*, 867–874.

Carver, C. S., Scheier, M. F., & Segerstrom, S. C. (2010). Optimism. *Clinical Psychology Review*, *30*(7), 879–889.

Connor K. M. & Davidson J. R. (2003). Development of a new resilience scale: The Connor-Davidson Resilience Scale (CD-RISC). *Depression and Anxiety*. *18*(2), 76–82.

Coughlan, B. & Carpenter, B. (2013). Mental health & emotional wellbeing in students with disabilities: understanding the complexities involved. *Special Education Resource Update, 23*(3), 1–4.

DelleFave, A. (2006). The impact of subjective experience on the quality of life: A central issue for health professionals. In M. Csikszentmihalyi & I. S. Csikszentmihalyi (Eds.), *A life worth living: Contributions to positive psychology* (pp. 165–181). New York: Oxford University Press.

Diener E. (1984). Subjective well-being. *Psychological Bulletin*. *95*(3), 542–575.

Diener E., Lucas R. E. & Oishi S. (2002). *Subjective well-being: The science of happiness and life satisfaction.* In (C. R. Snyder and S. J. Lopez, Eds.), *Handbook of positive psychology* (pp. 63–73). Oxford University Press: New York.

Diener, E, Sandvik E., & Pavot W. (1991). Happiness is the frequency, not the intensity, of positive versus negative affect. In F. Strack, M. Argyle, & N. Schwarz (Eds.), *Subjective well-being: An interdisciplinary perspective* (pp. 119–139). New York: Pergamon.

Diener, E., Suh, E. M., Lucas, R. E., & Smith, H. L. (1999). Subjective well-being: Three decades of progress. *Psychological Bulletin, 125*(2), 276–302.

Duckworth, A. L., Steen, T. A., & Seligman, M. E. P. (2005). Positive psychology in clinical practice. *Annual Reviews of Clinical Psychology*, *1*, 629–651.

Dunn D. S. & Dougherty S. B. (2005). Prospects for a positive psychology of rehabilitation. *Rehabilitation Psychology*, *50*(3), 305–311.

Dunn, D., Uswatte, G., & Elliot, T. (2009). Happiness, resilience, and positive growth following physical disability: Issues for understanding, research, and therapeutic intervention. In C. R. Snyder & S. Lopez (Eds.), *Oxford handbook of positive psychology* (2nd ed.) (pp. 651–664). New York: Oxford University Press.

Dunn, D. S., Uswatte, G., & Elliott, T. R. (2012). *Happiness, resilience and positive growth following physical disability: Issues for understanding, research and therapeutic intervention.* 10.1093/oxfordhb/9780195187243.013.0062

Elliott, T. R., Shewchuk, R. M., & Richards, J. S. (2001). Family caregiver social problem-solving abilities and adjustment during the initial year of the caregiver role. *Journal of Counseling Psychology, 48*, 223–232.

Faisal, T., & Mathai, S. M. (2017). Resilience and spirituality as correlates of psychological well-being among adolescents. *The International Journal of Indian Psychology, 4*(3), 23–29.

Faure, S. & Loxton, H. (2003). Anxiety, depression and self-efficacy levels of women undergoing first trimester abortion. *South African Journal of Psychology, 33*, 28–38.

Ganguly, R., Brownlow, C., Preez, J., & Graham, C. (2015). *Resilience/thriving in post-secondary students with disabilities: an exploratory study*. National Centre for Student Equity in Higher Education (NCSEHE), Perth: Curtin University.

Grotberg, E. H. (2003). What is resilience? How do you promote it? How do you use it? In E. H. Grotberg (Ed.), *Resilience for today: Gaining strength from adversity* (pp. 1–29). Westport, CT: Praeger.

James, William (1902). *The varieties of religious experience*: *A study in human nature*: Being the Gifford Lectures on Natural Religion delivered at Edinburgh in 1901–1902. Dover Publications.

Joshi, H. (2015). Happiness and quality of life. In U. Kumar, Archana & V. Parkash (Eds.), *Positive psychology: Applications in work, health and well-being* (pp. 65–78), Noida: Pearson India Education Services Pvt. Ltd.

Kashdan, T. B., Biswas-Diener, R., & King, L. (2008). Reconsidering happiness: the costs of distinguishing between hedonics and eudaimonia. *Journal of Positive Psychology, 3*, 219–233.

Kashdan, T. B. & Roberts, J. E. (2004). Social anxiety's impact on affect, curiosity, and social self-efficacy during a high self-focus social threat situation. *Cognitive Therapy and Research, 28*, 119–141.

Keyes C. L. M. (2002). The mental health continuum: from languishing to flourishing in life. *Journal of Health and Social Research*, 43, 207–222.

Keyes, C. L. M., Ryff, C. D., & Shmotkin, D. (2002). Optimizing well-being: the empirical encounter of two traditions. *Journal of Personality and Social Psychology, 82*(6), 1007–1022.

Khramtsova, I., Sarrino, D. A., Gordeeva, T., & Williams K. (2007). Happiness, life satisfaction, and depression in college students: Relations with student behaviors and attitudes. *American Journal of Psychological Research, 3*(1), 8–16.

Larsen R., Diener E. & Emmons R. (1985). An evaluation of subjective well-being measures. *Social Indicators Research, 17*(1), 1–17.

Lyubomirsky, S. (2001). Why are some people happier than others?: The role of cognition and motivational processes in well-being. *American Psychologist, 56*, 239–249.

Lyubomirsky S., Sheldon K. M., & Schkade, D. (2005). Pursuing happiness: the architecture of sustainable change. *Review of General Psychology*, 9(2), 111–131.

Magaletta, P. R. & Oliver, J. M. (1999). The hope construct, will, and ways: their relations with self-efficacy, optimism, and general well-being. *Journal of Clinical Psychology, 55*, 539–551.

Maslow, A. H. (1971). *The farther reaches of human nature*. New York: The Viking Press.

May, Rollo (1953). *Man's search for himself*. London: W W Norton & Co Inc.

McMillan J. & Jarvis, J. (2013). Mental health and students with disabilities: A review of literature. *Australian Journal of Guidance and Counselling, 23*(2), 236–251.

Mpofu, J., Sefotho, M. M. & Maree, J. G., (2017). Psychological well-being of adolescents with physical disabilities in Zimbabwean inclusive community settings: An exploratory study, *African Journal of Disability 6*(0), a325.

Olkin, R. (1999). *What psychotherapists should know about disability*. New York: Guilford.

Pajares, F., & Schunk, D. H. (2001). Self-beliefs and school success: Self-efficacy, self-concept, and school achievement. In R. Riding & S. Rayner (Eds.). *Self perception* (pp. 239–266). London: Ablex.

Ryff, C. D. (1989). Happiness is everything, or is it? Explorations on the meaning of psychological well-being. *Journal of Personality and Social Psychology*, *57*(6), 1069–1081.

Ryff, C. D., & Keyes, C. (1995). The structure of psychological well-being revisited. *Journal of Personality and Social Psychology*, *69*, 719–727.

Ryff, C. D., & Singer, B. (1996). Psychological well-being: Meaning, measurement, and implications for psychotherapy research. *Psychotherapy and Psychosomatics, 65*, 14–23.

Sagone, E., & Indiana, M. L. (2017). The relationship of positive affect with resilience and self-efficacy in life skills in Italian adolescents. *Psychology, 8*, 2226–2239.

Salami, S. O. (2010). Emotional intelligence, self-efficacy, psychological well-being and student's attitudes: implications for quality education. *European Journal of Educational Studies, 2*(3), 247–257.

Santos M. C. J., Magramo Jr. C. S., Oguan Jr., F., & Paat, J. N. J (2014). Establishing the relationship between general self efficacy and subjective well being among college students. *Asian Journal of Management Sciences and Education, 3*(1), 1–12.

Schwarzer R. (1994). Optimism, vulnerability and self-beliefs as health-related cognitions: A systematic overview. *Psychology and Health, 9*, 161–180.

Schwarzer, R. & Jerusalem, M. (1995). Generalized Self-Efficacy Scale. In Weinman, J, S. Wright, and M. Johnson (Ed.), *Measures in health psychology: A user's portfolio, causal and control beliefs* (pp. 35–37). Windsor England: NFER-NELSON.

Seligman, M. E. P. (2002). Positive psychology, positive prevention, and positive therapy. In C. R. Snyder & S. J. Lopez, *Handbook of positive psychology* (pp. 3–9) New York: Oxford University Press.

Seligman, M. E. P., & Csikszentmihalyi, M. (2000). Positive psychology: An introduction. *American Psychologist, 55*(1), 5–14.

Seligman, M. E. P., Steen, T. A., Park, N., & Peterson, C. (2005). Positive psychology progress: Empirical validation of interventions. *American Psychologist, 60*, 410–421.

Smith B. W., Dalen J., Wiggins K., Tooley E., Christopher P., & Bernard J. (2008). The brief resilience scale: assessing the ability to bounce back. *International Journal of Behavioral Medicine 15*(3), 194–200.

Souri, H. & Hasanirad, T. (2011). Relationship between resilience, optimism and psychological well-being in students of medicine, *Procedia—Social and Behavioral Sciences, 30,* 1541–1544.

Strobel, M., Tumasjan, A. & Spörrle, M. (2011). Be yourself, believe in yourself, and be happy: Self-efficacy as a mediator between personality factors and subjective well-being. *Scandinavian Journal of Psychology 52*, 43–48.

Veenhoven, R. (1989). Introduction. In R. Veenhoven (Ed.), How harmful is happiness? (pp. 1–6). Rotterdam, the Netherlands: Universitaire Perss, Rotterdam.

Watson, D., Clark, L. A., & Tellegen, A. (1988). Development and validation of brief measures of positive and negative affect: The PANAS scales. *Journal of Personality and Social Psychology, 54*(6), 1063–1070.

World Health Organisation (2007). *Mental health: Strengthening mental health promotion (fact sheet no. 220)* Geneva: World Health Organisation.

Wright, B. A. (1983). *Physical disability: A psychosocial approach,* 2nd edition New York: Harper & Row.

Yalom, I. D. (1980). *Existential psychotherapy*. USA: Basic Books.

Yazici, H., Seyis, S., & Altun, F. (2011). Emotional intelligence and self efficacy beliefs as predictors of academic achievement among high school students, *Procedia Social and Behavioral Sciences, 15*, 2319–2323.

Yee Ho, M., Cheung, F. M., & Cheung, S. F. (2010). The role of meaning in life and optimism in promoting well-being. *Personality and Individual Differences, 48*, 658–663.

Yu, X., & Zhang, J. (2007). Factor analysis and psychometric evaluation of the Conner-Davidson resilience scale (CD-RISC) with Chinese people. *Social Behavior and Personality: An International Journal, 35*(1), 19–30.

Zautra, A. J., Hall, J. S., & Murray, K. E. (2010). Resilience: A new definition of health for people and communities. In: Reich, J, W., Zautra. A. J., & Hall, J. S. (Eds.) *Handbook of adult resilience*. The Guilford Press: New York.

Zimmerman, B. J., Bandura, A., & Martinez-Pons, M. (1992). Self-motivation for academic attainment: The role of self-efficacy beliefs and personal goal setting. *American Educational Research Journal, 29*, 663– 676.

CHAPTER 11

Resilience, Coping, and Chronic Illnesses in the Workplace

MANISHA SAWHNEY[1*] and JULIJANA NEVLAND[2]

[1]*University of Mary, Bismarck, ND 58504, USA*

[2]*NEV Psychological Consulting, PLLC, Bismarck, ND 58504, USA*

[*]*Corresponding author. E-mail: msawhney@umary.edu*

ABSTRACT

Resilience is conceptualized in different ways, some researchers define resilience as the ability to withstand and rebound from serious life challenges, while others propose resilience is a process which involves positive adaptation after significant adversities. Traditional resilience research has been conducted for a long period of time with children, so earlier studies have focused more on developmental psychopathology. However, recently the focus of resilience research has shifted to the well-being of adults with chronic illnesses. This chapter will examine and synthesize how adults experiencing chronic illnesses like cardiovascular disease, cancer, and HIV, survive, manage, and thrive. Literature on chronic illnesses shows that effective coping styles are related to resilience. Resilient individuals tend to show lower levels of denial, and overall lower avoidant coping strategies, and higher active coping strategies. The resources that help these individuals adjust to the disease, cope and be resilient can be understood in the paradigm of health psychology, and is presented in this chapter. This information will help in working with all those who are diagnosed and are coping with chronic health conditions, and how it impacts their workplace attendance and efficiency.

11.1 INTRODUCTION

Gan et al. (2018) define resilience as a "dynamics process of positive adaptation when an individual successfully copes with stress, frustration, trauma, and other negative life events" (p. 595). Developing resilience is associated with lower stress levels, lower burnout rates, and improved employment retention (Curtis et al., 2005). Therefore, employers have much to gain by nurturing resilience in their workforce.

The construct of resilience is conceptualized in different ways. Some researchers define resilience as an ability to withstand and rebound from serious life challenges (Walsh, 2016), while others propose resilience is a process involving positive adaptation following adversity. In the construct of resilience, adversity refers to experiences or events which have a potential to disrupt the normal functioning of the individual. Examples of adversity, from a developmental perspective, include sexual abuse, maltreatment, or growing up in a war-torn region. Adversities are conceptualized as risk factors or characteristics which increase the probability of an undesirable outcome. Thus, there are two conditions required to assess for the presence of resilience in an individual's life (i) that a significant adversity has occurred and (ii) the functioning of the individual is within normal limits, despite the adversity.

Heart disease, cancer, and chronic obstructive pulmonary disease (COPD) are some of the leading causes of death, worldwide. Generally speaking, chronic diseases, such as cardiovascular disease, cancer, diabetes, and chronic respiratory diseases, account for 48% of global disability and 63% of all deaths (WHO, 2010). Chronic illness adversely impacts an individual's mental, physical, and social health, due to its generally long term and incurable nature. The stressors associated with chronic illnesses threaten the individual's overall welfare, including their future plans, financial stability, and capacity to successfully fulfill familial, social, and professional roles (Livneh & Antonak, 2005). Individuals who are diagnosed with chronic illnesses encounter new stressors for which usual resistance strategies may be ineffective, thus creating a need for additional coping strategies. By developing greater resilience and coping skills, an individual is able to reduce the negative effect of the stressors that accompany chronic illnesses. Moreover, employers who are proactive in accommodating chronically ill employees, through reduced work hours, training programs, role adjustment, and emotional support are more likely to retain their workforce and prevent absenteeism, early retirements, poor productivity, and a complete departure from the workplace.

11.2 COPING

Although various definitions and classification systems have been proposed to describe the concept of coping (Lazarus & Launier, 1978; Moos & Billings, 1982), a definition that encompasses many previous views is as follows—coping is a response aimed at diminishing the physical, emotional, and psychological burden that is linked to stressful life events and daily hassles (Snyder, 2000). Therefore, by this definition, coping strategies are those responses that effectively reduce an undesirable "load" (i.e., the psychological burden). Both personal and psychosocial factors (e.g., social support strategies) buffer individuals from the pernicious side effects of high stress and reduce any subsequent risk for psychological and physical disorders. Meisels and Shonkoff (2000) found the theoretical and empirical studies showed the most consistently enumerated components of effective coping are related to the activation of social support networks and the use of psychological resources. Social support networks are considered to be externally based coping mechanisms, while psychological resources are considered to be internally based coping mechanisms.

The contextual model of coping, which was heavily researched in the 1970s, emphasized risk factors for illness. The risk was found to be associated with exposure to life stressors and demands (Silver & Wortman, 1980). Stress was seen as any nonspecific result of mental or somatic demands placed on the body (Selye, 1976). Hinkle (1974) found that most people remain healthy even under high levels of stress.

Slowly, there was a fundamental change in the conceptualization of the stress process. Instead of envisioning stressors as resulting in the dysfunction, stress-resistant research emphasized one's capacity to remain healthy when stressors occurred.

Lazarus and Folkman (1984) moved the focus away from Selye's innate biochemical reactivity and directed attention toward the active appraisal processes. Likewise, researchers abandoned the previous emphasis on enduring traits and instead placed the spotlight on situational determinants, by conceptualizing coping as a cognitive transaction between the individual and the environment within a specific context. Lazarus's stress and coping model defined coping as "constantly changing cognitive, behavioral and emotional efforts to manage particularly external and/or internal demands that are appraised as taxing or exceeding the resources of the person" (Lazarus & Folkman, 1984). The significant point to highlight in Lazarus's (1984) definition relates to the fact that it conceptualized coping as process related rather than trait oriented; coping is interactive instead of automated, and relies on a developmental rather than a mastery model. The centerpiece of Lazarus and Folkman (1984) approach to understanding coping is the notion of how the person appraises the situation. The appraisal was posited to occur when the perceiver encounters situations that are interpreted as excessive relative to resources (i.e., stressors). When the event is perceived as harmful or threatening, the individual enters into secondary appraisal, wherein the available responses for coping are examined. Coping responses are thus initiated in an emotional environment, and often one of the first coping tasks is to down-regulate negative emotions that are stressful and may be interfering with instrumental forms of coping.

Moos (1984) proposed a framework that integrated findings on stress and coping processes as related to resource predictors of adaptation, stress resistance, and situational influences. He categorized stressful life circumstances as short-term stressors (e.g., meeting deadlines), longer-term stress (e.g., career concerns), and chronic life strains (e.g., conflict-ridden home environment). Moos (1984) defined the environmental system as encompassing all the characteristics of individuals in a setting, as well as the social climate factors. The personal system included such factors as an individual's sociodemographic characteristics and personal traits and resources, such as self-esteem, intellectual ability, health status, social network resources, and appraisal and coping resources. The model hypothesizes that life stressors, as well as the environmental and personal factors related to such stressors, can shape social network resources and the effectiveness of the coping responses.

Moos and Swindle (1990) conceptualized a model of the stress and coping process in which specific characteristics of a focal life crisis or transition, along with an individual's appraisal of the situation, provide a context for the selection of specific coping responses. Cognitive appraisal and coping responses are hypothesized to help a person avoid depressive symptomatology by mediating the potential effects the stressors have on functioning. Researchers have used two main conceptual approaches to classify coping responses. The first approach emphasizes the focus on coping, whether it is problem focused (approach coping) or emotions focused (avoidance coping). The second approach emphasizes the method of coping, whether it is cognitive or behavioral (Billings & Moos, 1984; Lazarus & Folkman, 1984). Severe stressors may initialize/mobilize more approach and/or avoidance coping responses. When severe stressors persist, individuals may gradually lessen their use of problem coping and increase their reliance on avoidance coping strategies. The research on the influence of coping responses on adjustment suggests that active, problem-oriented coping appears to moderate the adverse influences of negative life events on psychological functioning.

Consequently, the effectiveness of a coping strategy rests on an individuals' ability to reduce immediate distress, as well as to contribute to more long-term outcomes such as psychological well-being or chronic illness management. If coping is defined as an attempt to diminish the physical and psychological load, then both approach and avoidance coping may play a part in the response. Some examples of approach-focused coping include making a plan of action or concentrating on the next step. Examples of avoidance focused coping are engaging in distracting activities, using alcohol or drugs, or seeking emotional support. In sum, approach-focused coping is problem centered; it utilizes cognitive and behavioral efforts to master and resolve life stressors. In contrast, avoidance coping tends to be emotion-focused; it utilizes cognitive and behavioral attempts to avoid thinking about a stressor and its implications, or to manage the affect associated with it. Lastly, some researchers have also identified meaning-focused coping as a different type of coping in which cognitive strategies are used to manage the meaning of the situations. For instance, Pearlin and Scholar (1978) included the responses of positive comparison or selective ignoring in this category. The positive comparison includes engaging in positive comparisons with others or with one's past and selectively ignoring and minimizing the stressors.

11.3 RESILIENCE IN PATIENTS WITH CARDIOVASCULAR DISEASE (CAD) AND CANCER

Cardiovascular disease (CAD) is viewed as the number one health issue of the 21st century (Laaser & Brand, 2014). A large number of studies provide evidence that psychosocial factors are related to morbidity and mortality due to cardiovascular disease. When considering the resilience theory for chronic illness, there are two important components to consider: risk factors and protective factors. Research has shown that resilience is the balance between risk and protective factors. Risk factors are often caused by multiple stressful events or even a single traumatic event. Protective factors on the other hand, can reduce the negative influences of being at risk. Thus, both the interplay between these factors is associated with the development of resilience (Fraser et al., 2004).

Risk factors are biological, environmental, and psychosocial hazards, which increase the likelihood of maladaptive outcomes (Murray, 2003). Risk factors can negatively affect development and can heighten the likelihood of poor outcome status (Murray, 2003). Further, there are two kinds of factors that can increase the risk of developing CAD: modifiable risk factors or nonmodifiable risk factors such as age, male sex, family history of cardiovascular disease, and race ethnicity. Modifiable risk factors include characteristics such as high blood cholesterol, hypertension, diabetes, abdominal obesity, cigarette smoking, being overweight, low physical activity, dietary habits, alcohol consumption, and other psychosocial factors. Resilience is an important factor that impacts the prognosis of CAD, especially influencing the modifiable factors.

Protective factors are also important components of resilience theory and are related to positive adjustment when people face stressful events. According to Hasse (2004), protective factors are described as individual, family, social or contextual factors that promote resilience processes and outcomes. Protective factors can moderate the impact of risk exposure and can change the outcome of the disease. These factors may be working to counteract risk factors and can thus promote healing and positive long-term adjustment.

A large number of studies have examined the psychological effects of having CAD (Khayyam-Nekouei et al., 2013) but not many have explored the resilience factors. Studies have shown that higher levels of resilience may assist with chronic illness management (Becker & Newsom, 2005). Other studies have found that positive affect can be a source of resilience for individuals with fibromyalgia in chronic pain (Zautra et al., 2005). Higher levels of positive affect also predicted lower levels of pain. Overall, the literature indicates that personal characteristics like optimism, an active or adaptable coping style, and the ability to elicit social support, are associated with resilience in patients with chronic illnesses such as cancer and HIV/AIDS. However, little is known about the impact of resilience on the long-term prognosis of CAD.

11.4 SOURCES OF RESILIENCE

Overall, there are three main sources of resilience, the individual, the family, and the society. In patients with CAD, studies have shown that individual protective factors, such as high self-esteem, problem-solving abilities, and having high levels of activity, can reduce the risk of developing CAD later in life (Gallo et al., 2004). Factors such as positive affect, self-efficacy, self-mastery, optimism, and locus of control have also been found to be correlated with reduced distress and better health statuses for individuals with CAD. Chan et al. (2006) examined the relationship between resilience and illness outcomes in CAD patients. Resilience was operationalized as optimism, perceived control, and self-esteem. The study showed that Chinese patients with CAD who demonstrated higher levels of resilience had a better quality of life and physical health when compared to the low resilience group.

Family protection factors usually present a pattern of family interaction that is warm and supportive. The social or community protective factors are found outside the home (e.g., positive and supportive workplace, access to healthcare). These protective factors also play an important role in the development of resilience. Therefore, there are different sources of resilience for individuals with CAD, and Lim et al. (2015) demonstrated that individuals who are resilient are quick to reestablish physiological, psychological, and social equilibrium following stressful events. Therefore, resilient individuals with stable CAD show higher physical activity, greater medication adherence, and better overall treatment outcomes, as compared to individuals who are low on resilience.

11.5 CANCER: ITS DEVELOPMENTAL TRAJECTORIES, ECONOMIC BURDEN, AND ROLE OF RESILIENCE

Because cancer is widely considered as one of the most serious and traumatic illnesses, it can be emotionally disturbing (Jim & Jacobsen, 2008) and can lead to various mental health disorders, including anxiety and depression (Li et al., 2016; Park et al., 2010). Patients on the cancer continuum are exposed to a series of potentially traumatic events, beginning at the time of risk assessment and screening. Healthy individuals may perceive great stress from the potential of a life-threatening diagnosis. Recently diagnosed patients must cope with many life changes and continue to adapt throughout treatment. Survivors meet additional challenges with the adjustment to their "new normal" and the increased risk of poor physical and psychosocial outcomes. Patients who face death from their disease also encounter challenges as they struggle to maintain a positive outlook or find meaning at the end of life. Each phase of the cancer experience may have unique elements and shared aspects of resilience. For example, patients may draw on individual characteristics (e.g., hope) throughout the cancer trajectory; however, the stress and requisite adaptation associated with cancer screening is vastly different from the stress of a terminal diagnosis. Similarly, resilience may manifest at each time point with different clinical characteristics, and it may be fostered by various types of interventions or interactions. Thus, resilience is important during each stage of cancer.

In the last few decades there has been an increase in the overall survival ratio in cancer patients and research has shifted toward the emotional well-being of long-term cancer survivors (Woodgate, 1999) and routine screening for unmet needs and psychological distress (Carlson et al., 2012). A number of factors such as age, gender, education, and disease-related factors like the type of cancer, type of treatment, and stage of the disease have been found to be associated with different levels of emotional adjustment (Helgeson et al., 2004). However, some individuals diagnosed with the disease are more successful in processing and adapting to this life threat than others (Min et al., 2013). Studies show that these individuals score higher on standardized measures of resilience (Southwick et al., 2005) and thus are less likely to experience negative effects of chronic illness. It has been reported that low resilience increases the risk of feeling hopeless and that these feelings of hopelessness have an impact on brain functioning thus indicating the mind-body interaction. In fact, studies have highlighted that knowledge of the patient's resilience should be a prerequisite for planning individualized care for the cancer patient. A recently published study protocol is investigating the association between psychological resilience and biomolecular signatures in breast cancer patients, so treatment care can be more individualized.

Those who possess greater resilience are less likely to experience the negative effects of chronic illness, such as cancer. A recent study of 146 Chinese cancer patients found that the "low resilient" group benefited from a meaning-making process, while a highly resilient group did not appear to need meaning making (Gan et al., 2018). Meaning making refers to searching for meanings when faced with trauma, such as a cancer diagnosis. Therefore, meaning making is one of many coping skills that can be used to especially assist those who are deemed as less resilient. Moreover, Perez and Smith (2015) found that greater self-efficacy for coping with cancer predicted greater well-being for cancer patients. Thus, any action employers are able to undertake to foster self-efficacy (e.g., emotional support, reasonable workplace accommodations, and resilience/coping skills training for employees) should lead to improved employee mental health and workplace productivity.

Resilience has been proposed to serve as resistance to chronic illnesses like CAD and cancer. When patients decide the stressors are beyond their control, or that they cannot control what happens to them, their adaptive skills become restrictive and often ineffective. On the other hand, when patients believe they can manage their life events and outcomes, they eschew learned helplessness and actively seek coping strategies. This, in turn, aids in overcoming the aversive situation, opening the possibility of moving forward and achieving resilience. Studies have shown that by being resilient, individuals have the power to adjust, resist stress, and potentially thrive in the face of adversity (Munir et al., 2009).

11.6 COPD AND CHRONIC PAIN IN THE WORKPLACE

COPD is a lung disease that impairs normal breathing and is the third leading cause of death in the United States (Kochanek et al., 2011). COPD forces 1 in 5 employees to retire prematurely, due to loss in productivity and medically related complications (Nair et al., 2012). One study found that individuals with COPD had twice the rate of emergency room visits and were 1.5 times more likely to be absent from work (Nair et. al., 2012). The average age for a COPD diagnosis is 53 years, with half of COPD sufferers reporting their diagnosis negatively impacts their ability to work (American Lung Association, 2011; American Association for Respiratory Care, 2011). Moreover, COPD sufferers were more likely to utilize short-term and long-term disability, resulting in an approximately 50 billion annual direct and indirect cost in productivity to the US economy (National Heart Lung and Blood Institute, 2009). In order to minimize overall employer cost of hiring and retaining individuals with COPD, Nair et al. (2012) recommended employers invest in prevention-based smoking cessation programs.

A metanalysis of 91 studies related to chronic pain the workplace showed 13%–76% of individuals cited chronic pain as a cause for work disability (Patel et al., 2012). Individuals between 8% and 28% chose early retirement due to their chronic pain condition. Another study found that those who were able to cope more effectively with their chronic pain experienced a lesser intensity of the pain (Bergstrom et al., 2010). Numerous other studies have found a relationship between chronic pain and work absences, including sick leave (Lotters & Burdorf, 2006; Saastamoinen et al., 2009; Jeneweinet al., 2009). Extensive sick leave was found to be significantly associated with job loss (De Buck et al., 2006). Thus, it is critical that employees learn positive strategies for coping with chronic pain, to prevent negative consequences to themselves, their employers, and the society, as a whole. Patel et al. (2012) suggest employers identify workers who frequently utilize sick leave and provide interventions such as "ergonomic interventions, role-adjustment, and/or reduced working hours," to increase the likelihood of continued employment (p. 586).

11.7 BURDEN OF CHRONIC ILLNESSES IN THE WORKPLACE

Each year, chronic illness costs the US economy approximately one trillion dollars in lost worker productivity (Partners to Fight Chronic Disease, 2009). According to Koch et al. (2013), employers who engage in proactive problem solving and follow-up with employees struggling with chronic illness can successfully prevent employee turnover and loss in productivity. This is especially important considering those diagnosed with a chronic condition, such as diabetes, heart disease, stroke arthritis, and cancer, among others, are more likely to reduce their work activity (e.g., full-time to part-time) or depart entirely from the workforce. Furthermore, around 50% of those with chronic illnesses suffer from multiple illnesses, further complicating their ability to effectively cope and sustain long-term employment (Partners to Fight Chronic Disease, 2009).

The vast majority of Americans with a chronic illness are under the age of 65, meaning they are current members of the workforce (Allaire et al., 2005). Individuals with chronic illness may require ongoing medical treatment and experience difficulty with functional tasks of daily living, such as walking, showering, dressing, and driving. These realities can lead to greater stress levels, which tend to negatively impact one's mental health, job satisfaction, and ultimately work productivity (Partnership for Solutions, 2004). Certain chronic conditions may not manifest visible symptoms of discomfort, and a lack of disclosure by an employee will thus lead to a lack of support from supervisors and colleagues. It is not uncommon for employees to fail to disclose chronic diseases due to fear of office gossip, bullying, or being fired (Mount et al., 2005; Stewart et al., 2001).

Moreover, employees with chronic illness may experience overt discrimination or other forms of maltreatment from supervisors and colleagues (American Heart Association, 2012). Studies show that negative stigma and lack of education may cause employers to view employees struggling with chronic illness as a liability, leading to a lack of emotional support and workplace accommodations (Lightman et al., 2009). Individuals suffering from chronic illness may not be offered promotions, or may refuse them, due to physical limitations. Workplace bullying is another stressor which may adversely affect an individual's mental health, job commitment and satisfaction, and performance (Hugh et al., 2011; Nielsen & Einarsen, 2012; Hoel et al., 2011). Research shows that having a disabling chronic condition increases the likelihood of being bullied in the workplace (Van Cleave & Davis, 2006). In one study, 16% of employees with chronic conditions reported being victims of bullying, with vast majority of bullying behaving beginning with the onset of chronic illness (Fattori et al., 2015). Furthermore, a strong association was found between workplace bullying and greater absenteeism and the use of sick leave (Fattori et al., 2015). Due to fear of discrimination or negative reaction from employers and colleagues, some individuals with chronic illnesses may leave their place of employment without ever asking for flexibility or accommodation, which could have allowed them to continue their employment (Beatty & Joffe, 2006).

Employees with chronic illnesses are in a potentially burdensome position of proving they are ill and justifying their need for workplace accommodations (Gold et al., 2012). This negotiation process can be perceived as threatening and stressful, as employees are often asked to provide medical documentation, justify how a potentially costly accommodation will ultimately benefit both the employee and the employer. If an employee lacks assertive communication and advocacy skills, their request may be denied. Gold et al. (2012) caution employees against lying or exaggerating their medical conditions, as mutual trust is imperative to a positive working relationship. However, it is crucial that employers recognize that employees may struggle with disclosure due to feelings of shame about their disabilities, difficulty articulating their concerns, worry about being misunderstood and perceived as lazy, or fears of greater negative repercussions, such as public humiliation. Ironically, the added stress of communicating with an employer, especially one who is perceived as unsupportive, that may further aggravate certain chronic conditions, such as autoimmune diseases, causing symptom flare-ups and disruption of work duties (McGonagle & Barnes-Farrell, 2013).

11.8 EMPLOYER ACCOMMODATIONS TO CHRONICALLY ILL EMPLOYEES

Organizations can benefit from greater productivity and morale if appropriate supports and services are provided to individuals with chronic medical conditions. Koch et al. (2013) suggested rehabilitation counselors may assist by providing various interventions, including educating employers about chronic illness and teaching employees to utilize self-management strategies. For instance, employees should be encouraged to use medications as prescribed, manage discomfort by taking breaks, follow a balanced diet and exercise regiment, and stay socially engaged (Munir et al., 2009). Anticipatory coping skills may also assist employees in preparing for a future decline in functional or other abilities in the workplace (Gignac et al., 2006). Anticipatory coping skills may include behaviors such as periodic stretching, alternating rest with activity, and timing tasks based on the severity of one's pain (Koch et al., 2013). Consequently, if an employee can successfully anticipate future functional limitations, he or she is thus able to practice resilience, lowering stress levels, increasing self-efficacy, and ultimately productivity. However, it is worth noting that chronic illnesses tend to be unpredictable, with symptoms that could appear quickly and fluctuate from day to day, thus causing possible disruption in the workplace and requiring immediate accommodations (Beatty & Joffe, 2006). Periods of remission from symptoms can vary in duration, as no two individuals will experience a chronic condition in the same way.

Employers who are willing to provide reasonable accommodations, including emotional support, can exert a positive influence on their employees' well-being and productivity (Munir et. al., 2009). Teaching and modeling positive coping skills, such as mindfulness, active listening, stress management, and the use of humor, can further serve to decrease the employees' stress levels and lead to improved productivity. Stress has been shown to lead to absenteeism, reduced performance, and low employee morale (Bhui et al., 2012). Therefore, employers are well-advised to implement programs which foster employee resilience. Specialized training should also extend to employers and supervisors. Wong (2006) surveyed human resource professionals regarding their own training needs and the following were identified as most pertinent and desired: strategies for accommodating employees with chronic illness, methods for implement and evaluating these strategies, and tools for creating policies for employees who are returning to work after a prolonged illness related absence. Gold et al. (2012) further recommend skills training for employees, teaching tools for self-advocacy, how to collect the required medical evidence of chronic illness, and how best to negotiate with employers with confidence and respect, to improve the likelihood of receiving a necessary workplace accommodations. Jakobsen and Lillefjell (2014) reinforce the importance of chronically ill employees receiving training on stress management and effective communication skills with employers.

Of note, some chronically ill employees who are seeking workplace accommodations may struggle with the consequences of such accommodations. For instance, an employee who requires reduced working hours might suffer greater economic burdens and require additional coping strategies to manage elevated levels of financial and psychological stress (Jakobsen & Lillefjell, 2014). Some employees find that even with accommodation, the employers' performance expectations remain demanding and difficult to achieve, reinforcing the cycle of stress. Because an employee's personal life may also be negatively affected by chronic illness, relationship difficulties at home further exacerbate already elevated stress levels, negatively impacting an individual's self-esteem and productivity. Thus, employer support becomes essential and utilizing the expertise of occupational therapists could successfully facilitate the employer/employee support structure (Jakobsen & Lillefjell, 2014).

Simply accommodating chronically ill employees may not be enough. Varekamp et al. (2009) suggest that actively empowering such employees is a more desirable course of action. Empowering refers to encouraging others to develop knowledge and exercise decision making power in their workplace. Chronic illness forces an individual to redefine their identity (e.g., roles at home and at work) and accept physical limitations/ability (e.g., what they can and can no longer do), leading to a process of self-reinvention. Employers should view chronically ill employees as collaborators, and commit to understanding the employee's medical, physical, and psychological needs, so they may communicate openly and develop a plan that will satisfy mutual expectations (Varekamp et al., 2009).

Most importantly, practicing compassion with chronically ill employees is the greatest tool for retention and employee satisfaction. However, it is equally as important to avoid pitying the employee and using language which could be perceived as patronizing or disempowering. Chronically ill employees typically make the difficult decision to disclose their illness in order to negotiate duties, born from a desire to maintain their employment. Those who are unsuccessful in negotiating duties with their employer were more likely to apply for disability and permanently their jobs (Butler & Modaff, 2016). Thus, accommodating chronically ill employees is a win-win proposition, further supporting their ability to cope, in both personal and professional contexts, leading to improved well-being, job satisfaction, self-efficacy, and productivity (Siu et al., 2013).

11.9 SUMMARY

Resilience is a process of positive adaptation to stress, trauma, and frustration and is associated with improved mental well-being in all individuals, including patients with chronic illness (Gan et. al., 2018). Individuals who demonstrate greater resiliency tend to have more positive psychological adjustment and improved chronic illness treatment outcomes. Those who possess greater resilience are better able to manage and less likely to experience negative effects of chronic illness, such as cancer or cardiovascular disease. This is especially critical considering those diagnosed with a chronic condition, such as diabetes, heart disease, COPD, stroke arthritis, and cancer are more likely to reduce their work activity or depart entirely from the workforce (Curtis et al., 2005). The anticipated threat of employer discrimination, workplace bullying, negative response to requests for accommodation, and general stigma associated with chronic illness, results in many employees failing to disclose medical conditions, exhausting sick leave, accepting early retirement, or not giving their employer an opportunity to provide the needed supports. Fortunately, research shows that employers who are willing to provide reasonable accommodations, adjust their expectations, and provide emotional support, can exert a positive influence on their employees' well-being, productivity, and overall retention (Munir et. al., 2009).

KEYWORDS

- **resilience**
- **coping**
- **chronic illnesses**

REFERENCES

Allaire, S. H., Niu, J., & LaValley, M. P. (2005). Employment and satisfaction outcome from a job retention intervention delivered to persons with chronic diseases: *Rehabilitation Counseling Bulletin, 48*, 100–109.

American Association for Respiratory Therapy (2011). Confronting COPD in America: Executive summary. Retrieved from http://www.aarc.org/resources/confronting_copd/exesum.pdf

American Lunch Association (2011). Chronic obstructive pulmonary disease (COPD) fact sheet. Retrieved from http://www.lungusa.org/lung-disease/copd/resources/facts-figures/COPD-Fact-Sheet.html#note_01.

American Heart Association, (2012). Heart disease and stroke statistics update. Retrieved from http://www.heart.org/HEARTORG/General/Heart-ad-Stroke-Association-Statistics_UCM_319064_SubHomePage.jsp

Beatty, J. E., & Joffe, R. (2006). An overlooked dimension of diversity: The career effects of chronic illness. *Organizational Dynamics, 35*(2), 182–195.

Becker, G., & Newsom, E. (2005). Resilience in the face of serious illness among chronically ill African Americans in later life, *Journal of Gerontology Series B: Psychological Sciences and Social Sciences, 60*, S214–S223.

Bergstrom, G., Bergstrom, C., Hagberg, J., Bodin, L., & Jensen, I. (2010). A 7-year follow-up of multidisciplinary rehabilitation among chronic neck and back patients. Is sick leave outcome dependent on psychologically derived patient groups? *European Journal of Pain, 14*, 426–433.

Bhui, K. S., Dinos, S., Stansfeld, S. A., & White, P. D. (2012). A synthesis of the evidence of managing stress at work: A review of the reviews reporting on anxiety, depression, and absenteeism. *Journal of Environmental and Public Health, 2012*, 21.

Billings, A. G., & Moos, R. H. (1984). Coping, stress, and social resources among adults with unipolar depression. *Journal of Personality and Social Psychology*, 46, 877–891.

Butler, J. A., Modaff, D. P. (2016). Motivations to disclose chronic illness in the workplace. *Qualitative Research Reports in Communication, 17*(1), 77–84.

Carlson, L. E., Waller, A., & Mitchell, A. J. (2012). Screening for distress and unmet needs in patients with cancer: Review and recommendations. *Journal of Clinical Oncology*, 30, 1160–1177.

Chan, M. F., Wong, O. C., Chan, H. L., Fong, M. C., Lai, S. Y., Lo, C. W., Ho S. M., Ng S. Y. & Leung S. K. (2006). Effects of music on patients undergoing a C-clamp procedure after percutaneous coronary interventions. *Journal of Advanced Nursing 53*, 669–679.

Curtis, L., Moriarty, J., & Netton, A., (2005). The expected life of a social worker. *British Journal of Social Work, 40*, 1628–1643.

De Buck, P. D. M., De Bock, G. H., Van Dijk, F., Van den Hout, W. B., Vanderbroucke, J. P., & VlietVlieland, T. P. M. (2006). Sick leave as a predictor of job loss in patients with chronic arthritis. *International Archives of Occupational Environmental Health, 80*, 160–170.

Fattori, A., Neri, L., Aguglia, E., Bellomo, A., Bisogno, A., Camerino, D., Carpiniello, B., Cassin, A., Costa, G., De Fazio, P., Di Sciascio, G., Favaretto, G., Fraticelli, C., Giannelli, R., Leone, S., Maniscalso, T., Machesi, C., Mauri, M., Mencacci, C., Polselli, G., Quartesan, R., Risso, F., Sciaretta, A., Vaggi, M., Vender, S., & Viora, U. (2015). Estimating the impact of workplace bullying: Humanistic and economic burden among workers with chronic medical conditions. *BioMed Research International*. Retrieved from http://dx.doi.org/10.1155/2015/708908

Fraser, M. W., Kirby, L. D., & Smokowski, P. R. (2004). Risk and resilience in childhood. In M. W. Fraser(Ed.),

Risk and resilience in childhood (2nd ed., pp. 13–64). Washington, DC: NASW Press.

Gallo L. C., Ghaed, S. G., & Bracken, W. S. (2004). Emotions and cognitions in coronary heart disease: risk, resilience, and social context. *Cognitive Therapy & Research, 28*, 669–94.

Gan, Y., Zheng, L., Wang, Y., & Li, W. (2018). An extension of the meaning making model using data from Chinese cancer patients: The moderating effect of resilience. *Psychological Trauma: Theory, Research, Practice, and Policy, 10*(5), 594–601.

Gignac, M. A. M., Sutton, D., & Bradley, E. M. (2006). Reexamining the arthritis–employment interface: Perception of arthritis-work spillover among employed adults. *Arthritis & Rheumatism, 55*(2), 233–240.

Gold, P. B., Oire, A. N., Fabian, E. S., & Wewiorski, N. J. (2012). Negotiating reasonable workplace accommodations: Perspectives of employers, employees with disabilities, and rehabilitation service providers. *Journal of Vocational Rehabilitation, 37*, 25–37.

Haase, J. E. (2004). The Adolescent Resilience Model as a guide to interventions. *Journal of Pediatric Oncology Nursing, 21,* 289–299.

Helgeson, V. S., Snyder, P., & Seltman, H. (2004). Psychological and physical adjustment to breast cancer over 4 years: identifying distinct trajectories of change. *Health Psychology*, 23(1):3–15

Hinkle, L. E. (1974). The concept of "stress" in the biological and social sciences. *The International Journal of Psychiatry in Medicine*, *5*(4), 335–357.

Hoel, H., Sheehan, M. J., Cooper, C. L., & Einarsenin, S. (2011). Organizational effects of workplace bullying in Taylor & Francis. (Eds) *Bullying and harassment in the workplace: development in theory, research, and practice*.

Hogh, A., Mikkelsen, E. G., & Hansen, A. M. (2011). Individual consequences of workplace bullying/mobbing in Taylor and Francis. (Eds), In *Bullying and harassment in the workplace: development in theory, research, and practice*.

Jakobsen, K., & Lillefjell, M. (2014). Factors promoting a successful return to work: From an employer and employee perspective. *Scandinavian Journal of Occupational Therapy, 21*, 48–57.

Jenewein, J., Moergeli, H., Wittmann, I., Buchi, S., Kramer, B., & Schnyder, U. (2009). Development of chronic pain following severe accidental injury: Results of a 3-year follow-up study. *Journal of Psychosomatic Research, 66*, 119–126.

Jim, H. S., & Jacobsen, P. B. (2008). Posttraumatic stress and posttraumatic growth in cancer survivorship: a review. *The Cancer Journal*, 14(6): 414–419.

Khayyam-Nekouei, Z., Neshatdoost, H., Yousefy, A., Sadeghi, M., & Manshaee, G. (2013). Psychological factors and coronary heart disease. *ARYA Atherosclerosis, 9,* 102–11.

Koch, L. C., Rumrill, P. D., Conyers, L., & Wohlford, S. (2013). A narrative literature review regarding job retention strategies for people with chronic illnesses. *Work, 46*, 125–134.

Kochanek, K. D., Xu, J. Q., Murphy, S. L., Minino, A. M., & Kung, H. C. (2011). Deaths: Preliminary Data for 2009. *National Vital Statistics Reports, 59* (4), 1–51.

Laaser, U., & Brand, H. (2014). Global health in the 21st century. *Global Health Action, 7,* 10.3402/gha.v7.23694. http://doi.org/10.3402/gha.v7.23694

Lazarus, R. S. & Folkman, S. (1984). Stress, appraisal, and coping. New York: Springer Publishing.

Lazarus, R. S. and Launier, R. (1978). Stress-related transactions between person and environment. In L. A. Pervin and M. Lewis (Eds), (1978). Perspectives in interactional psychology (pp. 287–327). New York: Plenum.

Li, W., Miao, M., Gan, Y., Zhang, Z., & Chen, G. (2016). The relationship between meaning discrepancy and emotional distress among patients with cancer: The role of posttraumatic growth in a collectivistic culture. *European Journal of Cancer Care, 25*, 491–501.

Lightman, E., Vick, A., Herd, D., & Mitchell, A. (2009). 'Not disabled enough': Episodic disabilities and the Ontario Disability Support Program, *Disability Studies Quarterly, 29*(3), 1–16.

Lim, M. L., Lim, D., Gwee, X., Nyunt, M. S., Kumar, R., & Ng, T. P. (2015). Resilience, stressful life events, and depressive symptomatology among older Chinese adults, *Aging and Mental Health, 19* (11), 1005–1014.

Livneh H., & Antonak, R. F. (2005). Psychosocial adaptation to chronic illness and disability: A primer for counselors. *Journal of Counseling Development, 83*(1), 12–20.

Lotters, F., & Burdorf, A. (20060. A prognostic factors for duration of sickness absence due to musculoskeletal disorders. *Clinical Journal of Pain, 22*, 212–221.

McGonagle, A. K., Barnes-Farrell, J. L. (2013). Chronic illness in the workplace: Stigma, identity threat, and strain. *Stress Health (30)*, 310–321.

Meisels, S. J., & Shonkoff, J. P. (2000). Early childhood intervention: A continuing evolution. In J. P. Shonkoff & S. J. Meisels (Eds.) Handbook of early childhood intervention. (2d ed., pp. 3–33). New York: Cambridge University Press.

Min, J. -A., Yoon, S., Lee, C. -U., Chae, J. -H., Lee, C., Song, K. -Y. & Kim T. -S. (2013) Psychological resilience contributes to low emotional distress in cancer patients. *Supportive Care in Cancer*, 21, 2469–2476.

Moos, R. H. (1984). Context and coping: Toward a unifying conceptual framework. *American Journal of Community Psychology, 12*, 5–25.

Moos, R. H., & Billings, A. G. (1982). Conceptualizing and measuring coping resources and processes. In L. Goldberger and S. Breznitz (Eds.), Handbook of stress: Theoretical and clinical aspects (pp. 212–230). New York, NY: Free Press.

Moos, R. H., & Swindle, R. W. (1990). Stressful life circumstances: Concepts and measures. *Stress Medicine, 6*(3), 171–178.

Mount, D., Johnstone, B, White, C., & Sherman, A. (2005). Vocational outcomes: VR service determinants for persons with epilepsy. *Journal of Vocational Rehabilitation, 23*, 11–20.

Munir, F., Randall, R., Yarker, J., & Nielsen, K. (2009). The influence of employer support on employee management of chronic health conditions at work. *Journal of Occupational Rehabilitation, 19*, 333–344.

Murray, C. (2003). Risk factors, protective factors, vulnerability and resilience: a framework for understanding and supporting the adult transitions of youth with high-incidence disabilities. *Remedial and Special Education 24* (1), 16–26.

Nair, K., Ghushchyan, V., Van Den Bos, J., Halford, M. L., Tan, G., Frech-Tamas, F. G., & Doyle, J. (2012). Burden of chronic illness for an employed population with chronic obstructive pulmonary disease. *Population Health Management, 15*(5), 267–275.

National Heart Lung and Blood Institute (2009). Morbidity and mortality. *Chart book on cardiovascular, lung, and blood diseases*. Retrieved from http://www.nhlbi.nih.gov/resources/docs/cht-book.htm

Nielsen, M. B., & Einarsen, S. (2012). Outcomes of exposure to workplace bullying: A meta-analytic review. *Work and Stress, 26* (4), 309–332.

Patel, A. S., Farquaharson, R., Carroll, D., Moore, A., Phillips, C. J., Taylor, R. S., & Barden, J. (2012). The impact and burden of chronic pain in the workplace: A qualitative systematic review. *Pain Practices, 12* (7), 578–589.

Park, C. L., Chmielwski, J., & Blank, T. O. (2010). Post-traumatic growth: Finding positive meaning in cancer survivorship moderates the impact of intrusive thoughts on adjustment in younger adults. *Psycho-Oncology, 19*, 1139–1147.

Partnership for Solutions, (2004). Chronic conditions: Making the case for ongoing care: September 2004 update. Retrieved from http://www.partnershipforsolutions.org/DMS/files/chronicbook2004.pdf.

Partners to Fight Chronic Disease, (2009). Almanac of chronic disease. Retrieved from http://www.fightchronicdisease.org/resources/almanac-chronic-siseases-0.

Pearlin, L. I., & Schooler, C. (1978). The structure of coping. *Journal of Health & Social Behavior, 19*, 2–21.

Perez, J. E. & Smith, A. R. (2015). Intrinsic religiousness and well-being among cancer patients:

The mediating role of control-related religious coping and self-efficacy for coping with cancer. *Journal of Behavioral Medicine, 38*, 183–193.

Saastamoinen, P., Laaksonen, M., Lahelma, M., Leino-Arjas, P. (2009). The effect of pain on sickness absence among middle aged municipal employees. *Occupational Environmental Medicine, 66*, 131–136.

Selye, H. (1976). Stress in health and disease. Boston, MA: Butterworth's, Inc.

Silver, R. L., & Wortman, C. B. (1980). Coping with undesirable life events. In J. Garber & M. E. P. Seligman (Eds.), Human helplessness: Theory and applications (pp. 279–340). New York: Academic Press.

Siu, A. M. H., Hung, A., Lam, A. Y. L., & Cheng, A. (2013). Work limitations, workplace concerns, and job satisfaction of persons with chronic disease. *Work, 45*, 107–115. Doi 10.3233/WOR-121550

Snyder, C. R. (Ed.). (2000). *Handbook of hope: Theory, measures, and applications*. San Diego, CA, US: Academic Press.

Southwick, S. M., Vythilingam, M., Charney, D. S. (2005). The psychobiology of depression and resilience to stress: implications for prevention and treatment. *Annual Review of Clinical Psychology*, 1, 255–291.

Stewart, D. E., Cheung, A. M., Duff, S., Wong, F., McQuestion, M., & Cheng, T. (2001). Long-term breast cancer survivors: Confidentiality, disclosure, effects on work and insurance. *Psychoncoogyl, 10*, 259–263.

Van Cleave, J., & Davis, M. M. (2006). Bullying and peer victimization among children with special health care needs, *Pediatrics, 118* (4), 1212–1219.

Varekamp, In., Heutink, A., Landman, S., Koning, C. E. M., De Vries, G., & Van Dijk, F. J. H. (2009). Facilitating empowerment in employees with chronic disease: Qualitative Analysis of the process of change. *Journal of Occupational Rehabilitation, 19*, 398–408.

Walsh, F. (2016). *Strengthening family resilience* (3rd ed.). New York, NY, USA: Guilford Press.
Woodgate R. L. (1999). A review of the literature on resilience in the adolescent with cancer: Part II. *Journal of Pediatric Oncology Nursing, 16*(2), 78–89.
World Health Organization (2010). *Global burden of disease* [fact sheet]. Retrieved from http://www.who.int/news-room/fact-sheets/detail/depression.
Wong, L. (2006). A report on a survey of certified human resources professionals regarding episodic disabilities. Retrieved from www.cchra.ca/uploadedFiles/...29987-HR_report_English.pdf.
Zautra, A. J., Johnson, L. M., & Davis, M. C. (2005). Positive affect as a source for resilience for women in chronic pain. *Journal of Consulting and Clinical Psychology, 73,* 212– 220.

CHAPTER 12

Resilience and Suicide Risk

RENATA SARGON*, MARISSA N. EUSEBIO, RACHEL BACIGALUPI, KEVIN RODRIGUEZ, and BRUCE BONGAR

Department of Clinical Psychology, Palo Alto University, Palo Alto, CA, USA

Corresponding author. E-mail: rsargon@paloaltou.edu

ABSTRACT

Suicide continues to be a significant worldwide public health concern. Major efforts by theorists and researchers continue to increase our understanding of how to identify risk factors and what constitute the best evidence-based preventative measures. An important consideration in recent years is the relationship between resilience and suicide risk. With regard to this connection, a number of theories have been put forward. Some theorists see resilience as a personality feature characterized by an ability to adapt to change, cope with stress, and thrive in adverse conditions. Other researchers define resilience as a psychological construct where an individual possesses a set of positive beliefs, resources, and the perceived ability to overcome adverse situations. Deficits in how an individual tolerates internal conflict and develops adaptive coping skills have been posited to have an origin in negative life experiences that impact one's resilience. A number of researchers see resilience and risk as existing on a continuum. Deficits in resilience amplify risk, and enhanced resilience is considered a protective factor against suicide. Researchers have proposed that specific resilience factors exist as moderators and a significant protection against suicide. Furthermore, there is often a pattern of individuals who have a history of suicide attempts—which have a tendency to lower resiliency scores. Attempters can be compared to those who have never attempted suicide—thus, highlighting the importance of measuring resilience when assessing for suicidality. This chapter will further explore the role of resilience as a moderating variable, how resilience factors are measured and defined, and how this information can be used in suicide prevention and intervention.

12.1 INTRODUCTION

Suicide continues to be a significant worldwide public health concern. Significant efforts by theorists and researchers continue to increase our understanding of how to identify risk factors and establish evidence-based preventative measures. In recent years, there has been a movement toward understanding the relationship between resilience and suicide risk. Several theories have been put forward regarding this connection. Some theorists view resilience as a personality feature characterized by the ability to adapt to change, cope with stress, and thrive in adverse conditions. Other researchers conceptualize resilience as a psychological construct by which an individual possesses a set of positive beliefs, resources, and the perceived ability to overcome adverse situations. Deficits in how an individual tolerates internal conflict and develops adaptive coping skills have been theorized to having an origin in negative life experiences that impact one's resiliency. Numerous researchers see resilience and risk as existing on a continuum where deficits in resilience amplify risk, and enhanced resilience is considered a protective factor against suicide. Researchers have proposed that specific resilience factors exist as moderators and as protective factors against suicide. Furthermore, individuals that have a history of suicide attempts tend to score lower on resiliency measures than those who have never attempted suicide. This highlights the importance of measuring resilience when assessing for suicidality. This chapter will further explore the role of resilience as a moderating variable, how resilience factors are measured and defined, and how this information can be used in suicide prevention and intervention.

12.2 SUICIDE AS A PUBLIC HEALTH CONCERN

Suicide is a growing concern among health care and public health specialists alike. Worldwide, suicide has been found to be the tenth leading cause of death. However, this may be underestimated due to underreporting (Jenkins, 2002). Adding to the growing concern, rates of suicide are not projected to diminish in the near future, as projected by the World Health Organization (WHO), who estimates that suicide rates will increase by 10–20 times globally by 2020 (Bertolote & Fleischmann, 2015). With rates of suicide appearing alarmingly high with the projection of increase, researchers have begun to take action to better understand the prevalence of suicide throughout various regions and groups of individuals. This search to better understand the impact of suicide throughout the world is necessary to highlight the importance of both action and intervention. Additionally, better understanding the incidences of suicide across regions and demographics may help to develop appropriate programs where they are needed most.

12.2.1 SUICIDE IN THE UNITED STATES

Within the United States (US), suicide has been an increasing concern among mental health professionals and public health researchers. However, despite the increased attention in research and clinical work, suicide rates continue to climb in contrast to the decline in suicide rates observed by researchers in the 1990s (Hu et al., 2008). This decline may have been attributed to the beginnings of increased research attention and intervention on suicidality that occurred in the early 1990s (Knox et al., 2004). Nevertheless, rates of suicide began to increase by one percent from 1999 to 2006 and to two percent from 2006 to 2014 (Curtin et al., 2016). Collectively from 1999 to 2014, the rate of suicide has increased by 24% (Curtin et al., 2016). Within the United States, researchers have found that suicide rates in men were over three times greater than in women. Nevertheless, the largest increase observed in suicides within the United States occurred in females between the ages of 10–14 years old (Curtin et al., 2016). Although these demographics in recent suicide rates are exclusive to the United States, suicide appears to be a global public health concern and many countries have been found to be affected by this growing epidemic.

12.2.2 GLOBAL DIFFERENCES IN SUICIDE

Organizations, such as the WHO, have highlighted the reality of suicide as a significant public health problem around the world. Globally, suicide accounts for approximately one million deaths per year (WHO, 2010). Some researchers suggest that rates of suicide are highest in men throughout European countries (Bertolote & Fleischmann, 2015), yet, within European countries these rates can differ. Factors such as attitudes toward suicide may impact suicide rates and reports. For example, researchers have found a relationship between mental health stigma and suicide rates in different European countries (Schomerus et al., 2015). Other factors, such as socioeconomic differences have also shown to be related to the variation of suicide rates in different European countries (Fountoulakis et al., 2014). Similarly, in Asia, socioeconomic differences influence rates of suicide between countries (WHO, 2010).

Overall, more than 60% of suicides annually occur in Asia. Specifically, in China, there is a unique, substantially higher rate of death by suicide among women when compared to men (WHO, 2010). Additionally, in China, contrary to Western countries, larger rates of suicide are observed in rural areas rather than urban areas (Jenkins, 2002). While rates of suicide in China continue to be one of the highest globally, researchers and public health officials have observed a decrease over the past 10 years (Wang et al., 2014). One must consider patterns of reporting when considering suicide rates. Due to stigma, it is likely that many deaths are not reported as a suicide. This may lead to a large underestimation of the actual frequency of suicides in China as well as other countries. In response to this global epidemic, both the WHO and the United Nations (UN) have called for public health policy and national programs aimed at suicide prevention (Jenkins, 2002).

12.2.3 INDIVIDUAL DIFFERENCES IN SUICIDE

Multiple factors in individuals who die by suicide or attempt suicide have been studied in order to better understand and identify suicide patterns. The relationships between factors such as age and gender with suicide rates have been explored throughout psychological and public health studies. Adolescents have been identified as being a particularly high-risk population for suicide. In the United States, suicide is the second leading cause of death in young adults 15–19 years old (Shain, 2016). Moreover, the rate of suicide in young adults has continued to grow in the last 15 years (Curtin et al., 2016). Differences between adolescent girls and boys have also been a concern in public health research. For instance, girls display a higher frequency of suicide attempts, while rates of death by suicides are greater in boys (Shain, 2016), a trend that is mirrored in adult populations. While women have a higher number of suicide attempts, men engage in more lethal means of suicide which means that men have a higher rate of deaths by suicide (Mergl et al., 2015).

In addition to age and gender, public health professionals have expressed interest in the relationship between discrimination and minority stress and their impact on suicidality. Indeed, minority stress has been shown to be linked to suicide attempts in various marginalized racial and ethnic groups (Gomez et al., 2011). The consistent stressors faced by marginalized racial and ethnic groups may also be attributed to the increased prevalence of deaths by suicide among sexual and gender minorities. Internationally, research has shown that lesbian, gay, and bisexual individuals account for a greater number of suicide attempts than heterosexual individuals (Haas et al., 2010). Similarly, transgender and gender nonconforming individuals have shown substantially high rates of suicide attempts ranging from 30% to 60% (Haas et al., 2014). Despite the increased risk for suicide attempts among gender and sexual minorities, deaths by suicide within this population are not systematically recorded in the United States and thus there is a gap in research on the number of deaths by suicide among gender and sexual minorities (Haas et al., 2015). These alarmingly high rates of suicide in marginalized racial, ethnic, sexual, and gender minority groups constitute a growing public health concern.

An important element in the growing public health concern on suicide is its relationship with different mental health disorders. Being that 90% of individuals who died from suicide have been diagnosed with a mental health disorder, exploring this relationship is vital in understanding death by suicide (Bertolote & Fleischmann, 2002). Notably, the most prevalent diagnoses among individuals who died by suicide are mood disorders, specifically depression (Bertolote & Fleischmann, 2002). In the general public, substance-use related disorders were the second most common diagnosis in deaths by suicide. While, in clinical populations, the second most common mental health diagnosis was schizophrenia (Bertolote & Fleischmann, 2002). Given the prevalence of each of these disorders, it is important to be aware of the relationship between psychiatric disorders and suicide. Moreover, the widespread prevalence of suicide across varying clinical presentations, as well as individual demographics only highlights its significance as public health concerns across the globe.

12.3 RISK AND PROTECTIVE FACTORS OF SUICIDE

There are many factors that increase a person's risk for suicide. Symptoms of depression increase the risk for suicidal ideation, attempts, and suicide-related behaviors (Brown et al., 2000; Franklin et al., 2017). A person experiencing thoughts of suicide may include varying degrees of suicidal ideation, suicide plan, as well as passive suicidal ideation (Teismann et al., 2018). Psychic pain, or psychache, is a psychological construct that is defined as an internal experience of intense psychological pain that leads an individual to believe that there is no way to end their unbearable pain except for suicide (Tillman et al., 2017). Individuals with a history of suicidal ideation and/or attempts were found to have higher levels of psychic pain and fewer reasons for living (Tillman et al., 2017).

There are distal risk factors such as biopsychosocial variables and proximal, or trigger, risk factors that are increased by distal factors (Roy et al., 2007). Proximal risk factors tend to be more directly related to suicidal behaviors and include such things as life events, stress, acute or increased symptoms of a preexisting mental illness, and alcohol or substance abuse (Roy et al., 2007). Other risk factors include the presence of mental illness, gender, hopelessness, childhood trauma, health-related negative life events, and impulsivity (Bakhiyi et al., 2017; Bhatt et al., 2018; Mann et al., 1999).

There are also factors that protect and may prevent suicidal behavior. Some of these factors include marriage, employment, religion, and biopsychosocial factors (Roy et al., 2007). Positive mental health, life satisfaction, and social support have all been shown to help protect an individual from attempting suicide (Roy et al., 2007; Siegmann et al., 2018; Teismann et al., 2018). Positive mental health, for instance, does not only correlate negatively with suicidal ideation but also lessens the impact of depressive symptoms when positive factors are increased (Siegmann et al., 2018). An important way to understand about positive mental health is that it is not the same as psychopathology. Simply put, positive mental health is having a positive attitude toward the trials and tribulations of life (Siegmann et al., 2018; Teismann et al., 2018). Similarly, life satisfaction is an overall feeling of satisfaction toward life in general, even when an individual is experiencing a depressed mood (Siegmann et al., 2018). It is also important to consider that this is more prevalent in Western cultures than it is in Eastern cultures, as Western cultures tend to be more individualistic rather than collectivist (Siegmann et al., 2018). Religion is often also considered as a protective factor against suicide. Further study of this theory clarifies that it may not be a specific protective factor against suicide but lessens the impact of depressive symptoms (Teismann et al., 2017). It is not clear what aspect of being religious is a mitigating factor, but it is thought that trust in a higher power increases hopefulness, which in turn may decrease the risk that an individual will attempt suicide, even if there is some suicidal ideations present (Teismann et al., 2017).

Many factors are seen as existing on a continuum with regard to mitigating suicide risk. For example, social support is both a risk and protective factor against suicide. That is, decreased social support increases the risk of suicide, whereas increased social support decreases the risk of suicide (Teismann et al., 2017). Both risk and protective factors generally exist on a continuum, especially when considering the severity of an individual's history of suicidal severity (Tillman et al., 2017).

12.4 PREVENTATIVE METHODS AGAINST SUICIDE

An important consideration when discussing risk for suicide is that there are methods by which suicide can be addressed and prevented. Suicidal behavior involves a pattern of symptomatology that, when addressed, could reduce the presence and severity of suicidal behaviors. From a clinical perspective, therapists are highly encouraged to implement techniques that would reduce the manifestation of suicidal ideation and therefore prevent suicidal behavior.

12.4.1 PSYCHOTHERAPY

Therapists utilize talk therapy, commonly referred to as psychotherapy, in order to effectively reduce the risk of suicide. Two well-known evidenced-based treatment modalities that are used to help decrease suicidal ideation and behaviors are cognitive behavioral therapy (CBT) and dialectical behavioral therapy (DBT) (Linehan, et al., 2006; Tarrier et al., 2008). The principal component of CBT revolves around the modification of harmful emotions, behaviors, and thoughts in order to help individuals acquire new skills to learn how to cope with stressful circumstances (Tarrier et al., 2008). The overall goal of CBT is to assist individuals in recognizing maladaptive thought patterns and learn to replace these maladaptive thoughts, including suicidal ideation, with alternative thoughts and actions.

Tarrier and colleagues (2008) created a systematic meta-analysis analyzing the degree of effectiveness CBT has in reducing suicidal behavior across 28 research studies. The results of the meta-analysis suggest that CBT was more effective in reducing suicidal behavior through individual therapy in comparison to group therapy. Also, the implementation of CBT in reducing suicidal behavior was significant among adult samples, but not adolescents (Tarrier et al., 2008). A subcategory of CBT known as Cognitive-Behavioral Therapy for Suicide Prevention (CBT-SP) has been suggested as being significantly more effective in reducing suicidal behavior among adolescents at high risk for suicide (Stanley et al., 2009). Some of the components that encompass CBT-SP include the development of a safety plan and healthy coping skills, psychoeducation, relapse prevention, and family intervention.

DBT is a form of cognitive-behavioral psychotherapy which focuses on the psychosocial components of the individual through collaborative intervention, which manifests as the client having consistent access and support from their individual therapist as well as a DBT team (Linehan et al., 2006). One defining characteristic regarding DBT is that it has been suggested to be extremely beneficial among individuals with suicidal behavior, more specifically among those that have been diagnosed with Borderline Personality Disorder. Individuals who have a diagnosis of Borderline Personality Disorder are characterized by abrupt emotional lability and continuous interpersonal and personal crises (Linehan et al., 2006). Overall, DBT helps individuals in identifying healthy coping mechanisms in order to manage surges of intense emotional dysregulations and crises. Linehan et al. (2006) conducted a study in which they followed a sample of 100 individuals that were diagnosed with Borderline Personality Disorder within a two-year treatment period. The subjects who participated in DBT treatment had fewer psychiatric hospitalizations, fewer visits to the emergency department, and were less likely to stop treatment (Linehan et al., 2006).

Dialectical Behavior Therapy group skills training (DBT-ST) involves individuals meeting weekly to learn skills that help reduce suicidal behaviors, specifically nonsuicidal self-injury (NSSI), by using the four main components of DBT which are mindfulness, emotion regulation, distress tolerance, and interpersonal effectiveness (Krantz et al., 2018). DBT-ST was found to produce reductions in three risk factors for suicidal behavior which were impulsivity, affective instability, and psychoticism, as well as to decrease the presence of NSSI and suicide attempts than individual DBT treatment (Krantz, et al., 2018; Linehan et al., 2015; Soler et al., 2009). Additionally, three types of DBT treatments were compared across factors of suicidality and suicidal and self-injurious behaviors. The three types of DBT included in this study were standard DBT, which included individual- and group-skills training, DBT combined with case management without individual DBT (DBT-S), and DBT without skills training. With regard to a number of suicide attempts, suicidal ideation, use of crisis prevention and intervention services related to suicidality, and medical severity of self-injurious behaviors, all three types of DBT studied were shown to significantly reduce each of these factors (Linehan et al., 2015). Resilience in relation to DBT was studied by Ramaiyaet al. (2018) for a Nepalese population of women who exhibited suicidal behaviors, and it was shown that there was a reduction in suicidal ideation and improvement in resilience, suggesting that DBT may be effective across various cultures. This can facilitate further research cross-culturally in the use of DBT to reduce suicidality and increase resilience.

12.4.2 MEDICATION

Another preventive method toward the reduction of suicide behavior is medication. It is imperative for individuals that could benefit from medication to be compliant with their medication regimen. The psychiatrist and patient need to discuss whether combining medication with psychotherapy could be an important and more effective intervention. One example of a medication that has been proven effective in reducing suicidal behavior is Clozapine, an antipsychotic medication (Meltzer et al., 2003). One defining characteristic regarding Clozapine is that it is the only medication approved by the US Food and Drug Administration that specifically indicates a reduction of suicidal behavior in patients that have been diagnosed with Schizophrenia or Schizoaffective Disorder (Meltzer et al., 2003). A study by Meltzer et al. (2003) measured the effectiveness of Clozapine in reducing suicidal behavior in comparison to Olanzapine, another antipsychotic medication. Overall, the individuals that were treated with Clozapine had fewer suicide attempts, rescue interventions, and a reduction in the need to implement antidepressants to reduce suicidal behavior (Meltzer et al., 2003). It is significant for individuals that are consuming medications to reduce suicidal behavior to comply with medical recommendations and continuously monitor the effectiveness of the medication with their medical provider (Meltzer et al., 2003).

12.4.3 INDIVIDUAL FACTORS

There are also specific individual interventions that could be implemented to reduce the risk of suicide. One area of focus in reducing suicidal behavior is to work with a patient to combat the impact of self-stigma (Rimkeviciene et al., 2015). Fundamentally, the perceived stigma an individual may be experiencing regarding suicide and mental health could induce an increase in suicidal behavior due to reduced motivation to seek help, lower self-esteem, and the lack of social support. Rimkeviene et al. (2015) conducted a study with the aim to analyze the effects of personal stigma on suicidal behavior. The results suggested that there are four main themes correlated with perceived negative stigma: care, seriousness, badness, and avoidance. Results from this research study suggested that individuals that were in interpersonal relationships were more emotionally affected by perceived mental health stigmas (Rimkeviene et al., 2015). Substantially, combating negative stigma increases the likelihood of recognizing risk factors and creating preventive plans.

Another individual component that has the potential to decrease suicidal behavior is communal social support. This notion is particularly relevant in today's political and cultural climate given that certain populations have an increased likelihood of suicidal behavior (Heikkinen et al., 1994). Thus, it is imperative for community-based programs to implement tactics with the aim of reducing suicidal behavior. The literature strongly suggests that a poor sense of social support and weak social networks could potentially disintegrate protective factors and consequently increase the likelihood of suicide behavior (Heikkinen et al., 1994).

12.5 RESILIENCE AND SUICIDE RISK

12.5.1 RESILIENCE AS A PERSONALITY FACTOR

Resilience to stressful life events that increase the risk of an individual developing psychopathology can be influenced by genetic factors; however, resilience is also a personality feature (Amstader et al., 2016). Resilience as a personality feature is characterized by the ability to successfully adapt to changes, especially stressful or negative life changes, a high tolerance for stress and the adaptive ability to cope with it, and the ability to thrive in adverse conditions (Roy et al., 2007, 2011). Resilience has been found to play a role in moderating the potential severity of symptomology and personality characteristics that are connected to resilience (Craciun, 2016). In terms of the big five personality traits of openness, extraversion, conscientiousness, agreeability, and neuroticism, consciousness is the trait that correlates closest to resilience, which suggests that involvement and trustfulness is needed for an individual to be resilient (Craciun, 2016).

Resilience has been found to influence whether an individual has attempted suicide in the past. Individuals who have a history of suicide attempts have lower resiliency scores than those who have never attempted suicide, which highlights the importance of measuring resilience when assessing for suicidality (Roy et al., 2007). Specifically, resilience in regard to suicidality is the perceived or actual ability to regulate suicidal thoughts, an internal factor that defends against suicidal thoughts and behaviors, and positive thinking that helps decrease stress that in turn decrease suicidal thoughts and behaviors (Johnson et al., 2011).

Siegmann et al. (2018) stated that for a variable to be considered as conferring resilience, it should have three characteristics. First, it should possess both protective and risk qualities in order to determine whether or not it can increase or decrease the association between risk and outcome. Thus, the second characteristic is that it should exist on a continuum, that is, an inverse resilience factor amplifies risk such as an inverse risk factor is considered protective. Finally, it should be a psychological construct such as a perceived ability to overcome difficulties or a set of positive beliefs.

12.5.2 RESILIENCE AS A PSYCHOLOGICAL CONSTRUCT

Resilience as a psychological construct has been studied across a variety of fields of psychology. Recent research defines resilience as a psychological construct where an individual possesses a set of positive beliefs and resources and the perceived ability to overcome difficulties, including recovering from stressful or negative life events (Roy et al., 2007). Resilience can also be defined as "a construct connoting the maintenance of positive adaptation by individuals despite experiences of significant adversity" (Luthar et al., 2000, p. 543). Some of the earliest works on resiliency began with Smith and Werner (1982) in their longitudinal study of high-risk children living in Hawaii. Following Smith and Werner's research, resilience has been studied across various aspects of psychology such as low socioeconomic status (Brody et al., 2013; Garmezy, 1991; Rutter, 1979; Werner 1991), academics (Martin & Marsh, 2006), combat exposure (Aldwin et al., 1994; Gold et al., 2000; Youssef et al., 2013), and exposure to trauma (Alim et al., 2008; Wingo et al., 2010; Wright et al., 1997).

Considering that the study of resilience has increased in recent years, the purpose of studying resilience can help promote positive psychology and increase our understanding of how positive change can be created. Masten (2001) stated that resilience is innate in human adaptational systems and, aside from interference of these human adaptational systems; a child may flourish despite adversity. Recent research has also suggested an association between resilience and positive psychology. More specifically, Fredrickson (2004) has proposed the broaden-and-build theory of positive emotions, in which positive emotions work to broaden a person's momentary thought-action repertoire, increasing their resources to manage adverse events. This theory relates to resilience by suggesting that increasing a person's experience of positive emotions fuels psychological resilience (Fredrickson, 2004). Furthermore, Fredrickson (2004) hypothesized that positive emotions and resilience have a reciprocal relationship by which resilient people are more likely to utilize positive emotions in adverse situations, and positive emotions aid in fostering resilience. When analyzing the relationship between resilience and negative emotions, a study that focused on veterans found that resilience may be a protective factor against depressive symptoms and suicidality (Green et al., 2010; Yousef et al., 2013). These findings lend impetus to further psychological research regarding the utilization of resilience in a therapeutic setting.

A foundational method of understanding psychological constructs is the creation of various measures to determine the prevalence of a construct. In regard to resilience, several measures have been created and tested. Windle et al. (2011) recently conducted a methodological review of 15 measures of resilience. Each measure was evaluated based on content validity, criterion validity, internal consistency, construct validity, agreement, reliability, responsiveness, floor and ceiling effect, and interpretability (Windle et al., 2011). Results of this analysis concluded that of these 15 measures three of them were shown to have good psychometric properties, the Connor-Davidson Resilience Scale, the Resilience Scale for Adults, and the Brief Resilience Scale (Windle et al., 2011). Despite these findings, Windle et al. determined that all 15 of the resilience measures lacked some psychometric properties, suggesting that there continues to be a lack of a "gold standard" measure of resilience. The debate for determining a sound psychological measure for resilience was also highlighted by Friborg et al. (2006) who argued that a semantic version of the Resilience Scale for Adults was more psychometrically sound than the Likert version of the measure. More recently, Cassidy (2016) created the Academic Resilience Scale (ARS-30) that focuses on the process of resilience, in contrast with the outcome of resilience which is frequently the core of other resilience measures. Although the ARS-30 cannot be generalized to resilience outside of an academic setting, Cassidy (2016) proposes the use of cognitive-affective and behavioral responses and this preliminary study suggests good psychometric properties of the ARS-30.

Although the study of resilience is growing within the field of psychology, some researchers have reported being hesitant in the use of resilience as a psychological construct. Luthar et al. (2000) summarized the existing literature on resilience and argued that the main issues of utilizing resilience as a psychological construct is the lack of consensus of a definition and related terminology, the way in which resilience is conceptualized, theoretical concerns, and concerns regarding the measurement and prevalence of resilience. Similarly, Ungar (2011) argued that in order for the study of resilience to grow, a shift in focus from children to the environment in which a person experiences adversity and growth is needed.

12.5.3 RESILIENCE AND RISK CONTINUUM

One approach when studying resilience in relation to suicidality is to view resilience as existing on a separate dimension than that of risk. The Buffering Hypothesis, as described by Johnson et al. (2011) suggests that resilience is a separate construct from that of risk, each existing on its own continuum that has an influential relationship with each other. Johnson et al. (2011) describe this continuum as resilience being a factor that is activated when a person experiences a high level of stressors that may lead to suicidality, and thus resilience engages to reduce the risk of suicide. On the opposite end of the spectrum, resilience would not be activated when a person is not experiencing high levels of stress, and thus not at high risk for suicide. Depending on the type of stressors and circumstances a person experiences, the impact of resilience factors is altered, as resilience is not a fixed attribute (Rutter, 1987).

12.5.4 HOW SPECIFIC RESILIENCE FACTORS ACTING AS MODERATORS AND PROTECTION AGAINST SUICIDE ARE MEASURED AND DEFINED

When it comes to measuring how resilience acts as a moderator or protective factor against suicide, the first thing that must be understood is that resilience in this sense should be viewed as a factor that buffers the strength of the association between risk and suicidality (Johnson et al., 2011). When an individual is experiencing distress in terms of suicidal ideation, the buffering factor, resilience, should be active and work to counteract the likelihood that an individual will engage in suicidal behaviors like suicide attempts (Johnson et al., 2011). Alternatively, when an individual is not suicidal, resiliency is able to remain dormant and unused as it is not necessary at the moment to counteract any stressful life events (Johnson et al., 2011). In this case, resilience is seen as existing on a separate dimension than risk that interacts with risk when needed to decrease the likelihood of suicidality (Johnson et al., 2011). That is, when someone with high resilience levels is at high risk for suicide, their resilience will increase and act as a barrier against suicidal behavior, but when one is at high risk for suicide but has low levels of resilience, there is no barrier and therefore the individual is more likely to engage in suicidal behavior (Johnson et al., 2011).

There are many psychological factors that act as moderators and protection against suicidality. Namely, problem-solving ability and confidence, overall self-esteem, social support including familial support and support from a significant other, attachment style, and beliefs related to suicide were found to moderate and protect against suicide (Johnson et al., 2011). The factors that were most empirically supported to moderate the risk of suicide were attributional style, agency perfectionism, and hopelessness (Johnson et al., 2011). This view that resilience may be multifaceted encompasses a wider range of abilities and beliefs (Johnson et al., 2011).

Additionally, not all factors that have a negative correlation with suicide, for example optimism, have been found to be a resilience factor, or one that moderates or buffers against suicidality (Johnson et al., 2011). These findings provide further evidence for the theory that resilience should be seen as a moderating variable and cannot be viewed linearly, which provides directions for future research into the concept of resiliency as it relates to suicidality (Johnson et al., 2011). Therefore, it is best to consider resilience in the sense of being a moderating variable or buffer against suicide, a combination of many psychological constructs that account for a variety of beliefs and abilities that may also be associated with each other (Johnson et al., 2011).

12.5.5 SUICIDE AND RESILIENCE SCORES

Mental health professionals have at times relied on measures that operationalize resilience in order to observe trends in different populations. These measures can be particularly useful when trying to better understand the relationship between resilience and suicidal behavior. Various studies have used such measures in order to analyze this relationship in different groups of individuals. Different age cohorts, such as adolescents and older adults are particular groups of interest. Additionally, the relationship between resilience scores and suicidality in individuals diagnosed with psychiatric disorders is also of specific concern.

In regard to suicide, particular age groups have been identified to be at an increased risk. For example, adolescents and young adults display startlingly high rates of suicide (Shain, 2016). Additionally, there is a concern for elevated risk of suicide in older adults (Conwell et al., 2011). In order to better understand and treat suicide in these populations, one can observe patterns of resiliency as both possible protective and risk factors. In age groups overall, higher risk for suicide is associated with lower resilience scores on the Connor-Davidson Resilience Scale (CD-RISC) (Liu et al., 2014). The CD-RISC is a 25 item Likert scale measure of resilience. It is comprised of items that load on to five factors which are theorized to heavily influence resiliency. These five factors reflect aspects of tenacity, tolerance of stress or negative affect, secure relationships or acceptance of change, control, and spirituality. Factor analyses and studies in general population samples have shown the CD-RISC to be a valid, reliable, and psychometrically sound measure of resilience (Connor & Davidson, 2003).

With such a high prevalence of suicide in adolescent and young adult populations, identifying possible risk factors is imperative in order to prevent further deaths by suicide. Observing factors, such as resiliency, may serve to inform treatment and prevention of suicide in adolescents. For example, researchers have observed higher scores of resilience on the CD-RISC in depressed adolescents who had not attempted suicide than in those who had previously attempted suicide (Nrugham et al., 2010). This may further lead to evidence toward the connection between resilience and suicide in specific at-risk populations, such as adolescents.

Older adults have also been identified as being at risk for suicide and the relationship between resilience scores and suicide has also been studied in older adult populations. In older adults, resilience and meaning in life was shown to influence levels of suicidal ideation (Heisel & Flett, 2008). Yet, older adults have displayed greater levels of resilience than younger adults when measured by the Resilience Appraisal Scale (Gooding et al., 2012). These findings may highlight an opportunity to work with resilience as a protective factor against suicidal ideation, specifically in older adults.

While age may serve as a risk factor for suicide, other factors have also been observed. For example, the presence of a psychiatric disorder has been shown to put one at an increased risk for suicide. While depression is most common in those who die by suicide, substance-use disorders are also prevalent (Bertolote & Fleischmann, 2002). However, it is important to understand the risk factors within these populations. In individuals being treated for substance abuse, those who had attempted suicide displayed lower resilience scores than those who had not. Additionally, individuals who had attempted suicide who endorsed symptoms of depression showed lower scores of resilience (Roy et al., 2011). This further emphasizes the relationship between resilience and suicide in various groups at risk.

12.5.6 USING THIS INFORMATION IN SUICIDE PREVENTION AND INTERVENTION

Some researchers have argued the necessity to change the mental health care system in the United States in order to identify and treat mental illnesses in such a way that reduces the prevalence of deaths by suicide (Reiss & Dombeck, 2007). Approximately 50% of the individuals reported as having died by suicide and who had a mental health problem were not receiving adequate mental health treatment (Reiss & Dombeck, 2007). In addition, numerous individuals do not have the necessary access and means to appropriately receive mental health care—partially due to the deficient insurance policies and lack of accessible resources (Reiss & Dombeck, 2007). Although resilience is a trait that is often developed in individuals that have faced discrimination and adversity, the socioeconomic adversity often faced by many ethnic minorities could be related to mental illness and suicidal behavior, which has been shown to be particularly more prevalent among young people (Agerbo, 2002).

There is also the need to enhance and expand research regarding an understanding about suicide by analyzing statistical information and generating preventative methods (Reiss & Dombeck, 2007). With this in mind, there is a lack of funding from grant agencies and concerns by Institutional Review Boards due to liability—given that suicide is considered to be a highly sensitive topic (Reiss & Dombeck, 2007). Some researchers have suggested that, such limitations have constrained potential research such as long-term suicidality studies and best treatment options for specific conditions such as Bipolar Disorder (Reiss & Dombeck, 2007). With more potential research generated, agencies could implement accurate information to properly train health care providers. For example, 40% of the individuals who die by suicide are seen by a clinician within a month of their death (Reiss & Dombeck, 2007). This implies that perhaps the clinician did not assess for suicidality, perhaps because they were not properly trained. Another risk factor that has been identified among individuals that have engaged in suicidal behavior is that they had lower resilience scale scores (Windle et al., 2011). Therefore, further research is needed on the impact that implementation of coping mechanisms can have on the effort to develop resilience in individuals that are engaging in suicidal behavior. It is important to recognize that the overall concept of resilience and how resilience manifests itself is highly affected by culture (Harrod & Martin, 2013). Therefore, broader and more inclusive research that focuses on the correlation of resilience and suicidal behaviors needs to be conducted in order to generate more adequate preventative interventions for suicidal behaviors.

A common theme in recent research that focuses on the reduction of suicidal behavior is understanding the role of resilience. When considering patients who have experienced a recent major depressive episode, separated by those who made a suicide attempt in their lifetime and those who had not, higher presence of resilience factors were found in the nonsuicidal group (Rossetti et al., 2017). Furthermore, these results support the belief that resilience as a trait can modify and reduce a person's risk for suicide. Rossetti et al. (2017) urged an increase in research to focus on psychological skills to enhance resilience and therefore provide specific tools to be utilized in a clinical setting to reduce the risk of suicide. The Schematic Appraisals Model of Suicide, which postulates (a) the way in which a person appraises a situation and (b) self-appraisal plays a significant role in moderating suicidality, has been used to describe positive self-appraisal as a resilience factor in regards to suicide (Johnson et al., 2010). When evaluating the impact of positive self-appraisal in the face of stressful life events in adults, Johnson et al. (2010) describe self-appraisal as a resilience factor that in turn moderated the risk of suicidality. This work suggests the importance of understanding various aspects of resilience to mitigate the impact of suicidal ideation.

By properly implementing and prioritizing screening skills that focus on depression, substance abuse, and a more profound understanding of the effect of resilience on different cultural groups, this could potentially reduce the prevalence of suicidal behaviors and attempts. An increase in collaboration on resilience is needed between clinicians and researchers in order to reduce suicidal behavior. It is imperative to share the research generated with the general public by collaborating with the media and creating publicity that would deliver accurate information and advocate for suicide prevention across multiple media venues (Reiss & Dombeck, 2007). Consequently, creating a more supportive and mindful society on the effects of resilience on suicide behavior and its surrounding stigma.

12.6 CONCLUSION

Suicide is a public health concern worldwide and much research has been conducted to understand the nature of suicidality. In addition to identifying risk and protective factors, resilience has emerged as a key concept in relation to suicidality. Resilience has been defined as a personality factor, psychological construct, and is seen as existing on a continuum. In studying the relationship between an individual's score on resiliency measures, it was found that those who had lower resiliency scores tended to attempt suicide more frequently than those who scored high on resilience, highlighting the importance of considering resiliency when assessing for suicidality. This information can be used to further develop suicide prevention and intervention.

KEYWORDS

- **resilience**
- **suicide**
- **at risk**

REFERENCES

Agerbo, E. (2002). Familial, psychiatric, and socioeconomic risk factors for suicide in young people: Nested case-control study. *BMJ, 325*(7355), 74–74.

Aldwin, C. M., Levenson, M. R., & Spiro, A. (1994). Vulnerability and resilience to combat exposure: Can stress have lifelong effects? *Psychology and Aging, 9*(1), 34.

Alim, T. N., Feder, A., Graves, R. E., Wang, Y., Weaver, J., Westphal, M., Alonso, A., Aigbogun, N. U., Smith, B. W., Doucette, J. T., Meliman, T. A., Lawson, W. B., & Charney, D. S. Trauma, resilience, and recovery in a high-risk African–American population. *American Journal of Psychiatry, 165*(12), 1566–1575.

Amstder, A. B., Maes, H. H., Sheerin, C. M., Myers, J. M., & Kendler, K. S. (2016). The relationship between genetic and environmental influences on resilience and on common internalizing and externalizing psychiatric disorders. *Social Psychiatry and Psychiatric Epidemiology, 51*, 699–678.

Bakhiyi, C. L., Jaussent, I., Beziat, S., Cohen, R., Genty, C., Kahn, J.-P., Leboyer, M., Le Vaou, P., Guillaume, S., & Courtet, P. (2017). Positive and negative life events and reasons for living modulate suicidal ideation in a sample of patient with history of suicide attempts. *Journal of Psychiatric Research, 88*, 64–71.

Bertolote, J. M., & Fleischmann, A. (2002). Suicide and psychiatric diagnosis: a worldwide perspective. *World Psychiatry, 1*(3), 181.

Bertolote, J. M., & Fleischmann, A. (2015). A global perspective in the epidemiology of suicide. *Suicidologi, 7*(2), 6–8.

Bhatt, M., Perera, S., Zielinski, L., Eisen, R. B., Yeung, S., El-Sheikh, W., DeJesus, J., Rangarajan, S., Sholer, H., Iordan, E., Mackie, P., Islam, S., Dehghan, M., Thabane, L., & Samaan, Z. (2018). Profile of suicide attempts and risk factors among psychiatric patients: A case–control study. *PLoS One, 13*(2), 1–15.

Brody, G. H., Yu, T., Chen, E., Miller, G. E., Kogan, S. M., & Beach, S. R. (2013). Is resilience only skin deep? Rural African Americans' socioeconomic status-related risk and competence in preadolescence and psychological adjustment and allostatic load at age 19. *Psychological Science, 24*(7), 1285–1293.

Brown, G. K., Beck, A. T., Steer, R. A., & Grisham, J. R. (2000). Risk factors for suicide in psychiatric outpatients: A 20-year prospective study. *Journal of Consulting and Clinical Psychology, 68*, 371–377.

Cassidy, S. (2016). The Academic Resilience Scale (ARS-30): A new multidimensional construct measure. *Frontiers in Psychology, 7*, 1787.

Connor, K. M., & Davidson, J. R. (2003). Development of a new resilience scale: The Connor-Davidson resilience scale (CD-RISC). Depression and anxiety, *18*(2), 76–82.

Conwell, Y., Van Orden, K., & Caine, E. D. (2011). Suicide in older adults. *Psychiatric Clinics, 34*(2), 451–468.

Craciun., A. (2016). Psychological resilience in relation to potential factors and stress. *Roman Journal of Experimental Applied Psychology, Special Issue*, 410–414.

Curtin, S. C., Warner, M., & Hedegaard, H. (2016). Increase in suicide in the United States, 1999–2014.

Fountoulakis, K. N., Kawohl, W., Theodorakis, P. N., Kerkhof, A. J. F. M., Navickas, A., Höschl, C., Lecic-Tosevski, D., Sorel, E., Rancans, E., Palova, E., Juckel, G., Isacsson, G., Jagodic, H. K., Botezat-Antonescu, I., Warnke, I., Rybakowski, J., Azorin, J. M., Cookson, J., Waddington, J., & Juckel, G. (2014). Relationship of suicide rates to economic variables in Europe: 2000–2011. *The British Journal of Psychiatry, 205*(6), 486–496.

Franklin, J. C., Ribeiro, J. D., Fox, K. R., Bentley, K. H., Kleiman, E. M., Huang, X., Musacchio, K. M., Jaroszewski, A. C., Chang, B. P., & Nock, M. K. (2017). Risk factors for suicidal thoughts and behaviors. *Psychological Bulletin, 143*, 187–232.

Fredrickson, B. L. (2004). The broaden-and-build theory of positive emotions. *Philosophical Transactions of the Royal Society B: Biological Sciences, 359*(1449), 1367.

Friborg, O., Martinussen, M., & Rosenvinge, J. H. (2006). Likert-based vs. semantic differential-based scorings of positive psychological constructs: A psychometric comparison of two versions of a scale measuring resilience. *Personality and Individual Differences, 40*(5), 873–884.

Garmezy, N. (1991). Resilience in children's adaptation to negative life events and stressed environments. *Pediatric Annals, 20*(9), 459–466.

Gold, P. B., Engdahl, B. E., Eberly, R. E., Blake, R. J., Page, W. F., & Frueh, B. C. (2000). Trauma exposure, resilience, social support, and PTSD construct validity among former prisoners of war. *Social Psychiatry and Psychiatric Epidemiology, 35*(1), 36–42.

Gomez, J., Miranda, R., & Polanco, L. (2011). Acculturative stress, perceived discrimination, and vulnerability to suicide attempts among emerging adults. *Journal of Youth and Adolescence, 40*(11), 1465.

Gooding, P. A., Hurst, A., Johnson, J., & Tarrier, N. (2012). Psychological resilience in young and older adults. *International Journal of Geriatric Psychiatry*, *27*(3), 262–270.

Green, K. T., Calhoun, P. S., Dennis, M. F., & Beckham, J. C. (2010). Exploration of the resilience construct in posttraumatic stress disorder sensitivity and functional correlates in military combats veterans who have served since September 11, 2001. *The Journal of Clinical Psychiatry*, *71*(7), 823.

Haas, A. P., Eliason, M., Mays, V. M., Mathy, R. M., Cochran, S. D., D'Augelli, A. R., ... & Russell, S. T. (2010). Suicide and suicide risk in lesbian, gay, bisexual, and transgender populations: Review and recommendations. *Journal of Homosexuality*, *58*(1), 10–51.

Haas, A. P., Rodgers, P. L., & Herman, J. L. (2014). Suicide attempts among transgender and gender non-conforming adults. *Work*, *50*, 59.

Haas, A. P., Lane, A., & Working Group for Postmortem Identification of SO/GI. (2015). Collecting sexual orientation and gender identity data in suicide and other violent deaths: a step towards identifying and addressing LGBT mortality disparities. *LGBT Health*, *2*(1), 84–87.

Harrod, R. P., & Martin, D. L. (2013). Culture and resilience. *Bioarchaeology of climate change and violence, Springer briefs in anthropology*, 23–32.

Heikkinen, M., Aro, H., & Lönnqvist, J. (1994). Recent life events, social support and suicide. *Acta Psychiatrica Scandinavica*, *89*(S377), 65–72.

Heisel, M. J., & Flett, G. L. (2008). Psychological resilience to suicide ideation among older adults. *Clinical Gerontologist*, *31*(4), 51–70.

Hu, G., Wilcox, H. C., Wissow, L., & Baker, S. P. (2008). Mid-life suicide: an increasing problem in US Whites, 1999–2005. *American Journal of Preventive Medicine*, *35*(6), 589–593.

Jenkins, R. (2002). Addressing suicide as a public-health problem. *Lancet (London, England)*, *359*(9309), 813–814.

Johnson, J., Gooding, P. A., Wood, A. M., & Tarrier, N. (2010). Resilience as positive coping appraisals: Testing the schematic appraisals model of suicide (SAMS). *Behavior Research and Therapy*, *48*(3), 179–186.

Johnson, J., Wood, A. M., Gooding, P., Taylor, P. J., & Tarrier, N. (2011). Resilience to suicidality: The buffering hypothesis. *Clinical Psychology Review*, *31*, 563–591.

Knox, K. L., Conwell, Y., & Caine, E. D. (2004). If suicide is a public health problem, what are we doing to prevent it? *American Journal of Public Health*, *94*(1), 37–45.

Krantz, L. H., McMain S., Kuo, J. R. (2018). The unique contribution of acceptance without judgment in predicting nonsuicidal self-injury after 20-weeks of dialectical behavior therapy group skills training. *Behavior Research and Therapy*, *104*, 44–50.

Linehan, M. M., Comtois, K. A., Murray, A. M., Brown, M. Z., Gallop, R. J., Heard, H. L., Korslund, K. E., Tutek, D. A., Reynolds, S. K., & Lindenboim, N. (2006). Two-year randomized controlled trial and follow-up of dialectical behavior therapy vs therapy by experts for suicidal behaviors and borderline personality disorder. *Archives of General Psychiatry*, *63*(7), 757–766.

Linehan, M. M., Korslund, K. E., Harned, M. S. (2015) Dialectical behavior therapy for high suicide risk in individuals with borderline personality disorder: A randomized clinical trial and component analysis. *JAMA Psychiatry*, *72*(5), 475–482.

Liu, D. W., Fairweather-Schmidt, A. K., Roberts, R. M., Burns, R., & Anstey, K. J. (2014). Does resilience predict suicidality? A lifespan analysis. *Archives of Suicide Research*, *18*(4), 453–464.

Luthar, S. S., Cicchetti, D., & Becker, B. (2000). The construct of resilience: A critical evaluation and guidelines for future work. *Child Development*, *71*(3), 543–562.

Mann, J., Waternaux, C., Haas, G. L., & Malone, K. M. (1999). Toward a clinical model of suicidal behavior in psychiatric patients. *The American Journal of Psychiatry*, *156*, 181–189.

Martin, A. J., & Marsh, H. W. (2006). Academic resilience and its psychological and educational correlates: A construct validity approach. *Psychology in the Schools*, *43*(3), 267–281.

Masten, A. S. (2001). Ordinary magic: Resilience processes in development. *American Psychologist*, *56*(3), 227.

Meltzer, H. Y., Alphs, L., & Green, A. I. (2003). Clozapine treatment for suicidality in schizophrenia international Suicide Prevention Trial (InterSePT). *Archives of General Psychiatry*, *60*(1), 82.

Mergl, R., Koburger, N., Heinrichs, K., Székely, A., Tóth, M. D., Coyne, J., Quintão, S., Arensman, E., Coffey, C., Maxwell, M., Värnik, A., van Audenhove, C., McDaid, D., Sarchiapone, M., Schmidtke, A., Genz, A., Gusmão, R., & Hegerl, U. (2015). What are reasons for the large gender differences in the lethality of suicidal acts? An epidemiological analysis in four European countries. *PLoS One*, *10*(7), e0129062.

Miller, A. (2006). Faculty of 1000 evaluation for two-year randomized controlled trial and follow-up of dialectical behavior therapy vs therapy by experts for suicidal behaviors and borderline personality disorder. *F1000—Post-publication Peer Review of the Biomedical Literature*.

Nrugham, L., Holen, A., & Sund, A. M. (2010). Associations between attempted suicide, violent life events, depressive symptoms, and resilience in adolescents and young adults. *The Journal of Nervous and Mental Disease, 198*(2), 131–136.

Ramaiya, M. K., McLean, C., Regmi, U., Fiorillo, D., Robins, C., Kohrt, B. A. (2018). A dialectical behavior therapy skills intervention for women with suicidal behaviors in rural Nepal: A single-case experimental design series. *Journal of Clinical Psychology, 74*(7), 1071–1091.

Reiss, N. S., & Dombeck, M. (2007). Suicide prevention and societal, measures. Retrieved from https://www.mentalhelp.net/articles/suicide-prevention-and-societal-measures/

Rimkeviciene, J., Hawgood, J., Ogorman, J., & Leo, D. D. (2015). Personal stigma in suicide Attempters. *Death Studies, 39*(10), 592–599.

Rossetti, M. C., Tosone, A., Stratta, P., Collazzoni, A., Santarelli, V., Guadagni, E., Rossi, R., & Rossi, A. (2017). Different roles of resilience in depressive patients with history of suicide attempt and no history of suicide attempt. *Revista Brasileira de Psiquiatria, 39*(3), 216–219.

Roy, A., Carli, V., & Sarchiapone, M. (2011). Resilience mitigates the suicide risk associated with childhood trauma. *Journal of Affective Disorders, 133*(3), 591–594.

Roy, A., Sarchiapone, M., & Carli, V. (2007). Low resilience in suicide attempters. *Archives of Suicide Research, 11*, 265–269.

Rutter, M. (1979). Protective factors in children's responses to stress and disadvantage. *Annals of the Academy of Medicine, Singapore, 8*(3), 324.

Rutter, M. (1987). Psychosocial resilience and protective mechanisms. *American journal of orthopsychiatry, 57*(3), 316–331.

Schomerus, G., Evans-Lacko, S., Rüsch, N., Mojtabai, R., Angermeyer, M. C., & Thornicroft, G. (2015). Collective levels of stigma and national suicide rates in 25 European countries. *Epidemiology and Psychiatric Sciences, 24*(2), 166–171.

Siegmann, P., Teismann, T., Fritsch, N., Forkmann, T., Glaesmer, H., Zhang, X. C., Brailovskaia, J., & Margraf, J. (2018). Resilience to suicide ideation: A cross-cultural test of the buffering hypothesis. *Clinical Psychology & Psychotherapy, 25*(1), e1–e9.

Shain, B. (2016). Suicide and suicide attempts in adolescents. *Pediatrics, 138*(1), e20161420.

Smith, R. S., & Werner, E. E. (1982). *Vulnerable but invincible: A study of resilient children,* New York: Mc Graw-Hill.

Soler, J., Pascual, J. C., Tiana, T., Cebrià, A., Barrachina, J., Campins, M. J., Gich, I., Alvarez, E., & Pérez, V. (2009). Dialectical behavior therapy skills training compared to standard group therapy in borderline personality disorder: A 3-month randomised controlled clinical trial. *Behavior Research and Therapy, 47*, 353–358.

Stanley, B., Brown, G., Brent, D. A., Wells, K., Poling, K., Curry, J., Kennard, B. D., Wagner, A., Cwik, M. F., Klomek, A. B., Goldstein, T., Vitiello, B., Barnett, S., Daniel, S., & Hughes, J. (2009). Cognitive-behavioral therapy for suicide prevention (CBT-SP): Treatment model, feasibility, and acceptability. *Journal of the American Academy of Child & Adolescent Psychiatry, 48*(10), 1005–1013.

Tarrier, N., Taylor, K., & Gooding, P. (2008). Cognitive-behavioral interventions to reduce suicide behavior. *Behavior Modification, 32*(1), 77–108.

Tiesmann, T., Forkmann, T., Brailovskaia, J., Siegmann, P., Glaesmer, H., & Margraf, J. (2018). Positive mental health moderates the association between depression and suicide ideation: A longitudinal study. *International Journal of Clinical and Health Psychology, 18*, 1–7.

Tillman, J. G., Clemence, A. J., Cree, R., Lewis, K. C., Stevens, J. L., & Reiss, D. (2017). The persistent shadow of suicide ideation and attempts in a high-risk group of psychiatric patients: A focus for intervention. *Comprehensive Psychiatry, 77*, 20–26.

Ungar, M. (2011). The social ecology of resilience: addressing contextual and cultural ambiguity of a nascent construct. *American Journal of Orthopsychiatry, 81*(1), 1.

Wang, C. W., Chan, C. L., & Yip, P. S. (2014). Suicide rates in China from 2002 to 2011: An update. *Social Psychiatry and Psychiatric Epidemiology, 49*(6), 929–941.

Werner, E. E. (1991). High-risk children in young adulthood: A longitudinal study from birth to 32 years. In S. Chess & M. E. Hertzig (Eds.), *Annual progress in child psychiatry and child development, 1990.* (pp. 180–193).

Windle, G., Bennett, K. M., & Noyes, J. (2011). A methodological review of resilience measurement scales. *Health and quality of life outcomes, 9*(1), 8.

Wingo, A. P., Wrenn, G., Pelletier, T., Gutman, A. R., Bradley, B., & Ressler, K. J. (2010). Moderating effects of resilience on depression in individuals with a history of childhood abuse or trauma exposure. *Journal of Affective Disorders, 126*(3), 411–414.

Wright, M. O. D., Master, A. S., & Hubbard, J. J. (1997). Long-term effects of massive trauma: Developmental and psychobiological perspectives. In D. Cicchetti & S. L. Toth (Eds.), *Developmental perspectives on trauma: Theory, research, and intervention.* (pp. 181–225). University of Rochester Press.

WHO (World Health Organization) (2010). Towards evidence-based suicide prevention programmes. Manila: WHO Regional Office for the Western Pacific.

Youssef, N. A., Green, K. T., Dedert, E. A., Hertzberg, J. S., Calhoun, P. S., Dennis, M. F., Mid-Atlantic Mental Illness Research Education And Clinical Center Workgroup, & Beckham, J. C. (2013). Exploration of the influence of childhood trauma, combat exposure, and the resilience construct on depression and suicidal ideation among US Iraq/Afghanistan era military personnel and veterans. *Archives of Suicide Research, 17*(2), 106–122.

PART II

Resilience and Well-Being in Home, School, Community, and Sports Environments

CHAPTER 13

Systemic Analysis of Family Resilience

ALEXANDER V. MAKHNACH* and ANNA I. LAKTIONOVA

Institute of Psychology of the Russian Academy of Sciences, Moscow, Russia

**Corresponding author. E-mail: amak@inbox.ru*

ABSTRACT

This chapter presents analysis of family resilience by which we understand a systemic characteristic of a family capable of responding to stresses of various genesis using family resources inherent thereto, including the family's ability and willingness to cope, change, adjust and develop, form the point of A.A. Bogdanov's theory. The chapter gives a brief description of A.A. Bogdanov's organizational theory, consistently considering the notions of family resilience and its resources. It shows that the family is resilient when it is an organized system where the whole practically exceeds a mere total of its parts because its activity complexes match. In the event, the activities only partially sum up whereas partially represent mutual resistance, that is, subtract from the organizational point of view, the level of family resilience shall be determined by the correlation of risk factors and family resources. If the family members' activities fully counteract and destroy each other or mutually disorganize, we believe the family to be nonresilient since it is a disorganized system.

13.1 INTRODUCTION

The term "resilience" was first used in Russian philosophy by A.A. Bogdanov, the founder of the General Systems Theory (Bogdanov, 1989). His main work "Tektologia (Universal Organizational Science)" published in the 1910–1920 is a generally scientific concept of an organizational approach to studying any system in nature and in human activity[1]. It is believed that Ludwig von Bertalanffy first proposed a general theory describing the principles of the organization of living systems. However, even 20–30 years before he published the first works on his general theory of systems, Russian researcher, philosopher, and economist

[1] "The latter's fundamental and highly original work appeared in Russia under the title Tektologia, Vols. 1–3 (1912–1927), hence several years before von Bertalanffy's first system theoretic notions—at the time restricted to biology—were published (1928). Furthermore, a German version of the decisive first two volumes of Bogdanov's much more comprehensive presentation of systems organization was published in Bertalanffy's own mother tongue as early as 1926–1928, and was subsequently reviewed in the pertinent German literature. We are certainly not accusing Bertalanffy of plagiarism, and do not dispute his many genuine contributions; the history of science has shown repeatedly that ideas frequently mature in the minds of two or more scholars simultaneously. Von Bertalanffy may well have conceived his system ideas without any influence from Bogdanov" (Mattessich, 1982, p. 387).

Alexander Bogdanov developed an equally sophisticated and comprehensive system theory, which, unfortunately, is almost unknown outside of Russia (Capra, 1997). Bogdanov's Tektology preempted N. Wiener and W. Ashby's cybernetics, the general systems theory by L. von Bertalanffy and I. Prigogine's synergetics. The freshness of A.A. Bogdanov's ideas was in bringing together all human, biological and physical sciences understood as systems of interrelation and searching organizational principles on the basis of all types of systems. According to the fundamental prerequisites of tektology, the functioning of two and more elements incorporated in a single process can exceed under certain organization or, on the contrary, be inferior in efficiency to the functioning of the same elements on their own (A.A. Bogdanov also pondered upon a "neutral" way of interaction) (Loktionov, 2016). Bogdanov (1989) explained the development processes of nature and society in terms of the equilibrium principle borrowed from natural science. All developing objects of nature and society are holistic formations or systems consisting of many elements, according to Bogdanov. The starting point for his concept is that organization laws are one for all objects in which very heterogeneous phenomena (elements) are brought together by structural bonds. Hence the approach to studying any phenomenon should be based both on the study of any system as an interrelation of all its parts and its relationship as the whole with the environment, that is, with all external systems. A.A. Bogdanov was among the first in the world to introduce the notion of systematicity, specifying that an organization as a whole, which is greater than the sum of its parts. The author believed that to take an organizational point of view meant to study any system from the point of interrelation of all its parts as well as its relationship as the whole with the environment, that is, with all external materials. He developed the idea of a structural sustainability of a system and its conditions. He suggested two universal organizational principles (two types of pattern) for the system:

- The forming principle (patterns forming systems—and developments leading to the transition of the system into a different quality); and
- The regulating principle of "selection" (regulating, that is, functioning patterns that contribute to the stabilization of a given quality of the system).

The basis of the organizational mechanism forming any organizational system is in organizational actions—putting elements and complexes together and setting them apart. The process of formation of any organization is bringing elements together in one single body, the whole, but results attained with this can be different. The result depends on the properties of elements brought together, their congruity and correlation.

Any element has certain activity and resistance in relation to others or to the system. Resistance is also an activity but taken from a different point of view as if contraposed to another activity.

The consociation of complexes may go in three ways:

- Activities of the complexes coincide;
- Activities of one complex fully resist the other and paralyze it; and
- The most common case, activities of the components partially sum up and partially represent a mutual resistance to others, that is, organizationally subtract.

Three types of systems are possible to be formed in consequence of a continuous interaction:

- Organized (the whole practically exceeds a simple sum of its parts);
- Neutral (the whole is equal to the sum of its parts);
- Unorganized, disorganized (activities fully counteract and destroy each other, and get mutually disorganized).

The main peculiarity of tektological selection, according to Bogdanov, is its systemic nature: tektological selection affects the whole complex in its entirety, continuously transforming it in accordance with the environment. A tektological complex is capable of self-development and enhancement of self-discipline, characterized by such a level of instability that leads to a higher level of organization. A.A. Bogdanov believed the balanced state of a system to be not something given once and forever, but a "dynamic" or "transient" equilibrium. It acts as a continuous interaction of a progressively developing system with the environment, which gradually takes it out of equilibrium and subsequently leads to instability (crisis), another structural rearrangement which generates a new equilibrium and a new state of balance at a higher turn of its further development.

System instability is a source of its inner development and it takes the system to a higher level of organization, according to A.A. Bogdanov. His definition reads, "Any resilience is relative; it exists only in relation to this or that given environment; elements highly adjusted to one environment, highly sustainable under its influence, may turn (and this is very likely in the majority of cases) into poorly adjusted or unstable in a different environment or under substantially different impact" (Bogdanov, 1989, pp. 215–216).

Thus, we have very briefly described some aspects of A.A. Bogdanov's organizational theory. To analyze family resilience from this point of view we shall consistently consider the notion of family resilience and its resources.

13.2 FAMILY RESILIENCE

By family resilience, we understand a systemic characteristic of a family capable of responding to stresses of various genesis with the use of family resources (individual, belonging to the family as a system, and external environment resources) inherent only thereto, that envisages the family's ability and willingness to cope, change, adjust, and develop (Makhnach, 2016). A new interest toward the notion of "a resilient family" coincides, according to M. Worden, "with the change of the paradigm in the field of family psychotherapy. A resilient family is a notion that underlines strong sides of a family under stress" (Worden, 2002, p. 167). The paradigm based on a family pathology is replaced with the paradigm that focuses on competency and the family's strong aspects (Walsh, 2006). Family resilience is considered as a multifactor and dynamic factor. The family is a natural context of both growth and healing. One or another resilient family structure is necessary to fulfill the family's main objectives—to support individuality along with the creation of the feeling of belonging to the whole, according to Minuchin and Fishman (1981). McCubbin and McCubbin (1996) conceptualize "family resilience" with the help of two different but still interconnected processes continuously taking place in the family: (1) adjustment when the family leans on defense factors which helps it to preserve its integrity, functionality, and continue tackling tasks under risk conditions; and (2) adaptation, that is, bringing itself to "norm" with the account of a crisis situation in the family. They believe it is the success or failure to adapt after a trauma that helps to tell resilient families from nonresilient. Resilient families make their path through changes, can "cope with situational challenges, correct their life, adapt, and even thrive," in spite of these challenges. In contrast, nonresilient families easily become "exhausted" (McCubbin & McCubbin, 2005). G. Patterson assumed that there were four key processes that describe resilient families and conceptualize this notion. She identified the processes that facilitate family functioning, acting as defense factors under unfavorable life circumstances: (1) family cohesion—the balance of connectedness of family members and individualization thereof; (2) family flexibility—balance between willingness to accept changes and adjust to them, and perseverance in the preservation of earlier existing components of the family identity; (3) family ties—affective and instrumental; (4) family values, that is, how the family internally determines problems, in which context the family lives, and who are they as a family (Patterson, 2002). F. Walsh who was one of the first to turn to the theme of family resilience, described the main family processes that work as defense factors in her theoretical review. They include three main groups: (1) the system of opinions and convictions; (2) organizational processes; (3) communication processes. She described these processes in her fundamental book "Strengthening Family Resilience," having filled them with concrete content. She delves into the family as an organized unit with interdependent substructures; in its turn family resilience is determined by their interconnectedness (Walsh, 2006). The study of family resilience logically continued through the search of sources of family strength and resources in spite of all unfavorable factors of the external environment and internal family circumstances. From the position of its resilience, the family is not a system with dysfunctions and responses to external adverse impacts, but rather a system exposed to a stress impact. From this point of view, "proponents of the family assessment approach from the position of its resilience strive to understand what makes the family survive and even thrive under stress" (Worden, 2002, p. 206). Consequently, there appeared papers with the study of family resources which we class as a family protecting factor. Resources play a special protective compensatory role, contribute to adaptive behavior, and are indicators of family resilience as a whole (Windlea et al., 2008). Family defense factors can enhance pro-social behavior and resilience to negative consequences of a crisis or stress, ensuring stable but flexible and favorable environment which facilitates the development of "harmony and family members' growth" (McCubbin & McCubbin, 2005, p. 31). Studies of family resilience are carried out within the framework of various theoretical approaches. Our studies are undertaken from the point of family resources (Makhnach et al., 2015; Makhnach & Postylyakova, 2012; Laktionova, 2015, 2016; Makhnach, 2015, 2016), which allow us, first, to gain knowledge about family resources; and second, to assess family members' ability to use them for their own wellbeing.

13.3 FAMILY RESOURCES AS A SOURCE OF FAMILY RESILIENCE

The family as an integral unit has a number of strong features—resources—that are held in the family system itself (Balswick & Balswick, 1999; Hobfoll & Spilberger, 1992; Olson, 1991). As Walsh (2003) noted, some families that find themselves in adverse and critical life situations fall apart, whereas many demonstrate all symptoms of strength and potential in such instances. Family resources are valuable social, economic, psychological, emotional, and physical characteristics and qualities that family members can use to cope with stress. Such resources include flexibility of family boundaries, role flexibility, possibility to review family rules, clear communication, moderate family cohesion, and openness in world perception (Makhnach & Postylyakova, 2013). Since family resources as a whole develop from individual psychological resources of each family member and the resources that belong to the family as a system, the notion of individual resources is introduced by which we understand physiological, cognitive, personal, social and psychological qualities, and properties of a subject (which they are mainly aware of), that can be accumulated/disposed, developed or modified, to which the subject turns at times of stress and difficult life situations with the goal to cope therewith (Postylyakova, 2016). Along with individual resources, availability of formed family resources and the ability to use them facilitates both the family's adaptation to external conditions and coping with possible challenges, and further family development. Family resources are created, recognized and tested for reliability by various life situations. For clear reasons the family needs time—several years—to have resources generated in it. Family resources play a very important role in maintaining homeostasis in the family when responding to intrafamily and external requirements, challenges, and when coping with family stress. Resources can enhance pro-social behavior and resilience to negative consequences of crisis or stress, ensuring a stable but flexible and a favorable environment that facilitates the development of "stability, harmony, and growth of family members" (McCubbin & McCubbin, 2005, p. 31). For this reason, we should not only consider the family as a resource for an individual (for instance, as a source of social support), but rather as a resource potential of the family in its entirety, where individual resources of its members complement the family resources thus increasing their resilience (Walsh, 2003).

There are three main constituent components that exert considerable impact on family resilience: the system of family values, organizational patterns, and communication processes. We have correlated family systemic resources with these components, as well as with a number of stability factors of a family system (Varga, 2001; Voronina et al., 2011; Kovaleva, 2012; Kryukova et al., 2005; Kuftyak, 2010; Makhnach & Postylyakova, 2013; Tkachenko, 2014). Thus, we believe that communication processes are represented by family communication and family support; organizational patters—by cohesion, resource management, problem resolution skills in the family, family boundaries, family roles and rules, financial freedom of the family; family values—by physical health of family members, trust in the family, religion, interests, views, responsibility, children, and love. Under stress, the family will modify cohesion, adaptivity and flexibility, and adjust to the circumstances. In the sphere of family beliefs, resilient families are aware of the meaning of a crisis or problem, and they consider a crisis to be a problem common for the whole family. Such families find it very important to have a relationship and confidence shared by all its members that together they can enhance their ability to cope with a problem. By expanding their view of themselves as part of something larger than they are, the families keep a broader view of the crisis they are experiencing. The sphere of family organization and resources imply that resilient families have a more flexible structure which they will be able to change and adjust to their needs, rather than hold onto a rigid understanding of family roles and rules, which allows resilient families to get adjusted to changes that happen during the crisis. Such families are capable of reorganization in challenging times. Connectedness is an important characteristic of a resilient family. It reflects the family's conviction that family members can rely on each other in the times of crisis. Along with that resilient families become stronger, when all the members respect individual peculiarities and individual boundaries of each other. Such families know how to maintain a balance between the singularity of each member and general connectedness, which helps them to give a successful response to challenging situations (Makhnach et al., 2015). Noncontradicting clear messages play an important role in family communication. Direct, consistent, noncontradicting, honest communication in resilient families helps all the family members recognize and accept the crisis that the family has faced, and support each other, sharing feelings, and opinion of each family member. This type of communication facilitates a joint process of decision taking as to how the family will handle the problem. Open demonstration of emotions implies that a resilient family creates an atmosphere of trust and support to its members, sharing their feelings, expressing empathy, showing the sense of humor, etc. Such families have the tradition of a joint resolution of problems and looking for ways to handle it (Makhnach et al., 2015).

Individual psychological resources of people with a high level of resilience complement the skills of family resources management thus increasing the family's resilience (Postylyakova, 2016). Communication skills will help the family change if necessary its levels of intimacy and flexibility. A broad network of social support is of great importance to the family. Social support includes external and internal components: family members, friends, neighbors, colleagues, church, and public groups to whose help families may have recourse to if necessary (Balswick & Balswick, 1999; Boss, 1987; Helgeson & Lopez, 2010). Physical health is an important family resource. Chronically ill or disabled family members may cause dysfunction of a family system since the family finds itself in the situation of chronic stress. Meaningful efforts of family members to support and strengthen physical health enhance the family's resilience during stress (Boss, 1987). Researchers note that the family can successfully resolve one set of problems and experience difficulties while resolving another that it faces at different stages of its functioning. The studies of a family coping with stress on the basis of family resources also show the importance of not only the availability of resources but a developed skill to use them. For instance, it was noted that some families having many resources at hand use them so unskillfully that they tend to depreciate in the course of crisis, while the ability to handle resources is invaluable during stress (Balswick & Balswick, 1999). However, the problem is not only in the availability of resources in the family that ensure the progress of various intra-family processes and the ability to use them. Differences in family characteristics and tasks that the family faces at different stages of its life cycle also have their influence on the combination of family and individual resources that facilitates the support of normal family functioning and resolution of family problems. Thus, it is important to understand the peculiarities of resource combinations depending on: (a) structural characteristics of the family and stages of its life cycle; (b) peculiarities and differences that exist in individual and family resources of candidates to substituting parents.

All this will allow us to get a clearer picture which strong features characterize a particular family and which combination of individual and family resources can be considered as optimal. Rational and, what is particularly important, coordinated use of available resources by family members can ensure normal functioning of the family, regulate its behavior and facilitate problem resolution. Family resilience is not a mere result of an aggregated behavior of resilient family members, but rather it is a characteristic which the family develops after several years of existence when one or several components of individual resiliency of each family member join together with resiliency components of the other family members, and eventually the family becomes resilient (Postylyakova, 2015).

Thus, family resilience is a systemic characteristic of a family capable of responding to stresses of various geneses with the employment of family resources inherent to this particular family. Besides, the availability of resources in the family is important, as well as the ability to use these resources and peculiarities in combining these resources depending on structural characteristics of the family and stages of its life cycle. Which combination of resources promotes the enhancement of family resilience will subsequently be analyzed from the point of the organizational theory by A.A. Bogdanov.

13.4 ANALYSIS OF FAMILY RESILIENCE FROM THE POINT OF A.A. BOGDANOV'S THEORY

While coping with stress, the family interacts with such challenging situations whose resolution may be impossible through individual activity of each of its members acting separately; however, this is possible through joint activity since a number of resources exist only in a family activity (Kuftyak, 2010). Joint activity can be described as a process of mutual tuning and adjustment, including compensatory, of regulatory resources of partners who interact in the currently important life situation. The coordination of regulatory resources may have the nature of both direct positive accommodation and compensation. By compensation we understand redistribution of a resource when the implementation of a necessary behavior is carried out through the use of the partner's possibilities in the situation (Kovaleva, 2009).

Having undertaken the attempt to analyze family resilience from the organizational approach by A.A. Bogdanov, which as we said above, is applicable to the study of any system in nature and human activity, we consider the family both as a system and as a correlation of all its members (family members). The essence of considering the family as a system is that all processes unfolding in its different fields are interrelated and targeted at supporting the family integrity. The family's objectives are thus not only subordinated to the law of homeostasis under which the family maintains its own status, but also to the law of development, that is, the need to transit to the next stage of the life cycle with its specific objectives and functions. The transition is implemented as a result of experiencing a normal crisis by the family, the period when the old ways of achieving the goal become inefficient and the family has to develop new ones. This happens, for example, when children are born, when they grow up or start their independent life, when a family member retires or there is a need to contact new social institutions. Quite often, the family faces challenges of maintaining its own integrity under new circumstances. Then family goals begin to get implemented not by means of generating new ways of functioning but due to the so-called dysfunctional processes, that is, special interactions that allow the family to preserve its own integrity in a new situation using the old strategies. Dysfunctional processes include, for example, the change of internal and external psychological boundaries, the onset of psychosomatic symptoms, and transfer to the previous stage of the family's life cycle (regression) (Systemic Family Psychotherapy, 2002; Varga, 2001).

We can assume, leaning on A.A. Bogdanov's propositions, that the family as an organized system is resilient if the whole practically exceeds a mere sum of its parts through the match of activity complexes. This happens both on the basis of operating the already available resources by the family and, primarily, by generating new resources, formation of new patterns of family functioning which ensures necessary changes in the family, processes of coping, adaptation and development, that is, successful responses to new requirements that indicate the family's resilience. How well the family can mobilize its outer and inner resources determines its resilience at a certain period of time: when responding to a risk factor the available family resources get restructured each time. This happens so that the feedback response in the form of coping and adaptation allows the family to achieve its ultimate goal—problem resolution (Makhnach, 2016). Our study shows the importance of such family resources as availability of family support, physical health of family members, established skills of problem resolution, clear family roles and rules, warm emotional ties in the family, financial freedom (as an ability to manage financial resources), and family communication. Specifically it was identified that openness of family members to communication and their ability to discuss any issues are the family's key resources (Makhnach et al., 2014, 2015). Apparently, these resources help to achieve a joint coordinated activity of the family members that leads to higher family resilience. Our findings are that the total number of links between family resources in functional families is less than in dysfunctional. But although they are few in functional families, there are more links identified between individual and family resources. This fact means that functional family members are more involved in the life of their family, their mindset contributes to more openness and interest in other people and the world, ability to love, cooperate and be reliable, while their meaningful, rational and "energy-saving" use of resources speaks about established instrumental skills of dealing with them. These skills characterize the family efficiency and potential (Postylyakova, 2015).

The most common case is when activities partially sum up and partially represent mutual resistance, that is, subtract from the organizational point of view (Bogdanov, 1989). In this case, the level of family resilience is determined by the correlation of risk factors and family resources. Lately, there have been more and more studies dedicated to risk factors and defense factors in the context of family resilience (Boyden & Mann, 2005; Cowan et al., 1996; Flouri et al., 2010; Halford et al., 2001; Luthar & Suchman, 2000; Osofsky & Thompson, 2000; Steptoe & Marmot, 2003). However, the focal point of the studies on family resilience is quite often on the identification of strong sides of the family as challenges to ill-being, stress, etc. with the emphasis on family dysfunction and pathology (Walsh, 2006), whereas little attention is paid to the identification of family vulnerability as a result of its exposure to risks. The following factors may act as predictors of behavioral or functional disturbances in the family: abuse, incest (Daniel, 2006; Gaudin et al., 1993; Wright et al., 2013), early leaving of the family or loss of close relatives at a young age (Mannarino & Cohen, 2011), or problems in the mother's family history (Barnes et al., 2014; Collishaw et al., 2016). As a whole, it is considered that the main risk factors for the family are loss of employment, poverty, divorce, death, chronic diseases, infertility, and disposition of a family member to addictive behavior. Family vulnerability along with an individual's vulnerability has lately been viewed as a separate risk factor. Risk factors for a family can be found both in the external environment (economic, social, political, cultural, etc.) and in the family itself (illness of a family member, normative family crises, rigidity of the family system, poor communication, pursuit of goals and achievements contradicting the family system, etc.). Exerting impact on the family system, risk factors set requirements to the family's ability to respond to stresses of various origins using defense factors characteristic of this particular family which imply ability and willingness to change, cope, and adjust. For this the family employs its resources (individual and family), as well as the resources from the external environment. These resources in their turn ensure the processes of coping, adaptation, and recovery. We mentioned above that there are three main components that exert substantial influence on family resilience, that is, the system of family values, organizational patterns, and communication processes. Diverging activities intermediated by differences in the understanding of family values, communicative needs, and organizational patterns lead to mutual resistances and a lower level of family resources (potential). If we analyze, for instance, peculiarities of a young family, then its resilience is tested by such socially important processes as mutual adaptation of the spouses' personal qualities, internalization of most challenging parental roles, division of family responsibilities, and assertion of statuses in the family (Dzagkoev, 2000). If the response to a stress factor balances the risk and defense factors, or the latter exceed risk factors, such a family can be considered resilient.

If the activities completely counteract or destroy each other, or mutually disorganize, then we believe the family to be a disorganized system and consider it as a nonresilient one. Counter-organizational activity is a particular case of organizational activity. "If society, classes, or groups clash in a destructive manner, disorganizing each other, they do this exactly because each collective body strives to organize the world and the human race around itself in its own way. This is the result of singularity and detachment of organizing forces, because their unity or common harmonious organization has not been achieved yet. This is a fight of organizational forms" (Bogdanov, 1989, p. 210). The impossibility to coordinate own resources lays a regulatory foundation for the behavior that makes it impossible to tackle relevant tasks. The same results may happen because the excessively close troubled relationship, the absence of personal space, or merger does not allow the spouses to develop new joint strategies of behavior and its regulation in the current situation. We can only assume that the main condition for development—differentiation—does not work in this case, and no integration is possible without this as well as the generation of new forms of behavior (Aleksandrov, 2006; Chuprikova, 2009; Sergienko, 2009). The absence of a clear individual position of the partners on a situation does not allow them to lean on each other and correlate resources. When we look at the family from the child-centered perspective, we see that an excessively strong perception of family cohesion reduces the feeling of own social competence in adolescents. The similarity of values also has a different influence on them depending on whether they share the father's or the mother's values. For instance, the difference in opinion with the mother undermines their social competence. On the other hand, difference in opinion with the father that offers a possibility to challenge the family ties, virtually strengthen social competence of young people. The loss of maternal support may be an extremely powerful threat, whereas the father's behavior triggers independence (Craig & Baucum, 2001, p. 610). Werner (1993) showed that the development of a boy's resilience is under a positive influence of a clear role structure and rules in the family, open exchange of feelings, and a positive model of masculine behavior for identification. As for girls, it is very important to encourage independence in the family and a positive feminine model for identification.

13.5 CONCLUSIONS

The Organizational Theory by Alexander Bogdanov is a general academic concept of an organizational approach to the study of any system in nature and human activity. According to the fundamental assumptions of tektology, the functioning of two and more elements included in a single process may exceed under a certain organization (organizational structure) or, on the contrary, be inferior in efficiency to the functioning of the same elements taken separately. In that theory, any element is characterized by a certain activity and resistance in relation to others or the system. Resistance is, actually, the same activity but taken from a different perspective as set against the other activity. According to A.A. Bogdanov, three options are possible when complexes unite: the activities of complexes coincide; the activities of one complex fully turn into resistances to the other and paralyze it; most often the activities of components partially sum up and partially represent mutual resistance to others, that is, subtract from the organizational point of view.

Also three types of systems can be possibly formed as a result of a continuous interaction: as it can be organized (the whole practically exceeds a simple sum of its parts), neutral (the whole is equal to the sum of its parts), or unorganized, disorganized (activities fully counteract and destroy each other, and get mutually disorganized).

The Organizational Theory by A.A. Bogdanov as a general academic concept of an organizational approach to the study of any system makes it possible to analyze family resilience as its systemic characteristics, and with this we consider the family both as a system and correlation of all its parts (family members). Family resilience is a systemic characteristic of a family capable of responding to stresses of various origins with the employment of family resources inherent to this particular family, implying the family's ability and willingness to cope, change, adjust, and develop.

According to F. Walsh, the system of family values, organizational patterns, and communication processes are identified as three main constituent components that exert considerable impact on family resilience. Consideration of those components as the main family resources and availability of those to all family members, as well ability to use them at all stages of its life cycle play an important role in family resilience. The family as an organized system where the whole practically exceeds a mere sum of its parts through the match of activity complexes is resilient. This happens when the family both operates the already available resources and generates primarily new resources, forming new patterns of family functioning.

In the event the activities partially sum up and partially represent mutual resistance, that is, subtract from the organizational point of view, the level of family resilience shall be determined by the correlation of risk factors and family resources. If the response to a stress factor balances the risk and defense factors, or the latter exceed the risk factors, such a family can be considered as a resilient one. Diverging activities intermediated by differences in the understanding of family values, communicative needs, and organizational patterns lead to mutual resistances and a lower level of family resources (potential). If the activities completely counteract or destroy each other, or get mutually disorganized, we believe the family to be a disorganized system and consider it a nonresilient one. The failure to coordinate own resources lays a regulatory foundation for the behavior that makes it impossible to tackle relevant tasks and counteract stresses.

KEYWORDS

- **family resilience**
- **systemic approach**
- **Bogdanov's general system theory**
- **family resources**
- **systemic analysis of the family**

REFERENCES

Aleksandrov, I. O. (2006). Formation of the structure of individual knowledge. Moscow: Institute of Psychology of the Russian Academy of Sciences Publ. (in Russian).

Balswick, J. O., Balswick, J. K. (1999). The family: A Christian perspective of the contemporary home. Grand Rapids: Baker Books.

Barnes, J., Gardiner, J., Sutcliffe, A., Melhuish, E. (2014). The parenting of preschool children by older mothers in the United Kingdom. *European Journal of Developmental Psychology, 11*, 397–419.

Bogdanov, A. A. (1989). Tektology: General organizational science. Vol. 1. Moscow: Ekonomika (in Russian).

Boss, P. (1987). Family stress. In: M. Sussman & S. Steinmetz (Eds.). *Handbook of marriage and the family* (pp. 695–723). New York: Plenum.

Boyden, J., Mann, G. (2005). Children's risk, resilience, and coping in extreme situations. In: M. Ungar (Ed.). *Handbook for working with children and youth: Pathways to resilience across cultures and contexts* (pp. 3–26). Thousand Oaks, CA: Sage.

Capra, F. (1997). *The web of life: A new scientific understanding of living systems*. Doubleday: Anchor.

Chuprikova, N. I. (2009). The universal differentiation-integration law of development as the basis of an interdisciplinary paradigm theory of development. *Psychological Studies, 1*(3), 1–15. Available at: http://www.psystudy.ru (Accessed 28 May 2018). (in Russian)

Collishaw, S., Hammerton, G., Mahedy, L., Sellers, R., Owen, M. J., Craddock, N., Thapar, A. K., Harold, G. T., Rice, F., Thapar, A. (2016). Mental health resilience in the adolescent off spring of parents with depression: A prospective longitudinal study. *The Lancet Psychiatry, 3*(1), 49–57.

Craig, G. J., Baucum, D. (2001). *Human development*. (9th ed.). Upper Saddle River: Prentice Hall.

Cowan, P., Cowan, C., Schulz, M. (1996). Thinking about risk and resilience in families. In: E. M. Hetherington & E. A. Blechman (Eds.). *Stress, coping, and resiliency in children and families* (pp. 1–38). Mahwah: Lawrence Erlbaum.

Daniel, B. (2006). Operationalizing the concept of resilience in child neglect: Case study research. *Child: Care, Health and Development, 32*(3), 303–309.

Dzagkoev, A. K. (2000). Young family in modern conditions. Author's abstract of PhD dissertation. Moscow: Moscow State University Publ. (in Russian)

Flouri, E., Tzavidis, N., Kallis, C. (2010). Adverse life events, area socioeconomic disadvantage, and psychopathology and resilience in young children: The importance of risk factors' accumulation and protective factors' specificity. *European Child Adolescent Psychiatry, 19*, 535–546.

Gaudin, J. M., Polansky, N. A., Kilpatrick, A. C., Shilton, P. (1993). Loneliness, depression, stress, and social supports in neglectful families. *American Journal of Orthopsychiatry*, *63*(4), 597–605.

Halford, W. K., Sanders, M. R., Behrens, B. C. (2001). Can skills training prevent relationship problems in at-risk couples? Four-year effects of a behavioral relationship education program. *Journal of Family Psychology*, *15*, 750–768.

Helgeson, V. S., Lopez, L. (2010). Social support and growth following adversity. In: J. W. Reich, A. J. Zautra & J. S. Hall (Eds.). *Handbook of adult resilience* (pp. 309–330). New York: The Guilford.

Hobfoll, S. E., Spielberger, C. D. (1992). Family stress integrating theory and measurement. *Journal of Family Psychology*, *6*(2), 99–112.

Kovaleva, Yu. V. (2012). Joint regulation of spouses' behavior at different stages of the family life cycle. *Psikhologicheskij Zhurnal*, *33*(5), 50–70 (in Russian).

Kovaleva Yu. V. (2009). Study of subject regulation of behavior from the standpoint of the psychology of a collective subject. In: A. L. Zhuravlev, V. V. Znakov, Z. I. Ryabikina & E. A. Sergienko (Eds.). *Subjective approach in psychology* (pp. 466–482). Moscow: Institute of Psychology of the Russian Academy of Sciences Publ. (in Russian).

Kryukova, T. L., Saporovskaya, M. V., Kuftyak, E. V. (2005). Family psychology: Life difficulties and coping. Sankt Petersburg: Rech Publ. (in Russian).

Kuftyak, E. V. (2010). *Psychology of family coping*. Kostroma: Kostroma State University Publ. (in Russian).

Laktionova, A. I. (2015). Features of an effective foster family raising an orphan adolescent. In: A. V. Makhnach & K. B. Zuev (Eds.). *Family, marriage and parenthood in modern Russia*. Issue 2. (pp. 225–242). Moscow: Institute of Psychology of the Russian Academy of Sciences Publ. (in Russian).

Laktionova, A. I. (2016). Resilience development of the foster family in the process of its support. *Institute of Psychology of the Russian Academy of Sciences. Organizational Psychology and Psychology of Labor*, *1*(1), 254–271 (in Russian).

Loktionov, M. V. (2016). Alexander Bogdanov as a founder of the General Systems theory. *Philosophy of Science and Technology*, *21*(2), 80–96 (in Russian)

Luthar, S. S., Suchman, N. E. (2000). Relational psychotherapy mothers' group: A developmentally informed intervention for at-risk mothers. *Development and Psychopathology*, *12*, 235–253.

Makhnach, A. V. (2015). Psychopathological symptoms and family resources of candidates for replacement parents. In: A. V. Makhnach & K. B. Zuev. (Eds.). *Family, marriage and parenthood in modern Russia*. Issue 2. (pp. 249–264). Moscow: Institute of Psychology of the Russian Academy of Sciences Publ. (in Russian).

Makhnach, A. V. (2016). Family resilience: a new object and conceptual background to the study. In: A. L. Zhuravlev, E. A. Sergienko, N. V. Tarabrina & N. E. Kharlamenkova (Eds.). *Psychology of everyday and traumatic stress: Threats, consequences, and coping* (pp. 350–375). Moscow: Institute of Psychology of the Russian Academy of Sciences Publ. (in Russian).

Makhnach, A. V., Laktionova, A. I., Postylyakova, Yu. V. (2014). The program of psychological diagnostics of personal and family resources in the practice of selection, training and maintenance of foster parents. In: L. G. Dikaya, A. L. Zhuravlev & M. A. Kholodnaya (Eds.). *Methods of psychological support of professional activity and technology for the development of mental resources* (pp. 166–193). Moscow: Institute of Psychology of the Russian Academy of Sciences Publ. (in Russian).

Makhnach, A. V., Laktionova, A. I., Postylyakova, Yu. V. (2015). The role of the family resources in the assessment of foster parents' applicants. *Psikhologicheskij Zhurnal*, *36*(1), 108–122 (in Russian).

Makhnach, A. V., Postylyakova, Yu. V. (2012). Family resilience: Psychological resources as a protective factor of the family. In: A. L. Zhuravlev & E. A. Sergienko (Eds.). *Psychological problems of modern Russian society* (pp. 529–550). Moscow: Institute of Psychology of the Russian Academy of Sciences Publ. (in Russian).

Makhnach, A. V., Postylyakova, Yu. V. (2013). Model of family resilience. In: A. L. Zhuravlev & E. A. Sergienko (Eds.). *Psychological studies of the problems of modern Russian society* (pp. 438–460). Moscow: Institute of Psychology of the Russian Academy of Sciences Publ.

Mannarino A. P., Cohen J. A. (2011). Traumatic loss in children and adolescents. *Journal of Child & Adolescent Trauma*, *4*(1), 22–33.

Mattessich, R. (1982). The systems approach: Its variety of aspects. *Journal of the American Society for Information Science*, *33*(6), 383–394.

McCubbin, H. I., McCubbin, M. A. (1996). Resiliency in families: A conceptual model of family adjustment and adaptation in response to stress and crisis In: H. I. McCubbin, A. I. Thompson & M. A. McCubbin (Eds.). *Family assessment: Resiliency, coping and adaptation: Inventories for research and practice* (pp. 1–64). Madison: University of Wisconsin.

McCubbin, L. D., McCubbin, H. I. (2005). Culture and ethnic identity in family resilience: Dynamic processes in trauma and transformation of Indigenous people. In: M. Ungar (Ed.). *Handbook for working with children and youth: pathways to resilience across cultures and contexts* (pp. 27–44). Thousand Oaks: Sage.

Minuchin, S., Fishman, H. C. (1981). Family therapy techniques. Harvard: Harvard University Press.

Olson, D. H. (1991). Three-dimensional (3D) circumplex model and revised scoring of FACES III. *Family Process, 30*(1), 74–79.

Osofsky, J. D., Thompson, D. (2000). Adaptive and maladaptive parenting: Perspectives on risk and protective factors. In: J. P. Shonkoff & S. J. Meisels (Eds.). *Handbook of early childhood intervention* (pp. 54–75). Cambridge: Cambridge University Press.

Patterson, J. M. (2002). Integrating family resilience and family stress theory. *Journal of Marriage and Family, 64*(2), 349–360.

Postylyakova, Yu. V. (2015). Individual and family resources of foster parents' applicants. In: A. V. Makhnach, A. M. Prikhozhan & N. N. Tolstykh (Eds.). *The problem of orphanhood in modern Russia: The psychological aspect* (pp. 459–477). Moscow: Institute of Psychology of the Russian Academy of Sciences Publ. (in Russian)

Postylyakova, Yu. V. (2016). Family resources and individual viability of candidates for alternative parents. In; A. V. Makhnach & L. G. Dikaya (Eds.). *Human resilience: individual, professional and social aspects* (pp. 288–298). Moscow: Institute of Psychology of the Russian Academy of Sciences Publ. (in Russian)

Sergienko, E. A. (2009). Principles of differentiation—integration and continuity—discreteness of mental development. *Psychological Studies, 1*(3). Available at: http://www.psystudy.ru (Accessed 10 August 2018) (in Russian)

Steptoe, A., Marmot, M. (2003). Burden of psychosocial adversity and vulnerability in middle age: Associations with biobehavioral risk factors and quality of life. *Psychosomatic Medicine, 65*(6), 1029–1037.

Systemic Family Psychotherapy (2002). In: E. G. Eidemiller (Ed.). Sankt Petersburg: Piter Publ. (in Russian)

Tkachenko, I. V. (2014). Personal developmental resources of the family: Ethnic determination. *Family and Personality: Problems of Interaction, 1*, 173–178.

Varga, A. Y. (2001). Systemic family therapy. Sankt Petersburg: Rech Publ. (in Russian).

Voronina, T. D., Puchkina, Yu. A., Averina, E. A. (2011). Informal and institutional resources of the modern family in the context of the formation of the familial environment in the region: the possibility of using qualitative research methods. *Tomsk State University Bulletin, 1*(13), 32–43 (in Russian).

Walsh, F. (2003). Family resilience: a framework for clinical practice. *Family Process, 42*(1), 1–18.

Walsh, F. (2006). Strengthening family resilience. New York: The Guilford Press.

Werner, E. E. (1993). Risk, resilience, and recovery: Perspectives from the Kauai longitudinal study. *Development and Psychopathology, 5*, 503–515.

Windlea, G., Markland, D. A., Woods, R. T. (2008). Examination of a theoretical model of psychological resilience in older age. *Aging & Mental Health, 12*(3), 285–292.

Worden, M. (2002). Family psychotherapy basics (3rd edition). Cengage Learning.

Wright, E. M., Fagan, A. A., Pinchevsky, G. M. (2013). The effects of exposure to violence and victimization across life domains on adolescent substance use. *Child Abuse & Neglect, 37*(11), 899–909.

Declaration: This research has been made for State task: "Psychological problems of professional mentality in the conditions of organizational and technological innovations", No. 0159-2020-0001.

CHAPTER 14

Resilience and Successful Aging in the Elderly: Exploring the Indian Scenario

SHUBHANGI GOSWAMI and RABINDRA KUMAR PRADHAN

Department of Humanities and Social Sciences, Indian Institute of Technology Kharagpur, West Bengal, India

**Corresponding author. E-mail: alphashubhangi@gmail.com*

ABSTRACT

Most literature in the field of resilience focuses on studying resilience in children and youth. Comparatively, lesser studies emphasize the importance of resilience in old age. This chapter attempts to understand resilience from a developmental perspective and focuses on resilience in the old age as it can form an integral part of successful aging. When we talk about resilience, it is also imperative to discuss about the risks and vulnerabilities which this population faces. Following this discussion, the various approaches to successful aging and factors contributing to resilience among the elderly have been deliberated in this chapter. In the old age, as individuals reach the stage of retirement, they start being reflective of their lives. This phase of human life is just as crucial as any other developmental milestone, as the changes in lifestyle are profound. Changes in vocation, retirement, becoming grandparents, living off savings, losing spouse and/or friends and relatives—all such major happenings in life generally follow the retirement period. In many cases, without proper amenities and facilities, care of the elderly becomes an issue of major concern. At times, older adults depend on a caregiver, especially in Indian setting where a collectivistic social style is followed. Keeping this in view, the Indian scenario of the aging population has been discussed. Modernization, especially in the technological front has changed things rapidly. This chapter also tries to find out how the elderly population has coped with this change. Finally, few of the existing policies and laws which protect the interest of the elderly are discussed along with further ideas which might be implemented for the Indian elderly to help them lead a better life and age successfully.

14.1 INTRODUCTION

14.1.1 RESILIENCE AS A DEVELOPMENTAL PHENOMENON ACROSS LIFESPAN

The interest in the study of resilience has seen a massive increase in recent times. Yet, the literature on resilience lacks the clarity of definition of the term. There are multifaceted views that consider resilience from different perspectives. When one considers resilience as an *individual trait* (Wagnild & Young, 1993; Connor & Davidson, 2003; Ong et al., 2006), it indicates that resilience is a personality trait that helps individuals to cope with the adversity and fight its impact. Resilience is also considered to be a *dynamic process* (Luthar et al., 2000; Fergus & Zimmerman, 2005) which emphasizes on the fact that resilience is rather a process of adaptation to adversity followed by recovery. Another approach to understanding resilience is the *outcome approach* (Masten, 2001; Harvey & Delfabbro, 2004) which states that resilience is the behavioral outcome which helps individuals to cope with adversity. Whatever the definition or orientation to understanding resilience may be, the core idea which unites all these perspectives is, in the most common terms, "bouncing back" from adversities and growing thereafter. With time, experience, and learning, the strategies to cope with the problems might change. Hence, considering resilience as a dynamic process seems safe as it explains how an adverse situation can give rise to better adaptations and development of an individual.

One issue that arises while discussing the definition of resilience is the necessary presence of an adversity—whether resilience solely depends on encountering an adverse situation in life or can it develop otherwise too? In this regard Richardson (2002), in his *metatheory of resilience,* has encompassed both the aspects, that is, he considers resilience as a *quality* which every person possesses as well as a *process* of reintegrating with life disruptions. Resilience can also be considered to be the complex relationship between the risk factors, protective factors, adversity, and the positive behavioral responses (Antonovsky, 1974, 1983; Hayman et al., 2017). With perceptions varying from person to person, the meaning of adversity also varies and takes different forms and intensities for different individuals (Luthar et al., 2000). The nature of adversities and the behavioral response might also differ based on the context; hence resilience can be contextual (Hayman et al., 2017) in nature, with various life stages as important contexts. In this chapter, we focus on one stage of life—the advanced age, which has received less attention in terms of resilience. With vulnerabilities, health issues, and losses markedly increasing with age, advanced age is one of the most important contexts where resilience can play a major role in successful aging.

14.2 AGING

Aging is the process of growing old. Aging is inevitable in the normal course of life and everyone wants to age successfully. It has mostly been defined in terms of biological deterioration. Gerontologist Bernard Strehler (cited in Viña et al., 2007) defines aging in terms of four postulates—aging is *universal* (individuals of all species must go through the process), *intrinsic* (the factors associated with aging must be endogenous or personal), *progressive* (the changes associated with aging must occur throughout the lifespan), and *deleterious* (it will be considered aging only if there is physiological decline). This is also a biological model of aging and suggests that aging is essentially a continuous waning of physiological functioning. (The psychosocial theories are discussed later in Section 14.4). There are many other biological theories that have explained aging as a deterioration of our bodily functioning. In order to increase longevity and slow down the process of aging one needs to slow down the process of this decay and deterioration of biological functioning. Advances in the understanding of genes and other biological phenomena, and adapting to a healthy lifestyle have indeed added to the longevity of cells. The average lifespan of people has increased dramatically. Nevertheless, aging is still inevitable, no matter how much we delay the process.

14.3 RISKS AND VULNERABILITIES IN OLD AGE

Understanding risks and vulnerabilities for any particular age group or category of people gives us the insight into building up of the protective factors. It also greatly aids in policy making. All over the world, the elderly population is exposed to age-related vulnerabilities. The risks/vulnerabilities and domains of well-being go hand-in-hand, considering what constitutes as a risk factor, its contrary must be a determinant of well-being. With this frame in mind, we might discuss the vulnerabilities and the domains of well-being under the same head. As per the "Global Age Watch Index 2013," there can be four domains of well-being for older persons: *financial security, health status, employment and education,* and *enabling environments* (Zaidi, 2013). The vulnerabilities might also be explained using the same domains.

14.3.1 FINANCIAL SECURITY

We generally associate aging with retirement—a period of rest after long work life. But economic security poses a big question when it comes to deciding whether to continue to work or not. All jobs do not provide monetary security throughout life nor does every nation provide social security to its citizens. In terms of economic security, the developed and the developing nations are starkly different. While the developed nations provide substantial social security across lifespan, the developing countries, like India still struggle to provide employment to all. The elderly in the developed nations have economic security comprising basic needs and medical facilities, that is, the state takes care of its citizen even if there is no close relative or caretaker to do so. However, in a collectivistic society like ours, dependence on family and close relatives is very high. With increasing urbanization and nucleation of families, the aging population is finding it difficult to have caretakers and is being neglected. It is unfortunate, yet not very uncommon, to see old people abandoned and resorting to do menial jobs or even begging on the streets in India. People who can afford caretaker facilities for their parents often hire a caretaker or send their parents off to old-age homes. But the scenario is not the same everywhere. Although, many jobs do not provide a pension plan, yet it is encouraged nowadays to save from an early age. Plenty of life insurance and investment plans are available which gives the hope that sustenance in old age is possible without economic support from anyone else. These might sound like an individualistic society's approach while the culture of India calls for taking care of parents in their old age; but such services not only provide financial security, they also ensure the rights of the elderly as individuals. Nevertheless, even with rising consumerism, India has not embraced individualistic culture to the fullest (Alladi, 1994); yet with modernization, increasing urbanization and education, India is seeing a combination of individual and group embedded behavior in recent times (Sinha & Tripathi, 1994). Due to this, the elderly in the urban and educated society strive for self-sustenance in terms of the economy whereas the greater rural India still sees the elderly being monetarily dependent. Elderly women are more dependent in terms of finances, as the female workforce in India is still much lesser than that of the males. Financial security acts as a protective factor when it comes to thriving through adverse times. In general, health care, quality of life, mobility, and many more facilities can be availed if the finances are secure.

14.3.2 HEALTH STATUS

Being the second-most populous country, the percentage of India's older population is also the second largest and together, India and China account for over one-third of the global elderly population (World Population Ageing, United Nations Report, 2017). Today, India is facing an unprecedented burden of noncommunicable diseases like cardiovascular diseases, diabetes, cancers and respiratory illnesses, etc. In the past three decades, these chronic diseases have surpassed the number of perinatal, maternal, nutritional, and infectious diseases (Benziger et al., 2016). Developed countries with strong economies assure health-care services in lieu of mandatory health insurance.

Even though countries like India, Nigeria, Poland, and the Republic of Korea have displayed strong economic performance, yet it has not resulted in the improvement of the status of the well-being of old people. As already mentioned before, income security is largely dependent on pensions. Old people without any pension plan often become helpless. Similar is the case of health status where a lack of health-care services seems very problematic. In terms of life expectancy, health, education and income, India is quite behind the developed countries (Zaidi, 2014).

If we look at the world, we will find an uneven distribution in terms of longevity. According to the US Census Bureau (2015), in many countries life expectancy has increased (80 years or longer in like Japan, Macau, Singapore) while many countries have a life expectancy lower than 60 years, most of these countries being in Africa. But, increased longevity does not assure a healthy life. As life expectancy increases, the susceptibility to diseases and disabilities that come with old age also increases. In developed countries like the United States, mortality due to obesity (Alley et al., 2011; Crimmins et al., 2011) and smoking (Crimmins et al., 2011; Preston et al., 2014; Carter et al., 2015) has increased manifold. This reflects the inactive and stressful lifestyle in these countries leading to obesity, hypertension, and alcoholism, cardiovascular and respiratory diseases. In the case of developing countries, malnutrition is found to be very high.

Disability due to diseases of age-induced deterioration in functioning is common in old age. Along with this debilitating process, the mental health issue is also alarming, especially in developing countries, where society still stigmatizes anyone with a "mental illness." Loss of near and dear ones and abandonment are common in the elderly. When aging people, left with minimal social support, are psychologically still attached to the worldly things, it becomes very painful for them to live and lead an isolated life. In individualistic societies, however, this scenario is expected and the aging persons are prepared for this kind of life in the old age. In collectivist societies, however, the enmeshed relationship patterns and interpersonal dependence make it difficult to live a life in isolation unless one practices disengagement or detachment.

While the matter looks scary, the good news is that with proper knowledge and intervention, things can change. Taking care of health is not something which should start only after deterioration. As many diseases are attributed to lifestyle, an early intervention and adopting a healthy lifestyle can ensure health benefits. One need not wait to be old to take care of health. In fact, regular exercising, abstaining from drugs and alcohol, eating fiber-rich food, and modifying diet with age can ensure slowing of the aging process, consequently leading to meaningful longevity. Along with this, awareness about mental health and following mental hygiene is important. It has also been seen that engaging in physical activity, learning new things, social connectivity, and training the brain actively reduces the risks of Dementia, especially due to Alzheimer's disease (Farrow & Ellis, 2003). It is clearly conclusive that as health risks increase with age, health care forms a major protective factor when it comes to being resilient in the old age.

14.3.3 EMPLOYMENT AND EDUCATION

Aging does not equate with complete disability and deterioration. As long as a person is alive, he/she can be an active part of society, contributing to its development. But, post-retirement, the activity level or involvement of the individual suddenly lowers down due to lack of active engagement. On one hand, in the Eastern countries, the elderly are considered to be an asset and their knowledge and wisdom are respected, but on the other hand, the same people would consider the elderly to be physically useless and restrict their activity. When indoor mobility is restricted, chores are mundane and there is too much rest, it results in physical and mental inactivity. Most aging individuals are well equipped to carry out various activities. Although, every person deserves a break after a long term of service or work, the aging years need not be spent in isolation or inactivity. Nowadays, many people are continuing to work in their post-retirement age. This not only keeps them active but also provides for economic security. Promotion of employment for the elderly paves way for an "ageless society." In rural India where literacy rate is very low, adult education programs are run by the government. This gives them another chance at life. They can understand the pension plans, finances, and understand the world at large. Knowledge has no limits and there is no age to be educated. An educated elderly can develop a wider perspective of life and cope with problems in a broad-minded way.

14.3.4 ENABLING ENVIRONMENTS

The availability of a strong external support system is considered to be a keystone of healthy and resilient development (Masten & Garmezy, 1985; Rutter, 1987). An environment that provides support, understanding, and enables the continuity of development at an old age makes up a protective factor for resilience. An enabling environment would help an old person physically, socially, as well as mentally. A common scenario in India is a typical home where an elderly is restricted to minimal daily chores and spiritual activities. This decreases or restricts their participation in society. For a healthy and positive aging process, the opportunities for participation, health, and security should be continuous throughout life. The degree and type of the mentioned areas surely vary with age but their discontinuation will render a section of society useless. With an increasing percentage of the elderly population, it is imperative to give them an environment promoting healthy aging. Environments where family members live together or visit their elderly members regularly, where medical check-ups, exercise, and modification in diet according to the health status are encouraged, and where free expression, social exchange, education, and learning are promoted may be called enabling environments.

14.4 SUCCESSFUL AGING

It is difficult to streamline the determinants of successful aging. It may be said that "success" or "failure" is a value judgment (Baltes & Baltes, 1990, 2010). Rather than a biological determinism, success is mostly evaluated in social terms. Several models of successful aging have been proposed and refined over time. We have discussed here some significant researches and their relevance in successful aging.

The earliest psychosocial theory of aging is the *theory of disengagement*. This theory by Cumming and Henry (1961), states that aging is a natural and inevitable process in which an aging individual gradually disengages/withdraws from social interactions. Hence, successful aging, in this context, would be the desire and ability to disengage from personal and social relationships. This prepares them to detach from worldly things and prepare for a peaceful death. This theory is claimed to be culture-free, universal, and unidirectional and hence has attracted a lot of criticism as well.

Contrary to the aforementioned, the *activity theory (also known as implicit or lay theory)* proposed by Havighurst (1961) suggests that engagement, rather than disengagement is what determines healthy aging. It posits that as individuals get older, spending time in attaining pursuits and engaging in social and personal relationships lead to optimal aging. Later on, Havighurst (1963) added that life satisfaction and feeling happy about the past and present events in life lead to optimal aging.

Another psychosocial theory of aging is the *continuity theory*. As the name suggests, this theory posits that aging individuals attempt to continue and preserve behaviors which they did in their early and middle ages. This provides a continuity of psychological and social life, thus making continuity of behavior an adaptive strategy (Atchley, 1989).

Rowe and Kahn (1987, 1997), defined successful aging to be standing on three pillars, namely, *minimizing risk and disability, maximizing physical and mental abilities, and engagement in active life*. This model was set for intervention studies as it includes the biological as well as psychosocial aspects of aging. But something was still missing in the model until Crowther et al. (2002) added a forgotten factor to these three factors—*positive spirituality*. Positive spirituality is neither an individualistic concept like spirituality, nor a community-focused concept like religion; it is rather an amalgamation of the two and focuses on measurable positive outcomes (Crowther et al., 2002).

The aforementioned models of aging can be called criterion-based models. The recent shift in interest in the field of gerontology focuses on process orientation. In this regard, the Selective Optimization with Compensation model (SOC-model) developed by Baltes and Baltes (1990), is one of the most popular and well-accepted models of successful aging. According to the SOC model, successful aging comprises:

a) *selection* of functional domains on which to focus one's resources,
b) *optimizing* developmental potential (maximization of gains), and
c) *compensating* for losses—thus ensuring the maintenance of functioning and a minimization of losses.

This theory reinforces the importance of starting from an early age in order to enjoy successful aging. The process of selection and optimization starts from a young age. Focusing and selectively attending to the more important and functional areas of one's life gives a person a goal. Life is meaningless without a purpose or goal. In old age too, a goal or purpose is what keeps a person going. Selection can take place in two ways—one where the individual actively selects (elective selection) and another, where the individual is left with no choice (loss-based selection). The latter is common in old age where an elderly comes across several losses, be it illness, financial losses, and loss of a loved one—especially spouse which is a strong predictor of life satisfaction (Whisman et al., 2006). Accepting these losses and focusing on the "glass half full" leads to the optimization of the resources available. Loss of goal or purpose curbs the development of an individual. Therefore, even when a person is moving toward a declining pathway, a purpose makes this journey worthwhile. He/she will redirect all the resources and use them optimally to attain this goal, which in turn, promotes life satisfaction in the elderly (Baltes & Cartensen, 1996). Loss-based selection and acceptance of the loss lead to compensation. This behavior can be truly said to be the "growth" in the face of adversities. Left with no choice but to accept the losses, focus shifts from what is present rather than what is lost. This helps maintain positive aging.

14.5 FACTORS CONTRIBUTING TO RESILIENCE IN THE ELDERLY

As aging individuals keep on filling up their repertoire of experiences, they also learn to negotiate with and handle adversities better. When it comes to the elderly population, various factors determine their resilience. As discussed in Section 14.3, health, financial security, employment and education, and enabling environments can serve as important factors for resilience. Also, despite deteriorating cognitive or physical abilities, an elderly might be socially resilient. Another factor that comes into play while determining resilience is whether we are talking about it in terms of personal (micro level)—like self-efficacy, good health; or social (meso level)—like family, neighborhood and community or global—(macro level). The interplay among these different levels can be complementary or even cause hindrance in an individual's resilience. For example, an old person with deteriorating physical conditions (micro level) might have a caring family and an excellent neighborhood (meso level) and also the policies for elderly care might be facilitative of physical and mental health care (macro level). Among the various types of classification of determinants of resilience, the simplest classification may be "internal" and "external" determinants of resilience—internal factors are being attributed to an individual and external being attributed to other than the individual. The protective factors among the elderly are discussed below.

14.5.1 PROTECTIVE FACTORS OF RESILIENCE AMONG ELDERLY

Plenty of studies (Hatch, 2005; Lin & Peek, 1999, MacLeod et al., 2016, McClain et al., 2018) are indicative of the various factors associated with resilience in older adults. A summary of these factors is given below. Individual characteristics which might work as protective factors of resilience among elderly are:

a) *Adaptation to and readiness for change/plasticity*—The very definition of resilience calls for some form of flexibility to adapt to changes. Along with that flexibility, preparedness for the change is an added advantage. Accordingly, the resilient elderly prepare for subsequent changes in their life and modify their physical and psychological framework to suit the inevitable and imminent changes. While in children, plasticity involves maturation and learning, in the elderly plasticity involves being flexible in the face of stressors (Baltes, 1997; Fontes & Neri, 2015).
b) *Endurance capacity*—Increased endurance results in the potential to deal with stressors. It gives a person the stamina to go on. Stamina acts as an important buffer for dealing with stress and adversity (Garmezy, 1993, Luthar & Cichheti, 2000).
c) *Positive traits of gratitude*—Having a tendency to be thankful for what one has and expressing feelings of appreciation for something/someone is called gratitude. Although expressing gratitude is inculcated in us since childhood, still adults forget to feel "thankfulness," being busy and occupied in the day-to-day lives. Thankfulness for merely what exists and for little things in life helps focus one's attention to the positivity around us. Gratitude has been positively related to resilience and well-being in the elderly (Rawat & Gopal Jee, 2018; Killen & Macaskill, 2014).
d) *Having a sense of purpose*—Sense of purpose guides our decision making, goal setting and helps achieve a person the meaning of life. A life of fulfillment requires one to have a sense of purpose and achieve eudaimonic happiness. Sense of purpose is associated with a sense of coherence (Nygren et al., 2005), successful aging (Bothra & Dasgupta, 2011) and better performance in life as a whole (Pradhan et al., 2017). Strong sense of purpose has also been associated with health benefits such as low cardiovascular risk, glycemic control, and adequate levels of cholesterol and cortisol (Ryff et al., 2012).
e) *Self-efficacy*—Self-efficacy is one's belief in one's ability to succeed in a situation or accomplish a task. A plethora of research shows that self-efficacy is an integral part of building resilience. At every stage of life, one's belief drives an individual. If the person believes that he/she is capable to overcome adversities, start afresh, and cope with difficulties, it makes the path ahead easier (Windle et al., 2008; Ryff et al., 2012).
f) *Optimism*—Optimism or seeing the positive side of things is one of the psychological capitals which can help the elderly population overcome adverse situations and carry on with the journey of life (Ferguson & Goodwin, 2010; Vahia et al., 2011; Emlet et al., 2011; Rosado-Medina et al., 2012).
g) *Healthy lifestyle including balanced diet and exercise*—With ample motivation, a practical approach toward life and a will to be healthy, the elderly population which focuses on their health and being healthy tend to be more resilient than others. Balanced diet and exercises modified as per age and health status are essential parameters of well-being (Depp & Jeste, 2006; Wells, 2009; Grover, 2016).

h) *Spiritual practice*—Spiritual practice has been found to be positively correlated to resilience in the elderly (Vahia et al., 2011). It will be discussed in detail in forthcoming section.
i) *Being reflective about life: Life review*—Life review might be understood as an intersection of past and future. One reflects on the experiences of the past and iteration of the stories brings about wisdom. Life review also brings with it hope and ambition for the future (Merriam et al., 1995). Reiterating stories and experiences can be helpful for the guidance of the younger generation.
j) *Resistance to ageist norms*—Ageism refers to stereotyping and discriminating against the elderly. Historically, it has been observed that with changing times people who have resisted the ageist norms and considered age to be "just a number" have flourished (Marshall, 2015). Today we have elderly people in many professions like fashion modeling, wildlife photography, music, research, and so on. They are all well past their retirement age but are working perfectly fine and thriving, thus breaking the ageist stereotype.
k) *Humor*—Humor is an excellent coping resource. Researches show that using humor as a coping strategy and humorous communication can positively impact life satisfaction among older adults (Celso et al., 2003; Westberg, 2003; Wanzer et al., 2009).

Collective/social characteristics of resilience are:

a) *Well-designed homes*—Well-designed homes make it convenient for elderly people to navigate. Especially in the case of people with cognitive disabilities, a well-designed home that is accessible and has the amenities easily reachable can be very helpful in combating physical difficulties. A home is a place where a person is most comfortable and at ease. Hence, homes should be well lit, aerated, and well-designed, keeping the needs of the elderly in mind. Aging in familiar place rather than in residential care helps the elderly to participate in the community too (James & Saville-Smith, 2018).
b) *Accessibility to public amenities*—In order to keep the elderly people mobile and yet not put undue pressure on them, accessibility of day-to-day amenities is very important. In rural areas, a simple grocery shop is far away from residences. This makes accessibility to amenities difficult. Accessibility to public amenities, resting places like benches, and sidewalks also ensure adequate mobility and assurance (Turell et al., 2013).
c) *Safe neighborhood*—This is one of the most important areas of concern when it comes to the elderly population. More than a proactive neighborhood, a safe neighborhood is a must. Residential areas where elderly people (alone or couples) live by themselves are prone to theft and other crimes. A neighborhood watch can help achieve this purpose. Neighborhood characteristics (maintenance of the place, cohesion among the residents, community feeling, etc.) have been directly related to the process of resilience among the elderly (Bowling & Strafford, 2007; Fitzpatrick & LaGory, 2003; Ziersch et al., 2005; Clark & Gallagher, 2013; Turell et al., 2013)

d) *Strong social support system—supportive and caring family and friends*—Apart from a safe neighborhood, supportive family and friends probably serve as the greatest buffer to the elderly. This replenishes their emotional, informational, and social needs. With depleting number of contemporaries, the older adults gradually detach themselves and lead a solitary life but their life lights up when they are surrounded by family, especially children. Social support acts as a buffer and many studies have corroborated the fact that strong social support is positively correlated with well-being (Windsor et al., 2016; Schumm et al., 2006; Orthner et al., 2004; Walsh, 2003). It has also been associated with lower mortality rates (Brown et al., 2003) and a slower rate of cognitive decline (James et al., 2011; Ertel et al., 2008; Fratiglioni et al., 2000).

e) *Social services and policies*—Social services and policies have been discussed in the subsequent section. They form an integral part of care and support. If there is no one immediately to look after the elderly, it gives hope that the government is there and that it gives them the due importance. It allows them to live a life of dignity (Windle & Bennett, 2011; Windsor et al., 2016). Some policies for the elderly in the Indian scenario will be discussed in the forthcoming section.

f) *Inclusivity*—Finally, if society understands that the elderly population is as much a part of the society as are the children and the youth, it is a healthy society. Older members of society can be as productive as any other individual. When they are included in social events, they feel more productive, dignified, and can give back to the society a lot of wisdom, guidance, and even active service. As already mentioned, ageist norms often isolate the elderly population (Marshall, 2015) and defying these norms by both the old and the young can ensure inclusivity, resulting in a stronger social network and productivity (Windsor et al., 2016).

Many resilience frameworks have been suggested by researchers which give us a model to be implemented in societies, such as the strengths-based approach, focusing on positive attributes that contribute to an outcome (Zimmerman, 2013). A noteworthy ecological framework of resilience suggests that the various challenges across life-span are the antecedents of resilience while mental well-being and further challenges are consequences of it. The interesting part is the enmeshed structure of individual, society, and community, on which an individual depends during times of adversity, in order to be resilient. With respect to challenges across the lifespan, resilience comes from resources pertaining to the individual (like psychological strength, biological predispositions, gender, age, food habits, exercise, as well as material resources), pertaining to the community (like social support and strong network which can act as a buffer), and pertaining to social structure (which include health-care services, social services, policies, neighborhood, and economic conditions). It seems to be a comprehensive and useful framework. However, its cross-cultural validity needs to be empirically tested (Windle & Bennett, 2012).

14.6 SUCCESSFUL AGING AND RESILIENCE: THE INDIAN PERSPECTIVE

Let us rewind a little bit and go back in time, focus on the attitude toward aging in the Eastern context, especially the Indian scenario. We will find that here, older people have been considered to be an asset and not a liability. Yet, it is sad to see the number of old-age homes and old parents abandoned by their children today. From the perspective of developmental psychology, the Indian system of *varnashrama* divides our lives into four stages of equal span or *ashramas* namely, *brahmacharya* (celibate studentship); *grahastha* (household stage—preparation and fulfillment of a relational life); *vanaprastha* (retiring forest dweller stage where spiritual development takes place); and *sanyasa* (renunciation—loosening the bond and cremating all one's past and present relations). The last stage of life requires preparation—renunciation is preceded by a retired life (*vanaprastha*) wherein most of the time is focused on the spiritual enhancement and thriving on minimal amenities. In the retirement stage, an individual generally looks back in life to analyze and evaluate it. A satisfactory reconciliation with the past might lead to the renunciation of the worldly things with more ease. It helps in accepting the end of life gracefully. Talking about stages of life, Erikson (1950, 1993) proposed that life can be divided into eight stages where every stage is characterized by a crisis, a virtue point, and the quest to successfully overcoming the crisis. For example, in the context of old age, stages seven and eight are relatable. Although stage seven, which is *Generativity versus Stagnation*, mostly covers the 30s, 40s, and 50s of a person, it can also extend beyond that age. The virtue point at this stage is "care and production." The main theme is the feeling of connectedness to the next generations and to give them whatever an individual has learned or gained through life. The next stage, stage eight is characterized by Integrity versus Despair where the virtue point is "wisdom and renunciation." This final stage of life roughly relates to the "sanyasa" ashram of the Indian philosophy. This stage is characterized by total social renunciation. Thus, the varnashrama system seems to be a prescribed pathway for successful aging and roughly corresponds to the disengagement theory mentioned before. Roughly, because while the disengagement theory harps on the capability to prepare for loosening ties with the world and relationships, the sanyasa stage emphasizes on taking up a new role and provides with a goal—the spiritual goal. It means that while one disengages from one role, he/she is ready to enter a new role. Hence, there lies a continuity of behavior and purpose in life in this ancient Hindu system of "differential disengagement" (Hochschild, 1975). With changing times and modernization, there hardly seems to be a strict follow up of this system. It is not a common phenomenon today for the elderly to leave their homes, become hermits, and ultimately attain sanyasa. Instead, the total cessation of social activity is not a strict demand for spiritual growth. A certain degree of retreat in social and psychological terms is generally seen in old age but it can take place in the midst of society (Vatuk, 1980). Somewhere, this disengagement aims at continuity—the continuity of life. As one starts disengaging or losing attachment with the worldly and material possessions, it becomes easier for the person to pass it on to the next generation. From worldly possession to power, everything is gradually prescribed to be renounced step by step in order to finally meet one's destination of life. Probably, renunciation and disengagement helps to cope with worldly loss and make one more resilient to rapid changes and impending adversities of old age.

14.7 SPIRITUALITY AS A SOURCE OF RESILIENCE IN OLD AGE

Spirituality is often deeply connected with religion. These terms are not interchangeable as the former relates to rising above the material world while the latter defines a prescribed and largely accepted way of life. Nevertheless, these two are deeply connected as often religion becomes a way to attain spirituality. Every religious teaching aims to uncover the spiritual side of a person. During times of distress, most people "turn to God" to show them the way. According to Pargament (1997), "Religion offers a response to the problem of human insufficiency" and can complement nonreligious coping through offering solutions to "the limits of personal powers. The solutions may come in the form of spiritual support when other forms of social support are lacking, explanations when no other explanations seem convincing, and a sense of ultimate control through the sacred when life seems out of control, or new objects of significance when old ones are no longer compelling."

With increasing losses and having experienced traumas, wars, deaths of loved ones, and many such hardships, the elderly often turn to religion and spirituality and its importance increase with age. This indeed is a type of coping response that makes the elderly resilient to losses and enhances well-being (Koenig, 1995). Such coping can be termed as positive religious coping where a person's relationship with the divine is most secure, where a person can surrender without fear and derive meaning out of the most stressful situations (Pargament, 1997).

The Hindu religious philosophy looks at life and death as a cyclic process. It says that the body is merely like a cloth the soul wears. The soul never dies and simply moves on to another body, until this cycle is broken and *Moksha* or ultimate spiritual freedom is attained. Engaging in spiritual activities in the old age seems to make this process easier and accept death as a part of life's journey—the last stoppage. Very commonly, people in their old age in India are found to go for religious journeys or pilgrimages, whether be the four holy "dhams" for Hindus or the Hajj—pilgrimage to Mecca by the Muslims. They are considered to be pious and free people from their sins of this life. The Indian city of Varanasi is perhaps, one of the most interesting places to visit to witness the impact of religion and spirituality among the people of India. Located on the banks of the Ganges, the several ghats (banks) in the city are marked by religious offerings and burning pyres. From marriages to deaths, anything happening in Varanasi is considered to be holy. There are old age homes in this place where people come and wait to die. The Mukti Dham is one such place which is considered to be the final halt for the people bound to heaven. It has free entry and people of course, wait for their exit to the heavenly abode, having breathed their last on the holy land of Varanasi. Despite minimal facilities for survival, the head count at Mukti Dham is more than 14,000 (Mishra, 2009). The government also is under immense pressure to organize amenities for these people and to arrange for a safe, secure journey and stay for the pilgrimages such as Hajj and Amarnath Yatra. This is not a mere reflection of the strength of the religious belief of people in India but also an ever-increasing pressure to cater to the needs of the elderly. With changing times and increasing modernization, a stark change in the outlook toward living a life in the old age is seen. Although today's urban elderly are more active in society and in touch with the global development, yet the most of rural India still heavily relies on religion and spirituality as one of the biggest sources of support and positive coping.

14.8 AGING AND MODERNIZATION: BEING RESILIENT IN THE TECHNOLOGICAL ERA

With modernization, the idea of aging has transformed from a *natural order* (Callahan & Topinkova, 1998; Hayflick, 2007) to a *problem* (Cole & Winkler, 1994) and a *disease* (Bulterijs et al., 2015) which needs to be solved by medicine and science. With the rapid nucleation of families, the social order is changing and families have become less accommodative nowadays. With even more rapid technological developments, the elderly population might be at greater pressure for adaptability. Human generations and technological generations do not grow at the same pace, with the latter growing at a much faster pace. Adapting to these changes and learning to use these technologies quickly is relatively much easier for the younger generation than the old individuals. Yet, meeting the grandchildren on Skype who live miles away, or filing tax return on the internet or using a smartphone and many more such things are being learned by today's aging generation to keep pace with the demands of the society. India, as a whole, is facing challenges in meeting the demands of the changing demographics. With advanced medical services and technology, the number of people living past the age of 60 years has considerably increased over the past few decades and 60 is definitely not the average lifespan anymore. With a greater number of aged population in our country, the pressure on economic, social, and medical health-care systems has also increased. In order to cater to the needs of this stratum of society, we are still not as well equipped as we need to be. Despite the growth and development in the economy and health sector, the elderly are still marginalized in our country. In this line, there are already several policies worldwide and in India, but they need to be implemented on a large scale to provide the elderly with a graceful old age.

14.9 POLICIES: RETIREMENT, WORLD VIEW, INDIAN POLICIES

Taking into consideration the risks and vulnerabilities and the trajectory of life, Schröder-Butterfill and Marianti (2006) provide a framework for an empirical examination of "who is vulnerable" and for what reasons, and how different policy interventions can be put forward for an impact in these different domains. Schröder-Butterfill (2011) discussed in further detail four domains—exposure, threat, coping capacity, and bad outcomes in the context of old-age vulnerabilities. They suggest that exposure to threat makes an individual utilize his/coping capacities which leads to positive or negative outcomes. In order to ensure good outcomes or at least reduce bad outcomes, policy interventions at the three stages might be helpful. Firstly, reducing susceptibility at the level of exposure, followed by reducing the magnitude of threats at the level of exposure to threats and finally strengthening resilience at the time of coping can greatly impact the outcome in a favorable way.

The type of social support available to elderly can be further broken down into "social networks" and "formal support." Social network refers to the close circle of the elderly like the family, relatives, friends, and neighborhood—an informal network that they have formed. This informal group acts as a very important source of support and defenses against threats and challenges in old age (Ngan, 2011). With everyone engaged in their own worlds, providing support to the elderly in the family becomes an obligation which often burdens an individual, sometimes to the extent that it reduces their capacity to support themselves (Evandrou & Falkingham, 2004; Lloyd-Sherlock & Locke, 2008). For these reasons, Schröder-Butterfill (2011) stated that the availability and reach of formal support is important. Formal support comprises the various agencies and groups (both government and nongovernment organizations) which take care of the elderly, such as old age homes. The policies made by the government also take into consideration the various threats and challenges faced by the elderly and policies are designed accordingly. The challenging part is the implementation of the policies as the awareness regarding the same is often very less. Especially in the rural areas, people are not aware of the securities and convenience that the government provides them and hence they are unable to taste the benefits of these policies. Nevertheless, communicating this information through mass media and local methods is also a part of the strategies that the government takes up in order to make the policymaking a success. Knowing that the government is there to fall back on surely adds to the resilience of any citizen of the country.

The Ministry of Social Justice and Empowerment of the Indian Government has laid down plenty of policies and laws which protect the interest of the elderly in India. Discussing all of them is beyond the scope of this chapter. However, a few have been discussed below to throw some light on the areas which are taken care of by the government.

Unlike the Western countries, in India, social security system is very basic and caters to a smaller percentage of workforces. Also, central and state government pension funds indicated to change their defined benefit plan into a defined contribution plan. This shift paved the way to retirement planning, thus, people should decide how to allocate their savings to obtain the desired retirement state. In the budget of February 2018, exemption of interest income on deposits with banks and post offices were proposed to be increased from Rs. 10,000 to Rs. 50,000. A hike in deduction limit for the health insurance premium and/or medical expenditure from Rs. 30,000 to Rs. 50,000 was proposed. Another proposal was an increase in the deduction limit for medical expenditure for certain critical illnesses from Rs. 60,000 (in case of senior citizens) and from Rs. 80,000 (in case of very senior citizens) to Rs. 1 lakh for all senior citizens. Concessions will give extra tax benefit of Rs. 4000 crore to senior citizens. It is also proposed to extend the Pradhan Mantri Vaya Vandana Yojana up to March, 2020. The current investment limit is also proposed to be increased to Rs. 15 lakh from the existing limit of Rs. 7.5 lakh per senior citizen.

Keeping in view the recommendations made in the "National Policy on Older Persons" as well as the State's obligation under the "Maintenance and Welfare of Parents and Senior Citizens Act, 2007," the Ministry of Health and Family Welfare launched the "National Programme for the Health Care of Elderly" during the year 2010, in the 11th Plan period, to address various health-related problems of elderly people. The program has been initiated in 100 selected districts in 21 States. It is proposed to cover 225 more districts during the 12th Five-Year Plan in a phased manner and develop 12 additional Regional Geriatric Centers in selected Medical Colleges of the country. The main aims of this program are to focus on:

(a) Preventive and promotive care
(b) Management of illness
(c) Health man power development for geriatric services
(d) Medical rehabilitation and therapeutic intervention

Information, education, and communication—health education programs, mass-media education, etc.

14.10 SUMMARY

This chapter was an attempt to cover a rather huge area related to aging and resilience. The individual topics covered can be further elaborated. Nevertheless, to summarize, it can be said that we understood the concept of aging, delved into the various difficulties faced by the elderly population, and subsequently covered the two main areas of concern—successful aging and factors contributing to resilience in the elderly. While there are many theories of successful aging, the psychosocial theory by Rowe and Kahn and the SOC model by Baltes stand out. The factors contributing to resilience can be majorly categorized as internal factors (which pertain to the individual like self-efficacy, level of activity, lifestyle, coping strategy, spirituality, engagement, and more), and external factors (which pertain to the social and economic aspects like community, neighborhood, finances, policies, health care, design of homes, and more). Spirituality as a contributor of resilience was discussed. Spirituality has an added advantage of increasing social network too. Hence, it does not necessarily isolate a person from worldly ties; rather, religious and spiritual groups form a remarkable part of our society. Technological advancement as a challenge to the elderly was also discussed. The Indian perspective regarding aging was discussed in the light of the Vedic period and collectivistic society. It suggests that the Indian elderly population, when compared to the world population of the elderly is more dependent in terms of finances and emotional security, both. Nevertheless, some policies and government aid give us hope. Some of these policies were discussed in the last section which elaborates the various schemes for the elderly especially in the area of health care and pension plans.

14.11 CONCLUSION

Having discussed various aspects of the elderly life, the challenges and vulnerabilities of old age, and the things that make them bounce back after adversities, it can still be said that there are many more aspects of the old age not covered in the current chapter. A child looks forward to becoming an adult, an adult looks forward to the middle ages, and a middle-aged will plan for his/her old age. All these age groups are available in society to share their experiences and prepare other generations to face the same, equipping them in a better way. But an old person has no further stage in life to look forward to. Thus, an immense longing for "life" often creates a denial and makes the remaining period of life painful. How a person leads his/her life greatly affects the outlook in old age. Looking upon aging as a period of disease, disability, and declining cognitive and physical functions is what makes aging so scary. Hence, it becomes more important to emphasize on a life of dignity in the old age when an individual gradually goes down the plateau of physical and cognitive decline. A person well-prepared and having a positive attitude toward this inevitable stage of life shall be able to cope much better than a person full of regrets and worldly attachments. No matter how much society promotes being "young and beautiful," growing old has its own charm and aging gracefully is beautiful.

KEYWORDS

- **successful aging**
- **resilience**
- **old age**
- **Indian elderly**

REFERENCES

Alladi, V. (1994). India's changing consumer economy: A cultural perspective. In C. T. Allen & D. R. John (Eds), *NA—Advances in Consumer Research* (Vol. 21, pp. 323–328) Provo, UT: Association for Consumer Research.

Alley, D. E., Lloyd, J., & Shardell, M. (2011). Can obesity account for cross-national differences in life-expectancy trends? In E. M. Crimmins, S. H. Preston, and B. Cohen (Eds.), *International differences in mortality at older ages: Dimensions and sources.* Washington, DC: The National Academies Press.

Antonovsky, A. (1974). Conceptual and methodological problems in the study of resistance resources and stressful life events. In B. S. Dohrenwend & B. P. Dohrenwend (Eds.), *Stressful life events: Their nature and effects* (pp. 245–258). New York, NY: Wiley.

Antonovsky, A. (1983). The sense of coherence: Development of a research instrument. *Newsletter and Research Reports, 1,* 1–11

Atchley, R. (1989). A continuity theory of normal aging. *The Gerontologist, 29,* 183–190.

Baltes, P. B. (1997). On the incomplete architecture of human ontogeny: Selection, optimization, and compensation as a foundation of developmental theory. *American Psychologist 52,* 366–380.

Baltes, P., & Baltes, M. (Eds.). (1990). *Successful aging: Perspectives from the behavioral sciences.* New York: Cambridge University Press.

Baltes, P. and Baltes, M. (2010). *Successful aging.* Cambridge: Cambridge University Press.

Baltes, M., & Carstensen, L. L. (1996). The process of successful aging. *Aging and Society, 16,* 397–422.

Benziger, C. P., Roth, G. A. & Moran, A. E. (2016). The Global Burden of Disease study and the preventable burden of NCD. *Global Heart, 11*(4), 393–397.

Bothra, N., & Dasgupta, M. (2011). A comparative psychosocial study of selected elderly groups in the canvas of successful aging. *The Indian Journal of Gerontology, 21,* 1–24.

Bowling, A., & Stafford, M. (2007). How do objective and subjective assessments of neighborhood influence social and physical functioning in older age? Findings from a British survey of ageing. *Social Science & Medicine, 64*(12), 2533–2549.

Brown, S. L., Nesse, R. M., Vinokur, A. D., & Smith, D. M. (2003). Providing social support may be more beneficial than receiving it results from a prospective study of mortality. *Psychological Science, 14*(4), 320–327.

Bulterijs, S., Hull, R. S., Björk, V. C. E., & Roy, A. G. (2015). It is time to classify biological aging as a disease. *Frontiers in Genetics, 6,* 205.

Callahan D., Topinkova E. (1998). Is aging a preventable or curable disease. *Drugs & Aging 13,* 93–97.

Carter, B. D., Abnet, C. C., Feskanich, D., Freedman, N. D., Hartge, P., Lewis, C. E., Ockene, J. K., Prentice, R. L., Speizer, F. E., Thun, M. J. & Jacobs, E. J. (2015). Smoking and mortality-beyond established causes. *New England Journal of Medicine 372*(7), 631–640.

Celso, B. G., Ebener, D. J. & Burkhead, E. J. (2003). Humor coping, health status, and life satisfaction among older adults residing in assisted living facilities. *Aging & Mental Health, 7,* 438–445.

Clarke, P. & Gallagher, N.A. (2013). Optimizing mobility in later life: The role of the urban built environment for older adults aging in place. *Journal of Urban Health, 90*(6), 997– 1009.

Cole, T. R. & Winkler, M. G. (Eds.) (1994). *The Oxford book of aging: Reflections on the journey of life*, Oxford University Press, New York.

Connor, K. M., & Davidson, J. R. T. (2003). Development of a new resilience scale: The Connor–Davidson resilience scale (CD-RISC). *Depression and Anxiety, 18,* 76–82.

Crimmins, E. M., Preston, S. H. & Cohen, B (Eds.). (2011). *Explaining divergent levels of longevity in high-income countries.* National Research Council Panel on Understanding Divergent Trends in Longevity in High Income Countries. Committee on Population, Division of Behavioral and Social Sciences and Education. Washington, DC: The National Academies Press.

Crowther, M. R., Parker, M. W. & Achenbaum, W. A., Larimore, W. L. & Koenig, H. G. (2002). Rowe and Kahn's model of successful aging revisited positive spirituality—The forgotten factor. *The Gerontologist, 42,* 613–620.

Cumming, E., & Henry, W. E. (1961). *Growing old: The process of disengagement*. New York: Basic Books.

Depp, C. A, & Jeste, D. V. (2006). Definitions and predictors of successful aging: A comprehensive review of larger quantitative studies. *American Journal of Geriatric Psychiatry, 14,* 6–20.

Donnellan C. & O'Neill D. (2013). Baltes' SOC model of successful ageing as a potential framework for stroke rehabilitation. *Disability and Rehabilitation, 36*(5), 424–429.

Emlet, C. A., Tozay, S., & Raveis, V. H. (2011). "I'm not going to die from the AIDS": resilience in aging with HIV disease. *Gerontologist, 51*(1), 101–111.

Erikson, E. H. (1993) [1950]. *Childhood and society*. New York, NY: W.W. Norton & Company.

Ertel, K. A., Glymour, M. M., & Berkman, L. F. (2008). Effects of social integration on preserving memory function in a nationally representative US elderly population. *American Journal of Public Health, 98*(7), 1215.

Evandrou, M., & Falkingham, J. (2004). Demographic change in Europe: Implications for future family support for older people. In P. Kreager & E. Schröder-Butterfill (Eds.) *Ageing without children: European and Asian perspectives* (pp 176–197) Oxford: Berghahn.

Farrow M & Ellis K. (2013). *Physical activity for brain health and fighting dementia. Alzheimer's Australia Paper.* Canberra: Alzheimer's Australia.

Fergus, S., & Zimmerman, M. A. (2005). Adolescent resilience: A framework for understanding healthy development in the face of risk. *Annual Review of Public Health, 6,* 399–419.

Ferguson, S. J. & Goodwin, A. D. (2010). Optimism and well-being in older adults: The mediating role of social support and perceived control. *International Journal of Aging and Human Development, 71*(1), 43–68

Fitzpatrick, K. M., & LaGory, M. (2003)."Placing" health in an urban sociology: cities as mosaics of risk and protection. *City & Community, 2*(1), 33–46.

Fontes, A. P., & Neri, A. L. (2015). Resilience in aging: literature review. *Ciencia and Saude Coletiva, 20,* 1475–1495.

Fratiglioni, L., Wang, H. X., Ericsson, K., Maytan, M., & Winblad, B. (2000). Influence of social network on occurrence of dementia: A community-based longitudinal study. *The Lancet, 355*(9212), 1315–1319.

Garmezy, N. (1993). *Vulnerability and resilience.* In D. C. Funder, R. D. Parke, C. Tomlinson-Keasey, & K. Widaman (Eds.), *APA Science Vols. Studying lives through time: Personality and development* (pp. 377–398). American Psychological Association.

Grover S. (2016). Successful aging. *Journal of Geriatric Mental Health, 16*(3), 87–90.

Harvey, J., & Delfabbro, P. H. (2004). Psychological resilience in disadvantaged youth: A critical overview. *Australian Psychologist, 39,* 3–13.

Hatch, S. L. (2005). Conceptualizing and identifying cumulative adversity and protective resources: Implications for understanding health inequalities. *Journals of Gerontology: Series B:, 60B*, 130–134.

Havighurst, R. J. (1961). Successful aging. *The Gerontologist, 1,* 8–13.

Havighurst, R. J. (1963). Successful aging. In R. H. Williams, C. Tibbitts, & W. Donahue (Eds.), *Processes of aging* (pp. 299–320). New York: Atherton Press.

Hayflick, L. (2002). Anarchy in gerontological terminology. *Gerontologist 42*, 416–421

Hayman, K. J., Kerse, N., & Consedine, N. S. (2017). Resilience in context: The special case of advanced age. *Aging & Mental Health, 21*(6), 577–585

Hochschild, A. R. (1975). Disengagement theory: A critique and proposal. *American Sociological Review 40* (5), 553–569.

James, B. V. & Saville-Smith, K. (2018). Designing housing decision-support tools for resilient older people. *Architectural Science Review, 61*(5), 305–312.

James, B. D., Wilson, R. S., Barnes, L. L., & Bennett, D. A. (2011). Late-life social activity and cognitive decline in old age. *Journal of the International Neuropsychological Society, 17*(6), 998–1005.

Killen, A., & Macaskill, A. (2015). Using a gratitude intervention to enhance well-being in older adults. *Journal of Happiness Studies: An Interdisciplinary Forum on Subjective Well-Being, 16*(4), 947–964.

Koenig, H. G. (1995). *Research on religion and aging: An annotated bibliography.* Westport, CT: Greenwood Press.

Lin, N., & Peek, M. K. (1999). Social networks and mental health. In A. V. Horwitz & T. L. Scheid (Eds.), *A handbook for the study of mental health: Social contexts, theories, and systems* (pp. 241–258). New York: Cambridge University Press.

Lloyd-Sherlock, P., & C. Locke (2008). Vulnerable relations: lifecourse, wellbeing and social exclusion in Buenos Aires, Argentina. *Ageing and Society, 28*(6), 779–803.

Luthar, S. S., Cicchetti, D., & Becker, B. (2000). Research on resilience: Response to commentaries. *Child Development, 71,* 573–575.

MacLeod, S., Musich, S., Hawkins, K., Alsgaard, K., & Wicker, E.R. (2016).The impact of resilience among older adults. *Geriatric Nursing, 37*(4), 266–272.

Marshall, L. (2015). Thinking differently about aging: Changing attitudes through the humanities. *The Gerontologist, 55,* 519–525.

Masten, A. S. (2001). Ordinary magic: Resilience processes in development. *American Psychologist, 56,* 227–238.

Masten, A. & Garmezy, N. (1985). Risk, vulnerability, and protective factors in developmental psychopathology. In B. Lahey & A. Kazdin, (Eds). *Advances in clinical child psychology*. Vol. 8., New York: Plenum Press.

McClain, J., Gullatt, K., & Lee, C. (2018). Resilience and Protective Factors in Older Adults. *Graduate Master's Theses, Capstones, and Culminating Projects,* 296.

Merriam, S. B., Martin, P., Adkins, G., & Poon, L. (1995). Centenarians: Their memories and future ambitions. *The International Journal of Aging and Human Development 41*(2), 117–32.

Mishra, M. (2009, September 10). Mukti Dham, the final halt. *The Times of India.* Retrieved from https://timesofindia.indiatimes.com/city/lucknow/Mukti-Dham-the-final-halt/articleshow/4993061.cms

Ngan, R. (2011). Social care and older people. In I. Stuart-Hamilton (Ed.). *An introduction to gerontology* (pp. 126–158), Cambridge: Cambridge University Press.

Nygren, B., Alex, L., Jonsen, E., Gustafson, Y., Norberg, A., & Lundman, B. (2005). Resilience, sense of coherence, purpose in life and self-transcendence in relation to perceived physical and mental health among the oldest old. *Aging and Mental Health, 9,* 354–362.

Ong, A. D., Bergeman, C. S., Bisconti, T. L., & Wallace, K. A. (2006). Psychological resilience, positive emotions, and successful adaptation to stress in later life. *Journal of Personality and Social Psychology, 91,* 730–749.

Pargament, K. I. (1997). *The psychology of religion and coping: Theory research, and practice.* New York: The Guilford Press.

Pradhan, R. K., Panda, P., & Jena, L. K. (2017). Purpose, passion, and performance at the workplace: Exploring the nature, structure, and relationship. *The Psychologist-Manager Journal, 20*(4), 222–245.

Preston, S. H., Stokes, A., Mehta, N. K., & Cao, B. (2014). Projecting the effect of changes in smoking and obesity on future life expectancy in the United States. *Demography 51*(1), 27–49.

Rawat, R. & Gopal Jee, S. (2018) Gratitude and resilience among elderly and young people. *Research Review Journals, 3*(8), 259–262.

Richardson, G. E. (2002). The meta-theory of resilience and resiliency. *Journal of Clinical Psychology, 58,* 307–321.

Rosado-Medina, J. J., Rodríguez-Gómez, J. R., & Altieri Ramirez, G. (2012). Study on resilience internal factors in a sample of Puerto Rican centenarians. Boletin de la *Asociacion* Medica de *Puerto Rico, 104*(4), *17*–25.

Rowe, J. W., & Kahn, R. L. (1987). Human aging: Usual and successful. *Science, 237,* 143–149.

Rowe, J. W., & Kahn, R. L. (1997). Successful aging. *The Gerontologist*, 37, 433–440.

Rutter, M. (1987). Psychosocial resilience and protective mechanisms. *American Journal of Orthopsychiatry, 57,* 316–331.

Ryff, C. D., Friedman, E. M, Morozinki, J. Á. & Tsenkova, V. (2012). Psychological resilience in adulthood and later life: Implications for health. In B. Hayslip & G. Smith (Eds). *Annual review of gerontology and geriatrics. Emerging perspectives on resilience in adulthood and later life.* (Vol. 32, pp. 72–112). New York: Springer Publishing Company.

Schröder-Butterfill, E. (2011). *The concept of vulnerability and its relationship to frailty.* Online Version retrieved from http://www.academia.edu/12329314/The_Concept_of_Vulnerability_and_Its_Relationship_to_Frailty

Schröder-Butterfill, E., & Marianti, R. (2006). A framework for understanding old-age vulnerabilities. *Ageing and Society, 26*(1), 9–35.

Schumm, J. A., Briggs-Phillips, M., & Hobfoll, S. E. (2006). Cumulative interpersonal traumas and social support as risk and resiliency factors in predicting PTSD and depression among inner-city women. *Journal of Traumatic Stress, 19*(6), 825–836.

Sinha, D., & Tripathi, R. C. (1994). Individualism in a collectivist culture: A case of coexistence of opposites. In U. Kim, H. C. Triandis, C. Kagitcibasi, S. C. Choi, & G. Yoon (Eds.), *Individualism and collectivism: Theory, method, and application* (pp. 123–136). Thousand Oaks, CA: Sage.

Turrell, G., Haynes, M., Wilson, L.-A., & Giles-Corti, B. (2013). Can the built environment reduce health inequalities? A study of neighbourhood socioeconomic disadvantage and walking for transport. *Health & Place, 19,* 89–98.

U.N. Census Beaurau (2015). *An ageing world: 2015 international population reports.* Retrieved from https://www.census.gov/content/dam/Census/library/publications/2016/demo/p95-16-1.pdf

United Nations, Department of Economic and Social Affairs, Population Division (2017). *World population ageing 2017* (ST/ESA/SER.A/408).

Vahia, I. V., Deep, C. A., Palmer, B. W., Fellows, I., Golshan, S., Thompson, W., Allison, M., & Jeste, D. V. (2011). Correlates of spirituality in older women. *Aging and Mental Health, 15*(1), 97–102.

Vatuk, S. (1980). Withdrawal and disengagement as a cultural response to aging in India. In C. Fry (Ed.) *Aging in culture and society: Comparative viewpoints and strategies.* (pp. 126–148). New York: Praeger.

Viña, J., Borras, C., & Miquel, J., (2007). Theories of ageing. *Journal of the International Union of Biochemistry and Molecular Biology, 59*(4–5), 249–254

Wagnild, G. M., & Young, H. M. (1993). Development and psychometric evaluation of the resilience scale. *Journal of Nursing Measurement, 1,* 165–178.

Walsh, F. (2003). Family resilience: A framework for clinical practice. *Family Process, 42*(1), 1–18.

Wanzer, M. B., Sparks, L. & Frymier, A. B. (2009). Humorous communication within the lives of older adults: The relationships among humor, coping efficacy, age, and life satisfaction. *Health Communication, 24*(2), 128–136.

Wells N. (2009). Resilience in rural community-dwelling older adults. *Journal of Rural Health, 25*(4), 415–419.

Westberg, N. G. (2003). Hope, laughter, and humor in residents and staff at an assisted living facility. *Journal of Mental Health Counseling, 25,* 16–33.

Whisman, M. A., Uebelacker, L. A., Tolejko, N., Chatav, Y., & McKelvie, M. (2006): Marital discord and well-being in older adults: Is the association confounded by personality? *Psychology and Aging, 21*(3), 626–631.

Wiles, J. L. (2011). Reflections on being a recipient of care: Vexing the concept of vulnerability. *Social and Cultural Geography, 12*(6), 573–588.

Wiles, J. L., & Jayasinha, R. (2013). Care for place: The contributions older people make to their communities. *Journal of Aging Studies, 27*(2), 93–101.

Wiles, J. L., Wild, K. L., & Allen, R. E. S. (2012). Resilience from the point of view of older people: 'There's still life beyond a funny knee'. *Social Science and Medicine, 74*(3), 416–424.

Windle, G., & Bennett, K. M. (2012). Caring relationships: How to promote resilience in challenging times. In M. Ungar (Ed.), *The social ecology of resilience: A handbook of theory and practice* (pp. 219–231). New York, NY, USA: Springer Science + Business Media.

Windle, G. (2012). The contribution of resilience to healthy ageing. *Perspectives in Public Health, 132(4),* 159–160.

Windle, G., Markland, D. A., & Woods, R. T. (2008). Examination of a theoretical model of psychological resilience in older age. *Aging and Mental Health, 12*(3), 285–292.

Windsor, T. D., Hunter, M. L, & Browne-Yung, K. (2015). *Ageing well: Building resilience in individuals and communities.* Retrieved from Government of South Australia (SA) health website https://www.sahealth.sa.gov.au/wps/wcm/connect/e35948004ac667418a67de0b65544981/Ageing%2BWell%2BBuilding%2BResilience%2Bin%2BIndividuals%2Band%2BCommunities.pdf?MOD=AJPERES&CACHE=NONE&CONTENTCACHE=NONE

Windsor, T., Rioseco, P., Fiori, K., Curtis, R., & Booth, H. (2016). Structural and functional social network attributes moderate the association of self-rated health with mental health in midlife and older adults. *International Psychogeriatrics, 28*(1), 49–61.

Zaidi, A. (2013). *Global Age Watch Index 2013: Purpose, methodology and results.* Retrieved from HelpAge India website http://www.helpage.org/silo/files/global-agewatch-index-2013-purpose-methodology-and-results.pdf

Zaidi, A. (2014). *Life cycle transitions and vulnerabilities in old age: A review.* Occasional paper, UNDP Human Development Report Office, New York.

Ziersch, A. M., Baum, F. E., MacDougall, C., & Putland, C. (2005). Neighbourhood life and social capital: the implications for health. *Social Science & Medicine, 60(*1), 71–86.

Zimmerman, M. A. (2013). Resiliency theory: a strengths-based approach to research and practice for adolescent health. *Health Education & Behavior, 40*(4), 381–383.

CHAPTER 15

Making of a Resilient Student: Tracing Challenges and Interventions

SAMRIDHI AHUJA, AYESHA SAHA, and UPDESH KUMAR

Defence Institute of Psychological Research, Delhi, India

*Corresponding author. E-mail: samridhiahuja@gmail.com

ABSTRACT

Schools are an integral part of a child's learning and development. They are touted as premises that foster the student's many capabilities be it academic, social, or emotional. Schools endeavor to create an environment that aids in the positive development of the students; however, with the changing milieu, they are increasingly becoming a setting which harbors elements that can be detrimental for a student's growth and development. An increase in issues like academic pressure, body shaming, bullying to name a few, have led to deterioration in the psychosocial and behavioral adjustment of the students. Such an environment then no longer seems to play a protective role in the student's life. Acknowledging the influence of this environment as well as the factors that are at play would help us focus on the systematic as well as the individual change which is required. This would then help in drawing attention toward seeking pathways to change the environment for the students. An important factor that can help alleviate the difficulties that the students are going through is resilience. Promoting resilience in the students would empower them to develop mechanisms that would act as protective factors against experiences that could otherwise prove to be overwhelming for them. Espousing resilience in the school setting will enable the students to maintain a balance in their lives and thrive even when the circumstances seem tough. The present chapter aims at highlighting these pressing issues within the school environment that tend to impede the development of the students. The chapter aims to explore a whole-school approach of building resilience.

15.1 INTRODUCTION

> *Education is what remains after one has forgotten what one has learned in school.*
>
> *Albert Einstein (n.d.)*

Schools are an integral part of a child's learning and development. They herald a student's capabilities in academia, sports, and performing arts that shape them as an individual. Children and adolescents spend a large amount of their time in school. Thus, schools play a fundamental role in fostering capabilities that impacts several facets of a child's personality. Acknowledging the influence of this environment as well as the interplay of other factors brings to the forefront several underlying issues concerning student mental health. The competitive nature of academia, sports, and performing arts has often had a negative impact on students. This has not only led to cases of bullying, parental pressure, peer pressure, and rivalry but has drastically impacted the life of the student and their social support system.

The mental health of students has been a serious issue among researchers and policymakers alike. Schools endeavor to create an environment that would foster positive growth and development of the students. However, with the changing milieu, schools tend to harbor elements that lead to deterioration in the psychosocial and behavioral adjustment of the students.

Knowing the fundamental importance of school life, the increasing incidents of suicide amongst the student population worldwide suggest otherwise. These cases have brought to the forefront instances of academic stress, (cyber) bullying, body shaming, and peer pressure as key factors influencing student mental health. With an effort toward controlling this concern, schools have been involved in efforts to promote healthy behaviors among students.

Schools no longer are a harbinger of an environment that promotes education, socialization, and transmission of cultural norms and values. They have created an atmosphere of constant comparisons, rivalry, and detestation that have led to several risks that a child may face during the essential years of childhood and adolescence. Highlighting these grappling issues of academic pressure, (cyber) bullying, body shaming, peer pressure within schools that hamper positive growth and development, the chapter also discusses several protective factors that may change or transpose the negative outcomes associated with risk factors during the developmental phase of school-going children. Research has suggested that certain attributes of the family, school, and community may help in fostering resilience among children despite risks. Under adverse situations, caring and supportive relationships can cushion the child against negative experiences. Individual resilience is the interplay of personality characteristics as well as the external factors within the environment (Johnson & Thompson, 2008; Masten, 2011; Ungar, 2010). Adopting a whole-school approach is a collaborative and a cohesive effort toward building resilience in a school community. This strategically constructed initiative can facilitate the improvement of student learning, behavior and wellbeing, and foster conditions that provide support when faced with these challenges.

The chapter attempts to trace some of the significant challenges faced by the students during their school life, and talk about the support mechanisms that can be in place to facilitate the students in navigating their school life. Some of the challenges like academic pressure are at a surface level and tangible, while others like bullying, parental pressure, body shaming, and peer pressure take a subtle form and may not always be noticeable by the authorities.

15.2 ACADEMIC PRESSURE

Pranay had always been an above-average student, who took his studies seriously and tried to put his best efforts in scoring well. However, as the grade level and complexity of the subjects increased, Pranay had to enroll in a number of tuitions to keep up with the pace of his teachers and compete with his classmates. Attending so many classes left him drained and exhausted, he started experiencing burn out. During this time, absenteeism from school also increased.

Academic pressure is an outcome of myriad factors that often get clouded and classified as only a few. During our years in school, academic pressure is driven from our need for achievement, concern over grades, parental pressure, competition, sports, and a significantly competitive class distribution. Likewise, in a study by Kouzma and Kennedy (2004), it was reported that school-related situations—such as tests, grades, studying, self-imposed need to achieve, as well as that induced by others—are the main sources of stress for high school students. For most of this population, the intellectual, emotional, and physical strain caused by academic pressure is an outcome of poor time management and a tendency of procrastination.

Academic stressors may be environmental, emerging from the situation at home, school, or even through interpersonal relationships. Students fall prey to academic pressure not only due to factors stated above but also due to government laws and policies. Alarming situations have been cited globally. For instance, according to the Department of Education (2010), education policy within England has encouraged schools to maximize students' academic attainment, neglecting their broader well-being, personal growth, development, and health. Studies have also suggested that since subjects such as personal, social and health education are nonstatutory, schools spend less time in teaching them because of the continued focus on academic subjects (PSHE Association, 2013). Furthermore, recent reforms in the education policies of England have focused on mandating academic tests in the first year of primary school, and have proposed the early years' foundation stage profile (that had offered a holistic development) as nonstatutory (Department of Education, 2013). The background that underlines these flawed reforms is twofold. Firstly, it emphasizes promoting academic attainment on one hand, with personal health and development on the other, the outcome being a "zero-sum game" since more time gets allocated to health and well-being as opposed to academic development resulting in lower attainment. Secondly, that improvement in academic attainment is rarely crucial to rising economic competitiveness (Bonell et al., 2014). Contrary to the first assumption, research has suggested that students with better health have higher academic attainment. This outcome stands valid across schools in Finland, Australia, Singapore, and Sweden, where development and well-being are integrated into subject learning, accounting for better attainment. Thus, academic and broader development is not a "zero-sum game."

Academic success has therefore been directly linked with positive mental health. Adopting the approach of "teaching to the test" which is common in schools with a narrow focus on attainment can lead to poor mental health amongst students (Bonell et al., 2013). Overall research within this area has reiterated the need to integrate knowledge and skills that promote mental and physical health along with academic knowledge and cognitive skills that would benefit students in the long run.

Research has suggested that academic stress is a risk factor for psychopathology. Findings indicated a strong positive correlation between academic stress, parental pressure, and poor mental health. Examination anxiety has also been associated with maladaptive coping behaviors. The practice of drawing comparisons across academic performance amongst students creates anxiety and pressure to outperform other students. It is imperative to understand and judge the performance of a student with regard to their capabilities. With such differences being overlooked, the proneness to anxiety, depression, and fear of academic failure increases.

One of the leading newspapers in India reported parental pressure as a major determinant of academic stress. About 66% of the students admitted parental pressure as a driving factor to perform better academically (Mukherjee, 2010a). It is a common phenomenon for Indian parents to criticize their children's performance by comparing against the class topper. As a result of which, schools create an environment of rivalry and hatred instead of nurturance and friendship. The extent of comparisons is so high that some parents tend to disgrace the achievement of the class topper by stating it as a case of teacher favoritism.

The alarming number of incidences surrounding mental health is reported for school students appearing for their secondary (10th) and senior high (12th) grade final board examinations. Pressurized by parents to "excel" in extracurricular activities as well as academics, children grapple with the issues of being an "all-rounder" causing extreme mental demands. Children are unable to adapt to the situation and respond with maladaptive coping mechanisms. For some children, this stress may even cause the development of nervous disorders in early childhood (Mukherjee, 2010b). The thought behind parental pressure to excel is out of the concern for the child's welfare. More so, because of the competition for admission into reputed universities for a successful career ahead. However, what parents fail to realize is that the pressure that they exert for the welfare of the child emerges as one of the main reasons leading to mental health concerns.

The academic performance of a student does not exist in isolation; it is a part of the larger dynamics that are at play in a school set up. Incidents of bullying and peer pressure can also negatively impact the performance of the students' leading to increased pressure to prove themselves. Thus, the student is left struggling with a lot of issues at the same time.

15.3 BULLYING

Bullying is a widespread phenomenon, the manifestation of which changes from time to time. The term "bullying" refers to the victimization of a student or students when they are exposed repeatedly to harmful or negative acts on the part of one or group of students (Olweus, 1993). These acts are attempts to cause discomfort or humiliate the student. Examples of such negative acts are physical contact, verbal insults, intentional exclusion, and rumors. However, for these acts to be labeled as instances of bullying, an asymmetric power relationship between the victim and the bully must exist, such that the bullied child or the victim experiences difficulty in defending himself or herself (Pozzoli et al., 2012).

WHO (2010) defined bullying as a multifaceted form of abuse prevalent usually in school and sometimes in workplaces as well. It is marked by frequent exposure of an individual to physical/emotional aggression which may include name-calling, taunting, spreading rumors, social exclusion, etc. Olweus (1993, 1997), in his seminal work, talks about two types of bullying victims: passive and provocative. Passive victims tend to be cautious, sensitive, and react to the bullying episode by crying. Provocative victim on the other hand, experiences problems with concentration and is often hyperactive. Bullies are often aggressive toward parents, peers, and teachers. They are likely to use violence, are impulsive, and have a strong need to dominate others (Olweus, 1997). The impact of bullying is not just limited to the present functioning of the child, but also affects the future outcomes of the individual. The future outcomes also depend on whether the individual was a victim or a perpetrator. Victimization in the past may produce feelings of harassment and violence (Patchin & Hinduja, 2011), which can have adverse long-run consequences. These harmful consequences have been explained through the premise of the general strain theory (Agnew, 1992). According to this theory, the individual who experiences a strain may produce unfavorable emotions such as anger, irritation, anxiety, and frustration thereby leading the individual to take steps such as self-harm, suicide, etc. as a way to escape these emotions (Agnew, 1992).

The act of bullying was understood in a social network perspective by Faris and Felmlee (2011). From this perspective, bullying is interpreted as an act of aggression aimed at attaining and maintaining social status in one's peer group, as against the idea that it is a maladjusted reaction toward the marginalized group. Faris and Felmlee (2011) believe that an individual who is marginalized lacks the capability to bully and the individual who is at the apex of the social hierarchy does not have a reason to engage in such behaviors.

Bullying tends to negatively impact self-esteem, peer relations, as well as the academic performance of the individuals (Heckman, 2008; Waddell, 2006). Overcoming the emotional turmoil that bullying causes directs the investment of personal resources in other skills, while engaging in bullying behavior leads the perpetrator to direct the resources in counter-productive behavior and away from investing in beneficial skills for personal growth and development (Akerlof & Kranton, 2002). Over the years, variants of bullying behavior have been identified, and in the school setting both overt (physical such as kick, punches, pushes and verbal such as insult, blackmail, etc.) and covert forms (stealing, hiding and damaging material, spreading rumors, name-calling, etc.) of bullying have been prevalent.

Simran, a 16-year-old girl had been "skinny" since childhood. She had grown up listening to her friends, classmates, and family members make comments like "why are you so thin," "don't you eat?", "you look like a bag of bones," "you are so flat." Time and again she would breakdown and cry for being humiliated about her body weight. She would constantly worry about the type of clothes she should wear and her appearance. She would also hesitate while going out shopping with her friends, as she would find it difficult to find her size and always had to look for her size in the child section. This was always a point of ridicule for her.*

A debilitating manifestation of bullying that time and again rears its ugly head is "body shaming." The case example highlights the issue of body shaming and how it is experienced by the targeted individual. It is during puberty when adolescents tend to develop a heightened awareness of their bodies. Both girls and boys during this period are vulnerable toward any comments or criticisms about their appearances (Valutis et al., 2009). Every society intentionally or unintentionally idealizes a certain body image, which in turn can create a pressure, to work toward and maintain the said image. A concept that is related to body image is "body shame." It is a phenomenon that is used to describe individuals' negative and self-conscious emotional responses about oneself. It is when the individual feels that he/she is unable to meet this set ideal that they develop a sense of body shame (Fredrickson & Roberts, 1997). Body shaming on the other hand is when the individual is at the receiving end of inappropriate comments and statements about his/her weight and size. Women and men struggle with a sense of shame. Women who feel that they do not conform to the standards may experience a lot of negative self-directed emotions leading to a sense of shame (Fredrickson & Roberts, 1997). Men in society are often seduced by the idea of quintessential masculinity, failing to meet which they are often labeled as feminine (Grogan & Richards, 2002; Fisher et al., 2002). Body shaming is often found to be a predictor of depression among adolescent girls (Grabe et al., 2007). It is an amalgamation of the societal messages received by the individuals as well as what is retained and absorbed by them. The idea that meeting a certain body image standard can be life-altering leads to hopelessness and depression when the individual is unable to do so. Living with an innate sense of shame due to the messages received from various social media platforms and peers can be and is detrimental for an all-round development of the individual.

Over time, bullying has been recognized as a colossal issue affecting the social functioning and well-being of the school-aged children. Although relationships among peers are dictated by some amount of conflict, bullying presents a severe threat to healthy development (Wang et al., 2011). With the changing social milieu, students are also dealing with newer forms of bullying, which use different avenues.

15.3.1 CYBERBULLYING

Things changed overnight for Ayushi a class IX student of a reputed South Delhi School. With the advent of a new social networking site called Sarahah, which gave the users the opportunity to post comments anonymously, Ayushi found herself in the center of a social debacle. Using this social networking site, a group of students had managed to start a rumor that Ayushi was secretly in a relationship with her best friend's boyfriend. The next day, she realized that she had become an object of ridicule not just among her classmates but also among her seniors. The school counselor was called in when the class teacher had to break a fight in the girl's washroom, which had escalated into a physical fight. Ayushi was found huddled in one of the cubicles and refusing to leave the washroom.*

Cyberbullying is the use of digital communication platforms to humiliate and harass people or to spread defamatory information about them (Hinduja & Patchin, 2007). Using the cyberspace, online humiliation is carried out by posting derogatory messages and pictures of the targeted individual on various social networking sites and platforms. Cyberbullying tends to manifest in the form of leaving hate messages on chats or personal blogs, stalking, impersonation, social exclusion, slander, etc. (Slovak & Singer, 2011). This form of bullying is at a rise due to a lack of parental control and accountability to authority figures (Schneider et al., 2012). With almost every student having access to cell phones the damage that cyberbullying can do, if unchecked is monumental. Institutions in the past have been concerned only with bullying that takes place in the physical space, however with the increased use of mobiles and internet and rise in incidents of cyberbullying; the institutions are now directing their efforts to curtail such forms of bullying (Juvonen & Gross, 2008).

Research studies carried out in the area of cyberbullying have identified various negative effects on students who have been privy to this experience. Some of these include low academic performance, reduced school satisfaction, and withdrawn behavior. Reduced school satisfaction also tends to increase absenteeism without any specific reason. Cyberbullying has also been implicated in higher incidences of psychological harm and suicidal thoughts among the victims, leading to a lot of emotional distress (Schneider et al., 2012). Gender also seems to be an influencing factor in the type of bullying behavior. Boys as opposed to girls use physical and cyberbullying whereas girls are likely to rely on verbal means. Girls are however more likely to fall victim to cyberbullying than boys. Cyberbullying has also been linked with substance abuse, where alcohol consumption and drugs are seen as stress management strategies (Pozzoli et al., 2012).

Bullying, therefore, presents a huge challenge that if not addressed timely can be detrimental for the optimal functioning and positive growth of the student. Apart from bullying, another factor that poses a consistent challenge to the students is the peer pressure.

15.4 PEER PRESSURE

The social and emotional development of adolescents is influenced largely by peers (Allen et al., 2005). Although peer relations begin to develop from an early age, peers begin to play a prominent role during the teenage years. Peer pressure, as the name suggests, is the power that is exerted by the individual's peer group encouraging him/her to alter in their outlook, values, or behaviors in order to conform to the popular norms (Treynor, 2009). The impact of peer pressure can be positive as well as negative. Positive peer pressure aids in the growth and development of the individual, whereas negative peer pressure tends to be a hindrance. It may make the individual unhappy, uncomfortable, and unwell. Students are likely to give in to this pressure in order to feel accepted and fit into the larger group, giving them a sense of belongingness. Peer pressure is either implied or expressed. In its expressed form, the individual is challenged to abide by the existing group norms (Rubin et al., 2006). An example of expressed peer pressure is when the individual is challenged to partake in a risky activity that he/she may otherwise not be a part of. Studies reveal that in such a case, peers are prone to take a risk that they believe would garner a higher standing in the eyes of their peers and ensure acceptance in the group (Cotterell, 2007).

On the other hand, implied peer pressure is harder to combat as it takes a subtler approach. An example of an implied form of peer pressure would be when an individual is made fun of dressing the way he/she does, forcing them to comply with the perceived "acceptable" style of dressing. Many times, students who look, think, or dress a certain way are targeted and ousted from the social circle and may find themselves at the receiving end of undue criticism and ridicule. Such ridicule may cause the individual to feel rejected and depressed (Perkins, 2003).

Essentially, what peer pressure does is, it annihilates the sense of individuality, forcing the peers to be mare clones of each other. Dealing with peer pressure can take a toll on the individual, and if not dealt with in time can hamper the social and emotional development of the individual. Loss of individuality not just creates a void in student life but also impacts the adult life of the individual. It, therefore, becomes important for the students to realize that it is not important to fit into a mold that others seem to have set for them (Clausen, 1968). The school and teachers can play an important role in helping the students to be comfortable and proud of whom they are without feeling compelled to do and be what seems to be the fad for the rest of the peers.

15.5 INTERVENTIONS: PROVIDING THE NECESSARY TOOLS

The greatest glory in living lies not in never falling, but in rising every time we fall.
Nelson Mandela

Over the course of a decade, epidemiological research has reported escalating levels of mental health conditions among school children and youth. The reinforcement to develop preventive measures and appropriate intervention to counter mental health problems amongst students is a result of these studies (Twenge et al., 2010). With the time the students spend in the school premises, schools become a key setting to promote the well-being and resilience of the students.

Schools today are actively working toward improving the mental health of the students. One of the approaches that the schools across the country have adopted is to create awareness about challenges such as bullying and peer pressure and provide the students with the framework of dealing with these challenges thereby sending across the message that "it is okay" to discuss these issues. However, what this approach does is to just address the concerns that are at the surface level. An approach like this can be described as "psychological first aid," which happens usually after such an episode is reported. What necessitate attention are the vulnerabilities that make the students susceptible to these issues in the first place. Hence, an approach that focuses on building a resilient student is imperative.

The mechanism of successfully adapting to an adversity is defined as *Resilience*. It is the proclivity to bounce back and adapt to the demands of an exasperating situation (Block & Block, 1980; Block & Kremen, 1996; Lazarus, 1993). Resilience has been looked at as a protective factor, the presence of which can facilitate a good outcome even when there exist factors that can curb the growth of the individual (Anthony, 2011). The adaptation that takes place comprises of collaboration between risk as well as protective factors across the various domains of an individual's life (Olsson et al., 2003). It is the individual's interaction with the stressful life events that helps in fostering the ability of being resilient (Blum, 1998).

Whole-school approach toward building resilience warrants attention in the current scenario. A whole-school approach entails providing a secure and supportive learning environment, advocacy of positive peer relations as well as student–teacher relationships, designing a positive learning environment, promotion of social and emotional skill development, and involving the parents and students in the decision making process. Hence, all the stakeholders work in sync to achieve a shared vision and goal.

A positive learning and supportive social environment help in building a sense of connectedness to school and engagement in the learning process. Such a sense of belongingness to the school serves as a protective factor that facilitates in building student resilience (Resnick, 1997; Resnick et al., 1997). Student resilience has also been correlated with positive health and academic performance (Blum, 2005). The finding of several research studies indicate that students who experience a higher sense of connectedness with the school are less likely to engage in violence, substance abuse or report mental health problems or issues (McNeely et al., 2002; O'Brien & Bowels, 2013).

The key components of a whole-school approach toward building well-being and resilience of the students proposed by the National Health Promoting Schools Initiative (2000) are discussed in the following sections:

15.5.1 DEVELOPMENT OF A SOCIAL AND EMOTIONAL LEARNING CURRICULUM

A social and emotional learning curriculum helps in building a positive social behavior and restrains students from engaging in risky and disruptive behaviors (Durlak et al., 2011; Frydenberg, 2010). However, there seems to be a debate about the efficacy of such an explicit curriculum. Some researchers are of the opinion that modeling positive social behavior is of greater importance than a curriculum. Studies carried out to evaluate the efficacy of such a curriculum have revealed that these programs have a positive impact on both the cognitive and academic domains (Nelson et al., 2003).

In Australia, Geelong Grammar School adopted a continuous intervention to enhance character strengths, meaning, gratitude, flow, positive emotions, optimism, and positive relationships. The training was over a course of 4 years that involved 250 staff members participating in an intensive training with a team of senior trainer from the Positive Psychology Centre at the University of Pennsylvania. Results suggested improvement in conflict resolution between students, the emergence of a common strengths-based interaction between students and teachers, increased levels of resilience, and individual flourishing among seventh-grade students (Williams, 2011).

Studies focusing on "positive psychology" have identified individual factors that contribute toward youth well-being (Seligman et al., 2009). Factors such as positive emotion (Fredrickson, 2004), hope (Snyder et al., 2003), character strengths (Peterson & Seligman, 2004), goal setting (Locke & Latham, 2002), and gratitude (Froh et al., 2008) have revealed a strong correlation with subjective-well-being among the youth. Another intervention focused toward promoting resilience, positive emotions, a sense of purpose and meaning, and the use of character strengths in daily living among school children was empirically appraised at Strath Haven High School outside Philadelphia. The positive psychology intervention had 352 ninth grade students participating in approximately 25 lessons on related concepts of positive psychology. The findings suggested an increase in enjoyment and school engagement, improvement in social skills like empathy, cooperation, and self-control. However, the program was not successful in improving symptoms of anxiety and depression, character strengths, or increased participation in extracurricular activities (Seligman et al., 2009).

Furthermore, studies on adolescent well-being have suggested that interventions with an inherent theoretical underpinning of resilience and empowerment were a major determinant of positive mental health. In a study by Gillespie et al. (2007), a program was developed to enhance individual and interpersonal characteristics. The theoretical underpinnings of resilience factors such as self-efficacy, positive emotions, internal locus of control, optimistic world view, and social support were a significant contributor toward overall well-being. According to Fredrickson (2004), these personal resources vary from psychological to societal assets and act as protective factors that can be accessed to improve adaptive coping during adversities and crisis. Thus, these interventions can structure a platform of resilience for students to flourish.

15.5.2 PROMOTION OF POSITIVE STUDENT–TEACHER RELATIONSHIP

Promoting well-being also includes attention to cultivating strong student–teacher relationships. A meta-analysis of studies revealed positive linkages between student–teacher relationships and student engagement and achievement, whereas a negative relationship predicted poor student engagement and achievement (Roorda et al., 2011). Students who were identified to be "at-risk" were more susceptible to the quality of the relationships as compared to the normative students. Positive relationships between students and adults are also implicated in better educational outcomes (Anderson et al., 2004). The perception that a student holds about a teacher's authority and influence over the class as well as the belief that the teacher trusts and has positive regard for his/her students, leads to a higher school engagement (van Uden et al., 2014).

15.5.3 ENHANCING POSITIVE PEER RELATIONSHIPS

An essential component of the whole-school approach is the focus on building a healthy peer environment. The quality of the peer relationship that the students enjoy tend to impact their overall school experience (Holfve-Sabel, 2014). Healthy peer relationships can help in forming strong friendships as well as an atmosphere of support and inclusion. This is a component that is also linked to better academic performance as well as reduced risk-taking. Farrington and Ttofi's (2011) meta-analysis indicated that bullying and victimization instances reduced by at least 20% when programs targeting peer relationships were incorporated in the school setup.

Protective factors play an important role in facilitating the process of resilience building. Research has evidenced that even during adversity; individuals have been able to exhibit adaptive coping and functioning because of their strong social support (Masten, 2001). Even for children, strong social and family support plays an important role in their ability to be resilient in adverse circumstances. Positive supportive relationships are the foundation of resilience and positive mental health (Wang et al., 2003). Interventions targeted at promoting positive interpersonal relationships in the community are successful in the creation of a positive school environment. Strong social support plays an essential role in enhancing resilience among all individuals, especially youth.

15.5.4 ENHANCEMENT OF SCHOOL–HOME RELATIONSHIPS

The whole-school approach does not limit itself to the teachers or students; it goes beyond the school premises and includes parental participation as well. Various studies have indicated that parental engagement in school as well as the attitude toward learning at home influences the academic achievement of the students (Goodall, 2013). School's approach toward determining the specific ways in which the parents can be involved at a school level in the decision-making processes can ensure parental engagement for a longer duration. When the parents are involved in the decision-making processes they are also abreast of all the policy-level changes that the administration proposes to make, thereby helping the students in the transition.

15.6 CONCLUSION

Schools form an integral part of a student's development. It is an institution where the child not only advances academically but also grows exponentially in the social and emotional domains. However, school is not always a happy place for many children, due to issues such as bullying, peer pressure, or academic pressures. In such a scenario, school becomes an inhibiting environment for the student and navigation of such an environment requires the child to direct all his personal resources. Resilience building programs employing the whole-school approach addresses the student's distress at various levels be it peer relationship or student–teacher relationship. Apart from these components, this approach also focuses on the social and emotional learning of the student which fosters not only positive peer relations but also facilitates academic achievement. All these components come together to cultivate a resilient attitude in the students. Therefore, for uninterrupted progress to take place, it is important that the protective mechanisms that facilitate resilience building are in place so that even if the students were to face adverse circumstances they are capable of facing these challenges.

KEYWORDS

- **academic pressure**
- **bullying**
- **challenges**
- **interventions**
- **resilience**

REFERENCES

Agnew, R. (1992). Foundation for a general strain theory of crime and delinquency. *Criminology*, *30*(1), 47–88.

Akerlof, G. A., & Kranton, R. E. (2002). Identity and schooling: Some lessons for the economics of education. *Journal of Economic Literature*, *40*(4), 1167–1201.

Albert Einstein Quotes. (n.d.). BrainyQuote.com. Retrieved October 5, 2018, from BrainyQuote.com Web site: https://www.brainyquote.com/quotes/albert_einstein_ 108304

Allen, P, McGarland, M, & Elhaney, M. C. (2005). The two faces of adolescents success with peers: Adolescent popularity, social adaptation and deviant behavior. *Child Development*, *76*, 757–760.

Anderson, A. R., Christenson, S. L., Sinclair, M. F., & Lehr, C. A. (2004). Check & connect: The importance of relationships for promoting engagement with school. *Journal of School Psychology*, *42*(2), 95–113.

Anthony, E. K. (2011). Youth development tip sheet: Three questions about child and adolescent development. Retrieved from http://www.helpstartshere.org/kids-and families/youth-development/child-and-adolescent-resilience.html

Block, J. H., Block J. (1980). The role of ego-control and ego-resiliency in the origination of behavior. In *The Minnesota symposia on child psychology* (pp. 39–45). Hillsdale, NJ: Erlbaum.

Block, J., & Kremen, A. M. (1996). IQ and ego-resiliency: Conceptual and empirical connections and separateness. *Journal of Personality and Social Psychology, 70*, 349–361.

Blum, R. W. (2005). A case for school connectedness. *Educational Leadership*, *62*(7), 16–20.

Blum, R. W. (1998). Healthy youth development as a model for youth health promotion: A review. *Journal of Adolescent Health*, *22*, 368–375.

Bonell C., Humphrey N., Fletcher A., Moore L., Anderson R., Campbell R. (2014). Why schools should promote students' health and wellbeing. *BMJ, 7958*, 348–378.

Bonell, C., Parry, W., Wells, H., Jamal, F., Fletcher, A., Harden, A., Thomas, J., Campbell, R., Petticrew, M., Murphy, S., Moore, L., & Whitehead, M. (2013). The effects of the school environment on student health: a systematic review of multi-level studies. *Health & Place*, *21*, 180–191.

Clausen, J. A. (1968). *Socialization and society.* Boston: Little Brown and Company Publishers.

Cotterell, J. (2007). *Social Networks in Youth and Adolescence.* New York: Routledge.

Department for Education (2010). The importance of teaching. The schools white paper 2010. www.gov.uk/government/uploads/system/uploads/attachment_data/file/175429/CM-7980.pdf.

Department for Education (2013). Primary assessment and accountability under the new national curriculum. www.gov.uk/government/uploads/system/uploads/attachment_data/file/298568/Primary_assessment_and_accountability_under_the_new_curriculum_consultation_document.pdf

Durlak, J. A., Weissberg, R. P., Dymnicki, A. B., Taylor, R. D., & Schellinger, K. B. (2011). The impact of enhancing students' social and emotional learning: A meta-analysis of school-based universal interventions. *Child Development, 82*(1), 405–432.

Faris, R., & Felmlee, D. (2011). Status struggles: Network centrality and gender segregation in same- and cross-gender aggression. *American Sociological Review* 76, 48–73.

Farrington, D. P., & Ttofi, M. M. (2011). Bullying as a predictor of offending, violence and later life outcomes. *Criminal Behavior and Mental Health, 21*(2), 90–98.

Fisher, E., Dunn, M., & Thompson, J. K. (2002). Social comparison and body image: An investigation of body comparison processes using multidimensional scaling. *Journal of Social and Clinical Psychology, 21*(5), 566–579.

Fredrickson, B. L. (2004). Gratitude, like other positive emotions, broadens and builds. In R. A. Emmons & M. E. McCullough (Eds.), *The psychology of gratitude* (pp. 145–166). New York: Oxford University Press.

Fredrickson, B. L., & Roberts, T.-A. (1997). Objectification theory. *Psychology of Women Quarterly, 21*(2), 173–206.

Froh, J. J., Sefick, W. J., & Emmons, R. A. (2008). Counting blessings in early adolescents: An experimental study of gratitude and subjective well-being. *Journal of School Psychology, 46*(2), 213–233.

Frydenberg, E. (2010). *Think positively! A course for developing coping skills in adolescents.* London: Continuum International Publishing Group.

Gillespie, B. M., Chaboyer, W., & Wallis, M. (2007). Development of a theoretically derived model of resilience through concept analysis. *Contemporary Nurse, 25*(1–2), 124–135.

Goodall, J. (2013). Parental engagement to support children's learning: a six point model. *School Leadership & Management, 33*(2), 133–150.

Grabe, S., Hyde, S. J., & Lindberg, S. M. (2007). Body objectification and depression in adolescents: The role of gender, shame and rumination. *Psychology of Women Quarterly, 31*, 164–175.

Grogan, S., & Richards, H. (2002). Body image focus groups with boys and men. *Men and Masculinities, 4*(3), 219–232.

Heckman, J. J. (2008). Schools, skills, and synapses. *Economic Inquiry 46*, 289–324.

Hinduja, S., & Patchin, J. W. (2007). Offline consequences of online victimization: School violence and delinquency. *Journal of School Violence, 6*(3), 89–112.

Holfve-Sabel, M. A. (2014). Learning, interaction and relationships as components of student well-being: Differences between classes from student and teacher perspective. *Social Indicators Research, 119*(3), 1535–1555.

Johnson, H., & Thompson, A. (2008). The development and maintenance of posttraumatic stress disorder (PTSD) in civilian adult survivors of war trauma and torture: A review. *Clinical Psychology Review, 28*, 36–47.

Juvonen, J., & Gross, E. F. (2008). Extending the school grounds?—Bullying experiences in cyberspace. *Journal of School Health, 78*(9), 496–505.

Kouzma, N. M., & Kennedy, G. A. (2004). Self-reported sources of stress in senior high school students. *Psychological Reports, 94*, 314–316.

Lazarus, R. S. (1993). From psychological stress to the emotions: A history of changing outlooks. *Annual Review of Psychology, 44*, 1–21.

Locke, E. A., & Latham, G. P. (2002). Building a practically useful theory of goal setting and task motivation. A 35-year odyssey. *The American Psychologist, 57*(9), 705–717.

Masten, A. S. (2001). Ordinary magic: Resilience processes in development. *American Psychologist, 56*, 227–238.

Masten, A. (2011). Resilience in children threatened by extreme adversity: Frameworks for research, practice, and translational synergy. *Development and Psychopathology, 23*, 493–506.

McNeely, C. A., Nonnemaker, J. M., & Blum, R. W. (2002). Promoting school connectedness: Evidence from the national longitudinal study of adolescent health. *Journal of School Health, 72*(4), 138–146.

Mukherjee Pandey, J. (2010a). Schools junk ranks in pre-board tests. The Times of India.

Mukherjee Pandey, J. (2010b). Kids driven to breaking point; pushed by parents to be all rounders, the perform-or-perish pressure on school kids is turning them into nervous wrecks. The Times of India.

National Health Promoting Schools Initiative (2000). National framework for health promoting schools 2000–2003. Canberra: Commonwealth Department of Health and Family Services and Australian Health Promoting Schools Association.

Nelson, G., Westhues, A., & MacLeod, J. (2003). A meta-analysis of longitudinal research on preschool prevention programs for children. *Prevention & Treatment, 6*(1), 31a.

O'Brien, K. A., & Bowles, T. V. (2013). The importance of belonging for adolescents in secondary school settings. *The European Journal of Social & Behavioral Sciences, 5*(2), 976–985.

Olsson, C. A., Bond, L., Burns, J. M., Vella-Brodrick, D. A., & Sawyer, S. M. (2003). Adolescent resilience: A concept analysis. *Journal of Adolescence, 26*(1), 1–11.

Olweus, D. (1993). *Bullying at school: What we know and what we can do.* Malden, MA: Blackwell Publishing.

Olweus, D. (1997). Bully/victim problems in school: Facts and interventions. *European Journal of Psychology of Education, 12*, 495–510.

Patchin, J. W., & Hinduja, S. (2011), Traditional and non-traditional bullying among youth: A test of general strain theory. *Youth and Society, 43*, 727–751

Perkins, H. W. (Ed.). (2003). *The social norms approach to prevent school and college age substance abuse: A handbook for educators, counselors, and clinicians*. San Francisco: Jossey-Bass.

Peterson, C., & Seligman, M. E. P. (2004). Character strengths and virtues: A classification and handbook. New York/Washington, DC: Oxford University Press/American Psychological Association.

Pozzoli, T., Gini, G., & Vieno, A. (2012). Individual and class moral disengagement in bullying among elementary school children. *Aggressive Behavior*, *38*(5), 378–388.

PSHE Association. (2013). Comments on the national curriculum proposals published in February 2013 from the PSHE education strategic partners' group. file:///C:/Private/Downloads/PSHE-StratGroup-comments NC.pdf.

Resnick, M. D. (1997). Close ties to parents, school improve adolescent's lives. *Minnesita Medicine*, *80*(12), 24–26.

Resnick, M. D., Bearman, P. S., Blum, R. W., Bauman, K. E., Harris, K. M., Jones, J., Tabor, J., Beuhring, T., Seiving, R., Shew, M., Bearinger, L.,& Ireland, M. (1997). Protecting adolescents from harm: findings from the National Longitudinal Study on Adolescent Health. *JAMA*, *278*(10), 823–832.

Roorda, D. L., Koomen, H. M., Spilt, J. L., & Oort, F. J. (2011). The influence of affective teacher–student relationships on students' school engagement and achievement: A meta-analytic approach. *Review of Educational Research*, *81*(4), 493–529.

Rubin, K. H., Bukowski, W. M., & Parker, J. G. (2006). Peer interactions, relationships, and groups. In N. Eisenberg, W. Damon, & R. M. Lerner (Eds.), *Handbook of child psychology* Vol.3, (pp. 571–645). New Jersey: Wiley.

Schneider, S. K., O'donnell, L., Stueve, A., & Coulter, R. W. (2012). Cyberbullying, school bullying, and psychological distress: A regional census of high school students. *American Journal of Public Health*, *102*(1), 171–177.

Seligman, M. E., Ernst, R. M., Gillham, J., Reivich, K., & Linkins, M. (2009). Positive education: Positive psychology and classroom interventions. *Oxford Review of Education*, *35*(3), 293–311.

Slovak, K., & Singer, J. B. (2011). School social workers' perceptions of cyber bullying. *Children & Schools*, *33* (1), 5–16.

Snyder, C. R., Lopez, S. J., Shorey, H. S., Rand, K. L., & Feldman, D. B. (2003). Hope theory, measurements, and applications to school psychology. *School Psychology Quarterly*, *18*(2), 122–139.

Treynor, W. (2009). *Towards a General Theory of Social Psychology: Understanding Human Cruelty, Human Misery, and, Perhaps, a Remedy (a Theory of the Socialization Process)*. Redondo Beach: Euphoria Press.

Twenge, J. M., Gentile, B., DeWall, C. N., Ma, D., Lacefield, K., & Schurtz, D. R. (2010). Birth cohort increases in psychopathology among young Americans, 1938–2007: A cross-temporal meta-analysis of the MMPI. *Clinical Psychology Review*, *30*(2), 145–154.

Ungar, M. (2010). Cultural dimensions of resilience among adults. In J. W. Reich, A. J. Zautra, & J. Stuart Hall (Eds.) *Handbook of adult resilience* (pp. 404–426). New York: Guilford.

Valutis, S. A., Goreczny, A. J., Abdullah, L., Magee, E., & Wister, J. A. (2009). Weight preoccupation, body image dissatisfaction, and self-efficacy in female undergraduates. *Journal of Psychiatry, Psychology and Mental Health*, *3*(1), 1–11.

Van Uden, J. M., Ritzen, H., & Pieters, J. M. (2014). Engaging Students: The role of teacher beliefs and interpersonal teacher behavior in fostering student engagement in vocational education. *Teaching and Teacher Education*, *37*, 21–32.

Waddell, G. R. (2006). Labor-market consequences of poor attitude and low self-esteem in youth. *Economic Inquiry*, 44(1), 69–97.

Wang, J., Nansel, T. R., & Iannotti, R. J. (2011). Cyber and traditional bullying: Differential association with depression. *Journal of Adolescent Health*, *48*(4), 415–417.

Wang, H. H., Wu, S. Z., & Liu, Y. Y. (2003). Association between social support and health outcomes: A meta-analysis. *The Kaohsiung Journal of Medical Sciences*, *19*(7), 345–350.

Williams, P. (2011). Pathways to positive education at Geelong Grammar School. Integrating positive psychology and appreciative inquiry. *AI Practitioner*, *13*(2), 8–13.

WHO (2010). Prevention of bullying-related morbidity and mortality: a call for public health policies. Bulletin of the World Health Organization, *88*, 403–403.

CHAPTER 16

Teachers' Resilience: An Empirical Investigation Based on Demographic Profiles

G. N. SUMATHI, R. ARTHI, and FEBA THOMAS

Vellore Institute of Technology University, Vellore, Tamil Nadu, India

**Corresponding author. E-mail: sumathi.gn@vit.ac.in*

ABSTRACT

This chapter examined the level of resilience among teaching professionals based on demographic profiles. The study used descriptive research method. A cross-sectional survey was conducted among 65 teachers working in private and public schools and colleges in the selected towns of Vellore district. The teachers self-rated their level of resilience. Descriptive analysis, Mann–Whitney *U*-test and Kruskal Wallis H-test were used for analyzing the data. The finding of the study indicated high level of resilience among teachers. Also the resilience level differed across teachers' demographics. The future study could examine the role of mediators and moderators in the relationship of resilience with other outcome variables.

16.1 INTRODUCTION

In most of the literature, the term resilience is considered as an ability to bounce back from stressful situation (Rutter, 2006; *Rees et al.*, 2015). The American Psychological Association (2004) defined resilience as "the process of adapting well in the face of adversity, trauma, tragedy or even significant sources of stress." Perhaps, in reality resilience is likely to exist on a continuum in varying degrees across several realms of individual's life (Southwick et al., 2011). In order to develop resilience, it is important for oneself to resolve the problem promptly, shielding one's resistance at times of crisis, adopting new situations, coping with environmental problems and managing psychology during these circumstances (Kaner et al., 2011; Gurgan, 2006; Walsh, 2015; Luthar 2003). Such skill is highly necessary in a profession engaging in human relations and in professions where life conditions have a greater effect. Teaching profession is one such profession that often leads to exhaustion. Since teachers have to change his/her vision accordingly with daily changes, which critically needs resilience skill to overcome stress and burnout (Özbey et al., 2014). Increasing the resilience skill would help teachers to be more successful in their profession and also in turn help to improve their life satisfaction (Rees et al., 2015). Individuals who are considered to be more resilient seem to reflect more positive attitudes and emotions.

They can easily balance their negative emotions with positive attitude. Several studies in the area of psychology insisted the importance of resilience for employee effectiveness and productivity (Parker & Martin, 2009; Mayer 2006; Gu & Day 2007). Pretsch et al., (2012) claimed resilience as a foremost factor for well-being of the employee particularly in teaching context.

Understanding the role of emotions is indeed for teachers as it helps to shape their attitude and manage the stressful circumstances (Montgomery & Rupp, 2005). Resilient teachers are more likely to promote resilience among students in the classroom environment (Gu & Day, 2007). Further, the studies emphasized the investigation of teachers' resilience as a crucial factor in development of their self as well as for students' development (Howard & Johnson, 2000; Kyriacou, 2011). With such importance on the role of resilience, this study examines resilience among teaching professionals. Moreover, this study is important in a way that it emphasizes teacher's resilience skills that support them psychologically to handle different situations. Also, most of the studies adopted qualitative methods to study teachers' resilience, indicating that resilience is an important construct to predict teachers' behavior empirically. The upcoming section includes the theoretical review, literature review, methodology opted for the study, analyses, results and discussion, limitations of the study, and finally future direction of the study.

16.2 THEORETICAL BACKGROUND

16.2.1 RESILIENCE THEORY

Resilience as a term evolved over the past 70 to 80 years. Resilience theory addresses the adults, families, communities, workplaces, and children. Individual resilience includes research on children, locus of control, and self-efficacy. Family resilience includes family stress research and family strength research. Community resilience includes social support systems, resilience-based policy, and resilience theory of social work, which includes a historical review of social work (Van Breda, 2001). Resilience involves a transactional dynamic process, which encompasses an adaptation process of goodness of fit. Resilience is influenced by diversity, which includes race, gender, and age (Greene, 2002). The resilience theory is considered as a person–environment concept. There are external and internal factors related to resilience (Greene et al., 2004). The results of the qualitative research study conducted among three African–American female teachers indicated that the teachers had long careers in education subject to adversity (Taylor, 2013). The resilience theory indicates eight characteristics of resilience: flexible locus of control, commitment, autonomy, optimistic bias, religion, relationships, and education. Polidore's resilience theory (2004) focuses on eight characteristics of resilience as depicted in Figure 16.1.

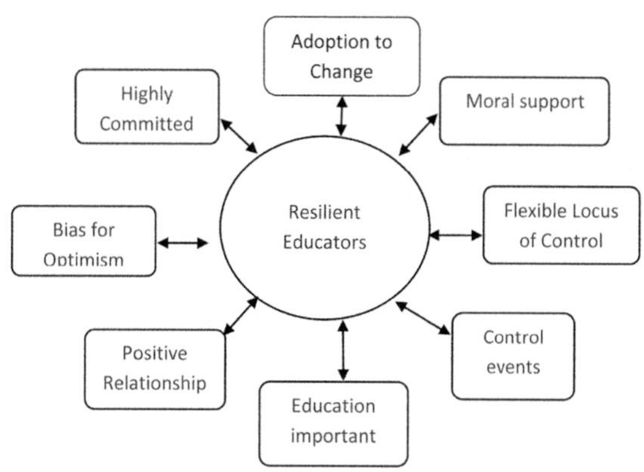

FIGURE 16.1 Graphical conceptualization of Polidore's resilience theory.

Resilience refers to the difficulty or negative contexts that individuals have the potential to disrupt or challenge from adapting to the functioning and development. A resilience framework offers a powerful tool for realizing the goals of positive psychology in the time of adversity. Resilience is dynamic as it emerges from many interactions within and between systems in a given cultural, developmental, and historical context that collectively influence the capacity of an individual system to adapt successfully to challenges (Yates et al., 2015). Resilience is conceptualized as a capacity of a dynamic system to adapt successfully to disturbances that threaten system function, viability, or development (Masten, 2014). According to Krovetz (2007), resilience theory is defined as protective factors among the schools, communities, and families that help children in their success. The schools and teachers help in building resilience among children. Resilience is built by providing intellectual stimulation, building relation among peers, and also providing job satisfaction among employees in the organization.

16.2.2 TEACHERS' RESILIENCE

Resilience is the capacity to exhibit the personal strengths needed to cope with difficulty or hardship and overcome the odds. It is the capacity to maintain competent functioning in the face of major life stressors (Kaplan et al., 1996). It is also referred to as the process of coping with stressful or challenging life events, coping with disruptive events, in a way that provides the individual with additional coping and protective skills that result from the event (Richardson et al., 1990). It represents one's capacity to cope successfully with stressors in the nonwork and work environment (Jex et al., 2013). It is defined as the "presence of factors that are protective such as personal, familial, social, and institutional safety which enable the individuals to resists life stress" (Kaplan et al., 1996). It can be developed over years through learning (Higgins, 1994). In association to this, Rutter (2006) and Benard (2002) stated the three most important aspects, such as individual characteristics, competences, and optimistic social environment, which are unique and combined in the process of resilience building. In line with this, empirical evidence clearly depicts that resilience is not influenced by a single factor rather it is influenced by the interaction of conditions in organization, social relationships, and individual characteristics (Gu, 2014). These distinct factors, such as individual characteristics, and organizational and social relationships are termed as promotive factors, which indeed operate in reversing the risk factors and developing positive views. Further, it is identified with two distinctions of these promotive factors as "assets and resources" (Zimmerman & Brenner, 2005). The asset represents the positive psychological factor of an individual such as self-efficacy, self-esteem, and motivation, which seems to reside inside the individual. In contrast, the resources represent the respective factors which are outside of an individual such as organizational support, family support, and societal support. Both these attributes are necessary for individual to become more resilient. For teaching profession, resilience is found to be the innate factor required to strive in adversity (Gu & Day, 2013). Resilience is an important characteristic needed for teachers as students see them as role models. Now teachers play extended roles as they are not only transmitting knowledge but also act as mentors and facilitators in the learning process of the students. The teaching role is rated to be one of the stressful professions (Nash, 2005). Therefore, their ability to bounce back to normal situation is critical both for them and for the students. The characteristics of resilience help to thrive, flourish, and sustain effectiveness in the teaching role. Broaden and build theory (Fredrickson, 2001) predicts that the positive emotions, such as joy, interest, and contentment over a period of time build resilience, as the positive emotions widen the individual's attention and thinking capabilities that enables creative thinking and adds further to the individual's ability to cope with stress. The emotional component is imbibed in the teaching role as a teacher brings not only competence to the classroom but also emotions such as anger, joy, love, depression, hope, and optimism. Resilience can fluctuate accordingly with individual as well as with professional circumstances (Day & Gu, 2007) and is a critical factor for teachers blooming in face of diversity (Ebersöhn, 2014). Resilience in general is an individual's attribute that helps one to cope with failures and conflicts, making them more responsible and capable of achieving self-efficacy and job satisfaction (Benight & Cieslak, 2011). Teachers' resilience is defined as a positive behavior that ensures teachers' long-term commitment toward their job in diverse educational environment (Beltman et al., 2011). Resilience is a relevant skill needed for teachers. It is needed for all professions worldwide (Boniwell & Ryan, 2012), since it allows individuals to adopt changes and regulate stressful situation by making use of the personal or environmental resources in adverse circumstances (Gibbs & Miller, 2014).

Some of the recent studies in the context of teachers have explored the factors that differentiate resilience and intention to quit. One such study by Hong (2012) identified the factors such as values, beliefs, emotional aspects, and self-efficacy that discriminates resilience and quitting behavior. According to Rep (2013), a study conducted among pre-school teachers in Tehran showed happiness and self-resilience as vital factors of teacher's self-efficacy. Since preschool is considered to be a critical phase for children, teachers should possess high self-efficacy which in turn helps them to be satisfied with their job and will also bring politeness among children through efficient communication.

Similarly, the resilience and retention among teachers depended on various influences. Student's behavior is commonly considered challenging and critical factor of stress for teacher's resilience. Resilience also influences teachers in dealing with stress, burnout, exhaustion, and fatigue (Chang 2009; Tait 2008; Bobek 2002). The consequences of burnout and stress are reported as exhaustion, reduced level of efficacy, loss of locus of control, and mismatch rewards (Maslach & Leiter, 2008). It is also identified that loss of resilience among teachers lead to consequences in three dimensions: first, to teachers as stress and burnout; second tends to create less managed learning atmosphere for students and third for management by imposing loss in perspectives of training cost (Gibbs & Miller, 2014). These in turn leads teachers toward leaving the profession (Hong, 2012; Tsouloupas et al., 2010).

Happy people seem to be more self-controlled and thereby prefer an effective stress coping mechanism in their work settings (Myers, 1992). Several studies explored the positive relationship between self-resilience and happiness (Rep, 2013; Meiklejohn et al., 2012; Chan 2009; Herman et al., 2015).

The teachers in the rural schools indicated a moderate to high level of happiness and resilience (De Jong, 2013). The three-dimensional model has been developed which includes life of pleasure, life of engagement, and life of meaning, and there was also a relationship between the subjective wellbeing and orientation for helping the teachers and reducing their stress and burnout (Chan, 2009). In order to develop resilience among teachers, it was suggested to overlook the challenges and adversities as a chance of self-growth (Wood et al., 2012). Few studies reported that individuals' core beliefs and values, peer relationship and support from their leaders, or all these collectively ensure development of resilience among them (Day et al., 2007, 2011). Teachers resilience has been conceptualized in several literatures based on the common idea and definitions as follows:

- "a quality that enables teachers to maintain their commitment to teaching and their teaching practices despite challenging conditions and recurring setbacks" (Brunetti 2006).
- "capacity to continue to 'bounce back', to recover strengths or spirit quickly and efficiently in face of adversity"; "a dynamic construct subject to influence by environmental, work-specific and personal contexts" (Sammons et al., 2007).

"Using energy productively to achieve school goals in the face of adverse conditions" (Patterson et al., 004).

- "a mode of interacting with events in the environment that is activated and nurtured in times of stress" (Tait, 2008).
- "capacity to overcome personal vulnerabilities and environmental stressors, to be able to 'bounce back' in the face of potential risks, and to maintain well-being" (Oswald et al., 2003).

To summarize, various definitions of resilience of teachers represent the dynamic process which can be developed through learning. The individual factors and contextual factors contribute more in building resilience. The individual factors include self-efficacy beliefs, coping skills, technical skills, and personal attributes such as patience, flexibility, intrinsic motivation, positive attitudes, perseverance, and emotional intelligence. The second factor related to contextual aspect includes organizational support, mentor support, peer support, support from family members, working with students, and support from friends. These factors are indeed necessary for fostering resilience among teachers.

16.2.3 TEACHERS RESILIENCE AND DEMOGRAPHICS

Previous research has examined the teachers' resilience across demographic characteristics, such as gender, age, additional qualification, specialization, and years of experience (Brouskeli et al., 2018). Most of the studies in the context of resilience revealed different results based on gender. Some studies showed that gender does not have any significant impact with respect to their resilience (Herman et al., 2015; Bowles & Arnup, 2016). In contrast, the experimental study carried by Gurgan (2006) and Özbey (2012) proved a meaningful difference in resilience with respect to gender. The female teachers are highly resilient compared to the male teachers. Age has been studied in the well-being and resilience research (King & Jex, 2014). About 78 percent of the female school teachers who were parents had work-to-family conflict and they had stress-related symptoms which had led the school teachers to resign (Kirchmeyer & Cohen, 1999). Gender and years of experience had a significant impact on the teachers' resilience. Resilience helps the majority of the teachers in building the teachers' self-perception and to withstand difficulty and also to bounce back at the time of adversity (Estaji & Rahimi, 2014). Gender and experience have an influence on the teachers' resilience among the vulnerable children. Female teachers have more resilience while dealing with vulnerable children (Silyvier & Nyandusi, 2015). Teachers' resilience has been studied across teachers' demographics: the years of experience, marital status, and the level of education. Teachers with children under the age of 18 years at their home reported higher resilience. Teachers who were married reported higher resilience than the single teachers (Morris, 2002). As a trait, resilience shows fluctuation. It is likely to be considered that teachers with more years of service and similarly with increasing years of age become more resilient (Gu, 2014; Morris, 2002). Alternatively, recent study on early career teachers (ECT) describes no significant differences based on the years of service and age related to resilience (Bowles & Arnup, 2016), Therefore based on the above studies, the following hypotheses were formulated.

H1: Female teachers will be more resilient than male teachers.

H2: There will be difference in resilience among different age groups of teachers.

H3: Teachers with more number of years of teaching experience will be more resilient than the teachers with less number of teaching experience.

H4: There will be difference in resilience among teachers working in schools and colleges.

16.3 METHODOLOGY

The study follows a descriptive research design. The sample for the study included 65 teachers (schools, $n = 29$, and colleges, $n = 36$) in Vellore district, Tamil Nadu. The respondents of the study are selected based on the purposive sampling technique.

16.3.1 INSTRUMENTS

16.3.1.1 THE RESILIENCE SCALE (RS-14)

The resilience of the teachers was measured using the resilience scale developed by Wagnild (2009). The scale comprised of 14 items (e.g., "I usually manage one way or another; I usually take things in stride; my belief in myself gets me through hard times; when I am in a difficult situation, I can usually find my way out of it"). The respondent rated each item using a 7-point Likert scale where 1 indicated "strongly disagree" and a score of 7 indicated "strongly agree." The reliability value of the resilience scale (RS-14) was measured using Cronbach's alpha, which was found to be 0.843. Also, the face validity of the scale used is confirmed through experts in the field of psychology. Various demographic factors such as age, gender, teaching experience, and type of working institution were included.

16.3.2 DATA ANALYSIS

Using SPSS 22, statistical analyses were carried out that included descriptive analysis, Mann–Whitney *U*-test, and Kruskal–Wallis *H*-test. Both Mann–Whitney *U*-test and Kruskal–Wallis *H*-test were used to identify the significant differences across demographic profile (age, gender, teaching experience, and type of working institution) with respect to the resilience level. The mean of resilience value is 3.67 (SD = 0.62).

16.4 RESULTS

The results of the study are described in this section. The results of demographic analyses are presented in Table 16.1 and the level of resilience is given in Table 16.2.

TABLE 16.1 Demographic Profile of the Respondents

Profile	Options	Mean	SD	N	Percentage
Gender	Male	1.46	0.50	35	53.8
	Female			30	46.2
Age	24–26 years	3.05	1.02	7	10.8
	27–29 years			11	16.9
	30–32 years			19	29.2
	Above 33 years			28	43.1
Teaching experience	1–3 years	2.98	1.08	7	10.8
	Above 3–5 years			12	18.5
	Above 5–7 years			21	32.3
	Above 7 years			25	38.5
Type of Institution	School	1.55	0.50	29	44.6
	College			36	55.4

TABLE 16.2 Level of Resilience of the Teachers

Level of Resilience	Percentage	Frequency
High	19	31
Moderate	26	40
Low	20	29

Table 16.1 indicates the demographic profile of the respondents. Among 65 respondents (teachers) in the sample, 54% were male and 46% were female. Majority of the respondents were above 33 years of age (43%, SD=1.02). With respect to experience, the majority of the respondents possess more than 7 years of teaching experience (39%, SD=1.08), whereas 32% possess above 5 years of experience. About 45% of respondents are teachers in schools, whereas 55% are teachers at colleges.

Based on the frequency distribution and percentage of overall 65 respondents (teachers) as mentioned in Table 16.2, 40% of respondents possess the moderate level of resilience which is comparatively higher than other levels. Hence, in this study, the majority of the teachers accounted for the moderate resilience level.

16.4.1 DIFFERENCE IN RESILIENCE LEVEL ACROSS TEACHERS' DEMOGRAPHICS

With regard to the major objective of the study, the differences among teacher's personal attributes (demographical profile) toward resilience are explored. The Kruskal–Wallis and Mann–Whitney *U*-tests (nonparametric analyses) were performed because of the violations in the homogeneity distribution of the data. The results obtained were discussed as follows:

16.4.1.1 GENDER AND RESILIENCE

The result of Mann–Whitney *U*-test showed that female (median = 56.00; mean rank = 33.25) scored higher on resilience than male (median =54.00; mean rank = 32.09). The Mann—Whitney *U*-value was found to be statistically insignificant $U = 514$ ($Z = -0.154$), $p > 0.05$ and the difference between male and female groups was small ($r = -0.01$). This result reflects that there exists no significant difference based on gender with resilience. Hence, Hypothesis (H1) is not accepted.

16.4.1.2 AGE AND RESILIENCE

The Kruskal–Wallis *H* test revealed that there exists a significant difference in the resilience level across the four diverse age groups, chi square (3, $n = 65$) = 19.43, $p = 0.000$. The age group (33–35) years reported the highest score (median = 56.00) and the lower score was reported by the age group of (27–29) years with median = 44.00. Among the remaining two age groups, 24–26 years age group reported the median value of 40.00 and 30–32 years age group scored a median value of 54.00, respectively. Thus, the hypothesis (H2) is accepted.

Since there exists a significant difference among the groups, further to examine which group differed from each other, a post-hoc test using pairwise comparisons was performed. The results from Table 16.3 showed that the age group (33–35 years) found significantly different from age group 27–29 years, ($p = 0.017$) and age group 24–26 years, ($p = 0.001$). Thus, the age group above 33 years shows greater difference with respect to resilience.

TABLE 16.3 Post-Hoc Test: Pairwise Comparison (Sample Average Rank of Age)

Group 1–Group 2	T-Statistics	Sig	Adj. Sig
27–29 years—above 33 years	–20.049	0.003	0.017**
24–26 years—30–32 years	–19.989	0.017	0.099
30–32 years—Above 33 years	–10.690	0.056	0.339
24–26 years—27–29 years	–10.630	0.244	1.000
27–29 years—30–32 years	–9.359	0.190	1.000

** The significance level is 0.05.

16.4.1.3 *TEACHING EXPERIENCE AND RESILIENCE*

The Kruskal–Wallis test based on the effect of teaching experience with the resilience level resulted in a statistically significant difference, chi-square (3, $n = 65$) = 20.03, $p = 0.000$ among the groups. The teaching experience of above 7 years recorded the highest median score (56.00) and those who served between 1 and 3 years reported the lowest level of resilience (median = 44.00). This result reflects that there exists significant difference based on teaching experience with the level of resilience. Therefore, the hypothesis (H3) is accepted.

Since there exists a significant difference among the groups, further to examine which group differed from each other, a post-hoc test using pairwise comparisons was performed. The results from Table 16.4 indicate that the group with an experience of above 7 years found significantly different to the group with 1-3 years of experience ($p = 0.000$) and group with 3-5 years of experience ($p = 0.014$). Thus, the group above 7 years of experience showed greater difference with respect to resilience.

TABLE 16.4 Post-Hoc Test: Pairwise Comparison (Sample Average Rank of Experience)

Group 1–Group 2	T-Statistics	Sig	Adj. Sig
1–3 years—above 3.5 years	–11.732	–1.308	1.000
1–3 years—above 5–7 years	–19.857	–2.413	0.095
1–3 years—above 7 years	31.917	–3.958	0.000**
Above 3–5—above 5–7 years	–8.125	–1.191	1.000
Above 3–5 years—above 7 years	–20.185	–3.048	0.014**
Above 5–7 years—above 7 years	–12.060	–2.161	0.184

** The significance level is 0.05.

16.4.1.4 *TYPE OF INSTITUTION AND RESILIENCE*

The results of the Mann—Whitney *U*-test indicated no significant difference between the teaching professionals who are working in schools and colleges; the median score showed 56.00 and 54.00, respectively (U = 489, Z = –0.318, p = 0.751). Since the obtained value of p >0.05, there is no significant difference with respect to type of working organization and resilience level. Hence, hypothesis (H4) is not accepted.

16.5 DISCUSSION

In line with the previous research findings, gender did not show a significant difference based on the resilient characteristics of teachers (Herman et al., 2015; Bowles & Arnup, 2016). Also, based on the mean score, it is identified that female teachers' level of resilience is found to be slightly higher than that of male teachers because resilience is the capability to bounce back to normal; individuals adopt coping strategies in order to cope with stressful situation. When encountered with stressful situation, female exhibits more of emotion-focused strategy and seeks greater social support than that of male (Melendez et al., 2012).

The years of teaching experience with respect to resilience characteristic of teachers showed significant difference. With increasing number of years, working individuals tend to be more resilient, as they know how to manage stressful situation. Thus, by practice, they can easily discriminate the coping strategies with respect to the situation encountered; this result is also supported by few studies in which experience has an influence over resilience (Gu, 2014).

Similarly, demographic characteristic based on teachers' age showed significant difference with respect to the resilience level. This may be possible because with the increasing age, individuals are capable to manage negative situation more in accordance with the prevailing conditions of the working environment (Herman, 2015; Morris, 2002). Furthermore, the type of working institution showed no significant difference in teachers' resilience. The findings were found to be consistent with the previous study (Bowles & Arnup, 2016; Gu, 2014).

Thus, the findings indicate that resilience among teachers was found to be different across the demographic characteristics, such as age and years of teaching experience. The teachers in the age group greater than 33-35 years and with more than 7 years of teaching experience reported the high level of resilience. However, the gender and the type of working institution did not have any influence over the resilience level.

16.6 CONCLUSION AND RECOMMENDATION

Resilience is considered as a critical characteristic among teaching professionals and is accounted to oscillate over time. There is a scant of research studies conducted in the area of resilience with respect to demographical characteristics of the teachers. Based on this view, the study was attempted with the major demographics (gender, age, experience, and type of education institute working) of teachers. The result of the study emphasized that the demographic characteristic particularly age and experience of the teachers influenced their resilience. Working consistently under stressful circumstances will likely pave way for the individuals in sustaining stressors. But in either way consistently working under such stressful situation should not make them to learn and develop the capabilities to sustain stressors. Thus, it is viable to make them engage in learning and practicing skills to become more resilient. Also, these results emphasize that teachers with less experience showed a lower level of resilience when compared with more experienced teachers. As the characteristics of resilient are dynamic and can be learnt, efforts need to be made from the organizational level to enhance the resilience level of new entry level and less experienced teachers. Despite the fact that the implication of this study reflect that age and teaching experience have significant difference on the level of resilience; a more detailed research through longitudinal studies should be carried out to establish other personal and organizational factors, which are necessary to develop the skill of resilience. In addition, indirect and direct effects can be analyzed using suitable mediators and moderators with respect to teachers' resilience.

KEYWORDS

- **resilience**
- **demographics**
- **teachers**
- **Vellore district**

REFERENCES

American Psychological Association. (2004). The road to resilience. *http://helping. apa. org/resilience/.*

Beltman, S., Mansfield, C., & Price, A. (2011). Thriving not just surviving: A review of research on teacher resilience. *Educational Research Review*, *6*, 185–207.

Benard, B. (2002). Applications of resilience. In *Resilience and Development* (pp. 269–277). Boston, MA: Springer.

Benight, C. C., & Cieslak, R. (2011). *Resilience and Mental Health: Challenges across lifespan: Cognitive factors and resilience: how self-efficacy contributes to coping with adversities*. Cambridge: Cambridge University Press.

Bobek, B. L. (2002). Teacher resiliency: A key to career longevity. *Clearing House 75*, 202–205.

Boniwell, I. & Ryan, L. (2012). *Personal well-being lessons for secondary schools: Positive psychology in action for 11 to 14 year-olds*. Oxford: McGraw-Hill.

Bowles, T., & Arnup, J. L. (2016). Early career teachers' resilience and positive adaptive change capabilities. *The Australian Educational Researcher*, *43*, 147–164.

Brouskeli, V., Kaltsi, V., & Loumakou, M. (2018). Resilience and occupational well-being of secondary education teachers in Greece. *Issues in Educational Research*, *28*, 43—60.

Brunetti, G. J. (2006). Resilience under fire: Perspectives on the work of experienced, inner city high school teachers in the United States. *Teaching and Teacher Education*, *22*, 812–825.

Tsouloupas C. N., Carson, R. L., Matthews, R., Grawitch, M. J., & Barber, L. K. (2010). Exploring the association between teachers' perceived student misbehavior and emotional exhaustion: The importance of teacher efficacy beliefs and emotion regulation. *Educational Psychology, 30*, 173–189.

Chan, D. W. (2009). Orientations to happiness and subjective well-being among Chinese prospective and in-service teachers in Hong Kong. *Educational Psychology*, *29*, 139–151.

Chang, M. (2009). An appraisal perspective of teacher burnout: Examining the emotional work of teachers. *Educational Psychology Review, 21*, 193–218.

Day, C., Edwards, A., Griffiths, A., & Gu, Q. (2011). Beyond survival: Teachers and resilience. In *Key Messages from an ESRC Funded Seminar Series,* Nottingham.

Day, C., & Gu, Q. (2007). Variations in the conditions for teachers' professional learning and development: Sustaining commitment and effectiveness over a career. *Oxford Review of Education*, *33,* 423–443.

Day, C., Sammons, P., Stobart, G., Kington, A., & Gu, Q. (2007). *Teachers matter: Connecting work, lives and effectiveness*. Buckingham, UK: Open University Press.

De Jong, H. (2013). *Measuring resilience, happiness and sense of coherence of teachers in rural schools* (Doctoral dissertation), University of Pretoria.

Ebersöhn, L. (2014) Teacher resilience: Theorizing resilience and poverty. *Teachers and Teaching, 20*, 568–594.

Estaji, M., & Rahimi, A. (2014). Examining the ESP teachers' perception of resilience. *Procedia-Social and Behavioral Sciences*, *98*, 453–457.

Fredrickson, B. L. (2001). The role of positive emotions in positive psychology: The broaden-and-build theory of positive emotions. *American Psychologist*, 56, 218.

Gibbs, A. S., & Miller, A. (2014). Teachers' resilience and well-being: A role for educational psychology. *Teachers and Teaching: Theory and Practice, 20*, 609–621.

Greene, R. R. (2002). *Resilience: Theory and research for social work practice.* Washington, DC: NASW Press.

Greene, R. R., Galambos, C., & Lee, Y. (2004). Resilience theory: Theoretical and professional conceptualizations. *Journal of Human Behavior in the Social Environment, 8*, 75–91.

Gu, Q. & Day, C. (2007) Teachers' resilience: A necessary condition for effectiveness. *Teaching and Teacher Education, 23*, 1302–1316.

Gu, Q. (2014). The role of relational resilience in teachers' career-long commitment and effectiveness. *Teachers and Teaching: Theory and Practice, 20*, 502–529.

Gu, Q., & Day, C. (2013). Challenges to teacher resilience: Conditions count. British *Educational Research Journal, 39*, 22–44.

Gurgan, U. (2006). *The Effect of the group program of resiliency level of university students.* (Unpublished Doctoral Dissertation), Ankara: Ankara University.

Herman, J. P., Maroun, M., & Richter-Levin, G. (2015). Good stress, bad stress and very bad stress. Stress: *The International Journal on the Biology of Stress, 18*, 267–268. doi: 10.3109/10253890.2015.1087091

Higgins, G.O.C. (1994). *Resilient adults: Overcoming a cruel past.* San Francisco, Jossey-Bass.

Hong, J. Y. (2012). Why do some beginning teachers leave the school, and others stay? Understanding teacher resilience through psychological lenses. *Teachers and Teaching, 18*, 417–440.

Howard, S., & Johnson, B. (2000). What makes the difference? Children and teachers talk about resilient outcomes for children 'at risk'. *Educational Studies, 26*, 321–337.

Jex, S. M., Kain, J., & Park, Y. (2013). Situational factors and resilience: facilitating adaptation to military stressors. *Building psychological resilience in military personnel: Theory and practice*, American Psychological Association, 67–84.

Kaner, S., Bayrakli, H., & Guzeller, C. O. (2011). Investigating perception of parental resilience in terms of some variables. *Ankara University Faculty of Educational Sciences Journal of Special Education, 12*, 63–81.

Kaplan, C. P., Turner, S., Norman, E., & Stillson, K. (1996). Promoting resilience strategies: A modified consultation model. *Social Work in Education, 18*, 158–168.

King, R., & Jex, S. (2014). Age, resilience, well-being, and positive work outcomes. In *The Role of Demographics in Occupational Stress and Well Being*. Emerald Group Publishing Limited.

Kirchmeyer, C., & Cohen, A. (1999). Different strategies for managing the work/non-work interface: A test for unique pathways to work outcomes. *Work and Stress, 13*, 59–73.

Krovetz, M. L. (2007). *Fostering resilience: Expecting all students to use their minds and hearts well.* Corwin Press.

Kyriacou, C. (2011). Teacher stress: From prevalence to resilience. *Handbook of stress in the occupations*, Edward Elgar Publishing, 161–173.

Luthar, S. S. (2003). *Resilience and vulnerability: Adaptation in the context of childhood adversities.* Cambridge University Press.

Maslach, C., & Leiter, M. P. (2008). Early predictors of job burnout and engagement. *Journal of Applied Psychology, 93*, 498–512.

Masten, A. S. (2014). Global perspectives on resilience in children and youth. *Child Development, 85*, 6–20.

Mayer, D. (2006). The changing face of the Australian teaching profession: New generations and new ways of working and learning. *Asia-Pacific Journal of Teacher Education, 34*, 57–71.

Meiklejohn, J., Phillips, C., Freedman, M. L., Griffin, M. L., Biegel, G., Roach, A., & Isberg, R. (2012). Integrating mindfulness training into K-12 education: Fostering the resilience of teachers and students. *Mindfulness, 3*, 291–307.

Melendez, J. C., Mayordomo, T., Sancho, P., & Tomas, J. M., (2012). Coping strategies: gender differences and development throughout life span, *The Spanish Journal of Psychology, 15*, 1089–1098.

Montgomery, C., & Rupp, A. A. (2005). Meta-Analysis for exploring the diverse causes and effects of stress in teachers. *Canadian Journal of Education 28*, 458–486.

Morris, B. A. (2002). *Measuring resilient characteristics of teachers.* (Unpublished doctoral dissertation), Athens: University of Georgia.

Myers, D. G. (1992). *The pursuit of happiness: discovering the pathway to fulfillment, well-being, and enduring personal joy.* Harper Collins Publishers.

Nash, P. (2005, April). Speech to work life support conference. In London Well Being Conference, London (Vol. 21).

Oswald, M., Johnson, B., & Howard, S. (2003). Quantifying and evaluating resilience-promoting factors: *Teachers' beliefs and perceived roles. Research in Education, 70*, 50–64.

Özbey, S. (2012). An investigation on candidate preschool teachers' communication skills and emotional intelligence.

New Trends on Global Education Conference (GEC-2012). Cyprus. 24–26 September.

Özbey, S., Büyüktanir, A., & Türkoglu, D. (2014). An Investigation of Preservice Pre-School Teachers' Resilience Skills. *Procedia-Social and Behavioral Sciences, 116*, 4040–4046.

Parker, P. D., & Martin, A. J. (2009). Coping and buoyancy in the workplace: Understanding their effects on teachers' work-related well-being and engagement. *Teaching and Teacher Education, 25*, 68–75.

Patterson, J. H., Collins, L., & Abbott, G. (2004). A study of teacher resilience in urban schools. *Journal of Instructional Psychology, 31*, 3–11.

Polidore, E. (2004). *The teaching experiences of Lucille Bradley, Maudester Hicks, and Algeno McPherson before, during, and after desegregation in the rural south: A theoretical model of adult resilience among three African American female educators* (Doctoral dissertation), Sam Houston State University, Huntsville.

Pretsch, J., Flunger, B., & Schmitt, M. (2012) Resilience predicts well-being in teachers, but not in non-teaching employees. *Social Psychology of Education, 15*, 321–336.

Rees, C. S., Breen, L. J., Cusack, L., & Hegney, D. (2015). Understanding individual resilience in the workplace: the international collaboration of workforce resilience model. *Frontiers in Psychology, 6*, 73.

Rep, A. S. (2013). Study of the relationship between happiness and self-efficacy with self-resilience of preschool teachers in Tehran. *Applied science Report, 2*, 10-16.

Richardson, G. E., Neiger, B. L., Jensen, S. & Kumpfer, K. L. (1990). The resiliency model. *Health Education, 21*, 33–39.

Rutter, M. (2006). Implications of resilience concepts for scientific understanding. *Annals of the New York Academy of Sciences, 1094*, 1–12.

Sammons, P., Day, C., Kington, A., Gu, Q., Stobart, G., & Smees, R. (2007). Exploring variations in teachers' work, lives and their effects on pupils: Key findings and implications from a longitudinal mixed-method study. *British Educational Research Journal, 33*, 681–701.

Silyvier, T., & Nyandusi, C. (2015). Factors influencing teachers' competence in developing resilience in vulnerable children in primary schools in Uasin Gishu County, Kenya. *Journal of Education and Practice, 6*, 172–182.

Sinclair, R. R., & Britt, T. W. (2013). *Building psychological resilience in military personnel: Theory and practice*. American Psychological Association.

Southwick, S. M., Pietrzak, R. H., & White, G. (2011). Interventions to enhance resilience and resilience-related constructs in adults. *Resilience and mental health: Challenges across the lifespan*, 289–306.

Tait, M. (2008). Resilience as a contributor to novice teacher success, commitment, and retention. *Teacher Education Quarterly, 35*, 57–75.

Taylor, J. L. (2013). The power of resilience: A theoretical model to empower, encourage and retain teachers. *The Qualitative Report, 18*, 1–25.

Van Breda, A. D. (2001). Resilience theory: A literature review. *Pretoria, South Africa: South African Military Health Service*.

Wagnild, G. M. (2009). *The resilience scale user's guide: for the US English version of the Resilience Scale TM and the 14-Item Resilience Scale TM (RS-14 TM)*. Resilience centre.

Walsh, F. (2015). *Strengthening family resilience*. Guilford Publications.

Wood, L., Ntaote, G. M., & Theron, L. (2012). Supporting Lesotho teachers to develop resilience in the face of the HIV and AIDS pandemic. *Teaching and Teacher Education, 28*, 428–439.

Yates, T. M., Tyrell, F.A., & Masten, A. S. (2015). Resilience theory and the practice of positive psychology from individuals to societies. *Positive psychology in practice: Promoting human flourishing in work, health, education, and everyday life*. John Wiley & Sons, pp. 773–788.

Zimmerman, M. A., & Brenner, A. B. (2010). Resilience in adolescence: Overcoming neighbourhood disadvantage. In: Reich J, Zautra AJ, Hall JS, (Eds.) *Handbook of adult resilience.* New York, NY: Guilford Press; 2010. pp. 283–308.

CHAPTER 17

Community Resilience and Place Identity

JHUMA MUKHOPADHYAY[1*] and SHUBHANGI GOSWAMI[2]

[1]*Asutosh College, University of Calcutta, Kolkata, West Bengal, India*

[2]*Indian Institute of Technology Kharagpur, Kharagpur, West Bengal, India*

[*]*Corresponding author. E-mail: dr.jhuma.m@gmail.com*

ABSTRACT

Community emphasizes the common ties and social interaction component. It is characterized not so much by locale as by a "high degree of personal intimacy, emotional depth, moral commitment, social cohesion and continuity in time." Permanent dwellers of a locality generally live together in such a way that they share the basic conditions of community life and develop a strong positive feeling for the locality which finally becomes a part of their respective identities. The place one lives in or the places one relates to, thus, form an extended identity of an individual. In difficult times, the place identity of an individual as well a community can also act as an agent contributing to coping and resilience (or bouncing back from adversities with a growth mindset). This chapter is an attempt to understand community resilience, and the contribution of place identity in it. Units of society—whether individuals or community, who are highly resilient can exhibit adaptive coping skill and deal with the environment more skillfully. When one has a sense of belongingness to an identified community, chances of anticipation of receiving resources from environment are more. But the intertwined nature of a community and the place they dwell in cannot be denied. Thus, the place identity becomes integral in community resilience.

17.1 INTRODUCTION

Resilience has been a much talked about concept lately. It has different connotations in different fields like physics, geosciences, telecommunications, social sciences, and more, making it a cross-disciplinary construct. In social sciences, resilience research is not merely concentrated on individuals, families, and communities, even cultural resilience is studied now (Fleming & Ledogar, 2008). Despite the vast and varied area of research, resilience is most easily and simply understood as the capacity of any system to bounce back despite adverse conditions. Our major thrust area in this chapter will be community resilience. To understand what community resilience means, perhaps it is better to start with understanding the word "community."

Community essentially refers to a unified collection with something common among them, which connects the units. The Merriam Webster online dictionary defines community from several standpoints. According to it, the community is a unified body of individuals with common interest living in an area (or broadly the area itself), or a group of people with a common characteristic or interest living together within a larger society, people with common professions, common history or common social, economic, and political interests. The community also refers to a social state or condition or the society at large or even the sense of caring among the members of a group—a community feeling (Merriam-Webster Online Dictionary, 2019). So we can see that the word community has multiple connotations and its meaning is pervasive in nature. Now, when we join these two concepts—community and resilience, it refers to a "community's ability to endure and recover from abrupt system shocks" (Summers et al., 2018). It includes both, dimensions of the natural ecosystems as well as the human communities, as both are intrinsically connected to each other (Summer et al., 2018). In the realm of environmental psychology, one is interested in studying this intricate and intertwined relationship between the environment and the human beings. Understanding a human being essentially requires studying the behavior in a context—here the context being the environment as a whole which comprises several levels like the immediate surroundings to the larger ecosystem. Ever since humanity thrived on Earth, humans evolved to be the only creatures that can significantly impact the environment. Our interaction with the environment has been continually shaping one another mutually. The Earth's and our history within it stands witness to evolution and change. This change has not been a mere mechanical manipulation of the physical setting around us. The responses to physical stimuli have been a result of the social nature of humans. Therefore, when we say environment, it not only refers to the physical aspects but also to the social components. Humans socialize, react to surroundings, and also perceive the symbolic meaning of the settings. In the broadest sense, we call the Earth our home whereas we all know that when enquired about our home in an everyday sense, we will most likely refer to our limited man-made residences. Thus, places around us have been explored and manipulated depending upon the meaning we have attached to them. Our dynamic responses also constitute the environment—the social environment embedded in the natural environment.

The relationships among people, people and nature, people and things, and relationships among things, are complex and multilayered, intimate, and reciprocal. The early hominids started their journey barefoot from Africa and today the world has become a global village as we have, and still are, changing our environment to suit the modernized and industrialized living. We can stop to think about the evolution of our need for space and place of living from caves to humble homes to skyscrapers, and from tribal colonies to residential buildings. Thus, within this vast natural environment, man finds a smaller unit that pertains to the physical and geographical dimensions of the environment being influenced by our social, political, and economic needs—it is called a "place." As we go further in the chapter, we will discuss this interrelationship among place, community, and resilience.

17.2 PLACE—ITS DEFINITION AND CONCEPT

A place can be understood as a unit of environmental experience (Canter, 1986), a convergence of how people perceive, understand and feel a place, and the behavior of the people who are experiencing them (Canter, 1991). The term place is very generic in nature and its use in different contexts shows the wide variety its meaning caters to. The place might convey a physical size, a tangible area, a symbolic entity, it might be known and experienced or unknown and novel. Whether be a natural open place like a landscape with mountains and rivers or a specific space like the cafeteria in the corner of a street, or an old market (the list can just go on), people form attachments with places and add meaning to it, sometimes these places serve as symbolic of something of importance (Williams & Vaske, 2003). The cafeteria might be a symbol of love where two lovers decided to get married while the river might be a fond memory of childhood picnic and good times or even a representation of the psyche of a person. Identification with natural places is not uncommon. When it comes to man-made spaces, today's architecture and city planning witnesses not merely fulfillment of the space requirement, but also an extension of personality, art, symbolism, and representation. Along with having definitive physical characteristics, places are changeable and acquire dynamic properties varying with social interaction and memory (Stokowski, 2002), thus making the place a fluid concept. Once people start residing in a place, these occupants give meaning to it; their personal, social, and cultural processes influence their definition of the place (Hummon, 1992). Hence, the place can be described in terms of multidimensional physical and psychological environmental attributes.

The concept of place has evolved from a location, to a place to dwell (the process of making a place a home), to the place having a spirit (e.g., Dilwalon ki Delhi, which roughly translates to Delhi being a place of whole-heartedness as it is where hearty people live), to ultimately establish a unique perception of architecture. Therefore, a place is beyond a physical space for dwelling and getting accustomed to it; it is also about the interaction with that place which gradually shapes our identity (Heidegger, 1962; Qazimi, 2014).

17.3 PLACE IDENTITY

Identity—the recognition of one's sameness as well as uniqueness (Jacobson-Widding, 1983), starts developing from an early age. As we grow, we associate ourselves with various social elements that render us our identity. Place identity is a substructure of the self-identity of a person, which is the broadly conceived cognition of the physical world. These cognitions are our perceptions, memories, attitude, values, preferences, meanings, etc., associated with a place. Starting from the room a child plays in the house, to the neighborhood, school, and so on; the place gets associated intrinsically with one's identity and becomes an inseparable setting in our memories.

Proshansky (1978), defined place identity as "the dimensions of an individual's self-identity in relation to his physical environment world by means of a complex pattern of conscious and unconscious ideals, beliefs, preferences, feeling, values, goals, and behavioral tendencies and skills which are relevant to his environment." In this sense, the setting offers an individual the opportunity to both express and affirm their identity.

Our biological, psychological, cognitive, aesthetic, and cultural needs are fulfilled by physical environment, for example, a place provides us with all—shelter, agrarian land, recreation, education, vocation, and so on. Since ages, this understanding of a place and its service to fulfill our needs has given us our "environmental past" (Proshansky, 1978, Proshansky et al., 1979; Proshansky & Kaminoff, 1982). With our environmental past and continued exploration of and settlement in places, place identity has become so important that we cannot express ourselves without taking into account the setting that surrounds us (Ryden, 1993).

Regardless of the various definitions and parameters to define "place," scientists converge at one point that place identity develops when there is some form of emotional bond/attachment to the place which is necessary for healthy psychological balance (Rowles, 1990; Lewicka, 2008). The emotional attachment to places makes people identify themselves in terms of the strong affective ties to house or home, neighborhood, community, and broader geographical units. A person's self-identity thus incorporates the conception of a place that forms a necessary and sufficient component for reference. If one understands race and ethnicities, it would not be difficult to point that they are our identities tracing back to our places of origin and how we are tied to a group of people, that is, a community. Surnames, especially in Indian culture, have been indicative of lineage and often the surnames have been direct names of either the occupation or the place of ancestral origin. Thus, we can see how place identity extends beyond one's lifetime and is transgenerational in nature.

17.4 PROBABLE FUNCTIONS OF PLACE IDENTITY

A sense of place is quite natural. It also has a subjective component. It starts with acquisition and access to natural resources of land and physical setting where the person lives. Whether be a perpetual mover like hunter, gatherer or bedouin, or having a fixed and permanent settlement, access to territory is a necessary condition for life and fundamental to human existence (Qazimi, 2014).

Formation of this relationship between place and identity feels like a beautiful meaning-making process where our thinking, reasoning, feeling, behaving, motives, curiosity—all work in unison with the places around us to give us, our identity, a shape. Objectively, every place has a temporal life. For instance, everyone can agree that a house or a residential alley would be recognizable even after ten to fifteen years, provided no sudden or acute damage has taken place, quite similar to meeting a person after many years. But for different individuals, the same place evokes so many different memories. What meaning one gives to a place and in turn how the individual's or a community's identity is shaped by these is central to the concept of place identity. Associations or bonds with places vary from no attachments to very high attachments. "Feeling at home" surely sounds like high attachment. Our images and memories of kindergarten, our homes, visit to a fair, first workplace or best workplace or worst workplace, or even a dream place—all have something to contribute to our identity, the way we are and the way we behave. But what purpose does place identity serve? Physical settings give an organism what it requires and in turn, the organism gives meaning to it. Different cognitive clusters aggregate to give rise to these meanings. They indicate what we can expect from the setting, how to behave in it, and how others might behave in it. These cognitive clusters, and further the meanings of places, enable a person to recognize a setting, to understand the intended purpose of the place, the activities, and thus sense an overall tone of the place in relation to its substantive properties like spatial design. Meaning about a place has the cognitive building blocks of attitude (Bem, 1970). These attitudes translate the physical setting into consideration and its attributes to beliefs, attitudes, and identity.

Place as self-identification incorporates the following aspects: personalization, individualism, appropriation, social and gender identification with home, identification as belonging, feeling in control, feeling habituated or adapted, and feeling that one has the freedom to do as one chooses. More personalized places and "ambiences" have some characteristics which reflect the personality of the occupant or architect. For instance, in case of home, Kron's (1983) Home-Psych research findings suggest that individuals invest a portion of their personalities into home decoration and homes can be built keeping their personalities in mind. Similarly, urban planning and design take into account the identities and characteristics of the residents. When we talk about home, city, and urban planning, we are focusing on "built" environment and shaping a place according to our needs rather than fitting ourselves into the place available. This concept of "built environment" reflects our associations with structured physical environment and beliefs about the urban life. The node that led to this development toward industrialization and urbanization is of course, life. Thus, environment and life are inseparable. The urban residents have a strong belief in this human-built environment system and attachment to this way of life which is commonly referred to as urban-identity. An array of physical settings and complex cognitive associations connect an individual's conception of the city with their self-identity. Such cognition of city dwellers extends to all the settings of life which are a part of that person's history and current life in the city. Dwellers of a place would seem to be far more involved with the "additive" consequences of positive and negative attachments due to variety, complexity, and diversity of specific physical settings in an urban context. The lifestyle that underlies urban-identity undoubtedly involves such general factors of the big city due to its size of space and number of people, complexity and dispersion of the visual scene, background noise level, the variety and the rapidity of technological change and still other general aspects of an urban existence (Proshansky et al., 1983). Thus, place identity holds an immensely important place in urban planning and design.

How organisms perceive and behave to urban physical settings depend on how individuals or groups are concern with problems of privacy, personal space, crowding, and territoriality. There is evidence that the spaces and places that young boys perceive to or have "adventures in" are quite different than those of young girls (Saegert & Hart, 1978). These inseparable ideas of different types of places etched in our memories during our formative years make place identity closely linked to self-identity and its other subsystems, in this case, the gender identity.

Place and its natural, historical, social, and political markers are inseparable. This readily tinges an individual's as well as communities' identity. The socio-geo-political conditions of a place render a community its position in the society. For instance, the social and political situations determine when a community can be called marginalized. Place identity can also act as an agent of change in that matter, by providing the community identity and a source of resilient action against oppression by more powerful communities (Smith & Bender, 2001). It is perhaps the strong place identity that forms the basis of patriotism. The political history of the world cannot be explained without the map or globe, pointing not merely to geographical locations, but to the strong characteristics of the communities that dwell in it, relocate, explore, cooperate, and compete.

Place identity can give a fair idea regarding the evolution of a place as well as social constructs and hence about the community as well. The phrase "When in Rome, be a Roman" must have some basis in that these few words reflect that people belonging to different places have unique characteristics. Place identity gets embedded in our lifestyle—our cuisine, clothing, language, diction, and many other aspects of our lives are intertwined with the place we associate with. This gives rise to stereotypes as well. But stereotypes can be considered as a cognitive shortcut to understanding someone. An Indian with a British accent gives us information about the ancestry and relocation of that individual along with racial origin. Based on these physical pieces of information one readily forms an impression about the individual, one of the most important being the characteristics pertaining to his/her community. Till now, it is quite clear that place identity is deeply connected with the community. Let us explore a little more of this and about community resilience.

17.5 COMMUNITY AND PLACE IDENTITY

Community emphasizes on common ties and social interaction. It is characterized not so much by location as by "a high degree of personal intimacy, emotional depth, moral commitment, social cohesion, and continuity in time" (Bernard, 1973). Permanent dwellers of a locality generally live together, so as to share the basic conditions of a community life and develop a strong positive feeling for the locality which finally becomes a part of their respective self-identity. Social ecology reflects variation in the social and cultural phenomena generated out of behavioral adjustment of the local community. It involves incidents along a psychological plane. It is always value-oriented and a conditional adaptation for the cause of the well-being and quality of life of the community member concerned. High place identity or attachment helps to develop in them strong sense of belongingness, unanimity, and a subculture. This is expressed in their social behaviors as reciprocal altruism. The dwellers of a local community are influenced by common sentiments, felt-needs, and perceived problems for survival. Accordingly, it may be assumed that high place attachments of community members contribute to generate common "perceived characteristics" of the locality concerned.

The social–environmental characteristics of place determine a sense of community; but with an urban living, community characteristics have become linked to residents' perceptions of physical characteristics of the built environment (Plas & Lewis, 1996). Sarason (1974) described the *sense of place* as "the extent to which a person feels part of a readily available, supportive, and dependable structured environment, in which one belongs." On the lines of belongingness, McMillan and Chavis (1986) assert that *sense of community* refers to "a feeling that members have of belonging, a feeling that members matter to one another and to the group, and a shared faith that members' needs will be met through their commitment to be together." The two aforementioned statements go hand in hand. Shared beliefs, common goals, and belongingness to a community make both, the people and place, dependable, supportive, and structured. Place attachment can influence the anticipation of receiving resources from a community as well as reciprocating with one's own resources. Other than a sense of belongingness and attachment with a place, another component that links communities to places is satisfaction about a place. The utilitarian value of place, which is all about meeting certain basic needs (Guest & Lee, 1983) ranging from physical characteristics of that place to the characteristic sociability of the place (Fried, 1982; Herting & Guest, 1983; John et al., 1986) determine the sense of satisfaction regarding a place.

Community and place are linked through generations of social behavior and culture. Adults in the community would often delineate their community sentiments and their judgments are considered very important, especially when it comes to place identity in rural or tribal areas (Goudy, 1990; Rowler, 1990; Stinner et al., 1990; O'Brien et al., 1994). Longer the residence period in a community, higher is the number of possible indicators of attachment and identity (Kasarda & Janowtiz, 1974; Sampson, 1989; Goudy, 1990; Rowler, 1990). The permanent and cross-generational dwellers grow and develop with the place and within the place and in turn with the community. In such cases, the place serves to maintain the sense of continuity of one's identity (Rubinstein & Parmelee, 1992).

A life-span perspective of community and place identity reveals the importance of the sense of place in the development of self-identity and its impact on the maturation processes (Proshansky et al., 1983; Sarbin, 1983; Proshansky & Fabian, 1987; Hummon, 1992; Giuliani & Feldman, 1993). Our relationship with the places we live in changes throughout the life span, but a sense of continuity in the community is indicative of well-being among the youth (Pretty et al., 1994, 1996; Chipuer et al, 1999). Elder et al. (1996) assert that the influence of place identity and attachment to their place and community on young people in rural communities is profound who move to other places, especially urban areas, looking for economic stability/future career when threatened by economic decline. Organizations have in-patriation, expatriation, and repatriation issues discussed extensively by the human resources as community and culture can be a big issue when it comes to adjustment to a new place. With the development of places and mankind, change is inevitable. Changes can sometimes be smooth and readily acceptable while at times they are adversities and cause turbulence. Communities have shown to perish or thrive in the face of adversities throughout history. Let us take a look at the concept of community resilience and further discuss how place identity can contribute to resilient behavior in communities.

17.6 COMMUNITY RESILIENCE

Context plays a vital role in understanding the process of resilience. The context here refers to broad social/environmental conditions such as socioeconomic status, geography, culture, and so on. The context shaped by these social/environmental conditions is an important determiner of protective and risk factors of resilience (Fleming & Ledogar, 2008). With a community as the context, community resilience can be defined as an unrelenting capability of a community to withstand, respond to, and recover from adverse situations by utilizing available resources (Bosher & Chmutina, 2017). In the case of ecological disequilibrium, natural disasters, economic fallout, political crisis, and any other adversity which affects communities pervasively, community resilience comes to rescue. It is imperative to sustain a community and helps to strengthen it.

A part of resilience research focuses on building frameworks and plans to build resilient communities. Several frameworks of resilience are available in the literature but empirical testing of these frameworks is not very prevalent. One of the frameworks of resilience in the context of community resilience is the Panarchy framework (Gunderson & Holling, 2002) which is considered to be one of the most ambitious structures of resilience (Cutter et al., 2008). This framework displays a nested hierarchical structure where the componential dynamic systems interact across space and time. When pushed to its limits, the system reorganizes itself where the smaller, faster, and nested levels of the community formulate and test, while the larger, slower levels stabilize and conserve the system. As we talk about the resilience of a community, the environment, or the setting automatically comes into picture.

Along with indicating the exigencies, the resilience of a place also considers long-term mitigation and adaptation strategies to confront social, economic, familial, and environmental problems. Resilient thinking adds a strategic meaning to a sense of safety, security, and protection from potential threats to society, economy, and environment. Magis (2010) contends that community resilience is an important indicator of social sustainability. She further elaborates some measurable dimensions of community resilience which are as follows:

a) Community Resources—Refers to the natural, human, cultural, social, financial, built, and political capital within the community and how the members of the community engage these resources.
b) Resource Development—Resources vary in their availability; they are dynamic in nature. To understand the availability of resources, their limits, their wise usage, and accordingly plan and develop strategies imply resource development. From choosing their political representatives to following civil rules, everything can contribute to resource development of a community.
c) Resource Engagement—Refers to active engagement of the resources toward a shared objective of the community. This strengthens the purpose and leads to a self-reinforcing cycle of identifying the resources, developing resources, and active engagement.
d) Active Agents—Needless to say, the community members are the active agents who can bring about the required change. People who can contribute to the well-being of the community and influential leaders play a vital role in this regard. Active communities have self-reliant members who can efficiently work in collective groups and are driven by community feeling.
e) Collective Action—As mentioned above, collective action is characteristic of a resilient community. Collective effort and common goals are the necessary preconditions to it.
f) Strategic Action—It might seem like an obvious consequence of collective effort but strategic action points out to the conscious and well-planned efforts made by the community members. A foresight is needed to take action. A good leader plays a vital role in this regard.
g) Equity—A resilient community looks into the entire community's benefit. Equity refers to the equal distribution of and access to all benefits—social or economic, to all the social groups within a community. This can ensure the optimum involvement of all members in order to fulfill all the aforementioned points.

h) Impact—It is what we see or experience after the crises. Resilience is not merely withstanding the odds and/or opportunities but also thriving. Communities undergo changes and how resilient actions impact their nature is an integral dimension of community resilience.

17.7 URBAN RESILIENCE

As discussed in the previous section, community resilience is more than responding to threats and adversities as it also marks the interdependence, cooperation, and strategic relationship between a community and the environment (Wilkinson, 2012). Cities within a resilience framework are seen as a complex adaptive system which helps consider the interdependence between communities and environment where culture and social relations play an important role. The relation between social and the ecological system is seen through the effectiveness of public policy and governance, the adaptive capacity of communities and the role of nature and the benefit it offers in terms of ecosystem service (De Groot et al., 2002; Walker & Salt, 2012). This aspect can help planners and policymakers to take into account, both, immediate and long term place-based effect on social, economic, and environmental changes. Urban places have experienced radical changes all over the world. The urban renaissance has strengthened the complex developments as the world is rapidly moving forward in technology with the ever-increasing needs of people. Between socialism and capitalism, reforms in the labor laws (Giddens, 1998) contributed in the radical shift in urbanization. Increasing industrialization, global businesses, and world policies prompted the development of many small and medium-sized towns and cities. New retail districts, public squares, high-rise apartments, and multinational companies—all these have promoted physical proximity of different social, economic, and ethnic groups in urban areas (Jones & Mean, 2010).

Urban resilience has, therefore, a different connotation as urban areas have their own challenges accommodating multicultural residents and catering to their demands. Sustainable city planning, thus, essentially takes into account the place identity of its residents in order to make the community more resilient.

17.8 SUMMING UP THE LINK BETWEEN PLACE IDENTITY AND COMMUNITY RESILIENCE

Throughout the chapter, the terms community and place identity have appeared sporadically in order to express the immediate links between them. The intertwined nature of these two concepts is hopefully well emphasized by now. So, summing up these connections, all the dimensions of community resilience (Magis, 2010) have a core feature—the feeling of oneness, belongingness to a group having some common objective. Without this core of identification with the community, being resilient is probably not possible. Thus, we can see the intricate relationship between a community and a place, place identity and community, and in turn the relationship between place identity and community resilience. Resilient communities have displayed a strong place identity and attachment to places. For instance, the farmers of Majuli Island in Assam, India display resilient actions. This island gets flooded every year but due to the lack of alternative resources of income and also a strong sense of belongingness to their place of dwelling, they continue to revive their lives every year. The relocated refugees carry a sense of their identity in the form of their place of origin. Marginalized communities fight back standing on the pedestal their place identity gives them. Numerous such examples of communities fighting back adverse situations based on their place identity can be cited throughout the history of humanity. Adversities can strike any time during the life span, but children growing up in adverse conditions and resilient communities must be experiencing something profound which shapes their identities.

17.9 RESILIENCE AND PLACE IDENTITY IN CHILDREN

The very concept of identity starts developing from childhood. The process of identity development is crucial to children's psychological development. The place children come in contact with has an immense contribution to this. The home, the neighborhood, interaction with nature, educational institute, and exposure to various natural and man-made places—all these shape the experience of children. For example, children growing up in hilly regions identify closely with hills and mountains while those who have grown up in coastal areas know the sea better. Children growing up in metropolitan cities can identify with the traffic and might be extremely comfortable with the dependency on technology. However, a village boy/girl might be able to identify very well with the ways and means of the countryside, be familiar with animals and farms. We can easily imagine that the resilience of children who grow up in flood-prone or earthquake-prone zones will be different from children who do not. Children growing up in politically volatile places will have a different approach in dealing with upheavals than the children who have watched these only on news channels. In this regard, place identity again might play a vital role as any memory they have of any significant adversity, must have had a geographical background or place associated with it. An association of a place with a strong community calls for a resilient attitude. A strong sense of community right from childhood can make for a vital support system affecting resilience throughout life.

17.10 CONCLUSION

One can say without a doubt that place—whether be tangible or intangible, forms a setting for any set. Place identity forms an integral part of our self-identity. On a global scale, our identity comes up as an Asian or American or Egyptian or Irish—in essence, related to a physical space. The place is also characterized by its community and the flavors that define the community. Community resilience can be easily seen in places that get afflicted by calamities very often, yet never fail to rebuild themselves. In urban planning, theory and practices, pertinently in the approach to spatial equilibrium, resilience has been an important point that has prevailed. Ordering the space is necessary when it comes to designing and transforming cities on geometrical and functional bases (Davoudi, 2012). So, in the modern world, when the architects, engineers, planners, and policymakers know the importance of creating a resilient place for a resilient community, the design of the place often reflects a state of equilibrium in all functions. It might be achieved by the power of planmaking, emphasizing on improving physical accessibility, and developing complimentary function (Davoudi, 2012). Therefore, when one goes beyond the psychological and social characteristics of a community, its resilience and resourcefulness is reflected by the way it would maintain its equilibrium as identified by environmental planners and policymakers.

KEYWORDS

- **community**
- **resilience**
- **place identity**

REFERENCES

Bosher, L. & Chmutina, K. (2017). *Disaster risk reduction for the built environment.* Wiley Blackwell.

Bem, D. J. (1970). *Belief, attitude and human affairs* (p. 246). Pacific Grove, CA: Brooks/Cole.

Bernard, J. (1973). *The sociology of community,* Glen View, IL: Scott, Foresman.

Canter, D. (1986). Putting situations in their place: Foundations for a bridge between social and environmental psychology. In A. Furnham (Ed.) *Social behavior in context* (pp. 215–230). London: Allyn and Bacon.

Canter, D. (1991). Understanding, assessing and acting in places: Is an integrative framework possible? In T. Garling and G. Evans (Eds.). *Environmental cognition and action: An integrated approach* (pp. 181–196). New York: Plenum.

Chipuer, H. M., Pretty, G. H., Delorey, E., Miller, M., Powers, T., Rumstein, O., Barners, A., Cordasic, N. & Laurent, L. (1999). The neighbourhood youth inventory: Development and validation. *Journal of Community and Applied Social Psychology, 9,* 355–368.

Community [Def. 1, 2, 3]. (2019). Merriam-Webster Online. In Merriam-Webster. Retrieved April 7, 2019, from http://www.merriam-webster.com/dictionary/community.

Cutter, S. L., Barnes, L., Berry, M., Burton, C., Evans, E., Tate, E., & Webb, J. (2008). A place-based model for understanding community resilience to natural disasters. *Global Environmental Change, 18,* 598–606.

Davoudi, S. (2012). The legacy of positivism and emergence of interpretive tradition in spatial planning, *Regional Studies, 46*(4), 429–441.

De Groot, R. S., Wilson, M. A. & Boumans, R. M. J. (2002). A typology for classification, description and valuation of ecosystem functions, goods and service. *Ecological Economics, 41*(3), 393–408.

Elder, G. H., King, V. & Conger, R. D. (1996). Attachment to place and migration prospects: A developmental perspective. *Journal of Research on Adolescence, 6,* 397–425.

Fleming, J., & Ledogar, R. J. (2008). Resilience, an evolving concept: A review of literature releva,,,,,,,,,,,,,,,,,,,,,,,,,,,,,
,,nt to aboriginal research. *Pimatisiwin, 6*(2), 7–23.

Fried, M. (1982). Residential attachment: Sources of residential and community satisfaction. *Journal of Social Issues, 38*(3), 107–120.

Giddens, A. (1998). *The third way: The renewal of social democracy.* Cambridge Polity Press.

Giuliani, M. V. & Feldman, R. (1993). Place attachment in a developmental and cultural context. *Journal of Environmental Psychology, 13,* 267–274.

Goudy, W. (1990). The ideal and the actual community: Evaluations from small town residents. *Journal of Community Psychology, 18,* 277–288.

Guest, A. M. & Lee, B. A. (1983). Sentiment and evaluation as ecological variables. *Sociological Perspectives, 26,* 159–184.

Gunderson, L. H., & Holling, C. S. (2002). *Panarchy: Understanding transformations in human and natural systems.* Washington, DC: Island Press.

Heidegger, M., 1962, *Being and time*. Oxford, Blackwell. [1927, *Sein and Zeit*].

Herting, J. R. & Guest, A. M. (1983). Components of satisfaction with local areas in the metropolis. *Sociological Quarterly, 26,* 99–115.

Hummon, D. M. (1992). Community attachment: local sentiment and sense of place. In I. Altman and S. Low (Eds.). *Place attachment,* New York, Plenum.

Jacobson-Widding, A. (1983). *Identity, personal and socio-cultural: A symposium.* Uppsala: Acta Universitatis Uppsaliensis.

John, C, Austin, D. M. & Baba, Y. (1986). The question of community attachment revisited. *Sociological Spectrum, 6,* 411–431.

Jones, S. & Mean, M. (2010). *Resilient place: Character and community in everyday heritage* (London: DEMOS Report).

Kasarda, J. & Janowitz, M. (1974). Community attachment in mass society. *American Sociological Review, 39,* 328–339.

Kron, J. (1983). *Home-psych: The social psychology of home and decoration.* New York: Potter.

Lewicka, M. (2008). Place attachment, place identity, and place memory: Restoring the forgotten city pas. *Journal of Environmental Psychology, 28*(3) 209–231.

Magis, K. (2010). Community resilience: An indicator of social sustainability. *Society and Natural Resources, 23*(5), 401–416.

McMillan, D. & Chavis, D. (1986). Sense of community: A definition and theory. *Journal of Community Psychology, 14*, 6–23.

O'Brien, D., Hassinger, E. & Dershem, L. (1994). Community attachment and depression among residents in two rural Midwestern Communities. *Rural Sociology, 59*, 255–265.

Plas, J. & Lewis, S. (1996). Environmental factors and sense of community in a planned town. *American Journal of Community Psychology, 24*, 109–133.

Pretty, G. H., Andrewes, L. & Collett, C. (1994). Exploring adolescent's sense of community and its relationship to loneliness. *Journal of Community Psychology, 22*, 346–358.

Pretty, G. M. H., Conroy, C, Dugay, J., Fowler, K. & Williams, D. (1996). Sense of community and its relevance to adolescents of all ages. *Journal of Community Psychology, 24*, 365–379.

Proshansky, H. M. (1978). *The city and self-identity. Environment and Behavior*, 10, 147–169.

Proshansky, H. M. & Fabian, A. K. (1987). The development of place identity in the child: In C. S. Weinstein and T. G. David (Eds.) *Spaces for Children.* New York: Plenum Press, pp. 134–141

Proshansky, H. M., Fabian, A. K. & Kaminoff, R. (1983). Place identity: Physical World socialisation of the self. *Journal of Environmental Psychology, 3*, 57–83.

Proshansky, H. M. & Kaminoff, R. D. (1982). The build environment of the young. In S. Messick (Ed). *Development in young adulthood: Characteristics and compencier Francisco, work and social life* (pp. 44–45). San Francisco: Jossey-Bass.

Proshansky, H. M., Nelson-Shulman, Y. & Kaminoff, R. D. (1979). The role of physical setting in life crisis experiences. In I. Sharon and C. Spielberger (Eds.). *Stress and anxiety* (Vol. 6, pp. 298–302). Washington, DC: Hemisphere Publishing Corporation,

Qazimi, S. (2014). Sense of place and place identity. *European Journal of Social Sciences, Education and Research, 1*(1), 306–310.

Rowler, G. (1990). Place attachment among the small town elderly. *Journal of Rural of Rural Community Psychology, 11*, 103–120.

Rowles, G.D. (1990). Place attachment among the small town elderly. *Journal of Rural Community Psychology, 1*, 103–120.

Rubinstein, R. & Parmelee, P. (1992). Attachment to place and the representation of the life course by the elderly. In I. Altman and S. M. Low (Eds.), *Place attachment* (pp. 132–149). New York: Plenum Press.

Ryden, K. C. (1993). *Mapping the invisible landscape: Folklore, writing, and the sense of place.* Iowa City: University of Iowa Press.

Saegert, S. & Hart, R. (1978). The development of environmental competence in girls and boys. In M. Salter (Ed.) *Play: Anthropological perspectives* (pp. 142–154). Cornwall NY: Leisure Press.

Sampson, R. (1989). Local friendship ties and community attachment in mass society: A multilevel systematic model. *American Sociological Review, 53*, 766–779.

Sarason, S. B. (1974). *The psychological sense of community: Perspective for community psychology* (pp. 127–139). San Francisco: Jossey-Bass.

Sarbin, T. R. (1983). Place identity as a component of self: An addendum. *Journal of Environmental Psychology, 3*, 337–342.

Smith, M. P. & Bender, T. (2001). *City and nation: Rethinking place and identity* (Vol. 7). Transaction Publishers.

Stinner, W., Vanloon, M., Chung, S. & Byun, Y. (1990). Community size, individual and social position and community attachment. *Rural Sociology, 55*, 494–521.

Stokowski, P. A. (2002). Language of place and discourses of power: Constructing new senses of place. *Journal of Leisure Research, 34*, 368–382.

Summers, J. K., Harwell, L. C., Smith, L. M., & Buck, K. D. (2018). Measuring community resilience to natural hazards: The Natural Hazard Resilience Screening Index (NaHRSI)—Development and application to the United States. *GeoHealth, 2*, 372–394

Walker, B. & Salt, D. (2012). *Resilience Thinking: Sustaining Ecosystem and People in a Changing World.* Washington, DC: Island Press.

Williams, D. R. & Vaske, J. J. (2003). The measurement of place attachment: Validity and generalizibility of a psychometric approach. *Forest Science, 49*, 830–840.

Wilkinson, C. (2012). Social-ecological resilience: Insight and issues for planning theory. *Planning Theory 11*(2), 148–169.

CHAPTER 18

Resilience and Well-Being in Sports

MOHITA JUNNARKAR,* SANJEEV P. SAHNI, and SHIVJOT GILL

Jindal Institute of Behavioral Sciences, O.P Jindal Global University, Sonipat, Haryana, India

**Corresponding author. E-mail: mjunnarkar@jgu.edu.in*

ABSTRACT

The field of sports psychology aims at assisting athletes to flourish in their sport-specific domain where the idea is to focus upon building resilience in athletes. Whereas the concept of well-being in sports further aims at developing life skills and promoting positive sporting experiences among athletes through sports participation and physical activities. However, research studies in both the domains have been context specific in nature and vulnerable to the conceptual ambiguity. Thus, demanding more explicit conceptual framework at global level. The present chapter discusses the concept and the application framework of resilience and well-being in the field of sports.

18.1 INTRODUCTION

Resilience and well-being are two prominent factors vigorously researched in the field of sports psychology. Due to the context specific (Luthar & Cicchetti, 2000) nature of resilience, research inclination is majorly on studying clinical population suffering with high traumatic circumstances and grave illness issues (Denz-Penhey & Murdoch, 2008) that however, does not address the challenges faced by sports achievers who by no chance can escape the forbidding conditions constraining sporting performance. As per the obligatory circumstances, calling for a certain amount of mental toughness is essential to sustain in the sport (Mguni et al., 2012). Such situations incite us to interrogate whether excelling in sporting performance is worth one's psychological well-being (Amirault & Orlick, 1999). However, finding the appropriate coping and stress preventive measures becomes essential for sporting professionals to help athletes deal with their distressed mental state.

The present chapter is categorized into four sections. The first section focuses on introducing the concept of well-being and resilience in context of sports; the second section focuses on the important theories in sports psychology that focus on resilience and well-being; third section emphasizes the different strategies that are widely employed by sports persons as part of mental skills training, and lastly the chapter concludes by suggesting prominent best globalized practices and way forward for future research in the context of sport psychology.

18.2 WELL-BEING AND RESILIENCE IN SPORTS: CONCEPTUAL FRAMEWORK

18.2.1 WELL-BEING IN SPORTS

Regular physical activity has proved to be beneficial in enhancing well-being of people (e.g., Mack et al., 2012; Wendel-Vos et al., 2004). In context of sports, the influence of regular physical activity and training may not always advantage the sports persons and can prove to be hazardous (Theberge, 2008). Conditions such as burnout, injury, illness, and so forth are common in sport and thus, such situations induce severe psychological challenges such as anxiety, low self-esteem, demotivation, etc. that enforce the risks of negatively affecting an athlete's well-being (Steffen et al., 2011). Thus, the most challenging task for an athlete is to protect his or her well-being in the situations that are highly demanding and challenging.

In sports psychology, there is a dearth of empirical studies to understand association between athletes' well-being and psychological and social environment (Lundqvist, 2011). The focus of earlier psychological literature was engaged in philosophical understanding of the concept of good life, health, and happiness (Ryan & Deci, 2001). In the past, the philosophers lead to the development of five well-being perspectives—the subjective state theories (Bentham, 1781), desire fulfillment theories (Griffin, 1993; Heathwood, 2014), life-satisfaction theories (Tiberius & Hall, 2010), objective list theories (Arneson, 2007; Gert, 1998), and nature fulfillment theories (Kraut, 2007; Norton, 1976). On the other hand, philosophers such as Norton (1976), Kraut (1979), and Haybron (2008) extensively worked on traditional concepts of well-being, that is, hedonic and eudemonic. It is only since 1960s that there is research in context of human flourishing and well-being has emerged tremendously and the interest has shifted to study nonclinical populations (Diener et al., 1999; Keyes, 2005).

As a concept of psychology, well-being in sports has evolved on the basis of two traditional approaches, that is, the hedonic and the eudemonic traditions (Ryan & Deci, 2001; Ryff et al., 2001). According to the hedonic well-being or subjective well-being, is conceptualized differently by various researchers such as an individual's aim in life is to achieve happiness and success by fulfilling essential life goals (Diener, 2009) or as a means to seek happiness by fulfilling the desires of success (Hobbes, 2012). Further, psychologists suggested that the concept of hedonism does not only comprise of human happiness but also involves human pleasures in relation to body and mind (Kubovy, 1999). In addition to this, Kahneman and colleagues (1999) defined hedonic psychology as a study of individual's enjoyable and hostile experiences. The concept of hedonism was also defined as subjective well-being which comprised of three components namely, life satisfaction, positive mood, and absence of negative mood and happiness (Ryan & Deci, 2000; Diener et al, 1999). The research in measuring subjective well-being components remains implicit, yet the overall theoretical development of the concept enables linking of human behavior and cognitive functioning smoothly (Peterson et al., 2005).

In the context of eudemonic well-being (psychological well-being), the term eudaimonia was initially popularized by Aristotle during the 4th century BC. The eudemonic tradition focuses on attaining optimal functioning and fulfillment of individual talents and capabilities (Ryan & Deci, 2001). Based on the various theoretical perspectives of developmental and humanistic psychology, Ryff (1989) and further Ryff and Keyes (1995) stated six components of psychological well-being namely purpose in life, positive relations with others, personal growth, environmental mastery, autonomy, and self-acceptance. Majority of the researchers (Diener et al. 1999; Kahneman et al., 1999; Keyes et al., 2002; Seligman, 2011) have adapted Ryff's model and claimed it to be sound and well-rooted in the previous research studies. However, Mehrotra (2013) reported a 20-items and 4-factor solution for Ryff's scales was valid for the Indian population. Keyes (1998) developed a cohesive Mental Health Continuum model comprising of the hedonic and eudemonic components of well-being. The hedonic component was related to the emotional well-being of individuals (feeling of happiness, satisfaction, and positive emotions) and the eudemonic is addressed through psychological and social well-being. The various subdimensions namely; self-acceptance, personal growth, purpose in life, positive relation with others, autonomy, and environmental mastery assess psychological well-being whereas social well-being is measured by social actualization, social acceptance, social contribution, social integration, and social coherence. Empirical evidence has demonstrated a significant enhancement in the social well-being of athletes through team support and team identification (Inoue et al., 2015). The progression in the theoretical framework of well-being has allowed the researchers to investigate beyond the general human differences and has led to the assessment of individual's goals, values, and social environment as a contributor toward individual's motives (Emmons, 1986; Little, 1989; Sheldon & Kasser, 1995; Ryan & Deci, 2001).

The practice-based research approach of well-being components in the field of sports psychology is observed as a stepping stone in the development of sports well-being literature (Lundqvist, 2011). However, lesser is known about the impact of sporting environment on the well-being of athletes. Thus, there is needed to empirically consider both the hedonic and eudemonic approaches in order to understand the overall well-being of athletes (Lundqvist, 2011).

18.2.2 RESILIENCE IN SPORTS

Resilience has been defined in various ways in the psychology research literature (Fletcher & Sarkar, 2013). However, the term resilience simply means to bounce back or to leap back (Fletcher & Sarkar, 2012). The word resilience has originated from the Latin verb "resilire" and has been mostly referred to the ability of an individual to resume the activity that one is engaged in (Strümpfer, 1999). Most often, resiliency is well-defined as a process in which an individual is optimistic regardless of the adversities or circumstances (Luthar & Cicchetti, 2000).

The Interest in conceptualizing resilience increased during the 1980s when development psychologists conducted various comparative studies between higher achiever individuals regardless of difficult circumstances then those who did not face any major crisis situations such as poverty, health issues, and monetary crisis (Rutter, 1985).

In addition to the forthcoming trend of research in developmental psychology, various researchers state that the definition and the conceptual clarity of resilience is ambiguous in its nature and it is difficult to measure resilience in sports and nonsports settings (e.g., Atkinson et al., 2009; Fletcher & Sarkar, 2013; Windle et al., 2011). However, various researchers suggest that resilience can be understood as per its contextual framework in a particular domain (Luthar & Cichetti, 2000; Galli & Gonzalez, 2015).

In the context of psychology, Fletcher and Sarkar (2012) defined psychological resilience as "the role of mental processes for promoting and protecting the individual from possible negative effect of stressors." The definition stated by Fletcher and Sarkar focuses on psychological resilience, that is, behavior and mental process of individual rather than other forms of resilience such as physical and structural resilience. In addition to it, the purpose of conceptualizing resilience was to understand how the mental and cognitive processes facilitate an individual to adapt to the difficult circumstances and obstacles they encounter (Connor & Davidson, 2003). Also, defining resilience and conducting advance research in a similar context aimed at exploring how resilient athletes can become in the tough environment (Mummery et al., 2004).

Specifically, in the context of sports, the number of obstacles is high and can be in the form of failure in performance (Mellalieu et al., 2009; Poczwardowski & Conroy, 2002; Tamminen et al., 2013), injury (Podlog & Eklund, 2006), organizational stressors (Fletcher & Hanton, 2003), and so forth. Studying resilience from the lens of sports provides an opportunity to examine the concept in a new light where other components such as intrapersonal factors of human behavior also play an evident role in addition to the general environmental factors in athletic settings (Galli & Gonzalez, 2015). Thus, future researchers can majorly contribute in theory building in the context of resilience and defining how it can act as a major protagonist in affecting an athlete's performance and well-being in sports settings.

18.3 THEORETICAL MODELS IN SPORTS PSYCHOLOGY

With the advancement in the field of sports psychology, various intervention methods and programs have evolved over a period of time to support athlete's performance goals as well as psychological well-being. These models stem from core positive psychology constructs. The most prominent and widely applied models are mentioned further:

18.3.1 PSYCHOLOGICAL SKILL TRAINING (PST)

Psychological skills are defined as the mental abilities that assist athletes in improved sports performance. The purpose of a sport psychologist is to support athletes by teaching them how to handle these skills in a competitive environment (Dosil & González-Oya, 2008). PST is a process comprising various psychological skills to reduce the mental barriers of athletes and enhancing the development of talent. According to Fredrickson (1998), the human being has a general tendency to focus on negative aspects of life and feelings which are discouraging in nature. Thus, for a holistic growth and improvement of an individual, intervention is required that would integrate certain aspects related to an individual's health and well-being (Vázquez et al., 2009). Thus, PST is one comprehensive package of various techniques that can be applied during training sessions as well as competition by athletes. Mostly, the implication of PST is integrated with the physical training program. There are two phases of PST namely general phase and specific phase (Castillo et al., 2007). General phase comprises of learning the basic psychological skill techniques such as imagery, self-talk, and so forth whereas, Specific phase is situation based and involves the application of the psychological skill in a particular situation. Thus, the implication of PST has proven to be 75%–80% beneficial for athletes in supporting their psychological preparation and sports performance (Vealey, 1994).

18.3.2 MINDFUL SPORT PERFORMANCE ENHANCEMENT (MSPE)

MSPE is a method of developing the quality of centering one's attention on a particular task and bringing control in one's actions, by being in the present moment with the ability of being nonjudgmental (Kaufman et al., 2009). These qualities promote a state of being in flow (Csikszentmihalyi & Robinson, 1990).

MSPE is a mindfulness-based approach that is different from the traditional, control-based approaches which can impair an individual's ability to develop meta-cognitive skills (Kaufman et al., 2009). MSPE is an extended version of the mindfulness-based stress reduction model (Kabat-zinn, 2000) and mindfulness-based cognitive therapy. The MSPE model involves a set of exercises such as sitting meditation, body scan, and so forth which involve focus on a specific thought. In addition to well-developed programs such as MSPE, numerous meditation exercises based on the mindfulness approach are available to enhance mindfulness skills for developing the skills of evaluating the situation or phenomena with no judgment attitude.

18.3.3 MODEL OF QUALITY OF LIFE (QOL) IN SPORTS

The quality of life model is based on the theoretical concept of hedonic well-being and aims to develop the physical, social, emotional, cognitive, and spiritual well-being of an individual contributing toward the overall quality of life (Gill et al., 2017). The model is divided into five categories naming social, physical, emotional, cognitive, and spiritual well-being. Further, physical well-being is divided into two parts namely physical health symptoms and physical functioning. The development of the model was based on multiple survey studies and applicable psychometric tests.

18.3.4 GROUNDED THEORY OF PSYCHOLOGICAL RESILIENCE AND SPORTS PERFORMANCE

Fletcher and Sarkar (2012) developed the grounded theory of psychological resilience and sports performance. The theory was developed using the grounded theory approach and aimed at studying mental resilience in Olympic champions for establishing stress–resilience–performance relationship. Hence, Fletcher and Sarkar (2012) defined resilience as the cognitive processes involving the protection of an individual from the potential stressors which can result in negative behavior. An individual when faced with stressors, strives to challenge the situation and involve oneself in the appraisal of the situation. This cognitive process further leads to facilitative responses as per the situation and as a result, an individual is successful in presenting the optimal sports performance. The theory states five psychological factors that contribute toward challenge appraisal and individual's meta-cognition namely positive personality, motivation, confidence, perceived social support, and focus.

Thus, the evolution of various theoretical models has contributed to the significant improvement in sporting performance and management of various psychological stressors intervening in the daily functioning of athletes. Thus, in the future, a more holistic approach can be proposed to address the sporting performance of athletes.

18.4 MENTAL SKILLS IN SPORTS

Traditionally, the focus of sports professionals has been on physical training of athletes but over the time the psychological aspects of the game have gained importance leading to the major changes in the field of sports psychology (Dosil & González-Oya, 2008). With the passage of time, the sports professionals such as sports psychologists, physiotherapists, and coaches have realized that psychology serves as a key component to attain optimal performance and better well-being to perform any sport. In the 21st century, the importance of physical, social, psychological, and tactical domains of sport performance stand equal by stating the possibility of the advent of sports specialists with expertise in the psychological fundamentals of a particular sport (Dosil & González-Oya, 2008). Today, the emphasis is more on integrating the theoretical and applied fundamentals of sports psychology in comparison to the past where lack of cross-disciplinary approaches in educating and training athletes lead to a very single-minded and narrow approach toward sports performance. Certainly, with the rapid changes and advancement over the years, the field of sports psychology has started to enquire areas such as diversity in sports, athlete's life skills, and cultural issues related to sports that have fuelled the growth of the field for the near future (Voelker et al., 2014).

In a sports setting, the emergence of application-based methods such as mental skill training and coping skills has made a great difference to the psychological state as well as the well-being of athletes (Brown & Fletcher, 2017). Various research studies state that the application of such skills and methods can contribute toward enhanced sports performance and athletes' well-being (Lauer & Zakrajsek, 2017). Application of mental skills and coping strategies such as goal setting, self-talk, imagery, and relaxation has been a major part of the various psychological training programs and is widely used by sports psychologists (Vealey, 2007). The application of mental skill techniques and coping strategies was an earliest attempt of Greenspan and Feltz (1989). Further intervention studies were conducted by Vealey (1994) and Weinberg and Comar (1994).

It was after 20th century that the focus of sports psychology had shifted toward applying the mental skills programs on a single subject and conducting in-depth and detailed experiments with a sample of few people (Martin et al., 2004). In the context of resilience, various research studies have shown that implication of the coping skills and mental training techniques can contribute significantly toward building athletes' confidence, increasing confidence, and contribute toward coping with the stressors (Burton & Raedeke, 2008; Omar-Fauzee et al., 2010; Williams, 2001; Vealey, 2007). The implementation of these techniques can contribute toward making an athlete resilient and can contribute toward executing the finest sports performance and mental capabilities (Bertollo et al., 2009).

Mental training skills can be classified as follows:

18.4.1 SELF-TALK

It is defined as the process which involves inner voice, silent speech, or inner dialogue with the self which can help in bringing oneself to an optimal level of improving and enhancing sporting performance by gaining control on negative emotions and thoughts (Shannon et al., 2012). Self-talk as a mental skill has proven to be very effective as per various research studies where athletes implement self-instruction and usage of positive statements as a tool for preparing themselves to perform (Van Raalte et al., 1995).

Hatzigeorgiad and colleagues (2011) stated three significant types of self-talk namely the content, overtness, and selection. The content can be divided into instructional and motivational cues. Further, overtness can be divided into internal and external. In the case of selection, recent studies are based on the selection of the self-talk cues by participants either from a designated list or out of their own choice (Hatzigeorgiadis et al., 2011). Thus, self-talk can be considered as vital as human language and cognition is an important process which is majorly responsible for negative thinking pattern of the human mind and which further can be tackled with the implication of self-talk strategies and by being more mindful about oneself (Bervoets, 2013).

18.4.2 MENTAL IMAGERY

It can be defined as the rehearsal of an action or experience cognitively which can be easily utilized for improving sports performance (Gill et al., 2017). Various sports professionals have stated that by implication of imagery the athlete experiences the competing or the action experiences in the mind which are similar to the real sporting experiences (Gammage et al., 2000; Vealey & Greenleaf, 2001). Athletes often imagine themselves performing their sport which is similar to imagining oneself performing the skill. Various research studies show that imagery has a very positive impact on an athlete's real-time actions (Gammage, et al., 2000; Vealey, 2007) and also enhances performance and tactical skills (Guillot et al., 2009). The most common question about imagery is that whether it leads to better performance or not (Morriser et al., 2005). Various research studies have been conducted to answer this question and it is found that implication of imagery can often contribute toward improving improve athletes' self-confidence and enhance attention while performing the sports task (Farrow & Kemp, 2003). Hence, the application of imagery in sports is prominently recommended by sports professionals and has been able to yield a positive effect in improving the task performance.

18.4.3 GOAL SETTING

It is defined as the process that involves the development of the action plan comprising of subjective and objective goals (Biddle & Ekkekikakis, 2005). Scholars in the field of sports have differentiated between three types of goals, that is, outcome goals, performance goals, and process goals where outcome goals are associated with the final result of the performance, performance goals are related to the athlete's goals in the context of training interval (time spent on practice), accuracy in executing certain sport specific skills and so forth and process goals are based on how a certain sporting skill will be executed and what sort of preparation is required to execute the same (Weinberg et al., 2017).

In recent years, the concept of goal setting has received some consideration as it is the most widely used technique by sports professionals, physical trainers, and so forth (Williams, 2001). Locke et al. (1981) specified the effect of goal setting on an athlete's performance. Studies related to goal setting have a major role to play in order to perform the task with accuracy. In sports, short-term goals play a major role as they can support the achievement of long-term goals (Bandura & Simon, 1977; Locke et al., 1970). Further, goal setting is only successful if the goals are definite and meaningful in nature (Locke et al., 1978).

Thus, goal setting is not only a method but a way to organize one's priorities in life. Goal setting is very much relevant at all ages and majorly impacts cognitive structuring of an individual across their lifespan.

18.4.4 RELAXATION METHOD: THE METHOD OF PROGRESSIVE MUSCULAR RELAXATION (PMR)

It is advantageous to athletes who are usually high in arousal or conscious during the competition which can result in a mental block such as negative thoughts, poor decision-making skills, and so forth. Simply application of relaxation methods will not demonstrate to be effective. However, the systematic application of relaxation methods can prove to be useful. One of the most pertinent methods is PMR. In PMR, an individual gradually relaxes and contracts their muscles through the tension and relaxation process, develops awareness about the particular muscle, and is able to control their arousal. PMR implication for relaxing the whole body usually takes about 30 minutes whereas, relaxing a specific target muscle can take approximately 15 minutes (Ost, 1988). With the implication of instructional cues such as "contract" or "relax," the time can be reduced to 5–7 min. Thus, the constant application of PMR can be highly effective and can be reduced to few minutes for the application during pre-performance routine.

In the field of sport psychology, it can be stated that the implication of various mental skills such as imagery, self-talk, and visualization can play a significant role in reducing performance anxiety levels and support in achieving a high level of psychosomatic congruency (Behncke, 2004). These conventional techniques can be beneficial if integrated with existing training programs and can offer a way to upsurge self-awareness and self-monitoring among athletes to regulate their sporting performance and behavior. Thus, application of such mental skills is essential as without their application, the improvement in sporting performance will be limited.

18.5 GLOBAL BEST PRACTICES

Globally sports psychology is a budding field and it has emerged radically in recent years, as many people now are able to recognize the importance of psychological training in sports. Over the years, various skills and procedures have been utilized for the well-being of athletes. This section attempts to highlight a few of the best practices utilized globally.

Numerous strategies are applied to address difficulties faced by athletes and other sports participants. Some of the primary areas for training, as given by the American Psychological Association (2019) are:

- **Cognitive and behavioral skills training for performance enhancement**: In this area, professionals in the field of sports are imparted training about various traditional mental training practices such as goal setting, imagery, and concentration. Self-confidence development, self-regulation techniques, leadership skills, and emotional management.
- **Counseling and clinical interventions**: These interventions focus on motivating Athletes and assisting them if they are undergoing any kind of severe psychological issue such as eating disorders, depression, burnout athletic injury and rehabilitation, and career transitions.
- **Consultation and training**: This area focuses on team building and cohesion;

parent–athlete relationship interventions, guiding the coaches, and educating them regarding various psychological skills; talent development.

Sports psychology has evolved over the years. The first World Congress in Sports Psychology was held in 1965 in Rome. Twenty years later, it was held in Scandinavia, that is 6th World Congress was held June 23–27 in Copenhagen, Denmark. In Congress, 400 delegates from different parts of the world were congregated to celebrate the 20th anniversary of International Sports Psychology (Uneståhl, 1985). Finest practices and methods in sports psychology were discoursed. Following are a few of these techniques:

1. *Mental training*: Mental training (Uneståhl, 1982) refers to the use of psychological techniques that aim at controlling and transform an individual's external and internal behavior and experience. The concept of mental training is based on the philosophy that the "mental strength" of athletes can be handled and enhanced in precisely the same way as physical strength. Mental training can be categorized as general or area-specific in nature where general training can be categorized into three steps:

TABLE 18.1 Table Depicting Steps of Mental Training

Step 1: Mental Conditioning	Step 2: Mental Technique Training	Step 3: Mental Strength Training
1. Muscular and mental relaxation	1. Direct indirect and posthypnotic suggestibility	1. Skills—attainment, detection, and development
2. Self-hypnotic induction training	2. Short-term and long-term images	2. Motivational training
		3. Attitudinal training

2. *Inner mental training*: It is an organized program for mental training, and was formulated during mid-seventies and been tested on 12,000 Swedish athletes (Uneståhl, 2013). The purpose of inner mental training is to increase the quality of Ideal Performance State (IPS) by developing and enhancing Ideal performance-related mental skills, increase the ability to reproduce, induce and control IPS, and increase the commitment toward physical training by: (a) refining the quality of training and (b) improving the capacity for rest/recovery between the training sessions (Uneståhl, 2013). The technique is self-instructional, concrete, and emphasizes on acquisition, enhancement, and automatization of mental skills to proceed from general training to more sport-specific and competition directed training.
3. *Competitive training (model training)*: Competitive training model can be divided into two parts—the general model training and the specific model training (Uneståhl, 1985).
   A. General model-training includes focusing on
      1. External conditions such as
         a. Time limitations, task-difficulty.
         b. Optimum speed and efforts.
      2. Internal conditions such as
         a. Psychological simulation.
         b. In vivo simulation.

B. Definite/specific model-training includes
   a. Time period of the year or days
   b. Weather.
   c. Form of competition. "Mental atmosphere."

   The model uses the technique of Mental Rehearsal. The main aim of the program is to reboot the "ideal performing feeling" and to acquaint body and mind with the upcoming competition (goal-orientation). The program is used as per the type of sport.

4. *The confidence cycle: prepare to be confident*: In this technique, athletes learn to understand that they have control over their preparation, which can in turn benefit their confidence and performance (Heil et al., 2013).

   It focuses on setting up a preparation routine that is consistent and specific for each athlete. Mental techniques such as breathing and positive affirmations assist athletes to focus on the process and on what they control.

5. *Priming the real me: rational thinking and strengths-based techniques in youth sport*: This technique involves working the athlete on identifying four key components of themselves and of competitive situations, so that they are better prepared for competition and can engage competition with a more approach-oriented mindset (Heil et al., 2013).

The technique involves collaborating with the athlete to draw out the "rational" rules of engagement (e.g., What rights do you and your opponent have in this competition? What challenges and adversities are you going to be ready for and accept? What will you learn and take away from competition every time regardless of the outcome?), drawing out the athlete's apparent strengths—technical, tactical, physical, psychological, social (in the context of team sports), discussing the definite mental tools or skills that the athlete will use in competition situation (pre and during the performance), indulging the athlete to discuss about their "real self" and their personal identity different from being a sport person (e.g., what qualities make you the real you? How are you different from what you are as a sports person?), and drawing together the poster overleaf, refining the lists and approving upon what the athlete feels within themselves every time they step out into competition. Finally, the athlete has to assign a coded name to themselves as an individual and as a sportsperson.

18.6 WAY FORWARD AND CONCLUSION

Sports psychology has emerged to be an important stepping stone in the life of athletes. Empirical evidence indicates that the evolving techniques have enhanced the well-being of athletes thus resulting in enhanced sporting performance. Although, sports psychology has come a long way, there is still a huge gap as many athletes are still not receiving adequate psychological training. Future research should focus on the dissemination of empirical-based techniques and trainings, which have proven to enhance well-being, resilience, and performance among athletes, in various settings to a greater number of athletes. Furthermore, global techniques should be adapted according to the cultural context as well as the sport.

KEYWORDS

- **well-being**
- **resilience**
- **sports**

REFERENCES

Amirault, K., & Orlick, T. (1999). Finding balance within excellence. *Journal of Excellence, 2*, 37–52.

Arneson R (2007). *Desire formation and human good.* In S. Olsaretti (Ed.) *Preferences and well-being*. Royal Institute of Philosophy, supp. vol. 59. Cambridge: Cambridge University Press, pp. 9–32.

Atkinson, P. A., Martin, C. R., & Rankin, J. (2009). Resilience revisited. *Journal of Psychiatric & Mental Health Nursing, 16*, 137–145.

Bandura, A., & Simon, K. M. (1977). The role of proximal intentions in self-regulation of refractory behavior. *Cognitive Therapy and Research, 1*(3), 177–193.

Behncke, L. (2004). Mental skills training for sports: A brief review. *Online Journal of Sport Psychology, 6*(1).

Bentham, J. (1781). *An introduction to the principles of morals and legislation*. McMaster University Archive for the History of Economic Thought.

Bertollo, M., Saltarelli, B., & Robazza, C. (2009). Mental preparation strategies of elite modern pentathletes. *Psychology of Sport and Exercise, 10*(2), 244–254.

Bervoets J., 2013. *Exploring the relationships between flow, mindfulness, and self-talk: A correlational study*. Master's thesis in sport and exercise psychology (p. 45). Department of Sport Sciences, University of Jyväskylä.

Biddle, S. J., & Ekkekakis, P. (2005). Physically active lifestyles and well-being. *The Science of Well-Being, 140*, 168.

Brown, D. J., & Fletcher, D. (2017). Effects of psychological and psychosocial interventions on sport performance: A meta-analysis. *Sports Medicine, 47*(1), 77–99.

Burton, D., & Raedeke, T. D. (2008). *Sports Psychology for Coaches*. Campaign, IL: Human Kinetics.

Castillo, I., Balaguer, I., & Garcia-Merita, M. (2007). Effects of physical activity and sport on healthy lifestyles in adolescence according to gender. *Revista De Psicologia Del Deporte, 16(2)*, 201–210.

Connor, K. M., Davidson, J. R., & Lee, L. C. (2003). Spirituality, resilience, and anger in survivors of violent trauma: A community survey. *Journal of Traumatic Stress, 16*(5), 487–494.

Csikszentmihalyi, M., & Robinson, R. E. (1990). *The art of seeing: An interpretation of the aesthetic encounter*. Getty Publications.

Denz-Penhey, H., & Murdoch, C. (2008). Personal resiliency: Serious diagnosis and prognosis with unexpected quality outcomes. *Qualitative Health Research, 18*(3), 391–404.

Diener, E. (2009). Subjective well-being. In The science of well-being (pp. 11–58). Dordrecht: Springer.

Diener, E., Suh, E. M., Lucas, R. E., & Smith, H. L. (1999). Subjective well-being: Three decades of progress. *Psychological Bulletin, 125*(2), 276.

Dosil, J., & González-Oya, J. (2008). *Eating disorders and the athlete's environment*. In J. Dosil (Ed.), Eating disorders in athletes (pp. 41–63). Chichester, England: John Wiley & Sons.

Emmons, R. A. (1986). Personal strivings: An approach to personality and subjective well-being. *Journal of Personality and Social Psychology, 51*(5), 1058.

Farrow, D., & Kemp, J. (2003). *Run like you stole something: The science behind the score line.* Allen & Unwin.

Fletcher, D., & Hanton, S. (2003). Sources of organizational stress in elite sports performers. *The Sport Psychologist, 17*(2), 175–195.

Fletcher, D., & Sarkar, M. (2012). A grounded theory of psychological resilience in Olympic champions. *Psychology of Sport and Exercise, 13*(5), 669–678.

Fletcher, D., & Sarkar, M. (2013). Psychological resilience: A review and critique of definitions, concepts, and theory. *European Psychologist, 18*(1), 12.

Fredrickson, B. L. (1998). Cultivated emotions: Parental socialization of positive emotions and self-conscious emotions. *Psychological Inquiry, 9*(4), 279–281.

Galli, N., & Gonzalez, S. P. (2015). Psychological resilience in sport: A review of the literature and implications for research and practice. *International Journal of Sport and Exercise Psychology, 13*(3), 243–257.

Gammage, K. L., Hall, C. R., & Rodgers, W. M. (2000). More about exercise imagery. *The Sport Psychologist, 14*(4), 348–359.

Gert, B. (1998). *Morality: Its nature and justification.* Oxford, UK: Oxford University Press

Gill, D. L., Williams, L., & Reifsteck, E. (2017). *Psychological dynamics of sport and exercise*. USA: Human Kinetics.

Greenspan, M. J. & Feltz, D. L. (1989). Psychological interventions with athletes in competitive situations: A review. *The Sports Psychologist, 3*, 219–236.

Griffin, P. (1993). Homophobia in Sport: Addressing the Needs of Lesbian and Gay High School Athletes. *The High School Journal, 77(1/2)*, 80–87.

Guillot, A., Nadrowska, E., & Collet, C. (2009). Using motor imagery to learn tactical movements in basketball. *Journal of Sport Behavior, 32*(2), 189–206.

Hatzigeorgiadis, A., Zourbanos, N., Galanis, E., & Theodorakis, Y. (2011). Self-talk and sports performance: A meta-analysis. *Perspectives on Psychological Science, 6*(4), 348–356.

Haybron, D. (2008). Happiness, the self and human flourishing. *Utilitas, 20*, 21–49.

Heathwood, C. (2014). Subjective theories of well-being. *The Cambridge Companion to Utilitarisnism*, 199–219. https://doi.org/10.1017/CCO9781139096737.011.

Heil, J., Bean, E., Morgan T., & Hardwood, C. (2013). *Signature sport psychology techniques*. Los Angeles: Association for Applied Sport Psychology in the United States of America, Champaign, IL: Human Kinetics. 387.

Hobbes, T. (2012). *Leviathan*. N. Malcolm (Ed.) (2012) (3 Vols, p. 1141). Oxford: Oxford University Press.

Inoue, Y., Funk, D. C., Wann, D. L., Yoshida, M., & Nakazawa, M. (2015). Team identification and post disaster social well-being: The mediating role of social support. *Group Dynamics: Theory, Research, and Practice, 19(1)*, 31–44.

Kabat-Zinn, J. (2000). Indra's net at work: The mainstreaming of Dharma practice in society. In G. Watson, S. Batchelor, & G. Claxton (Eds.), *The psychology of awakening: Buddhism, science, and our day-to-day lives* (pp. 225–249). York Beach, ME, US: Samuel Weiser.

Kahneman, D., Diener, E., & Schwarz, N. (Eds.). (1999). *Well-being: Foundations of hedonic psychology*. Russell Sage Foundation.

Kaufman, K. A., Glass, C. R., & Arnkoff, D. B. (2009). Evaluation of Mindful Sport Performance Enhancement (MSPE): A new approach to promote flow in athletes. *Journal of Clinical Sport Psychology, 3*(4), 334–356.

Keyes, C. L. M. (1998). Social well-being. *Social Psychology Quarterly, 61(2)*, 121–140.

Keyes, C. L. M. (2005). Mental illness and/or mental health? Investigating axioms of the complete state model of health. *Journal of Consulting and Clinical Psychology, 73*, 539–548.

Keyes, C. L., Shmotkin, D., & Ryff, C. D. (2002). Optimizing well-being: the empirical encounter of two traditions. *Journal of Personality and Social Psychology, 82* (6), 1007.

Kraut, R. (1979). Two conceptions of happiness. *Philosophical Review, 88(2)*, 167–197.

Kraut, R. (2007). *What is good and why: The ethics of well-being*. Cambridge, MA: Harvard University Press.

Kubovy, M. (1999). *On the pleasures of the mind*. In Kahneman, D., Diener, E. & Schwarz, N. (1999). *Well-being: the foundations of hedonic psychology*, New York: Russell Sage Foundation, 134–154.

Lauer, E. E., Zakrajsek, R. A., & Lauer, L. (2017). *The Role of Sport Psychology for Young Athletes* (pp. 9–20). In Knight, C. J., Harwood, C. G. & Gould, D. (Ed) Sport psychology for young athletes. Routledge.

Little B. R. (1989). *Personal projects analysis: Trivial pursuits, magnificent obsessions, and the search for coherence*. In D. M. Buss & Cantor N. (Eds.) Personality psychology. New York, NY: Springer.

Locke, E. A., Cartledge, N., & Knerr, C. S. (1970). Studies of the relationship between satisfaction, goal setting, and performance. *Organizational Behavior and Human Performance, 5(*2), 135–158.

Locke, E. A., Mento, A. J., & Katcher, B. L. (1978). The interaction of ability and motivation in performance: An exploration of the meaning of Moderators. *Personnel Psychology, 31*(2), 269–280.

Locke, E. A., Shaw, K. N., Saari, L. M., & Latham, G. P. (1981). Goal setting and task performance: 1969–1980. *Psychological Bulletin, 90(*1), 125.

Lundqvist, C. (2011). Well-being in competitive sports—The feel-good factor? A review of conceptual considerations of well-being. *International Review of Sport and Exercise Psychology, 4*(2), 109–127.

Luthar, S. S., Cicchetti, D., & Becker, B. (2000). The construct of resilience: A critical evaluation and guidelines for future work. *Child Development, 71*(3), 543–562.

Mack, D. E., Wilson, P. M., Gunnell, K. E., Gilchrist, J. D., Kowalski, K. C., & Crocker, P. R. (2012). Health-enhancing physical activity: associations with markers of well-being. *Applied Psychology: Health and Well-Being, 4*(2), 127–150.

Martin, G. L., Thompson, K., & Regehr, K. (2004). Studies using single-subject designs in sport psychology: 30 years of research. *The Behavior Analyst, 27*(2), 263–280.

Mehrotra, S. (2013). Feeling good and doing well? Testing efficacy of a mental health promotive intervention program for Indian Youth. *International Journal of Psychological Studies, 5*(3), 28.

Mellalieu, S. D., Neil, R., Hanton, S., & Fletcher, D. (2009). Competition stress in sport performers: Stressors experienced in the competition environment. *Journal of Sports Sciences, 27(7)*, 729–744.

Mguni, N., Bacon, N., & Brown, J. F. (2012). The well-being and resilience paradox. The Young Foundation, London.

Morris, T., Spittle, M., & Watt, A. P. (2005). *Imagery in sport.* Champaign, IL: Human Kinetics.

Mummery, W. K., Schofield, G., & Perry, C. (2004). Bouncing back: The role of coping style, social support and self-concept in resilience of sport performance. *Athletic Insight*, 6(3), 1–15.

Norton, D. L. (1976). *Personal destinies: A philosophy of ethical individualism. Princeton*, NJ: Princeton University Press.

Omar-Fauzee, M. S., Sadeghi, H. S., Jamalis, M., & Abd-Latif, R. (2010). The mental skills training of university soccer players. *International Education Studies 3*(2), 81–90.

Öst, L. G. (1988). Applied relaxation vs progressive relaxation in the treatment of panic disorder. *Behavior Research and Therapy, 26(*1), 13–22.

Peterson, C., Park, N., & Seligman, M. E. P. (2005). Orientations to happiness and life satisfaction: The full life versus the empty life. *Journal of Happiness Studies, 6*, 25–41.

Poczwardowski, A., & Conroy, D. E. (2002). Coping responses to failure and success among elite athletes and performing artists. *Journal of Applied Sport Psychology, 14*(4), 313–329.

Podlog, L., & Eklund, R. C. (2006). A longitudinal investigation of competitive athletes' return to sport following serious injury. *Journal of Applied Sport Psychology, 18*(1), 44–68.

Rutter, M. (1985). Resilience in the face of adversity: Protective factors and resistance to psychiatric disorder. *The British Journal of Psychiatry*, *147*(6), 598–611.

Ryan, R. M., & Deci, E. L. (2000). Self-Determination Theory and the facilitation of intrinsic motivation, social development, and well-being. *American Psychologist, 55,* 68–78.

Ryan, R. M., & Deci, E. L. (2001). On happiness and human potentials: A review of research on hedonic and eudaimonic well-being. *Annual Review of Psychology, 52*(1), 141–166.

Ryff, C. D. (1989). Happiness is everything, or is it? Explorations on the meaning of psychological well-being. *Journal of Personality and Social Psychology, 57*, 1069–1081.

Ryff, C. D., & Keyes, C. L. M. (1995). The structure of psychological well-being revisited. *Journal of Personality and Social Psychology, 69*, 719–727.

Ryff, C. D., Singer, B. H., Wing, E., & Love, G. D. (2001). Elective affinities and uninvited agonies: Mapping emotion with significant others onto health. In: C. D. Ryff, & B. H. Singer (Eds). *Emotion, social relationships, and health: Third annual Wisconsin symposium on emotion* (pp. 133–175). Oxford University Press; New York.

Seligman, M. (2011). *What you can change and what you can't: the complete guide to successful Self-Improvement*. UK: Hachette.

Shannon, V., Gentner, N. B., Patel, A., & Muccio, D. (2012). Striking gold: mental techniques and preparation strategies used by Olympic gold medallists. *Athletic Insight: The Online Journal of Sport Psychology, 4*, 1–11.

Sheldon, K. M., & Kasser, T. (1995). Coherence and congruence: Two aspects of personality integration. *Journal of Personality and Social Psychology, 68(3)*, 531–543.

Steffen, K., Soligard, T., & Engebretsen, L. (2011). The IOC's endeavor to protect the health of the athlete continues. *British Journal of Sports Medicine, 45*, 551–552.

Strümpfer, D. J. W. (1999). Psychosocial resilience in adults. *Studia Psychologica, 41(2)*, 89–103.

Tamminen, K. A., Holt, N. L., & Neely, K. C. (2013). Exploring adversity and the potential for growth among elite female athletes. *Psychology of Sport and Exercise, 14*(1), 28–36.

Theberge, N. (2008). "Just a normal bad part of what I do": Elite athletes' accounts of the relationship between health and sport. *Sociology of Sport Journal, 25*(2), 206–222.

Tiberius, V., & Hall, A. (2010). Normative theory and psychological research: Hedonism, eudaimonism, and why it matters. *The Journal of Positive Psychology, 5*(3), 212–225.

Unestahl, L. E. (1982). Inner mental training for sport. *Mental training for coaches and athletes*, 135–140.

Uneståhl, L. E. (Ed.) (1985). *Sport psychology in theory and practice*. Örebro, Sweden: VEJE International Publ.

Uneståhl. L. E. (2013). 50 years with Integrated Mental Training. *Swedish Journal of Sport Research, 1*, 82–108.

Van Raalte, J. L., Brewer, B. W., Lewis, B. P., & Linder, D. E. (1995). Cork! The effects of positive and negative self-talk on dart throwing performance. *Journal of Sport Behavior, 18*(1), 50.

Vázquez, C., Hervás, G., Rahona, J. J., & Gómez, D. (2009). Psychological well-being and health. Contributions of positive psychology. *Annuary of Clinical and Health Psychology, 5*, 15–27.

Vealey, R. S. (1994). Current status and prominent issues in sport psychology interventions. *Medicine & Science in Sports & Exercise, 26, 495–502.*

Vealey, R. S. (2007). Mental skills training in sport. In G. Tenenbaum, & R. C. Eklund (Eds.), *Handbook of sport psychology* (3rd ed.). Hoboken, NJ: Wiley.

Vealey, R. S., & Greenleaf, C. A. (2001). Seeing is believing: Understanding and using imagery in sport. Applied sport psychology: Personal growth to peak performance, *Applied Sport Psychology: Personal Growth to Peak Performance, 4*, 247–272.

Voelker, D. K., Gould, D., & Reel, J. J. (2014). Prevalence and correlates of disordered eating in female figure skaters. *Psychology of Sport and Exercise, 15*(6), 696–704.

Weinberg, R., Butt, J., Mellano, K., & Harmsion, R. (2017). The stability of mental toughness across situations: taking a social-cognitive approach. *International Journal of Sport Psychology, 48*(3), 280–302.

Weinberg, R. S. & Comar, W. (1994). The effectiveness of psychological interventions in competitive sport. *Sports Medicine, 18*(6), 406–418.

Wendel-Vos, G. W., Schuit, A. J., Tijhuis, M. A. R., & Kromhout, D. (2004). Leisure time physical activity and health-related quality of life: Cross-sectional and longitudinal associations. *Quality of Life Research, 13*(3), 667–677.

Williams, P. T. (2001). Physical fitness and activity as separate heart disease risk factors: A meta-analysis. *Medicine and Science in Sports and Exercise, 33*(5), 754.

Windle, G., Bennett, K. M., & Noyes, J. (2011). A methodological review of resilience measurement scales. *Health and Quality of Life Outcomes, 9*(1), 8.

PART III

Emotion, Resilience, and Well-Being in Work and Organizational Context

CHAPTER 19

Positive Emotions at Workplace

ARCHANA and UPDESH KUMAR

Defence Institute of Psychological Research, Defence Research and Development Organisation, Timarpur, Delhi 110054, India

Corresponding author. E-mail: archanasingh2410@gmail.com

ABSTRACT

In recent years, workplace emotions have gained immense attention in organizational research. Emotions are ever-present in the workplace as they are the integral part of human lives. Workplace experience involves a blend of positive and negative emotions that plays a vital role in shaping the mental health and well-being of the employees. Positive emotions tend to be extremely beneficial toward productive work environment, whereas negative emotions are considered to be harmful and unproductive. Lack of positive emotions makes an individual more vulnerable to stressful events. The presence of positive emotion strengthens the personal and social resources of an individual at the workplace. Employees are often found to be more satisfied with their jobs where positive emotions are seen as a valuable asset. Positive emotions at the workplace are characterized by the feelings of happiness, joy, enjoyment, enthusiasm, optimism, gratitude, and wellness that buffer against stress. Positive employees display higher level of competence, greater sense of pride, good interpersonal skills, creativity, and resilient behavior. Varied researches in the area of positive emotions address emotional processes that help in enhancing cognitive functions, promoting constructive behavior, and attaining desired organizational goals. Organizations propose an array of mechanisms for counteracting emotions at the work place. This chapter deals with understanding how positive emotions stimulate people to perform effectively in an organization. It also focuses on identifying the factors promoting positive emotions that serve as a trademark of positive health and happiness at the workplace.

19.1 POSITIVE EMOTIONS AT WORKPLACE

Stress within the work environment has become a growing cause of concern for the organizations as well as their employees. Stress is an interface between the individual and the demands placed by the environment. An individual reacts in multiple ways while experiencing stress. These reactions are directly related to an interaction between the individual traits and the resources available for dealing with the stressors. Inability to cope with these stressors often creates emotional complications. According to Lazarus (2000), stress occurs at three different levels: acute, episodic, and chronic. Acute stress takes place when individuals are not able to adjust to the demands and pressures imposed on them. These demands are above the threshold level of adaptability of an individual. The demands occur due to unrealistic expectations and work overload that causes disappointment and dissatisfaction with the work environment. Acute stress is reflected in the form of frustration, anxiety, fatigue, increased blood pressure, headache, inability to take decisions, and so on. As compared to acute stress, episodic stress occurs more frequently and consistently in multiple episodes. Symptoms of episodic stress include hostility, low levels of patience, and a sense of time urgency. Chronic stress illustrates stressors that persist and are long lasting. These stressors are generally caused due to job stress, family problems, and prolonged illness. According to Khosla (2001), both positive and negative effects co-occur during stressful conditions. Individuals with positive effects rely more on problem-focused coping strategies as compared to emotion-focused coping strategies (Khosla & Hangal, 2004).

Scholars have established that individuals working in toxic work environment, characterized by persistent demands and intense pressures, are prone to experience stress (Macklem, 2005). Murphy (1995) presented five different types of work stress that occur when there is a disagreement between the employees and the pressures of job placed upon them. The first type refers to *factors unique to the job* that includes working hours, level of freedom in the work setting, meaningfulness of the job, physical conditions prevailing within the work environment, and workload. The second type relates to the *employee's role in the organization*. Stress occurs when employees are not able to perform multiple roles simultaneously. This may result due to role ambiguity, wherein there is a lack of clarity about the responsibilities of the employees and the performance expected from them. The third type is concerned with the *career development*, which focuses on the issues related to over/underpromotion and opportunities for career development. Stress occurs when an employee feels that there is no opportunity for advancement. The fourth type focuses on *interpersonal relationships within the workplace*. The troubled interpersonal relationship between the leaders and the subordinates is a main cause of concern in the organization that often results in workplace stress. The fifth type of workplace stress relates to *organizational structure and climate*. It focuses on the communication linkages, management style, and active contribution in decision making.

Until recently, stress has been cast as a negative energy. If not dealt with effectively, the prolonged exposure of this negative force can lead to a number of problems. However, if handled well, it can result in achieving good health and high performance at workplace. The positive psychology movement has emerged as an innovative approach toward understanding human dynamics within the work setting. This approach builds a foundation for developing wellness over time and focuses on cultivating strengths in day-to-day life that plays a prominent role toward enhancing positive emotions and optimizing well-being.

19.2 CONCEPTUAL UNDERSTANDING OF EMOTIONS

Emotions play a significant role in individual's behavior and well-being. An individual's behavior and opinion are strongly influenced by both cognition and emotion. Mostly, emotional reactions are instinctive in nature and often occur without one's awareness. Emotions are represented through changes in one's behavior that includes changes in facial expression, pitch, tone, increase in heart rate, blood pressure, and feelings of anger, fear, frustration, and anxiety.

Research on emotions has gained a wide significance in the domain of positive psychology. Fredrickson and Branigan (2005) define emotions as short-lived experiences that produce changes in physiology, thoughts, and behaviors. Emotions are adaptive in nature and have an extensive range of implications at the work setting. The biggest challenge for the employees at the workplace is to manage their emotions. Employees are often required to display one emotion at a time while experiencing other emotions simultaneously. These discrepancies of emotions are fake in nature and are often termed as emotional dissonance that takes a serious toll on an individual's health (Ekman et al., 1997). Emotional dissonance has a negative impact on the quality of life and well-being of the employees. The emotions at the organizational front are often reflected by the feelings that are either felt or displayed (Hochschild, 1979). The greater the discrepancy between the felt and the displayed emotions, the more the employees tend to experience job stress and work absenteeism. The employees in the organization use different emotion regulation strategies in order to deal with the challenges of emotional dissonance. These strategies can take two forms: surface acting and deep acting. Surface acting relates to hiding one's inner feelings, which means modifying one's behavior to be consistent with the required emotions, but at the same time, the individual still persists to hold different internal feelings. It focuses on the displayed emotions. Deep acting refers to changing true emotions in order to match the required emotions. It focuses on felt emotions (Brotheridge & Lee, 2003). Employees adopting a strategy of deep acting are less likely to experience emotional dissonance (Grandey, 2003). The employees who express their true feelings while interacting with others display a healthier life style with a better quality of life.

19.2.1 EMOTION WORK

Understanding the essence of emotion work is valuable to any organization as it is through this emotion work that employees are able to meet the organizational obligations. Employees at the workplace need to perform emotion work so that they are able to regulate inappropriate emotions in their day-to-day life while dealing with others. According to Grandey (2000), emotion work is a process of adjusting emotions (feelings and expression) in order to meet organizational demands. In order to comply with the emotional standards of the organizations, employees are required to modulate their feelings and emotions. The emotion work has both good and bad sides. The good side of the emotion work is reflected in those individuals who are emotionally more adaptable to the prevailing conditions of the organizations. These individuals display higher levels of job satisfaction, personal accomplishment, and less burnout (Brotheridge & Grandey, 2002). Emotion work can also be a source of work stress for employees who lack in their abilities to deal effectively with the work conditions, thus affecting their well-being (Grandey, 2003).

Various approaches to emotion work suggest that emotions are handled at work setting so as to meet organizational rules. Ashforth and Humphrey (1993) laid more emphasis on emotion work as an expressive behavior. They focused on the association between visible manifestation and task performance. According to them, if employees are not able to express their feelings genuinely, then emotion work becomes negative, thus affecting the organizational functioning. Grandey (2000) described emotion work as the process of regulating feelings and expression in order to fulfill the expectations of the organization. According to him, individual differences exist in emotion work with regard to gender, emotional intelligence, self-monitoring, and emotional expressivity. The work environment of an organization depends upon the level and type of emotion work displayed by the employees. Emotions play an active role in workplace learning (Wall et al., 2017). Learning is more effective when people are more positively engaged in work (Hazelton, 2014).

19.2.2 EMOTION TYPES

Emotions are affective states occurring at both conscious and unconscious levels. Work incorporates a wide range of positive and negative emotions. At the organizational level, individuals express positive emotions when they are able to meet their goals. By cultivating positive emotions, people are able to improve their health and well-being. Negative emotions are unpleasant states displayed through the feelings of sadness, fear, anxiety, worry, and disappointment. They are experienced when people fail in their abilities to achieve the objectives and express inefficient emotional management. Positive emotions serve as a protective factor in guarding individual from negative emotions. Zautra et al. (2001) proposed the dynamic model of affect. The model suggests that under normal conditions, positive and negative emotions are fairly independent. During stressful situations, an inverse relationship increases between negative and positive emotions. Enhancing positive emotions during stress help in reducing the effect of negative emotions. Lack of positive emotions makes an individual more vulnerable to the effects of stress. Positive emotions can undo the effects of negative emotions by generating quicker returns to baseline levels of cardiovascular activation (Fredrickson et al., 2000). Well-being in the organization is characterized by an employee's positive emotional states. According to Fredrickson and Losade (2005), the ratio of positive and negative emotions is highly beneficial for understanding the well-being of an individual. In order to maintain well-being, higher rate of positive emotions is required to subside the effects of negative emotion.

19.3 POSITIVE EMOTIONS AT WORK LIFE

Positive emotions occur when people are in a safe and satisfied state and not in any sort of life-threatening situation. According to Fredrickson (1998), joy, interest, pride, and contentment are some of the basic positive emotions that bring positivity in the work environment. Joy relates to the drive to play, being innovative and display logical and artistic behavior; interest involves an urge to discover, acquire new information, and develop the self in the process; pride takes into account the achievements and sense of visualizing even greater future achievement; and contentment refers to the drive to appreciate present life conditions and incorporate these conditions into novel outlook of self and the humanity. Extensive literature in the area of positive emotions has focused on strong association of positive emotions with happiness (Szczygiela & Mikolajczakb, 2017), thriving at work (Paterson et al., 2014), positive orientations (Alessandri et al., 2012), and psychological capital (Newman et al., 2014).

Research has shown that positive emotions have an extreme range of impact on individuals and organizations. Positive emotions reduce the negative effects of emotions and contribute toward the optimal functioning of the organizations by enhancing the quality of people's life. Positive emotions are momentary in nature and not a function of predetermined, stable personality traits (Fredrickson, 2009). Positive emotions are also related to upward spirals of enhanced well-being that broadens an individual's scope of attention (Fredrickson & Joiner, 2002). The existence of positive emotions in the organization facilitates adaptive coping and enhances tendencies toward developing good interpersonal relations with respect and dignity. Positive emotions lead to greater affective and cognitive flexibility resulting in enhanced adaptive capacities (Rothman & Melwani, 2017).

Empirical research on positive emotion has shown a rich record in the area of positive psychology. This research has illustrated that positive emotions are strongly related to favorable organizational goals and valuable human outcomes. Employees who experience pleasant emotions at the workplace are more likely to set higher goals for themselves that enhances the productivity of the organization. Positive emotions promote helping behavior and cooperation, trim down aggression against the people and the organization, and enhance creativity (Madjar et al., 2002). It increases one's efficiency at the time of difficult decision making and also results in more collaborative approaches during crisis intervention (George et al., 1996). It has been empirically established that positive emotions expand cognitive processes and improve human performance (Fredrickson, 1998). It shapes the behavior of an individual by addressing their life challenges. According to Aarrestad et al. (2015), positive emotion intensifies collaboration, help-seeking and help-giving behavior, and sensory motor connectivity in knowledge creation.

Positive emotions capture less attention as compared to negative emotions. Expressing negative emotions is not always deleterious. At times, negative emotions lead to positive consequences. For example, anger as a negative emotion serves as a driving force that leads to counterproductive work behaviors. Positive emotions have contributed significantly toward enhanced companionship, elevated incomes, and improved physical health (Waugh & Fredrickson, 2006). Pioneers such as Lyubomirsky et al. (2005) have asserted that positive emotions have a casual influence on outcomes pertaining to social interactions, work productivity, health behavior, and psychological capital. People with positive emotions flourish in lives, since they get the boost out of everyday pleasant experiences. These pleasant events are generally observed in the form of altruistic behavior and the ways how an individual interacts with others and gets spiritually connected. Positive emotions not only focus on making people good in present, but also enhance the possibility on making them good in future.

Positive emotions benefit people at mental, social, psychological, and physical levels. At mental level, positive emotions influence the way one's intellect works and enhance divergent thinking that leads to greater creativity. At social level, positive emotions help in connecting to others that results in personal growth, better exchange of information, and less inconsistency in relationships. At psychological level, positivity helps in building psychological resources, enhancing well-being and resilience (Lyubomirsky et al., 2005). This, in turn, increases enthusiasm, helps in establishing goals, and promotes adaptive coping skills. At physical level, positive people are healthy and do not show any severe physical disability. Positive emotions at the workplace have a huge impact on the employee well-being, health, job satisfaction, and personal and organizational resilience (Quoidbach et al., 2015).

19.3.1 BROADEN-AND-BUILD THEORY OF POSITIVE EMOTIONS

The broaden-and-build theory of positive emotions conceptualized by Fredrickson (1998) suggests that positive emotions equip individuals and organizations toward future success. Broaden-and-build theory is based on two hypotheses. The broaden hypothesis states that positive emotions broaden an individual's attention and thinking, widen their repertoires of desired behavior, and enhance their openness to innovative experiences (Fredrickson & Branigan, 2005). The build hypothesis suggests that positive emotions set people on trajectories of growth, which, over a period of time, build consequential personal resources. Research has summarized that individuals who express positive emotions more often than others are more likely to display optimism and enhanced quality of interpersonal relationships (Waugh & Fredrickson, 2006).

The broaden-and-build theory envisages positive emotions. It can renovate workplace by broadening people's habitual modes of thinking. In this process, it makes the employees of the organization more flexible and innovative in their approach. The positive characteristics of employees help in enhancing organizational functioning, thus leading to stronger social networks and healthy organizational culture and climate. Fostering positive emotions help others in finding positive meaning in their day-to-day work life. In broadening an individual's momentary thought action repertoires, positive emotions play a pivotal role in building an individual's intellectual and social resources (Isen, 1987). Intellectual resources relate to developing cognitive maps for finding ways. Positive emotions build intellectual resources by facilitating learning and performance. Social resources relate to enduring social relationships that involve finding someone to turn to for seeking help.

19.4 FOSTERING POSITIVE EMOTIONS AT WORKPLACE

Sustaining positivity requires persistent efforts, since human beings have negativity bias and are more prone to remember negative incidents of life as compared to the positive ones. According to Fredrickson (2009), negativity cannot be avoided; however, it is important to build a positive-to-negative-emotion ratio of 3:1 in order to enhance positivity. To understand positivity, it is essential to know the factors leading to positive emotions. According to Lyubomirsky (2007), positivity is determined by genes and certain life events that include health and income. However, at the same time, positivity is also determined by an individual's thoughts and actions.

Emotional management requires an understanding about the nature of emotions, identifying positive factors for developing self-adjustment and increasing organizational effectiveness. At workplace, emotions are managed at two levels: individual and organizational. Managing emotions at individual level relates to the skills required on the part of the individual to have self-control over their emotional state and the capability to manage resources within the organization. At organizational level, there are certain strategies required on the part of the leaders that help in optimizing the organizational functioning and enhancing the quality of life of the employees. In an organization, the leader's emotions influence the emotions of their employees as emotions flow from top to bottom. The foremost criteria for the leaders to be effective are to manage their own emotions prior to the emotion of their subordinates.

Savoring and gratitude are effective approaches toward increasing the intensity of positive emotions. According to Bryant (2003), savoring refers to behavior for increasing the influence of positive emotions. A positive relationship exists between savoring and positive emotions, as savoring correlates positively with well-being and negatively with distress. Savoring relates to past, present, and future. Savoring in the past indicates recalling about the past successes. Savoring in present relates to conscious focusing on pleasurable experiences. Savoring in future focuses on predicting future incidents and building exciting plans (Bryant & Veroff, 2007). Positive emotions can also be enhanced by cultivating gratitude. According to McCullough et al. (2002), gratitude relates to a thankful gesture of admiration that results from the kindness of others. Gratitude serves as an effective strategy for enhancing positive emotions, thus leading to upward spirals of well-being. High levels of gratitude result in higher life satisfaction and enhanced well-being (Froh et al., 2010).

19.4.1 EMOTIONAL INTELLIGENCE

Emotions play a vital role in managing stress. The impact of stress varies from person to person. Some people remain calm under extreme stressful condition, and some go to pieces even under trivial frustration. Under these circumstances, the role of emotional intelligence comes into play that helps an individual to cope with stressful situations. Managing stress to a great extent depends upon bringing emotional balance between the probable stress conditions and one's reaction to those conditions. Emotional intelligence relates to one's ability to identify, understand, and deal with emotional cues. It suggests keeping oneself away from pessimistic feelings (such as anger, anxiety, worry, tension, and irritation) and encouraging positive and constructive feelings (such as self-efficacy, empathy, psychological flexibility, and psychological capital). Emotional intelligence can be learned, developed, and improved over a period of time. Emotional intelligence relates to an individual's ability to recognize, express, and regulate emotions. Emotionally intelligent people are in touch with their emotions and are capable of reading other's emotions.

Chadha and Singh (2001) stated emotional intelligence as the capability of an individual to react to huge array of stimuli drawn from one's nature and surroundings. These researchers have developed a tool on emotional intelligence that has been widely accepted across the country. Another Indian perspective suggests emotional intelligence to be rooted in our traditional values and philosophical framework. Familial and societal values play a vital role in shaping one's emotions. Social skills expressed in the form of altruism and concern for others are found to be associated with the goal accomplishment (Dalal et al., 1988). Expressing emotions ascertains positive connections with others thereby producing effective performance, positive result, and meaningful change in the organization. A closely connected aspect related to emotional intelligence is emotional competence that focuses on regulation and utilization of effect. Emotional competence is a learned capability based on emotional intelligence that enhances the efficiency of the employees at the organizational front. Emotional competence involves reading the feelings of others and managing those feelings skillfully. At work setting, high emotional intelligence implies that the employees have a great potential to learn emotional competencies. According to Goleman (1998), emotional intelligence takes into account five elements: self-awareness (being away of one's emotions, internal states, performances, motives, and intentions), self-regulation (managing one's internal states, impulses, and resources), motivation (emotional tendencies that assists in reaching set targets), social competence (deals with the ways of handling relationships) and empathy (awareness of other's feelings and concerns). An individual requires high degree of emotional intelligence to manage all these elements. Emotional intelligence plays an essential role in controlling the activated emotions, which generally arise when an individual is dealing with organizational conflicts (Bodtker & Jameson, 2001). Positive emotions boost emotional intelligence by enhancing one's capability to deal with adversity, overcoming negative emotional state, and enabling individuals to be more resilient and performing strong social networks. Employees with high level of emotional intelligence have better ability to deal with their emotions under challenging conditions, whereas employees with low emotional intelligence display negative tendency in managing emotions.

Extensive literature on emotional intelligence shows the role of emotional intelligence at work place and its impact on work success. In a study carried out by Pradhan and Nath (2012), emotional intelligence at organizational work setting was found to be associated with entrepreneurial orientation and contributed toward creativity, innovation, and intuitiveness. Emotional intelligence upholds performance of an individual leading to positive output at the work front (Pradhan et al., 2005). Research has also revealed an inverse relationship between emotional intelligence and perceived stress (Naidoon & Pau, 2008). Emotional intelligence helps in minimizing the negative effects of stress and is considered as an important component in defining success in life and ascertaining psychological well-being. Emotional intelligence plays a significant role in shaping the interaction between the individual and their work environment. Low emotional intelligence is found to be related to worry states and avoidance coping. It not only results in perceived occupational stress, but also results in negative health outcomes (Oginska-Bulk, 2005). Individuals with high emotional intelligence display more capability in making independent decisions and mobilizing human resources for its optimum utilization (Pradhan, 2003). A study carried out by Singh (2007) on software employees' highlighted emotional intelligence to be positively related to leadership effectiveness.

19.4.2 CHARISMATIC LEADERSHIP

Emotion serves as an effective powerful tool for leader's effectiveness. Emotion flows from top level to bottom level, that is, from leaders to followers. Leadership theories based on charismatic and transformational leadership placed emphasis on affective and emotional aspects of leader's influence. Varied research studies reveal a strong association between employee satisfaction and leadership behavior. Transformational leadership is found to be associated with job satisfaction and motivation of followers (Judge & Piccolo, 2004). Charismatic leadership is linked to the employee cooperation and task performance at the work place (De Cremer & van Knippenberg, 2002). Positive emotions affect the leadership process in an organization. Positive group mood is found to have strong association with better cooperation and reduced interpersonal differences among group members (Barsade, 2002). Positive affect also leaves a huge impact on the level of motivation of the employees. According to Conger and Kanungo (1998), leaders who are capable of extracting emotional responses from their subordinates are more likely to achieve positive outcomes.

Positive emotional responses of leaders result in heightened energetic arousal that affects the performance of the employees in the organization. Individuals experiencing positive affect set higher targets for themselves. Fostering positive emotions in organization results in several kinds of behavioral outcomes related to positive moods, effectiveness of leaders, job satisfaction, well-being, and enhanced performance (Diener et al., 2003). Charismatic leaders interpret, integrate, and paint positive outlook of the world. The personality of the leader is important in understanding the association between positive emotions and effective leadership. Influential leaders are observed to be more extroverts and display cheerful disposition toward their employees (Ployhart et al., 2001).

Besides positive emotions, negative emotions are also influential in leader's effectiveness. It has been examined that negative emotions leave a much greater impact with long-lasting effects than positive emotions. Bucy (2000) has established that leader's negative emotional displays are perceived with greater intensity than their positive emotional displays. That means the energetic and optimistic leaders energize their followers, whereas leaders who display negative emotional expressions are likely to activate their followers negatively. The emotion and moods of the leaders get transferred to their subordinates. This is a process called emotional contagion, wherein the followers get influenced by the leader's emotions (Hatfield et al., 1994). Due to greater emotionally expressive behavior, extroverts are more effective and influential senders of emotions (Friedman & Riggio, 1981).

19.4.3 RESILIENCE BUILDING

Resilience has become important in today's fast paced world. Resilience is the capability of an individual to adapt successfully to the challenging conditions. According to Luthans (2002), resilience is the ability to bounce back from the adverse and difficult life situations in order to deal with the stressors of life. During adversities, people experience wide range of emotions ranging from anxiety to worry, sadness, and grief. There are individual differences in psychological resilience, wherein some people are able to bounce back from almost any adversity and move on with their lives, while for others, even a small hindrance may bring a major setback in their lives. Resilience is required in the organization in order to deal and adapt well with the stressors of work life. Stress does not occur from the situation itself; rather, it arises from an individual's perception about their inability to deal with the situation.

Positive emotion stimulates psychological resilience. Research suggests that resilience strengthens an individual's resistance to stress. High-resilient individuals display quicker emotional recovery from stressful events, thus leading to adaptive coping (Ong et al., 2006). Resilience is a steady personality trait that enables an individual to rebound and cope with negative life events. Positive emotions are essential components of resilience and serves as an adaptive behavior in coping with stress. It not only strengthens resistance, but also provides respite from ongoing stresses (Zautra et al., 2005). In a study conducted by Fredrickson et al. (2003), it was found that high-resilient individuals show faster cardiovascular recovery from stressors along with lower levels of depression. Nath and Pradhan (2013) highlighted the importance of positive emotional experiences in building psychological resilience. According to these researchers, during life-threatening conditions, positive emotions help in managing emotional resources in a flexible manner that contribute toward dealing with life adversities and promoting psychological resilience. Positive emotions increase resilience, which further results in improved life satisfaction. Individuals who are optimistic about their abilities are likely to exhibit more positive perceptions of their work along with life satisfaction (Judge et al., 1998). Self-efficacy is one of the important variables that promote resilience by reducing the impact of stressors prevailing at work place. Confident employees find ways and means for managing their higher workloads and overcoming the feelings of distress (Jex et al., 2001). Resilience at workplace requires its employees to have willpower, strong vigor, and resourcefulness to cope with the organizational climate. Maddi and Khoshaba (2005) have opined that resilient individuals turn hardship into growth experience by enhancing their work life and living conditions.

Positive emotions enhance resilience and act as a force toward self-flourishing. The broaden-and-build model of Fredrickson (2001) emphasizes the role of emotions in building resilience. According to this model, positive and negative emotions have different impacts on people dealing with stress. Positive emotions build personal resources by broadening the thought action repertoires and enhancing well-being.

19.4.4 EMPLOYEE ENGAGEMENT

Engagement is characterized by high levels of energy, interest, and passion about the work. Work engagement relates to an optimistic and satisfying state of mind that involves vigor, dedication, and absorption (Schaufeli & Bakker, 2004). Vigor relates to an increased level of dynamism and hardiness that are required in order to function well in the organization. Dedication focuses on a sense of commitment and perseverance that is essential for carrying out any set of work. Absorption refers to keeping oneself fully engrossed in the work. Engaged employees are highly self-efficient people who hold a positive attitude toward the events that occur in their day-to-day life. These people are not workaholics. They maintain their pleasant state of mind by enjoying all kinds of work they undertake. An engaged employee is well aware of the organizational climate and maintain a good interpersonal relationship with their colleagues. This culture benefits the organization by enhancing its functioning and performances at the job front.

Employee engagement is associated with varied psychological concepts prevailing at workplace that includes organizational scholarship, organizational commitment, organizational citizenship behavior, and positive employee behavior (Organ & Paine, 1999). A study by Mohapatra and Sharma (2010) focused on identifying the predictors of organizational commitment. The findings of the study showed that pay, job content, and objectivity emerged as the most significant predictors of employee engagement. Work engagement focuses on two kinds of resources that are associated with work engagement: job resources and personal resources. Job resources focus on the physical, social, and organizational characteristics of the job that helps in fostering the growth and development of an individual in the organization. Personal resources relate to an individual's capability to keep a control over their environment. Engaged employees hold a positive self-evaluation with a belief that they will experience desirable outcomes in their life. They exhibit an optimistic approach by being resilient in adapting to challenging life events. Resilience tends to show a positive relationship with work engagement under tough job conditions (Bakker & Demerouti 2007).

Varied researchers suggest that engaged employees mostly experience positive emotions, which enable them to be more constructive at their work place (Schaufeli & Salanova, 2007). According to Fredrickson (2001), positive emotions help in building physical, psychological, and social resources. All these resources are associated with better employee engagement and health outcomes, thus facilitating positive change in the organizational setup. Positive emotions strengthen physical resources by enhancing the functioning of immune system and protecting the body against any disease. Psychological resources relate to an individual's ability to deal with the stressors. Positive emotions contribute to psychological resources by cultivating resilience that buffers against stress. Social resources build on the quality of relationship one shares with others. It boosts an individual's social resources by fostering social relationships that serve to be supportive at the time of crisis. Higher levels of positive emotions are found to be associated with higher levels of employee engagement and organizational citizenship behavior (Wright & Staw, 1999).

The organizational culture plays a significant role in creating positive work environment by means of (i) promoting positive thinking that includes instilling hope and optimism among the employees and developing a culture of gratitude and forgiveness, (ii) building positive relationships through trust, respect, recognition, and generosity, (iii) enhancing strengths by making employees identify their potentials and by providing feedback and developmental opportunities, and (iv) adding a sense of fulfillment, happiness, and prosperity.

19.5 CONCLUSION

Positive emotions are essential and valuable ingredients of positive psychology. Positive emotions relate to the pleasant experiences that lead to contentment and broaden one's range of ideas and innovation. Such emotions at work result in increased levels of engagement, organizational citizenship behavior, and job satisfaction. Positivity in employees helps in making better decisions, enhancing positivity and creating healthier interpersonal skills. Positive emotional state contributes toward making an employee more flexible, altruistic, and adaptable and strengthens the resources within the workplace so as to make the functioning of the organizations more effective. Positive emotions are more likely to open healthy interpersonal relationships with others, fuel one's mental health, and foster adaptive coping under challenging circumstances. Positive emotions in an organization can be enhanced through various strategies that include cultivating emotional intelligence, building charismatic leadership, fostering resilience, and keeping employees engaged in their work. These strategies contribute toward transforming people in finding a positive meaning in their life and empowering them to flourish in challenging environments. Enhancing positive emotions under stressful conditions will definitely pay off toward the growth and development of the individual as well as organization as a whole. Positive emotions need to be seen as an important tool for producing meaningful change and enhancing the organizational effectiveness.

KEYWORDS

- **positive emotions**
- **negative emotions**
- **organizational productivity**
- **mental health**

REFERENCES

Aarrestad, M., Brondbo, M. T., & Carlsen, A. (2015). When stakes are high and guards are low: High-quality connections in knowledge creation. *Knowledge and Process Management, 22*(2), 88–98.

Alessandri, G., Vecchione, M., Tisak, J., Deiana, G., & Caprara, G. V. (2012). The utility of positive orientation in predicting job performance and organizational citizenship behaviors. *Applied Psychology, 61*(4), 669–698.

Ashforth, B. E., & Humphrey, R. H. (1993). Emotional labor in service roles: The influence of identity. *Academy of Management Review, 18*(1), 88–115.

Bakker, A. B., & Demerouti, E. (2007). The job demands resources model: State of the art. *Journal of Managerial Psychology, 22*(3), 309–328.

Barsade, S. (2002). The ripple effect: Emotional contagion and its influence on group behavior. *Administrative Science Quarterly, 47*(4), 644–675.

Bodtker, A., & Jameson, J. (2001). Emotion in conflict formation and its transformation: Application to organizational conflict management. *International Journal of Conflict Management, 12*(3), 259–275.

Brotheridge, C. M., & Grandey, A. A (2002). Emotional labor and burnout: Comparing two perspectives of people work. *Journal of Vocational Behavior, 60*(1), 17–39.

Brotheridge, C. M., & Lee, R. T. (2003). Development and validation of the emotional labour scale. *Journal of Occupational & Organizational Psychology, 76*(3), 365–379.

Bryant, F. B. (2003). A scale for measuring beliefs about savoring. *Journal of Mental Health, 12*(2), 175–196.

Bryant, F. B., & Veroff, J. (2007). Savoring: A new model of positive experience. Mahwah, NJ, USA: Lawrence Erlbaum Associates.

Bucy, E. P. (2000). Emotional and evaluative consequences of inappropriate leader displays. *Communication Research, 27*(2), 194–226.

Chadha, N. K., & Singh, D. (2001). How to measure your EQ? In D. Singh (Ed.), *Emotional Intelligence at Work: A Professional Guide* (pp. 210–218). New Delhi, India: Response Books.

Conger, J. A., & Kanungo, R. N. (1998). *Charismatic leadership in organizations.* Thousand Oaks, CA, USA: Sage Publications.

Dalal, A. K., Singh, A. K., & Misra, G. (1988). Reconceptualisation of achievement behavior: A cognitive approach. In A. K. Dalal (Ed.), *Attribution theory and research* (pp. 82–97). New Delhi, India: Wiley Eastern.

De Cremer, D., & van Knippenberg, D. (2002). How do leadership promote cooperation? The effects of charisma and procedural fairness. *Journal of Applied Psychology, 87*(5), 858–866.

Diener, E., Oishi, S., & Lucas, R. E. (2003). Personality, culture, and subjective well-being: Emotional and cognitive evaluations of life. *Annual Review of Psychology, 54*(1), 403–425.

Ekman, P., Friesen, W. V., & O'Sullivan, M. (1997). Smiles when lying. In P. Ekman & E. L. Rosenberg (Eds.), *What the face reveals: Basic and applied studies of spontaneous expression using the facial action coding system* (pp. 210–216). London, UK: Oxford University Press.

Fredrickson, B. L. (1998). What good are positive emotions? *Review of General Psychology, 2*(3), 300–319.

Fredrickson, B. L. (2001). The role of positive emotions in positive psychology: The broaden-and-build theory of positive emotions. *American Psychologist, 56*(3), 218–226.

Fredrickson B. L. (2009). *Positivity*. New York, NY, USA: Random House.

Fredrickson, B. L., & Branigan, C. (2005). Positive emotions broaden the scope of attention and thought-action repertoires. *Cognition and Emotion, 19*(3), 313–332.

Fredrickson, B. L., & Joiner, T. (2002). Positive emotions trigger upward spirals toward emotional well-being. *Psychological Science, 13*(2), 172–175.

Fredrickson, B. L., & Losada, M. F. (2005). Positive affect and the complex dynamics of human flourishing. *American Psychologist, 60*(7), 678–686.

Fredrickson, B. L., Mancuso, R. A., Branigan, C., & Tugade, M. M. (2000). The undoing effect of positive emotions. *Motivation and Emotion, 24*(4), 237–258.

Fredrickson, B. L., Tugade, M. M., Waugh, C. E., & Larkin, G. (2003). What good are positive emotions in crises? A prospective study of resilience and emotions following the terrorist attacks on the United States on September 11th, 2001. *Journal of Personality and Social Psychology, 84*(2), 365–376.

Friedman, H. S., & Riggio, R. E. (1981). Effect of individual differences in nonverbal expressiveness on transmission of emotion. *Journal of Nonverbal Behavior, 6*(2), 96–104.

Froh, J. J., Bono, G., & Emmons, R. (2010). Being grateful is beyond good manners: Gratitude and motivation to contribute to society among early adolescents. *Motivation and Emotion, 34*(2), 144–157.

George, J. M., Brief, A. P., & Motowildo, S. J. (1996). Motivational agendas in the workplace: The effects of feelings on focus of attention and work motivation. In L. Cummings & B. W. Staw (Eds.), *Research in Organizational Behavior* (pp. 75–100). Greenwich, CT, USA: JAI Press.

Goleman, D. (1998). *Working with emotional intelligence.* New York, NY, USA: Bantam.

Grandey, A. A. (2000). Emotion regulation in the workplace: A new way to conceptualize emotional labor. *Journal of Occupational Health Psychology, 5*(1), 95–110.

Grandey, A. A. (2003). When the show must go on: Surface acting and deep acting as determinants of emotional exhaustion and peer-rated service delivery. *Academy of Management Journal, 46*(1), 86–96.

Hatfield, E., Cacioppo, J. T., & Rapson, R. L. (1994). *Emotional contagion.* New York, NY, USA: Cambridge University Press.

Hazelton, S. (2014). Positive emotions boost employee engagement: Making work fun brings individual and organizational success. *Human Resource Management International Digest, 22*(1), 34–37.

Hochschild, A. R. (1979). Emotion work, feeling rules, and social structure. *American Journal of Sociology, 85*(3), 551–575.

Isen, A. M. (1987). Positive affect, cognitive processes, and social behavior. In L. Berkowitz (Ed.), *Advances in experimental social psychology,* (pp. 203–253). New York, NY, USA: Academic Press.

Jex, S. M., Bliese, P. D., Buzzell, S., & Primeau, J. (2001). The impact of self-efficacy on stressor–strain relations: Coping style as an explanatory mechanism. *Journal of Applied Psychology, 86*(3), 401–409.

Judge, T. A., Locke, E. A., Durham, C. C., & Kluger, A. N. (1998). Dispositional effects on job and life satisfaction: The role of core evaluations. *Journal of Applied Psychology, 83*(1), 17–34.

Judge, T. A., & Piccolo, R. F. (2004). Transformational and transactional leadership: A meta-analytic test of their relative validity. *Journal of Applied Psychology, 89*(5), 755–768.

Khosla, M. (2001). Gender differences in coping with stress. *Journal of Research and Applications in Clinical Psychology, 4*(1–2), 63–72.

Khosla, M., & Hangal, E. (2004). Role of optimism and pessimism in coping with stress. *Journal of Personality and Clinical Studies, 20*(1–2), 71–78.

Lazarus, J. (2000). *Stress relief & relaxation techniques.* Lincolnwood, IL, USA: Keats Publishing.

Luthans, F. (2002). The need for and meaning of positive organizational behavior. *Journal of Organizational Behavior, 23*(6), 695–706.

Lyubomirsky, S. (2007). *The how of happiness: A scientific approach to getting the life you want.* New York, NY, USA: The Penguin Press.

Lyubomirsky, S., King, L., & Diener, E. (2005). The benefits of frequent positive affect: Does happiness lead to success? *Psychological Bulletin, 131*(6), 803–855.

Macklem, K. (2005). The toxic workplace. *Maclean's, 118* (5), 34.

Maddi, S. R., & Khoshaba, D. M. (2005). *Resilience at work: How to succeed no matter what life throws at you.* New York, NY, USA: AMACOM.

Madjar, N., Oldham, G. R., & Pratt, M. G. (2002). There's no place like home: The contributions of work and non-work creativity support to employees' creative performance. *Academy of Management Journal, 45(*4), 757–767.

McCullough, M. E., Emmons, R. A., & Tsang, J. A. (2002). The grateful disposition: A conceptual and empirical topography. *Journal of Personality and Social Psychology, 82*(1), 112–127.

Mohapatra, M., & Sharma, B. R. (2010). Study on employee engagement and its predictors in an Indian Public Sector Undertaking. *Global Business Review, 11*(2), 281–301.

Murphy, L. R. (1995). Occupational stress management: Current status and future direction. *Trends in Organizational Behavior, 2*, 1–14.

Naidoon, S., & Pau, A. (2008). Emotional intelligence and perceived stress. *South African Medical Journal, 63*(3), 148–151.

Nath, P., & Pradhan, R. K. (2013). *Implications of Positive Emotions in Growing Out of Adversity: Examining the Role of Individual Differences.* Paper presented at 3rd Indian Academy of Management at Indian Institute of Management, Ahmedabad, India. December 12–14, 2013.

Newman, A., Ucbasaran, D., Zhu, F., & Hirst, G. (2014). Psychological capital: A review and synthesis. *Journal of Organizational Behavior*, *35*(1), 120–138.

Oginska-Bulik, N. (2005). Emotional intelligence in the workplace: Exploring its effects on occupational stress and health outcomes in human service workers. *International Journal Occupational Medicine and Environmental Health, 18*(2), 167–175.

Ong, A. D., Bergeman, C. S., Bisconti, T. L., & Wallace, K. (2006). Psychological resilience, positive emotions, and successful adaptation to stress in later life. *Journal of Personality and Social Psychology, 91*(4), 730–749.

Organ, D. W., & Paine, J. B. (1999). A new kind of performance for industrial and organizational psychology: Recent contributions to the study of organizational citizenship behavior. *International Review of Industrial and Organizational Psychology, 14*, 337–368.

Paterson, T. A., Luthans, F., & Jeung, W. (2014). Thriving at work: Impact of psychological capital and supervisor support. *Journal of Organizational Behavior, 35*(3), 434–446.

Ployhart, R. E., Lim, B. C., & Chan, K. Y. (2001). Exploring relations between typical and maximum performance ratings and the five-factor model of personality. *Personnel Psychology, 54*(4), 809–844.

Pradhan, R. K. (2003). Emotional intelligence as tool for managing workforce diversity. *Prestige Journal of Management and Research*, *7*(2), 240–249.

Pradhan, R. K., Awasthy, S., Kumar, U., & Pattanayak, S. P. (2005). Role of emotional intelligence on organizational commitment and conflict management. In S. K. Srivastava (Ed.), *Applied and Community Psychology* (pp. 25–43). New Delhi, India: Sarup & Sons.

Pradhan, R. K., & Nath, P. (2012). Perception of entrepreneurial orientation and emotional intelligence: A study on India's future techno-managers. *Global Business Review, 13*(1), 89–108.

Quoidbach, J., Mikolajczak, M. & Gross, J. J. (2015). Positive interventions: An emotion regulation perspective. *Psychological Bulletin, 141*(3), 655–693.

Rothman, N. B. & Melwani, S. (2017). Feeling mixed, ambivalent, and in flux: The social functions of emotional complexity for leaders. *Academy of Management Review, 42*(2), 259–282.

Schaufeli, W., & Bakker, A. B. (2004). Job demands, job resources and their relationship with burnout and engagement. *Journal of Organizational Behavior*, *25*(3), 293–315.

Schaufeli, W. B. & Salanova, M. (2007). Efficacy or inefficacy, that's the question: Burnout and engagement, and their relationships with efficacy beliefs. *Anxiety, Coping & Stress, 20*(2), 177–196.

Singh, S. K. (2007). Emotional intelligence and organizational leadership: A gender study in Indian context. *International Journal of Indian Culture and Business Management, 1*(1), 48–63.

Szczygieła, D. & Mikolajczakb, M. (2017). Why are people high in emotional intelligence happier? They make the most of their positive emotions. *Personality and Individual Differences, 117,* 177–181.

Wall, T., Tran, L. T. & Soejatminah, S. (2017). Inequalities and agencies in workplace learning experiences: International student perspectives. *Vocations and Learning, 10*(2), 141–156.

Waugh, C. E., & Fredrickson, B. L. (2006). Nice to know you: Positive emotions, self-other overlap, and complex understanding in the formation of new relationships. *Journal of Positive Psychology, 1*(2), 93–106.

Wright, T. A., & Staw, B. M. (1999). Affect and favorable work outcomes: Two longitudinal tests of the happy productive worker thesis. *Journal of Organizational Behavior*, *2*(1), 1–23.

Zautra, A. J., Johnson, L. M., & Davis, M. C. (2005). Positive affect as a source of resilience for women in chronic pain. *Journal of Consulting and Clinical Psychology, 73*(2), 212–220.

Zautra, A., Smith, B., Affleck, G., & Tennen, H. (2001). Examinations of chronic pain and affect relationships: Applications of a dynamic model of affect. *Journal of Consulting and Clinical Psychology, 69*(5), 786–795.

CHAPTER 20

Navigating Through the Thunderstorm: Understanding and Developing Leadership Resilience at Workplace

SURENDRA KUMAR SIA[1*] and K. ALPHONSA JOSE[2]

[1]*Department of Applied Psychology, Pondicherry University, Pondicherry 605014, India*

[2]*Department of Psychology, BGR Campus, CHRIST (Deemed to be University), Bangalore, India*

[*]*Corresponding author. E-mail: surendra.sia@gmail.com*

ABSTRACT

A leader is considered best when he or she remains invisible till the aim is fulfilled, and the team members feel that it was all done by themselves. But leading the organization through the downpour requires special abilities. In addition to the traditional leadership characteristics and skills, being a resilient leader is found highly influential in the context dealing with unforeseen adversities and challenges. Recently, empirical researches focusing on resilient leadership is on the rise. The method used in the present chapter was review of relevant literatures identified from refereed papers published in scientific journals or books and other electronic sources. This chapter tries to understand resilient leadership and the various factors that contribute to resilient leadership. This also discusses the different strategies to foster leadership resilience for a better and productive organization.

20.1 INTRODUCTION

Resilience is of wider application as noted by Diane Couto in Harvard Business Review. It says that, "more than education, more than experience, more than training, a person's level of resilience will determine who succeeds and who fails. That's true in the cancer ward, it's true in the Olympics, and it's true in the boardroom" (Couto, 2002).

Applying positive psychology and its principles to various walks of life are found important and highly effective. Applying positive psychology researches exclusively to workplace resulted in the comparatively new field of Positive Organizational Behavior (POB) which aimed at improving one's performance at the workplace by measuring, developing and managing positive human resource strengths, and psychological capacities (Luthans, 2002). Psychological Capital or PsyCap is a widely studied variable in the field of POB (Luthans et al., 2006a,b). PsyCap is a meta-component in POB, consisting of hope, optimism, resilience, and self-confidence/efficacy (Luthans et al., 2006b). It is proposed analogous to the concept of financial capital and hence retains the possibilities of investing, developing, and a return (Luthans et al., 2004, 2007). Initial researches focused more on optimism, hope, and self-confidence, rather than on resilience in the workplace. Luthans et al. (2006b) pointed out this lack of evidence and put forth the importance of studying resilience. Instead of considering resilience as a "trait," they proposed it as possessing a "state" like characteristics (Luthans et al., 2006b; Youssef & Luthans, 2005; Masten & Reed, 2002), with ample chances of development and investing (Luthans et al., 2006b). Later on, the additions and observations also confirmed this possibility of developing resilience. The book "Option B" written by Sandberg and Grant (2017) described resilience as follows: "Resilience is the strength and speed of our response to adversity—and we can build it. It is not about having a backbone. It is about strengthening the muscles around our backbone." In sum, resilience is identified to be a characteristic, viable for training and development, and highly efficient at enhancing the work outcome. Luthans et al. (2006b) proposed the need to develop resilience among the employees and in the organization in general. Navigating the organization through the turbulent times require special abilities (Adhikari, 2014). Being lead by a resilient leader during the turbulent times helps the organization to bounce back from the situation, and also to thrive in the face of adversities (Jackson et al., 2007). Broaden and Built model by Fredrickson (2001) also supports the inherent value of positive emotions to yield a fuller life. While considering the model in the workplace it becomes obvious that fostering leadership resilience and resilient work culture will definitely yield better results. This chapter particularly tries to address the need of resilient leaders at the workplace.

20.2 RESILIENCE AND LEADERSHIP RESILIENCE

A definition given by Masten and Reed (2002, p. 75) says that resiliency as "a class of phenomena characterized by patterns of positive adaptation in the context of significant adversity or risk." Initial discussions of resilience considered it to be a rare personality trait that inoculates one against vulnerable situations (Connor & Davidson, 2003; Ong et al., 2006). Studies on children of schizophrenic mothers showed that resilience has state-like characteristics and it can protect the children from later developing mental health issues (Garmezy, 1974). Outcome-oriented approach of resilience considers it as a behavioral outcome that helps the individual to get better from the harder experiences (Harvey & Delfabbro, 2004; Masten, 2001). The state-like characteristics of resilience (open to development) is supported by many empirical evidences (e.g., Coutu, 2002; Maddi & Koshaba, 2005; Hu et al., 2015). Resilience is found negatively related to negative mental health outcomes and positively correlated to positive outcomes (Hu et al., 2015; Tugade & Fredrickson, 2004). Resilience yields mental health via reducing harm, protecting from harm, and promoting the individual ability to deal with the adversities; postulates the Resilience system mechanism, put forwarded by Davydov et al. (2010). Resilience is identified by some researchers as the interaction with psychological, social, biological, and cultural factors (Ungar,

2013; Bonanno et al., 2007). It may change over time as a function of development and one's interaction with these environmental factors (Kim-Cohen & Turkewitz, 2012).

The stories of many world-famous leaders remind one thing "triumph or failure, the leader should not be much touched. Being euphoric or elated or being depressed and down is not acceptable." A leader is considered resilient when he or she can balance the emotions and communicate optimism and positivity to the followers, rather than merely exerting power and authority. Globalization and technological improvements together created a job market that is comparatively wider but chaotic and volatile. This volatility creates the emotional highs and lows very easily and to be unaffected in the face of these ups and downs, the leader should be flexible, resourceful, and resilient. In broad terms resilient leadership demands the leader to be personally effective to consider the challenges and adverse situations as the change catalysts for better performance and he or she should be influential to organization toward yielding positive results during the harder times (Adhikari, 2014).

Resilient leaders are also portrayed to possess strategic or pro-active thinking, ability to learn from past experiences, change oriented, with high emotional intelligence (Afsheen et al., 2011), and flexibility (Clayton, 2012; Patel, 2010). Today's "wicked problems" do require collective actions of the people of the organization rather than command and control-heroic leadership styles. So researches point out the need of the leader who is flexible in the roles. Resilient organizations are expected to work by establishing two complementary subsystems, with domains of performance and adaptation (Robb, 2000). Performance system bureaucratic in nature is responsible for activities aiming for immediate survival and achieving the current goals and tasks. The sustainability of the organization is achieved mainly through an adaptation system that constantly generates possible futures. The volatile nature of the market always keeps the organization at the demand of flexible leadership and the resilient leaders are identified best to cater to the need (Robb, 2000).

Coutu (2002) describes the basic three characteristics of resilient individuals on the work "How Resilience Works," as follows: Accepting the reality as it is; possessing a clear sense of purpose and meaning, and possessing extreme ability to improvise. Other determinants include grit, resilient attitude, PsyCap, transformational coping, and social support. Grit is the personal trait consisting of perseverance and passion for achieving long-term goals. People with grit possess greater self-control and can delay short term gratification for achieving better long-term success. Constant passion for learning and development holds them rooted through the difficult times. Resilient leaders were also found possessing high levels of grit (Duckworth et al., 2007; Duckworth & Duckworth, 2016). Resilience is a component of PsyCap, but it is also predicted by the other components of PsyCap. Psycap researches show that a leader, possessing an optimistic attitude, hope, and confidence can easily showcase resiliency in his leadership. Luthans et al. (2006b) even proposed that other PsyCap factors may converge to act as a pathway to resilience.

20.3 RESILIENT ATTITUDE—3CS OF RESILIENT LEADERS

Resilience attitude or hardiness is best described using the three personality traits of resilient leaders, known as 3Cs, Commitment, Control, and Challenge (Maddi & Khoshaba, 2005).

Attitude of Commitment: Possessing a commitment toward the work and outcomes, keeps people fully engaged in work and life. Once being committed to something gives a better focus and meaning over the challenging situation. Possessing a genuine interest in people around one and world, or to specific situations forms the basis for evaluation and thereby to decision making.

Attitude of Control: Attitude of control or simply will-power plays a major role in promoting resilience. Believing in one's power to influence the adverse situations and its outcome helps in coping. Control attitude facilitates the coping efforts in the face of adversity by increasing the estimation of the probability of carrying in a positive change. Control attitude enhances the commitment attitude also, resulting in a dedication to solving the task.

Attitude of Challenge: The third hardiness attitude is the attitude of challenge. It is about approving the change, and about seeking new "normal," against the known, previous, comfortable normalcy. This attitude keeps one motivated in the face of unexpected changes.

These three hardiness attitudes together can result in a personality of courage and motivation, along with these hardiness attitudes, two skills are also identified to be important. They are transformational coping skills and social support.

Transformational Coping: It is about getting a deeper understanding of the situation, and using those situations to one's own benefit, instead of breaking down while facing adverse situations. Recent Meghalaya polls and the coalition government formation can be considered an expression of transformational coping. Though the Bharatiya Janata Party (BJP) could yield only two seats against the parties, and no party had the simple majority to form the government alone. BJP had leaders strategic enough to move forward fast to form a coalition government with other parties, and this movement overthrew the competitors from the victory that was so near to them (Nagshbandi, 2018). Transformational coping requires broader viewpoints, understanding that the adversity is not something unique, and it is happening to many others, making it comparatively tolerable and viable to well-planned problem-solving actions, instead of breaking down facing it.

Social Support: It is age-old wisdom that social support can alleviate stress, resilience studies are proving it further that resilient workplace encouraged constructive interaction among employees in the face of difficulties. Also believed that problems opened a path to interact with others and also the expertise of the team members can be further used to encourage achieving better solutions.

20.4 RESILIENCE—PROACTIVE APPROACHES AND REACTIVE APPROACHES

20.4.1 PROACTIVE APPROACHES (RISK-ASSET-PROCESS FOCUSED STRATEGIES)

These approaches are about sculpting the organization and leaders in the probability of upcoming stresses and challenges. It is about developing leadership resilience through various techniques in the anticipation of challenges and keeping them prepared to deal with the situations as it arises. Luthans and colleagues put forward the three possible strategies to build a resilient organization, and they are risk-focused strategies, asset-focused strategies, and process-focused strategies. *Risk-focused strategies* articulate the importance of developing and maintaining organizational culture and ethics, which prepare the employees to be prepared for any unforeseen incidents. *Asset-focused strategies* generally aim to enhance the employability of the employees and they focus more on providing training and skill development to the employees to make them prepared to deal with the adversities. The final *process-focused strategies* are focused majorly on the way in which one interprets the events and experiences (Luthans et al., 2006b; Madden & Reed, 2002). It is mainly about fostering self-efficacy in the work contexts to induce proactive resiliency. Avoilo et al. (2006c) describe efficacy as a major contributor to resilience. Efficient employees are likely to show openness to development (Bandura, 2009) and high work performance (Stajkovic & Luthans, 1998).

20.4.2 REACTIVE APPROACHES

Reactive approaches of resiliency building draw heavily from Frederickson's Broaden and Build model (2001) and the clinical psychology studies. It focuses upon the need to remind people while facing challenges that various strategies such as meaning-making of the event, positive thinking, etc. will help them to yield through the adversities. Broaden and build model proposes the possibility of people to broaden the momentary thought-action repertoires in response to traumatic circumstances. These broadened thought-action repertoires are even powerful enough to leave anatomical changes in the brain, influencing the synaptic activity. Many empirical studies also support this possibility (Garland et al., 2010; Peterson et al., 2008). Other benefits of promoting positive emotions in the face of change and adversities are its effect in undoing the negative emotions, also helps in fostering social support reception, which in turn can increase the positive emotions (Fredrickson & Levenson, 1998). Other put forth strategies are hardiness strategies, attribution strategies, and self-enhancement strategies to enhance resilience (Luthans et al., 2006b).

Both reactive and proactive strategies are important for yielding better resilient leadership. The following section deals with the methods to develop leadership resilience.

20.5 DEVELOPING RESILIENT LEADERSHIP

Leaders are expected to face certain setbacks, failures, and mistakes inevitably (Farson & Keyes, 2003; Gladwell, 2008; Heifetz et al., 2009; Griswell & Jennings, 2009; Sanaghan, 2016). How effectively they deal with all these negative experiences determine the effectiveness of the leader, this way of thinking was called Stockdale Paradox (Collins, 2014). Therefore, the methods to build and promote resilient leadership to face adversities are unavoidable. Human resource development strategies should look into these possibilities seriously. Following a state-based approach to leadership resilience, it becomes obvious that it can be developed using different methods. It is important to remember at this point that resilience is a psychological capital viable for investment and development. The possibility of investing and developing leadership resilience hence requires attention. The following session discusses the various techniques that can be used to induce and enhance resilient leadership. State-based approaches toward resilience leave the scope of investing upon and developing the resilience.

Although many of the individual and organizational resilience fostering techniques can be used even in developing leadership resilience, this session solely focus upon the various techniques that can be utilized to foster leadership resilience. American Psychological Association's guide titled the road to resilience (2018) points out the important factors in resilience and those factors are having supportive relationships that act as role models or provide encouragement and reassurance in and out the family relations. In a similar line Sanaghan (2016) also support this notion and says that the relationships should be vivid as follows; thought partners, confidants, and therapists. Personal resilience researches also support this notion that people with good social support are equipped to deal with the thunderstorms of challenges than people with no known, explicit support system (Sanaghan, 2016; Snyder 2013; Zolli & Healy, 2013; Pulley & Wakefield, 2001; Goleman, 2001). Other suggested skills include planning and executive skills, efficacy believes, communication and problem-solving skills, and emotion regulation capacity (APA, 2018).

20.5.1 MASTERING RESILIENCE THROUGH LESSONS OF LIFE

Earlier experiences of trauma and successful resolution of trauma can act as the best teacher in building later resilience (Walsh, 2007; Carver, 1998). Another scope is the vicarious learning of resilience. Learning the stories of successful as well as resilient leaders familiarizes the adversities and the ways through which they resolved it (Sanaghan, 2016). As Dr. APJ Abdul Kalam, Former Indian President, and world-famous scientist of Indian Space Research Organization, rightly said "Don't read success stories, you will get only message, read failure stories, you will get some ideas." Reading the stories of resilient leaders leave the reader with some ideas and suggestions to bounce back from the adversities; therefore, it can be considered a simple technique to foster resilience (Maddi & Khoshaba, 2005).

By providing a schedule with reflection and review time another widely suggested method to build leadership resilience (Sanaghan, 2016; Pulley & Wakefield, 2001). Thoughts on questions regarding the various aspect of leadership *(e.g., What are the consistent challenges?, Who are the people confiding to me?, Who do I talk with?, How are my people feeling? What makes me anxious?, What I learned from the experience?, etc.)* can be either recorded as videos or journal entries or using various mobile apps. This reflection time mostly reveals the important and strategic information. Taking the time to reflect will often reveal important, even strategic, information that might not be apparent at first glance (Sanaghan, 2016). Reviewing with the help of the confidents or the significant supporters will enhance the effectiveness of the review process (Sanaghan, 2016).

20.5.2 PROACTIVE STRATEGY OF BUILDING RESILIENCE

Risk-asset-process strategy of building resilience preparedness is applicable even in the case of resilient leadership. It is about thinking how to act in a difficult situation and being prepared intellectually and experientially. Working with unfamiliar people, placing one in unfamiliar situations, and all can prepare oneself to deal with unpredicted challenges. As early mentioned it is about being healthy enough to deal with the harsh conditions, it is about strengthening the resilience muscle.

20.5.3 VARIETY OF EXPERIENCE MOBILITY AND JOB ROTATION

Leadership requires variety and the person trapped in the same career path may find it difficult to become the leader of everyone, who works in different–different situations. Moreover, repeated experiences may create mindsets that block creative thinking and problem solving. Evidences do support job rotation as an accepted practice (Fernandez-Aroaz et al., 2011; Sanghan, 2016). Job rotation enables the leader to experience and deal with another group and their culture. Also provides chance to become familiarized with the difficulties, nature of work, and the possible and familiar ways to deal with the various situations. Along with providing the leader with richer experience, job rotation also gives closer interpersonal ties and increased support to the leader within the organization. This can act as the necessary investment to effectively working together as a team during later periods. Job rotation also enables the person to spread out their perspectives and insights. Most of the time the person ends up meeting somebody within the organization, who works with a different value and attitude or perspectives, thereby the job rotation assignment can act as the source of expanded perspectives, excellent problem-solving, creativity, and resilience skills.

20.5.4 ENHANCE THE SKILL SETS THROUGH DEVELOPMENTAL ASSIGNMENTS

As we have seen earlier in pro-active resilience-building strategies, keeping the person active in various assignments and divisions helps in building the skill sets and hence enhances the employability of the person. Assignments that utilize the unique skills of the person to the best, also avails the opportunity to deal with others who are proficient, thereby perks up the chances of learning and progressing on the particular skill. Facilitating team work, building up teams, reviewing the works and coming up with brighter solution spreads out the leadership capacity, and dealing with the new, unfamiliar situations may promote the resiliency.

Cultural traveler is another strategy put forwarded by Sanaghan. Cultural travelers are those employees; who are highly resilient, carry great reputation, get along well with others, and work harder for getting things done in favor of the organization. He suggests that every resilient leader should practice cultural travelling and becoming cultural traveler requires effort and time. They possess skills such as listening, keeping up the commitments—without being over-promising or under-promising but being committed to what is achievable. Other characteristics include deep trust and confident with others, orientation toward accomplishing things, serve for the best of the organization, and building connections across different groups. Deciding to be a traveler enhances the leadership resilience (Sanaghan et al., 2001; Sanaghan, 2014, 2016).

20.5.5 BUILDING UP FAILURE TOLERANCE

The discussion is not about toxic failures, that incapacitates the growth. But it is mainly about considering failures as the growth opportunities, to ascertain novel directions, solutions, and approaches. Farson and Keyes (2003) book *Whoever Makes The Most Mistake Wins* proposes failure as the essential ingredient of learning. Rather than trying to ignore or to deviate from failures, it is advised to consider failure as a learning opportunity. A leader cannot be fearful about failures, but should be ready to show acceptance, review, and reflect the possible reasons and possible way out. Dealing with failure and moving ahead is the best hands-on training that any leader can receive in developing resilience. A nonjudgmental attitude toward failure is the key.

20.5.6 EMBRACING CONFUSION AND REFLECTIVE INQUIRY AND ACTION

Lost leader syndrome as given by Jentz and Murphy (2005) on the monograph titled "embracing confusion: what leaders do when they don't know what to do" discusses the confusion experience of leaders. Certain situations arise in every body's life, where the challenges be ambiguous, and may not make much sense. These situations, surpass the leader's capacity to detect a clear path forward, and may leave them lost, confused and all at sea despite their intelligence and experiences. The right attitude to take up is to admit that they are confused, rather than trying to take control of the situation, or covering it up. This confused state can be considered a fruitful situation, facilitating the reality checking and applicability evaluation of older assumptions. Jentz and Murphy consider confusion as potter's clay rather than as quick sand. Hidden confusion may result in hasty decisions that later induce regret rather than resilience. It is admissible and acceptable to express the concern and confusion, in a healthy manner, if the leader finds the situation as ambiguous or chaotic even at the expense of all the available resources. Sharing real feeling promote mutual trust and respect. And the act of sharing the confusion itself requires a great deal of courage.

Sharing the confusion consists of five stages and these can be together called a reflective inquiry and action plan (RIA). Major postulates of RIA include; embracing the confusion, asserting the need to make sense, structuring the conversation, reflective listening and learning, and loud response processing. By acknowledging oneself about confusion opens the opportunity for dialogue and invention. Asserting the need to make sense enhances one's chance of yielding better understanding and better solutions. Structuring the conversation in the manner goes well with the leader, but clearly stating the need for help to make sense of the situation is a critical step. Caution should be applied when selecting the right words to express the concern. After posing the need for clarity, leader should also take caution to listen reflectively and understand the situation better. Every communication becomes successful when there is feedback, so the leader can also build the habit of explaining the decisions made to come out of the confusion to the team and thereby create clarity and focus on the team. This skill of embracing the confusion, and dealing with it require a lot of skills. Practicing those skills at an acceptable place is heavily helpful in building resilient leadership (Sanaghan, 2016).

20.5.7 AFTER ACTION REVIEWS (AARS)

Picking up the cues left behind by both the success and failure leaves the team to learn lessons from the past and to apply it in the future. AARs are indebted to the US army's intensive review to learn from the past; at times it can be called a "Post mortem" (Morrison & Meliza, 1999). Mainly four review questions are utilized. These questions are; "What did we set out to do? (What was our intention?), What actually happened? (What did we do?), What caused our results? (Why did it happen?), and

What will we sustain/improve? (What can we do better next time?)." If conducted rigorously, with the help of a facilitator and the participation of all the important parties surely yield a wealth of knowledge to prepare better for upcoming events and issues. These endeavors become rigorous and resilient promoting only when the organizational culture is supportive of honesty, feedback, openness, and transparency. Initial AARs can focus on the successes rather than on failures, and later on, the shift can be moved toward failures, once everyone gets familiar with it. Ideas derived or lessons learned out of the AARs can be circulated among the organization, in order to receive proper feedback and different perspectives, which is quite useful in the face of a crisis (Sanaghan, 2016).

20.5.8 METAPHORICAL PROBLEM SOLVING

It is about rethinking a problem in a metaphorical manner. Metaphorical manner uses seemingly different things, but in fact connected, to explain the condition at hand. Initially, in this manner we create a metaphorical problem, and later understand the underlying connections and further utilize this metaphorical connection to yield solution to the real problem. While using metaphors to describe the problem, the scope of the problem is reduced to one important piece, so working for a solution to this piece becomes easy against trying to solve everything at once.

20.5.9 INTERACTING WITH EDGY-CREATIVE THINKERS

Interaction with the creative thinkers, who view the world through a quite different glass than normal people can stimulate the thinking of leaders. It can improve the intelligence resilience. Sanaghan uses the term "Edgy thinkers," they may be the normal employees of the organization, but on their personal life, they may be living on the edge. Identify the edgy thinkers, build a connection with them, stimulate the intellectual resilience by having a stimulating conversation, this is the strategy given by Sanaghan (2016). Hamel and Valikangas, (2003), Ho (2012) and Zolli and Healy (2013) also supported the need of leaders to find and establish a connection between the different and creative employees of the organization. Edgy thinkers can easily identify the "heralds" of change, well ahead in time than many other people, this foreseeing capacity can help the leader to be prepared for the future.

20.6 CONCLUSION

This chapter tried to focus upon the resilient leadership and the technique to improve resilience in the organization as well as in the leaders. Our discussions focused on the state-like character of resilience rather than on the trait-like characteristics. State-like characters give the scope of investing and developing. Necessary strategies for enhancing the resilience of the organization like reactive resilience strategies and risk–asset–process-based proactive strategies are all discussed in the current chapter. Deeper discussions on leadership resilience and techniques to foster it are also given.

It is identified that resilience is the outcome of hard work and it can be learned. It requires commitment, challenge and control in face of adversities (Hardiness), grit, and risk-taking tendencies. While the negative emotions fear, anger, and confusion leaves the leader paralyzed.

Having close-knit ties with others is essential, as relationships act as the key in fostering resilience. Social support and sense of community can trigger resilience. Keeping thought partners at the workplace with whom one can discuss or brainstorm for creative and resilient outcomes, confidants, to whom the leader can share the secrets, and a good therapist to assist one in the process of becoming resilient are highly recommended at the workplace for developing resilient leadership. On the other hand family, support is also highly influential.

Keeping the goals short and easily achievable rather than trying to scale the mountain all at once can trigger efficacy feeling, this in turn can contribute to resilience. In the face of uncertain and disastrous situations one can feel easily overthrown and confused, but keeping the shorter goals and the accomplishment of these goals will enhance the resiliency. And resiliency requires a lot of improvising, leader can consider improvising by taking risks and tackling the unfamiliar situations through creative thinking. Identifying one's non-negotiable values in life helps the leader to be sustaining in difficult times. Values determine one's character strength which is the forerunner of resilience. Therefore identifying one's lived, non-negotiable values and using it for the purpose of meaning making in life is a required skill of the leader. And a resilient leader should be failure tolerant and should perceive challenges, rather than the blocks. It is about bouncing back from the pit rather than climbing out (Kanter, 2013). The final attitude a resilient leader should always possess is the attitude that being resilient is a choice. It is about choosing to move forward in the face of challenges and setbacks, accepting the challenges rather than denying it. Leaders should always learn to develop constructive responses rather than just reacting to the situations.

KEYWORDS

- **resilient leadership**
- **adversities**
- **work place**
- **resilient attitude**

REFERENCES

Adhikari, S. (2014). What it takes to build resilient leaders. People matters. Retrieved from https://www.peoplematters.in/article/hr-insights/what-it-takes-to-build-resilient-leaders-7042 on 07 May 2018.

Afsheen, F., Rabia, I. and Sajid, H. A. (2011). Emotional intelligence and transformational leadership: finding gender differences, *World Applied Sciences Journal*, *14* (11), 1734–1743.

American Psychological Association (2018). The road to resilience. Retrieved from http://www.apa.org/helpcenter/road-resilience.aspx on 12 May 2018.

Bandura, A. (2009). Cultivate self-efficacy for personal and organizational effectiveness. In E.A. Locke (Ed). *Handbook of Principles of Organization Behavior* (2nd ed.), (pp.179–200), Vol. 2, pp. 0011–0021.

Bonanno, G. A., Galea, S., Bucciarelli, A., & Vlahov, D. (2007). What predicts psychological resilience after disaster? The role of demographics, resources, and life stress. *Journal of Consulting and Clinical Psychology*, *75*(5), 671–682.

Carver, C. S. (1998). Resilience and thriving: Issues, models, and linkages. *Journal of Social Issues*, *54*(2), 245–266.

Clayton, H. (2012), The changing leadership landscape, *Strategic HR Review, 11*(2), 78–83.

Collins, J. (2014). The Stockdale Paradox. *The brutal facts*.

Connor, K. M., & Davidson, J. R. (2003). Development of a new resilience scale: The Connor-Davidson resilience scale (CD-RISC). *Depression and Anxiety*, *18*(2), 76–82.

Coutu, D. (2002). How resilience works? *Harvard Business Review.* 46–55. Retrieved from https://hbr.org/2002/05/how-resilience-works on 08 May 2018.

Davydov, D. M., Stewart, R., Ritchie, K., & Chaudieu, I. (2010). Resilience and mental health. *Clinical Psychology Review*, *30*(5), 479–495. doi:10.1016.j.cpr.2010.03.003.

Duckworth, A. L., Peterson, C., Matthews, M. D., & Kelly, D. R. (2007). Grit: Perseverance and passion for long-term goals. *Journal of Personality and Social Psychology, 92*(6), 1087.

Duckworth, A., & Duckworth, A. (2016). *Grit: The power of passion and perseverance*. Findaway World LLC.

Farson, R., & Keyes, R. (2003). *Whoever makes the most mistakes wins*. Simon and Schuster.

Fernández-Aráoz, C., Groysberg, B., & Nohria, N. (2011). How to hang on to your high potentials. *Harvard Business Review, 89*(10), 76–83.

Fredrickson, B. L. (2001). The role of positive emotions in positive psychology: The broaden-and-build theory of positive emotions. *American Psychologist, 56*(3), 218–226.

Fredrickson, B. L., & Levenson, R. W. (1998). Positive emotions speed recovery from the cardiovascular sequelae of negative emotions. *Cognition & Emotion, 12*(2), 191–220.

Garland, E. L., Fredrickson, B., Kring, A. M., Johnson, D. P., Meyer, P. S., & Penn, D. L. (2010). Upward spirals of positive emotions counter downward spirals of negativity: Insights from the broaden-and-build theory and affective neuroscience on the treatment of emotion dysfunctions and deficits in psychopathology. *Clinical Psychology Review, 30*(7), 849–864.

Garmezy, N. (1974). The study of competence in children at risk for severe psychopathology. In E. J. Anthony & C. Koupernik (Eds.), *The child in his family: Vol. 3, children at psychiatric risk* (pp. 77–97). New York: Wiley.

Gladwell, M. (2008). *Outliers: The story of success*. Hachette UK.

Goleman, D. (2001). Emotional intelligence: Issues in paradigm building. *The emotionally intelligent workplace, 13*, 26.

Griswell, J. B., & Jennings, B. (2009). The adversity paradox. *The Adversity Paradox: An Unconventional Guide to Achieving Uncommon Business Success.* New York: St. Martin's Press.

Hamel, G., & Valikangas, L. (2003). The quest for resilience. Harvard Business Review. Retrieved from https://hbr.org/2003/09/the-quest-for-resilience on 10 May 2018.

Harvey, J., & Delfabbro, P. H. (2004). Psychological resilience in disadvantaged youth: A critical overview. *Australian Psychologist, 39*(1), 3–13.

Heifetz, R. A., Grashow, A., & Linsky, M. (2009). *The practice of adaptive leadership: Tools and tactics for changing your organization and the world.* Harvard Business Press.

Ho, P. (2012). Coping with complexity. McKinsey & Company. Retrieved from https://www.mckinsey.com/industries/public-sector/our-insights/coping-with-complexity on 11 May 2018.

Hu, T., Zhang, D., & Wang, J. (2015). A meta-analysis of the trait resilience and mental health. *Personality and Individual Differences, 76*, 18–27. DOI: 10.1016/j.paid.2014.11.039.

Jackson, D., Firtko, A., & Edenborough, M. (2007). Personal resilience as a strategy for surviving and thriving in the face of workplace adversity: A literature review. *Journal of Advanced Nursing, 60*(1), 1–9.

Jentz, B. C., & Murphy, J. T. (2005). Embracing confusion: What leaders do when they don't know what to do. *Phi Delta Kappan, 86*(5), 358–366.

Kanter, R. M. (2013). Surprises are the new normal; resilience is the new skill. *Harvard Business Review*. Retrieved from https://hbr.org/2013/07/surprises-are-the-new-normal-r on 13 May 2018.

Kim-Cohen, J., & Turkewitz, R. (2012). Resilience and measured gene–environment interactions. *Development and Psychopathology, 24*(4), 1297–1306.

Luthans, F. (2002). Positive organizational behavior: Developing and managing psychological strengths. *The Academy of Management Executive, 16*(1), 57–72.

Luthans, F., Avey, J. B., Avolio, B. J., Norman, S. M., & Combs, G. M. (2006a). Psychological capital development: Toward a micro-intervention. *Journal of Organizational Behavior, 27*(3), 387–393.

Luthans, F., Luthans, K. W., & Luthans, B. C. (2004). Positive psychological capital: Beyond human and social capital. *Business Horizons, 47*(1), 45–50. doi 10.1016/j.bushor.2003.11.007.

Luthans, F., Vogelgesang, G. R., & Lester, P. B. (2006b). Developing the psychological capital of resiliency. *Human Resource Development Review, 5*(1), 25–44. doi 10.1177115344843052853 35

Luthans, F., Youssef, C. M., & Avolio, B. J. (2007). Psychological capital: Investing and developing positive organizational behavior. *Positive Organizational Behavior, 1*(2), 9–24.

Luthans, F., Zhu, W., & Avolio, B. J. (2006c). The impact of efficacy on work attitudes across cultures. *Journal of World Business, 41*(2), 121–132.

Maddi, S. R., & Khoshaba, D. M. (2005). *Resilience at work: How to succeed no matter what life throws at you*. AMACOM Div American Management Association.

Masten, A. S. (2001). Ordinary magic: resilience processes in development. *American Psychologist, 56*(3), 227–238. http://dx.doi.org/10.1037/0003–066X.56.3.227

Masten, A. S., & Reed, M.-G. J. (2002). Resilience in development. In C. R. Snyder and S. Lopez (Eds.),

Handbook of positive psychology (pp. 74–88). Oxford: Oxford University Press.

Morrison, J. E., & Meliza, L. L. (1999). *Foundations of the after action review process* (No. IDA/HQ-D2332). Alexandria, VA, USA: Institute for Defense Analyses.

Nagshbandi, A (2018) Meghalaya govt formation: How BJP outwitted Congress, again. *Hindustan Times. Retrieved from* https://www.hindustantimes.com/india-news/meghalaya-govt-formation-how-bjp-outwitted-congress-again/story-GUpTDT79N7ePFUcRt0TcVP.html On 15 May 2018.

Ong, A. D., Bergeman, C. S., Bisconti, T. L., & Wallace, K. A. (2006). Psychological resilience, positive emotions, and successful adaptation to stress in later life. *Journal of Personality and Social Psychology*, *91*(4), 730.

Patel, B. (2010), *The importance of resilience in leadership*, Clore Social Leadership Programme.

Peterson, S. J., Balthazard, P. A., Waldman, D. A., & Thatcher, R. W. (2008). Neuroscientific implications of psychological capital: Are the brains of optimistic, hopeful, confident, and resilient leaders different? *Organizational Dynamics*, *37*(4), 342–353.

Pulley, M. L., & Wakefield, M. (2001). *Building resiliency: How to thrive in times of change*. Center for Creative Leadership.

Robb, D. (2000), Building resilient organisations, *OD Practitioner*, *32*(3), 27–32.

Sanaghan, P. (2016). Building leadership resilience in higher education. Higher Ed Impact. Academic Impressions.

Sanaghan, P., Goldstein, L., & Jurow, S. (2001). A learning agenda for chief business officers. *Business Officer*, 42–49.

Sandberg, S., & Grant, A. (2017). *Option B: Facing adversity, building resilience, and finding joy.* London: WH Allen.

Sanaghan, P., & Kimberly, E. (2014). *Creating the exceptional team: A practitioner's guide*. HRD Press.

Snyder, S. (2013). *Leadership and the art of struggle*. Berrett-Koehler Publishers.

Stajkovic, A. D., & Luthans, F. (1998). Self-efficacy and work-related performance: A meta-analysis. *Psychological Bulletin*, *124*(2), 240.

Tugade, M. M., & Fredrickson, B. L. (2004). Resilient individuals use positive emotions to bounce back from negative emotional experiences. *Journal of Personality and Social Psychology*, *86*(2), 320–333. doi:10.1037/0022-3514.86.2.320

Ungar, M. (2013). Resilience, trauma, context, and culture. *Trauma, Violence, & Abuse*, *14*(3), 255–266.

Walsh, F. (2007). Traumatic loss and major disasters: Strengthening family and community resilience. *Family Process*, *46*(2), 207–227.

Youssef, C. M., & Luthans, F. (2005). Resiliency development of organizations, leaders & employees: Multi-level theory building for sustained performance. In W. Gardner, B. J. Avolio & F. O. Walumbwa (Eds.), *Authentic leadership theory and practice. Origins, effects, and development*. Oxford, UK: Elsevier.

Zolli, A., & Healy, A. M. (2013). *Resilience: Why things bounce back*. Simon and Schuster.

CHAPTER 21

Promoting Workplace Health: A Positive Occupational Health Psychology Perspective

AISHWARYA JAISWAL, YOGESH KUMAR ARYA, and TUSHAR SINGH

Department of Psychology, BHU, Varanasi, India

**Corresponding author. E-mail: tusharsinghalld@gmail.com*

ABSTRACT

Work, in addition to providing financial rewards, has a pronounced role in ensuring workers' healthiness and well-being. Based on their employment type, workers pervasively get exposed to varying occupational hazards at the workplace that not only affect them but also tax the entire occupational hygiene. Tremendous changes (such as the prominence of precarious contracts, contingent workforces, globalization, work intensification, etc.) underwent by the world of work in recent decades have intensified the already prevalent hazardous psychosocial factors in the work environment, thereby professing fresh challenges to the organizations' occupational wellness. Prominently arising out of the work design, context, organization, and management, these factors are considerable contributors to employee's counterproductive physiological and/or psychological health outcomes.

However, the promotion of psychological capital among employees, that incorporates resilience amongst its prime components, can be an expedient protective factor against these counterproductive outcomes. Enhanced resilience improves health, well-being, and life quality, along with bearing a positive association with industrious work outcomes such as greater job satisfaction, better performance and productivity levels, etc. When weighed against the standard workplace hazard and risk management approaches, positive interventions exhibit superior receptivity at contemporary workplaces.

The concept of positive organizational behavior (POB) is of growing interest among positive psychology, management, and health researchers as an object of study of occupational health. This chapter reviews the salient contemporary psychosocial hazards pervading across workplaces and introduces the POB concept of psychological resilience, as a means to address an array of detrimental outcomes resulting from these hazards. The role that fostering worker resilience has in improving workplace health has been highlighted. Also, an overview of the interventions and strategies to be adopted for building up a resilient industrial workforce has been provided.

21.1 INTRODUCTION

Employment is vital not only for obtaining economic resources but also for ensuring one's material well-being and fuller social participation. Work provides the individual with a sense of identity, boosts self-confidence, enhances self-esteem, improves self-efficacy, facilitates socialization, and has a significant role in one's overall personal development. While the traditional approaches to occupational health and safety viewed work as a potential hazard adversely impacting workers' health outcomes, contemporary perspective regards work to be the best form of welfare. A safe, healthy working environment reportedly enhances the employee's self-esteem and their sense of well-being, in addition to boosting their productivity levels, work engagement, and job satisfaction. Besides, such an environment also countersigns reduced absenteeism resulting from physical or mental illness/injury. However, apparent socioeconomic and technological changes along with increased trends in restructuring and downsizing, have considerably changed the world of work today, thereby making the maintenance of healthy working environments difficult.

The 21st-century work sphere has witnessed the prominence of globalization, work intensification, contracting-out, hiring of contingent workers, lean productions, multicultural working classes, aging workforces, heightened emotional demands of tasks, and the inability of workers to effectively manage both personal and professional lives. Alongside these comprehensive changes, the demands placed upon employees are also changing. All these have significantly altered the modern work scenario and have led to—besides physical, chemical, and biohazards—the advent of what have been termed as "psychosocial hazards."

Psychosocial hazards mainly prevail out of the inadequacies in the design and specification of the job, improper task and role arrangement therein, uneducated job roles and responsibilities, and the faulty integration of work processes and procedures within the organization. Apart from detrimentally impacting the workers' physiological, psychological, and social health and the organizations' health indices, they also levy an additional burden on the existentially high global GDP (4%–5%) spent on work-related health and productivity losses (Takala, 2002).

This chapter, rather than focusing on outcomes, focuses on workplace situation as a primary area of risk assessment and prevention. A positive occupational health psychology perspective, an approach that examines the role of positive phenomena in addressing the employee welfare issues at the workplace, has been adopted to address the issue of rising hazardous psychosocial conditions in the work environment. The human resource authorities have been suggested to outplay the psychosocial hazards-linked detrimental outcomes via investing in the expansion of employee's psychological capital of resilience (Luthans, 2002; Luthans et al., 2004, 2007).

21.2 WORKPLACE HEALTH

Workplace health seeks to ensure the promotion and maintenance of paramount well-being among workers, across occupations, by—preventing health departures, controlling risks, and facilitating workers' work adaptation. It aims at preventing work-related illness and injury, by encouraging safe working practices and ergonomics, monitoring workforce health, and supporting the management of sickness absence.

Workplace health promotion enhances employee's self-esteem, reduces job stress, improves work morale, increases job satisfaction, and leads to improved health and well-being. It also portrays a positive and caring image of the organization, thereby contributing to valued organizational outcomes such as lowered staff turnover, reduced absenteeism and health care/insurance costs, improved worker employability, raised productivity, and promotion of sustainable development.

However, presently, even the informal sectors and small-scale industries that are considered to be the key arenas in poverty alleviation are not intact to workplace health risks. Thus, for organizations to ensure their occupational wellness, addressing the occupational risks and/or hazards prominent in the workplace is of inevitable importance.

21.3 PSYCHOSOCIAL HAZARD

Workplace hazards are hazards experienced in the workplace. It can be chemical, biological, psychosocial, or physical by nature. The Occupational Health Psychology (OHP) framework looks at psychosocial hazards as hazards stemming from work and the work environment. Leka et al. (2003) highlighted workload, work pace, and schedule of work; control over job tasks, work processes and procedures; job environment, and equipment; organizational culture and function; interpersonal relations at work; role in organizations; career development; and homework interface, to be the focal within-job context factors triggering psychosocial hazards. EU-OSHA (2007), on the other hand, held that globalization and the trending new forms of precarious employment contracts that add to employee vulnerability and job insecurity shall be regarded as key contributors to the rising psychosocial hazards. These hazards are a major occupational health and safety challenge, which if not effectively managed can negatively impact valued individual and organizational factors. The most prominent of these hazards, witnessed across most businesses, include occupational stress, work-related fatigue, and workplace bullying, harassment, and violence.

21.3.1 OCCUPATIONAL STRESS

Occupational/job stress encompasses psychophysiological responses (physiological, cognitive, and emotional) of workers, and surfaces in instances when there is an ample perceived discrepancy between the employee's work demands and their available coping resources. Individual's perception of their work characteristics forms an integral part of their stress equation. The perception of a stressor (as positive/negative) depends upon the type of demand and the amount of situational control and support availability one possesses. Unresolved stress by instilling a state of continual bodily stimulation leads to physiological and/or psychological ailments. Probable stress-linked deleterious health outcomes incorporate cardiovascular dysfunction, gastrointestinal disorders, problematic musculoskeletal conditions, immunodeficiency diseases, and mental disorders. Stress also results in familial problems, loss of productivity, work absenteeism, faulty health behaviors, and cancerous conditions (Cooper & Cartwright, 1994).

21.3.2 WORK-RELATED FATIGUE

Fatigue, whether acute or chronic, is a state characterized by extreme tiredness, weariness or sleepiness, and is conventionally an upshot of deficit sleep, continued physical or mental labor, and/or prolonged stress or anxiety. While acute fatigue resulting from short-term sleep loss or short periods of heavy physical/mental workload is reversible; chronic fatigue syndrome is a constant, severe state of tiredness not relieved by having rest. Owing to difficulties in measuring/quantifying fatigue levels, it is tough to estimate the intensity and frequency of injuries or accidents resulting exclusively from fatigue. However, there are well-documented pieces of evidence that it leads to reduced mental and physical functioning, decreased productivity or performance, impaired communication, judgment errors, lack of concentration, lowered motivation, reduced reaction time, increased risk-taking tendency, and an enhanced propensity toward accidents.

21.3.3 WORKPLACE BULLYING

Workplace bullying and harassment are recognized as core psychosocial hazards under the Occupational Health and Safety laws across the globe. Bullying is a persistent, offensive, intimidating, humiliating behavior (verbal, nonverbal, psychological, or physical) attempting to undermine an individual or group of employees. It might include: abusing, insulting, making offensive comments, unjust criticizing/complaining, deliberate workplace exclusion, spreading malicious rumors, withholding essential information, etc. It leads to physical and mental health problems, somatization, burnout, post-traumatic stress, lowered satisfaction, dropped work commitment, elevated absenteeism, higher intentions to leave, and degraded performance.

21.3.4 WORKPLACE HARASSMENT

Harassment is an unwanted and offensive intentional/unintentional conduct such as disrespecting, demeaning or showing unacceptability toward a person or a group that affects the dignity of the targeted individual(s). It could be prejudice-based, such as racial and sexual harassment, or gender-, disability-, age-, religion-, nationality-, or sexual-orientation based. It leads to outcomes similar to bullying and counts amongst the most serious health issues in modern workplaces.

21.3.5 OCCUPATIONAL/WORKPLACE VIOLENCE (WPV)

WPV is an acute global health concern involving behaviors ranging from verbal threats and abuse to physical assaults and homicidal acts directed against other staff, patrons, clients, and/or regular visitors. Matters concerning the acts of violence in the work environment have been a subject of extensive study in OPH literature (Beale et al., 1999; Chappell & Di Martino, 2000; Cox & Leather, 1994). Violence whether external/intrusive, consumer-related, or relationship-related could be prevalent in any workplace and could involve any type of worker. Fatal risks are highest for workers in sales, protective services, and transportation; while not-fatal for those in healthcare and social assistance services. Exposure to violence in the workplace could lead to experiences of emotional exhaustion, depersonalization, flashbacks, sleeplessness, depression, post-traumatic stress disorder, etc., and thereby, detrimentally impact workers' psychological and physical health. It also adds to lowered performance and productivity at work.

21.4 THE IMPACT OF PSYCHOSOCIAL HAZARDS

When the psychosocial conditions prevailing in the work setting are hazardous in nature, then the certain counterproductive employee and organizational outcomes are likely. Table 21.1 enlists the physiological, psychological, behavioral,

TABLE 21.1 The Physiological, Psychological, Behavioral, and Organizational Impact of Psychosocial Hazards

Physiological	Psychological	Behavioral	Organization
• Cardiovascular disease (CVD) • Musculoskeletal disorders (MSD) • Metabolic syndrome (MetS) and diabetes	• Poor mental health – Anxiety – Depression • Sleep disorders • Burnout syndrome • Suicide	• Alcoholism • Smoking • Physical inactivity • Poor diet • Poor sleep • Injuries • Accidents	• Absenteeism • Presenteeism • Turnover intentions • Poor interpersonal relations • Poor customer relations • Reduced performance • Loss of productivity • Low job satisfaction

and organizational outcomes resulting from psychosocial hazards at work.

21.4.1 PHYSIOLOGICAL IMPACT

Cox et al. (2000) found high blood pressure, heart diseases, wound-healing, gastrointestinal diseases, and debilitated immunocompetence to be the most commonly experienced physiological effects of psychosocial risks, particularly work-related stress. The other perceptible effects included:

- **Cardiovascular Disease:** CVD is the foremost cause of death globally. Epidemiological researches (Belkic et al., 2000) suggest close linkages of psychosocial hazards with the CVD prominence and the CVD-allied morbidity and mortality (Everson-Rose & Lewis, 2005; Landsbergis et al., 2001). Employees with high perceived stress are at an elevated risk of fatal and nonfatal CV events.
- **Musculoskeletal Disorders:** Epidemiological studies (Devereux et al., 2002; Rugulies & Krause, 2008) around different work sectors (from office work to manual work) have found significant linkages between psychosocial hazards and MSD. Lack of social support, low levels of job satisfaction, poor organization of job tasks, and low job content are a few psychosocial factors identified as hazardous to the work environment, and that subsequent into the development of musculoskeletal disorders (De Beeck & Hermans, 2000). Randall et al. (2002) noticed that the worker's musculoskeletal complaints were recognizably related with their job demands; the availability of work equipment; their job status; the provision of managerial support; their job participation; the workplace environment; and their perceived adequacy of organizational management's practices.
- **Metabolic Syndrome and Diabetes:** Based on a longitudinal study, Raikkonen et al. (2007) suggested psychosocial factors to be having a causative role in leading to MetS. Chandola et al. (2006) noticed an exposure–response relation between work-related stress and the MetS. Heraclides et al. (2009) concluded psychosocial work stress to be an independent predictor of type 2 diabetes.

21.4.2 PSYCHOLOGICAL IMPACT

Poor psychosocial working conditions detrimentally impact workers cognitive-affective processing and might lead to impaired decision-making, anxiety, depression, distress, and even burnout (Cox et al., 2000). The salient psychological impacts of hazardous psychosocial conditions include:

- **Poor Mental Health-Anxiety and Depression:** Mental health is an indicator of one's overall psychological well-being. Stress at work significantly relates to poor mental health functioning (Kopp et al., 2008). Studies (Cox et al., 2000, 2005; Middeldorp et al., 2006) have shown that exposure to psychosocial hazards at work leads to experiences of job stress, depression, and anxiety. The psychosocial risk factors associated with depression are poor interpersonal relations at work, poor organizational justice, and compromised work–life balance.
- **Sleep Problems:** Exposure to psychosocial risks at work causes significant alterations in workers' circadian rhythms. Job-related strain, excessive job demands, disparity and imbalance in effort and reward at work, and extended work hours together with shift work fall amongst the foremost work conditions linked to sleep disturbances (Linton, 2004).
- **Burnout Syndrome:** Burnout is a result of persistent response or prolonged exposure to overwhelming emotional and interpersonal risk factors at work. It is characterized by emotional exhaustion, cynicism, depersonalization, lack of involvement at work, lower levels of personal accomplishment, and ensuing job inefficiency. Some chief contributors to this psychological syndrome include ambiguous job roles, unmanageable workload, organizational change and transformation, low employee satisfaction, personal unfulfillment, work–life conflict, poor interpersonal relations, lack of perceived managerial/organizational support, and the prevalence of occupational violence, involving harassment and bullying. It results in specific/nonspecific health issues such as hypertension, myocardial infarction, feeling tired, irritation, emotional instability, and rigid social relations.
- **Suicide:** With the rise in hazardous psychosocial working conditions, an increment in the rate of work suicide has been observed not only in the Eastern world (e.g., Japan, China, India, and Taiwan) but also in the Western world nations (e.g., the United States, Australia, and France). Byun et al. (2016) found job stress to be significantly related to suicide and suicide ideation. This postulation has been confirmed by numerous other studies as well (Feskanich et al., 2002; Ostry et al., 2007; Kivimaki et al., 2010).

21.4.3 BEHAVIORAL IMPACT

Exposure to psychosocial hazards leads to a wide array of unhealthy behaviors (Kouvonen et al., 2006) such as alcoholism (Head et al., 2004), smoking (Radi et al., 2007), physical inactivity (Wemme & Rosvall, 2005), poor dietary habits, and reduced sleep quality (Cox et al., 2000). These hazards have been found to develop cognitive failure propensity among workers, thereby making them even more accident-prone. Swaen et al. (2003) established fatigue to be an independent risk factor in relation to work accidents.

21.4.4 THE IMPACT ON ORGANIZATION

At the organizational level, the disadvantageous impact cast by psychosocial hazards include: increased absenteeism rate, increased intentions to leave and turnover rates, presenteeism, poor interpersonal relations, unsatisfactory customer relations, increased workplace injuries and accidents, elevated workers' compensation claims, and an enhanced disability rate (e.g., Michie, 2002; Spurgeon et al., 1997; Vahtera et al., 2004; van den Berg et al., 2009).

A consideration of these detrimental consequences necessitates the need for the implementation of effective interventions to prevent psychosocial hazards at work. The next section focuses on the role positive psychological concepts could have in addressing the issue.

21.5 POSITIVE PSYCHOLOGY AT WORK

POB implicates an investigation into and implementation of positively-positioned human capital strengths and psychological capacities toward improving work performance (Luthans, 2002). Psychological capital (PsyCap), a construct of essence within POB, has an important role in addressing organizations manifold behavioral issues. PsyCap refers to an individual's positive and developmental psychological state and is characterized by the possession of high levels self-efficacy (confidence to commence and successfully complete challenging tasks), optimism (positive acknowledgment about the present and future success), hope (positive motivational state based on a sense of successful goal-directed determination and planning to meet these goals (Snyder et al., 1991), and encountering (the ability to bounce back with even more vigor for the achievement of success, when encountered challenges and adversity) (Luthans et al., 2007).

This section introduces the psychological capital of individual resilience toward providing an effective solution in preventing or mitigating the disadvantageous consequences of workplace psychosocial hazards.

21.6 RESILIENCE

Resilience is a widely applied concept in research and practice across varied disciplines such as sociology, psychology, nursing, medicine, business, and ecology. As per Luthans (2002), it is the positive psychological capacity that enables the individual to "bounce back" not only from adverse or uncertain instances, conflict or failure, but also from occurrences involving positive changes, progresses, and those commanding escalated responsibilities. It is not a rare phenomenon, rather is a dynamic process encompassing positive adaptation even in the presence of significant adversity, and thus enables the individual to continue leading a healthy and functional life. It focuses on the preemptive assessment of the risks and resources (personal) concerning employee outcomes.

21.7 CHARACTERISTICS OF RESILIENCE

As per Wagnild and Young (1990), the individual's possessing resilience display the following characteristics: awareness of their purpose in life, persistence in spite of adverse situations and discouragement, self-belief, a clear-cut acknowledgment of their competences and confines, a balanced life outlook, openness to experiences, recognition of individual uniqueness, and an understanding of the worth of self-experiences.

Robertson Cooper Model proposes that resilience encompasses confidence, purposefulness, adaptability, and social support. This implies that being resilient involves being proficient, having a goal-directed life with clarity of purpose, adopting effective coping skills, bearing a strong sense of self-esteem, having flexibility and adaptability to those situational changes that are not under one's control, and persisting and achieving regardless of setbacks. A review of over 270 research articles by Windle et al. (2011) found self-esteem, self-efficacy, optimism, hardiness, humor, creativity, initiative, problem-solving skills together with spiritual influences, personal and social competence, social resources, and family cohesion to be the constituting components of resilience.

The individuals high on resilience exhibit adaptive behaviors like revising goals on encountering adversity, displaying positive organizational behaviors, contributing to the maintenance of a positive work climate and/or culture, attaining higher productivity levels, displaying superior innovativeness, use of adequate coping skills and strategies, and enhanced job satisfaction. Contrary to this, those low on resilience allege higher perceived stress, burnout, and lowered work engagement; make frequent physiological complaints; display affective responses like helplessness, hurt, guilt, and fear in distressing situations; and largely practice avoidance strategies.

21.8 FACTORS AFFECTING RESILIENCE

There are certain personal, situational, and contextual factors affecting individual resilience that should be considered for the planning and effective implementation of the varied resilience-based intervention or training, in order to yield the desired outcomes. Some such factors have been enlisted in the Table 21.2.

21.9 RESILIENCE AND OTHER PSYCAP FACTORS

Resilience is differentiated from the other PsyCap concepts (viz., efficacy, hope, and optimism) in the following ways:

TABLE 21.2 The Individual Characteristics and the Individual-, Group-, and Organization-Level Disruptions Affecting Employees' Psychological Resilience at Work

Personal Characteristics	Situational/Contextual Characteristics		
	Individual	Group	Organizational
• Personality traits • Ego-resilience • Emotional intelligence • Coping skills • Locus of control • Problem-solving skills • Flexibility • Self-esteem	• Switching over to a new job • Changed work role • Redundancy • Physical abuse • Psychological abuse	• Changed team structure • Changed team dynamics • Inexperience of employees • Insufficient resources • Ineffective leadership • Stakeholder management issues	• Corporate restructuring • Acquisitions and mergers • Disasters • Terror instances • Global financial crisis • Global health crisis

- **Resilience and Self-efficacy:** While self-efficacy refers to the possession of confidence in one's own selves such that one is capable of undertaking challenging tasks and succeeding in their pursuit of those tasks; resilience as a PsyCap factor promotes individuals to learn to reframe adversity and challenges as opportunities for learning and development. Seen this way, efficacy impacts an individual's "resilience to adversity" (Bandura, 1997). The more confidence an individual displays toward accomplishing a task, the more prospectively they develop a path to resilience. Resilience enables individuals to keep attempting and restores their sense of efficacy even in the face of threatening situations.
- **Resilience and Hope:** In positive psychology, resilience is viewed as the one's capacity to retort constructively and prosper in both the negative and/or the positive stressful circumstances, while hope is seen as the presence of positive expectancies, having specified goals, and the possession of alternate pathways toward the goal attainment in the face of failures. This shows that both resilience and hope incorporate flexibility as an essential component, but hope, unlike resilience, does not encompass a troublesome or bothersome happening that activates the progression of resilience.
- **Resilience and Optimism:** Optimism is about having a generalized positive expectancy of future, that is, one will have positive experiences and outcomes in life, and should therefore persistently strive toward one's goals. It is a positive attribution about succeeding continually. Contrary to resilience, an exhibit of optimism does not require the triggering of any challenges or adverse situations. While optimists have a positive attributional style and avoid probing the actual meaning of adversity, resilient individuals adopt a planned, realistic, and practical approach to deal with stress and adversity. Owing to this difference, those high on resilience are better at adapting to stress and overcoming challenges than ones with high optimism.

21.10 RESILIENCE AND THE PSYCHOSOCIAL RISKS

The past decade saw a shift in focus of resilience researches conducted in organizational contexts, from assessing the impact that protective, as well as risk factors, have on employee's well-being to investigating the role that transformative processes had in this regard. In line with this, the current researches are probing the utility of resilience interventions in the promotion of individual and organizational level positive functioning.

Numerous studies have established that resilience is amenable to change (e.g., Arnetz et al., 2009; Sood et al., 2011). The organizational populace researched so far, in regard to resilience, largely encompasses public school teachers, university administrative staffs, resource sector company employees, sales managers from industrial organizations, human service professionals, public sector middle managers, public health service agency executives and senior managers, nurses and tertiary care physicians, civil servants, police officers, and military personnel.

Lu et al. (2014) and Mak et al. (2011) have revealed that resilience has a sizeable inverse association with mental health outcomes of anxiety, depression, secondary traumatic stress, and burnout. While those high on resilience demonstrated reduced anxiety, depression, post-traumatic stress disorder, and burnout; those low on resilience had poor psychological functioning. McCarthy et al. (2002), in their study, found resilience to be an important aiding factor in handling trauma resulting from workplace bullying. Following resilience-based interventions, police officers reported an enhanced personal resilience (Sood et al., 2011), higher antithrombin levels, an improved observed behavioral performance (Arnetz et al., 2009) along with reduced distress, negative emotions, and depression (McCraty & Atkinson, 2012). Similar findings exist in relation to civil servants (e.g., Liossis et al., 2009) and nurses (Pipe et al., 2012). Physicians too reported significantly reduced stress and anxiety following the resilience intervention (Sood et al., 2011). The healthcare workers even beheld improved locus of control, bettered self-esteem and interpersonal relations, and perception of purposefulness in life (Waite & Richardson, 2004). Further, enhancement of resilience results in lowered emotional exhaustion (Carvalho et al., 2006), higher job satisfaction (Matos et al., 2010), and an increased organizational commitment and happiness (Youssef & Luthans, 2007).

Psychological resilience is also a partial mediating mechanism in the association betwixt dispositional optimism and subjective well-being (He et al., 2013) and a moderator in the inverse relationship that exists between work-linked strain and employee satisfaction (Earvolino-Ramirez, 2007). This is suggestive of resilience being a protective factor against the job strain in demanding work settings.

In sum, resilience relates significantly to *employee's job-related attitudes* like work commitment, employee satisfaction, and flexibility; their *job-related behaviors* like organizational citizenship behavior, performance and effectiveness; their *psychological experiences* of self-awareness, self-esteem, and work happiness; and their *organizational attributes* of competitive advantage, vitality, strategic agility and leadership. Simultaneously, all of the aforementioned constructs exhibit a negative relationship with hazardous psychosocial working conditions.

Above discussions lead us to following elucidations: (1) psychosocial hazards lead to the counterproductive individual (physiological, psychological, behavioral), and organizational outcomes, (2) there exists an inverse relationship between resilience and the intensity of psychosocial risk factors, and (3) resilience relates positively to the valued individual and organizational outcomes and negatively to the counterproductive outcomes resulting from psychosocial risk factors. This implies that the presence of psychosocial hazards leads to disadvantageous consequences at both individual and organization level; however, the presence of resilience can hamper the occurrence of such psychosocial conditions that are injurious to the work environment and reduce the intensity of occurrence of associated hazards. Moreover, worker resilience could substantially reduce the counterproductive outcomes resulting from these conditions.

Thus, fostering psychological resilience among employees holds immense potential not only for mitigating hazardous psychosocial conditions but also for improving the organizations' overall occupational health.

21.11 BUILDING RESILIENCE AT WORK

Resilience building has also undergone a shift in focus in the 21st-century workplaces. Contrary to being viewed as a remedial or protective measure that's good enough only in reducing anxiety and stress, there has been a heightened realization that resilience is a strength-building capacity that operates across, and impacts, each organizational level (individual, group, and organization) and leads to the display of superior endurance on the part of employees even on occasions of persistent challenges and difficulties. This section briefs about the interventions to be adopted for building/boosting workplace resilience.

Most resilience-based interventions are based on a combination of psychotherapeutic procedures/methods of stress inoculation training, acceptance and commitment therapy, mindfulness-based attention and interpretation therapy, and problem-solving therapy. The prime modes of intervention delivery are web-based training, one-to-one training, group-based sessions, and a combination of these. The intervention duration may vary from a single few hours session to the one ranging between 3 and 12 weeks, depending upon the content and the characteristics of participants.

The American Psychological Association (2014) suggests that individual resilience could be built by: maintaining good interpersonal and familial relationships, not viewing stressful events as unbearable or critical, having acceptance for experiences irrespective of their emotional valence, setting up realistic goals and working for their realization, taking decisive actions even in unfavorable situations, looking for prospects of self-discovery even after struggling with any loss, being confident of one's own selves, bearing a long-run future perspective, appraising stressful events with a broadened context, being optimistic, doing regular exercise, learning to care about one's own selves, and developing self-awareness.

Leaders and managers could also have a role in developing individual employees' resilience. They could do so by minimizing the impact of situational factors, building a facilitating organizational climate, modeling resilient behavior at the workplace, ensuring holiday pay and flexible work arrangements, and enforcing proper policies in regard to bullying and harassment such that the overall employee well-being remains ensured.

Table 21.3 enlists the practical strategies to be adopted at each organizational level for building/boosting/facilitating the formation of a resilient workforce.

21.12 CONCLUSION

Numerous psychosocial hazards at the workplace like occupational stress, work fatigue, workplace bullying, harassment, and violence lead to counterproductive physiological, psychological, behavioral, and organizational outcomes. These detrimental outcomes could, however, be overcome and managed by the facilitation of a strong psychological capital among workers, in particular by building up a resilient workforce. Enhancing resilience reframes the individual cognition in ways such that the earlier adversities and challenges start to appear as coeval prospects of learning and self-enhancement, and the workers become able enough to proficiently adapt to the multifarious changes that the leading-edge ultramodern organizations of the present-day undergo. This chapter, thus, elucidates the role of workplace resilience in ensuring the individual worker's health and well-being, in addition to highlighting the immense potential that resilience holds for the creation of a productive, sustainable, safe, and healthy workplace.

TABLE 21.3 Practical Strategies for Building Individual, Group and Organizational Resilience

Individual	Group	Organizational
• Introducing resilience training interventions grounded in evidence-based approaches • Promoting, greater than before, involvement of workers in organizational decision-making • Annexing the employee's sense of control over their job design and practices • Providing stress management training to employees • Organizing physical activity program for employees • Promoting access to electronic health initiatives such as self-help websites • Appraising and appreciating good work	• Training the managers in such skills that can be of aid and assistance to the group members, as and when needed • Gaining top management commitment toward the maintenance of mentally sound workplace • Organizing workshops/training sessions on handling interpersonal conflict • Facilitating routine intercommunications among the group members • Formulating peer support programs • Hosting group-level social events for fostering coworker cooperation • Educating the employees on appropriate means to describe, and respond to the mental health issues they encounter	• Display of active involvement on the part of senior management • Framing up of a mental health policy • Forming a working group for monitoring the implementation of psychological resilience strategies • Ensuring the maintenance of high levels of organizational justice • Adopting policies and programs aimed at reducing mental health stigma • Routine assessment and addressal of the mental health risk factors, using appropriate measures

KEYWORDS

- **psychosocial hazards**
- **psychological resilience**
- **workplace health**

REFERENCE

American Psychological Association. (2014). *The road to resilience.* Retrieved from http://www.apa.org/helpcenter/road-resilience.aspx

Arnetz, B. B., Nevedal, D. C., Lumley, M. A., Backman, L., & Lublin, A. (2009). Trauma resilience training for police: Psychophysiological and performance effects. *Journal of Police and Criminal Psychology*, *24*(1), 1–9. DOI: 10.1007/s11896-008-9030-y

Bandura, A. (1997). *Self-efficacy: The exercise of control.* New York: Freeman.

Beale, D., Clarke, D., Cox, T., Leather, P., & Lawrence, C. (1999). System memory in violent incidents: Evidence from patterns of reoccurrence. *Journal of Occupational Health Psychology*, *4*(3), 233–244.

Belkic, K., Landsbergis, P., Schnall, P., Baker, D., Theorell, T., Siegrist, J., Peter, R., & Karasek, R. (2000). Psychosocial factors: Review of the empirical data among men. *Occupational Medicine, 15*, 24–46.

Byun, J., Kim, H. -R., Lee, H. -E., Kim, S. -E., & Lee, J. (2016). Factors associated with suicide ideation among subway drivers in Korea. *Annals of Occupational and Environmental Medicine, 28*, 31.

Carvalho, M. V. A., Calvo, F. B., Martin, H. L., Campos, R. F., & Castillo, C. I. (2006). Resilience and the burnout-engagement model in formal caregivers of the elderly. *Journal of Employee Counselling, 18*, 791–796.

Chandola, T., Brunner, E., & Marmot, M. (2006). Chronic stress at work and the metabolic syndrome: Prospective study. *British Medical Journal, 332*(7540), 521–525.

Chappell, D., & Di Martino, V. (2000). *Violence at work.* Geneva: International Labour Office.

Cooper, C. L., & Cartwright, S. (1994). Healthy mind, healthy organisation: A proactive approach to occupational stress. *Human Relations*, *47*, 455–470.

Cox, T., Griffiths, A., & Leka, S. (2005). Work organization and work-related stress. In K. Gardiner and J. M. Harrington (Eds.), *Occupational hygiene* (3rd ed.). Oxford: Blackwell Publishing.

Cox, T., & Leather, P. (1994). The prevention of violence at work: Application of a cognitive behavioral theory. In C. L. Cooper & I. Robertson (Eds.), *International review of industrial & organizational psychology*, Chichester: Wiley & Sons.

Cox, T., Griffiths, A., & Rial-Gonzalez, E. (2000). *Research on work related stress.* Luxembourg: Office for Official Publications of the European Communities.

De Beeck, R. O., & Hermans, V. (2000). *Research on work-related low back disorders (No. 204).* Luxembourg: Office for Official Publications of the European Communities.

Devereux, J. J., Vlachonikolis, I. G., & Buckle, P. W. (2002). Epidemiological study to investigate potential interaction between physical and psychosocial factors at work that may increase the risk of symptoms of musculoskeletal disorder of the neck and upper limb. *Occupational & Environmental Medicine, 59*(4), 269–277.

Earvolino-Ramirez, M. (2007). *Resilience: A concept analysis. In Nursing forum* (pp. 73–82). doi: 10.1111/j.1744–6198. 2007.00070.

EU-OSHA (2007). *Expert forecast on emerging psychosocial risks related to occupational safety and health.* Luxembourg: Office for Official Publications of the European Communities.

Everson-Rose, S. A. & Lewis, T. T. (2005). Psychosocial factors and cardiovascular diseases. *Annual Review of Public Health, 26*, 469–500.

Feskanich, D., Hastrup, J. L., Marshall, J. R., Colditz, G. A., Stampfer, M. J., Willett, W. C. & Kawachi, I. (2002). Stress and suicide in the Nurses' Health Study, *Journal of Epidemiology and Community Health, 56*, 95–98.

He, F., Cao, R., Feng, Z., Guan, H., & Peng, J. (2013). The impacts of dispositional optimism and psychological resilience on the subjective well-being of burn patients: A structural equation modelling analysis. *PLoS One, 8*(12): e82939. https://doi.org/10.1371/journal.pone.0082939.

Head, J., Stansfeld, S. A., & Siegrist, J. (2004). The psychosocial work environment and alcohol dependence: A prospective study. *Occupational & Environmental Medicine, 61*, 219–224.

Heraclides, A., Chandola, T., Witte, D. R., & Brunner, E. J. (2009). Psychosocial stress at work doubles the risk of type 2 diabetes in middle-aged women: Evidence from the Whitehall II study. *Diabetes Care, 32*(12), 2230–2235.

Kivimaki, M., Hotopf, M., & Henderson, M. (2010). Do stressful working conditions cause psychiatric disorders? *Occupational Medicine (London), 60*, 86–87.

Kopp, M. S., Stauder, A., Purebl, G., Janszky, I., & Skrabski, A. (2008). Work stress and mental health in a changing society. *European Journal of Public Health, 18*(3), 238–244.

Kouvonen, A., Kivimaki, M., Elovainio, M., Pentti, J., Linna, A., Virtanen, M., & Vahtera, J. (2006). Effort-reward imbalance and sedentary lifestyle: An observational study in a large occupational cohort. *Occupational & Environmental Medicine, 63*(6), 422–427.

Landsbergis, P. A., Schnall, P. L., Belkic, K. L., Baker, D., Schwartz, J., & Pickering, T. G. (2001). Work stressors and cardiovascular disease. *Work, 17*, 191–208.

Leka, S., Griffiths, A., & Cox, T. (2003). *Work Organization and Stress.* Geneva: World Health Organization.

Linton, S. J. (2004). Does work stress predict insomnia? A prospective study. *British Journal of Health Psychology, 9*, 127–136.

Liossis, P. L., Shochet, I. M., Millear, P. M., & Biggs, H. (2009). The Promoting Adult Resilience (PAR) program: The effectiveness of the second, shorter pilot of a workplace prevention program. *Behavior Change, 26*, 97–112.

Lu, W., Wang, Z., & Zhang, Y. L. (2014). Resilience as a mediator between extraversion, neuroticism, and happiness, PA and NA. *Personality and Individual Differences, 63*, 128–133.

Luthans, F. (2002). The need for and meaning of positive organizational behavior. *Journal of Organizational Behavior, 23*(6), 695–706.

Luthans, F., Luthans, K. W., & Luthans, B. C. (2004). Positive psychological capital: Going beyond human and social capital. *Business Horizons, 47*(1), 45–50.

Luthans, F., Youssef, C. M., & Avolio, B. J. (2007). Psychological capital: Developing the human competitive edge. Oxford, UK: Oxford University Press.

Mak, W. W. S., Ng, I. S. W., & Wong, C. C. Y. (2011). Resilience: Enhancing well-being through positive cognitive triad. *Journal of Counselling Psychology, 58*, 610–617.

Matos, P. S., Neushotz, L. A., Griffin, M. T. Q, & Fitzpatrick, J. J. (2010). An exploratory study of resilience and job satisfaction among psychiatric nurses working in inpatient units. *International Journal of Mental Health Nursing, 19*, 307–312.

McCarthy, P., Henderson, M., Sheehan, M., & Barker, M. (2002). Workplace bullying: Its management and prevention. In Australian master *OHS and environment guide* 2003 (pp. 519–549). Sydney: CCH Australia Limited.

McCraty, R., & Atkinson, M. (2012). Resilience training program reduces physiological and psychological stress in police officers. *Global Advances in Health and Medicine, 1*, 44–66. doi:10.7453/gahmj.2012.1.5.013

Michie, S. (2002). Causes and management of stress at work. *Occupational & Environmental Medicine, 59*, 67–72.

Middeldorp, C. M., Cath, D. C., & Boomsma, D. I. (2006). A twin-family study of the association between employment, burnout and anxious depression. *Journal of Affective Disorders, 90,* 163–169.

Ostry, A., Maggi, S., Tansey, J., Dunn, J., Hershler, R., Chen, L., Louie, A. M., & Hertzman, C. (2007). The impact of psychosocial work conditions on attempted and completed suicide among western Canadian sawmill workers, *Scandinavian Journal of Public Health, 35,* 265–271.

Pipe, T. B., Buchda, V. L., Launder, S., Hudak, B., Hulvey, L., Karns, K. E., & Pendergast, D. (2012). Building personal and professional resources of resilience and agility in the healthcare workplace. *Stress and Health, 28*, 11–22. doi: 10.1002/smi.1396

Radi, S., Ostry, A., & Lamontagne, A. D. (2007). Job stress and other working conditions: Relationships with smoking behaviors in a representative sample of working Australians. *American Journal of Industrial Medicine, 50*(8), 584–596.

Raikkonen, K., Matthews, K. A., & Kuller, L. H. (2007). Depressive symptoms and stressful life events predict metabolic syndrome among middle-aged women. *Diabetes Care, 30*, 872–877.

Randall, R., Griffiths, A., Cox, T., & Welsh, C. (2002). The activation of mechanisms linking judgments of work design and management with musculoskeletal pain. *Ergonomics, 45,* 13–31.

Rugulies, R, & Krause, N. (2008). Effort-reward imbalance and incidence of low back and neck injuries in San Francisco transit operators. *Occupational and Environmental Medicine*, *65,* 525–533.

Snyder, C.R., Irving, L., & Anderson, J.R. (1991). Hope and Health: Measuring the will and the ways. In C.R. Snyder and D.R. Forsyth (Eds.) *Handbook of social and clinical psychology: The health perspective* (pp. 285–305). Elmsford, New York: Pergamon Press. Cited in Snyder, C.R. (2000). Hypothesis: There is hope. In C.R. Snyder (Ed.), *Handbook of hope theory, measures and applications* (pp. 3–21). San Diego: Academic Press.

Sood, A., Prasad, K., Schroeder, D., & Varkey, P. (2011). Stress management and resilience training among Department of Medicine Faculty: A pilot randomized clinical trial. *Journal of General Internal Medicine, 26*(8), 858–861.

Spurgeon, A., Harrington, J. M., & Cooper, C. L. (1997). Health and safety problems associated with long working hours: A review of the current position. *Occupational & Environmental Medicine, 54*(6), 367–375.

Swaen, G., van Amelsvoort, L. G. P. M., Bultmann, U., & Kant, I. (2003). Fatigue as a risk factor for being injured in an occupational accident: Results from the Maastricht Cohort Study. *Occupational and Environmental Medicine, 60*(Suppl1), 88–92. http://doi.org/10.1136/oem.60.suppl_1.i88

Takala, J. (2002). *Introductory report: Decent work—safe work.* XVIth World Congress on Safety and Health at Work Vienna: ILO.

Vahtera, J., Pentti, J., & Kivimäki, M. (2004). Sickness absence as a predictor of mortality among male and female employees. *Journal of Epidemiology & Community Health, 58*(4), 321–326.

van den Berg, T. I. J., Elders, L. A. M., de Zwart, B. C. H., & Burdorf, A. (2009). The effects of work-related and individual factors on the Work Ability Index: A systematic review. *Occupational & Environmental Medicine, 66*, 211–220.

Wagnild, G., & Young, H. (1990). Resilience among older women. *Journal of Nursing Scholarship, 22*(4), 252–255.

Waite, J. P., & Richardson, E. G. (2004). Determining the efficacy of resiliency training in the work site. *Journal of Allied Health, 33*(3), 178–183.

Wemme, K. M., & Rosvall, M. (2005). Work related and non-work related stress in relation to low leisure time physical activity in a Swedish population. *Journal of Epidemiology & Community Health, 59*(5), 377–379.

Windle, G., Bennett, K. M., & Noyes, J. (2011). A methodological review of resilience measurement scales. *Health and Quality of Life Outcomes, 9*(8). doi:10.1186/1477-7525-9-8

Youssef, C. M., & Luthans, F. (2005). Resiliency development of organizations, leaders and employees: Multilevel theory building for sustained performance. In W. Gardner, B. Avolio, and F. Walumbwa (Eds.), *Authentic leadership theory and practice: Origins, effects and development* (Vol. 3, 303–343). Oxford, UK: Elsevier.

CHAPTER 22

Linkage Among Work-Life Balance, Psychological Capital, and Life Satisfaction: Testing a Spillover-Crossover Model Among Indian Dual-Working Couples

NIDHI BANSAL* and UPASNA A. AGARWAL

National Institute of Industrial Engineering, Vihar Lake Road, Powai, Mumbai, Maharashtra 400087, India

**Corresponding author. E-mail: Nidhi.kunalbansal@gmail.com*

ABSTRACT

Drawing upon the actor–partner interdependence model (APIM), spillover and crossover effects among psychological capital, work–life balance (WLB) perception, and life satisfaction were examined among dual-working couples in India. Matched data were collected from 213 dual-working Indian couples. The APIM_SEM app was used to analyze the dyadic data. The results of APIM_SEM analysis show the following: (1) psychological capital positively predicts the WLB, and enhanced WLB leads to increased life satisfaction of husbands as well as their wives (spillover effects); and (2) crossover of psychological capital on WLB was significant for husband to wife, but insignificant for wife to husband. Crossover of WLB on life satisfaction was found significant for wife to husband but not for husband to wife. To the best of authors' knowledge, this is the first attempt to understand the crossover effects among dual-working couples in India. This study suggests that employers should adequately address work–life issues of employees, because these issues impact not only their well-being but also their partners' well-being at home.

22.1 INTRODUCTION

Employees' most important domains are their work and family (Michel et al., 2011; Zhang et al., 2013). Voluminous past studies examined the predictors and outcomes of work–family interactions and how employees deal with the interplay of work–family domains (Amstad et al., 2011; Eby et al., 2005). However, individuals live in a larger social system consisting of their family at home and work with their colleagues at the workplace. They are part of this system consisting of their family, friends, colleagues, and society. All these may influence their behaviors and their experiences at the work–life interface. Thus, crossover is likely to occur. Crossover is described as a dyadic or interindividual transmission across individuals, whereby experiences crossover between closely related persons (Westman, 2001). The crossover process in the work–family interface is particularly frequent and influential in couples and, above all, in parents, because they need to highly integrate their work and family domains to accommodate their work–family demands. Past work–family research less investigated the crossover phenomenon as compared with the spillover process. Yet, there is evidence for growing research considering the crossover effects and the transmission of work–life experiences between husband and wives (Park & Fritz, 2015; Sprung & Jex, 2017; Tian et al., 2017). However, these efforts are limited and absent in the Indian context. Therefore, more research is needed to understand how dual-working couples fulfill their work–family responsibility, as current society has an increasing number of dual-working couples. This is also important to know how a spouse affects other spouse and is affected by his/her spouse's work and family experiences. Accordingly, several theoretical perspectives have suggested that adopting a couple perspective enhances our understanding of the work–family interface (Parasuraman & Greenhaus, 2002).

However, despite voluminous research examining factors related to the work–life interface, most of the researchers have focused only on the individual as the unit of analysis (Lu et al., 2016). This is a significant gap in work–family research (Westman & Etzion, 2005). Furthermore, most of the crossover studies were conducted mainly in western contexts (Tian et al., 2017). There is an urgent need for understanding work–family issues among couples in different contexts (Kossek et al., 2011), because there is evidence that work–family experiences of individuals and couples are different in different cultures (Yang et al., 2000; Spector et al., 2004). Therefore, a crossover study in the non-Western context, such as Indian, is needed to understand the phenomena clearly. Considering these research gaps, this study extends the work–family literature by studying the spillover–crossover effect among dual-working couples.

The predominant objective of this research is to understand the spillover and crossover of psychological capital on work–life balance (WLB) perceptions of employees and of WLB on life satisfaction at a dyadic level (with couple data) in a non-Western (i.e., Indian) context. The spillover–crossover model of Bakker and Demerouti (2013) posits that one spouse's work-related experiences first spill over to the home (intraindividually), where they then crossover to the other spouse (interindividually), through marital interactions. Therefore, the spillover–crossover model shows how an individual's experience passes to another domain, namely through intraindividual spillover and to his/her partner through crossover. For investigating spillover–crossover effects, the study used the actor–partner interdependence model (APIM) to understand the spillover–crossover process. Kashy and Kenny (2000) developed the APIM to study the dyadic relationships. Interdependence of two persons in close relationships is the basic tenet of this model. The APIM helps researchers to understand "how an individual's response for a particular independent variable affects his or her own responses to a dependent variable (i.e., an actor effect), as well as on his or her partner's responses to a dependent variable measure (i.e., a partner effect)" (Campbell & Kashy, 2002).

This research contributed to the work–family literature in different manners. First, this study explores spillover and crossover effects among dual-working couples in the Indian context. Second, more enquires were required to understand the gender role in the crossover process (Zhang et al., 2013); a gender asymmetry in the crossover process was found for psychological capital–WLB and WLB–life satisfaction relationships. Third, most of the previous studies examined the crossover of work–family conflict and enrichment, crossover of WLB scarcely examined. Fourth, this is the first study to examine the crossover effects of psychological capital and WLB in the Indian context. Finally, the findings of this study have important organizational implications in human resource management because it is evident in the extant literature that psychological capital and WLB are associated with various important organizational outcomes such as job satisfaction, organizational commitment, turnover intention, and job performance (Adame-Sánchez et al., 2016; Haar et al., 2014; Mishra et al., 2017).

The issue of WLB among Indian population has increasingly attracted researchers' attention (Korabik et al., 2017). However, work–family research in Indian samples is relatively limited (Kalliath et al., 2018). This study focuses on the WLB among Indian dual-working couples because the number of dual-working couples is still at their growing stage in India (Munn & Chaudhuri, 2016). Indians face greater difficulties in managing work–life-related issues than the couples from different parts of the world due to contextual differences, family structures and values, and societal norms (Buddhapriya, 2009). Hence, examining the crossover–spillover effects (psychological capital, WLB, and life satisfaction) among Indian dual-working couples would benefit work–family research in a significant manner.

22.2 LITERATURE REVIEW AND HYPOTHESIS DEVELOPMENT

22.2.1 WORK–LIFE BALANCE

"Work–life balance is the individual's perception that work and nonwork activities are compatible and promote growth by an individual's current life priorities" (Kalliath & Brough, 2008). Agreeing with this definition, recently, Casper et al. (2018) have defined *work–nonwork balance* as: "employees' evaluation of the favorability of their combination of work and nonwork roles, arising from the degree to which their affective experiences and their perceived involvement and effectiveness in work and nonwork roles are commensurate with the value they attach to these roles." Researchers are giving increasing attention toward WLB due to its association with several important individual and organizational outcomes (Beauregard & Henry, 2009; Greenhaus et al., 2014; Greenhaus & Allen, 2011; Greenhaus et al., 2012). However, WLB construct is still under focus in the area of work–family literature (Casper et al., 2018; Greenhaus & Allen, 2011; Valcour, 2007).

The aforementioned definition is based on the conservation of resources (COR) theory given by Hobfoll (1989). According to the COR theory, "an individual motivates to create, conserve, and protect the quality of their resources" (Hobfoll, 1989). Drawing on the theory, it is argued that difficulty in balancing work–life roles leads to struggle in managing resources such as time or energy. In such a situation, employees must be motivated enough to save their energy and resources. Therefore, this leads to negative individual and organizational outcomes. For example, Fisher-McAuley et al. (2003) suggested that experiencing work–life imbalance leads to work-related stress, which further results into undesirable outcomes. Recently, Kraak et al. (2018) have also mentioned that WLB leads to employee's involvement in development activities due to availability of enough resources.

22.2.2 HOW INDIANS PERCEIVE THEIR WORK AND FAMILY ROLES: CULTURAL REALITIES

Managing employees at the workplace is a culture-specific phenomenon (Budhwar & Khatri, 2001); Powell et al. (2009) mentioned that the work–life interface of employees is influenced by contextual variables. A study suggested that work–life discourses are influenced by the societal norms and emergent of a specific place, time, and context (Rajan-Rankin, 2016), Therefore, understanding the cultural aspects is important to deeply know the work–life interface of Indian dual-working couples.

The Indian context differs from the Western context in many ways, for example, for Indians, their relationships with their family and community are very important (Hofstede, 2007). In a collectivist society like India, family plays an important role in individual lives. Indians show loyalty and interdependence for their family members. They give priority to the interests of their family members in taking important decisions regarding their marriage and job. Research shows that Indian dual-working couples experience an entirely different set of problems than the couples from the West (Rajadhyaksha et al., 2013). Research shows that Indian working women experiences different work–family dilemmas from those reported by Western women because of cultural values and social expectations (Sekaran, 1992). For example, Indian men usually do not share domestic work and childcare (Ramu, 1987; Sekaran, 1992). Therefore, working women in India prefer to get paid help for household work and childcare instead of delegating these tasks to their husbands. They do so to avoid any conflict in the family and maintaining equilibrium between their work–family roles (Rout et al., 1999). Paid help can preserve inequalities between men and women because it reduces the women's burden but does not change the men's role at home (Lewis, 1992). India is considered as low in gender egalitarianism, where women's primary role is being a "wife" and a "mother" and men's role as breadwinner (Chhokar et al., 2007). This perceived gender division affects the division of household, child-caring responsibilities, and employees' attitude and behavior regarding work–life. A global survey suggests that most Indians (64%) view women as good mothers and wives. According to them, women's priority should be home (The Ipsos Global Trends Survey, 2017).

In the societies where both men and women work, most of them struggle to fulfill the responsibilities of their work and family roles. Male employees face additional responsibilities and uncertain expectations regarding their family roles due increasing egalitarian culture in the society. Female employees face dual expectations, performing work duties along with childcare and household work. Therefore, the work and family domains of dual-working couples are closely intervened for both husbands and wives.

22.2.3 THE SPILLOVER EFFECT

"PsyCap refers to one's positive developmental state, where positive psychological resources enhance the employees' ability to cope with challenges, stress, or strain" (Luthans et al., 2006). "The PsyCap is a second-order construct, comprising four positive psychological resources: (a) self-efficacy, (b) present and future optimism, (c) goal perseverance (hope), and (d) resiliency to problems and adversity to attain goals" (Luthans et al., 2006, 2007). A better predictor of psychological capital of an individual than the four separate dimensions is found (Luthans et al., 2007a).

This study argues that PsyCap would positively predict WLB, because individuals with high self-efficacy have more personal resources (Bandura, 1997) and, with high self-confidence, are prone to be happier with their lives (Lyubomirsky et al., 2005). In a similar way, individuals hope to contribute to the enhanced WLB perceptions. For instance, individuals with high hope tend to have good both mental and psychological health, have good faith on their achievement, and are happy with their surroundings (Snyder, 2000). Resilient people handle unexpected problems or setbacks in a better manner and reverse the problems more willingly (Masten & Reed, 2002). Optimism helps individuals to increase or maintain their motivation level, energies, and productivity (Avey et al., 2009). These individuals experience more positive emotions and well-being (Diener, 2000). The COR theory (Hobfoll, 1989) helps understand the relationship between psychological capital and WLB. According to the COR theory, "PsyCap is a personal characteristic resource that enables individuals to cope with stress" (Avey et al., 2009) and have positive outcomes (Hobfoll et al., 2018). Thus, employees with high PsyCap experience more balance. Alarcon et al. (2013) also reported that PsyCap is an individual's characteristic, which acts as personal resource contributing to the employees' well-being. Thus, based on the given discussion, we propose the following:

Hypothesis 1: PsyCap will be positively related to WLB.

Life satisfaction is defined as "an overall assessment of feelings and attitudes about one's life at a particular point in time ranging from negative to positive" (Diener et al., 1985). Some authors believe that life satisfaction can be accessed by evaluating an individual's happiness and state of well-being (Benz & Frey, 2008; Binder & Coad, 2010; Coad & Binder, 2014). Life satisfaction is an important variable because satisfaction level of individuals indicates their social progress (Skeppner et al., 2008). In addition, life satisfaction is important for well-functioning and happiness of human being (Ryff et al., 2017). It is a measure of success (Wach et al., 2016) and is positively related to individuals' productivity (Lyubomirsky et al., 2005).

This study believes that individuals who all experience WLB are more satisfied with their lives (Brough et al., 2014; Haar et al., 2014). Haar et al. (2014) found a significant WLB–life satisfaction relationship. They also found that in individualistic culture, the relationship between WLB and life satisfaction was stronger as compared to collectivistic culture. It is also found that the relationship is stronger in gender egalitarian culture than the traditional culture. This suggests further study of the WLB–life satisfaction relationship in different contexts such as India.

The broaden-and-build theory by Fredrickson (2001) provides the support for the WLB–life satisfaction relationship. The theory predicts that positive emotions "broaden people's momentary thought–action repertoires, widening the array of the thoughts and actions that come to mind." Previous studies found that positive emotions are positively related to various success measures and well-being (Keyes et al., 2002; Lyubomirsky et al., 2005). These positive emotions may give strength to the employees to cope with adverse work–life situations (Baumeister et al., 2006). Therefore, we argue that experiencing increased balance leads to positive emotions, and these elevated levels of positivity will contribute to the enhanced life satisfaction. Thus, drawing on the above discussion, we propose the following hypothesis:

Hypothesis 2: WLB will be positively related to life satisfaction.

22.2.4 THE CROSSOVER EFFECT

"The process whereby the psychological stress or well-being experienced by one person affects the level of well-being of another person is referred to as crossover" (Westman, 2001; Westman & Vinokur, 1998). The crossover process is a dyadic phenomenon; in this process, interindividual transmission of stress and well-being occurs between the persons in the close relationship. This phenomenon occurs within a particular domain, such as organization or home. Crossover occurs because spouses identify and are concerned for each other; they divide most of their responsibilities related to their family life. They react emphatically toward each other and thereby emphasize each other's experiences. For instance, if the wife comes home from work tensed, she is likely to trigger an empathetic reaction in her husband, thereby "infecting" him with strain. Indirect crossover occurs through social interactions between spouses: If the wife comes home from work tensed, she is likely to interact with her husband in a more hostile or aggressive manner, thereby increasing the husband's level of distress. Conversely, if she comes home in a good mood because she experienced something positive at work, she is likely to interact with her husband in a positive manner, thereby increasing her husband's well-being.

This study believes that in married life, employees often exchange their feelings with their spouses and they try to understand each other. For example, when person is stressed, he/she might share his/her feelings to the spouse and then develop same level of stress into the spouse. The extant literature also suggests the direct crossover of work–family balance (Ferguson et al., 2015; Ferguson et al., 2012).

The ecological systems theory (Bronfenbrenner, 1986) provides support for crossover effect; the theory asserts that individuals interact with and are influenced by several environmental systems, including systems in which the individual himself or herself may play or not play an active role, such as the spouse's personality and attitudes. Both husband and wife, especially dual-earner parents, actively synchronize and coordinate their lives in the work and family domains to manage their work–life issues effectively (Parasuraman & Greenhaus, 2002). This is apparent not only in normal days but also in unforeseeable or unusual situations, for example, when a child is not well, and family function. Therefore, the personality and attitude of a spouse have influence on the other spouse. Thus, building on the discussion, we develop the following hypotheses.

Hypothesis 3: There will be a positive relationship between one spouse's psychological capital and other spouse's WLB.

Hypothesis 4: There will be a positive relationship between one spouse's WLB and other spouse's life satisfaction.

22.3 METHODS

22.3.1 SAMPLE AND PROCEDURE

The data for this study were collected from 213 dual-working couples from various Indian organizations. Diversified data were collected through sectors such as education, banking, IT, and aviation. Although more than 300 couples were contacted, among them a few employees were reluctant to participate due to their work pressure. There were 213 males and 213 female respondents, and all participants were married and had living with working spouses. All the participants have at least one child; all are dual-working parents. Of all respondents, 314 were middle-level employees and 112 were senior-level employees. The respondents were 23–55 years old. The average age of respondents was 36 years. The data were collected from participants by a self-administered, self-reported survey questionnaire from October 2017 to May 2018. Data were collected by using a snowball sampling technique from different sources (organizations, acquaintances, schools, and daycare centers,). This procedure for collecting data from dual-working couples is similar to those in previous work–family studies (Demerouti, 2012; Delina & Raya, 2016; Radcliffe & Cassell, 2014; Schooreel & Verbruggen, 2016).

22.3.2 *MEASURES*

The measures of the survey questionnaire were adopted from the published scales of respective constructs. All scale items were in English language. The details of the measures are as follows:

22.3.2.1 *PSYCHOLOGICAL CAPITAL*

A 12-item scale developed by Luthans *et al.* (2007) was used to measure the construct, (five-point scale, ranging from 1—strongly disagree to 5—strongly agree). This scale is composite of four separate dimensions, each with three items: hope, optimism, and self-efficacy and resilience. A sample item is "I feel confident in representing my work area in meetings with management." Cronbach's alpha was 0.85.

22.3.2.2 *WORK-LIFE BALANCE*

WLB was measured by a four-item scale developed by Brough et al. (2014) (four-item, five-point Likert scale, ranging from 1—strongly disagree to 5—strongly agree). A sample item is "overall, I believe that my work and nonwork life is balanced." Cronbach's alpha was 0.81.

22.3.2.3 *LIFE SATISFACTION*

Diener et al.'s (1985) scale of life satisfaction was used. It is a five-item, seven-point Likert scale, ranging from 1—strongly disagree to 7—strongly agree. A sample question is "In most ways my life is close to ideal." Cronbach's alpha was 0.88.

22.3.2.4 *DATA ANALYSIS*

Exploratory factor analysis and confirmatory factor analysis in statistical package for the social sciences were used to ensure the reliability and validity of the survey questionnaire. All scales were found reliable and validate. The spillover and crossover effects were analyzed using the APIM_SEM app using the APIM approach (Stas et al., 2018).

22.4 RESULTS

22.4.1 *DESCRIPTIVE STATISTICS*

Descriptive statistics of the study variables are shown in Table 22.1. Correlation analysis

TABLE 22.1 Descriptive Statistics and Correlations

	Variable	Mean	SD	1	2	3	4	5	6
1	WLBw	3.63	0.700	1					
2	WLBm	3.44	0.709	0.161*	1				
3	LSw	4.72	0.688	0.348**	0.175*	1			
4	LSm	4.51	0.758	0.199**	0.430**	0.347**	1		
5	PsyCapw	4.01	1.03	0.282**	0.010	0.356**	0.191**	1	
6	PsyCapm	3.96	1.24	0.284**	0.510**	0.280**	0.501**	0.248**	1

Note: *(N = 213 couples), WLBw = work–life balance of women, WLBm = work–life balance of men, LSw = life satisfaction of women, LSm = life satisfaction of men, PsyCapw = psychological capital of women, and PsyCapm = psychological capital of men.*

**p < 0.05; **p < 0.01; ***p < 0.001.*

indicates that the correlations of all the variables are significant in the proposed direction except for the PsyCapw–WLBm relationship.

22.4.2 APIM RESULTS

The analysis results from (Stas et al., 2018) revealed that "the actor effect for women is equal to 0.313 ($p < 0.001$, 95% CI [0.13, 0.49]). The overall standardized effect for women is 0.248 (partial $r = 0.227$). The actor effect for men is equal to 0.625 ($p < 0.001$, 95% CI [0.49, 0.76]), and the overall standardized actor effect for men is 0.494 (partial $r = 0.496$). When tested if the two actor effects are equal, the difference was found to be statistically significant, $p = 0.007$, 95% CI [–0.54, –0.08]. The overall actor effect is equal to 0.469 and is statistically significant ($p < 0.001$, 95% CI [0.36, 0.58])."

"The partner effect from men to women is equal to 0.261, which is statistically significant ($p < 0.001$, 95% CI [0.11, 0.41]), and its overall standardized effect is 0.207 (partial $r = 0.283$). The partner effect from women to men is equal to –0.167 and is not statistically significant ($p = 0.051$, 95% CI [–0.33, 0]) and its overall standardized partner effect is –0.132 (partial $r = –0.140$). When tested if the two partner effects are equal, the difference was found to be statistically significant ($p < 0.001$, 95% CI [0.2, 0.65]). The overall partner effect is equal to 0.047 is not statistically significant ($p = 0.404$, 95% CI [–0.06, 0.16])." Figs. 22.1 and 22.2 show the results of APIM analysis of PsyCap–WLB relationship. Table 22.2 shows the estimates and significance of the actor and partner effects of PsyCap on WLB (Stas et al., 2018).

"The actor effect for women is equal to 0.487 ($p < 0.001$, 95% CI [0.3, 0.67]). The overall standardized effect for women is 0.300 (partial $r = 0.363$). The actor effect for men is equal to 0.716 ($p < 0.001$, 95% CI [0.5, 0.93]), and the overall standardized actor effect for men is 0.440 (partial $r = 0.442$). When tested if the two actor effects are equal, the difference was found not to be statistically significant, $p = 0.121$, 95% CI [–0.52, 0.06]. The overall actor effect is equal to 0.602 and is statistically significant ($p < 0.001$, 95% CI [0.46, 0.74])" (Stas et al., 2018).

"The partner effect from men to women is equal to 0.179, which is not statistically significant ($p = 0.058$, 95% CI [–0.01, 0.36]), and

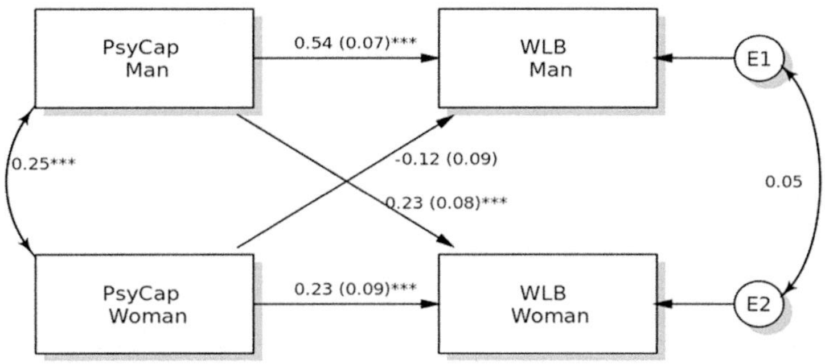

FIGURE 22.1 APIM results of PsyCap on WLB.

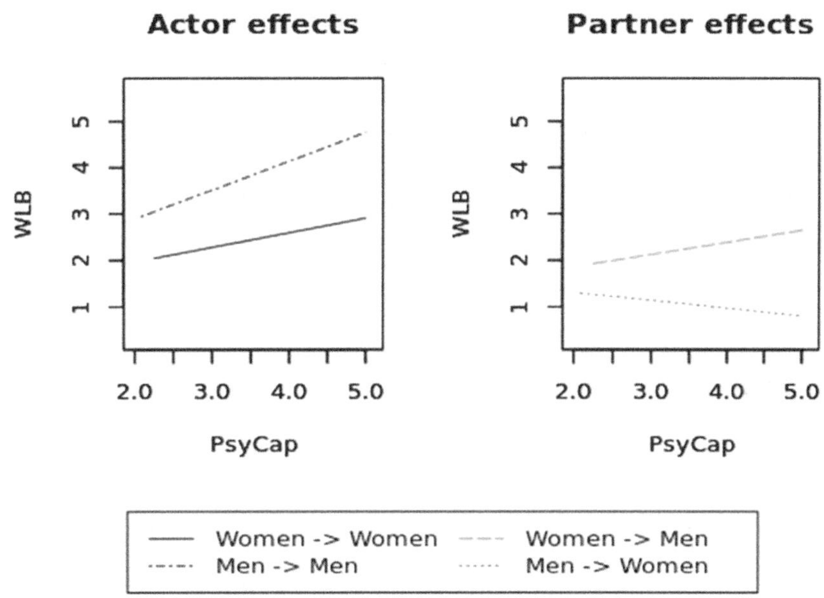

FIGURE 22.2 Actor and partner effects of PsyCap on WLB.

TABLE 22.2 APIM Results for PsyCap–WLB Relationships

Effect	Role Estimate	Lower 95% CI	Upper CI	*p* Value
Women				
Actor	0.313	0.133	0.493	0.000
Partner	0.261	0.113	0.409	0.000
Men				
Actor	0.625	0.488	0.762	0.000
Partner	–0.167	–0.334	0.000	0.051

its overall standardized effect is 0.110 (partial $r = 0.085$). The partner effect from women to men is equal to 0.236 and is statistically significant ($p = 0.032$, 95% CI [0.02, 0.45]), and its overall standardized partner effect is 0.145 (partial $r = 0.129$). When tested if the two partner effects are equal, the difference was found not to be statistically significant ($p = 0.702$, 95% CI [–0.35, 0.23]). The overall partner effect is equal to 0.207, which is statistically significant

(p = 0.003, 95% CI [0.07, 0.35])." Figs. 22.3 and 22.4 show the results of APIM analysis of PsyCap–WLB relationship. Table 22.3 shows the estimates and significance of the actor and partner effects of PsyCap on WLB" (Stas et al., 2018).

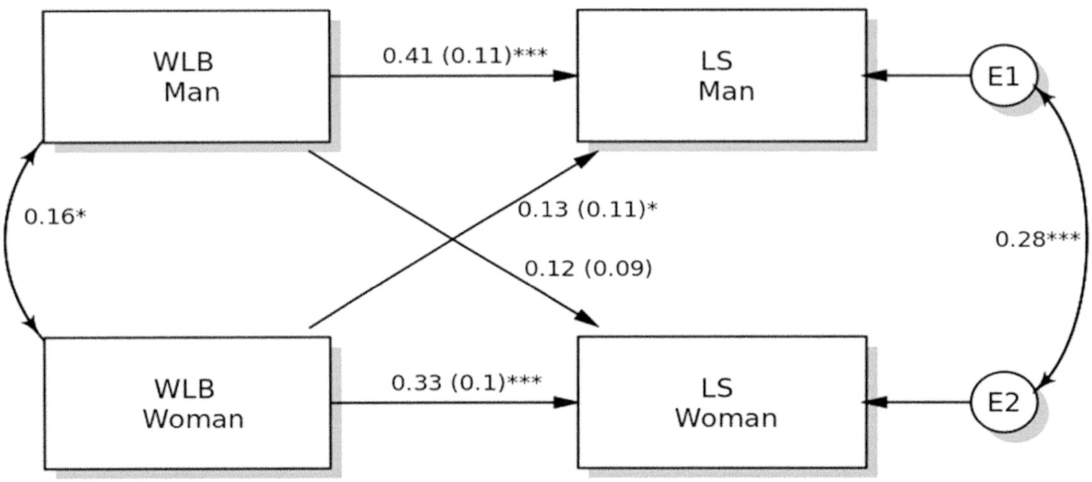

FIGURE 22.3 APIM results of WLB–life satisfaction.

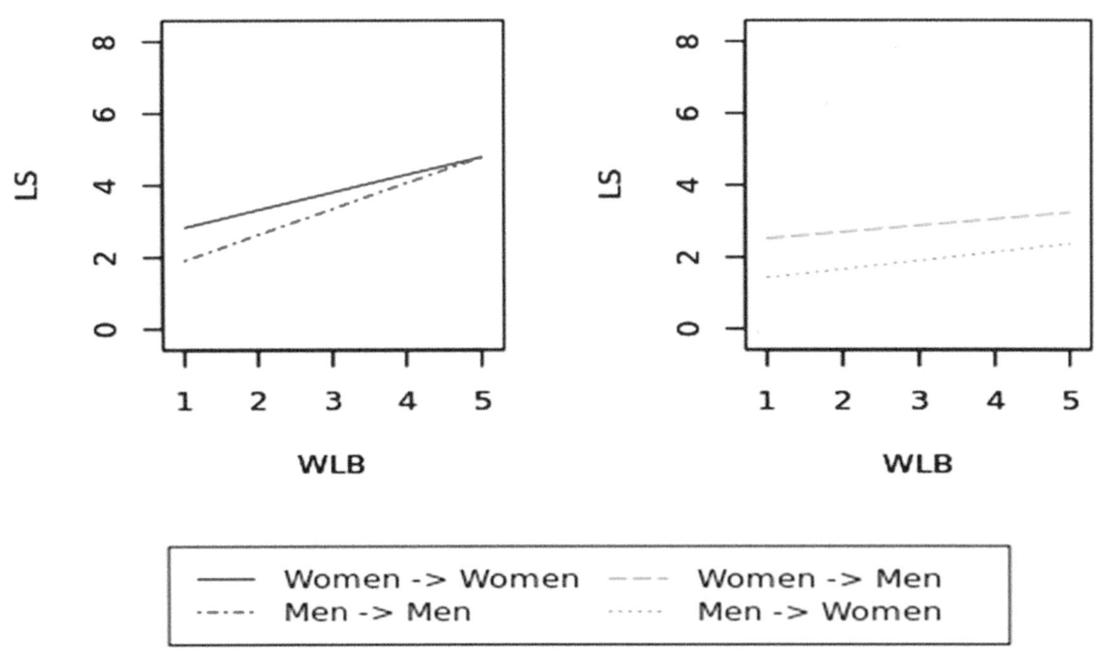

FIGURE 22.4 Actor and partner effects of WLB–life satisfaction.

TABLE 22.3 APIM Results for WLB–Life Satisfaction Relationships

Effect	Role Estimate	Lower 95% CI	Upper CI	p Value
Women				
Actor	0.487	0.299	0.675	0.000
Partner	0.179	–0.006	0.364	0.058
Men				
Actor	0.716	0.503	0.929	0.000
Partner	0.236	0.020	0.451	0.032

22.5 DISCUSSION

This study focused on the spillover and crossover effects in the work–life interface of dual-working employees. To understand the spillover–crossover process, the APIM approach was adopted. This study examined that PsyCap has spillover and crossover effects on WLB, and WLB has spillover and crossover effects on life satisfaction for employees and their spouses. Most previous studies have focused on the negative crossover among couples; individual factors were largely ignored in crossover studies. Therefore, we decided to investigate the spillover–crossover effects of a personality variable (PsyCap) and positive experience (WLB) among dual-working couples. The results of APIM_SEM analyses clearly showed that PsyCap and WLB have spillover effect as hypothesized. Furthermore, results supported the hypothesis only in single direction for both crossover hypotheses. Results suggested that the crossover of PsyCap on WLB was significant for husband to wife, but insignificant for wife to husband. Crossover of WLB on life satisfaction was found significant for wife to husband but not for husband to wife. The reason for these asymmetrical results lies in cultural norms and role of perceived gender role ideology. Furthermore, the contributions and the implications of this study are discussed.

22.5.1 IMPLICATIONS FOR RESEARCH AND PRACTICE

This study significantly contributed toward the work–family literature. It has mainly two important contributions. First, it highlighted the spillover and crossover process of underfocused constructs; the results suggest the situations under which individuals are affected by their spouses. Second, this study first examines the crossover effects among Indian dual-working couples and extends the crossover research in the non-Western context. The findings of the study suggest that the crossover process also applies well in Indian culture. Furthermore, the study got significant results for the unidirectional crossover effect (husbands to wives for PsyCap–WLB and from wives to husband for WLB–life satisfaction). Gender differences in the crossover process may be found because of differences in societal gender roles, gender role attitudes of individuals, as well as the influence of traditional values of Indian family cultures.

The findings of this study also provide various fruitful implications to the individuals and organizations. Crossover effects of positive variables provide useful benefits to the organizations and employees. These findings may encourage the organizations to focus on employees' work–life issues through designing the appropriate WLB practices and policies (such as childcare and flexible working). It is also well known that work-related attitudinal and behavioral outcomes are influenced by their partners (Brough & O'Driscoll, 2010; Westman & Etizon, 2005), especially in collectivist societies such as India (Shukla & Purani, 2012). Therefore, it would be beneficial for employers to focus on employees' work–family demands by implementing WLB policies, to reduce the negative outcomes and increase the positive outcomes. Organizations should also enhance the training to improve the psychological capital of the employees.

22.5.2 LIMITATIONS AND FUTURE DIRECTIONS

However, despite these contributions to the literature, few limitations of this study should be acknowledged. First, the study adopted self-reported questionnaire to collect study data; therefore, there is a possibility of a common method bias in this study. However, suitable techniques were used to avoid the common method bias. Second, cross-sectional design of this study may limit the establishing causal relationships. In future research, adopting longitudinal research design or diary study can be adopted to understand the causal conditions of spillover–crossover effects in a better way. Third, the sample for this study was drawn from managerial-level full-time employees. This raises question to the applicability of the findings of this study in different contexts. This study can be replicated by considering other population such as self-employed people and less educated workers in the future. This study examined the crossover effects only on the WLB perceptions and life satisfaction. Crossover effects of WLB on other important outcomes such as turnover intentions, organizational commitment, and job satisfaction should be considered in future research. In the future, the authors may consider the influence of gender role ideology in the crossover process.

22.6 CONCLUSION

This study finds that high level of PsyCap predicts high level of balance, which results in increased life satisfaction for individuals. The current positive crossover findings clearly indicate that dual-working couples also affect well-being in the form of WLB and life satisfaction of each other. Asymmetric results in the crossover process highlight the role of socially prescribed and perceived gender roles for men and women in experiencing balance and satisfaction. Furthermore, we should acknowledge the importance of work–life practices to promote increased life satisfaction through enhanced WLB for both partners. Individuals and organizations should do efforts to increase the PsyCap by various interventions for enhanced well-being. Nonetheless, the work–life interface of dual-working couples is complicated, and further research is required to enhance the knowledge about the intervening relationships between partners.

KEYWORDS

- **work–life balance**
- **psychological capital**
- **life satisfaction**
- **crossover**
- **spillover**
- **dual-working couples**
- **India**

REFERENCES

Adame-Sánchez, C., González-Cruz, T. F., & Martínez-Fuentes, C. (2016). Do firms implement work–life balance policies to benefit their workers or themselves? *Journal of Business Research, 69*(11), 5519–5523.

Alarcon, G. M., Bowling, N. A., & Khazon, S. (2013). Great expectations: A meta-analytic examination of optimism and hope. *Personality and Individual Differences*, *54*(7), 821–827.

Amstad, F. T., Meier, L. L., Fasel, U., Elfering, A., & Semmer, N. K. (2011). A meta-analysis of work–family conflict and various outcomes with a special emphasis on cross-domain versus matching-domain relations. *Journal of Occupational Health Psychology*, *16*(2), 151.

Avey, J. B., Luthans, F., & Jensen, S. M. (2009). Psychological capital: A positive resource for combating employee stress and turnover. *Human Resource Management*, *48*(5), 677–693.

Bakker, A. B., & Demerouti, E. (2013). The spillover-crossover model. In *New frontiers in work and family research* (pp. 54–70). Hove, UK: Psychology Press.

Bandura, A. (1997). *Self-efficacy: The exercise of control*. New York, NY, USA: Macmillan.

Baumeister, R. F., Gailliot, M., DeWall, C. N., & Oaten, M. (2006). Self-regulation and personality: How interventions increase regulatory success, and how depletion moderates the effects of traits on behavior. *Journal of Personality*, *74*(6), 1773–1802.

Beauregard, T. A., & Henry, L. C. (2009). Making the link between work-life balance practices and organizational performance. *Human Resource Management Review*, *19*(1), 9–22.

Benz, M., & Frey, B. S. (2008). Being independent is a great thing: Subjective evaluations of self-employment and hierarchy. *Economica*, *75*(298), 362–383.

Binder, M., & Coad, A. (2010). An examination of the dynamics of well-being and life events using vector autoregressions. *Journal of Economic Behavior & Organization*, *76*(2), 352–371.

Bronfenbrenner, U. (1986). Ecology of the family as a context for human development: Research perspectives. *Developmental Psychology*, *22*(6), 723.

Brough, P., & O'Driscoll, M. P. (2010). Organizational interventions for balancing work and home demands: An overview. *Work and Stress*, *24*(3), 280–297.

Brough, P., Timms, C., O'Driscoll, M. P., Kalliath, T., Siu, O.-L., Sit, C., & Lo, D. (2014). Work–life balance: A longitudinal evaluation of a new measure across Australia and New Zealand workers. *The International Journal of Human Resource Management*, *25*(19), 2724–2744.

Buddhapriya, S. (2009). Work-family challenges and their impact on career decisions: A study of Indian women professionals. *Vikalpa*, *34*(1), 31–46.

Budhwar, P. S., & Khatri, N. (2001). A comparative study of HR practices in Britain and India. *International Journal of Human Resource Management*, *12*(5), 800–826.

Campbell, L., & Kashy, D. A. (2002). Estimating actor, partner, and interaction effects for dyadic data using PROC MIXED and HLM: A user-friendly guide. *Personal Relationships*, *9*(3), 327–342.

Casper, W. J., Vaziri, H., Wayne, J. H., DeHauw, S., & Greenhaus, J. (2018). The jingle-jangle of work–nonwork balance: A comprehensive and meta-analytic review of its meaning and measurement. *Journal of Applied Psychology*, *103*(2), 182.

Chhokar, J. S., Brodbeck, F. C., & House, R. J. (2007). India: Diversity and complexity in action. *Culture and Leadership across the World. The GLOBE Book of in-Depth Studies of*, *25*, 1004–1005.

Coad, A., & Binder, M. (2014). Causal linkages between work and life satisfaction and their determinants in a structural VAR approach. *Economics Letters*, *124*(2), 263–268.

Diener, E. (2000). Subjective well-being: The science of happiness and a proposal for a national index. American Psychologist, *55*(1), 34.

Delina, G., & Raya, R. P. (2016). Dilemma of work-life balance in dual-career couples—A study from the Indian perspective. *International Journal of Indian Culture and Business Management*, *12*(1), 1–27.

Demerouti, E. (2012). The spillover and crossover of resources among partners: The role of work–self and family–self facilitation. *Journal of Occupational Health Psychology*, *17*(2), 184.

Diener, E. D., Emmons, R. A., Larsen, R. J., & Griffin, S. (1985). The satisfaction with life scale. *Journal of Personality Assessment*, *49*(1), 71–75.

Eby, L. T., Casper, W. J., Lockwood, A., Bordeaux, C., & Brinley, A. (2005). Work and family research in IO/OB: Content analysis and review of the literature (1980–2002). *Journal of Vocational Behavior*, *66*(1), 124–197.

Ferguson, M., Carlson, D., & Kacmar, K. M. (2015). Flexing work boundaries: The spillover and crossover of workplace support. *Personnel Psychology*, *68*(3), 581–614.

Ferguson, M., Carlson, D., Zivnuska, S., & Whitten, D. (2012). Support at work and home: The path to satisfaction through balance. *Journal of Vocational Behavior*, *80*(2), 299–307.

Fisher-McAuley, G., Stanton, J., Jolton, J., & Gavin, J. (2003). Modelling the relationship between work life balance and organisational outcomes. In *Annual Conference of the Society for Industrial-Organisational Psychology* (pp. 1–26).

Fredrickson, B. L. (2001). The role of positive emotions in positive psychology: The broaden-and-build theory of positive emotions. *American Psychologist, 56*(3), 218.

Greenhaus, J. H., & Allen, T. D. (2011). Work–family balance: A review and extension of the literature. In 2nd ed., *Handbook of occupational health psychology* (pp. 265–183). Washington, DC, USA: American Psychological Association

Greenhaus, J. H., Peng, A. C., & Allen, T. D. (2012). Relations of work identity, family identity, situational demands, and sex with employee work hours. *Journal of Vocational Behavior, 80*(1), 27–37.

Greenhaus, J. H., Ziegert, J. C., & Allen, T. D. (2014). Relationships of work-family conflict and enrichment with work-family balance. In *Contribution to the symposium "What is balance* (pp. 14–17).

Haar, J. M., Russo, M., Suñe, A., & Ollier-Malaterre, A. (2014). Outcomes of work–life balance on job satisfaction, life satisfaction and mental health: A study across seven cultures. *Journal of Vocational Behavior, 85*(3), 361–373.

Hobfoll, S. E. (1989). Conservation of resources: A new attempt at conceptualizing stress. *American Psychologist, 44*(3), 513.

Hobfoll, S. E., Halbesleben, J., Neveu, J.-P., & Westman, M. (2018). Conservation of resources in the organizational context: The reality of resources and their consequences. *Annual Review of Organizational Psychology and Organizational Behavior, 5*, 103–128.

Hofstede, G. (2007). Asian management in the 21st century. *Asia Pacific Journal of Management, 24*(4), 411–420.

Ipsos Global Trends. (2017). Ipsos Global Trends Survey. Retrieved from https://www.ipsosglobaltrends.com

Kalliath, T., & Brough, P. (2008). Work–life balance: A review of the meaning of the balance construct. *Journal of Management & Organization, 14*(3), 323–327.

Kalliath, P., Kalliath, T., Chan, X. W., & Chan, C. (2019). Linking work–family enrichment to job satisfaction through job well-being and family support: A moderated mediation analysis of social workers across India. *The British Journal of Social Work, 49*(1), 234–255.

Kashy, D. A., & Kenny, D. A. (2000). The analysis of data from dyads and groups. In *Handbook of research methods in social and personality psychology* (pp. 451–477). Cambridge, UK: Cambridge University Press.

Keyes, C. L., Shmotkin, D., & Ryff, C. D. (2002). Optimizing well-being: The empirical encounter of two traditions. *Journal of Personality and Social Psychology, 82*(6), 1007.

Korabik, K., Aycan, Z., & Ayman, R. (2017). *The work–family interface in global context*. Abingdon, UK: Taylor & Francis.

Kossek, E. E., Pichler, S., Bodner, T., & Hammer, L. B. (2011). Workplace social support and work–family conflict: A meta-analysis clarifying the influence of general and work–family-specific supervisor and organizational support. *Personnel Psychology, 64*(2), 289–313.

Kraak, J. M., Russo, M., & Jiménez, A. (2018). Work-life balance psychological contract perceptions for older workers. *Personnel Review, 47*(6), 1198–1214.

Lewis, J. (1992). Gender and the development of welfare regimes. *Journal of European Social Policy, 2*(3), 159–173.

Lu, C., Lu, J.-J., Du, D., & Brough, P. (2016). Crossover effects of work-family conflict among Chinese couples. *Journal of Managerial Psychology, 31*(1), 235–250.

Luthans, F., Avolio, B. J., Avey, J. B., & Norman, S. M. (2007a). Positive psychological capital: Measurement and relationship with performance and satisfaction. *Personnel Psychology, 60*(3), 541–572.

Luthans, F., Avolio, B. J., Avey, J. B., & Norman, S. M. (2007b). Positive psychological capital: Measurement and relationship with performance and satisfaction. *Personnel Psychology, 60*(3), 541–572.

Luthans, F., Vogelgesang, G. R., & Lester, P. B. (2006). Developing the psychological capital of resiliency. *Human Resource Development Review, 5*(1), 25–44.

Luthans, F., Youssef, C. M., & Avolio, B. J. (2007). Psychological capital: Investing and developing positive organizational behavior. *Positive Organizational Behavior, 1*(2), 9–24.

Lyubomirsky, S., King, L., & Diener, E. (2005). The benefits of frequent positive affect: Does happiness lead to success? *Psychological Bulletin, 131*(6), 803.

Masten, A. S., & Reed, M.-G. J. (2002). Resilience in development. In *Handbook of positive psychology* (pp. 74–88). Oxford, UK: Oxford University Press.

Michel, J. S., Kotrba, L. M., Mitchelson, J. K., Clark, M. A., & Baltes, B. B. (2011). Antecedents of work–family conflict: A meta-analytic review. *Journal of Organizational Behavior, 32*(5), 689–725.

Mishra, P., Bhatnagar, J., Gupta, R., & Wadsworth, S. M. (2017). How work–family enrichment influence innovative work behavior: Role of psychological capital and supervisory support. *Journal of Management & Organization, 25*(1), 58–80.

Munn, S. L., & Chaudhuri, S. (2016). Work–life balance: A cross-cultural review of dual-earner couples in India and the United States. *Advances in Developing Human Resources*, *18*(1), 54–68.

Parasuraman, S., & Greenhaus, J. H. (2002). Toward reducing some critical gaps in work–family research. *Human Resource Management Review*, *12*(3), 299–312.

Park, Y., & Fritz, C. (2015). Spousal recovery support, recovery experiences, and life satisfaction crossover among dual-earner couples. *Journal of Applied Psychology*, *100*(2), 557.

Powell, G. N., Francesco, A. M., & Ling, Y. (2009). Toward culture-sensitive theories of the work–family interface. *Journal of Organizational Behavior*, *30*(5), 597–616.

Radcliffe, L. S., & Cassell, C. (2014). Resolving couples' work–family conflicts: The complexity of decision making and the introduction of a new framework. *Human Relations*, *67*(7), 793–819.

Rajadhyaksha, U., Ramadoss, K., Pestonjee, D. M., & Pandey, S. (2013). Work-family conflict in India: test of a causal model. In *Stress and work: Perspectives on understanding and managing stress*, (129), Thousand Oaks, CA, USA: SAGE.

Rajan-Rankin, S. (2016). Paternalism and the paradox of work–life balance: Discourse and practice. *Community, Work & Family*, *19*(2), 227–241. [Online]. Available: https://doi.org/10.1080/13668803.2016.1134131

Ramu, G. N. (1987). Indian husbands: Their role perceptions and performance in single-and dual-earner families. *Journal of Marriage and the Family*, *49*(4), 903–915.

Rout, U. R., Lewis, S., & Kagan, C. (1999). Work and family roles: Indian career women in India and the West. *Bulletin (Centre for Women's Development Studies)*, *6*(1), 91–105.

Ryff, C., Almeida, D. M., Ayanian, J., Carr, D. S., Cleary, P. D., Coe, C., & Williams, D. (2017). Midlife in the United States (MIDUS 2), 2004-2006. *Ann Arbor, MI, USA: Inter-University Consortium for Political and Social Research [Distributor]*, 11–20.

Schooreel, T., & Verbruggen, M. (2016). Use of family-friendly work arrangements and work–family conflict: Crossover effects in dual-earner couples. *Journal of Occupational Health Psychology*, *21*(1), 119.

Sekaran, U. (1992). Middle-class dual-earner families and their support systems in urban India.

Shukla, P., & Purani, K. (2012). Comparing the importance of luxury value perceptions in cross-national contexts. *Journal of Business Research*, *65*(10), 1417–1424.

Skeppner, E., Windahl, T., Andersson, S.-O., & Fugl-Meyer, K. S. (2008). Treatment-seeking, aspects of sexual activity and life satisfaction in men with laser-treated penile carcinoma. *European Urology*, *54*(3), 631–639.

Snyder, C. R. (2000). The past and possible futures of hope. *Journal of Social and Clinical Psychology*, *19*(1), 11–28.

Spector, P. E., Cooper, C. L., Poelmans, S., Allen, T. D., O'driscoll, M., Sanchez, Sui O. L., Dewe P., Hart P., & Lu, L. (2004). A cross-national comparative study of work-family stressors, working hours, and well-being: China and Latin America versus the Anglo world. *Personnel Psychology*, *57*(1), 119–142.

Sprung, J. M., & Jex, S. M. (2017). All in the family: Work–family enrichment and crossover among farm couples. *Journal of Occupational Health Psychology*, *22*(2), 218.

Stas, L., Kenny, D. A., Mayer, A., & Loeys, T. (2018). Giving dyadic data analysis away: A user-friendly app for actor–partner interdependence models. *Personal Relationships*, *25*(1), 103–119.

Tian, L., Chen, H., Zhu, L., Tang, D., Huebner, E. S., Yang, Y., & Yang, H. (2017). Crossover of weekly work engagement among dual-working couples. *Journal of Business and Psychology*, *32*(4), 441–453.

Valcour, M. (2007). Work-based resources as moderators of the relationship between work hours and satisfaction with work-family balance. *Journal of Applied Psychology*, *92*(6), 1512.

Wach, D., Stephan, U., & Gorgievski, M. (2016). More than money: Developing an integrative multi-factorial measure of entrepreneurial success. *International Small Business Journal*, *34*(8), 1098–1121.

Westman, M. (2001). Stress and strain crossover. *Human Relations*, *54*(6), 717–751.

Westman, M., & Etzion, D. L. (2005). The crossover of work-family conflict from one spouse to the other 1. *Journal of Applied Social Psychology*, *35*(9), 1936–1957.

Westman, M., & Vinokur, A. D. (1998). Unraveling the relationship of distress levels within couples: Common stressors, empathic reactions, or crossover via social interaction? *Human Relations*, *51*(2), 137–156.

Yang, N., Chen, C. C., Choi, J., & Zou, Y. (2000). Sources of work-family conflict: A Sino-US comparison of the effects of work and family demands. *Academy of Management Journal*, *43*(1), 113–123.

Zhang, M., Foley, S., & Yang, B. (2013). Work–family conflict among Chinese married couples: Testing spillover and crossover effects. *The International Journal of Human Resource Management*, *24*(17), 3213–3231.

CHAPTER 23

Validation of Resilience at Work (RAW) Scale: An Insight from an Indian Sample

POOJA GARG[1*] and HAN KI-SOON[2]

[1]*Department of Humanities and Social Sciences, Indian Institute of Technology Roorkee, Roorkee, Uttarakhand, India*

[2]*Incheon National University, Incheon, South Korea*

[*]*Corresponding author. E-mail: pooja.garg@hs.iitr.ac.in or gargpdhs@gmail.com*

ABSTRACT

The term "resilience" has its roots from positive psychology and in true essence, signifies "good life," for which every individual desires, while surpassing every change. Nevertheless, the concept of resilience at work (RAW) is not novel. But exploring RAW in Indian context would carve a niche for Indian managers to cope up with the adversities of the environment and emerge to be more flexible and adaptable. The present research aims to validate resilience at work scale developed by Winwood and colleagues. The validation process has been conducted on a sample of 300 employees working in manufacturing industries. The data were subjected within the framework of Exploratory and Confirmatory Factor Analysis. The analysis yielded five dimensions in contrast to the original seven dimensions. The study provides a breakthrough evidence for the psychometric properties of resilience at workplace in Indian work setting.

23.1 INTRODUCTION

The present work environment has become increasingly challenging and uncertain due to technological advancements, competition, and globalization. Researchers have massively explored novel ways to sustain at global level with positive organizational outcomes and effectiveness. Still, it can be laid upon that a paradigm shift is required to enrich work environment while gauging and utilizing strengths of employees to perform effectively in volatile environment and also employees evolve as an optimal person.

To withstand such challenging and threatening situations, much more emphasis has to be paid on the human strengths in order to achieve optimal organizational functioning. In terms of human strengths, *resilience* is one such construct which has to be well researched in workplaces. As being, resilient make employees capable to overcome changes and adversity in workplace (Rossi et al., 2013).

The term resilience has been well explored in developmental and clinical psychology with enormous utility. And now research evidences are showing its promising implications for organizations as well (Paul & Garg, 2012a). It is understood that in the complex interaction between individual and the environment, resilience not only provides for the positive psychological capacity to rebound from occupational stress and crisis, but also for building an attitude to deal with the realities of business world. To be more explicit, to meet the global challenges, employees invest long working hours, with mounting pressure to do more with less and constant change. This yields to higher levels of stress, maximization of work errors, and diminished mental health (insomnia, depression, anxiety, and severe headaches). Although, it is a laid fact that no job is free from stress and workload. But this becomes imperative to gauge the strengths which transform employees with positive psychological capacities and help them not only to survive but also thrive and make positive contribution to organizations.

Nevertheless, it becomes essential to develop awareness among employees of their own capacities which are valuable in terms of human strengths and at the same time organizations also represent their concerns to accentuate employees' strengths, while increasing employee contribution and create human capital.

The school of thought can be extended with an opinion that employees do have certain strengths and it is a dire need of the present scenario to help employees make a conscious effort to gauge, develop, and manifest their own strengths and facilitate themselves to recover from adversity at work.

23.1.1 WHY RESILIENCE IS IMPORTANT AT WORK

Building on definitions of resilience, it can be deduced that resilience can be considered as a conduit to positively cope, adapt, and thrive in response to changing work circumstances and recover a state of equilibrium. Considerable attention has been offered to cope up with global crisis at organizational level. Such crisis has underpinned employees under tremendous work pressure and employees have involved themselves in a rigorous phase of transformation to recover from demands and challenges of work in a consistent and effective ways. In this rigorous phase of coping, adaptation and transformation from the paradoxes of workplace, employees have accrued a mechanical image for themselves and as a consequence have become completely disoriented with their personal identity, strengths, and weaknesses.

Against this background, it has become all-important for the organizations to facilitate resilience with a proactive approach. This entails not only to enhance executive functioning but also help them gauge, utilize, and manifest their strengths and weaknesses for their personal growth and well-being. As stated by Winwood and colleagues (2013), focus must be maintained on identifying elements of resilience that can be *consciously and deliberately considered and modified through appropriate skill training, that is, behaviors and strategies that are not limited by fixed genetic and personality factors, rather as a specific capacity within the work environment.*

23.1.2 DEFINING RESILIENCE

Resilience is a complex concept with numerous definitions. The term resilience has resulted in a plethora of conceptual and operational definitions. Central to most of these is the notion that resilience involves being able to withstand and overcome adversity and unpleasant or difficult events successfully and to be able to adapt to change and uncertainty. Interestingly, the term itself is derived from the Latin word *resilere* which means *jump*, or *bounce-back* and also incorporates some sense of *personal growth* as we use each adverse experience not just to enhance our coping but also to better manage the next one. Most definitions of resilience outline one of two perspectives on resilience as: (a) *reactive recovery* and (b) *stability after traumatic events,* such as natural disaster or bereavement (Maguire & Cartwright, 2008). Generally, resilience can be defined as, "*an individual's response and methods used to allow them to successfully navigate through or past an event perceived to be stressful".*

The concept of resilience has been explored in varied domains; however, it is largely explored in clinical and developmental psychology. There are myriad of studies conducted on resilience with psychopathological orientation. Early research has majorly studied resilience among children in the context of high-risk circumstances (McCubbin et al., 1985), parental mental illness (Garmezy, 1974), socioeconomic disadvantage (Rutter, 1971), maltreatment (Crittenden, 1985), survivors of disasters (Moskovitz, 1985), devastating life events (Brown et al., 1986), and patients suffering from chronic diseases like climacteric women (Gerdes, 1980). Very recently organizational research on resilience has been initiated which can be broadly categorized on the basis of the *capability* and *nature*. On the basis of *capability level,* resilience is studied either as a capacity of an individual (Caverley, 2005; Knight, 2007) or as the ability of an organization (Hamel & Välikangas, 2003; Bolton, 2004; Lengnick-Hall and Beck, 2005). However, two broad categories exist based on the *nature* of resilience—*trait* (Block and Block, 1980; Wagnild & Young, 1993; Block & Kremen, 1996; Luthar et al., 2000; Fredrickson et al., 2003) or *state* (Youssef & Luthans, 2007; Luthans et al., 2008; Luthans & Avolio, 2009; Toor & Ofor', 2010).

Wagnild and Young (1990), in their grounded-theory research, identified five characteristics which serve as the conceptual foundation for Resilience Scale (RS)-14. These characteristics includes: (1) *meaningfulness of life* or the realization that life has a purpose and recognition that; there is something for which to live; (2) *perseverance* or the act of persistence despite adversity or discouragement; (3) *self-reliance* or belief in oneself with a clear understanding of own capabilities and limitations; (4) *equanimity* is a balanced perspective for life and experiences, and might be viewed as sitting loose and taking what comes; thus moderating the extreme responses to adversity; and (5) *existential aloneness* is the realization that each person is unique and that while some experiences can be shared, others must be faced alone (Wagnild & Young, 1990, 1993). Wagnild and Young (1990) termed these five characteristics as resilience core.

Tugade et al. (2004) defined resilience as, "the flexibility in response to changing situational demands, and the ability to bounce back from negative emotional experiences." According to Neenan (2010), "resilience is a set of flexible cognitive, behavioral and emotional responses to acute or chronic adversities which can be unusual or common place." Mowbray (2010) defined the term resilience as, "the capacity to mobilise personal features that enable individuals, groups and communities (including controlled communities such as a workforce) to prevent, tolerate, overcome and be enhanced by adverse events and experiences."

The definition of resilience proposed by Luthans suggests that, "it is a recovery process in which one returns back to one's original state of equilibrium." Further, defining resiliency at work confers to behaviors leading to skill development, autonomy, and encouragement to stretch goals, which map on the underlying facets of, learning, adaptability, and networking.

Although in recent years, researches on resilience have expanded its scope from clinical and developmental foci to applications in occupational settings, and has already been reflected in the studies by Avey et al. (2009), King and Rothstein (2010), Lee et al. (2011), Lengnick-Hall and Beck (2011), and Luthans (2002). Recent work suggests that resilience comprises a more fruitful construct in organizational research, if conceptualized as an individual capability that can be developed though interactions between people and their environment (Fletcher & Sarkar, 2013; Lengnick-Hall et al., 2011; Luthans & Avolio, 2003).

Cárdenas-Jiménez and López-Díaz (2011) explained that resilience can be defined in at least four ways: (a) *personal ability or capacity*, (b) *positive characteristics of personality*, (c) *transactional process*, and/or (d) *coping capacity*.

Winwood et al. (2013), defined resilience as "an individual's capacity to manage the everyday stress of work and remain healthy, rebound and learn from unexpected setbacks and prepare for future challenges proactively." Key assets and resources underpinning resilience have been identified as strong social support networks, flexible coping styles, maintaining perspective, moving toward your goals and staying healthy (American Psychological Association [APA], 2010; Friborg et al., 2003; Grant & Kinman, 2012; Winwood et al., 2013).

Appendix A represents a sampling of some definitions of resilience. Some definitions of resilience refer to the construct as a capacity residing within individuals (Masten & Narayan, 2012) to maintain stable functioning in the face of highly stressful and traumatic events (Bonanno, 2012), and reflecting growth and positive changes after an adverse event (Maguen et al., 2006). In addition, highlighting adaptation versus growth in defining resilience, these definitions also reflect the distinction between resilience represented as a capacity residing within the individual (or team or organization) and resilience represented as the demonstration of adaptation on an important outcome.

The definition given by Winwood et al. (2013) comprises a proactive element which transmits a significant message to employees to be mindful on identifying and overcoming future challenges through conscious choice and decision, while facilitating recovery from occupational stress and losses. This demands a high degree of self-awareness regarding one's own personal values and strengths that we can in turn use to position ourselves as positively as we can in our employment. A whole-of-life approach is advocated with the strategies applied to both work and personal lives. Approaching issues only at work or at home makes no sense as problems can spill over from home to work or vice versa. Thus, investment in all of life's domains pays dividends.

Therefore, we can state that while some employees may have more resilient personal characteristics such as optimism and flexibility; employees can be encouraged to become aware, gauge, learn, and develop the thoughts and behaviors that underpin resilience.

23.1.3 MEASURES OF RESILIENCE

There are different scales available to measure resilience with different perspectives. Connor–Davidson Resilience Scale (CD-RISC); which was originally developed by Connor and

Davidson (2003) is used to measure resilience within the Post-Traumatic Stress Disorder and measure factors of resilience as: *personal competence, acceptance of change and secure relationships, trust/tolerance/strengthening effects of stress, control and spiritual influences.* Resilience Scale for Adults (RSA), developed by Friborg et al. (2003, 2005), examines the intra and interpersonal protective factors that promote adaptation to adversity. These factors are: *personal competence, social competence, social support, and family coherence.* The Brief Resilience Scale (BRS) developed by Smith et al. (2008) provides some key insights for individuals with health-related stress. There are different scales available to measure resilience for different age groups. Resiliency Scales for Children and Adolescents is used to measure resilience of children and youth from 9 to 18 years (Prince-Embury, 2008). Child and Youth Resilience Measure is available for youth aged 12 to 23 years old (Ungar & Liebenberg, 2009). Similarly, Assessing Developmental Strengths Questionnaires is also an instrument to measure resilience for the same age group (Donnon & Hammond, 2007). Paul and Garg (2012a) validated the RS-14 scale by Wagnild and Young (1990) and revealed the four factor structure of resilience as: *self-reliance, perseverance, meaningfulness of life, and self-efficacy* for working population in India. The items of the scale indicate clinical and developmental orientations; therefore it is pertinent to have a valid and reliable tool to measure this psychological capacity in context of working individuals. Appendix B outlines a set of resilience scales of varying popularity, applicability, and validity.

23.1.4 AIM OF THE STUDY

Literature suggests that there are evidences to measure resilience from different perspectives and which leads to positive workplace outcomes. A lacuna has been observed in the management discourse, which reflects to consider and measure the levels of resilience within working environment and help employees to evolve as a *good human being* and lead *a good life.* It is pertinent to have a valid and reliable tool to measure this psychological capacity in the context of working individuals. The study delineates the crucial gap in the current literature and aims to study the factor structure of RAW Scale, developed by Winwood et al. (2013), with working population in Indian context.

23.2 METHOD

23.2.1 SAMPLE AND PROCEDURE

The participants in the study were employees working in a manufacturing industry in India. Questionnaires and the accompanying cover letter stating the voluntary purpose of the study and an assurance of the confidentiality of responses were distributed to a total of 450 employees. Out of 450 questionnaires, 400 were returned, giving a response rate of 66.6 percent. Of these 100 incomplete questionnaires were rejected, leaving 300 for further analysis. The participants were chosen using convenient sampling. The demographic characteristics of the study participants are shown in Table 23.1. The respondents were encouraged to participate in the survey within two weeks. The sample comprised 72.7% males and 27.3% females; 77.3% were married and 22.7% were unmarried; 30.3% of employees were in the age bracket of 25–40, 56.7% of employees were in the bracket of 41–56 years of age and 13% of employees were in the bracket of 57–72 years of age.

23.2.2 MEASURES

To measure resilience, *RAW Scale* has been used which is developed by Winwood et al. (2013). It is a 20-item self-report instrument that measures individual workplace resilience. It has seven subscale components, all of which contribute to overall individual resilience. The subscales are:

1. ***Living Authentically:*** Knowing and holding onto personal values, deploying your strengths and having a good level of emotional awareness and regulation. *Sample item as: I know my personal strengths and make sure I use them regularly in my work.*
2. ***Finding Your Calling:*** Seeking work that has purpose, gives a sense of belonging and fits well with one's core values and beliefs. *Sample item as: The work that I do fits well with my personal values and beliefs.*
3. ***Maintaining Perspective:*** Having the capacity to reframe setbacks, maintain a solution-focus and manage negativity. *Sample item as: Nothing at work ever really fazes me for long.*
4. ***Managing Stress:*** Employing work and life routines that help manage everyday stressors, maintain work-life balance and ensure time for relaxation. *Sample item as: I have developed some reliable ways to deal with the stress of challenging events at work.*
5. ***Interacting Cooperatively:*** Seeking feedback, advice and support and providing support to others. *Sample item as: I believe in giving help to my work colleagues, as well as asking for it.*
6. ***Staying Healthy:*** Maintaining a good level of physical fitness and a healthy diet. *Sample item as: I have a good level of physical fitness.*
7. ***Building Networks:*** Developing and maintaining personal support networks. *Sample item as: I have a strong and reliable network of supportive colleagues at work.*

Further, all the items relating to the subdimensions were measured on a seven-point scale ranging from *1= Strongly Disagree to 7= Strongly Agree*. Cronbach alpha coefficient of reliability for the RAW scale total has been reported to be 0.86. Also for the subscales the values vary from 0.60 to 0.87.

TABLE 23.1 Demographic Characteristics of the Study Participants

Demographic Variables	Frequency	Percent
Gender		
• Male	218	72.7
• Female	82	27.3
Marital Status		
• Married	232	77.3
• Unmarried	68	22.7
Work Experience		
• Up to 1–15 years	143	47.7
• Up to 16–30 years	133	44.3
• Up to 31–46	23	7.7
Age		
• Up to 25–35 years	91	30.3
• Up to 36–46 years	170	56.7
• Up to 47–57 years	39	13.0

23.3 ANALYSIS

The validation of the scale has been done within the framework of using SPSS-AMOS v. 20. The core analysis consisted of descriptive statistics of the study variables (Table 23.2), exploratory factor analysis, and confirmatory factor analysis (CFA). Further, the analysis has been extended to convergent and discriminant validity of RAW scale.

TABLE 23.2 Descriptive Statistics of the Study Variables ($n = 300$)

Variables	Sum	Mean	SD	SEM
LA	2864.00	9.54	3.80	0.21
FOC	3766.00	12.55	4.85	0.28
MP	2795.00	9.31	3.70	0.21
MS	3888.00	12.96	5.01	0.28
IC	1905.00	6.35	2.66	0.15
SH	1894.00	6.31	2.47	0.14
BN	1944.00	6.48	2.57	0.14

Note: LA–Living authentically; FOC–Finding one's calling; MP–Managing Perspective;

MS—Managing Stress; IC—Interacting cooperatively; SH—staying healthy; BN—Building network.

The Keiser–Meyer–Olkin test of sampling adequacy has been reported to be acceptable as 0.95. The sample is adequate if the value of KMO is greater than 0.5 (Field, 2009). The KMO for the present study is 0.746 which exceeds the recommended value of 0.60 (Ryan and Caltabiano, 2009) and shows the adequacy of the sample size. Whereas, Bartlett Test of Sphericity gave a χ^2 value of 5849.68 ($p < 0.0001$), df = 190. The 20-item RAW scale was subjected to principal component analysis with Kaiser's Varimax Rotation technique with orthogonal rotation.

23.3.1 PRINCIPAL COMPONENT ANALYSIS (PCA)

Since the resilience scale has been factor analyzed within the framework of exploratory factor analysis with oblique (direct oblimin) rotation that set the inter-factor correlations to be high. The rotation extracted seven factors and explained a total of 85.37% of variance of the RAW scale (Table 23.3). Item-factor loadings below 0.40 were excluded from the scale and thus, only 15 items have been retained. To mention, when items do not load on a dimension (that is, has loadings less than 0.40) on all dimensions or factors, it may indicate that the item/s is unreliable and as a result may be a candidate for deletion. The item loadings for the extracted factors, eigenvalue, and variance explained by each factor are presented in Table 23.4. Communalities that give the proportion of variance for each of the variables are also entered in the last column of the matrix in Table 23.3. Communalities for all the 20 items ranged between 0.37 and 0.66. The RAW scale demonstrated good internal consistency, with the cronbach alpha value for overall scale as 0.90. Whereas, for the subdimensions, the reliability ranged from 0.85 to 0.90, respectively.

23.3.2 CONFIRMATORY FACTOR ANALYSIS (CFA)

The model of resilience arising from the exploratory factor analysis was further examined with CFA, using AMOS v. 20 software. CFA for testing factorial validity was evaluated by using the CMIN/df statistics and several other recommended practical fit indices (Byrne, 2001) such as: Goodness of Fit Index (GFI), Root Mean Squared Error of Approximation (RMSEA), Normed Fit Index (NFI), and Comparative Fit Index (CFI).

CFA was used in order to compare the newly obtained five factor structure (15 items) model of RAW scale with the original seven factor structure (20 items) model. It was found that the five factor structure (15 items) model showed better fit in comparison to the seven factor structure (20 items) model (Figures 23.1

TABLE 23.3 Principal Component Analysis

Items	Components							
	1	2	3	4	5	6	7	h2
LA1—I have important core values that I hold fast to in my work life	0.088	**0.553**	−0.040	0.019	0.068	0.017	−0.552	0.37
LA2—I know my personal strengths and make sure I use them regularly in my work	−0.030	**0.476**	0.012	0.474	−0.004	0.354	0.229	0.53
LA3—I am able to change my mood at work when I need to	−0.042	**0.781**	0.059	0.071	−0.132	0.088	0.010	0.64
FYC1—The work that I do helps to fulfill my sense of purpose in life	−0.355	**0.467**	0.239	−0.202	−0.069	−0.071	−0.015	0.45
FYC2—My work place is somewhere where I feel that I belong.	0.133	**0.788**	0.113	−0.094	0.017	−0.045	0.028	0.66
FYC3—The work that I do fits well with my personal values and beliefs	−0.158	.333	**0.530**	0.097	−0.043	0.105	−0.135	0.39
FYC4—Generally I appreciate what I have in my work environment	−0.141	−0.257	**0.447**	0.143	−0.523	0.234	0.040	0.63
MP1—When things go wrong at work, it usually tends to overshadow the other parts of my life	−0.144	0.131	−0.010	−0.625	0.184	0.015	.045	0.46
MP2—Nothing at work ever really "fazes me" for long	−0.011	0.121	−0.385	−0.149	0.117	−0.103	−0.089	0.58
MP3—Negative people at work tend to pull me down	0.032	−0.084	−0.509	0.116	−0.119	0.117	.081	0.66
MS1—I make sure I take breaks to maintain my strength and energy when I am working hard	**0.532**	−0.041	−0.120	−0.326	−0.115	0.380	.109	0.57
MS2—I have developed some reliable ways to relax when I am under pressure at work	**0.670**	0.048	.230	0.147	0.114	0.030	.098	0.54
MS3—I have developed some reliable ways to deal with the stress of challenging events at work	0.404	0.308	−0.228	0.163	−0.146	−0.003	.011	0.26
MS4—I am careful to ensure my work does not dominate my personal life	**0.610**	−0.149	−0.121	0.005	−0.023	−0.055	−0.026	0.41
IC1—I often ask for feedback so that I can improve my work performance	**0.661**	0.025	−0.017	−0.662	−0.010	0.109	0.155	0.49
IC2—I believe in giving help to my work colleagues, as well as asking for it	**0.466**	−0.133	−0.030	−0.044	0.313	0.156	0.307	0.45
SH1—I have a good level of physical fitness	−0.118	−0.101	0.027	−0.025	−0.448	−0.507	0.340	0.59
SH2 —I am careful about eating well and healthily	−0.084	−0.154	0.068	−0.038	0.339	−0.062	0.017	0.58
BN1—I have friends at work I can rely on to support me when I need it	0.147	0.017	−0.061	0.−449	0.272	−0.165	**0.441**	0.40
BN2—I have a strong and reliable network of supportive colleagues at work.	0.203	0.245	−0.058	0.138	0.059	−0.030	**0.709**	0.63
Eigenvalues	12.6	1.27	0.72	0.68	0.66	0.54	0.49	
Percent of total variance explained	63.41	6.37	3.63	3.44	3.31	2.71	2.47	

Note: The significance of bold in the table indicates that the items have been retained for further analysis.

and 23.2). On the basis of Table 23.4, the five-factor model, reported the values of certain practical fit indices as: CMIN/df = 4.20, $p < 0.01$, RMR= 0.07, GFI = 0.86, CFI= 0.92, NFI= 0.90, TLI = 0.90, RMSEA = 0.08). In addition, for seven-factor model the values reported were as: CMIN/df = 5.40, $p < 0.01$, RMR = 0.09, GFI= 0.77, CFI= 0.84, NFI= 0.72, TLI = 0.72, RMSEA = 0.01), for Indian population. Therefore, the newly obtained five-factor structure model of RAW scale demonstrated superiority over the seven-factor model.

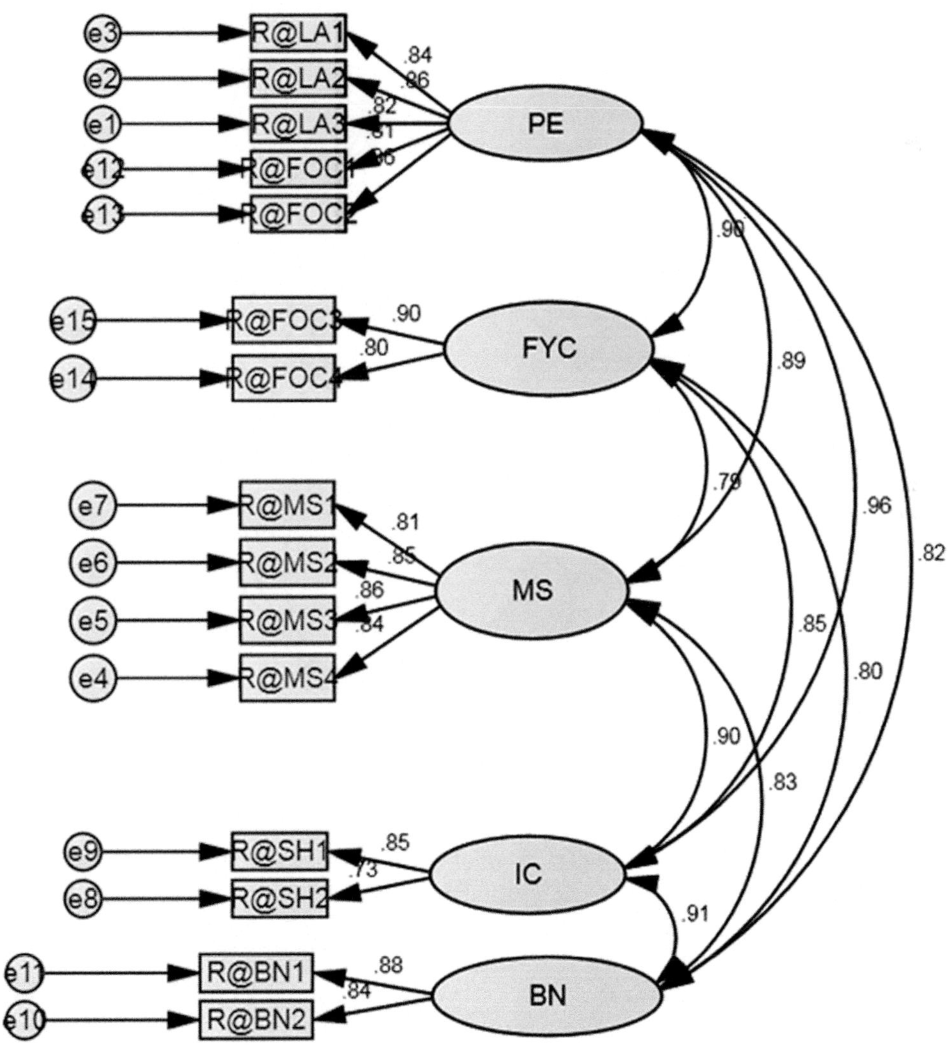

FIGURE 23.1 Confirmatory factor analysis for 15-item scale.

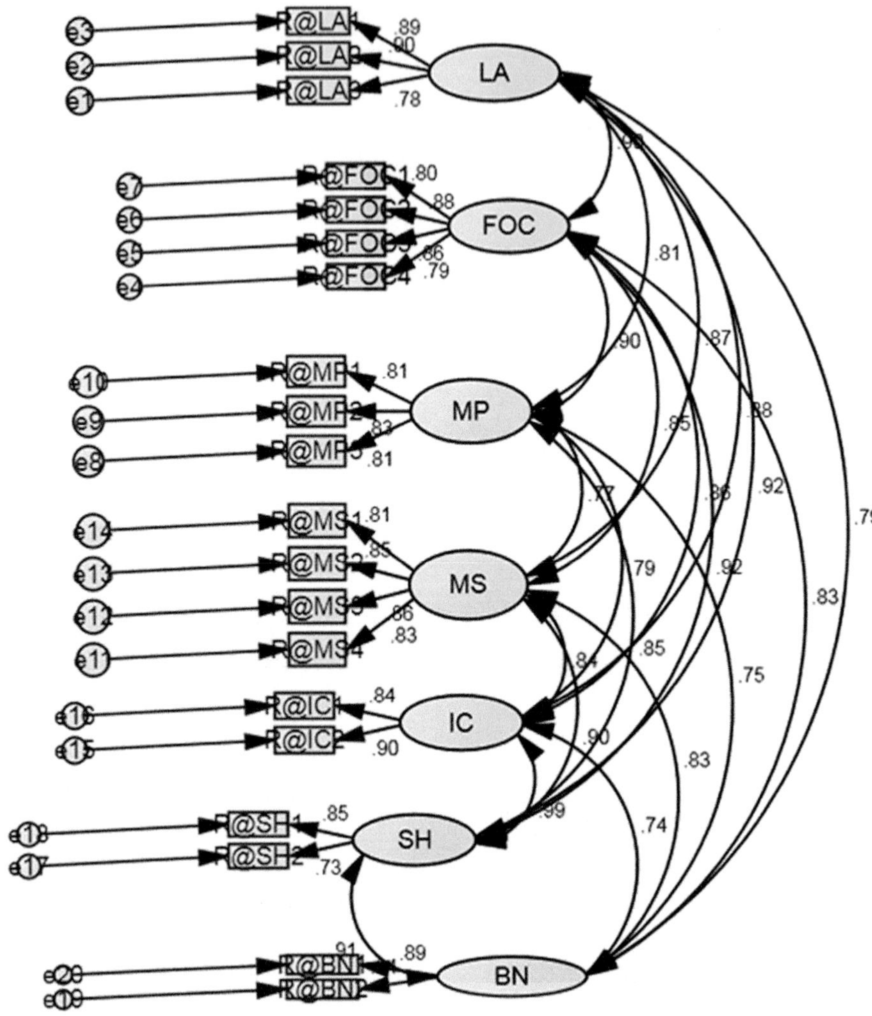

FIGURE 23.2 Confirmatory factor analysis for 20-item scale.

TABLE 23.4 Confirmatory Factor Analysis in Order to Determine the Structure of RAW Scale

Scale	CMIN/ df	RMR	GFI	CFI	NFI	TLI	RMSEA
*M1	5.40	0.09	0.77	0.84	0.72	0.72	0.01
**M2	4.20	0.07	0.86	0.92	0.90	0.90	0.08

Notes: *M1-20-item RAW Scale; **M2-15-item RAW Scale

TABLE 23.5 Results of Convergent and Discriminant Validity of RAW Scale

Values	IC	LA	FOC	MS	BN
CR	0.86	0.92	0.84	0.90	0.85
AVE	0.81	0.81	0.81	0.79	0.74
MSV	0.75	0.70	0.72	0.70	0.69

Note: CR: Composite reliability, AVE: Average variance extracted, MSV: Maximum shared variance

Moreover, the analyses were performed for convergent and discriminant validity for the 15-factor solution. The composite reliabilities (CR), the average variance extracted (AVE), and the maximum shared variance (MSV) are listed in Table 23.5, indicating adequate reliability and convergent validity of the measurement model.

23.4 DISCUSSION

The analysis of factor composition of RAW scale revealed a seven-factor matrix. In the newly obtained five-factor structure, only 15 items have been retained. Closer examination of the factor descriptors outlined that Factor *1* included item *number MS1—I make sure I take breaks to maintain my strength and energy when I'm working hard, MS2—I have developed some reliable ways to relax when I'm under pressure at work, MS3—I have developed some reliable ways to deal with the stress of challenging events at work and MS4—I am careful to ensure my work does not dominate my personal life for managing stress* and item number *IC1—I often ask for feedback so that I can improve my work performance and IC2—I believe in giving help to my work colleagues, as well as asking for it.* These items connotes to dealing with work stress in an adaptable fashion, while reframing stressful situations in a positive way and seeking and providing support to others. This corresponds to developing capacity to resolve intra-personal and interpersonal conflicts in workplace and develop strength to work effectively in team. This also construes to accumulate personal support via positive feedback mechanism, which enable employees to maintain a solution focus for a smoother functioning. The items also corresponds to the original dimensions of scale; hence no alteration is required in the labeling the respective dimensions.

Factor 2 is loaded with items *LA1—I have important core values that I hold fast to in my work-life, LA2—I know my personal strengths and make sure I use them, LA3—I am able to change my mood at work when I need to*, for living authentically and items *FYC1—The work that I do helps to fulfill my sense of purpose in life and FYC2—My work place is somewhere where I feel that I belong*, for finding your calling. It was found that the items are related to taking control of one's own life by making positive choices and setting goals. And developing self-awareness, that is, understanding of one's own strengths and weaknesses and knowing one's own limitations. Based on these reflections, the five items have been conjoined and has been named as p*ersonal empowerment*. As per the structure matrix derived from PCA, personal empowerment can be attained by having a good level of emotional regulation and awareness. Personal empowerment when gauged fosters confident sense making (Mallak and Yildz, 2016); implying that employees possess the ability to quickly filter out unnecessary signals and information and focus on the relevant stimuli for appropriate decision making and action. This extends to the notion of *bricolage,* (as used by the US Army and outdoor survival agencies to communicate the need for being mindful in the face of danger) and manifesting RAW benefits from fashioning solutions creatively to address the situation. And when confronted with chaotic, extreme, and dangerous situations, the resilient individual takes intelligent risks and realizes there is time to stop, think, observe, and plan. (Mallak and Yildz, 2016).

Factor 3 includes items *FYC3—The work that I do fits well with my personal values and beliefs and FYC4—Generally I appreciate what I have in my work environment*, for finding your calling. This entails respectful interactions with oneself, rooted in trust, honesty and self-respect, and also appreciating one's own perspective. This would further augment mindfulness, along with increased capacity of self-regulation and positive self-core evaluations; employee's subconscious and fundamental evaluations about themselves, and their own abilities and their control. In essence, building and capitalizing on one's own strengths to foster effectiveness. This further helps to exhibit intuitive response to any situation which can create options rather than constraints and invest in personal integration and behavior especially in the face of unexpected events through a broad repertoire of perceptions and for sense making and decision making.

Table 23.4 also indicates that no items have been loaded on *factor 5 and 6*, for the items having loadings for less than 0.40 have been subjected to deletion for the analysis. Turner et al. (2016) adapted the RAW scale for use in university setting. The scale proved to be valid and reliable measure of student resilience, with the major grouping of the interacting cooperatively (IC) and living authentically (LA) subscales into one factor. Whereas, *factor 7* include items *BN1—I have friends at work I can rely on to support me when I need it* and *BN2—I have a strong and reliable network of supportive colleagues at work,* for building network. This indicates about being effective and satisfied in social environment, being able to communicate adequately, especially when dealing with or preventing conflicts. Further, this also reflects an employee's systematic understanding of sharing organizational goals and performing team roles effectively and in turn guiding and mentoring actions and decisions of co-workers amicably. Such foundation for exchanges endures and grows into mutually beneficial and multifaceted long-term partnerships. Finally, enable employees to build bridges that cross conventional internal and external boundaries and forge a network of support and resources.

23.5 CONCLUSION

The result of this study differs with the original factor analysis of RAW scale. The items loaded on seven factors are in contrast to the underlying factor as described by Winwood et al. (2013). Based on the results, the factor structure of RAW scale implies that resilient workers invest in their general health, self-care, and stress management techniques. They are also adaptable when unexpected changes and setbacks occur, responding in a way that focuses on growth and solutions rather than noncoping. They are proactive regarding possible future challenges through understanding their values and strengths and knowing how to position themselves at work to maximize personal performance and mental well-being and their coping ability.

The study indicates that resilience at workplace can be considered as augmentation of a conscious self-regulation that brings about the possibility of further flourishing. This entails that an individual can be mindful about prioritizing those behaviors, thoughts, and feelings that facilitate contentment and an adequate pathway to a good life, encompassing psychological strengths such as meanings and virtues.

23.5.1 LIMITATIONS AND FUTURE IMPLICATIONS

The primary limitation of this chapter relates to the sample. All the participants in the study belong to the manufacturing industry. Consequently, it limits the conclusion for the wider Indian context, as different results may be obtained when used with other services like, IT, or any other industry. Further, the study sample had a greater representation of males to females which also limit the generalization of the results. Also, a social desirability bias may have had an impact on responses of the participants.

Despite these limitations, the present study offers significant implications for both resilience research and practice. Since very limited studies have talked about resilience in the context of Indian population. This study provides better insights into the factor structure of RAW scale and empirical evidences for its future use.

In practice, organizations can pay attention toward building resilience of its workforce through well-designed training and intervention programs focused on the revealed five factors of resilience. Human resource interventions can be applied at recruitment stage itself to get more resilient employees on board by making it as a recruitment criterion and using RAW scale to assess the resilience levels of prospective employees. This could also help organizations to retain their employees in those domains where the job is not very lucrative in terms of working environment (like manufacturing) or the job is very demanding (like sales and time-bound projects). It may also be helpful for the industries marked by high attrition rate like IT, ITES, and call centers. The validated RAW scale has the ability to demonstrate a more comprehensive and integrated model of resiliency. In sum, the scale can be used to earmark the improvements in productivity, performance, employee well-being, organizational commitment, and employee retention, and other desirable organizational outcomes. The scale can be used to develop enduring resilient approaches, such as: self- management and self-awareness, to recognize stress triggers or warning signs and take active steps to minimize their impact at workplace. The scale can also be used to emphasize the need for sustained emotional resilience and the development of creative, evidence-based strategies to enhance the competencies of employees that underpin it at all stages of professional development and value professional identity.

The scale can also be used as a key moderator along with resilience-building programs, and are likely to be most effective and provide important practical implications for conducting these programs in organizational settings in the future. For example, those considering implementing resilience-building programs should attempt to identify who will benefit from the development of protective factors (such as increased social support, enhanced self-efficacy, and more positive appraisals of potentially stressful experiences.) and carefully consider designing programs in order to produce optimal and lasting program effects. The scale can also have a strong implication in military setting in conjunction with resilience-building programs for military personnel and their family members and identify areas of improvement (Vanhove et al., 2015). The scale can be used to encourage employees to identify their own coping strategies by raising their consciousness and can be taught how to deal with the stresses of deployment and how families can be taught coping skills to help them decompress at home. In nutshell, the scale can be used as a *toolkit* for various specific factors in military settings such as: at *individual level* (dealing with positive coping and positive effect, realism, improved emotional regulation and altruism in the context of post-traumatic experiences), at *family level* (enhancing emotional ties, communication, based *on socio-emotional culture* and nurturing military families adaptable based on *strength-based culture*. The scale can also have future implication with reference to *unit level factors* as: positive command climate, teamwork and unit cohesion and last, improve well-being index of military personnel based on enhancing their community living on the basis of belongingness, cohesion, connectedness and collective efficacy (Meredith et al., 2011).

Based on the above discussion, we can say that the one of the primary purposes of this study is to provide a foundation for moving the organizational resilience literature forward in a unified and practical manner, specifically in the Indian context. The scale has the potential to be a new mandate for HRM as an agent of continuous transformation, shaping processes and a culture that together improve an organization's capacity for change.

KEYWORDS

- **resilience**
- **scale**
- **measurement**
- **validation**

REFERENCES

American Psychological Association (APA). (2010). The road to resilience. American Psychological Association. Retrieved October 12, 2015, from http://www.apa.org/helpcenter/road-resilience.aspx

Avey, J. B., Luthans, F., & Jensen, S. (2009). Psychological capital: A positive resource for combating stress and turnover. *Human Resource Management, 48*, 677–693.

Bartone, P. (2007). Test–retest reliability of the dispositional resilience scale-15, a brief hardiness scale. *Psychological Reports, 101*, 943–944.

Block, J. H., & Block, J. (1980). The role of ego-control and ego-resiliency in the organization of behavior. In W. A. Collins (Ed.), *Development of cognition, affect and social relations: The Minnesota symposia on child psychology* (Vol. 13, pp. 39-101). Hillsdale, NJ: Erlbaum.

Block, J., & Kremen A M. (1996). IQ and Ego-resiliency: Conceptual and empirical connections and separateness. *Journal of Personality and Social Psychology, 70*, 349–361.

Bolton, D. (2004). Change, coping and context in the resilient organizations, *Mt Eliza Business Review, 7*(1), 56–66.

Bonanno, G. A. (2004). Loss, trauma, and human resilience: Have we underestimated the human capacity to thrive after extremely aversive events? *American Psychologist, 59*, 20–28.

Bonanno, G. A. (2012). Uses and abuses of the resilience construct: Loss, trauma, and health-related adversities. *Social Science & Medicine, 74*, 753–756.

Britt, T. W., Robert, W. S., Sinclair, R. R., Grossman, M. R., & Klieger, D. M. (2016). How much do we really know about employee resilience? *Industrial and Organizational Psychology, 9*(2), 378–404.

Brown, G. W., Harris, T. O., & Bifulco, A. (1986). The long term effects of early loss of parent. In M. Rutter, C. E. Izard and P. B. Read (Eds.), *Depression in Young People*, New York: Guilford Press.

Byrne, B. M. (2001). *Multivariate applications book series. Structural equation modeling with AMOS: Basic concepts, applications, and programming.* Lawrence Erlbaum Associates Publishers.

Cárdenas-Jiménez, A., & López-Díaz, A. (2011). Resiliencia en la vejez. *Revista de salud pública, 13*(3), 528–540.

Caverley, N. (2005). Civil service resilience and coping. *International Journal of Public Sector Management, 18*(5), 410–413.

Connor, K. M., & Davidson, J. R. (2003). Development of a new resilience scale: The Connor–Davidson resilience scale (CD-RISC). *Depression and Anxiety, 18*(2), 76–82.

Crittenden, P. M. (1985). Maltreated infants: Vulnerability and resilience. *Journal Child Psychology and Psychiatry and Allied Disciplines, 26*(1), 85–96.

Donnon, T., & Hammond, W. (2007). A psychometric assessment of the self-reported youth resiliency: Assessing developmental strengths questionnaire. *Psychological Reports, 100*, 963–978.

Field, A. (2009). *Discovering Statistics using SPSS* (3rd ed.), London, Thousand Oaks: New Delhi, Singapore: Sage Publications.

Fletcher, D., & Sarkar, M. (2013). Psychological resilience: A review and critique of definitions, concepts, and theory. *European Psychologist, 18*(1), 12–23.

Fredrickson, B. L., Tugade, M. M., Waugh, C. E., & Larkin, G. R. (2003). What good are positive emotions in crises? A prospective study of resilience and emotions following the terrorist attacks on the United States on September 11, 2001. *Journal of Personality and Social Psychology, 84*(2), 365–376.

Friborg, O., Barlaug, D., Martinussen, M., Rosenvinge, J., & Hjemdal, O. (2005). Resilience in relation to personality and intelligence. *International Journal of Methods in Psychiatric Research, 14*(1), 29–42.

Friborg, O., Hjemdal, O., Rosenvinge, J., & Martinussen, M. (2003). A new rating scale for adult resilience: What are the central prospective resources behind healthy adjustment?. *International Journal of Methods in Psychiatric Research, 12*(2), 65–76.

Garmezy, N. (1974). *The study of competence in children at risk for severe psychopathology.* In E. J. Anthony & C. Koupernik (Eds.) *The child in his family: Children at psychiatric risk.* John Wiley & Sons.

Gerdes, L. C. (1980). Vulnerability and Resilience in Climacteric Women. *South African Medical Journal, 58* (19), 763–766.

Grant, L., & Kinman, G. (2012). Enhancing wellbeing in social work students: Building resilience in the next generation. *Social Work Education: The International Journal, 31*(5), 605–621.

Hamel, G., & Välikangas, L. (2003). The quest for resilience. *Harvard Business Review, 81* (9), 52–56.

King, G. A. & Rothstein, M. G. (2010). Resilience and leadership: The self-management of failure. In M. G. Rothstein & R. J. Burke (Eds.), *Self-management and leadership development* (pp. 361–394). Cheltenham, UK: Edward Elgar.

Klohnen, E. C. (1996). Conceptual analysis and measurement of the construct of ego-resiliency. *Journal of Personality and Social Psychology, 70*(5), 1067–1079.

Knight, C. (2007). A resilience framework: Perspectives for educators. *Health Education, 107*(6), 543–555.

Lee, Jennifer E. C., Sudom K. A., & McCreary D. R. (2011). High order model of resilience in the Canadian forces. *Canadian Journal of Behavioral Science 43*(3), 222–234.

Lengnick-Hall, C. A., & Beck, T. E. (2005). Adaptive fit versus robust transformation: How organizations respond to environmental change. *Journal of Management, 31*, 738–757.

Lengnick-Hall, C. A., Beck, T. E., & Lengnick-Hall, M. L. (2011). Developing a capacity for organizational resilience through strategic human resource management. *Human Resource Management Review, 21*(3), 243–255.

Luthans, F. (2002). The need for and meaning of positive organizational behavior. *Journal of Organizational Behavior, 23,* 695–706.

Luthans, F., & Avolio, B. J. (2003). Authentic leadership development. In K. S. Cameron, J. E. Dutton, & R. E. Quinn (Eds.), *Positive organizational scholarship* (pp. 241–258) San Francisco: Berrett-Koehler.

Luthans, F., Avey, J. B., Avolio, B. J., Norman, S. M., & Combs, G. J.(2006). Psychological capital development: Toward a micro-intervention. *Journal of Organizational Behavior, 27,* 387–393.

Luthans, F., Avey, J. B., & Patera, J. L. (2008). Experimental analysis of a web based training intervention to develop positive psychological capital. *Academy of Management Learning & Education, 7*(2), 209–221.

Luthans, F., Avolio, B. J. (2009). The "point" of positive organizational behavior. *Journal of Organizational Behavior, 30*, 291–307.

Luthar, S., Cicchetti, D., & Becker, B. (2000). The construct of resilience: A critical evaluation and guidelines for future work. *Child Development, 71*(3), 543–562.

Maguen, S., Vogt, S. D. S., King, L. A., King, D. A. & Litz, B. T. (2006). Posttraumatic growth among Gulf War I veterans: The predictive role of deployment- related experiences and background characteristics. *Journal of Loss and Trauma: International Perspectives on Stress & Coping, 11*, 373–388.

McCubbin H. I., Needle R. H., & Wilson, M. (1985). Adolescent health risk behaviors: Family stress and adolescent coping as critical factors. *Family Relations, 34*, 51–62.

Maguire, B. & Cartwright, S. (2008). *Assessing a community's capacity to manage change: a resilience approach to social assessment.* Canberra, ACT: Bureau of Rural Sciences.

Mallak, L. A., & Yildiz, M. (2016). The development of a workplace resilience instrument. *WORK: A Journal of Prevention, Assessment and Rehabilitation. 54*(2), 241–253.

Masten, A. S. (2001). Ordinary magic: Resilience processes in development. *American Psychologist, 56*(3), 227–238.

Masten, A. S., & Narayan, A. J. (2012). Child development in the context of disaster, war, and terrorism: Pathways of risk and resilience. *Annual Reviews of Psychology, 63*, 227–257.

Matos, S., Neushotz, L. A., Griffin,M. Q., & Fitzpatrick, J. J. (2010). An exploratory study of resilience and job satisfaction among psychiatric nurses working in inpatient units. *International Journal of Mental Health Nursing, 19*(5), 307–312.

McEwen K. (2011). *Building Resilience at Work.* Australian Academic Press. Brisbane, Australia.

McLarnon, M. J., & Rothstein, M. G. (2013). Development and initial validation of the workplace resilience inventory. *Journal of Personnel Psychology, 12*(2), 63–73.

Meredith, L. S., Cathy, D., Sherbourne, S. J. G., Lydia H., Hans V. R., Andrew, P., & Glenda, W.(2011). *Promoting psychological resilience in the U.S. military.* Santa Monica, CA: RAND Corporation.

Moskovitz, S. (1985). Longitudinal follow-up of child survivors of the holocaust. *Journal of the American Academy of Child Psychiatry, 24*(4), 401–407.

Mowbray (2010). http://www.personneltoday.com/hr/understanding-the-importance-of-resilience/

Neenan, M. (2010). *Developing resilience: A cognitive behavioral approach.* New York, NY: Routledge.

Paul, H. & Garg, P. (2012a). Revisiting resilience in Indian context. In Conference Proceeding of International Conference on Management in *New World Order: Concepts and Practices from India* (pp. 380–397). New Delhi: Tata McGraw-Hill.

Prince-Embury, S. (2008). The resiliency scales for children and adolescents, psychological symptoms and clinical status in adolescents. *Canadian Journal of School Psychology, 23*, 41–56.

Rossi, A. M., Meurs, J. A., & Perrewé, P. L. (2013). *Improving employee health and well being.* United States of America: Information Age Publishing Inc.

Rutter, M. (1971). Parent–child separation: Psychological effects on the children. *Journal of Child Psychology and Psychiatry, 12*, 233–260.

Ryan, L., & Caltabiano, M. L. (2009). Development of a new resilience scale: the eesilience in midlife scale (RIM). *Asian Social* Science, *5*(11), 39–51.

Sinclair, V. G. & Wallston, K. A. (2004). The development and psychometric evaluation of the brief resilient coping scale. *Assessment, 11*, 94–101.

Smith, B., Dalen, J., Wiggins, K., Tooley, E., Christopher, P., & Bernard, J. (2008). The brief resilience scale: Assessing the ability to bounce back. *International Journal of Behavioral Medicine, 15*, 194–200.

Tedeschi, R. G., & Kilmer, R. P. (2005). Assessing strengths, resilience, and growth to guide clinical interventions. *Professional Psychology Research and Practice, 36*(3), 230–237.

Toor, S., & Ofori, G. (2010). Positive psychological capital as a source of sustainable competitive advantage for organizations. *Journal of Construction Engineering and Management, 136*(3), 341–352.

Tugade, M. M., Fredrickson, B. L., & Feldman Barrett, L. (2004). Psychological resilience and positive emotional granularity: Examining the benefits of positive emotions on coping and health. *Journal of Personality, 72*, 1161–1190.

Turner, M., Scott-Young, C. M., & Holdsworth, S. (2016). *Bouncing back to move forward: Resilience of students in the built environment.* Association of Researchers in Construction Management's Annual ARCOM Conference, Manchester, UK, 589–598.

Tymon, W. G., & Thomas, K. W. (1997). Stress resiliency profile: A measure of interpretive styles that contribute to stress. In C. P. Zalaquett & R. J. Wood (Eds.), *Evaluating Stress: A Book of Resources* (pp. 343–352), Lanham, MD: Scarecrow Press.

Ungar, M., & Liebenberg, L. (2009). Cross-cultural consultation leading to the development of a valid measure of youth resilience. The International Resilience Project, *Studia Psychologica, 51*(2–3), 259–269.

Vanhove, A, J., Herian, M., Perez, A. L. U., Harms, P. D., and Lester, P. B. (2015). Can resilience be developed at work? A meta-analytic review of resilience-building programme effectiveness. *P.D. Harms Publications, 11*, 1–30.

Wagnild, G., & Young, H. (1990). Resilience among older women. *Journal of Nursing Scholarship, 22*, 252–255.

Wagnild, G., & Young H. (1993). Development and psychometric evaluation of the resilience scale. *Journal of Nursing Measurement, 1*(2), 165–178.

Windle, G., Markland, D. A., & Woods, B. (2008). Examination of a theoretical model of psychological resilience in older age. *Aging & Mental Health, 12*(3), 285–292.

Windle, G., Bennett, K. M., Noyes, J. (2011). A methodological review of resilience measurement scales. *Health and Quality of Life Outcomes, 9*(8), 1–18.

Winwood, P. C., Colon, R., & McEwen, K. (2013). A practical measure of workplace resilience. *Journal of Occupational and Environmental Medicine, 55*(10), 1205–1212. doi:10.1097/jom.0b013e3182a2a60a

Youssef, C. M., & Luthans, F. (2007). Positive organizational behavior in the workplace: The impact of hope, optimism, and resilience. *Journal of Management, 33*(5), 774–800.

APPENDIX A

Sr. No.	Author/s	Definitions
1.	Bonanno et al., 2004	"the ability of adults in otherwise normal circumstances who are exposed to an isolated/and potentially highly disruptive event....to maintain relatively stable, healthy levels of psychological functioning"
2.	Fredrickson et al., 2003	"a relatively stable personality trait characterized by the ability to bounce back from negative experience and by flexible adaptation to the ever-changing demands of life"
3.	Luthans et al., 2006	"when beset by problems and adversity, sustaining and bouncing back and even beyond (resiliency) to attain success"
4.	Luhans, 2002	"the positive psychological capacity to rebound, to "bounce back" from adversity, uncertainty, conflict, failure, or even positive change, progress and increased responsibility"
5.	Maguen et al., 2006	"growth and positive life changes that may result from exposure to traumatic incidents"
6.	Masten, 2001	"a class of phenomena characterized by good outcomes n spite of serious threats to adaptation or development"
7.	Masten and Narayan, 2012	"Resilience can be defined as the capacity of a dynamic system to withstand or recover from significant challenges that threaten its stability, viability, or development"
8.	Matos et al., 2010	"a positive outcome resulting from the experience of adversity"
9.	Tedeschi and Kilmer, 2005	"effective coping and adaptation in the face of major life stress"
10.	Winwood et al., 2013	"[Resilience is] the process of negotiating, managing, and adapting to significant sources of stress or trauma. Assets and resources within the individual, their life and environment facilitate the capacity for adaptation and "bouncing back" in the face of adversity"

Appendix A has been adapted from the original work with Permission from Corresponding author (Britt et al., 2016).

APPENDIX B

Scale	Author (s)	Target Population	Mode of Completion	Factors Measured	Comments
The dispositional resilience scale	Bartone et al. (2007)	Adults	Self-report	Originally designed to measure psychological hardiness	Derives from the literature on hardiness, confused with other constructs and often proposed as a measure of psychological resilience
The Connor–Davidson resilience scale (CD-RISC)	Connor & Davidson (2003)	Adults	Self-report	Has its origins in clinical practice, popular measure in the context of stress coping ability	Perspective is that resilience is personal quality or characteristics. Widely used in medicine
The resilience scale for adults (RSA)	Friborg et al. (2003, 2005)	Adults & young adults	Self-report	Interpersonal and intrapersonal factors, personal strength, social competence, family cohesion	Focuses on the key features of resilient people, multilevel nature of the questionnaire is consistent with resilience as a dynamic process
The brief resilience scale (BRS)	Smith et al.(2008)	Adults	Self-report	Ability to recover from stress	Developed with a particular focus on stress and the ability to "bounce back"
The resilience scale (RS)	Wagnild & Young (1993)	Adults	Self-report	Confident optimism, productive and autonomous activity, interpersonal warmth and insight, skilled expressiveness	Individual level measure, developed in qualitative research and has been widely used
Psychological resilience	Windle, et al. (2008)	Older adults	Self-report	Personal competence, self-esteem and interpersonal control	Contains items from established scales with strong empirical backing and consequently provides a good basis for generalization
Ego resiliency	Klohnen (1996)	Adults	Self-report	Confident optimism. Productive and autonomous activity	Based on a previous measure Block and Block (1993) with items drawn from the California Psychological Inventory (Gough 1987)
The brief resilient coping scale	Sinclair & Wallston (2004)	Adults	Self-report	Optimism, helplessness, self-efficacy	Clinical base, very brief, possible use in identifying at risk individuals to situate interventions
Stress resiliency profile	.Tymon & Thomas (1997)	Adults	Self-report	Deficiency focusing, low skill recognition, necessitating	Designed to give clients new insights into ways they may be unintentionally raising their stress level and reducing their job satisfaction and engagement. By understanding the thought patterns that can cause stress, clients can improve their ability to effectively interpret situations and manage stress.

APPENDIX B *(Continued)*

Scale	Author (s)	Target Population	Mode of Completion	Factors Measured	Comments
Workplace Resilience Inventory	McCarnon & Rothstein (2013)	Adults	Self-report	Individual's personal characteristics, social support network, initial response to significant and life changing events, self-regulatory process	A more comprehensive measure was developed based on the theoretical model of King and Rothstein (2010)
Workplace resilience instrument	Mallak & Yildiz	Adults	Self-report	The instrument identified four dimensions of workplace resilience: Active Problem-Solving, Team Efficacy, Confident Sense-Making, and Bricolage.	Focus is on more effective actions taken in a crisis, reduced stress, higher quality decision making, decreased use of sick days, and higher job satisfaction
RAW	McEwen (2011)	Adults	Self-report	Living authentically, finding your calling, Maintaining perspective, managing stress, interacting cooperatively, staying healthy, building networks	Compares performance to adult Australian population, new untested, currently being investigates further.

Adapted from Windle et al. (2011), but the content has been modified by the present author as per the requirement of the book chapter.

CHAPTER 24

Can Meaningful Work Mitigate the Impact of Technostress Creators? Evidence from India

MONALISA MAHAPATRA[1*], SURYA PRAKASH PATI[1], and RABINDRA KUMAR PRADHAN[2]

[1]*Organizational Behavior and Human Resources Management Group, Indian Institute of Management Kozikhode, Kunnamangalam, Kerala 673570, India*

[2]*Department of Humanities and Social Sciences, Indian Institute of Technology Kharagpur, Kharagpur, West Bengal 721302, India*

[*]*Corresponding author. E-mail: mahapatra.monalisa16@gmail.com*

ABSTRACT

Although prior research has examined the influence of technostress (TS) creators on various job outcomes, insights into the importance of meaningful work on the perceptions of individual TS creators and their impacts on job outcomes are somewhat limited. In this research, using the lens of the job demand-resource framework, we investigate the mitigating influence of meaningful work on the relationship between TS creators and burnout. We also examine if burnout mediates the relationship between TS creators and turnover intention. We situate the research in India, one of the fastest growing economies in the world. Analyzing data collected through an online survey that was administered to professionals who regularly use information and communication technologies for their work, we found encouraging pieces of evidence for our hypotheses. Specifically, our findings revealed that burnout mediated the relationship between techno-insecurity and turnover intention, while meaningful work dampens the positive relationship between techno-overload and techno-uncertainty with burnout. The contributions of this study to theory and practice are also discussed.

24.1 INTRODUCTION

> *Technology is a double-edged sword. When it comes to technological innovation and its effects on our lives and culture, both the luddites and the evangelists have a point.*
>
> *The Guardian, December 2012*

This study owes its origin to many informal conversations we had with our friends working in the Indian IT Industry. Many of them had expressed their fear of possible job loss and had enrolled in various executive/certificate programs for acquiring newer technological skills. Their concerns are supported by numerous reports in the print media of many organizations axing employment contracts of their middle and top management personnel. For example, Cognizant Technology Solutions announced the voluntary separation of some 1000 top executives (Directors, Associate Vice Presidents, and Senior Vice Presidents) in October 2017. Similarly, Infosys Limited had announced that nearly 1000 employees in job level 6 and above (i.e., project managers, group project managers, senior architects, etc.) would be asked to leave. The workforce at these roles was also asked to identify the bottom 10% of their reportees, in terms of performance (Times of India, 2017). Furthermore, TCS announced its intention of trimming 1%–2% of its employees, which largely includes engineers with at least 10 years of experience (The Hindu Business Line, 2015). Reacting to such a move by the IT organizations, Mr. Abid Ali Neemuchwala, the CEO of Wipro Limited, had remarked: "Today the performance report is not on what you are doing till yesterday, but whether you are ready for tomorrow" (Times of India, 2017). Toward this, Mr. R. Chandrasekhar, the President of NASSCOM, had remarked that reskilling and upskilling is necessary for employee retention, with every employee not paying attention to these, facing the risk of being redundant (Economic Times, 2017). The IT industry is concerned about its mid-level managers, who seem no longer in tune with their coding abilities and have done precious little to upgrade themselves to newer technologies (The Hindu Business Line, 2015).

We extend this concern of inadequately skilled personnel to all the industries; for information and communication technologies (ICTs), adoption and diffusion are widely prevalent in almost all sectors in India, including healthcare, service industries, education systems, construction industry, etc. (Ahuja et al, 2009; Budhwar et al., 2006; Keniston, 2002; Noir & Walsham, 2007; Snehi, 2009). According to NASSCOM, India's IT industry is projected to grow by 8% in FY2017 to USD 154 billion (NASSCOM, 2017). IT-BPM exports from the country were expected to reach USD 117 million during FY 2016–17, a 7.6% year-on-year growth. Additionally, e-commerce is expected to generate revenue of USD 33 billion (Ministry of Electronics and Information Technology, Government of India, Annual Report 2016–2017). Furthermore, the Government of India is also attempting to transform India into a digitally empowered society through the ambitious e-government initiative named Digital India[1] (2015). It is needless to elaborate that the sustainability and success of these endeavors are largely contingent on technological adeptness of the personnel and citizenry.

Therefore, it is essential that employees need to reskill themselves, for the juggernaut of technological growth will continue to roll with an accelerated pace. Organizations will continue to embrace the latest ICTs for increasing their productivity, effectiveness, and efficiency (Bharadwaj, 2000; Melville et al., 2004) and will expect their employees to realize its full potential (Tsai et al., 2007). However, this need for constant learning may subject employees to perpetual stress. For example, "coping with rapid change in IT" (Benamati & Lederer, 2001) has been acknowledged as an ongoing challenge in the IT profession. Gallivan (2003), while examining the antecedents and consequences of IT professionals' perceptions of job stress in a financial services and chemical engineering firm, provides the following striking anecdote on a female Australian IT manager:

---

[1] The Digital India program is a flagship program of the Government of India with a vision to transform India into a digitally empowered society and knowledge economy. For more details, refer to the official web page. URL: http://www.digitalindia.gov.in/content/about-programme

> *Despite Australian maternity leave policy that enables a woman to protect her job for one year, Cynthia was reluctant to take more than three months off after having her baby. "The IT field changes too quickly," she said.... "it is too difficult for a woman to keep up learning and never stop because women have two full time jobs: the IT career and motherhood (p. 18)*

In fact, Ang and Slaughter (2000) remark that changes in information technology is "competency destroying" for employees. Thanks to the ubiquity of ICTs, the skills of knowledge workers are in constant erosion over time (Tsai et al., 2007). Thus, it is of little surprise that professionals using ICTs for their day-to-day work feel the stress of being antiquated, thereby subjecting themselves to continuous retooling on top of their already demanding workload (Mak & Sockel, 2001). Accordingly, in this study, we propose to identify specific technology-associated stressors or technostress creators (Tarafdar et al., 2007) that contribute toward enhancing stress and related outcomes among the Indian workforce.

In addition, we will research the possibility of meaningful work mitigating the influence of TS creators on individuals. Meaningful work is acknowledged to be the principal mode of bringing meaning to one's life (Peters & Waterman, 2015), which constitutes a major part of one's well-being (Allan et al., 2016). The construct has been found to relate very strongly with work adjustment ($r = 0.73$, $p < 0.001$; Fairlie, 2011), which provides us encouragement to consider it as a possible moderator in this study. Alternatively, it has been found to relate positively with work engagement, life satisfaction, life meaning, positive affect, organizational commitment, and job satisfaction (see Allan et al. (2016) and Fairlie (2011) for citations and evidence). Further meaningful work relates negatively to disengagement, exhaustion, and turnover intention (Fairlie, 2011). Our findings may be useful for organizations seeking knowledge for designing interventions to help their workforce cope with the ever-changing technological space.

We organize the rest of this manuscript as follows. First, we discuss the variables to be used in this study. Second, employing the job demand-resources (JD-R; Schaufeli & Taris, 2014) framework, we establish the hypotheses. Third, we elaborate our research method, analyses, and the findings. Finally, we discuss the results deliberating on the contributions to theory and practice.

24.2 THEORY AND HYPOTHESES

The JD-R model (Bakker et al., 2003; Demerouti et al., 2001) postulates that although every occupation may have its own distinct factors associated with job stress, yet these can be separated into two broad groups: job demands and job resources. Broadly speaking, job demands represent aspects of job that require effort and skill, whereas job resources are job aspects that assist in meeting job demands. Together, these make an expansive framework that could be employed across workplace settings, despite the specificity of demands and resources involved. It is in this sense that the JD-R model is "flexible and rigorous" (Xanthopoulou et al., 2007). This strength of generalizability of the JD-R model motivated us to utilize it in this research.

24.2.1 RELATIONSHIP BETWEEN TS CREATORS AND JOB BURNOUT

Job demands indicate those physical, psychological, social, or organizational aspects of the job that necessitate sustained physical and/or psychological (cognitive and emotional) effort or skills and are, therefore, associated with certain physiological and/or psychological costs (Xanthopoulou et al., 2007). A few illustrations of such job demands in occupational contexts include emotionally demanding interaction with clients, an unfavorable physical environment, high work pressure (Bakker & Demerouti, 2007), heavy lifting, work overload, interpersonal conflict, and job insecurity (Schaufeli & Taris, 2014). While job demands need not be negative, they may transform to job stressors when they call upon a greater degree of effort from individuals (Bakker & Demerouti, 2007). Such exertion may result in irritability and fatigue from which individuals may recover by taking a break, switching tasks, or performing less demanding activities (Schaufeli & Taris, 2014). However, when such recovery is insufficient, the individual is in a continued state of activation that gradually exhausts her/him physically and/or mentally leading to burnout (Schaufeli & Taris, 2014). Burnout is often understood to be a psychological ailment that emerges as a prolonged response to persistent interpersonal stressors on the job. Conceptually, it manifests as three distinct dimensions: overwhelming exhaustion, feelings of cynicism and detachment from the job, and a sense of ineffectiveness and lack of accomplishment (Maslach & Leiter, 2016). However, off late, a general consensus has emerged among scholars that emotional exhaustion constitutes the core of burnout (Seidler et al., 2014). Burnout has been associated with numerous negative work outcomes such as job dissatisfaction, low organizational commitment, high absenteeism, and higher turnover intention (see Maslach & Leiter, 2016, for a detailed review). In summary, sustained job demands with scant opportunity for recovery are likely to lead to burnout.

We introduce *TS creators* as job demands in this study. Technostress is understood to be "a modern disease of adaptation caused by an inability to cope with new computer technologies in a healthy manner" (Brod, 1984). It is the stress experienced by the end users of ICTs (Ragu-Nathan et al., 2008). Similarly, Tarafdar et al. (2010) have elucidated it as the "stress caused by an inability to cope with the demands of organizational computer usage." TS creators are stressors (stressful contexts) associated with the technology being used (Srivastava et al., 2015).[2] Tarafdar et al. (2007) empirically distinguished five subfactors of TS creators: *Techno-overload, Techno-invasion, Techno-complexity, Techno-insecurity, and Techno-uncertainty.* They define *Techno-overload* as a state where ICTs force people to work more and faster, while *techno-invasion* refers to a situation where the employees feel eternally available, thanks to the ICTs, can be reached anywhere and anytime, and feel the need to be unceasingly connected. Thus, the regular workday is extended with office work done at all possible hours with the employees finding impossible to "cut away" (Srivastava et al., 2015). The third subfactor, *techno-complexity,* relates to complexities associated with technologies, which makes employees feel less competent as far as their skills are concerned. This, in

[2] There is immense confusion with the way TS creators have been defined in the literature. Originally defined by Tarafdar et al. (2007), they were variously labeled by the authors in their study. On p. 314, they identified them collectively (second-order factor) as "TS creators"; however, as part of the structural model of their study, the "five subfactors were modeled into a second-order construct" called technostress (p. 315). Future researchers have contributed in propagating this confusion. For example, Jena and Mahanti (2014) in their study involving a sample of Indian academicians collectively identified these five subfactors as "technostress" (pp. 1–2) as well as "TS" (p. 3) in separate pages of their manuscript. However, some scholars such as Srivastava et al. (2015) have recognized the label "TS creators" exclusively to identify these five subfactors. We concur with them thanks to the intuitive appeal of this label over "technostress."

turn, forces them to expend time and effort in learning and comprehending the manifold attributes of ICTs (Tarafdar et al., 2007). *Techno-insecurity,* on the other hand, may be understood as a situation that heralds a threat of job loss in employees, as they perceive new ICTs replacing them. Techno-insecurity may also result from perceived redundancy, when employees perceive that people with better understanding of the ICT may replace them (Tarafdar et al., 2007). Finally, *techno-uncertainty* indicates a condition where the rapid and continuous change of ICTs results in uncertainty for employees, so as to force them to constantly learn and educate themselves on the new technologies (Tarafdar et al., 2007).

Thus, in line with the JD-R theory, we argue that some of these TS creators (job demands) may require more effort from the employees than they can cope with. The escalation of such demands coupled with little opportunity for recovery may most likely result in burnout of employees (Demerouti et al., 2001). The literature provides encouraging evidence in support of such relationships between TS creators and burnout. For example, Srivastava and colleagues (2015) found a positive association between aggregated TS creators and burnout in their study involving senior-level organizational managers who regularly use ICTs to accomplish their tasks. Similarly, Salanova et al. (2013) stated that intensive ICT users experience fatigue on account of work overload (analogous to techno-overload) as well as role ambiguity (corresponding to techno-invasion). Hence, we hypothesize:

Hypothesis 1 (H1): TS creators are positively associated with burnout. Specifically:

Hypothesis 1a (H1a): Techno-overload will be positively associated with burnout.
Hypothesis 1b (H1b): Techno-invasion will be positively associated with burnout.
Hypothesis 1c (H1c): Techno-complexity will be positively associated with burnout.
Hypothesis 1d (H1d): Techno-insecurity will be positively associated with burnout.
Hypothesis 1e (H1e): Techno-uncertainty will be positively associated with burnout.

24.2.2 JOB BURNOUT AS MEDIATOR BETWEEN TS CREATORS AND TURNOVER INTENTION

Lately, scholars have enlarged the scope of the JD-R model to investigate the degree to which burnout predicts specific organizational outcomes such as performance, citizenship behaviors, and absenteeism (Nahrgang et al., 2011; Schaufeli et al., 2009). Specifically, a meta-analytic study by Nahrgang et al. (2011) found evidence that burnout is positively associated with accidents and adverse events. Similarly, Schaufeli et al. (2009), through a longitudinal study, established that burnout affects the duration of absence from work. All these findings could be easily explained through the JD-R model. An exhausted individual will have little energy left to meet additional work demands (Nahrgang et al., 2011). According to health impairment process of JD-R (Schaufeli & Taris, 2014), this may result in negative health outcomes for the employee. Hence, in order to limit such occurrence as well as to protect themselves, employees may choose to quit the organization.

Consequently, employees who feel exhausted and frustrated owing to the demands of continuous adaptation to new ICTs in their workplaces are more likely to have higher turnover. Our reasoning is also supported by previous research. For example, Cordes and Dougherty (1993) in their review of the literature on burnout provide citations of studies that have reported evidence of a positive association between the burnout and turnover in various populations. Similarly, Kim and Stoner (2008), involving a sample of social workers from the United States, also provides evidence for a positive association between burnout and turnover intention. Hence, we hypothesize:

Hypothesis 2 (H2): Burnout will be positively associated with turnover intention

Furthermore, since TS creators have been argued to cause burnout in employees adapting to ICTs, and burnout is known to be associated with many organizational outcomes, we argue that burnout is the likely mediator between TS creators and turnover intention. We suggest that mentally and physically spent employees may most likely exit the organization when they perceive their resources to be inadequate to cope with the TS creators. Burnout has been found to be the intervening mechanism in similar studies previously. For example, Koeske and Koeske (1991) reported that indirect effects of student stress on intention to quit school through student burnout were substantial. Similarly, Huang et al. (2003) reported that burnout partially mediated the relationship between organizational politics (a work stressor) and turnover intention in a sample of civil servants from Taiwan. Accordingly, we hypothesize:

Hypothesis 3 (H3): Burnout will mediate the relationship between TS creators and turnover intention.

24.2.3 MEANINGFUL WORK AS MODERATOR OF RELATIONSHIP BETWEEN TS CREATORS AND JOB BURNOUT

Job resources in the JD-R model refer to those physical, psychological, social, or organizational aspects of the job that may assist in any of the following: (a) be instrumental in achieving work goals, (b) reduce job demands and the associated physiological and psychological costs, and (c) stimulate personal growth and development (Demerouti et al., 2001). The JD-R model advocates that job resources buffer the negative effect of job demands on job burnout (Schaufeli & Taris, 2014). According to Bakker et al. (2005), job resources may bring about such buffer through any of the following ways: reducing the tendency of workplace properties to generate specific stressors, "coloring" the perceptions and cognitions evoked by such stressors, sobering the responses that result from the appraisal process, or limiting the occurrence of detrimental health outcomes associated with such responses.

In this chapter, we introduce meaningful work as a job resource and argue that it moderates the relationship between TS creators and burnout. Meaningful work may be understood as a pursuit of meaning in life through work. It is defined as "job and other workplace characteristics that facilitate the attainment or maintenance of one or more dimensions of meaning" (Fairlie, 2011). According to Steger et al. (2012), the construct consists of three fundamental dimensions: (1) positive meaning at work, that is, the subjective sense that one's work has personal significance; (2) meaning making through work, that is, the degree to which work helps people make sense of the world and contributes to personal growth; and (3) greater good motivation, that is, the belief that one's work positively impacts others. Meaningful work has a rich history in the industrial psychology literature. For example, Maslow (1987) mentioned that a musician will create music, a writer will write, and a painter will paint, even if there is little material gain from the endeavor. Similarly, McGregor (1960) advocated for work that permits expression of imagination, ingenuity, and creativity in individuals. Even

Locke and Dunnette (1976) asserted that job satisfaction results from engaging in what is personally valued. It will be erroneous to believe that people are just instrumentally related to their work. Instead, they actively seek to do work that holds high significance for them (Fagely & Adler, 2012). In fact, Fagley and Adler (2012) assert that staying true to one's meaning, and purpose makes life worth living.

It has been argued previously that TS creators may require more effort from the employees than they can cope with. However, if employees perceive that their work contributes to personal growth as well as helps them make significant contribution to their organizations/society, they may interpret the resultant stress owing to such adaptation to new ICTs positively. Employees may delineate this as a necessary state in their journey toward self-transcendence, an important component of meaning in life (Fairlie, 2011), thus making them cope better. Accordingly, we reason that people perceiving their work as highly meaningful will shield themselves from the negative impact of the TS creators.

Our argument is also supported by the pieces of evidence furnished by many scholars on the moderating effect of meaningful work on the relationship between negative job characteristics and their outcomes. For example, in their study involving individuals of multiple races/ethnicities, Allan et al. (2016) found meaningful work to moderate the relationship between work stress and search for meaning in life. Similarly, drawing a sample from an automotive company, Harris et al. (2007) found that meaningful work moderated the relationship between abusive leadership and job performance. In fact, their evidence suggested that the negative relationship between abuse and performance was strongest for those employees reporting higher meaningful work. Thus, we hypothesize:

Hypothesis 4: Meaningful work will positively moderate the relationship between TS creators and burnout such that this relationship will be weaker for individuals high in meaningful work compared to individuals low in it.

Hypothesis 4a (H4a): Meaningful work will positively moderate the relationship between techno-overload and burnout such that this relationship will be weaker for individuals high in meaningful work compared to individuals low in it.

Hypothesis 4b (H4b): Meaningful work will positively moderate the relationship between techno-invasion and burnout such that this relationship will be weaker for individuals high in meaningful work compared to individuals low in it.

Hypothesis 4c (H4c): Meaningful work will positively moderate the relationship between techno-complexity and burnout such that this relationship will be weaker for individuals high in meaningful work compared to individuals low in it.

Hypothesis 4d (H4d): Meaningful work will positively moderate the relationship between techno-insecurity and burnout such that this relationship will be weaker for individuals high in meaningful work compared to individuals low in it.

Hypothesis 4e (H4e): Meaningful work will positively moderate the relationship between techno-uncertainty and burnout such that this relationship will be weaker for individuals high in meaningful work compared to individuals low in it.

Figure 24.1 summarizes the study hypotheses.

24.3 METHODOLOGY

24.3.1 PARTICIPANTS AND PROCEDURE

Executive education participants (N = 168) enrolled in a two-year executive MBA program in a public management institute located in southern India formed the principal respondents for the study. We contacted them over an email explaining the purpose of this research

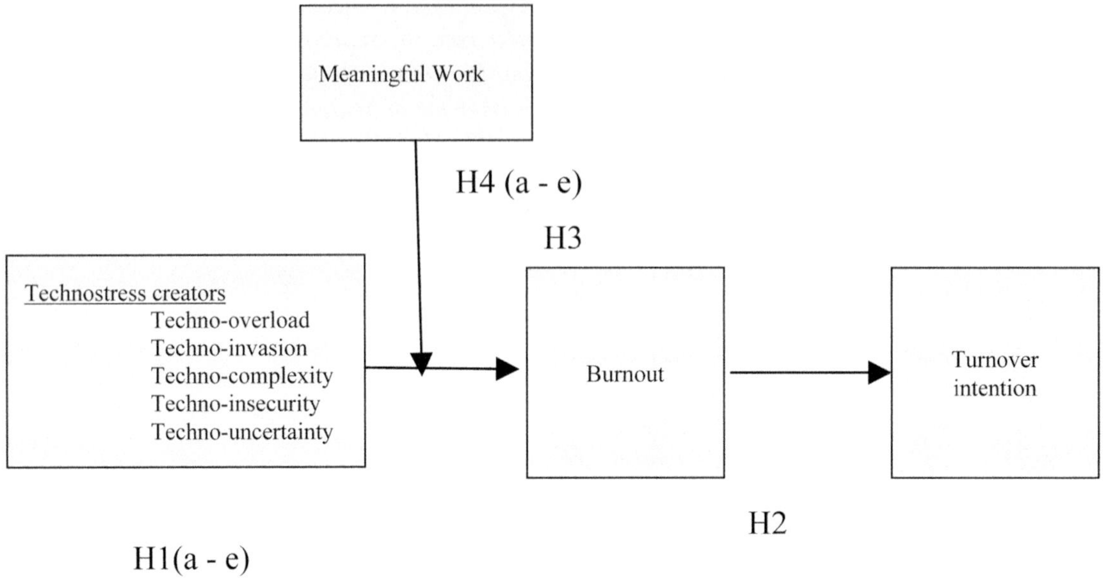

FIGURE 24.1 Research framework.

and invited them to respond to an online survey in case they regularly use ICTs for their professional work. Toward limiting the effect of common method bias, we explicitly assured the participants on the confidentiality of their responses (Podsakoff et al., 2003). We also encouraged the participants to remain anonymous, if they wished to (Podsakoff et al., 2003). We received a total of 133 usable responses.

To increase the sample size, we also forwarded the survey to employees across sectors through our acquaintances. However, we could manage to receive only 21 usable responses, which we combined with the executive education participants' data. In summary, 154 responses were used for the study.

Our respondents are employed in the following sectors: IT/ITES/BPO/KPO (49.3%), Banking and Financial Services (8.44%), Public Sector/Government (5.84%), Telecom (5.84%), Healthcare (3.24%), Manufacturing (3.24%), Automobile (3.24%), Energy (2.59%), FMCG (2.59%), Construction (1.94%), Education (1.94%), Mass Media (1.29), Agriculture (0.64%), Chemical (0.64%), Hospitality (0.64%), Non-IT Consulting (0.64%), Aerospace (0.64%), and Transport (0.64%). The rest reported being employed in "other" sector. Approximately, 70% of the respondents were undergraduates, while 28% of them had a postgraduate degree. The rest had other professional qualifications. The average work experience of the sample was calculated to be about 8.81 years (S.D. = 5.57). Nearly 52% of the respondents self-identified themselves to be working in a managerial position, while 43% had technical roles. The rest checked themselves into "other" roles. The mean age of the sample was determined to be 32.27 years (S.D. = 5.56) with nearly 80% of the respondents being males. Finally, the average ICT use for professional work (as reported by the respondents) was estimated to be 27.97 h per week (S.D. = 19.31).

24.3.2 MEASUREMENT

We used validated scales from the existing literature to assess the study variables (see Annexure A for the list of items used in this study).

TS creators. To measure the five dimensions of TS creators, we borrowed items from Srivastava et al. (2015). The respondents were requested to respond on a seven-point Likert continuum (1 = Strongly disagree to 7 = Strongly agree). A sample item for techno-overload (five items) is "I am forced by ICTs to work much faster," while a sample item for techno-invasion (four items) is "Because of ICTs, I spend less time with my family." Similarly, sample items for techno-complexity (five items) and techno-insecurity (five items) are "I often find it too complex for me to understand and use new ICTs" and "Because of new ICTs, I feel constant threat to my job security," respectively. Finally, a sample item for techno-uncertainty (four items) is, "In my organization, there are always constant changes in ICT software."

Burnout. Burnout was assessed using a 10-item short scale developed by Malach-Pines (2005). The participants were to respond on a seven-point Likert scale (1 = Never to 7 = Always). A sample item is "hopeless."

Turnover intention. Three items were borrowed from Gatling et al. (2016) to assess turnover intention. The responses were received on a seven-point Likert scale (1 = Strongly disagree to 7 = Strongly agree). A sample item is "I scan sources for prospective jobs."

Meaningful work. A six-item scale was used to measure meaningful work (Ashmos & Duchon, 2000), seeking responses on a seven-point Likert scale (1 = Strongly disagree to 7 = Strongly agree). A sample item is "My work is connected to what I think is important in life."

The control variables used in the analysis of these data include age, gender, qualification, career work experience, and extent of ICT use (measured as the number of average hours of ICT use per week). These variables were collected as part of the survey.

24.4 ANALYSES AND FINDINGS

The statistical software SPSS 21.0 was used for data analysis. In the following, we elaborate each step of the analysis.

24.4.1 COMMON METHOD BIAS

Prior to performing any regression analysis, we first checked for possible common method bias. Common method variance can either inflate or deflate observed relationships between constructs (Podsakoff et al., 2003). While we had taken necessary precautions during the design of the questionnaire to limit its influence, it is still important to statistically determine the existence of the effect. Therefore, we performed Harman's one factor test (Podsakoff & Organ, 1986). All study items were made to load on a single factor. The single factor explained only 24.93% of the total variance, indicating that common method bias does not affect the data and by implication the results.

24.4.2 VALIDITY OF THE MEASURES

We checked for two types of validity for the measures used in the study: convergent validity and discriminant validity. Convergent validity ascertains whether the measures for a construct are more correlated with one another than with the measures of another construct (Petter et al., 2007). Toward that, we estimated the factor loadings for each item, which indicates the strength of the correlation between each item and the corresponding construct. An item-factor loading of 0.7 or more is desirable (Morris & Venkatesh, 2010).

Factor analyzing the data[3] revealed many weak items that either exhibited an item-factor loading < 0.7 or cross-loaded with other constructs. These items were excluded from further analyses (see Annexure A for the final list of items for each construct retained after factor analysis). The construct of techno-insecurity was the most affected with many of the items cross loading with other factors. Finally, based on face validity, we retained only one item (i.e., "Because of new ICTs, I feel constant threat to my job security") to represent the construct. Results in Annexure A shows that there is a strong association between each of the indicators and the corresponding constructs. As the loadings within the construct are higher than those across constructs, this demonstrates convergent validity (Fornell & Larcker, 1981).

We adhered to the recommendation of Fornell and Larcker (1981) to investigate discriminant validity. Accordingly, the average variance explained (AVE) for each factor was compared with the squared correlation (SQ-COREL) between that factor and another factor (except Techno Insecurity). Table 24.1 reveals that for each factor, AVE > SQ-COREAL. Hence, the condition of discriminant validity is satisfied.

24.4.3 DESCRIPTIVE AND CORRELATION ANALYSIS

Table 24.2 presents the means, standard deviations, correlations, and Cronbach's alpha (internal consistency reliability estimate) of all the study variables. The Cronbach's alpha for all the measures, barring techno-insecurity, was calculated to be greater than 0.7 (Nunnally, 1978). Turnover intention reported a moderate

TABLE 24.1 Analysis for Discriminant Validity

	1	2	3	4	5	6	7
1. Techno-overload	(0.647)	0.164	0.035	0.077	0.0004	0.062	0.071
2. Techno invasion		(0.676)	0.020	0.110	0.055	0.091	0.030
3. Techno complexity			(0.833)	0.004	0.004	0.045	0.002
4. Techno-uncertainty				(0.725)	0.006	0.034	0.038
5. Meaningful work					(0.724)	0.078	0.105
6. Burnout						(0.614)	0.134
7. Turnover intention							(0.742)

Note: The values in the brackets are the AVE for the corresponding factor. The other values are SQ-COREL among the factors.

[1]The Kaiser–Meyer–Olkin measure for sampling adequacy was calculated to be 0.826, and the Bartlett's test for sphericity was found to be significant.

TABLE 24.2 Descriptive Statistics and Intercorrelations

	Mean	Std. Dev.	1	2	3	4	5	6	7	8
1. Techno-overload	4.57	1.36	(0.858)	0.405**	0.189*	0.279**	0.232**	–0.021	0.249**	0.267**
2. Techno invasion	4.51	1.64		(0.867)	0.142	0.332**	0.365**	–0.236**	0.302**	0.175*
3. Techno complexity	3.14	1.49			(0.856)	0.069	0.540**	–0.068	0.214**	0.049
4. Techno-uncertainty	4.41	1.38				(0.895)	0.292**	0.079	0.185*	0.197*
5. Techno insecurity	3.32	1.84					N/A	–0.209**	0.385**	0.252**
6. Meaningful work	5.05	1.23						(0.932)	–0.280**	–0.325**
7. Burnout	3.08	1.37							(0.912)	0.367**
8. Turnover intention	4.63	1.88								(0.875)

Note: *$p < 0.05$; **$p < 0.01$; numbers in diagonals are Cronbach's alpha.

relationship with techno-overload (r = 0.267, p < 0.01) and techno-insecurity respectively (r = 0.252, p < 0.01), while associating weakly with techno-invasion (r = 0.175, p < 0.05) and techno-uncertainty (r = 0.197, p < 0.05). Furthermore, the TS creators were found to share a weak to moderate relationship between themselves (r = 0.189 - 0.540, p < 0.05), thus removing any suspicion of multicollinearity effect affecting the study findings. Burnout too correlated positively with techno-overload (r = 0.249; p < 0.01), techno-invasion (r = 0.302; p < 0.01), techno-complexity (r = 0.214, p < 0.01), techno-uncertainty (r = 0.185, p < 0. 05), and techno-insecurity (r = 0.385, p < 0.01). This provided preliminary evidence in support of H1. However, the most important finding was that burnout related positively with turnover intention (r = 0.367, p < 0.01), thereby lending support to H2.

24.4.4 BURNOUT AS A MEDIATOR

Hierarchical multiple regression analyses were conducted to investigate evidence concerning the mediating role of burnout in the TS creators–turnover intention relationship. As reported in Table 24.3, the control variables (i.e., age, gender, work experience, qualifications, and ICT use/week) were entered in the first step, individual TS creators in the second step, and finally burnout in the third step. It can be observed that only techno-overload (β = 0.209, p <0.05) and techno-insecurity (β = 0.236, p <0.05) reported a significant association with turnover intention in Step 2 (see Table 24.3). The third mediation requirement from Kenny et al. (1998) is that the mediator variable (burnout) should be associated with the outcome variable (turnover intention), while controlling for the causal variables (individual TS creators). Furthermore, the relationship between exogenous variables (individual TS creators) and the outcome (turnover intention) reduced when the mediator (burnout) was entered the model. It may be noted from Step 3 (Table 24.3) that burnout positively associated with turnover intention (β = 0.281, p < 0.05), while the relationship between techno-insecurity and turnover intention became insignificant (β = 0.158). Furthermore, it may be observed that the regression coefficient between techno-overload and turnover intention has reduced in comparison to Step 2, although significant (β = 0.186, p < 0.05). We, therefore, conclude that burnout fully mediates the relationship between techno-insecurity and turnover intention, while partially mediating the relationship between techno-overload and turnover intention. H3 is thus partially supported.

TABLE 24.3 Hierarchical Regression Analysis of Turnover Intention on Technostress Creators and Burnout

Variable	β	ΔR^2
Step 1		
Gender	0.090	0.026
Age	0.029	
Qualification	–0.097	
Work experience	–0.042	
ICT use per week	0.089	
Step 2		
Gender	0.099	0.123**
Age	0.049	
Qualification	–0.078	
Work experience	–0.056	
ICT use per week	0.072	
Techno-overload	0.209*	
Techno-invasion	–0.010	
Techno-complexity	–0.111	
Techno-uncertainty	0.092	
Techno-insecurity	0.236*	
Step 3		
Gender	0.039	0.060**
Age	0.073	
Qualification	–0.089	
Work experience	–0.077	
ICT use per week	0.065	
Techno-overload	0.186*	
Techno-invasion	–0.061	
Techno-complexity	–0.122	
Techno-uncertainty	0.078	
Techno-insecurity	0.158	
Burnout	0.281**	

Note: †$p < 0.1$; *$p < 0.05$; **$p < 0.01$; dependent variable: turnover intention.

24.4.5 MEANINGFUL WORK AS A MODERATOR

Hierarchical moderated multiple regression analysis (Cohen et al., 2003) was conducted to examine the interaction effects of meaningful work and TS creators on burnout. The control variables such as gender, age, qualification, work experience, and ICT usage (in hours/week) were entered in the first step. In the second step, individual TS creators and meaningful work were entered. In the third step, the interaction terms (i.e., *Meaningful Work × Techno-overload; Meaningful Work × Techno-invasion; Meaningful Work × Techno-complexity; Meaningful Work × Techno-uncertainty; and Meaningful Work × Techno-insecurity*) were entered. All predictors were standardized. We expected that in the third step, only the interaction terms should predict burnout.

It may be observed from Table 24.4 that burnout was only predicted by techno-uncertainty (β = 0.152, $p < 0.10$) and techno-insecurity (β = 0.244, $p < 0.05$). Thus, H1 was partially supported. Furthermore, it was found that meaningful work moderated the relationship between techno-overload and burnout (β = –0.160, $p < 0.05$) as well as techno-insecurity and burnout (β = –0.171, $p < 0.10$), thus lending partial support to H4. Both the interactions are depicted graphically in Figures 24.2 and 24.3.

24.5 DISCUSSION

This study examined two major objectives. First, we sought to identify the contribution of TS creators to burnout and through burnout to turnover intention. Second, we also examined empirically the possibility of meaningful work moderating the influence of TS creators

TABLE 24.4 Interaction Effect of Meaningful Work on the Relationship between Technostress Creators and Burnout

Variable	β	ΔR^2
Step 1		
Gender	0.183	0.039
Age	–0.069	
Qualification	–0.029	
Work experience	0.057	
ICT use per week	0.005	
Step 2		
Gender	0.208	0.236**
Age	–0.044	
Qualification	0.045	
Work experience	0.060	
ICT use per week	0.028	
Techno-overload	0.097	
Techno-invasion	0.126	
Techno-complexity	0.052	
Techno-uncertainty	0.088	
Techno-insecurity	0.236*	
Meaningful work	–0.194*	
Step 3		
Gender	0.226**	0.058*
Age	–0.048	
Qualification	0.065	
Work Exp.	0.078	
ICT Use per Week	0.004	
Techno-overload	0.081	
Techno-invasion	0.081	
Techno-complexity	0.052	
Techno-uncertainty	0.152†	
Techno-insecurity	0.244*	
Meaningful work	–0.269**	
Meaningful work × Techno-overload	–0.160*	
Meaningful work × Techno-invasion	0.073	
Meaningful work × Techno-complexity	0.011	
Meaningful work × Techno-uncertainty	–0.171†	
Meaningful work × Techno-insecurity	0.139	

Note: †$p < 0.1$; *$p < 0.05$; **$p < 0.01$; dependent variable: Burnout.

on burnout. We used the theoretical lens of the JD-R model in this study for the development of specific hypotheses (Schaufeli & Taris, 2014). It posited individual TS creators as job demands that contribute to burnout among the employees, which, in turn, increases their turnover intention. Specifically, our results suggest that techno-uncertainty and techno-insecurity lead to burnout. It also indicates that meaningful work dampens the positive relationship between techno-overload and techno-uncertainty with burnout. Furthermore, our findings also reveal that burnout fully mediates the relationship between techno-insecurity and turnover intention, while partially mediating the relationship between techno-overload and turnover intention. In the following, we list the contributions from our study as well as elaborate implications for research and practice.

Our study contributes substantially to the understanding of TS creators and their implications. The literature on organizational stress is rich with evidence suggesting a significant relationship between job stressors and undesirable job outcomes. Situated within this stream of literature, researchers had investigated the role of TS creators toward heralding unwanted organizational outcomes such as diminished job satisfaction (Ragu-Nathan et al., 2008), lessened productivity, and innovation (Tarafdar et al., 2010). Furthermore, Srivastava et al. (2015) provide evidence supporting the positive relationship between TS creators and burnout. Our study findings are thus in tune with the literature. However, to the best of our knowledge, previous research treated TS creator as a second-order factor (see Ragu-Nathan et al., 2008; Tarafdar et al., 2010; Srivastava et al., 2015), which we believe reduced the explanatory power of the construct. Accordingly, we positioned each of the TS creators (i.e., techno-overload,

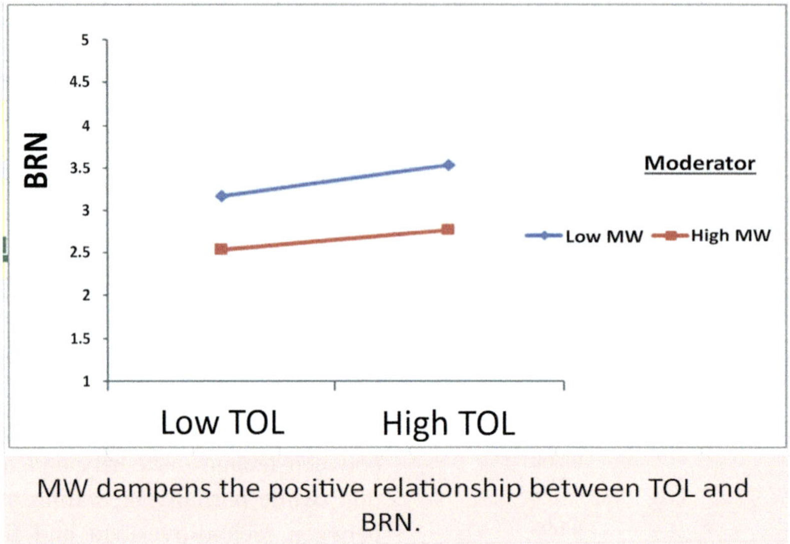

Note: BURN is Burnout; TOL is Techno-overload; MW is Meaningful Work.

FIGURE 24.2 Interaction effect of meaningful work on techno-overload–burnout relationship. *Note:* BURN is burnout, TOL is techno-overload, and MW is meaningful work.

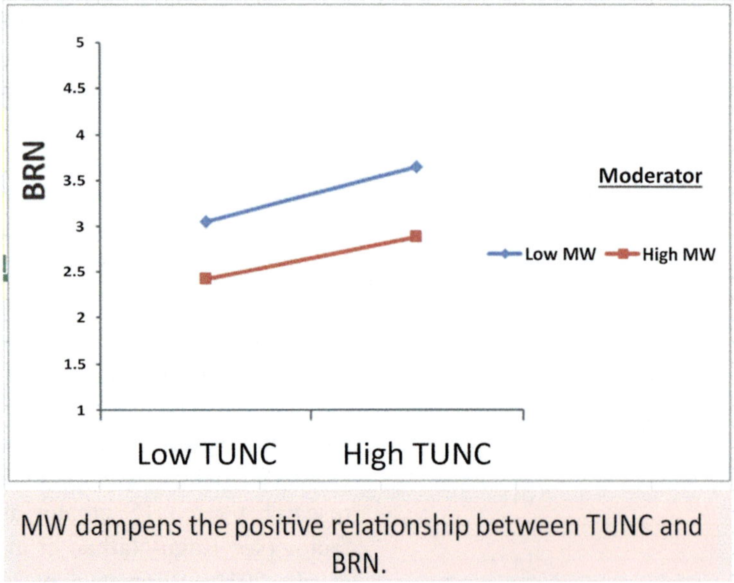

Note: BURN is Burnout; TUNC is Techno-uncertainty; MW is Meaningful Work.

FIGURE 24.3 Interaction effect of meaningful work on techno-uncertainty–burnout relationship. *Note:* BURN is burnout, TUNC is techno-uncertainty, and MW is meaningful work.

techno-invasion, techno-complexity, techno-insecurity, and techno-uncertainty) as a distinct causal variable, to understand their unique contribution toward burnout. This is the first contribution of our study. Second, the inclusion of TS creators also helped expand the list of job demands (see Schaufeli & Taris, 2014) for a comprehensive list of job demands identified in literature). These job demands are reflective of job characteristics in contemporary workplaces, whose association with other organizational outcomes may be studied. Third, our finding that techno-uncertainty and techno-insecurity explain maximum variance in burnout (in comparison to other TS creators) is a contribution to the burnout literature. Finally, ours is probably the first study that integrated meaningful work with the technostress literature. Meaningful work had been studied in the context of burnout literature previously (see, e.g., Allan et al., 2016; Harris et al., 2007), and our findings extended the scope and application of the construct. Previous research had identified computer self-efficacy (Shu et al., 2011), personality (Srivastava et al., 2015), user involvement, and innovation support (Tarafdar et al., 2010) as possible moderators of the relationship between TS creators and their outcomes. The inclusion of meaningful work enriches this list and opens up several avenues for research in the information systems domain.

Our study findings present significant insights for the Indian organizations in the digital economy. We began this study highlighting the organizational demand for frequent reskilling of the employees in new and emergent ICTs. While this is necessary, organizations must learn to manage such demands from the employee. Else these may result in burnout and attrition, which, in turn, can burgeon their costs. Since burnout is associated with job dissatisfaction, low organizational commitment, high absenteeism, higher turnover intention, etc. (see Maslach & Leiter, 2016), organizations would find it extremely difficult to sustain the motivation of their workforce if they refuse to acknowledge the harm resulting from technostress. On a brighter side, our study provides evidence for a coping strategy that could be embraced by the organizations. Since meaningful work moderates the relationship between a few techno-stressors and burnout, organizations must engage in conversations with their employees at every level, helping them appreciate the benefit they incur by learning new technologies. Drawing inspiration from the items of the meaningful work construct, we propose that organizations must also attempt to relate the learning of new technology to social good. Doing so will make the employees reconstruct technostress as greater good behind the great pain of frequent learning.

24.5.1 LIMITATIONS AND FUTURE RESEARCH DIRECTIONS

Like every research, our work too has certain limitations. First, the data were self-reported and, thus, may be subject to respondents' personal memory and biases while answering the questions. In addition, this study was cross-sectional, and the respondents' perceptions and intentions were measured at a single point in time. These limitations can lead to common method bias. Although we had conducted statistical tests and inferred that common method bias did not affect the study findings, yet in future, a longitudinal study with similar objectives may be designed to complement the findings from this research. Second, this study was conducted employing a sample of the Indian workforce.

Hence, researchers must exercise caution in generalizing the study findings. Accordingly, future researchers may only take the insights from this study, while investigating evidence for their support in different occupational and national contexts. Third, many items of the TS creator variables poorly loaded or cross loaded on their corresponding constructs. The problem was acute for techno-insecurity, where we had to satisfy ourselves by taking a single item for statistical analysis. Future studies may be designed to improve the items of TS creators. Finally, this study has a limited focus on two job-related outcomes (job burnout and turnover intention) in line with JD-R framework. Future studies may incorporate more or different outcomes for research.

KEYWORDS

- **meaningful work**
- **technostress**
- **information and communication technologies (ICTs)**
- **India**

REFERENCES

Ahuja, V., Yang, J., & Shankar, R. (2009). Study of ICT adoption for building project management in the Indian construction industry. *Automation in Construction*, *18*(4), 415–423.

Allan, B. A., Douglass, R. P., Duffy, R. D., McCarty, R. J. (2016). Meaningful work as a moderator of the relation between work stress and meaning in life. *Journal of Career Assessment*, *24*(3): 429–440

Ang, S., & Slaughter, S. (2000). The missing context of information technology personnel: A review and future directions for research. In R. W. Zmud (Ed.). *Framing the Domain of IT Management* (pp. 305–327). Cincinnati, OH, USA: Pinnaflex Educational Resources.

Ashmos, D. and Duchon, D. (2000). Spirituality at work. *Journal of Management Inquiry*, *9*(2), 134–145.

Bakker, A. B., & Demerouti, E. (2007). The job demands-resources model: State of the art. *Journal of Managerial Psychology*, *22*(3), 309–328.

Bakker, A. B., Demerouti, E., De Boer, E., & Schaufeli, W. B. (2003). Job demands and job resources as predictors of absence duration and frequency. *Journal of Vocational Behavior*, *62*(2), 341–356.

Bakker, A. B., Demerouti, E., & Euwema, M. C. (2005). Job resources buffer the impact of job demands on burnout. *Journal of Occupational Health Psychology*, *10*(2), 170.

Benamati, J., & Lederer, A. L. (2001). Coping with rapid changes in IT. *Communications of the ACM*, *44*(8), 83–88.

Bharadwaj, A. S. (2000). A resource-based perspective on information technology capability and firm performance: An empirical investigation. *MIS Quarterly*, *24*(1), 169–196.

Brod, C. (1984). Managing technostress: Optimizing the use of computer technology. *Personnel Journal*, *61*(10), 753–757.

Budhwar, P. S., Luthar, H. K., & Bhatnagar, J. (2006). The dynamics of HRM systems in Indian BPO firms. *Journal of Labor Research*, *27*(3), 339.

Cohen, J., Cohen, P., West, S. G., & Aiken, L. S. (2003). *Applied multiple regression/correlation analysis for the behavioral sciences* (3rd ed.) Mahwah, NJ: Lawrence Erlbaum.

Cordes, C. L., & Dougherty, T. W. (1993). A review and an integration of research on job burnout. *Academy of Management Review*, *18*(4), 621–656.

Demerouti, E., Bakker, A. B., Nachreiner, F., & Schaufeli, W. B. (2001). The job demands-resources model of burnout. *Journal of Applied Psychology*, *86*(3), 499.

Fagley, N. S., & Adler, M. G. (2012). Appreciation: A spiritual path to finding value and meaning in the workplace. *Journal of Management, Spirituality & Religion*, *9*(2), 167–187.

Fairlie, P. (2011). Meaningful work, employee engagement, and other key employee outcomes: Implications for human resource development. *Advances in Developing Human Resources*, *13*(4), 508–525.

Fornell, C., & Larcker, D. F. (1981). Evaluating structural equation models with unobservable variables and measurement error. *Journal of Marketing Research*, *18* (1), 39–50.

Gallivan, M. (2003). Examining gender differences in IT professionals' perceptions of job stress in response to technological change. In Proceedings of the *2003 SIGMIS Conference on Computer Personnel Research: Freedom in Philadelphia—Leveraging Differences and Diversity in the IT Workforce* (pp. 10–23). ACM.

Gatling, A., Gatling, A., Kim, J., Kim, J., Milliman, J., & Milliman, J. (2016). The relationship between workplace spirituality and hospitality supervisors' work attitudes: A self-determination theory perspective. *International Journal of Contemporary Hospitality Management*, *28*(3), 471–489.

Harris, K. J., Kacmar, K. M., & Zivnuska, S. (2007). An investigation of abusive supervision as a predictor of performance and the meaning of work as a moderator of the relationship. *The Leadership Quarterly*, *18*(3), 252–263.

Huang, I. C., Chuang, C. H. J., & Lin, H. C. (2003). The role of burnout in the relationship between Perceptions of organizational politics and turnover intention. *Public Personnel Management*, *32*(4), 519–531.

Jena, R. K., & Mahanti, P. K. (2014). An empirical study of technostress among Indian academicians. *International Journal of Education and Learning*, *3*(2), 1–10.

Keniston, K. (2002). Grassroots ICT projects in India: Some preliminary hypotheses. *ASCI Journal of Management*, *31*(1), 2.

Kenny, D. A., Kashy, D. A. & Bolger, N. (1998). Data analysis in social psychology. In D. Gilbert, S. T. Fiske & G. Lindzey (Eds.), *Handbook of Social Psychology* (4th ed. Vol. 1) (pp. 233–265). New York, NY, USA: McGraw-Hill.

Kim, H., & Stoner, M. (2008). Burnout and turnover intention among social workers: Effects of role stress, job autonomy and social support. *Administration in Social Work*, *32*(3), 5–25.

Koeske, G. F., & Koeske, R. D. (1991). Student "burnout" as a mediator of the stress-outcome relationship. *Research in Higher Education*, *32*(4), 415–431.

Locke, E. A., & Dunnette, M. D. (1976). The nature and course of job satisfaction. In *Handbook of Industrial and Organizational Psychology*: 1297–1343. Chicago, IL, USA: Rand McNally

Mak, B. L., & Sockel, H. (2001). A confirmatory factor analysis of IS employee motivation and retention. *Information & Management*, *38*(5), 265–276.

Malach-Pines, A. (2005). The burnout measure, short version. *International Journal of Stress Management*, *12*(1), 78.

Maslach, C., & Leiter, M. P. (2016). Understanding the burnout experience: Recent research and its implications for psychiatry. *World Psychiatry*, *15*(2), 103–111.

Maslow, A. H. (1987). *Motivation and Personality*. New Delhi, India: Pearson.

McGregor, D. (1960). *The Human Side of Enterprise*. New York, NY, USA: McGraw-Hill.

Melville, N., Kraemer, K., & Gurbaxani, V. (2004). Information technology and organizational performance: An integrative model of IT business value. *MIS Quarterly*, *28*(2), 283–322.

Ministry of Electronics & Information Technology (2017). *Annual Report 2016–17*. Available at http://meity.gov.in/writereaddata/files/AR2016-17_English.pdf [accessed 04.01.2018]

Morris, M. G., & Venkatesh, V. (2010). Job characteristics and job satisfaction: Understanding the role of enterprise resource planning system implementation. *MIS Quarterly*, *34*(1), 143–161.

Nahrgang, J. D., Morgeson, F. P., & Hofmann, D. A. (2011). Safety at work: A meta-analytic investigation of the link between job demands, job resources, burnout, engagement, and safety outcomes. *Journal of Applied Psychology*, *96*(1), 71–94.

NASSCOM (2017). *The IT-BPM Sector in India: Strategic Review 2017*. Available at http://www.nasscom.in/knowledge-center/publications/it-bpm-industry-india-2017-strategic-review [accessed 04.01.2018]

Noir, C., & Walsham, G. (2007). The great legitimizer: ICT as myth and ceremony in the Indian healthcare sector. *Information Technology & People*, *20*(4), 313–333.

Nunnally, J. C. (1978). *Psychometric Theory* (2nd ed.). New York, NY, USA: McGraw-Hill

Petter, S., Straub, D., & Rai, A. (2007). Specifying formative constructs in information systems research. *MIS Quarterly*, *31*(4), 623–656.

Peters, T. & Waterman R. H. (2015). *In Search of Excellence: Lessons from America's Best-run Companies*. In Aggarwal. Y. & Singh, S. (Eds.) (2016). Antecedents and consequences of work significance in Indian organizations. *Journal of Management, & Religion*. *14*(4), 318–342.

Podsakoff, P. M., MacKenzie, S. B., Lee, J. Y., & Podsakoff, N. P. (2003). Common method biases in behavioral research: A critical review of the literature and recommended remedies. *Journal of Applied Psychology*, *88*(5), 879.

Podsakoff, P. M., & Organ, D. W. (1986). Self-reports in organizational research: Problems and prospects. *Journal of Management*, *12*(4), 531–544.

Ragu-Nathan, T. S., Tarafdar, M., Ragu-Nathan, B. S., & Tu, Q. (2008). The consequences of technostress for end users in organizations: Conceptual development and empirical validation. *Information Systems Research*, *19*(4), 417–433.

Salanova, M., Llorens, S. & Cifre, E. (2013). The dark side of technologies: Technostress among users of information and communication technologies. *International Journal of Psychology*, *48*(3), 422–436.

Schaufeli, W. B., Bakker, A. B., & Van Rhenen, W. (2009). How changes in job demands and resources predict burnout, work engagement, and sickness absenteeism. *Journal of Organizational Behavior*, *30*(7), 893–917.

Schaufeli, W. B., & Taris, T. W. (2014). A critical review of the Job Demands-Resources Model: Implications for improving work and health. *Bridging Occupational, Organizational and Public health*, 43–68. Dordrecht, The Netherlands: Springer.

Seidler, A., Thinschmidt, M., Deckert, S., Then, F., Hegewald, J., Nieuwenhuijsen, K., & Riedel-Heller, S. G. (2014). The role of psychosocial working conditions on burnout and its core component emotional exhaustion—A systematic review. *Journal of Occupational Medicine and Toxicology*, *9*(1), 10.

Shu, Q., Tu, Q., & Wang, K. (2011). The impact of computer self-efficacy and technology dependence on computer-related technostress: A social cognitive theory perspective. *International Journal of Human-Computer Interaction*, *27*(10), 923–939.

Snehi, N. (2009). ICT in Indian Universities and Colleges: Opportunities and Challenges. *Management & Change*, *13*(2), 231–244.

Srivastava, S. C., Chandra, S., & Shirish, A. (2015). Technostress creators and job outcomes: Theorising the moderating influence of personality traits. *Information Systems Journal*, *25*(4), 355–401.

Steger, M. F., Dik, B. J., & Duffy, R. D. (2012). Measuring meaningful work: The work and meaning inventory (WAMI). *Journal of Career Assessment*, *20*(3), 322–337.

Tarafdar, M., Tu, Q., Ragu-Nathan, B. S., & Ragu-Nathan, T. S. (2007). The impact of technostress on role stress and productivity. *Journal of Management Information Systems*, *24*(1), 301–328.

Tarafdar, M., Tu, Q., & Ragu-Nathan, T. S. (2010). Impact of technostress on end-user satisfaction and performance. *Journal of Management Information Systems*, *27*(3), 303–334.

The Economic Times (2017). Stress at work: Pink slips giving employees the blues. Available at https://economictimes.indiatimes.com/jobs/stress-at-work-pink-slips-giving-employees-the-blues/articleshow/59376858.cms[accessed on 04.01.2018]

The Hindu Business Line (2015). Mid-level IT engineers face re-skilling challenge. Available at http://www.thehindubusinessline.com/info-tech/midlevel-it-engineers-face-reskilling-challenge/article6887725.ece [accessed on 04.01.2018]

The Times of India (2017). IT industry slows, Indian companies ready for large layoffs. Available at https://timesofindia.indiatimes.com/business/india-business/technology-giants-prepare-for-layoffs/articleshow/58585806.cms [accessed on 04.01.2018]

Tsai, H. Y., Compeau, D., & Haggerty, N. (2007). Of races to run and battles to be won: Technical skill updating, stress, and coping of IT professionals. *Human Resource Management*, *46*(3), 395–409.

Xanthopoulou, D., Bakker, A. B., Demerouti, E., & Schaufeli, W. B. (2007). The role of personal resources in the job demands-resources model. *International Journal of Stress Management*, *14*(2), 121.

ANNEXURE A

Scale Items with Factor Loadings

Technostress Creators	Factor Loading
Techno-Overload	
I am forced by ICTs to work much faster.	0.793
I am forced by ICTs to do more work than I can handle.	0.862
I am forced by ICTs to work with very tight time schedules.	0.850
I am forced to change my work habits to adapt to new technologies.	
I am forced by ICTs to handle higher workload because of increased technological complexity.	0.703
Techno Invasion	
Because of ICTs, I spend less time with my family.	
Because of ICTs, I have to be in touch with my work even during my vacation.	0.863
Because of ICTs, I have to sacrifice my vacation and weekend time to keep current on new ICTs.	0.803
Because of ICTs, I feel my personal life is being invaded.	0.800
Techno Complexity	
I do not know enough about the new ICTs to handle my job satisfactorily.	
I do not find enough time to study and upgrade my ICT skills.	
I need a long time to understand and use new ICTs.	0.919
I find new recruits to this organization know more about ICTs than I do.	
I often find it too complex for me to understand and use new ICTs.	0.907
Techno Insecurity	
Because of new ICTs, I feel constant threat to my job security*	
Because of new ICTs, I feel constant need to update my skills to avoid being replaced.	
Because of new ICTs, I feel constant threat by co-workers with newer ICT skills.	
For fear of being replaced, I do not share my knowledge with my co-workers.	
For fear of being replaced, I feel there is less sharing of knowledge amongst co-workers.	
Techno-Uncertainty	
In my organization, there are always new developments in the ICTs we use.	0.838
In my organization, there are always constant changes in ICT software.	0.902
In my organization, there are always constant changes in ICT hardware.	0.840
In my organization, there are always frequent upgrades in ICT networks.	0.824
Meaningful Work	
I experience joy in work.	0.872
My spirit is energized by work.	0.925
My work is connected to what I think is important in life.	0.835
I look forward to coming to work.	0.860
I see a connection between work and social good.	0.800

Annexure A *(Continued)*

Technostress Creators	**Factor Loading**
I understand what gives my work personal meaning.	0.809
Turnover Intention	
I will probably look for a new job next year.	0.865
I scan sources for prospective jobs.	0.858
I frequently think of quitting my job.	
Burnout	
When you think about your work overall, how often do you feel the following:	
Tired	
Disappointed with people	
Hopeless	0.783
Trapped	0.765
Helpless	0.809
Depressed	0.833
Physically weak/sickly	0.760
Worthless/like a failure	0.799
Difficulties sleeping	
"I've had it"	0.736

Note: Factor loading of items included in final scale only provided. For techno-insecurity, the included item is indicated by (*).

CHAPTER 25

Antecedents and Consequences of Employee Well-Being: A Theoretical Framework

LOPAMUDRA HATI*, and RABINDRA KUMAR PRADHAN

Department of Humanities and Social Sciences, Indian Institute Technology Kharagpur, Kharagpur, West Bengal 721302, India

**Corresponding author. E-mail: lopamudrahati@gmail.com*

ABSTRACT

In the era of globalization, organizations often face volatility, uncertainty, complexity, ambiguity, and cut-throat competition. In order to survive under this kind of uncertainty and complex environment, organizations need to put their best efforts to maximize their resources for sustaining healthy competition and ensuring their safe existence. Under this uncertainty and ambiguity, organizations resort to exploit the employees' physical and mental powers to gain competitive advantage. In the process, employees experience unrest, discontentment, and frustration at large that hamper their productive potentialities. This is one of the topmost perennial problems that disturb the congenial atmosphere of the workplace, which is essential for achievement of employee well-being (EWB) and happiness. Thus, organizations need to protect the employees' well-being and happiness by adapting healthy measures that could generate more efficient manpower for producing better results creating bright future of the organizations. Keeping this in view, this paper tries to examine the antecedents and consequences of EWB by proposing a theoretical framework for Indian organizations. It also explores various factors at individual, interpersonal, and organizational levels that contribute to EWB. This study tries to establish a strong link between EWB and their performance. In addition to this, this study also delves to find the future implications of EWB with regard to effective human resource management.

25.1 INTRODUCTION

In recent years, lots of research activities have been taken up in the concept of well-being (Diener et al., 1999; Keyes et al., 2002; Stratham & Chase, 2010; Seligman, 2011). However, its conceptual clarification and definition still remain largely unresolved (Forgeard et al., 2011). Well-being is a multidimensional concept, which covers psychological, emotional, social, and physical aspects of an individual. In this regard, well-being, as defined by Warr (1994), is the all-embracing quality of an employee's experience and functions in both physical and psychological dimensions, whereas in their integrative approach to the concept of well-being, Ryan and Deci (2000) are of the opinion that well-being of an individual takes into account the feelings, functioning, and evaluation criterion of that individual. The same can be discussed in three discrete approaches—hedonic, eudaimonic, and evaluative. Hedonic well-being is the superficial sense of pleasure, which originates from an individual's intrinsic sense of feelings and emotions. Eudaimonic well-being stems from the belief of an individual when he/she has a purpose of life and has a general feeling of leading "a life well lived." Evaluative well-being, however, deals with an individual's own evaluation of their lives with regard to their appraisal of life and achievements.

The notion that the productivity of an employee increases with the increase in well-being of an employee is well known in the professional and academic world. People spend a substantial amount of their day's hours at work, and in order to be healthy and happy, people not only have to find purpose and meaning in life, but also have to work in organizations that develop and recognize the contribution of work and focus on individual health and happiness (Gavin & Mason, 2004).

Academicians have pointed out several factors that have a bearing on an employee's well-being, such as psychological capital, emotional intelligence (EI), fitness, support structure of family and friends, and human resource management (HRM) practices. The effect of employee well-being (EWB) and its lack therof leading to increased healthcare costs has several financial impacts on the bottom line of an organization, and same has been studied very comprehensively (Matthew et al., 2006). EWB accelerates productivity at individual and organizational levels, whereas in the absence of it, organization may face cumulative financial as well as nonfinancial loss.

This study delves into various concepts of employee well-being in a more comprehensive manner and clearly outlines its antecedents and consequences. This chapter has three main objectives. First, it aims at exploring different factors that contribute to generating EWB at different levels such as individual, interpersonal, and organizational. The second objective is to establish a link between EWB and employee productivity, and the third one is to develop a conceptual framework that takes the available knowledge into consideration and provides additional lucidity to the concept of EWB.

In the current business scenario, organizations are facing a lot of competition and challenges from their rival for their safe existence. Thus, they usually ignore the well-being of their employees in an effort to develop competitive advantage. The study has become a necessity to probe into the complexities, problems, and issues related to sustainability of both the employees and the organization (Economic and Social Research Council, 2006). It has also become indispensible to know the problems of the employees and how to generate their well-being by striking a balance between the employees and expectations nurtured by the organizations. It is well known that a mentally and physically healthy employee is an asset and provides excellence in qualitative and quantitative service. Therefore, it is of paramount importance that the organizations focus on well-being of their employees. It should be binding on the part of the organizers to implement healthy practices and moral code of conduct to emphatically consider and solve problems of employees, which will contribute to the prosperity of the organization. Despite the significance of such initiatives, there is a scarcity in the number of empirical research studies that have explored the relationship among HRM practices, EWB, and performance (Guest, 2002;

Kersley et al., 2006; Peccei, 2004). Therefore, this study focuses on the overall factors at individual, interpersonal, and organizational levels that have an impact on EWB and its implications in the context of Indian service industry.

25.2 LITERATURE REVIEW

25.2.1 WHAT IS EWB?

Well-being, as defined by New Oxford Advanced Learner's Dictionary (7th ed., revised 2005), is a state of being comfortable, healthy, or happy. Warr (1999, 2007) has defined EWB from a perspective of employee's experiences and understanding of various life events. According to the researcher, there exist three sets of contrary dimensions that have been used by majority of the studies as dependent variables. These are (i) displeasure/pleasure, (ii) anxiety/comfort, and (iii) depression/enthusiasm. Displeasure/pleasure is one of the affective dimensions, which captures the positive and negative emotions related to the workplace in general. With respect to anxiety/comfort dimension, he adds that while anxiety blends minimal pleasure with maximum mental arousal, comfort is maximum pleasure with minimal mental arousal. The third dimension, depression/enthusiasm, oscillates between the extremes of dolefulness and eagerness.

Ryan and Deci (2000), in their study, have mentioned that the hedonism and eudemonism are the building blocks of EWB. While hedonism encompasses the feelings of an individual like joy and delight, inquisitiveness, and satisfaction, eudemonism revolves around an individual's functioning such as having good relationships, having control on life's events, or having a purpose in life and its reflection in day-to-day work. Most often EWB has been defined as employees' total well-being that they perceive to be determined primarily by work and can be influenced by workplace interventions (Juniper et al., 2011; Siegrist et al., 2006). World Health Organization (2013) has provided definition of it as "a state of every individual employee to realize his or her own potential, to cope with the normal stresses of life, to work productively, and to be able to make a contribution to her/his community."

Multiple definitions of EWB exist across literature as each researcher has tried to focus on one of the many facets of EWB, and there is a lack of a study that takes into account all the perspectives in a single study. This has made it difficult to gauge the exact level of understanding of this concept. This problem has been compounded by the fact that there exists an abundance of terms and expressions that refer to similar concepts as EWB. For example, Van Laar et al. (2007), Sirgy et al. (2001), and Warr (1999) refer to quality of work life (QWL) or quality of working life; Page and Vella-Brodrick (2009) report on workplace well-being and employee mental health; and Wright and Cropanzano (2007) refer to psychological well-being in their studies. Interestingly, Sirgy (2012), observes the concept of EWB using various terms such as "work well-being," "QWL," "quality of working life," "work quality of life," and "work well-being."

25.2.2 ELEMENTS OF EWB

Though well-being is a vast concept, it has two crucial elements: feeling good and functioning well. Feelings of happiness, contentment, enjoyment, curiosity, and engagement are characteristics of someone who has a positive experience in his life. Having a sense of purpose, experiencing positive and amiable relationships with others as well as having some control over one's life are all important attributes of well-being (Aked et al., 2008). This integrated "well-being" overture encompasses various components:

(1) Physical health (improvement in one's physical fitness);
(2) Psychological/emotional health (effective management of day to day life and issues like stress);
(3) Financial health (amount of savings, financial freedom, etc.);
(4) Spiritual health (one's belief system on the purpose of life and actions leading to sense of peace).

Ryff and Keyes (1995) conducted an analysis on dimensions of well-being and proposed a multidimensional model, which is shown in Table 25.1. Page and Vella-Brodrick (2009) have expounded and argued in favor of EWB having three major components: (i) psychological well-being, (ii) workplace well-being, and (iii) subjective well-being. According to Huhtala and Parzefall (2007), EWB can be approached from two dissimilar aspects, from a positive well-being and negative unwell-being approaches. However, the significance of mental health as well as positive well-being has been accepted for a long time. There are also many studies that have focused on employee "unwell-being," that is, burnout or occupational stress.

25.2.3 BENEFITS OF EWB

Previous research confirms that well-being of an employee is a vital area of matter for both employees and organizations. EWB has a substantial influence on the organizational productivity, as it controls different expenses associated with employee illness (Grawitch et al., 2006) and reduces absenteeism and turnover (Spector, 1997). EWB accelerates performance at both individual and organizational levels, whereas in the absence of it, organizations may face cumulative financial as well as nonfinancial loss (Wright, 2010; Wright & Cropanzano, 2007). EWB is the source of several benefits, such as the following:

TABLE 25.1 Elements of Well-being

Elements of Well-being		
Psychological Well-being	**Social Well-being**	**Emotional Well-being**
Self-acceptance	Social acceptance	Positive affect
Personal growth	Social actualization	Negative affect
Purpose in life	Social contribution	Life satisfaction
Environmental mastery	Social integration	Happiness
Positive relations with others		

Source: Lopez et al., 2013.

1. *Job performance/productivity:* A contented employee can go extra miles in his job (Bryson et al., 2017; Judge & Church, 2000; Wright, 2010; Bevan, 2010; Wright & Cropanzano, 2000).
2. *Work attendance:* A physically fit and happy employee tries to put his best efforts by ensuring 100% attendance (Wright, 2010).
3. *Employee turnover (Bevan, 2010; Wright, 2010):* Well-being has direct influence on productivity at both individual and organizational levels, which ultimately result in better recognition and fair payment of employee; as a result, employees feel self-satisfied, which ultimately reduces employee turnover.

4. *Acceleration of personal resources (Wright, 2010):* Employees who experience psychological well-being have a sense of positive motivation resulting in out of the box thinking and better social interactions. They also tend to better remember the positive events and take into stride the ambiguous events in a less negative way. Positive thought process also plays an important role in encouraging employees to boost up their personal resources such as physical, emotional, intellectual, and social resources. Hence, this positivity can help an employee for higher job performance, higher job involvement, and more commitment and dedication to the organization and work attendance.

25.3 ANTECEDENTS OF EWB

25.3.1 FACTORS AT INDIVIDUAL LEVEL

25.3.1.1 PERSONALITY

Extensive literature exists that explores the relationships between personality and well-being. According to Judge and Larsen (2001), there are many studies that suggest that employees' disposition to job satisfaction plays a major part in the determination of well-being.

Affective disposition reflects the emotional style and feelings that one has for other things, objects, and activities. It presents itself in two separate personalities: positive affectivity and negative affectivity. From numerous research studies that have been conducted in this area, it is quite evident that employee's affective disposition is one of the major contributors in the measurement of EWB (Sirgy, 2012). Positive and negative personality traits of individuals were found to be the major predictors of affective well-being at work.

An individual's personality traits can predict positively affective experiences (Authayarat & Umemuro, 2012). According to Judge and colleagues (2002), three separate personality traits have effect on EWB: neuroticism, extraversion, and conscientiousness. Neuroticism has been inversely related to satisfaction at work. Positive relation to job satisfaction has been associated with extraversion and conscientiousness traits. In a similar study conducted by Goodwin and Engstrom (2002), extraversion, conscientiousness, and openness have been associated with better perception of health. Meanwhile, in the same study, poor perceived health has been related to neuroticism trait.

25.3.1.2 AGE

Age and its relation to job-specific well-being was the focus area for the study conducted by Sirgy (2012). Evidence from the study points out that employees who are older have higher perception of workplace well-being than their younger counterparts. There can be two reasons behind this perception difference: one is that older employees are comparatively better placed in an organization's hierarchy and other explanation could be that the older workforce has less reward expectations than the younger workforce (Warr, 1999)

25.3.1.3 PHYSICAL FITNESS

It is well known that employees can reach its full potential only when they have best possible fitness, both mentally and physically. A fit employee will not only have the ability to function in an efficient and effective way, but also will adapt in a better way to the demanding challenges faced by them. Employees have their own optimum levels of fitness levels, which allow them to work at their peak. The same is determined by a combination of multiple factors such as their habits, internal and external stimuli, behavior and emotional conditioning, and socio-financial environment (C3 Collaborating for Health, 2011). As evidenced through multiple studies, involvement in any sort of physical activity has a positive impact on an employee's well-being. It not only helps the body to remain healthy medically, but also helps to reduce job-related stress and anxiety (Frank & Jason, 2005; C3 Collaborating for Health, 2011). Daily physical activities in any form and shape relieve the stress and anxiety faced by employees of all ages. Being a regular in physical activities also delays the onset of age-related disease and complications (Pruyne, 2011; Paradise, 2016).

25.3.1.4 EMOTIONAL INTELLIGENCE

Emotions are an important part of being human, and they affect our actions and motivations for behaviors and practices (Stanley & Burrows, 2005), which, in turn, affect our perceptions of psychological health (Slaski & Cartwright, 2003). Mayer and Salovey (1997) argued that EI refers to "the ability to perceive access and generate emotions so as to assist thoughts, to understand emotions and emotional knowledge, and to reflectively regulate emotions so as to promote emotional and intellectual growth." According to Emma and Frank (2004), EI in an individual can also be explained on the basis of one's adaptation and reaction to stress, self-management, conscientiousness, and social stimuli such as conflicts and leadership issues. Salovey and Mayer (1990) in their study pointed out that individuals with high EI will have higher perception of psychological well-being and vice versa. Individuals with high EI are able to handle demanding situations in a better way by effectively recognizing, generating, regulating, and promoting their emotions (Abraham et al., 2009).

In addition to the studies conducted by Mayer and Salovey (1997), the study conducted by Shaheen and Shaheen (2016) also found empirical evidence that individuals with higher EI have higher levels of life satisfaction, self-acceptance, and self-esteem, thus pointing to the possibility that EI may be playing a significant role in predicting both life satisfaction and self-esteem. High levels of EI in individuals also help them achieve better psychophysiological well-being (e.g., passive coping, rumination, and blood pressure) when exposed to different laboratory-based stressors.

25.3.1.5 WORK-RELATED BEHAVIORS

Work-related behaviors refer to constructs such as job performance, absenteeism, turnover, discretionary activities, burnout, and workaholism. According to the study conducted by Bakker and Oerlemans (2011), burnout and workaholism are indicators of low EWB at work. Exposure to stress for a prolonged period of time leads to burnout (Innanen et al., 2014), and it has been found in multiple studies that burnout can lead to lower job satisfaction and decreased fitness levels (Bakker et al., 2005; Hakanen et al., 2006). Workaholism, on the other hand, is marked by an inability of an employee to disengage from work. Employees often work harder than what is required and hence are not able to contribute to their other social responsibilities (Innanen et al., 2014). The study conducted by Berglass (2013) also points out that workaholics often do not have any meaningful intimate relationships and hence make work as an excuse to avoid social gatherings. This leads to reduced interpersonal relationship quality (Bakker et al., 2009) and poor life satisfaction (Hakanen & Schaufeli, 2012).

25.3.2 FACTORS AT INTERPERSONAL LEVEL

25.3.2.1 WORK-SUPPORTIVE FAMILY MEMBERS

According to the empirical evidence provided in the study conducted by Lynda and colleagues (2006), the support structure of an employee from friends and family provides emotional and instrumental sustenance to face tough situations at workplace. Emotional sustenance involves the act of listening to the issues faced by the employee and providing them emotional support to overcome those challenges. This helps the employee to relax at home and feel secure and calm. The energy thus saved is diverted to the employee's workplace, making them much more efficient and productive. Having a strong family support structure helps the employees to minimize strain-based personal life to interfere in their work. Families that support the employees by reducing their responsibilities and workload at home help them reduce the time-based family interference with work and thus provide them with instrumental sustenance. In addition to this, with fewer issues to worry about, the employee can focus their energy to achieve excellence at workplace. In a similar study conducted by Voydanoff (2005), it was observed that resources of a family can be classified into two distinct types: enabling resources and psychological rewards. Enabling resources helps employees to increase their competency and capacity to complete their tasks. Along with enabling resources, the psychological rewards also help the employees achieve a sense of respect for the unpaid work performed by them at home. All of these factors play an important role in the well-being of an employee.

25.3.2.2 EFFECT OF SOCIAL SUPPORT

Workplace relationships and social support are significant determinants of overall EWB. Hobfoll (1989) and Ryff and Singer (2000) viewed social support as a resource to the extent that such support provides or facilitates the preservation of valued resources in the form of time and energy. Positive social interactions among the managers and their subordinates go a long way in the development of an employee's well-being and also enhance their performance. Studies suggest that happiness is one of the most essential elements of EWB (Diener & Seligman, 2002) and helps the employees to flourish in their surrounding environment (Berscheid, 1994). In a similar study, Laverack (2006) has found evidence that positive social support leads to happiness and, in turn, plays an important role to enhance an employee's well-being. It has also been evidenced that more than the positive event in itself, sharing and communicating it with others is what makes it much more beneficial to an employee (Gabel et al., 2004). This also helps the employees to connect with other people and strengthen their relationships and created additional social resources. Furthermore, these additional resources created impetus for aiding EWB. As explained by Wayne et al. (2007), positive interactions and support from managers help employees not only in the work-related aspects, but also in learning new skills to face situations faced by them in their lives outside of office walls.

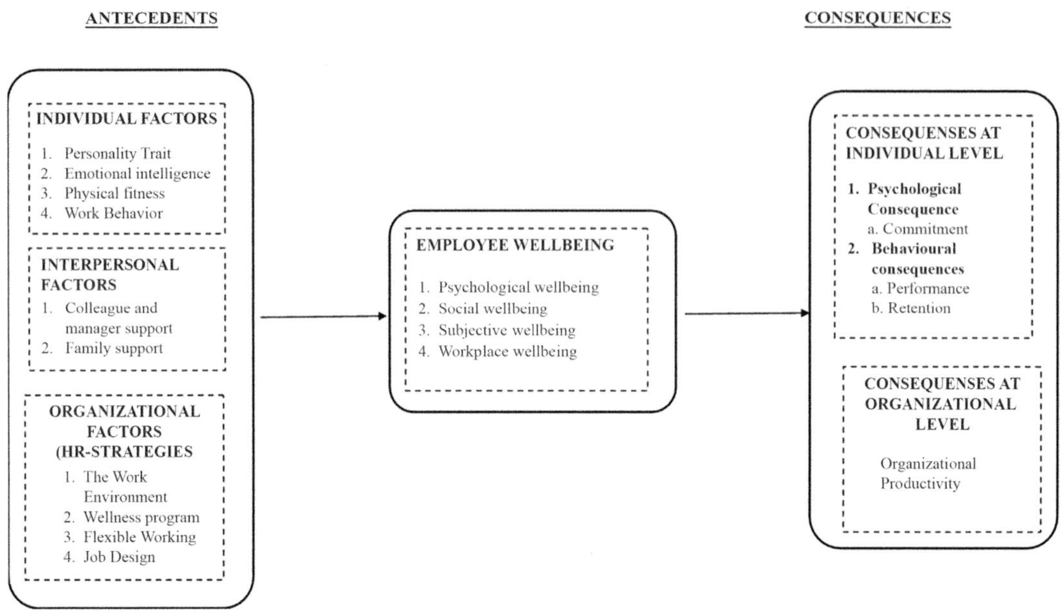

FIGURE 25.1 Framework of antecedents and consequences of EWB.

25.3.3 FACTORS AT ORGANIZATIONAL LEVEL

25.3.3.1 THE WORK ENVIRONMENT

Work environment has a marked effect on EWB (Arnold, 2015). According to Grawitch et al. (2006), work environment constitutes three separate factors that have an effect on EWB: employee growth and development, employee recognition, and employee involvement. Similarly, Warr (2007) identified nine aspects of the environment that can affect EWB and personal happiness: (i) opportunity for personal control, (ii) opportunity for skill use, (iii) externally generated goals, (iv) variety, (v) environmental clarity, (vi) contact with others, (vii) availability of money, (viii) physical security, and (ix) valued social position. The employees who have an optimal exposure of these environments will have adequate job satisfaction and personal happiness. Too much or too little may contribute to employee ill-being and not well-being. According to the study conducted by Kossek et al. (2012), following are the necessary work environments required for EWB:

1. Adequate pay meeting the market standards;
2. Healthy work environment;
3. Sufficient staffing to ensure that distributed sustainable workloads;
4. Overall training;
5. Consideration for work life balance.

25.3.3.2 FLEXIBLE WORKING

Flexibility at work helps ensures work life balance and initiatives that help achieve this enable the employees to provide their attention to all the necessary duties within work and outside of it, in a timely manner (Jamieson &

O'Mara, 1991). Flexible working is generally considered to be a positive working practice, as it allows people with family or other important commitments to work around other responsibilities. It is a way to give employees a sense of control over their work life and goes a long way in increasing employees' commitment and loyalty toward the organizations they work for. Flexible working has also been linked to psychological well-being (Stansfeld, 2002).

According to Ellen et al. (2012) found that flexible working practices offered by organizations are flexi-time, work from home, job sharing, and the ability to change hours, work condensed hours or work patterns, encouraged use of vacation time, flexible schedules, community involvement programs, child-care assistance, elder-care assistance, care giving assistance, etc.

In the recent times, many initiatives have been introduced by organizations to provide flexible working environment to their employees. Compressed work weeks help the employees to compress full-time work in a short period, and flexi-time arrangement allows the employees to start and end each day at their convenience (Ellen et al., 2012). Telework (Karina et al., 2011) allows an employee to choose the place of work and avoid unnecessary commute time. There are also few initiatives, in which employees can choose to have a reduced work load period and adjust their pay accordingly. All these initiatives allow the employees to devote time outside of office on the other priorities of life such as personal health (Ellen et al., 2012), family, eldercare, and education.

25.3.3.3 WELLNESS PROGRAM

During the past decade, many employers have implemented programs targeted at wellness of their employees. These programs not only increase the employee's well-being of an organization, but also reduce their resource attrition rate. According to the "World at Work" report, following are the top six offerings included in the wellness programs:

- Employee assistance program.
- Workplace safety (workplace environment).
- Health-related programs: immunizations, physical fitness, mental/behavioral health coverage, diet and nutrition, smoking cessation, disease management, health risk assessments (biometrics, physical fitness, etc.), and behavioral/mental health-risk assessment.
- Stress-related programs: employee assistance program, yoga, resiliency training, and meditation.
- Retirement-related programs: financial education and financial counseling.
- Skill-building and education programs: wellness coaching, stress management, time management, healthy workplace relationships, behavioral modification, and healthy personal relationships.

Sirgy and others (2008) proposed a conceptual framework linking the QWL programs with the subjective well-being. QWL affects subjective well-being in the following six ways: (a) providing suitable work resources to meet the prospects of employee role identities, (b) reducing role conflict in both work as well as nonwork life, (c) reducing role demands (d) increasing different role identities, (e) reducing stress, and (f) increasing the value of role identity.

25.3.3.4 JOB DESIGN

Job design, in its broader terms, encompasses all the constituents of a job and describes the tasks and activities related to a job structure and their enactment. The impact of these structures and enactments on individual, group, and organizational outcomes (Arnold, 2015) is also studied as part of job design.

Job design has been the focal area of research from a long time, and in recent years, researchers have focused more on the characteristic approach to job design (Garg & Rastogi, 2006). Jonge and Schaufeli (1998) in their study have mentioned that the Job Characteristics Model (Hackman & Oldham, 1975) and the Demand-Control-Support Model (Jason & Steve, 2010) have emerged as the most significant job design models. A well-designed job may foster EWB. Warr's (2007) Vitamin Model shows that design of a job can have a profound impact on EWB. Work-related stress and exposure to unsafe work environments are often a result of ill-conceived jobs. Although certain characteristics are necessary to have a positive outcome from a job, some characteristics if present can lead to negative effects on job performance and hence result in decreased EWB. Warr (2011) in his study has introduced the concept of "work vitamins" or a set of 12 characteristics that define a job. According to him, six work vitamins belong to the category of additional decrement (AD) and the remaining six work vitamins belong to the category of constant effect (CE). The characteristics under the AD category, if in excess, leads to negative effects on EWB. They are (a) job demands and related goals, (b) job autonomy, (c) social support and interpersonal connect, (d) skill utilization, (e) skill variety, and (f) task clarity along with appropriate feedback. The job characteristics under the CE category do not have any negative effect on EWB even if they are in excess. The characteristics that fall into this category are (a) safety including physical security, (b) salary and compensation, (c) job significance, (d) supportive supervision, (e) career outlook, and (f) equity.

25.4 CONSEQUENCES OF EWB

25.4.1 CONSEQUENCES OF EWB AT INDIVIDUAL LEVEL

25.4.1.1 EMPLOYEE COMMITMENT

A healthy employee usually shows a better and enhanced level of commitment as compared to the employee having poor level of mental and physical well-being (Bevan, 2010). It is evident from research findings that EWB has direct impact on commitment (Efraty & Sirgy, 1990; Rode et al., 2007; Wright, 2010). Employee with good state of well-being found to be connected and more committed to the organization not only mentally but also emotionally, as they understand the benefits of staying with the organizations who have well-being-oriented practices (Jain et al., 2009). A number of research activities have been carried to support the direct influence of physical and mental fitness and healthy organizational practices on employee commitment. Few research studies suggest that both affective and normative commitments are the results of healthy and positive mental condition, whereas the continuance commitment is related to the cognitive domain of an individual's personality. Employees with good level of commitment toward organization are the real assets for the organization as they are capable enough to identify and judge the objectives and values of their organization to work for it as well as they are proud for the same (Bevan 2010).

25.4.1.2 EMPLOYEE RETENTION

Previous research shows a strong connection between EWB and employee retention. EWB plays an important role of predictor and has strong influence on employee's decision of whether he/she should leave or stay in their current job rather than other job-related factors such as job satisfaction or job commitment (Wright, 2006; Harter et al., 2002; 2003). Rationales used by human resource (HR) stakeholders in seeking to acknowledge and satisfy EWB may achieve higher employee commitment toward work for organization, enhanced business performance, and a happier and stable workforce (Renwick, 2003).

There are a number of research activities that support an existence of strong connection between psychological well-being and employee retention. Employees who found their work stressful, and who are demotivated and not engaged and have poor interpersonal relationship at workplace, may feel that their job and organization is not worthwhile. This will ultimately lead to more periods of absence and mental disturbance, and they are more likely to resign their job, especially if they feel that an alternative job would be better. These elements of psychological well-being will ultimately affect the commitment and attachment of employees toward their organization (Bevan, 2010).

25.4.1.3 EWB AT WORK IMPACTS POSITIVELY ON PERFORMANCE

Matthew et al. (2006) linked EWB with performance and productivity. EWB stimulates high morale, job satisfaction, healthy practices, and wellness programs, reduces burnout and stress, and establishes supportive workplace climate, which ultimately results in enhanced sustainability and competitive advantage, better performance both at individual and organizational levels, better productivity along with service quality, customer satisfaction, and reduced turnover and absenteeism along with decreased healthcare cost, which ultimately increases individual and organizational productivity. It is evident from research that there is a strong bivariate correlation between well-being and job performance. Employees with high level of psychological well-being are found to be perfect decision makers, having good social relationships, and perform high level of productivity. Employee health status has a direct influence on the job performance, employee work behavior, and work attendance. Hence, developing healthier employees will result in a more productive workforce.

HR professionals are well aware of that human capital is the most important resource of an organization; it is the heart of any successful enterprise. An organization could achieve competitive advantage only if it has good and improved employee health status and well-being.

25.4.2 CONSEQUENCES OF EWB AT ORGANIZATIONAL LEVEL

25.4.2.1 EWB AND EMPLOYEE PRODUCTIVITY

Previous research reports that EWB is a crucial factor in determining an organization's long-term profitability (Lewis et al., 2012). A happy, healthy workforce is an integral element of a successful business. The more mentally and physically healthy the employee are, the more benefit for the organization in terms of lower sickness and absenteeism levels, attraction and retention of talented people, and more satisfied employees as well as customers. Employees with higher levels of well-being work better live longer and have happier lives. Nowadays, not only global organizations but also small organizations are also adopting healthy workplace practices, employee assistance programs, and flexible working practices in order to improve productivity, commitment, and work attendance (Bevan, 2010). Research shows that organizations giving importance to employee wellness practices are benefited by means of enhancing sustainability, enhanced profitability, improved competitive advantage, and better productivity and performance (Grawitch et al., 2006).

25.5 CONCLUSION

Lately, EWB has gained significant traction as the focal area of research. Although many studies have pointed out that EWB is critical to increasing an organization's productivity and builds its competitive advantage, there is dearth of empirical evidence to substantiate these claims. Keeping this in view, this study tries to examine the antecedents and consequences of EWB by proposing a theoretical framework. The literature review has revealed that EWB has been studied from various perspectives, and no single definition has been universally accepted. However, it has been well established that well-being is a multifaceted construct (Zheng et al., 2015; Juniper et.al, 2011).The results suggest that personality traits, especially neuroticism, extraversion, and conscientiousness (Judge et al., 2002), along with age, physical fitness, EI, and work-related behaviors such as workaholism and stress are the individual-level factors, which act as significant predictors of EWB. From literature survey, it is also evident that positive social interaction and presence of solid family support structure plays an important role in EWB. This paper also revealed that different organizational practices such as wellness programs and quality improvement programs help maintain a healthy work environment and, along with flexible working environment, act as important factors of EWB. This study has also tried to provide evidence on the link between EWB and productivity.

While this study is based on the review of literature, the proposed conceptual framework in this paper requires further empirical validation. Thus, future research can be done accordingly. This study adds to the existing literature on EWB and specifically is intended to inspire managers to come up with various interventions on how to enhance well-being in their organizations by identifying its factors. As a summary, this paper focuses on EWB as a concept, its contributing factors, and the impact on employee productivity.

KEYWORDS

- **employee well-being**
- **volatility uncertainty complexity and ambiguity**
- **productivity**
- **happiness**

REFERENCES

Abraham, C., Meyrav, Y., & Jacob W. (2009). The relationship between emotional intelligence and psychological well-being. *Journal of Managerial Psychology, 24*(1),66–78.

Aked, J., Marks, N., Cordon, C., & Thompson, S. (2008). *Five Ways to Wellbeing: The Evidence*. Centre for Well-Being, NEF (The New Economics Foundation). Retrieved from https://b.3cdn.net/nefoundati on/8984c5089d5c2285ee_t4m6bhqq5.pdf.(Accessed 20 August 2016).

Arnold B. B. (2015). Towards a multilevel approach of employee well-being. *European Journal of Work and Organizational Psychology*, *24*(6), 839–843.

Authayarat, W., & Umemuro, H. (2012). Workplace Environment Characteristics as Antecedents of Affective Well-being in the Workplace. *International Journal of Business Research and Management*, *3*(6), 307–324.

Bakker, A. B., Demerouti, E., & Burke, R. (2009). Workaholism and relationship quality: A spillover–crossover perspective. *Journal of Occupational Health Psychology*, *14*(1), 23–33.

Bakker, A. B., Demerouti, E., & Euwema, M. C. (2005). Job resources buffer the impact of job demands on burnout. *Journal of Occupational Health Psychology*, *10*(2), 170–180.

Bakker, A. B., & Oerlemans, W. G. M. (2011). Subjective wellbeing at work in organizations. InK. Cameron, & G. Spreitzer (Eds.), Handbook of positive organizational scholarship (pp. 178–189). Oxford, UK: Oxford University Press.

Berglass, S. (2013). *Handbook of Addictive Disorders: A Practical Guide to Diagnosis and Treatment* (383–407). Hoboken, NJ, USA: John Wiley & Sons.

Berscheid, E. (1994). Interpersonal relationships. *Annual Review of Psychology*, 45, 79–129.

Bevan, S. (2010). *The Business Case for Employee Health and Wellbeing*. The Work Foundation, London, UK. Retrieved from: http://www.theworkfoundation.com/downloadpublication/rep ort/245_245_iip270410.pdf. (Accessed 19 July 2016).

Bryson, A, Forth, J & Stokes, L. (2017). Does employees' subjective well-being affect workplace performance? *Human Relations*, *70*(8) 1017–1037.

C3 Collaborating for Health. (2011). *The Benefits of Physical Activity for Health and Well-Being*. Retrieved from: file:///C:/Users/shree/Desktop/PA%20n%20EWB%201.pdf.(Accessed 21 July 2016).

Diener, E. & Seligman, M. (2002). Very happy people. *Psychological Science*, *13*(1), 81–84.

Diener, E., Suh, E., Lucas, R., & Smith, H. (1999). Subjective well-being: Three decades of research. *Psychological Bulletin*, *125*(2), 276–302.

Ellen, E. K., Thomas, K., & Parveen, K. (2012). Achieving employee well-being in a changing work environment. *International Journal of Manpower*, *33*(7),738–753.

Economic and Social Research Council. (2006). *Health and Wellbeing at Work of Working Age People*. Seminar Series: Mapping the Public Policy Landscape, Economic and Social Research Council, Swindon, UK. Retrieved from: http://www.hcc.uce.ac.uk/craigjackson/ESRCWorkingPeople.pdf. (Accessed 10 August 2016).

Efraty, D., & Sirgy, M. J. (1990). The effects of quality of working life (QWL) on employee behavioral responses. *Social Indicators Research*, *22*(1), 31–47.

Emma, J., Donaldson, Frank, W. (2004), The relative importance of psychological acceptance and emotional intelligence to workplace well-being. *British Journal of Guidance & Counselling*, *32*(2), 187–203.

Forgeard, M. J. C., Jayawickreme, E., Kern, M., & Seligman, M. E. P. (2011). Doing the right thing: Measuring well-being for public policy. *International Journal of Well-being*, *1*(1), 79–106.

Frank, J., & Jason, R. (2005). Exercise and well-being: A review of mental and physical health benefits associated with physical activity. *Current Opinion in Psychiatry*, *18*(2),189–193.

Gabel, S. L., Reis, H. T., Impett, E., & Asher, E. R. (2004). What do you do when things go right? The intrapersonal and interpersonal benefits of sharing positive events. *Journal of Personality and Social Psychology*, 87, 228–245.

Garg, P., & Rastogi, R. (2006). New model of job design: Motivating employees' performance. *Journal of Management Development*, *25*(6), 572–587.

Gavin, J. H., & Mason, R. O. (2004). The virtuous organization: The value of happiness in the workplace. *Organizational Dynamics*, *33*(4), 379–392.

Goodwin R., & Engstrom G. (2002). Personality and the perception of health in the general population. *Psychological Medicine*, *32*(2), 325–332.

Guest, D. (2002). Human resource management, corporate performance and employee well-being: building the worker into HRM. *Journal of Industrial Relations*, *44* (3), 335–358.

Grawitch, M. J., Gottschalk, M., & Munz, D. C. (2006). The path to a healthy workplace: A critical review linking healthy workplace practices, employee well-being, and organizational improvements. *Consulting Psychology Journal*, *58*(3), 129–147.

Hackman, J. R., & Oldham, G. R. (1975). Development of the job diagnostic survey. *Journal of Applied Psychology*, *60*(2), 159–170.

Hakanen, J. J., Bakker, A. B., & Schaufeli, W. B. (2006). Burnout and work engagement among teachers. *Journal of School Psychology*, *43*, 495–513.

Hakanen, J. J., & Schaufeli, W. B. (2012). Do burnout and work engagement predict depressive symptoms and life satisfaction? A three-wave seven-year prospective study. *Journal of Affective Disorders*, *13*(2–3), 415–424.

Harter, J., Schmidt, F., & Hayes, T. (2002). Business-unit-level relationship between employee satisfaction, employee engagement, and business outcomes: A meta-analysis. *Journal of Applied Psychology*, *87*(2), 268–279.

Harter, J., Schmidt, F., & Hayes, T. (2003). Wellbeing in the workplace and its relationship to business outcomes: A review of the Gallup studies. *In Flourishing: Positive Psychology and the Life Well Lived*, C. Keyes and J. Haidt, Eds. (181–196). Oxford, UK: Oxford University Press.

Hobfoll, S. E. (1989). Conservation of resources: A new attempt at conceptualizing stress. *American Psychologist*, *44*(3), 513–524.

Huhtala, H., & Parzefall, M. R. (2007). A review of employee well-being and ınnovativeness: An opportunity for mutual benefit. *Creativity and Innovation Management*, *16*(3), 299–306.

Innanen, H., Tolvanenb, A., & Salmela, K. (2014). Burnout, work engagement and workaholism among highly educated employees: Profiles, antecedents and outcomes. *Burnout Research*, 1(2014), 38–49.

Jain, A., Giga, S., & Cooper, C. (2009). Employee wellbeing, control and organizational commitment. *Leadership & Organization Development Journal*, *30*(3), 256–273.

Jamieson, D., & O'Mara, J. (1991): *Managing Workforce 2000: Gaining the Diversity Advantage*. San Francisco, CA, USA: Jossey-Bass.

Jason, K., & Steve, J. (2010), Karasek's (1979) job demands-control model: A summary of current issues and recommendations for future research, in Pamela L. Perrewe, Daniel C. Ganster (Eds.) *New Developments in Theoretical and Conceptual Approaches to Job Stress* (Research in Occupational Stress and Well-Being, Volume 8 (237–268). Bingley, UK: Emerald Group Publishing Limited.

Jonge, J., & Schaufeli, W. B. (1998). Job characteristics and employee well-being: A test of Warr's Vitamin Model in health care workers using structural equation modelling. *Journal of Organizational Behavior*, 19, 387–407.

Judge, T. A., & Church, A. H. (2000). Job satisfaction: Research and practice. In C. L. Cooper &: E. A Locke (Eds.), Industrial and Organizational Psychology: Linking Theory with Practice. (166–198). Oxford, UK: Blackwell.

Judge, T. A., Heller, D., & Mount, M. K. (2002). Five-factor model of personality and job satisfaction: A meta-analysis, *Journal of Applied Psychology*, *87*(3), 530–541.

Judge, T. A., & Larsen, R. J. (2001). Dispositional affect and job satisfaction: A review and theoretical extension. *Organizational Behavior and Human Decision Processes*, *86*(1), 67–98.

Juniper, B. A., Bellamy, P., & White, N. (2011). Testing the performance of a new approach to measuring employee well-being. *Leadership & Organization Development Journal*, *25*(4), 344–357.

Karina. M., Maria, P., & Izel. R. (2011). Work–home interference and the relationship with job characteristics and well-being: A South African study among employees in the construction industry. *Stress and Health*, *27*(3), 238–251.

Keyes, C. L. M., Shmotkin, D., & Ryff, C. D. (2002). Optimizing well-being: The empirical encounter of two traditions. *Journal of Personality and Social Psychology*, *82*(6), 1007–1022.

Kersley, B., Alpin, C., Forth, J., Bryson, A., Bewley, H., Dix, G. & Oxenbridge, S. (2006). *Inside the Workplace: Findings from the 2004 Workplace Employment Relations Survey*. Abingdon, UK: Routledge, Taylor & Francis Group.UK.

Kossek, E. E., Kalliath, T., & Kalliath, P. (2012). Achieving employee wellbeing in a changing work environment. *International Journal of Manpower*, *33*(7),738–753.

Laverack, G. (2006). Improving health outcomes through community empowerment: A review of the literature. *Journal of Health, Population and Nutrition*, *24*(1), 113–120.

Lewis, R., Donaldson-Feilder, E., & Tharani, T. (2012). *Managing for Sustainable Employee Engagement: Developing a Behavioral Framework*. CIPD, London, UK. Retrieved from: http://engageforsuccess.org/wp-content/uploads/2017/05/Managing-for-sustainable-employee-engagement-Developing-a-behavioral-framework.pdf.

Lopez, S. J., Pedrotti, J. T., & Snyder, C. R. (2013). *Positive psychology: The scientific and practical explorations of human strengths*. California, USA: Sage Publications,133–160.

Lynda, A. K, Mattimore, K. L., Daniel W. K. & Adams. (2006). Family Support Inventory for Workers: A new measure of perceived social support from family members. *Journal of Organizational Behavior*, *16*(3), 235–258.

Mayer, J. D., & Salovey, P. (1997). What is emotional intelligence: implications for educators. In Salovey, P. and Sluyter, D. (Eds.), *Emotional Development, Emotional Literacy, and Emotional Intelligence* (3–31). New York, NY, USA: Basic Books.

Matthew J., Grawitch. M. G., & Munz. D. (2006). The path to a healthy workplace a critical review linking healthy workplace practices, employee well-being, and organizational Improvements. *Consulting Psychology Journal: Practice and Research, 58*(3), 129–147.

New Oxford Advanced Learner's Dictionary. (2005). *Well-being*, 7th ed., Lucknow, India: Oxford University Press.

Paradise, C. (2016). *Measuring User Satisfaction: Health and Wellbeing*. Retrieved from https://www.bsria.co.uk/download/asset/soft-landings-conference-well-being.pdf.(Accessed 19 June 2016).

Page, K. M., & Vella-Brodrick, D. A. (2009). The 'what', 'why' and 'how' of employee well-being: A new model. *Social Indicators Research, 90*(3), 441–458.

Peccei, R. E. (2004). Human resource management and the search for the happy workplace. *ERIM Inaugural Address Series Research in Management*. Retrieved from http://hdl.handle.net/1765/1108.

Pruyne, E. (2011). *Corporate investment in Employee Well-being the emerging strategic imperative*. Ashridge Business School, Nuffield Health. Retrieved from https://www.nuffieldhealth.com/downloads/corporate-investment-in-employeewell-being.(Accessed 14 July 2016).

Renwick, D. (2003). Line manager involvement in HRM: An inside view. *Employee Relations, 25*(3), 262–280.

Rode, J. C., Rehg, M. T., Near, J. P., & Underhill, J. R. (2007). The effect of work/family conflict on intention to quit: The mediating roles of job and life satisfaction. *Applied Research in Quality of Life*, 2, 64–82.

Ryan, R. M., & Deci, E. L. (2000). Self-determination theory and the facilitation of intrinsic motivation, social development, and well-being. *American Psychologist, 55*(1), 68–78.

Ryff, C. D. & Keyes, C. L. (1995). The structure of psychological well-being. *Journal of Personality and Social Psychology, 69*(4), 719–27.

Ryff, C. D., & Singer, B. H. (2000). Interpersonal flourishing: A positive health agenda for the new millennium. *Personality and Social Psychology Review*, 4(1 Special Issue: Personality and Social Psychology at the Interface: New Directions for Interdisciplinary Research), 30–44.

Salovey, P. & Mayer, J. D. (1990). Emotional intelligence. *Imagination, Cognition and Personality, 9*(3), 185–211.

Seligman, M. E. P. (2011). *Flourish: A New Understanding of Happiness and Well-Being and How to Achieve Them* (14–6). London, UK: Nicholas Brealey Publishing.

Shaheen, S., & Shaheen, H. (2016). Emotional ıntelligence ın relation to psychological well-being among students. *The International Journal of Indian Psychology, 3*(4), 206–213.

Siegrist, J., Wahrendorf, M., Knesebeck, O., Jurges, H., & Borsch-Supan, A. (2006). Quality of work, well-being, and intended early retirement of older employees: Baseline results from the Share Study. *European Journal of Public Health, 17*(1), 62–68.

Sirgy, M. J. (2012). *The Psychology of Quality of Life*, 50 (269–274). New York, NY, USA: Springer.

Sirgy, M. J., Efraty, D., Siegel, P., & Lee, D. J. (2001). A new measure of quality of work life (QWL) based on need satisfaction and spill over theories. *Social Indicators Research, 55*(3), 241–302.

Sirgy, M. J., Reilly, N. P., Wu, J., & Efraty, D. (2008). A work-life identity model of well-being: Towards a research agenda linking quality of work life (QWL) programs with quality of life (QOL). *Applied Research in Quality of Life, 3*(3), 181–202.

Slaski, M., & Cartwright, S. (2003). Emotional intelligence training and its implications for stress, health and performance. *Stress and Health, 19*(4), 233–239.

Spector, P. E. (1997). *Job Satisfaction: Application, Assessment, Causes and Consequences* (1–104). Thousand Oaks, CA, USA: Sage.

Stanley, R. O., & Burrows, G. D. (2005). The role of stress in mental illness: The practice. In C. L. Cooper (Eds.), *Handbook of Stress Medicine and Health*, London, UK: CRC Press.

Stansfeld, S. (2002). Work, personality and mental health. *The British Journal of Psychiatry, 181*(2), 96–98.

Stratham, J., & Chase, E. (2010). *Childhood Wellbeing—A Brief Overview*. Loughborough, UK: Childhood Wellbeing Research Centre. Retrieved from: http://cite seerx.ist.psu.edu/viewdoc/ download? doi=10.1.1.453.9199&rep=rep1&type=pdf. (Accessed 3 September 2016).

Van Laar, D., Edwards, J. A. & Easton, S. (2007). The work-related quality of life scale for healthcare workers. *Journal of Advanced Nursing, 60*(3), 325–333.

Voydanoff, P. (2005). Toward a conceptualization of perceived work-family fit and balance: A demands and resources approach. *Journal of Marriage and Family, 67*(4), 822–836.

Warr, P. (1994). A conceptual framework for the study of work and mental health. *Work & Stress, 8*(2), 84–97.

Warr, P. (1999). Well-being and the workplace. In Kahneman, D., Diener, E. and Schwarz, N. (Eds.): *Well-Being: The Foundations of Hedonic Psychology*, New York, NY, USA: Russell Sage Foundation.

Warr, P. (2007). *Work, Happiness and Unhappiness*. New York, NY, USA: Lawrence Erlbaum Associates.

Warr, P. (2011). *Jobs and Job-Holders: Two Sources of Happiness and Unhappiness Happiness and Organizations* (1–21). Oxford, UK: Oxford University Press.

Wilkinson, A. (1998). Empowerment: Theory and practice. *Personnel Review*, *27*(1), 40–56.

Wright, T. A. (2006). To be or not to be [happy]: The role of employee well-being. *The Academy of Management Perspective*, *20*(3), 118–125.

Wright, T. A. (2010). More than meets the eye: The role of employee well-being in organizational research. In P. A. Linley, S. Harrington, & N. Garcea (Eds.), *Oxford Handbook of Positive Psychology and Work* (143–154). Oxford, UK: Oxford University Press.

Wright, T.A. & Cropanzano, R. (2000). Psychological well-being and job satisfaction as predictors of job performance. *Journal of Occupational Health Psychology*, *5*(1), 84–94.

Wright, T. A., & Cropanzano, R. (2007). The happy/productive worker. In Marocchio, J. (Ed.): *Research in Personnel and Human Resource Management*, 26 (269–313). Amsterdam, The Netherlands: Elsevier.

Wayne, J. H., Grywacz, J. G., Carlson, D. S., & Kacmar, K. M. (2007). Work-family facilitation: A theoretical explanation and model of primary antecedents and consequences. *Human Resource Management Review, 17*(1), 63–76.

World Health Organization. (2013). *Mental Health: A State of Wellbeing*. Retrieved from: http://www.who.int/features/factfiles/mental_health/en/. (Accessed 11 April 2017).

Zheng, X., Zhu, W., Zhao, H., & Chi, Z. (2015). Employee well-being in organizations: Theoretical model, scale development, and cross-cultural validation. *Journal of Organizational Behavior*, *36*(5), 621–644.

PART IV

Resilience in Extreme Environment

CHAPTER 26

Psychological Flexibility: An Overview and Avenues for Application in the Military

SWATI MUKHERJEE* and UPDESH KUMAR

Mental Health Division, Defense Institute of Psychological Research, DRDO, Ministry of Defence, Timarpur, Delhi 110054, India

Corresponding author. E-mail: swatimukherjee@dipr.drdo.in

ABSTRACT

Psychological flexibility is an important but relatively neglected construct in the domain of positive psychology. Though inflexibility of thought, emotion, and behavior often provides significant markers for diagnosing problems with mental health and well-being, a mere absence of psychopathology does not necessarily indicate beneficial impact of psychological flexibility. Psychological flexibility is a multidimensional construct that has been defined to include adaptation to fluctuating situational demands, reconfiguration of mental resources, shifting of perspectives, and balancing of competing desires, needs and life domains. In this sense, psychological flexibility concerns more with interactions of the individuals with their environment, rather than inherent personality traits or dispositions. This chapter conceptualizes psychological flexibility as consisting of two major interdependent domains, namely, cognitive flexibility and emotional adaptability. Providing a brief overview of the existing research in the area, this chapter proposes that there is a need to constructively engage with the concept of psychological flexibility and systematically assess its role in social, emotional and behavioral adaptability that often contributes significantly to resilient outcomes under difficult circumstances. In the applied domain, the chapter proposes that training military personnel for developing psychological flexibility can enhance their preparedness for facing difficult circumstances, as well as act as a buffer for dealing with day to day stressors inherent in the military life.

26.1 INTRODUCTION

Inflexibility of thought, emotion, and behavior is a well-accepted and significant criterion for diagnosis of psychopathology (Berman et al., 2010); however, the construct of psychological flexibility is yet to be explored to reveal its true potential in contributing toward human well-being, resilience, and flourishing. Kashdan and Rottenberg (2010) remark that even if psychological inflexibility is a hallmark of many psychological disorders, it cannot be assumed that psychological flexibility necessarily confers good health. Kashdan and Rottenberg (2010) also provide a detailed review of the health benefits of psychological flexibility, dividing the studies in two domains, namely mental and behavioral response shifts, and balance among important life domains. Mental and behavioral response shifts denote the ability of an individual to modify the response in consonance with the situation at hand, and balance among life domains is denoted by ability to shift perspectives in order to satisfy various identity domains in one's life. This chapter begins with an overview of the various ways psychological flexibility has been defined, the varied dimensions that are included within the construct, and behavioral indicators of psychological flexibility. The chapter posits that it might be useful to view the components of psychological flexibility under two interdependent domains: cognitive flexibility and emotional adaptability. Reviewing the research in these two domains, the chapter emphasizes that psychological flexibility when defined as an ability (rather than a trait) provides a potent tool for enhancing individual well-being by improving adaptation to varied situations, emotions, and time perspectives. Placing this discussion in the applied contexts the chapter concludes with probable applications of psychological flexibility for the soldiers, and the benefits it accrues for the military organization as a whole.

26.2 DEFINITION AND DIMENSIONS

Psychological flexibility has been considered a difficult construct to define. Kashdan and Rottenberg (2010) comment that the construct has been largely pushed to the periphery by positive psychology researchers, and most of the existing research has been fragmented lacking integration. Different researchers have used different labels for the construct, for example, ego-resiliency (Block, 1961), executive control (Posner & Rothbart, 1998), response modulation (Patterson & Newman, 1993), and self-regulation (Carver & Scheier, 1998; Muraven & Baumeister, 2000). It is easier to identify the markers of psychological flexibility than arrive at a unitary definition. Kashdan and Rottenberg (2010) list the major markers of psychological flexibility in an individual in how he/she (i) adapts to fluctuating situational demands, (ii) reconfigures mental resources, (iii) shifts perspective, and (iv) balances competing desires, needs, and life domains.

Hayes et al. (2006) define psychological flexibility as the ability to contact the present moment more fully as a conscious human being, and to change or persist in behavior when this serves valued ends. A more elaborative definition is provided by Kashdan and Rottenberg (2010, p. 866) that describes it through its behavioral indicators. According to them psychological flexibility is a person's ability to (a) adapt to fluctuating situational demands, (b) reconfigure mental resources, (c) shift perspectives, and (d) balance competing desires, needs, and life domains. Psychological flexibility has also been defined as a mindfulness-based construct and individuals higher in dispositional mindfulness have been reported to be high on psychological flexibility too (Silberstein et al., 2012). A major contribution to the elaboration of the construct of psychological flexibility in the recent years has been done by the literature on the "acceptance and commitment therapy" (ACT), the third and latest wave of cognitive behavioral therapy, that emphasizes the significance of context in behavioral adaptability (Hayes, 2004). Though, initiated in the clinical context of cognitive behavior therapy (CBT), the ACT has found widespread applications in organizational behavior and interpersonal contexts too. Being one of the contextualized approaches to CBT, the ACT conceptualizes psychological flexibility as a transdiagnostic process that cuts across and influences the manifestations across many psychological disorders.

The conceptualization of psychological flexibility juxtaposed to psychological inflexibility by the ACT makes it easier to comprehend the components and dynamics involved. Psychological inflexibility is defined as a pattern of behavior that is excessively controlled by internal thoughts, feelings, and other internal processes (either for fulfillment or for avoidance of these), in disregard for the contextually relevant, meaningful and effective behavioral actions (Bond et al., 2011). An individual displays psychological inflexibility when their behavior is guided by internal psychological processes, not taking in consideration the situational contingencies or the preferred value positions. Hayes et al. (2011) place the root of psychological inflexibility in the tendency to avoid painful or unwanted internal thoughts, memories or emotions. Such experiential avoidance follows two pathways: first is the inability to tolerate aversive internal processes, and the second is narrowing of adaptive behavioral responding (Hayes et al., 1999). Another related process that has been conceptualized to aid psychological inflexibility is cognitive fusion (Hayes et al., 2005). Cognitive fusion is defined as the process when an individual's thoughts regulate the overt behavior in ineffective ways due to excessive focus on the content of thoughts at the cost of neglecting the context (Hayes et al., 2006; Pierson et al., 2004).

Ciarrochi et al. (2010) have described the processes involved in developing psychological flexibility using the ACT model. The ACT identifies six core processes that lead to developing psychological flexibility. These six processes have been divided into two major domains, first being acceptance and mindfulness processes, and the second being commitment and behavioral change processes. The two domains are overlapping. Acceptance and mindfulness processes include acceptance, defusion, the present moment, and a transcendent sense of self. Commitment and behavior change processes include a transcendent sense of self, values, and committed action. A brief description of these six subprocesses is in order for gaining a better understanding of psychological flexibility.

Acceptance refers to accepting the internal experiences rather than trying to control or avoid these. This includes acceptance of unwanted memories, aversive emotions, or painful thoughts. Whiting et al. (2017) define acceptance as, "The opening up and making room for distressing thoughts, emotions or experiences so that there is no longer an ongoing struggle." This implies learning to perceive and acknowledge whatever negative emotions or pain one is experiencing, instead of struggling to deny or suppress these. For example, if an individual feels jealous of a colleague's achievements, acceptance requires acknowledging the feeling, owning it up, rather than denying or negating the emotion.

The second process is defusion, which is the process of weakening the cognitive fusion. Cognitive fusion denotes "the process of creating some separation from distressing thoughts, emotions or experiences and changing the function of the thought rather than the content" (Whiting et al., 2017). Fusion is the process that causes the individual to emphasize the content of the thought, at the cost of becoming oblivious to the context in which they occur. Excessive emphasis on certain verbal content (e.g., "This is how a 'good' person should behave"; "My family would want me to value 'Y'") results in a rigid inflexible system of beliefs, which then dictate all action. Defusion, thus, involves learning to view one's thoughts as a neutral observer and understanding how use of certain parameters of language promotes fusion. Perceiving thoughts as resulting from one's experiences rather than as objective reality promotes defusion, breaking the thought–behavior link that might be causing ineffective behavioral response.

The third subprocess is a significant contributor to both mindfulness processes as well as the behavioral processes. Labeled as "getting in contact with the present moment," this involves developing awareness for the internal processes and being fully aware of what is happening in the present moment. Whiting and colleagues (2017) define it as, "being in the here and now, being more behaviorally flexible and consciously connecting with what is happening in that moment." This requires learning to view thoughts and emotions as passing events while the individual remains a neutral observer, rather than objective binding truths dictating behavior. Such awareness is a hallmark of mindfulness, and also connects to the behavioral domain by influencing the behavioral choices one makes in full awareness of one's preferred values and the situational context.

The fourth subprocess involves learning to view "self-as-context" or the observant self. This emphasizes the notion that certain aspect of self always remains the same, despite a constantly changing flow of thoughts and emotions. Closely related to learning to be mindful, viewing self as a context also involves developing a certain level of detachment to the self, and learning to distinguish between various self-evaluations and the self that is at the core of one's existence. Pierson et al. (2004) describe it as learning to let go of the unhelpful self-evaluations and retaining the sense of self.

The fifth subprocess is about keeping the values one holds in the center field of one's awareness when deciding upon taking action. Values are personally relevant principals that are unique to the individual (Whiting et al., 2017). When one learns to set goals that are consistent with one's values and molds one's responses so that they are not inconsistent with one's values, this leads to flexible and purposeful behavior. Values provide directions to behavior, but when individual is overwhelmed due to cognitive fusion or experiential avoidance, the behavior becomes inconsistent failing to achieve values outcomes. The last subprocess labelled "committed action" involves consistent engagement and pursuance of valued goal despite internal discomfort or distress. It is about learning to rise again after failures and setbacks, accepting difficult feelings and thoughts and continuing to move in the values direction despite these. Also implied in the process is the readiness to change or modify the chosen behavioral action when it becomes evidently inconsistent with the valued goals. The dynamic combination of determined pursuance of goals along with the flexibility and readiness to change paths if required is the hallmark of psychological flexibility.

These six core integrated processes have been found to be involved in both human adaptability and suffering (Hayes et al., 2012). These processes are also reported as potent indicators of psychological flexibility, though Ciarrochi et al. (2010) have argued for the need for further psychometric research for determining the specifics. The elaboration of the super-arching construct of psychological flexibility through these six processes is in consonance with the theorization that a psychologically flexible individual is less likely to respond to unwanted or negative private events with control or avoidance-based behaviors and will be more likely to respond in a way that is congruent with his or her personal values (Hayes, 2004; Hayes et al., 2006).

26.3 PSYCHOLOGICAL FLEXIBILITY AS COGNITIVE FLEXIBILITY

As compared to the literature on the construct of psychological flexibility that primarily derives from research in clinical and therapeutic domains, cognitive flexibility literature is rooted in cognition and information processing approaches. Cognitive flexibility is defined as the human ability to adapt the cognitive processing strategies to face new and unexpected conditions in the environment (Cañas et al., 2003). Cognitive flexibility refers to the individual's ability to transition from thinking about one concept to another, and to process multiple concepts at a time. A related construct is cognitive shifting, that involves adapting one's behavior and thoughts to new, changing, unexpected events, in order to make appropriate changes to respond adequately to situational demands. Cognitive flexibility has been recognized as a component of executive functioning, and highly related to many of the other components of executive functioning, like memory, inhibition, emotional stability, planning, and organization (Miyake et al., 2000).

The definition by Cañas and colleagues highlights certain important points. First, it conceptualizes cognitive flexibility as an ability that can be acquired through learning; second it focuses on acquiring the cognitive strategies or sequence of operations that lead to effective problem solving; third it conceptualizes flexibility as a complex behavior repertoire, rather than a single discreet response; and fourthly it highlights the core process of adaptation to novel situations and changing environment.

Ionescu (2012) considers cognitive flexibility as a property of the cognitive system, rather than a cognitive skill. Described as "the unified framework of cognitive flexibility" the model proposed by Ionescu (2012) highlights the potential usefulness of cognitive flexibility in problem solving and creativity. Ionescu (2017, p. 6) asserts that flexibility is present "when abilities are well tuned to changing demands," implying that being responsive to the context is a core factor in being flexible. Flexibility is often achieved after a prolonged period of stability in thought and other executive functions. Ionescu (2017) proposes a variability–stability–flexibility model that is used for explaining the nonlinear pattern in which stability, variability and flexibility follow each other in a seemingly random fashion.

Several explanatory propositions have been forwarded for the construct of cognitive flexibility. Cañas et al. (2006) discuss some of these. They propose that engaging in cognitively flexible behavior while performing a complex task puts a twofold demand on the individual. The first is of maintaining a conscious awareness of the situational complexity, which rarely if ever remains unchanged. Along with monitoring the changing dynamics of the situation, the individual also needs to curb their automatic prelearned responses. The second demand is placed by constant restructuring required of their knowledge in order to effectively deal with the situation at hand by coming up with new strategically appropriate solutions. They name these two processes as attentional processes and knowledge representation. Spiro and Jehng (1990) in proposing the Cognitive Flexibility Theory assert that people who develop a multiperspective representation of the task are able to easily interpret situational changes in the environment, and quickly restructure their knowledge in consonance with it. Both the knowledge representation and attentional process approaches have found application in the training for cognitive flexibility. The knowledge representation approaches to training have focused on developing a knowledge base through exposure to variable situational demands, as contrasted with practicing a fixed set of procedural steps for dealing with the situation (Antolí et al., 2005). However, as it might not be always feasible to train people through exposure to variable environmental conditions, the second approach to training focuses on the attentional processes. Such training involves instructing the trainees to shift their attention from one task to the other, or to different subparts of a task at hand in accordance with the instruction. The idea behind such training is to enable the participants to take control of their attentional processes and dedicate these to specific focus areas in accordance with the situational demands.

Attentional disengagement and variability in the behavioral repertoire obtained through a flexible knowledge representation can be two interrelated processes helping the development of cognitive flexibility. It is relevant, in the context that cognitive flexibility is a conscious process of shifting the attentional resources from one aspect to the other and utilizing the relevant knowledge representation. When happening at the unconscious level, it is termed as "task switching," and might involve some "cost" in terms of attention loss.

26.4 PSYCHOLOGICAL FLEXIBILITY AS EMOTIONAL ADAPTABILITY

Emotional adaptability or emotion regulation is the ability of an individual to be aware of one's emotions and regulate them in consonance with the demands of the situation. Drawing a parallel from the cognitive flexibility model, emotional adaptability may be conceptualized as the capability to manage one's own emotions and understand the emotional content of the context (other people's emotions), and modify one's behavioral response in accordance with the valued goals. This also implies not resorting to defensive strategies or emotion suppression for managing overt behavioral response, rather being able to consciously make a choice for functional emotional experience and expression. In this sense, emotional adaptability appears to be skill closely related to the construct of emotional intelligence. Goleman (1998) has conceptualized emotional intelligence as a cluster of 25 competencies, grouped in five clusters, namely self-awareness cluster, self-regulation cluster, empathy cluster, motivation cluster, and social skills cluster. On the conceptual level, some of these competencies, like self-awareness, self-assessment, self-control, and understanding others' emotions can be utilized to develop the construct of emotional adaptability.

Maladaptive emotional regulation is increasingly being recognized and included in various models of psychopathology (Berenbaum et al., 2003; Greenberg, 2002; Kring & Bachorowski, 1999; Mennin & Farach, 2007). Many strategies for emotion regulation have been identified that act as protective factors against psychopathology. Two of these strategies that find application across many intervention approaches are reappraisal and problem solving (Aldao et al., 2010). Reappraisal is a skill or strategy that involves eliciting positive or gentle interpretations of a stressful situation in order to reduce the distress and generate an adaptive response (Gross, 1998). Problem solving is a conscious attempt or orientation to reduce the distress by seeking to bring in a change in the stressful situation through taking some action. Though it does not directly focus on emotions, it reduces distress by changing the stressor itself. Enhancing problem solving skills is a component of many therapeutic approaches for reducing emotional distress. The third adaptive strategy for emotion regulation focuses on mindful awareness and nonjudgmental acceptance of one's emotions. Mindfulness-based therapies focus on this detached self-awareness of thought and emotion, in order to break the conditioned responses elicited by the thought–emotion–behavior linkage.

Many therapeutic approaches such as dialectical behavioral therapy (Linehan, 1993), emotion focused therapy (Greenberg, 2002), emotion-regulation therapy (Mennin & Fresco, 2009) include some form of emotion regulation training for facilitating adaptive behavior. The construct of emotional adaptability is also in consonance with the cognitive therapies that facilitate cognitive reappraisal of emotions in place of expressive regulation, and mindfulness-based therapies focusing on awareness and acceptance (Chambers et al., 2009).

Adaptive emotion regulation, as opposed to maladaptive regulation using avoidance, suppression, or rumination can be conceptualized as one of the major contributors toward psychological flexibility, the other being cognitive flexibility. Flexibility of cognition and adaptive regulation of emotional experience and expression have interconnected and overlapping processes and influence psychological flexibility in a dynamic manner. The purpose behind operationalizing the construct of psychological flexibility in terms of these two components is primarily their trainability. Academic as well as clinical research literature has amply demonstrated that both flexible cognitive processing as well as adaptive regulation of emotion are competencies or skills that can be learnt through experience and use of specific strategies.

To sum up, psychological flexibility connotes an ability to live and fully function in the present moment with conscious awareness and acceptance of the situational context on one hand and of one's inner thoughts and emotions on the other. This ability is then translated into purposive action that is in consonance with the goals one finds valuable. The two distinct yet interrelated components of psychological flexibility, thus, are mindful acceptance and committed action. In a similar vein, Kashdan and Rottenberg (2010) delineate three core components of psychological flexibility. The first component is the cognitive component that is reflected in one's ability to see novel solutions in problem situations, cognitive shifting, and creative thought. It is essentially defined by the executive functions of memory, attention, problem solving, and the like. The second component is reflected in the individual's ability to achieve a balance or psychological equilibrium when under stress. The third component is defined by the underlying personality traits, such as neuroticism, positive affect, openness to experience, and self-control.

When considering the construct of psychological flexibility in the context of structuring the training program for the soldiers, both cognitive as well as the emotional adaptability components of psychological flexibility appear to be crucial.

26.5 TRAINING FOR PSYCHOLOGICAL FLEXIBILITY: A CASE FOR THE MILITARY

Military has been described as a greedy institution (Coser, 1974; Segal, 1986). A greedy institution is one that puts excessive demand on the time and energy resources of the individual, leaving little or no resources for committing to other aspects of life. Military as a greedy institution demands total commitment from the service personnel seeking total conformity with the service rules and demands of the job per se. Along with the stresses associated with combat deployment, the soldier has to face innumerable routine stressors with minimal support. Moreover, given the inherent nature of the military job, the threshold of the day-to-day stressors faced by the soldiers is higher as compared to their civilian counterparts. With the changing nature of warfare on one hand, and changing structure of the society on the other, the service personnel face demanding situations on multiple fronts. Given the enormous literature available on the impact of day-to-day stressors on the well-being of the soldier, it has become imperative to design specific preventive strategies.

In the recent years, soldier well-being and mental health have been receiving increasing attention. While the earlier strategies focused on identification of the vulnerable and designing intervention strategies for those who broke down under stress, recent years have seen a shift from intervention to prevention, and further toward identifying and developing soldier strengths as a pre-emptive strategy. The pre-emptive strategies need to be deployed at three levels: at the level of the individual soldier, at the interpersonal level, and at the organizational level. A training for inculcating psychological flexibility operates both at the intrapersonal as well as the interpersonal levels. It might appear paradoxical and antithetical to propagate flexibility of thought and emotion in an organization characterized by rigid hierarchies and rules; however, inculcation of psychological flexibility through the ranks and files of the organization can be foreseen to bring in positive benefits for the individual soldier as well as for the organizational goals.

The two major components of the training for psychological flexibility are conceptualized as (a) developing flexible thought and (b) developing emotional adaptability. Developing flexible thought makes use of the two underlying mechanisms for training the soldiers. First is training the soldier for regulating their attentional processes in a conscious manner, developing varied perspectives to the problem at hand, and deliberately shifting attention from one subtask to the other. The second aspect focuses on encouraging the trainees to take stock of their knowledge base, be aware of their biases and stereotypical ways of responding. Along with upgrading their knowledge they are trained to employ self-monitoring strategies for checking the appropriateness of the behavioral response in dealing with the situation at hand.

Along with skills training to take conscious control of the cognitive processes, it is also important to develop emotional adaptability through adaptive emotion regulation. Failing to regulate one's emotions might lead to inflexibility of behavior, even if the cognitive processes are flexible enough. Based on a comprehensive review of literature, the training for emotional adaptability focuses on three aspects: (a) facilitating emotional awareness and acceptance (mindfulness based approaches), (b) facilitating emotional expression rather than suppression, shifting from nonfunctional to functional emotions (cognitive-therapy-based approaches), and (c) skills training for enhancing problem solving capability.

Psychological flexibility has been shown to enhance well-being and positive mental health outcomes for an individual. Kashdan and Rottenberg (2010) provide a detailed review of empirical research that documents the health benefits of psychological flexibility. Not disputing that the benefits of positive emotions, strengths, positive traits, or satisfaction of the basic needs for belonging, competence and autonomy are important, they assert for the complementary and significant role played by psychological flexibility in dealing with the complexity and fluctuation of the daily stressors. Psychological flexibility provides the individual with a dynamic way to accept, acknowledge, and adaptively deal with the challenges presented by the change induced by the situational context. In this sense, the processes involved in psychological flexibility are akin to the processes of resilient adaptation to adversity. Zautra and colleagues have defined resilience as "an adaptive response to adversity through the three processes of recovery, sustainability, and growth" (Zautra et al., 2010). This model highlights three aspects of resilience: the ability to sustain under stress and adversity, the propensity to quickly recover from the stress-response in order to regain prestress baseline state, and a possibility for post-traumatic growth. Given the conceptualization of psychological flexibility as a skill trainable through exposure and learning, strategies can be designed for helping individuals enhance this capability. A positive relationship can reasonably be hypothesized between enhanced flexibility of cognition and emotional adaptability on one hand, and resilient adaptation on the other.

Training the soldier in cognitive flexibility skills and in adaptive emotion regulation requires a systematic and regular schedule of practice. The training program thus proposes a short and focused training workshop with skilled trainers for learning the basic skills. This then needs to be followed with regular practice on various exercises enhancing creative problem solving skills, perspective taking, and finding novel solutions (the cognitive component); and acceptance of inner thoughts and emotions as well as external contexts despite the pain and discomfort involved, structuring the behavioral action in a manner that it is in consonance with one's values. The training program for enhancing psychological flexibility among Indian soldiers is being developed including both these focus areas. It ensures ease of implementation, given the busy schedule and all-pervasive nature of military assignments. The program can be implemented as a structured training course in a graded manner in the training institutions as well as in the form of an intensive training capsule for the soldiers on deployment in peace and field locations. The efficacy of the program, however, has been found to be dependent on two factors. First is the requirement of trained instructors, who have a conceptual understanding of the issues and are able to utilize appropriate training methodologies for training the soldiers. The second issue pertains to the reflective nature of the exercises designed for development of psychological flexibility, which requires some effort and willingness on the part of the trainee to intellectually engage with the concepts. However, there is ample literature to support the efficacy of training in enhancing cognitive flexibility and adaptive emotion regulation, the two main components of flexibility training.

26.6 CONCLUDING COMMENT

Flexibility, especially cognitive flexibility has been a construct of interest since long for the cognitive psychologist and the neuropsychologists (Berg, 1948). However, recent developments have seen a broadening of the construct to include adaptive emotion, and the incorporation of both cognitive and emotion aspects within the larger construct of psychological flexibility. Whiting and colleagues (2017) highlight the definitional commonalities between the two conceptualizations emphasizing that both perspectives find loss or impairment of flexibility a major factor underlying many forms of psychopathology. Though, they also emphasize that the relationship between the two perspectives needs further exploration, the present work has focused on the applied aspects. Hypothesizing cognitive flexibility as a subset of the skills that define psychological flexibility, it has focused on developing a training program for the soldiers that would have specific benefits in the military environment. The major domains where positive outcomes are expected as a result of training in psychological flexibility are as follows: (i) an enhanced capability to search for novel solutions, and also readiness to change perspectives on problems in order to arrive at workable solutions (creative problem solving), (ii) accepting and acknowledging one's painful inner feelings and thoughts (acceptance), instead of suppression or denial (that often leads to maladaptive behaviors), (iii) managing balanced behavioral outcomes through inculcating certain level of detachment from one's thoughts and emotions (mindfulness), (iv) clarity about what one values in life and a nonjudgmental approach toward molding behavior in pursuance of the valued goals (value), and (v) purposive and goal directed behavior despite difficult circumstances or persistent stress (committed action).

The acute stressors like war and violence that are an imminent part of military life get exacerbated by the chronic stressors faced by the soldier like, prolonged separation from family, and persistent threat of violence. The daily life stressors like financial concerns, provision of basic facilities, though of a lower intensity contribute significantly to the soldiers' overall stress. In the circumstances, it is a good strategy to pre-emptively equip the soldier with effective strategies that not only help him manage his stress, but in a proactive manner propel him toward achieving his true potentialities by being resilient in the face of adversities. A resilient soldier functioning with psychological flexibility in dealing with organizational requirements as well as in dealing with his emotional issues is likely to contribute significantly toward enhancing overall organizational resilience.

KEYWORDS

- **cognitive flexibility**
- **emotional stability**
- **training for flexibility**

REFERENCES

Aldao, A., Nolen-Hoeksema, S., & Schweizer, S. (2010). Emotion-regulation strategies across psychopathology: A meta-analytic review. *Clinical Psychology Review*, *30*(2), 217–237.

Antolí, A., Cañas, J. J., Fajardo, I., & Salmerón, L. (2005). Cognitive flexibility of the strategies in dynamic complex problem solving tasks: Effects of different types of training. *Theoretical Issues in Ergonomics Science*, *6*(1), 95–108.

Berenbaum, H., Raghavan, C., Le, H. N., Vernon, L. L., & Gomez, J. J. (2003). A taxonomy of emotional disturbances. *Clinical Psychology: Science and Practice*, *10*(2), 206–226.

Berg, E. A. (1948). A simple objective technique for measuring flexibility in thinking. *The Journal of General Psychology*, *39*(1), 15–22.

Berman, N. C., Wheaton, M. G., McGrath, P., & Abramowitz, J. S. (2010). Predicting anxiety: The role of experiential avoidance and anxiety sensitivity. *Journal of Anxiety Disorders*, *24*(1), 109–113.

Block, J. (1961). *The Q-sort method in personality assessment and psychiatric research.* Springfield, IL: Charles C Thomas.

Bond, F. W., Hayes, S. C., Baer, R. A., Carpenter, K. M., Guenole, N., Orcutt, H. K., Waltz, T. & Zettle, R. D. (2011). Preliminary psychometric properties of the Acceptance and Action Questionnaire-II: A revised measure of psychological flexibility and acceptance. *Behavior Therapy*, *42*(4), 676–688.

Cañas, J. J., Fajardo, I., & Salmerón, L. (2006). Cognitive flexibility. In Waldemar Karwowski (Ed.), *International encyclopedia of ergonomics and human factors* (pp. 297–300). Boca Raton, FL: CRC Press.

Cañas, J. J., Quesada, J. F., Antolí, A., & Fajardo, I. (2003). Cognitive flexibility and adaptability to environmental changes in dynamic complex problem-solving tasks. *Ergonomics*, *46*(5), 482–501.

Carver, C. C., & Scheier, M. F. (1998). *On the self-regulation of behavior*. New York: Cambridge University Press.

Chambers, R., Gullone, E., & Allen, N. B. (2009). Mindful emotion regulation: An integrative review. *Clinical Psychology Review*, *29*(6), 560–572.

Ciarrochi, J., Bilich, L., & Godsel, C. (2010). Psychological flexibility as a mechanism of change in acceptance and commitment therapy. In Ruth Baer's (Ed.), *Assessing mindfulness and acceptance: Illuminating the processes of change* (pp. 51–76). Oakland, CA: New Harbinger Publications, Inc.

Coser, L. A. (1974). *Greedy institutions: Patterns of undivided commitment*. New York: Free Press.

Goleman, D. (1998). *Working with emotional intelligence*. New York: Bantam.

Greenberg, L. S. (2002). *Emotion-focused therapy: Coaching clients to work through their feelings*. Washington, DC: APA.

Gross, J. J. (1998). The emerging field of emotion regulation: An integrative review. *Review of General Psychology*, *2*(3), 271–299.

Hayes, S. C. (2004). Acceptance and commitment therapy, relational frame theory, and the third wave of behavioral and cognitive therapies. *Behavior Therapy*, *35*(4), 639–665.

Hayes, S. C., Luoma, J. B., Bond, F. W., Masuda, A., & Lillis, J. (2006). Acceptance and commitment therapy: Model, processes and outcomes. *Behavior Research and Therapy*, *44*(1), 1–25.

Hayes, S. C., Strosahl, K., Bunting, K., Twohig, M., & Wilson, K. G. (2005). What is acceptance and commitment therapy? In S. C. Hayes & K. D. Strosahl (Eds.), *A practical guide to acceptance and commitment therapy* (pp. 3–29). New York: Springer Science + Business Media.

Hayes, S. C., Strosahl, K., & Wilson, K. G. (1999). *Acceptance and commitment therapy: An experiential approach to behavior change*. New York: The Guilford Press.

Hayes, S. C., Villate, M., Levin, M., & Hildebrandt, M. (2011). Open, aware, and active: Contextual approaches as an emerging trend in the behavioral and cognitive therapies. *Annual Review of Clinical Psychology*, *7*, 141–168.

Hayes, S. C., Strosahl, K. D., & Wilson, K. G. (2012). *Acceptance and commitment therapy: The process and practice of mindful change*. New York: Guilford.

Ionescu, T. (2012). Exploring the nature of cognitive flexibility. *New Ideas in Psychology*, *30*(2), 190–200.

Ionescu, T. (2017). The variability–stability–flexibility pattern: A possible key to understanding the flexibility of the human mind. *Review of General Psychology*, *21*(2), 123–131.

Kashdan, T. B., & Rottenberg, J. (2010). Psychological flexibility as a fundamental aspect of health. *Clinical Psychology Review*, *30*(7), 865–878.

Kring, A. M., & Bachorowski, J. A. (1999). Emotions and psychopathology. *Cognition and Emotion*, *13*(5), 575–599.

Linehan, M. M. (1993). *Cognitive-behavioral treatment for borderline personality disorder*. New York: Guilford Press.

Mennin, D. S., & Farach, F. J. (2007). Emotion and evolving treatments for adult psychopathology. *Clinical Psychology: Science and Practice*, *14*, 329–352.

Mennin, D. S., & Fresco, D. M. (2009). Emotion regulation as a framework for understanding and treating anxiety pathology. In A. M. Kring & D. M. Sloan (Eds.), *Emotion regulation in psychopathology* (pp. 356–379). New York: Guilford Press.

Miyake, A., Friedman, N. P., Emerson, M. J., Witzki, A. H., Howerter, A., & Wagner, T. (2000). The unity and diversity of executive functions and their contributions to complex "frontal lobe" tasks: A latent variable analysis. *Cognitive Psychology*, *41*(1), 49–100.

Muraven, M., & Baumeister, R. F. (2000). Self-regulation and depletion of limited resources: Does self-control resemble a muscle? *Psychological Bulletin*, *126*, 247–259.

Patterson, C. M., & Newman, J. P. (1993). Reflectivity and learning from aversive events: Toward a psychological mechanism for the syndromes of disinhibition. *Psychological Review*, *100*, 716–736.

Pierson, H., Gifford, E. V., Smith, A. A., Bunting, K., & Hayes, S. C. (2004). *Functional acceptance and commitment therapy scale*. Unpublished manuscript.

Posner, M. I., & Rothbart, M. K. (1998). Attention, self-regulation, and consciousness. *Philosophical Transactions of the Royal Society of London*, *353*, 1915–1927.

Segal, M. W. (1986). The Military and the family as greedy institutions. *Armed Forces & Society*, *13*(1), 9–38.

Silberstein, L. R., Tirch, D., Leahy, R. L., & McGinn, L. (2012). Mindfulness, psychological flexibility and emotional schemas. *International Journal of Cognitive Therapy* *5*(4), 406–419.

Spiro, R. J., & Jehng, J. C. (1990). Cognitive flexibility and hypertext: Theory and technology for the nonlinear and multidimensional traversal of complex subject matter. In D. Nix & R. J. Spiro (Eds.), *Cognition, education, and multimedia: Exploring ideas in high technology* (pp. 163–205). Hillsdale: Lawrence Erlbaum Associates.

Whiting, D. L., Deane, F. P., Simpson, G. K., McLeod, H. J., & Ciarrochi, J. (2017). Cognitive and psychological flexibility after a traumatic brain injury and the implications for treatment in acceptance-based therapies: A conceptual review. *Neuropsychological Rehabilitation*, *27*(2), 263–299.

Zautra, A. J., Arewasikporn, A., & Davis, M. C. (2010). Resilience: Promoting well-being through recovery, sustainability, and growth. *Research in Human Development*, *7*(3), 221–238.

CHAPTER 27

Psychological Resources of Resilience to Terrorist Threat

NADEZHDA N. KAZYMOVA, NADEZHDA V. TARABRINA, and JULIA V. BYKHOVETS*

Institute of Psychology, Russian Academy of Sciences, Moscow, Russia

**Corresponding author. E-mail: bykhovets@yandex.ru*

ABSTRACT

This chapter deals with the problem of determining the positive and negative factors of psychological resilience to the experience of the terrorist threat. The risk factors are psychopathological symptoms and signs of post-traumatic stress. Their presence presumably increases susceptibility to the terrorist threat. Resource factors are the components of psychological well-being and features of life prospects. The entire sample showed a negative relationship between the experience of the terrorist threat and positive factors and a positive relationship with risk factors. People, who are resistant to the threat of terrorist attacks, report a lower level of psychopathological symptoms and a higher estimate of their psychological well-being. They are more positive about their social relations, strive for greater autonomy, easier to cope with everyday tasks, have life goals, and strive for self-development. Women experience more intense terrorist threat than men, they are less resilient to information about terrorism, anticipate terrorist attacks, and are therefore more stressed. A high level of anxiety is manifested in the motivational content of the life perspective: in women it is associated with concern for loved ones, and in men—with concern for their own lives and well-being. Age specificity was studied: respondents, older than 35 years, experience more intense terrorist threat, have more pronounced signs of stress, and their level of psychological well-being is lower than that of younger age groups.

27.1 INTRODUCTION

A lot of data about the stressful nature of the terrorist threat have been accumulated in psychology. Understanding the clinical and psychological consequences of experiencing the terrorist threat is necessary to develop effective strategies to reduce the negative effects of the threat of terrorism on the mental health. The most common psychological consequences of experience of this threat are stress reactions, depression, sleep disorders, increased consumption of psychoactive substances. The results of longitudinal studies of the consequences of terrorist attacks show that the intensity of post-traumatic symptoms caused by information about terrorist attacks, gradually reduced for most of society and remains in a special, vulnerable part of society (Caramanica et al., 2014; Neria et al., 2013; Thoresen et al., 2014; Welch et al., 2016). These studies have demonstrated the ability of people to maintain optimal levels of functioning in extreme conditions, as well as to recover from traumatic events. The search and study of resources that allow a person to cope with life's difficulties, to function effectively, and succeed is an urgent task of psychology, since the life activity of modern man is forced to unfold in the conditions of the constant threat of terrorist attacks. Social and psychological resources (e.g., social support, social skills, level of education, and income) and individual psychological characteristics should be considered.

The experience of the terrorist threat can be considered as a model of a stressor, whose action is based not on the direct impact on a person, but on the perception of information about its presence (Tarabrina et al., 2007). To date, in Western psychology, concepts of psychological resilience are being actively developed in contrast to the study of vulnerability factors and negative psychological consequences of experiencing the threat of terrorist attacks. In Russian psychology, the term "resilience" was introduced into a wide scientific use by V. A. Makhnach in 2003. According to his definition, resilience is "the ability of a person to overcome adverse life circumstances with the ability to recover and use all possible internal and external resources, the ability to live in all its manifestations, based on the will to it" (Makhnach & Dikaya, 2016, p. 13). This definition is consonant with foreign concepts of resilience, according to which it means that a person has the ability to overcome negative circumstances, adapt to a situation of injury, to realize themselves and reach well-being despite the potential stress of living conditions (Mukherjee & Kumar, 2017). The theoretical significance of the concept of resilience in psychology is the generalization and understanding of empirical data on the adaptive functioning of a person in stressful circumstances of threatening influence.

Currently, most researchers tend to view resilience as a process rather than a stable personality trait (Mukherjee & Kumar, 2017). It is noted that the action of risk or protective factors alone is not sufficient to enhance or reduce a positive outcome. The specificity of resilience as process is its dynamism, which consists in balancing between risk factors and protective factors. Risk factors do not automatically lead to negative consequences; their presence only exposes the person to the circumstances associated with a high probability of such consequences. Protective factors have positive potential, regardless of risk status. Protective factors mediate the impact of risk factors, reducing their impact on the individual in a situation of threat and helping him adapt successfully (Mukherjee & Kumar, 2017).

The construct of resilience still lacks a single definition. The difficulties of operationalization of resilience lead to the lack of a unified method for its study. As a deductive construct, resilience cannot be observed or measured directly, but a judgment of the level of resilience can be derived from the observed behavior of an individual or a living system that has experienced extreme impacts.

From our point of view, resilience can be assessed using the construct of psychological well-being, which was proposed by C. Ryff in the 1980s to denote the set of psychological characteristics that act as determinants of health and happiness. The theoretical model and its main components originate in the concept of positive mental health, formulated in the mainstream of humanistic psychology.

In our study, we consider the presence of psychopathological symptoms and signs of post-traumatic stress as risk factors that increase sensitivity to the terrorist threat. The components of psychological well-being can be considered as resources that support the resilience of the individual and help to overcome the negative consequences of the experiencing of terroristic threat.

The aim of this study is to examine the relationship of the terrorist threat experience with the components of the psychological well-being of the individual on one hand and with the presence of psychopathological symptoms and signs of post-traumatic stress on the other.

27.2 METHOD

The following questionnaires have been used to attain the aim of the research:

1. The scales of psychological well-being (SPW) (Ryff, 1989; adapted for Russia by Shevelenkova & Fesenko, 2005) is designed to assess the actual psychological well-being of a person. SPW consists of six dimensions: autonomy, environmental mastery, personal growth, positive relations with others, purpose in life, and self-acceptance. The summation of points of all scales allows to get a general index of psychological well-being. The SPW is a widely used 84-item self-report measure of psychological well-being with well-established psychometric properties. Each item is rated on a 6-point Likert-type scale ranging from 1 to 6. The SPW has demonstrated excellent internal and test–retest reliability, in addition to strong convergent validity (Shevelenkova & Fesenko, 2005).
2. A scale of "method of motivational induction (MMI)" (Nuttin, 2004; adapted for Russia by Tolstyh, 2005) is aimed at studying the characteristics of the time perspective of the future. The blank contains 30 unfinished sentences (inductors: 20 sentences with positive wording ("I want...," "I dream...," etc.) and 10 negative sentences ("I don't want...," "I'm afraid that...," etc.). Inductors encourage people to write about their plans, goals, desires, intentions, etc. The answers of respondents are coded according to the system of codes: the code of content and temporal code. Ten categories are proposed for the analysis of the content of motivational objects: S (self), SR (self-realization), R (realization), C (contact), E (exploration), T (transcendental), P (possessions), L (leisure), Tt (test), U (unclassified). Temporal codes: categories of calendar time periods (T (test), D (day), W (week), M (month), Y (year), categories of human social life periods (E (education), a (adult), O (old)), additional categories whose temporal localization is difficult to establish (L (life), l ("open present"), x ("historical future"). This questionnaire does not provide normative indicators and is used for qualitative analysis of the content and length of life prospects.
3. A questionnaire of "prospect of my life" (Misko & Tarabrina, 2004) allows to get the data of respondents' views about the future life: emotional attitude, certainty and duration, social welfare, and the role of personality about the future. The method includes 15 pairs of definitions of life prospects, including emotional, cognitive, goal setting, etc.,

components. The indicator is obtained by summing up the scores of all points of the questionnaire. The extreme negative pole was assigned a value of 1, and the extreme positive—7. The overall score reflects a tendency to assess the life perspective as a whole as pessimistic or optimistic (Misko & Tarabrina, 2004).

4. The Mississippi scale (MS) (Kean, 1987, 1988) focuses on assessment of the severity of the symptoms of post-traumatic stress. The behavioral reactions and emotional experiences described in 39 items and grouped into several groups: obsessive memories, affective lability, various personal problems, etc. Each item is rated on a 5-point Likert-type scale ranging from 1 to 5. The scale has one final indicator. It gives an overall assessment of the severity of post-traumatic reactions. The MS has good reliability and validity. The Cronbach's alpha coefficient of the MS is 0.76 (Tarabrina et al., 2007).
5. Symptom check list-90-revised (SCL-90-R) (Derogatis et al., 1976) consists of 90 statements reflecting the presence of certain somatic and psychological problems. Each item is rated on a 5-point Likert-type scale ranging from 0 to 4. Evaluation and interpretation of the results are carried out on nine main symptomatic subscales: somatization, obsessive-compulsive, interpersonal sensitivity, depression, anxiety, hostility, phobic anxiety, paranoid ideation, and psychoticism. There are three global scores: the Global Severity Index, Positive Symptom Distress Index, and Positive Symptom Total. The SCL-90-R has demonstrated high internal and test-retest reliability (Tarabrina et al., 2007).
6. The questionnaire of the terrorist threat (QTT) (Byhovets & Tarabrina, 2010) assesses the intensity of the terrorist threat experience. The questionnaire is based on the three-component structure of the terrorist threat experience and includes the following scales: anticipation, resistance, and post-traumatic stress. It is important to note that the scale of resistance is the reverse: the higher the value, the lower the resistance. The QTT consists of 50 items, which have 5-point variable from 1—"absolutely true" to 5—"absolutely not." The global index is obtained by summing up the scores of all points of the questionnaire and expresses the intensity of the subjective experience of the terrorist threat. Psychometric properties of the QTT are on satisfactory level (Byhovets & Tarabrina, 2010).

27.2.1 SAMPLE

The participants of research were 301 residents (182 women and 109 men) of Moscow and the Moscow region aged 17–60 years. The sample was divided into three age groups. The youth group included 133 people aged 17–21 years—80 females (mean age: 19.67) and 53 males (mean age: 19.6). The respondents of this group were the students of higher educational institutes. The early adult group consists of people aged 22–35 years and included 58 females (mean age: 26.72) and 50 males (mean age: 27.44). The third group included 60 respondents aged 36–60 years—44 females (mean age: 48.07) and 16 males (mean age: 51.94). The participants of the study older than 22 years were people of different professional backgrounds and different social status, who expressed their consent to participate in an individual survey. The study participants were never directly affected by the terrorist attack and had no relatives or close friends who were injured or killed in the terrorist attack.

27.2.2 DATA ANALYSIS

Mathematical data processing was carried out using the computer program Statistica 8.0 and included the procedures of descriptive statistics, calculation of the correlation coefficients *r*-Spearman and Pearson-*r*, comparative analysis of groups of respondents using the criteria *t*-student (in the case of normal distribution) and *U*-Mann–Whitney (for distributions other than normal).

27.3 RESULTS

27.3.1 STUDY OF TERRORIST THREAT EXPERIENCE, PSYCHOLOGICAL WELL-BEING AND PSYCHOPATHOLOGICAL SYMPTOMS IN GROUPS OF RESPONDENTS OF DIFFERENT AGE AND GENDER

The main constructs used in the study were analyzed in groups of respondents with different age and sex.

27.3.1.1 THE EXPERIENCE OF TERRORIST THREAT

Statistical analysis of age differences in QTT indicators was carried out using the *t*-test for independent samples. The group of respondents of early adult age statistically significantly differed on the scale of "stability" from the group of young adults ($M_1 = 33.64$, $M_2 = 31.03$, $p = 0.008$) and the group of middle-aged adults ($M_3 = 34.88$, $p = 0.002$), while the differences on this scale are not observed between the groups of early and middle-aged adults ($p = 0.308$). Thus, resistance to information about terrorist attacks is higher in the group of early adulthood (22–35 age). With the exception of this scale, the youth and early adult groups have no statistically significant differences in QTT. The group of middle adult age is the most different in all indicators of the QTT ($p < 0.05$). Despite the fact that people of this age expect the new terrorist attacks in the future about the same as other age groups, they experience more symptoms of PTS ($M_1 = 29.59$, $M_2 = 27.73$, $M_3 = 32.60$, $p < 0.05$) and are more intensely worried about the terrorist threat in general ($M_1 = 120.24$, $M_2 = 116.81$, $M_3 = 131.00$, $p < 0.05$).

Thus, the experience of the terrorist threat is the most intense among older respondents. Compared to the group of early adulthood, the older group, in which women predominate in our study, may no longer feel the self-confidence that is inherent in young adults. In addition, such specific stressors associated with age, deteriorating physical condition, health problems, and other circumstances, which limit opportunities for older age, can increase sensitivity to threats.

Analysis of gender differences showed that all components of the terrorist threat experience have a higher level in the group of women (differences in the *t*-criterion are reliable at the level of significance $p = 0.000$ on all scales). Thus, women seem to experience a terrorist threat more intensively, which is expressed in the alarming expectation of new terrorist attacks, greater sensitivity to information about terrorist attacks and a higher level of post-traumatic stress reactions in connection with it.

27.3.1.2 PSYCHOLOGICAL WELL-BEING

A comparative analysis of age groups on psychological well-being using the *t*-criterion showed that the youth and early adult groups did not differ statistically in the level of expression of the components of psychological well-being ($p > 0.05$), except for the scale of "goals in life." This scale is higher in the early adult group ($M_1 = 63.21$, $M_2 = 65.44$, $p = 0.034$). The senior group of respondents whose indicators of psychological well-being are significantly different from other groups in the direction of its decrease ($p < 0.05$) is singled out.

Sexual differences were found in the expression of the component of psychological well-being "environmental mastery" (the average value in the group of women $M_1 = 56.97$, in the group of men $M_2 = 59.35$, $p = 0.015$). Since the overall psychological well-being index does not differ significantly between men and women, it can be assumed that psychological well-being can be achieved in different ways. Thus, higher values on the scale of "environmental mastery" in men may indicate a greater importance of the business sphere of life, the importance of competence in the performance of activities.

27.3.1.3 LIFE PERSPECTIVES

Table 27.1 presents data on the overall outlook for the future in the three age groups.

In Table 27.1, it is possible to trace a tendency to decrease of total score of PML with age. However, if the youth and early adult group differ little from each other in assessing their own life prospects, the older group has significant differences from them in the direction of reducing this indicator (according to the Mann–Whitney criterion, $p < 0.05$). Nevertheless, it should be noted that the values of the method in all age groups are at a fairly high level.

TABLE 27.1 Mean Values and Standard Deviations of Total Score of PML in Three Age Groups

	Youth Group		Early Adult Group		Middle Adult Group	
	M	**σ**	**M**	**σ**	**M**	**σ**
PML	87.74	9.41	85.71	13.57	78.87	13.22

Note. The sign "*" indicates statistically significant correlation coefficients. M: mean and σ: standard deviation.

27.3.1.4 PSYCHOPATHOLOGICAL SYMPTOMS

Psychopathological symptoms were examined in our study using the methods of SCL-90-R and the MS, which assesses the severity of post-traumatic stress. Similar to the previous results of comparison in the study of psychopathological symptoms of the study participants, a group of older respondents found differences from the other two groups in the level of severity of signs of post-traumatic stress ($M_1 = 73.77$, $M_2 = 76.55$, $M_3 = 82.88$, $p < 0.05$). In addition, the severity of signs of PTSD is higher in women than in men ($M_1 = 79.25$, $M_2 = 72.19$, $p = 0.000$).

Comparison of groups by SCL-90-R indicators was carried out using the Mann–Whitney *U*-test and did not reveal significant differences between groups. However, the data allow us to trace the general trend of lower values for all symptomatic scales of SCL-90-R in respondents of early adult age (from 22 to 35 years). In addition, the tendency of higher assessments of the components of psychological well-being was noted earlier. All this suggests that the group of early adulthood is the most resource in our study. The age of early adulthood allows a person to lead an active lifestyle, while young adults may already have a fairly high social status, providing them with various opportunities for the realization of their goals and desires, stable earnings, arranged personal and family life, etc., giving confidence in their abilities and the future.

The group of middle-aged people (from 36 to 60 years) has a lower psychological well-being compared to other age groups, as well as a tendency toward higher rates of negative psychopathological status and post-traumatic stress. However, it should be noted that in the interpretation of the results for this group, there may be some limitations related to the uneven representation of men and women. Thus, the prevalence of women in this group, almost three times, does not allow to draw a conclusion about the differences of this group, only due to age. The distribution of indicators is likely to be influenced by sexual characteristics in the formation of psychological reactions to the terrorist threat, in the structure of psychological well-being, as well as in the level of stress reactions.

27.3.2 CORRELATION ANALYSIS OF TERRORIST THREAT EXPERIENCE, PSYCHOPATHOLOGICAL SYMPTOMS AND COMPONENTS OF PSYCHOLOGICAL WELL-BEING

A correlation was established between the experience of terroristic threat and most components of psychological well-being. The results are presented in Table 27.2.

Table 27.2 shows that there is a significant inverse relationship between the experience of the terrorist threat and almost all components of psychological well-being: "autonomy" ($r = -0.26$), "environmental mastery" ($r = -0.24$), "personal growth" ($r = -0.12$), "goals in life" ($r = -0.12$), "self-acceptance" ($r = -0.13$). The inverse relationship with the general index of psychological well-being ($r = -0.21$) is also established.

TABLE 27.2 Values of the Pearson-r Correlation Coefficients Between the Indicators of SPW and QTT Methods in the Whole Sample ($n = 301$)

Scales of Psychological Well-being	**QTT**	
	r	***p*-Level**
Positive relationships with others	–0.11	0.071
Autonomy	–0.26*	0.000
Environmental mastery	–0.24*	0.000
Personal growth	–0.12*	0.044
Purpose in life	–0.12*	0.038
Self-acceptance	–0.13*	0.022
Index of PW	–0.21*	0.000

Note: The sign "*" indicates statistically significant correlation coefficients.

On the other hand, we obtained data on the association of experiencing the threat of terrorist attacks with negative risk factors presented in our study by indicators of PTSD and other psychopathological symptoms. It is shown that the intense experience of the terrorist threat is associated with all the indicators of the SCL-90-R (the values of the r_s coefficients are in the range from 0.27 to 0.35 at the significance level of $p = 0.0000$) and a high level of PTS ($r_s = 0.44$, $p = 0.0000$).

Thus, the overall sample showed the conjugation of resource factors on the one hand, and negative risk factors on the other. It is shown that the higher the level of psychological well-being, the lower the intensity of the terrorist threat and the lower the level of psychopathological symptoms and post-traumatic stress.

The assessment of the future prospects was not related to the intensity of the experience of the threat of terrorist attacks, but is positively related to all components of psychological well-being (the values of r_s coefficients are in the range from 0.18 to 0.51 at the level of significance $p < 0.002$). It can be assumed that the assessment of their own prospects is not a resource in coping with stressors but is the result of assessing the current level of well-being and its extrapolation into the future.

Further analysis took into account the age and sex characteristics of the respondents. In all age groups, the correlation analysis of association between the total score of QTT and components of psychological well-being was carried out. The results of the analysis for the youth and early adult groups are presented in Table 27.3.

TABLE 27.3 Values of *r*-Spearman Correlation Coefficients Between Indicators of WBS and QTT Methods in Youth ($n = 133$) and Early Adult Groups ($n = 108$)

Scales of Psychological Well-being	Total Score QTT			
	Youth Group		Early Adult Group	
	r_s	*p*-Level	r_s	*p*-Level
Positive relationships with others	–0.13	0.1333	–0.10	0.2940
Autonomy	–0.25*	0.0042	–0.33*	0.0005
Environmental mastery	–0.17	0.0523	–0.37*	0.0001
Personal growth	–0.09	0.2934	–0.01	0.8953
Purpose in life	–0.09	0.3127	–0.11	0.2430
Self-acceptance	–0.16	0.0645	–0.18	0.0643
Psychological well-being index	–0.18*	0.0400	–0.24*	0.0116

Note: The sign "*" indicates statistically significant correlation coefficients.

In both groups, the experience of the terrorist threat has a statistically significant negative relationship with the general index of psychological well-being and the scale of "autonomy." In the group of early adulthood, the index QTT is also associated with a scale of "environmental mastery." It is possible that people of this age have a great responsibility, and a sense of competence and the ability to plan their own daily lives is more important for them to cope with various external threats.

In the older group, none of the parameters of psychological well-being is interrelated with the experience of the threat of terrorist attacks ($p > 0.05$). Thus, in this group, the intensity of the terrorist threat experience is not related to the actual status of psychological well-being, i.e. to the general feeling of satisfaction/dissatisfaction with one's own life. However, respondents with a low level of psychological well-being are more likely to suffer from signs of post-traumatic stress, compared with psychologically prosperous representatives of this age. The result may be due to the fact that the older group in our study is the smallest and is represented mainly by women. In our previous studies (Tarabrina & Bykhovets, 2014), it has been shown that women are more worried about the terrorist threat and have higher levels of posttraumatic stress, which in turn can cause low level of psychological well-being. Thus, the results of comparison of different age groups may be influenced by sexual differences.

To test this assumption, a correlation analysis of the variables studied in the groups of women and men was carried out. The results are presented in Table 27.4.

In the group of women, the total score of QTT has a greater number of significant links with the components of psychological well-being. For both men and women, the threat of terrorist attacks is inversely linked to autonomy and environmental management. Apparently, in resilience to the experience of the terrorist threat in women is also more important the presence of warm relations with relatives and self-acceptance, while for men such importance is the scale of "personal growth."

TABLE 27.4 Values of r-Spearman Correlation Coefficients Between Indicators of WBS and QTT Methods in Groups of Women (n = 182) and Men (n = 119)

Scales of Psychological Well-being	QTT			
	Women		Men	
	r_s	*p*-Level	r_s	*p*-Level
Positive relationships with others	−0.17*	0.0236	−0.16	0.0782
Autonomy	−0.28*	0.0002	−0.27*	0.0029
Environmental mastery	−0.23*	0.0015	−0.27*	0.0032
Personal growth	−0.08	0.2700	−0.20*	0.0343
Purpose in life	−0.12	0.0969	−0.15	0.1124
Self-acceptance	−0.24*	0.0011	−0.16	0.0907
Psychological well-being index	−0.25*	0.0008	−0.22*	0.0202

Note. The sign * indicates statistically significant correlation coefficients.

27.3.3 *COMPARATIVE ANALYSIS OF THE STRUCTURE OF PSYCHOLOGICAL WELL-BEING AND SIGNS OF STRESS IN GROUPS OF RESPONDENTS WITH DIFFERENT INTENSITY OF EXPERIENCE OF TERRORIST THREAT*

Further analysis consisted in comparing the subgroups contrasting in terms of the intensity of the terrorist threat experience. Low and high scores on the total score of QTT were used as a criterion for the selection of subgroups. With this division, the group with a low intensity of experiencing the threat of terrorist attacks included those participants of the study, whose values of the total score of QTT are not less than ½ of the standard deviation below the average value, and in the group with a high intensity—those participants whose values of the total score of QTT exceed the average value of not less than ½ of the standard deviation. Thus, the group with a low level of experience of the threat of terrorist attacks included 96 people who scored by the QTT method from 53 to 108 points (Me = 95.00). The group with high intensity of experience of terrorist threat is represented by 96 respondents who scored 134–188 points in the QTT (Me = 148.50). Differences in the variables studied between the subgroups were analyzed using the Mann–Whitney *U*-test. The results are presented in Table 27.5.

TABLE 27.5 Comparative Analysis of the Components of Psychological Well-being in Groups with Different Intensity of Experience of Terrorist Threat

Scales of Psychological Well-being	Me_1	Me_2	U	p-Level
Positive relationships with others	62.50	58.00	3719.500	0.0210*
Autonomy	61.00	54.00	2939.500	0.0000*
Environmental mastery	60.00	55.00	2988.500	0.0000*
Personal growth	63.00	60.00	3795.500	0.0348*
Purpose in life	65.00	62.00	3616.000	0.0010*
Self-acceptance	61.00	56.00	3275.500	0.0005*
Psychological well-being index	371.00	348.50	3119.000	0.0001*

Notes. Me_1: the median value in the group with a low level of terroristic threat experience, Me_2: the median value in the group with a high level of terroristic threat experience. The sign * indicates statistically significant differences.

Table 27.5 shows that respondents with a high level of terrorist threat experience have lower scores on all SPW: positive relations with others, autonomy, environmental mastery, personal growth, goals in life, and self-acceptance. Their psychological well-being is generally lower than that of respondents with a low level of experience of terrorist attacks.

It should be noted that the values given in both subgroups correspond to the norms proposed by the authors of the method (Shevelenkova & Fesenko, 2005). Nevertheless, the results of the comparative analysis allow us to describe the differences between the subgroups. For example, respondents with a high level of experience of the terrorist threat are less capable of establishing close relations, more focused on the opinion of others in decision-making, lower assess their ability to resist social pressure. They are harder to perform daily activities and they often feel an inability to change or improve the conditions of life, and they are more likely to experience a feeling of powerlessness in managing the environment. These respondents are less satisfied with themselves, underestimate their strength and ability to overcome life's obstacles. In general, they are less satisfied with their lives than respondents with a low level of terrorist threat.

On the contrary, respondents who are more resistant to the terrorist threat demonstrate significantly higher rates of psychological well-being. These people are more satisfied with their relationships with others, more independent, and they are able to control the circumstances and create conditions for achieving goals. They are more open to new experiences, positive about their past, and have more goals for the future, on the way to which they feel a sense of continuous development and realization.

A comparative analysis of the intensity of psychopathological symptoms in the groups of respondents with high and low intensity of experience of the terrorist threat showed significant differences between them in all indicators of symptoms used in our study: they are higher in the group with a high level of experience of the threat of terrorist attacks. The results are presented in Table 27.6.

TABLE 27.6 Comparative Analysis of Signs of Psychopathological Symptoms in Groups with Different Intensity of Terrorist Threat Experience

		Me_1	Me_2	*U*	*p*-Level
SCL-90-R	Somatization	0.42	0.67	2990.500	0.0000*
	Obsessive-compulsive	0.40	0.90	2605.500	0.0000*
	Interpersonal sensitivity	0.56	0.89	3171.500	0.0003*
	Depression	0.33	0.69	3042.500	0.0001*
	Anxiety	0.30	0.70	2685.500	0.0000*
	Hostility	0.33	0.67	3211.000	0.0004*
	Phobic anxiety	0.00	0.29	2956.000	0.0000*
	Paranoid ideation	0.33	0.83	2807.500	0.0000*
	Psychoticism	0.10	0.40	2855.500	0.0000*
	Additional items	2.00	4.00	2722.000	0.0000*
MS		69.00	85.00	2093.500	0.0000*

Notes. Me_1: the median value in the group with a low level of terrorist threat experience, Me_2: the median value in the group with a high level of terrorist threat experience. The sign * indicates statistically significant differences.

27.3.4 STUDY OF THE MOTIVATIONAL AND TEMPORAL CHARACTERISTICS OF THE LIFE PROSPECTS

Protocols of 223 respondents aged from 17 to 35 years (the average age of the sample: 23 years) were analyzed in the study of the features of life prospects. Participants were divided into subgroups according to age and gender. The youth group (17–21 years) included 121 people (average age: 19.66 years): 72 girls (average age: 19.74 years) and 49 boys (average age: 19.54). A group of early age adults included 102 respondents aged 22–35 years (mean age of 26.90 years): 57 women (mean age: 26.72) and 45 men (mean age: 27.14). The smaller number of participants than in the previous stage is explained by the fact that some respondents refused to fill in the MIM methodology.

Table 27.7 presents the data of descriptive statistics of motivational and temporal categories of MIM (n = 223). The most common categories in the test subjects' responses were motives related to self (S), social contacts (C), self-realization (SR), productive activity (R), and motives and goals formulated for other people (C3). Nuttin (2004) refers to the empirical study, according to which the most frequently mentioned categories in the content of motivation greater significance in the life of the subjects (on a group level). The goals, desires, motives reported by the study participants are most often planned by them for the period of 1–2 years (Y) and for the first period of adult life (A1).

In empirical studies, the dynamics of the content of motivation and the length of the time perspective of the future in time were shown. Each stage of human development makes a qualitatively unique contribution to the development of the time perspective of the individual. Thus, it is shown that the content of time perspective differs in different age and sex groups (Tolstykh, 2010). This material was the basis for a more detailed analysis of the data obtained in our study, taking into account the age and gender of the respondents. Using the Mann–Whitney U-test, a comparative analysis of the studied parameters in different age and sex groups was carried out.

Comparison of age groups in all categories of responses in MIM did not show significant differences ($p > 0.05$): there is similar frequency of mention of motivational categories in both groups.

Statistically significant differences were found only for temporal parameters of future prospects. In the youth group, such plans and goals are more often indicated, the implementation of which should take place in the next 1–2 years (category Y) ($M_1 = 8.53$, $M_2 = 1.68$, $U = 1225.50$, $p = 0.000$). In the group of respondents of early adult age, plans relating to the period of the first half of professional career or adult life are more often indicated ($M_1 = 2.10$, $M_2 = 9.98$, $U = 1093.00$, $p = 0.000$). This result is due to the fact that for each age group this period is a relevant period of life: for the youth group this period is 1–2 years, which corresponds to the time of study in higher educational institutions (their average age is 19.66). The age of respondents in the middle group is 22–35 years, which corresponds to the beginning of adult life, designated by category A1. Thus, despite the statistical differences, we can say that there are no age differences: the periods of social life that are relevant for each age are more motivational.

Further analysis of the data was carried out without taking into account the age of the respondents. A comparative analysis of the responses to the MMI points of men (n = 94) and women (n = 129) revealed differences in the frequency of mentioning the following motivational and temporal categories of the questionnaire. So, women are more likely to write about the motives related to self-preservation ($M_1 = 1.09$, $M_2 = 0.82$, $U = 4966.50$, $p = 0.014$), achievement of personal autonomy ($M_1 = 1.89$, $M_2 = 1.23$, $U = 4317.00$, $p = 0.000$), self-realization ($M_1 = 3.47$, $M_2 = 2.51$, $U = 4856.00$, $p = 0.010$), productive activity ($M_1 = 3.83$, $M_2 = 3.32$, $U = 5063.00$, $p = 0.034$), social contacts ($M_1 = 5.92$, $M_2 = 4.02$, $U = 3885.00$, $p = 0.000$), wishes to others ($M_1 = 2.62$, $M_2 = 1.51$, $U = 4232.00$, $p = 0.000$), and knowledge ($M_1 = 1.07$, $M_2 = 0.66$, $U = 4896.50$, $p = 0.008$). Among the time categories of MMI women are more likely to meet the category of the beginning of adult life ($M_1 = 6.33$, $M_2 = 4.84$, $U =$

TABLE 27.7 Descriptive Statistics of Motivational and Temporal Categories of MIM in the Whole Sample ($n = 223$)

	Code	Name of Category	M	min	max	σ
Categories of motivation content	*S*	Self	6.16	0	20	3.38
	Spre	Self-preservation	0.98	0	6	1.21
	Saut	Self-autonomy	1.62	0	14	1.89
	SR	Self-realization	3.06	0	12	2.54
	R	Productive activity	3.61	0	19	2.76
	C	Social contact	5.12	0	19	3.44
	C3	Motives toward other people	2.15	0	12	2.08
	C3H	Wishes to all mankind	0.79	0	6	1.27
	E	Education	0.89	0	7	1.22
	Tx	Existential motives	0.21	0	4	0.52
	P	Consumer motivation	1.88	0	12	2.14
	L	Rest, entertainment	1.96	0	27	2.65
	Tt	Motives related to the testing procedure	0.09	0	5	0.48
Temporal categories	TDWM	Near future (up to 1 month)	2.04	0	28	3.05
	Y	1–2 years	5.39	0	24	5.19
	A1	The first period of adulthood	5.70	0	21	5.36
	A, AO, O	Maturity, old age	2.07	0	10	2.40
	L	All life ahead	7.43	0	25	4.36
	L	Open present	3.15	0	16	2.89
	X	Historical future	0.54	0	5	0.98
	X	The period after the death	0.03	0	1	0.16

5019.00, $p = 0.027$), "open present" ($M_1 = 4.36$, $M_2 = 1.49$, $U = 2435.00$, $p = 0.000$) and the historical future ($M_1 = 0.63$, $M_2 = 0.43$, $U = 5243.00$, $p = 0.038$). Higher mean values for all indicators (higher frequency) in women can be explained by the greater openness and willingness of women to share their internal goals and desires compared to men. Indeed, women were more willing to participate in the study and expressed greater interest in answering MIM questions.

Thus, we found sexual differences and did not find age differences in the features of life prospects. In accordance with the research objectives, we carried out further analysis among respondents with high and low levels of experience of the terrorist threat separately in a group of women and in a group of men.

In each age group, subgroups with a high and low level of experience of the threat of terrorist attacks were identified in terms of the QTT methodology. In this division, the group with low values included those participants of the study whose values of the total score of QTT are not less than ½ of the standard deviation below the average value, and in the group with high values—those whose values of the total score of QTT exceed the average value of not less than ½ of the standard deviation. Protocols of respondents with an average level of experience of the threat of terrorist attacks were excluded from the analysis. Differences in the variables between the subgroups were analyzed using the Mann–Whitney *U*-test.

It is shown that women with a high level of terrorist threat experience ($n = 43$) are more likely than women with a low level ($n = 41$) to use category C3 in their responses ($M_1 = 1.98$, $M_2 = 3.00$, $U = 617.00$, $p = 0.017$). This category includes motivation in which the subject expresses his or her hopes, desires or concerns about others. The goals and motives formulated for other people contain a whole complex of experiences, feelings, desires, which the subject, for some reason, is difficult to talk about in the first person. In such statements, respondents often indicate anxiety and excitement about relatives and friends (e.g., wishes such as "that my family was not sick," "that my children did not know grief," etc.). Where there is no real danger, such fears may mask anxiety and worry about one's own well-being. Statements of this type have the character of an attributive projection and are a consequence of the subject's perception of other people by analogy with themselves, attributing to them those qualities that are peculiar to him (Shlyapnikova, 2005). Thus, more frequent mention of this category in the responses of respondents with a high level of experience of the terrorist threat can be considered as a projection of their own concerns to other people due to the lack of personal reflection of their own experiences.

Comparison of men with high ($n = 33$) and low ($n = 27$) level of experience of terrorist threat made it possible to find a difference in the frequency of mentioning the category of self-preservation: men with a high score in the QTT often report motivation associated with self-preservation ($M_1 = 0.56$, $M_2 = 1.21$, $U = 323.00$, $p = 0.046$). At the same time, the motives for the protection and preservation of the "self" can refer to both the physical level (not to get sick, recover, stay healthy, etc.) and the psychological level—to protect or ensure internal balance, stability, psychological well-being. People who are prone to neurotic manifestations are afraid of anxiety, tension, and want to be calm and relaxed. They want to be protected from criticism and neglect, or they are afraid that their plans and aspirations will not come true.

27.4 CONCLUSION

Modern scientific psychology has accumulated extensive theoretical and empirical material that testifies to the stressful nature of the information impact of terrorism on society. Therefore, a natural direction of research development was the theoretical and empirical study of the factors contributing to the increase of people's resistance to the extreme impact of the terrorist threat. This is consistent with the overall interdisciplinary trends in psychology to the treatment of the attention of researchers on the positive aspects of life, the development of models, including resource and protective factors, in contrast to the studies of vulnerability.

Modern social, political, economic, etc., circumstances force a person to make significant efforts to protect themselves and their loved ones, to achieve stability in everyday life, and at the same time maintain a sense of harmony, well-being, and happiness. The terrorist threat, of course, contributes to the creation of a general negative background of uncertainty for the future, but it is in such conditions that the life of an ordinary citizen of a modern metropolis is realized. This is confirmed by the data obtained in our study that regardless of the level of experience of the terrorist threat, people tend to be optimistic about their future. In other words, most people find the strength to cope with every day and intense stress and continue to live, plan their future and hope for the best.

It has already been shown that increased anxiety is the predictor of intense experience of the threat of terrorist attacks. In the context of this study, anxiety was considered as one of the characteristics of the emotional sphere of the person, which determines his or her ability to adequately assess the reality of the terrorist threat. Accordingly, the experience of the terrorist threat was seen as a specific anxiety (Bykhovets & Tarabrina, 2010).

The level of psychological well-being (or distress) may indicate a degree of satisfaction with the need for security. "The theoretical and empirical study of the phenomenon of psychological well-being shows that it can be rightly attributed to such integrative characteristics that have close links with the subject's signs of negative affectivity and, consequently, affect the level of mental health" (Zhuravlev & Tarabrina, 2012, p. 9). In our study, it was shown that the experience of the terrorist threat is negatively correlated with most of the components of psychological well-being (positive relationships with others, autonomy, environmental mastery, personal growth, goals in life, self-acceptance). However, this experience is associated with a high level of psychopathological symptoms and signs of post-traumatic stress. Thus, the reduction or absence of internal protective resources negatively affects the overall resilience of the individual, increasing his sensitivity to various threats. The experience of the terrorist threat in this case can be considered as a model of a special kind of stressors invisibly presented in the life of society and forcing people to correlate their life activity with a subjective assessment of the probability of suffering from this stressor.

The data obtained in previous studies (Tarabrina & Bykhovets, 2014) on the age and sexual specifics of the terrorist threat experience were confirmed in the present work. The present study shows that the group of respondents of early adulthood (from 22 to 35 years) seems to be the most optimally functioning in terms of resilience. This group has a higher psychological well-being, less psychopathological symptoms and less intense experience of the terrorist threat than other age groups.

We also obtained data on the differences in the content of motivation of respondents with a high and low level of experience of the threat of terrorist attacks. In this regard, it is interesting to study the linguistic predictors of PTSD symptoms after the events of September 11. The authors of this study showed that the use of certain words by the participants of the study (who were not directly affected in the terrorist attacks), in describing the memories of how they first heard about terrorist attacks, affects the subsequent psychopathological symptoms. The use of more emotional (r = 0.63), religious (r = 0.60), and death-related words (r = 0.41) in stories about their memories after a short period of time (about 1 week) was significantly associated with the onset of PTSD symptoms 5 months later. The use of the words for his own anxiety, at the first examination was associated with lower levels of PTSD when re-examination. Thus, people who are open to discussing their anxiety provoked by terrorist attacks can thereby reduce their symptoms (D'Andrea-Merrins et al., 2006).

The projective mechanisms of protection were examined in the study of identifying with the victims of the sniper attacks in Washington. In this work, the relationship of the signs of PTSD and symptoms of depression with different types of identification with the victims were studied (identification of ourselves as victims, identification of friends or family members). The authors of the study believe that, in the wake of a negative event, people may begin to emphasize their dissimilarity with those who have experienced the event, allowing them to establish a psychological distance and maintain a sense of control and invulnerability to future threats. Identifying yourself with the victim of a traumatic event can, on the contrary, increase the sense of vulnerability, which will serve as a trigger for increased anxiety and sense of threat. The results of this study showed that the overall score of identification is positively associated with the symptoms of post-traumatic stress and depression. Self-identification with victims is only associated with increased levels of post-traumatic stress, which may reflect the experience of a personal threat. Identification of friends and relatives with the victims of the attack is associated with symptoms of posttraumatic stress and depression symptoms, which can speak about the experience of threat or anticipation of the loss of the significant other (Mash et al., 2016).

Thus, to understand the consequences of experiencing a terrorist threat, it is not enough to measure the level of negative psychopathological symptoms. In the practice of providing psychological assistance to people who are concerned about terrorist attacks, it is also important what form and content this concern takes. Also, in this regard, the results obtained in our study may be useful to improve the effectiveness of psychotherapeutic assistance to the population.

KEYWORDS

- **terrorist threat**
- **experience of terrorist threat**
- **resilience**
- **psychological well-being**
- **life perspective**
- **motivation**
- **psychopathological symptoms**
- **post-traumatic stress**

REFERENCES

Bykhovets, J. V., & Tarabrina, N. V. (2010). *Psychological estimation of perception of the terrorist threat: Guidance.* Moscow: Institute of Psychology RAS. [Bykhovets, J. V., & Tarabrina, N. V. (2010). Psikhologicheskaya otsenka vospriyatiya terroristicheskoy ugrozy: Rukovodstvo. M.: IP RAN] (in Russian).

Caramanica, K., Brackbill, R. M., Liao T., & Stellman, S. D. (2014). Comorbidity of 9/11-related PTSD and depression in the World Trade Center health register 10–11 years post disaster. *Journal of Traumatic Stress, 27*(1), 680–688.

D'Andrea-Merrins,W., Chiu, P., King-Casas, B., & Deldin, P. (2006). *Linguistic predictors of PTSD symptoms following the 9/11 attacks*. Paper presented at the 22nd Annual Meeting of International Society for Traumatic Stress Studies, Los Angeles, CA (p. 91).

Derogatis, L.R., Rickels, K., & Rock, A. (1976). The SCL-90-R and MMPI: A step in the validation of new self-report scale. *The British Journal of Psychiatry, 128*, 280–289.

Keane, T.M., Wolf, J., & Taylor, K.L. (1987). PTSD: Evidence for diagnostic validity and methods of psychological assessment. *Journal of Clinical Psychology, 43*, 32–43.

Keane, T.M., Caddell J. M., & Taylor, K. L. (1988). Mississippi scale for combat-related PTSD: Three studies in reliability and validity. *Journal of Consulting and Clinical Psychology, 56*(1), 85–90.

Makhnach, A. V., & Dikaya, L. G. (2016). Interdisciplinary and inter-system studies of resilience: Current state and prospects (instead of introduction). In A. V. Makhnach

& L. G. Dikaya (Eds.), *Human resilience:* Individual, professional and social aspects (pp. 7–28). Moscow: Institute of Psychology RAS. [Makhnach, A. V., & Dikaya, L. G. (2016). Mezhdistsiplinarnyye i mezhsistemnyye issledovaniya ustoychivosti: tekushcheye sostoyaniye i perspektivy (vmesto vvedeniya). V knige: A. V. Makhnach, & L. G. Dikaya (Red.). *Zhiznesposobnost' cheloveka: individual'nye, professional' nyeisocial' nyeaspekty.* M.: IP RAN] (in Russian).

Mash, H. B. H., Ursano, R. J., Benevides, K. N., & Fullerton, C. S. (2016). Identification with terrorist victims of the Washington, DC sniper attacks: Posttraumatic stress and depression. *Journal of Traumatic Stress*, *29*(1), 41–48.

Misko, E. A., & Tarabrina, N. V. (2004). Peculiarities of life perspective in veterans of Afghanistan war and liquidators of accident at Chernobyl atomic power station. *Psihologicheskiyzhurnal*, *25*(3), 44–52. [Misko, E. A., & Tarabrina, N. V. (2004). Osobennosti zhiznennoy perspektivy u veteranov voyny v Afganistane i likvidatorov avarii na Chernobyl'skoy AES. *Psihologicheskiyzhurnal*, *25*(3), 44–52] (in Russian).

Mukherjee, S., & Kumar, U. (2017). Psychological resilience. A conceptual review of theory and research. In U. Kumar (Ed.), *The Routledge international handbook of psychological resilience* (pp. 3–12). London, New York: Taylor & Francis Group.

Neria, Y., Wickramaratne, P., Olfson, M., Gameroff, M. J., Pilowsky, D. J., Lantigua, R., et al. (2013). Mental and physical health consequences of the September 11, 2001 (9/11) attacks in primary care: a longitudinal study. *Journal of Traumatic Stress*, *26*, 45–55.

Nuttin, J. (2004). *Motivation, action and future prospects.* Moscow: Smysl. [Nuttin, J. (2004). Motivatsiya, deystviya i budushchiye perspektivy. M.: Smysl] (in Russian).

Ryff, C. D. (1989). Happiness is everything, or is It? Explorations on the meaning of psychological well-being. *Journal of Personality and Social Psychology, 57(6)*, 1069–1081.

Shevelenkova, T. D., & Fesenko, P. P. (2005). Psychological well-being of personality (a review of basic concepts and methods of research). *Psihologicheskaya Diagnostica, 3*, 95–129. [Shevelenkova, T. D., & Fesenko, P. P. (2005). Psikhologicheskoye blagopoluchiye lichnosti (obzor osnovnykh ponyatiy i metodov issledovaniya. *Psihologicheskaya Diagnostica, 3*, 95–129] (in Russian).

Shlyapnikova, I. A. (2005). *Projective methods of psychodiagnostics (textbook).* Chelyabinsk: Izdatel'stvo YUU rGU. [Shlyapnikova, I. A. (2005). Proyektivnyye metody psikhodiagnostiki (uchebnik). Chelyabinsk: Izdatel'stvo YUUrGU] (in Russian).

Tarabrina, N. V., Agarkov, V. A., Bykhovets, J. V., Kalmykova, E. S., Makarchuk, A. V., Padun, M. A., et al. (2007). *The guide to the psychology of post-traumatic stress. Part 1. Theory and methods.* Moscow: Izdatel'stvo "Cogito-centre". [Tarabrina, N. V., Agarkov, V. A., Bykhovets, J. V., Kalmykova, E. S., Makarchuk, A. V., Padun, M. A., et al. (2007). *Prakticheskoe rukovodstvo po psihologii posttravmaticheskogo stressa. Ch. 1. Teoriyaimetody.* M.: Izdatel'stvo «Cogito-centre»] (in Russian).

Tarabrina, N. V., & Bykhovets, J. V. (2014). *Terrorist threat: Theoretical and empirical research.* Moscow: Institute of Psychology RAS. [Tarabrina, N. V., & Bykhovets, J. V. (2014). Terroristicheskaya ugroza: teoreticheskiye i empiricheskiye issledovaniya. M.: IP RAN] (in Russian).

Thoresen, S., Jensen, T. K., & Dyb, G. (2014). Media participation and mental health in terrorist attack survivors. *Journal of Traumatic Stress*, *27*, 639–646.

Tolstykh, N. N. (2005). Using the method of motivational induction to study motivation and the time perspective of the future. *Psihologicheskaya Diagnostica, 3*, 77–94. [Tolstykh, N. N. (2005) Ispol'zovaniye metoda motivatsionnoy induktsii dlya izucheniya motivatsii i vremennoy perspektivy budushchego. Psihologicheskaya Diagnostica, 3, 77–94] (in Russian).

Tolstykh, N. N. (2010). *Chronotope: Culture and ontogenesis.* Smolensk-Moscow: Universum. [Tolstykh, N. N. (2010). Hronotop: kul'tura i ontogenez. Smolensk-M.: Universum] (in Russian).

Welch, A. E., Caramanica, K., Maslow, C. B., Brackbill, R. M., Stellman, S. D., & Farfel, M. R. (2016). Trajectories of PTSD among Lower Manhattan residents and area workers following the 2001 World Trade Center disaster, 2003–2012. *Journal of Traumatic Stress*, *29*, 158–166.

Zhuravlev, A. L., & Tarabrina, N. V. (2012). Psychological safety: Towards an integrated interdisciplinary research (instead of the preface). In A. L. Zhuravlev & N. V. Tarabrina (Eds.), *Problems of psychological safety* (pp. 5–21). Moscow: Institute of Psychology RAS. [Zhuravlev, A. L., & Tarabrina, N. V. (2012). Psikhologicheskaya bezopasnost': na puti k kompleksnomu mezhdistsiplinarnomu issledovaniyu (vmesto predisloviya). In A. L. Zhuravlev, & N. V. Tarabrina (Eds.), Problemy psikhologicheskoy bezopasnosti (pp. 5–21). M.: IP RAN] (in Russian).

CHAPTER 28

Terrorism, Emotion, and Well-Being

CATALINA UDANI[1*], and JONATHAN MATUSITZ[2]

[1]*Department of Political Science, University of Pennsylvania, Philadelphia, PA, USA*

[2]*Nicholson School of Communication and Media, University of Central Florida, USA*

[*]*Corresponding author. E-mail: cudani@sas.upenn.edu*

ABSTRACT

Developed by Bormann, symbolic convergence theory rests on the premise that ideological themes and fantasies are shared through a group, strengthening or creating common bonds. Groups create meaning by interpreting shared experiences into symbols, and these symbols, in turn, reinforce the group's ideas. This occurs through commonplace conversation, stories, and rituals in which group members participate. Fantasies are "creative and imaginative interpretation of events that fulfills a psychological or rhetorical need." SCT essentially addresses how and why group identities are constructed and how the resultant shared consciousness assigns meaning within the group. Core principles of SCT are reflected in ISIS's attempts to shape reality through its rituals as well as its online publications. SCT is the synthesis of a thematic narrative through communication, particularly within groups. The group can affect the reality that ISIS recruits experience. Likewise, these recruits can affect the shared reality of the terrorist group. Ultimately, ISIS recruits become more resilient for waging and upholding jihad (holy war). Those shared group fantasies give them direction and a sense of belonging to the jihadist world.

28.1 INTRODUCTION

The presence of emotion in terrorism is visible at all levels of terrorist action, from the propaganda that terrorist groups diffuse to the violent actions that they commit, and the emotions present may range from very positive to intensely negative. Although white supremacist terrorists, for example, may express vitriolic hatred toward minority groups, such as Jews and blacks, they may also exhibit positive emotions toward what they perceive as traditional values (Freilich et al., 2009). Their negative emotions divide them from the groups they target, while their positive emotions draw them toward radical groups. Similarly, jihadist terrorist propaganda material published by the Islamic State of Iraq and Sham (ISIS) describes intense hatred for Western civilization and immense zeal for militant Islam and for ISIS itself, praising the group's own constituents as brothers, martyrs, and heroes.

Clearly, the role of emotion in terrorism is threefold: First, strong negative emotions (such as hatred, desperation, and cruelty) generally fuel violent actions on the part of individual terrorists. Second, intense positive shared emotions (including solidarity, belonging, and ideological or religious zeal) within a terrorist group usually contribute to group cohesion, the well-being of group members, a shared rhetorical vision, group identity, and organizational durability. Third, certain instances of emotion within terrorism may exhibit overlap between the two functions of driving terrorist action and promoting group durability. Instances of tragedy and great hardship, for example, may evoke negative emotions but shared suffering within a group may promote group cohesion as much as a shared victory through bolstering group camaraderie.

This chapter analyzes the impact of emotion on the well-being of terrorist groups by applying two theories of group communication, symbolic convergence theory (SCT) and social identity theory (SIT), to the rhetoric of well-known jihadist groups Al Qaeda and ISIS. By understanding how the groups' narratives engage their constituents emotionally, we aim to attain a better understanding of terrorist behavior.

28.1.1 TERRORISM: A DEFINITION

To begin, terrorism has become a highly salient issue in international relations. The attacks on September 11th, 2001, and the ensuing endeavors by the United States to develop an overarching antiterrorism coalition, have elicited the question of what constitutes terrorism (Jenkins, 2015). The United States Department of State (2001) has published a description according to which terrorism is the intentional use of violence against noncombatants (both civilians or noncivilians). From this vantage point, terrorism is violence against unarmed targets or in war against noncombatants. White (2001) extends this definition by calling terrorism the use of deliberately indiscriminate violence to cause fear and terror among the masses in order to accomplish a financial, political, religious, or ideological objective.

Another important aspect of terrorism is its ideological or political motivation (Ruby, 2002). Indeed, terrorism is often politically motivated. It is directed toward aims that are political; terrorist actions, then, serve to steer or influence governmental policy. Extreme acts such as theft, homicide, and kidnapping, perpetrated in the furtherance of personal or criminal goals, do not constitute terrorism. This principle stresses that the social and psychological pasts of personally or criminally driven violence differ from the antecedents of terrorist violence (Matusitz, 2012). Terrorism sends a political message; its ends rise above damaging an enemy's material resources. The direct and indirect victims of terrorist attacks do not matter as much as the larger human audience whose reaction the terrorists pursue (Crenshaw, 1981).

Third, Hoffman (2006) proposes that, by differentiating terrorists from other categories of criminals and terrorism from other types of crime, we understand that terrorism is (1) often political in aims and motives, (2) violent or threatening the use of violence, (3) intended to have sweeping psychological ramifications beyond the direct victims or targets, (4) committed either by an organization with a pyramidal chain of command or a flatter cell structure (like a terrorist cell or network) or by people or a "franchise" of people clearly inspired by the ideological aims of a terrorist movement, and (5) executed by a subnational or nonstate entity.

In line with these contentions, terrorists cannot declare themselves to be "freedom fighters" by sole virtue of fighting for national liberation or some other "noble" goal. Even if its proclaimed objectives are legitimate, an organization that calculatingly strikes civilians is a terrorist organization. There is no worth or remission in fighting for the freedom of one larger group if the rights and lives of another population are destroyed in the process. By extension, the terms "terrorism" and "guerrilla warfare" are often used interchangeably. However, the former contains a much more negative connotation, arguably requiring one to take an extreme position, whereas the latter has been perceived, in the media, as neutral and carrying a more positive connotation (Ganor, 2002).

28.2 TERRORISM AND EMOTIONS

Acts of terror are often fueled by intense emotion on the part of the radicalized individuals who take part in them, and organizations that perpetuate terrorist violence utilize emotions to enhance group cohesion through shared fantasies and an overarching rhetorical vision. Emotions are both driving forces of terrorist action and products of shared group identity. Individuals cannot adopt and adhere to radical ideologies without strong positive emotion toward them, and members of a group share emotional narratives that bind them together.

28.2.1 EMOTIONS

The concept of emotions is defined as any experience of intense, conscious mental activity with some perception of a level of pleasure or displeasure (Cabanac, 2002; Damasio, 1998). Within individuals, emotions may be observed through factors such as mood, temperament, personality, disposition, and motivation (Clark et al., 1994; Elliot & Thrash, 2002). Emotions are present at all levels of society and are embedded in the narratives of nations, institutions, cultures, and groups (Hoggett, 2009).

As a unit of analysis, emotions are vital, as they are central to human decision-making, social understanding, and the strengthening or weakening of social bonds. Individuals use emotions to make decisions and judgements about past experiences, current situations, and future possibilities. Emotions are also particularly salient in the social formation of morality and ideology and, thus, are necessary in the formation of a political identity (Hogget, 2009). For politically motivated groups, emotions may influence the way their constituents perceive injustice within aspects of society. Emotions of faithfulness and gratitude, for example, are necessary for individuals to establish trust with one another—that is, for the purpose of creating organizations and institutions (Simmel, 1950). Groups often coalesce naturally around strong mutual emotions over certain issues; within the jihadist movement, strong positive emotion over a fundamentalist, Salafist interpretation of Islam may draw individuals to a jihadist group. Psychologically, Ruby (2002) states that terrorists require sufficient ideological or political motivation to enact radical deeds. Strong emotions allow terrorists to adhere to their ideologies.

Beyond ideology, emotions can also take individuals to emotional states in which they are capable of believing in concepts that may be perceived as fantastic, unrealistic, or even irrational. For example, Durkheim's (1965) study examined the emotions experienced by members of an Australian aboriginal group during totemic rituals. Within the group's totemic ritual practices, individuals reported feelings of transportation, giving them strong positive sensations of awe, as if they were in the presence of a higher power embedded in the sacred objects they worshipped. Durkheim (1965) attributed these sensations to the emotions of exaltation that emerged during the group's worship. Similar emotions of exhilaration—rational and irrational—are observable in all variations of belief systems, and are even seen in secular systems, such as the one expressed by sports fans celebrating the victories. By and large, overwhelming emotions can be shared within all types of groups and are often positive sources of group cohesion and collective well-being.

However, such overwhelming emotions can also be twisted negatively into impetus for irrational hatred, social hostility, and even violent terrorist action. Post (2005) examined the psychological and cultural foundations of the Black September attack on the Israeli team in the 1972 Munich Olympics and members of the Palestinian Fatah movement. The results of Post's study identified intense hatred of all Israelis, despite the fact that the few individuals that Post interviewed actually reported having any contact with Israeli individuals. According to Post, the members of Black September projected their hatred for what they believed Israel represented to the individual members of the Israeli Olympic team. The emotion of intense hatred that drove the Munich massacre is echoed in other terrorist attacks through history, such as Omar Mateen's 2016 Pulse nightclub shooting; in Mateen's case, his jihadist ideology was strengthened by his irrational hatred for the LGBT and Latino communities (Stults et al., 2017). Likewise, San Bernardino terrorist attack that took place in December 2015 reflected the same level of hateful jihadist ideology—expressed by Syed Rizwan Farook and Tashfeen Malik—toward the American people in general. Hence, most of the 14 victims were Americans (D'Orazio & Salehyan, 2018).

28.3 A PERSPECTIVE FROM SYMBOLIC CONVERGENCE THEORY (SCT)

Tyrannical or extreme ideologies are the perfect storm for radicalization. Groups that espouse such ideologies become more popular when they feel endangered by powerful outsiders (Perliger & Pedahzur, 2016). To gain a better understanding of the emotional appeal and social benefits that terrorists gain from membership in a terrorist group, we use multiple theoretical frameworks to analyze terrorist group narratives and social identities. Terrorist groups create a shared rhetorical vision that capitalizes on the emotions of its constituents, positive and negative; this is visible through the theory of symbolic convergence.

Developed by Bormann (1972), SCT describes how groups share common ideological themes and fantasies, creating and reinforcing bonds within the collective. Rhetorically, groups create meaning by interpreting shared experiences, assigning symbols to these experiences, and utilizing these symbols in ways that reinforce the group's ideology (Deetz, 2001). This process of interpretation occurs during ordinary interactions within the group, including communicative storytelling and participatory rituals, spanning the individual level to the group (or even societal) level entirely. "Symbolic convergence" refers to the creation of a collective consciousness within a group; this consciousness includes common beliefs in the forms of symbols and fantasies. SCT describes the process of group identity construction and how the resultant shared identity continues to create and ascribe meaning within any group (Cragan & Shields, 1992).

This process occurs in terrorist groups, providing its members with a radical, emotional ideology and a rhetorical vision to align with. Symbolic convergence strengthens shared group identity and may be a tool for group cohesion. Through assigning meaning to symbols that all the members of a group can understand, group members can better identify with a single group ideology, connecting them with one another through their shared vision.

A fantasy within symbolic convergence refers to a "creative and imaginative interpretation of events that fulfills a psychological or rhetorical need" (Griffin, 2006, p. 38). For example, within the jihadist terrorist movement, the overarching apocalyptic theme of war between the righteous caliphate and a morally indefensible West provides a key psychological need for terrorists; it interprets the events of Islamist history through a mythic frame and provides them with a just movement they can align with.

SCT consists of two essential premises: First, communication creates the reality that people experience. Second, through communication, individuals can share their perceived reality throughout their group. This results in the convergence of meaning and the creation of a shared consciousness and a collective identity. Within a terrorist group, a shared consciousness and identity is necessary for to motivate individual terrorists to join the group, fight for the group, and even die for the group (Vasquez, 1993).

28.3.1 PRINCIPLES OF SCT

In their 1996 work, Bormann, Cragan, and Shields outlined three principles that allow for the development of a shared rhetorical consciousness within a group: novelty, explanatory power, and imitation. The principle of novelty states that shared rhetorical visions that are already established within a group may fail to attract potential recruits. Elements of novelty introduced over time are necessary additions to shared group fantasies for a group to grow and to remain durable. For modern-day terrorist groups, a necessary element of novelty is the usage of digital and social media. Most new recruits to jihadist groups are young men, and groups like ISIS rely increasingly on expatriate foreign fighters and funding from overseas (Hoffman, 2006). Through digital avenues, terrorist groups can more efficiently reach out to the youth and to overseas populations. By effectively incorporating the novelty of social media into their recruitment strategies, groups like ISIS successfully can secure the loyalties of foreign fighters and sources of external funding.

Explanatory power describes how symbolic fantasies provide satisfying explanations of certain experiences for members of a group. These experiences often include traumatic events. Through symbolic convergence, a group can collectively rationalize an event and understand how the event may affect functionally rather than experientially (Bormann et al., 1996). The explanatory power of symbolic convergence gives groups the tools to understand how to move forward from an event.

One example of explanatory power is in the movement ISIS from a short-term, tangible victory of its Caliphate over the Western world to a long-term, distant victory after its defeats in 2016, visible in the change of its English-language magazine title from *Dabiq* to *Rumiyah*. Dabiq is a town in Syria, about which a well-known *hadith*, an Islamic saying attributed to the Prophet Muhammad, states that the apocalyptic "Great Battle" between the Caliphate and the West will occur, resulting in an Islamic victory over the control of the entire world. Issue 3 of

Dabiq addresses this, noting that "the name of our magazine was taken from the area named Dābiq in the northern countryside of Halab, due to the significant role it will play during the events of al-Malhamah al-Kubrā (The Grand Battle) against the crusaders" ("Dabiq Issue 3," 2014, p. 15). Dabiq had been under ISIS control since 2014, but by March 2016, the group lost 22% of their territory within Iraq and Syria (Wignell et al., 2017).

The decline of ISIS territory was marked by the loss of Dabiq roughly a month after the first issue of *Rumiyah* was released. *Rumiyah*, the Arabic word for Rome (i.e., the seat of the Roman Catholic Church and an immensely significant city for Western civilization) implies that though a battle at Dabiq may not be in the near future, due to its loss to Western-allied forces, the far future includes Islamist dominance over Rome and the West at large. This name change suggests that ISIS anticipated the loss of Dabiq and decided to shift their group's primary goal from the tangible to the ideological. The fantasy of a distant apocalypse in favor of an Islamic caliphate helped the group explain its material losses; figuratively, though they lost Dabiq today, they will defeat Rome tomorrow (Wignell et al., 2017).

Finally, the principle of imitation within symbolic convergence states that individuals within a group may apply familiar symbols and fantasies to their own issues and personal events, imitating the wider symbolic convergence of a group by applying it to individual experiences (Bormann et al., 1996). The principle of imitation showcases how symbolic convergence allows for the sustained existence and salience of collective fantasies and group symbols. Jihadist groups utilize the principle of imitation through applying well-known Quranic verses and *hadiths*—which may refer to fantasies common to the broader Islamic community—to their present situations. These groups often repeat Medinan verses of the Quran to justify their violent attacks; the Medinan verse 2:191 commands, "Slay them wherever you find People of the Book," referring to Jews and Christians. By imitating the sentiment of these verses (i.e., the acceptable murder of disbelievers) during their acts, terrorist groups apply the familiar symbols of the Quran, written in the seventh century, to present-day actions.

28.3.2 TOOLS OF SCT

Shared rhetorical visions developed by symbolic convergence results in a collective understanding of mutual ideology within a group; the group's ideology frames its experiences. According to Bormann (1972), groups attribute meaning to fantasies and signs, allowing group members to understand their shared experiences through an overarching rhetorical movement. To achieve this, Bormann lists four tools of symbolic convergence: fantasy themes, symbolic cues, fantasy types, and sagas. Bormann (1973) describes fantasy themes as narrative-driven dramatizations of past, current, and hypothetical future events. These themes are concepts denoted by individual words, phrases, and statements, and they may contribute to the group's central rhetorical movement. The narratives that groups portray through fantasy themes are described through three different thematic factors: setting, character, and action.

Setting themes refer to the locations the narratives take place in (Bormann, 1972). For example, a prominent setting theme present in the jihadist fantasy theme of an apocalyptic

Great Battle, *Al-Malhama Al-Kubra*, is the town of Dabiq (Pregill, 2016). Reportedly, the Prophet Muhammad stated in a *hadith* that the final battle would be set in Dabiq, giving the town thematic significance in ISIS rhetoric as a setting. Character themes describe motivations and characteristics of actors within the fantasy; group consciousness often labels heroes and villains within shared fantasies (Bormann, 1972). Heroes are lauded in a rhetorical vision for their actions, embodying the group's causes, while villains are reviled by the group for their opposition to the group's causes. For jihadist terrorist groups, their deceased fighters are glorified as martyred heroes, while political leaders who oppose their actions are their clear villains. Finally, action themes are the plotlines of the fantasy's narrative (Bormann, 1972). Referring once more to the jihadist apocalyptic fantasy of a Great Battle, the plotline commonly accepted by jihadist groups is that the final battle between the West and the caliphate in Dabiq will result in the Caliphate's complete dominance over the world and the coming of Armageddon (Pregill, 2016). Through rhetorically and critically analyzing fantasies shared by groups, one can evaluate the values of the wider group consciousness (Bormann, 1985).

Symbolic cues are slogans and signs that evoke the group's fantasies; their repeated usage builds group cohesion. They may include words, phrases, nonverbal gestures, and images. These cues act as heuristics, reminding group members of their shared ideology, emotions, and past experiences. Jokes are examples of common symbolic cues, representative of shared group humor; when one "gets the joke," he or she may feel a sense of belonging within a group, while listeners unaware of the joke's symbolic value may respond to it with confusion. This holds true for all forms of symbolic cues; individuals who are not members of the group may not understand the cues' significance (Cragan & Shields, 1992).

One symbolic cue used by members of ISIS is a single raised index finger. This is more widely known as the symbol for the number one, but it is also used as a symbolic cue of the jihadist movement. In the broader jihadist movement, the gesture refers to the Islamic principle of the *tawhid*, meaning the unity of the *ummah*, the worldwide Muslim community, and the oneness of Allah. Within ISIS specifically, the gesture refers to the goal of a unilateral destruction of pluralism, individualism, and the Western civilization that values the will of individuals. It also symbolizes ISIS's ideology of complete, unified world domination under its Caliphate (Novenario, 2016).

Another tool of SCT is the fantasy type. Fantasy types are recurring fantasy themes, which groups apply to common issues, situations, and events (Bormann, 1985). By applying relevant fantasy types to concerns shared by group members, groups can use fantasy types as references to understand and assign meaning to new experiences. The usage of fantasy types showcases the principle of imitation within symbolic convergence. Groups imitate recurrent fantasy themes, apply them to shared experiences through fantasy types, and create new symbols through a dramatized version of the experience. An example of a prevalent fantasy type is "fetching good out of evil," a concept often used to explain the outcomes to adversity (Bormann, 1973). This fantasy type is commonly used after calamities and disasters; individuals may find beneficial experiences within difficult situations, allowing them to rationalize grave issues.

Sagas are the retellings of significant events relevant to a group, its members, and its ideology (Cragan & Shields, 1992). Bormann (1985) states that sagas describe the group's achievements by recounting the life of a person, community, organization, or state. These retellings represent the group's shared symbolic consciousness by constructing emotional narratives. These narratives become part of the group's rhetorical vision, and may be manipulated, appended, and repeated in order to reflect the group's changing principles. Sagas are reflective of the group's shared consciousness and its rhetorical community and are particularly relevant to physical organizations, such as centrally organized terrorist groups (Cragan & Shields, 1992). ISIS commonly recounts the lives of the caliphs of Islamic history, such as the Sunni-recognized caliph Abu Bakr; Abu Bakr al-Baghdadi announced himself the new caliph in 2014, calling himself a successor of caliphs Muhammad and Abu Bakr (Hashim, 2014). By recounting the lives of historical caliphs and claiming succession, ISIS leaders create a narrative saga painting ISIS as the direct heir of the prophet Muhammad's original vision.

These aspects of symbolic convergence develop an overarching rhetorical vision: the structure of communication within a group through which shared group identity coalesces (Bormann, 1985). The group's rhetorical vision may affect the perspectives of reality group members perceive.

28.4 A PERSPECTIVE FROM SOCIAL IDENTITY THEORY (SIT)

Developed by Tajfel (1970), SIT rests on the premise that individuals are defined in terms of their membership to groups (i.e., ingroups) and compared with individuals from other groups (i.e., outgroups). Social identities, then, influence tendencies toward supporting members of one's own group and believing in outgroup derogation or hostility (Brown, 1984; Turner, 1979). The problem with considering ingroup identity is that it often amounts to an "imagined community" (Anderson, 1983). Such imagined communities generate prototypical images of "good" and "evil," which Post (2005) refers to as "hatred bred in the bone" (p. 615).

Terrorist groups often obtain their membership from disenfranchised, alienated, and inadequate people from troubled backgrounds. For many, membership in a terrorist group constitutes a first real occasion to truly belong to any group. Committed to resist (and even fight) the outgroup overall (i.e., society), the ingroup is now regarded as all good and the outside society as all bad. This divergence is essential to the group's ideology and is the underlying principle for committing violent terrorist attacks. If they frame the outgroup as the most significant danger to the ingroup itself or to its core values, it will be easier for terrorist leaders to radicalize new recruits and increase the probability of terrorist violence. The shift to terrorism is facilitated through such close-knit social networks (Post, 1986).

Terrorists generally perceive the outgroup as responsible for some level of egregious moral violation, whether that be decadent behavior, invasion of disputed territory, or abasement of culture or religion (Barbalet, 2006). They regard their own violent deeds as a morally righteous response to the outgroup's egregious acts. Terrorist interactions within their groups reinforce their ingroup commitment and nurture a mindset that enables people to learn and resort to violence. Eventually, the would-be terrorists express strong identification with the ingroup's values and extreme alienation toward the outgroup (Perliger & Pedahzur, 2016). Rituals play significant roles during ingroup interactions. They bolster social cohesion, the belief in the superiority of the ingroup's way of life, and the inclination to put the group's interests above those of the individual members (Lacey, 2005; Sosis, 2009; Wessinger, 2009; Whitehouse, 2013).

28.5 DISCUSSION

Through the frameworks of SCT and SIT, one can observe how terrorist narratives encourage emotion within its constituents; these emotions drive violent action, strengthen group cohesion, and contribute to group member well-being. This discussion examined the narrative themes present within jihadist terrorist groups, the emotions terrorist rhetoric may evoke for terrorist group members, and how the latter use these negative emotions for the benefit of the group. Within the imagined communities drawn together through powerful, deep-seated terrorist rhetoric, shared narratives encourage both negative and positive emotion, both directions fostering ingroup identity and outgroup detestation. Generally, positive emotions reinforce the moral righteousness of the ingroup and fulfill the psychological needs of group members. Parallel to positive emotions, the negative emotions generated by terrorist groups' narratives encourage hatred and anger toward the perceived disenfranchisement and oppression that group members face from society at large. Together, both positive and negative emotions fostered by collective terrorist group rhetoric produce a mindset that allows terrorist group members to commit violent and militant action for the sake of the group.

28.5.1 POSITIVE EMOTION

Positive emotions instilled in group members through terrorist narratives are primarily concerned with terrorist identity, group cohesion, and moral rectitude. Through ideological themes, terrorist groups fulfill the psychological needs of their members by providing them with fantasy narratives of victory and righteousness. As groups with dichotomous worldviews, which describe saintly believers and evil disbelievers, terrorist groups cast their own members as heroes of their imagined communities through their shared group narratives and social identities. For example, positive reinforcement of terrorist moral rectitude is clearly visible within the ideology of Al Qaeda. Quiggin (2010) describes one of Al Qaeda's key ideological principles as "only Al Qaeda and its followers are fighting the oppressors of Islam"; as Al Qaeda considers itself the sole champion of the correct form of Islam, it sees itself as "the vanguard of the uprising of the oppressed" (p. 20). As a group, Al Qaeda maintains its perception of its own social identity as morally and religiously righteous, appealing to the positive emotions of justice and heroism of its constituency.

ISIS, too, paints itself as the rightful, divinely sanctioned champion of Allah and of Islam, further fostering positive emotions of righteousness for its members by using phrases like "the cause of Allah," and "Allah is with us" throughout its propaganda magazines *Dabiq* and *Rumiyah*. Issue 3 of *Rumiyah* (in 2016) also states that ISIS is "supporting the religion of Allah and making His word supreme." By communicating to its members that the group they are loyal to is heroic, moral, and religiously correct, ISIS can evoke similar positive emotions within its constituency, further encouraging cohesion and loyalty to the group.

It is important to note the importance of the positive ideals of world cohesion and purity under fundamentalist Islam promoted by Al Qaeda. Al Qaeda leader Ayman al Zawahiri prioritized the goal of achieving a genuine, pure Islamic state, unified worldwide. Through the narrative themes of purity and worldwide unity combined with justice and righteousness, Al Qaeda appeals to the positive emotion of heroism in an imagined community that uses a black-and-white system of morality. Group members can derive the positive emotions of righteousness and pride in their group membership from the themes Al Qaeda forwards, thus contributing to overall group cohesion.

Another theme common across jihadist terrorist groups is martyrdom, and its usage within terrorist rhetoric also is evocative of positive emotion. Becoming a *shaheed*, or martyr, through a suicide attack is celebrated within jihadist terrorist groups as part of a true Muslim's path to heaven; suicide attacks are viewed positively as acts of devotion to Allah and are essential to the terrorist identity (Quiggin, 2010). Groups like Al Qaeda and ISIS encourage their constituents to find positive emotions (i.e., happiness, satisfaction, and pride) in a violent suicide attack in the name of the terrorist group. ISIS propaganda materials routinely publish exaltations of deceased fighters who died in acts of self-professed martyrdom for the sake of the caliphate, encouraging group identity and loyalty. Similar to their promotion of suicide, terrorist groups are capable of finding positive emotions even in occurrences that affect the group negatively through the explanatory power of fantasy themes. ISIS used a shared group fantasy found in a *hadith*. The aim was to maintain the group's well-being despite its loss of territory, particularly the town of Dabiq.

The positive emotions encouraged within terrorist narratives through themes of heroism, righteousness, and martyrdom encourage group cohesion by forwarding a shared group consciousness—a like-minded community in which group members truly belong. Disenfranchised individuals may find the positive emotion of belonging for the first time within a terrorist group, fulfilling the psychological human desire to be a valued member of an imagined community. This, thereby, strengthens the social identity of the members of the terrorist group, encouraging group growth, member loyalty, stronger interpersonal relationships within the group, and group durability over time.

28.5.2 NEGATIVE EMOTION

While the positive emotions forwarded by terrorist narratives are valuable for the group's shared rhetoric and identity, the negative emotions present in terrorist rhetoric clarify the group's goals of enacting violence. To accomplish acts of violent terror, terrorist narratives cast the group's enemies as inhuman villains. According to Quiggin (2010), jihadist terrorist groups use narratives of oppression, poverty, and exploitation to evoke hatred of perceived suppressors (i.e., Christians, Jews, so-called *takfiri* or apostate Muslims, democratic governments, and Western nations). Within the jihadist movement, the commonly used theme *Al Wala' Wal Bara* can be roughly translated to "us versus them"; *Al Wara* means "those to whom we are loyal," while *Al Bara* refers to enemies and disbelievers (Quiggin, 2010). Although this theme is a concept present throughout the greater Islamic community without violent connotations, groups like Al Qaeda use it to strengthen their collective social identity and encapsulate their dichotomous worldview, encouraging negative emotions of hatred, disgust, and anger against *Al Bara*. This is where the value of the main tenets of SIT can be observed. The dichotomous worldview—i.e., often referred to as the "us vs. them" or "ingroup vs. outgroup" perspective—present in jihadist groups serves to incite recruits to take up arms against a hated enemy. For terrorist groups, negative emotions are clearly tools of radicalization.

Al Qaeda is not the only jihadist group that underscores the immorality of the outgroup, ISIS paints so-called disbelievers in a starkly unforgiving light in its publications. The text of the Issue 3 of its magazine *Rumiyah* states that "Jews, Christians, atheists, Rafidah, apostates, and all the nations of disbelief" are "the enemies of Allah" who have "dedicated their media, wealth, armies, and vehicles to wage war against the Muslims" ("Rumiyah Issue 3," 2016, p. 5). ISIS not only vilifies non-Muslims and apostate Muslims; it also demonizes them by labelling them the enemies of Allah. The magazines also describe villainous deeds done by the outgroup, who "extinguish the light of Allah, in enmity toward His religion and methodology on the earth" ("Rumiyah Issue 3," 2016, p. 4). Issue 3 of Rumiyah states that the armies of the disbelievers "raid your lands under the pretext of waging war against the Islamic State, then they don't depart until they've... killed your men and taken your women and children prisoner" ("Rumiyah Issue 3," 2016, p. 5). These harsh descriptions of the enemies of ISIS are clearly intended to stoke the anger of the members of terrorist groups, giving them more incentive to take up arms against the outgroup and commit acts of violent terrorism.

Other Islamist themes present in the jihadist community are also used by terrorist groups to incite negative emotions among their constituents. The Wahhabi/Islamist theme of *takfir* further encourages negative emotions against nonradical Muslims, alienating terrorists from perceived apostates. *Takfir* is the action of accusing others of being nonbelievers, unfaithful to Islam, and according to fundamentalist Wahabi scholars, "permissible to kill" (Firro, 2013). The established theme of *takfir* within the jihadist movement incites the hatred and suspicion of the outgroup and of nonradical Islamic communities. By driving a wedge between the members of the terrorist group and the greater Islamic community, groups can use the notion of *takfir* to encourage violence even between terrorists and those who share their faith.

Beyond the hatred of disbelievers, terrorist groups also appeal to the visceral negative emotions of cruelty and sadism to encourage their radicalized members to enact extreme abuse against their enemies. In its publications, ISIS explicitly uses action themes of brutality and militancy to encourage violent negative emotions; words and phrases like "harvesting," "smashing," "crushing," "blood bath," "carnage," "destruction," and "reaping casualties" are repeated in ISIS magazines, referencing ISIS's targets ("Rumiyah Issue 3," 2016, p. 10–12). Unambiguous calls for violence against disbelievers are also present throughout ISIS's publications. In reference to Christians in the Middle East, multiple instances of the phrases "kill the *mushrikin* wherever you find them" and "the blood of *mushrikin* is that it is permissible to shed" appear, repetitively appealing to the negative emotion of sadism ("Rumiyah Issue 9," 2017, p. 5).

The three primary negative emotions that terrorist groups evoke within their shared group rhetoric—hatred, anger, and sadism—showcase how terrorists can militarize emotion to fuel their acts of violence. By encouraging a dichotomous worldview and hatred against a perceived enemy, terrorist groups incite anger among their constituency for nonbelievers, giving group members a reason to take up arms. Finally, by inspiring sadism among their members, terrorist groups explicitly encourage violent terrorist action.

28.6 CONCLUSION

By crafting shared symbols, a collective rhetorical vision, and a group social identity, terrorist groups encourage both strong positive and negative emotions among their constituents. The groups evoke positive emotions through their narratives of moral righteousness, divine sanction, and heroism, allowing group members to perceive themselves as the just members of an organization that champions the causes of Allah. As many recruits of terrorist groups come from vulnerable situations and disempowered communities, the belonging and righteousness that they may feel as members of the terrorist ingroup can contribute to their sense of well-being and their loyalty and cohesion to their group (i.e., SIT). Negative emotions, meanwhile, are used within terrorist rhetoric to encourage alienation from and hatred toward the outgroup. Terrorist groups even operationalize the hatred and anger they incite by further fostering the emotion of sadism; by encouraging cruelty among group members against their targeted victims, group rhetoric fuels acts of terrorism through the intensity of the emotions of their constituents.

Emotions are clearly key aspects of a terrorist group's rhetorical success. By successfully instilling within their members genuine positive emotions toward the ingroup and negative emotions toward the outgroup, terrorist narratives strengthen the group's ideological identity and drive the group's operations (i.e., SCT). From a counterterrorism perspective, understanding the emotional effect terrorist rhetoric has on the individual terrorist is necessary to fight terrorism from the inside out. According to Post (2005), the best counterterrorism policy prevents the recruitment of potential terrorists; the emotional control the terrorist group has over an individual's psychology should not be underestimated. Understanding both the positive and negative emotional power of terrorist narratives is necessary to prevent vulnerable individuals from seeking out positive emotional reinforcement from terrorist groups and from acting upon the negative emotions terrorist narratives evoke.

KEYWORDS

- **symbolic convergence**
- **resilience**
- **ISIS terrorist**

REFERENCES

Anderson, B. (1983). *Imagined communities*. London: Verso.

Barbalet, J. M. (2006). Emotions in politics: From the ballot box to suicide terrorism. In S. Clarke, P. Hoggett, & S. Thompson (Eds.), *Emotions, politics and society* (pp. 34–52). New York: Palgrave Macmillan.

Bormann, E. G. (1972). Fantasy and rhetorical vision: The rhetorical criticism of social reality. *Quarterly Journal of Speech, 58*(1), 396–407.

Bormann, E. G. (1973). The Eagleton affair: A fantasy theme analysis. *Quarterly Journal of Speech, 59*(2), 143–159.

Bormann, E. G. (1985). Symbolic convergence theory: A communication formulation. *Journal of Communication, 35*(4), 128–138.

Bormann, E. G., Cragan, J. F., & Shields, D. C. (1996). An expansion of the rhetorical vision component of the symbolic convergence theory: The Cold War paradigm case. *Communication Monographs*, *63*(1), 1–25.

Brown, R. J. (1984). The role of similarity in intergroup relations. In H. Tajfel (Ed.), *The social dimension* (pp. 603–623). Cambridge: Cambridge University Press.

Cabanac, M. (2002). What is emotion? *Behavioral Processes*, *60*(2), 69–83.

Clark, L. A., Watson, D., & Mineka, S. (1994). Temperament, personality, and the mood and anxiety disorders. *Journal of Abnormal Psychology*, *103*(1), 103–116.

Cragan, J. F., & Shields, D. C. (1992). The use of symbolic convergence theory in corporate strategic planning: A case study. *Journal of Applied Communication Research*, *20*(2), 199–218.

Crenshaw, M. (1981). The causes of terrorism. *Comparative Politics*, *13*(4), 379–399.

Dabiq (2014). A call to Hijrah. *Dabiq*, *3*.

Damasio, A. R. (1998). Emotion in the perspective of an integrated nervous system. *Brain Research Reviews*, *26* (2), 83–86.

Deetz, S. (2001). Alternative perspectives in organizational communication studies. In F. Jablin & L. Putnam (Eds.), *The new handbook of organizational communication* (pp. 3–46). Thousand Oaks, CA: SAGE.

D'Orazio, V., & Salehyan, I. (2018). Who is a terrorist? Ethnicity, group affiliation, and understandings of political violence. *International Interactions*, *44*(6), 1017–1039.

Durkheim, E. (1965). *The elementary forms of the religious life*. New York: Free Press.

Elliot, A. J., & Thrash, T. M. (2002). Approach-avoidance motivation in personality: Approach and avoidance temperaments and goals. *Journal of Personality and Social Psychology*, *82*(5), 804–818.

Firro, T. K. (2013). The political context of early Wahhabi discourse of *takfir*. *Middle Eastern Studies*, *49*(5), 770–789.

Freilich, J. D., Chermak, S. M., & Caspi, D. (2009). Critical events in the life trajectories of domestic extremist white supremacist groups. *Criminology & Public Policy*, *8*(3), 497–530.

Ganor, B. (2002). Defining terrorism: Is one man's terrorist another man's freedom fighter? *Police Practice and Research*, *3*(4), 287–304.

Griffin, E. (2006). *A first look at communication theory* (6th ed.). Boston: McGraw-Hill.

Hashim, A. S. (2014). The Islamic State: From Al Qaeda affiliate to caliphate. *Middle East Policy*, *21*(4), 69–83.

Hoffman, B. (2006). *Inside terrorism*. New York: Columbia University Press.

Hoggett, P. (2009). *Politics, identity, and emotion*. London: Paradigm Publishers.

Jenkins, B. M. (2015). International terrorism: A new mode of conflict. In D. Carlton & C. Schaerf (Eds.), *International terrorism and world security* (pp. 13–49). New York: Routledge.

Lacey, A. (2005). Networked communities: Social centres and activist spaces in contemporary Britain. *Space and Culture*, *8*(3), 286–301.

Matusitz, J. (2012). *Terrorism & communication: A critical introduction*. Thousand Oaks, CA: SAGE.

Novenario, C. M. I. (2016). Differentiating Al Qaeda and the Islamic State through strategies publicized in jihadist magazines. *Studies in Conflict & Terrorism*, *39*(11), 953–967.

Perliger, A., & Pedahzur, A. (2016). Counter cultures, group dynamics and religious terrorism. *Political Studies*, *64*(2) 297–314.

Post, J. M. (2005). When hatred is bred in the bone: Psycho-cultural foundations of contemporary terrorism. *Political Psychology*, *26*(4), 615–636.

Post, J. M. (1986). Hostilité, conformité, fraternité: The group dynamics of terrorist behavior. *International Journal of Group Psychotherapy*, *36*(2), 211–224.

Pregill, M. (2016). ISIS, eschatology, and exegesis: The Propaganda of Dabiq and the sectarian rhetoric of militant Shi'ism. *Mizan: Journal of Interdisciplinary Approaches to Muslim Societies and Civilizations*, *1*(2), 1–36.

Rumiyah (2017). *Rumiyah, 9*

Quiggin, T. (2010). Understanding al-Qaeda's ideology for counter-narrative work. *Perspectives on Terrorism*, *3*(2), 10–21.

Ruby, C. L. (2002). The definition of terrorism. *Analyses of Social Issues and Public Policy*, *2*(1), 9–14.

Rumiyah (2016). Shawwal 1438. *Rumiyah*, *3*.

Simmel, G. (1950). Faithfulness and gratitude. In K. Wolff (Ed.), *The sociology of Georg Simmel* (pp. 379–395). New York: The Free Press.

Sosis, R. (2009). The adaptationist-byproduct debate on the evolution of religion: Five misunderstanding of the adaptionist program. *Journal of Cognition and Culture*, *9*(3), 315–332.

Stults, C. B., Kupprat, S. A., Krause, K. D., Kapadia, F., & Halkitis, P. N. (2017). Perceptions of safety among LGBTQ people following the 2016 Pulse nightclub shooting. *Psychology of Sexual Orientation and Gender Diversity*, *4*(3), 251–256.

Tajfel, H. (1970). Experiments in intergroup discrimination. *Scientific American*, *223*(5), 96–103.

Turner, J. C. (1979). Social categorization and social discrimination in the minimal group paradigm. In H. Tajfel (Ed.), *Differentiation between social groups: Studies in the social psychology of intergroup relations* (pp. 101–140). London: Academic Press.

United States Department of State (2001). *Patterns of global terrorism 2000*. Washington, D.C.: United States Department of State.

Vasquez, G. M. (1993). A homo narrans paradigm for public relations: Combining Bormann's symbolic convergence theory and Grunig's situational theory of publics. *Journal of Public Relations Research*, *5*(3), 201–216.

Wessinger, C. (2009). Deaths in the fire at the Branch Davidians' Mount Carmel: Who bears responsibility? *Nova Religio*, *13*(2), 25–60.

White, J. (2001). *Terrorism: An introduction* (3rd ed.). Boston: Wadsworth Publishing.

Whitehouse, H. (2013). Religion, cohesion and hostility. In S. Clarke, R. Powell, & J. Savulescu (Eds.), *Religion, intolerance and conflict: A scientific and conceptual investigation* (pp. 36–47). Oxford: Oxford University Press.

Wignell, P., Tan, S., O'Halloran, K. L., & Lange, R. (2017). A mixed methods empirical examination of changes in emphasis and style in the extremist magazines *Dabiq* and *Rumiyah*. *Perspectives on Terrorism*, *11*(2), 2–20.

CHAPTER 29

Post-Traumatic Growth: Positive Transformation beyond Resilience

N. D. S. NAGA SEEMA

Centre for Health Psychology, University of Hyderabad, Hyderabad 500046, India
E-mail: nagaseemavsp@gmail.com

ABSTRACT

The chapter outlines the divergence of resilience and post-traumatic growth, as these two constructs have conceptual confusion and ostensible overlapping. The chapter presents the meaning of trauma, post-traumatic reactions, and the potency of post-traumatic growth. It also explains the biopsychosocial outcomes of post-traumatic growth. The content of the chapter focuses on the cultural meanings attached to post-traumatic growth, religious coping for the development and fostering of post-traumatic growth, and post-traumatic growth as an outcome as well as an ongoing positive transformation process. The chapter not only draws attention on post-traumatic growth as an approach to trauma treatment and care, but also paves a pathway for preparing the person for post-traumatic growth as a "life-changing" psychological shift toward well-being.

29.1 INTRODUCTION

The aim of the chapter is to focus on the divergence of resilience from post-traumatic growth. It is pertinent to bring clarity as these two constructs have conceptual confusion and ostensible overlapping due to the presence of numerous similar ideas and their co-occurrence. The chapter clarifies the discrepancy of resilience and post-traumatic growth in trauma survivors. This chapter contributes to the understanding of post-traumatic growth as a process, outcome, and as an intervention warranting cultural and religious coping following trauma and its trailing effect.

The term "resilience" is originated from the Latin word *resilire*, which means "to spring back." The concept of resilience, initially used in physics, when adopted in the field of psychology, it brings issues and challenges as the properties of object and person are different, even if the core dynamics of resilience remain the same. In context of psychology, resilience refers to the ability of a person to withstand and respond to a stressor event, traumatic exposure, or crisis in a constructive way and bounce back from adverse experiences and circumstances quickly and effectively. Walsh (2003) explained it as the reparative potential and capacity for recovery and sustainability which may develop in individuals or systems as the product of the interaction among developmental processes and environmental factors that confirms resilience as the product of the continual interaction between personal, social, and environmental factors. Social resources, such as effective parenting and social support, are the most important environmental factors that enhance the probability of manifestation of resilience. Another view expounded resilience as a difference rather than as growth (Norris et al., 2008). A recently proposed model of resilience, called synergy model of resilience, postulates that resilience is a unique function of adversity, operating factors, and resistance across a time continuum that brings and sustains positive reflection in forms of achievement and flourishing in the person's life (Hariharan & Rana, 2016).

Resilience has been conceptualized encompassing three abilities, that appear to be simple, but the process of acquisition of those abilities may be complex. These three are—(i) the personal abilities, (ii) the ability for coping with highly challenging situations, and (iii) the ability for recovering from traumatic experience (Masten et al., 1990). Keeping this in mind, a comprehensive model of resilience was developed by Tedeschi and Calhoun (1996), who coined the term "post-traumatic growth" based on their experience with adults who had become physically handicapped and with older women who had lost their spouses. The origins of post-traumatic growth can be found in the ancient writings and contemporary religious, philosophical, and psychological lineages as observed by Calhoun and Tedeschi (2013).

Post-traumatic growth refers to the perceived positive changes following the encounter and struggle with highly stressful events, that is, the promotion of gaining new meaning, developing new perspectives and directions, and a sense of learning valuable lessons after the initial distress (Calhoun & Tedeschi, 2006). The terminology related to post-traumatic growth and its related concepts have been illustrated by Richard and Lawrence (2004). Perceived benefits, positive aspects, and the transformation of trauma are the terms that have been used much before the introduction of the term post-traumatic growth in the literature of psychology. Many other terms having the parallel meaning of post-traumatic growth are strength conversion, positive psychological changes, perceived benefits or construing benefits, stress-related growth, flourishing, positive by-products, discovery of meaning, positive emotions, and thriving (Bower et al., 1998; Calhoun & Tedeschi, 1991; Finkel, 1974, 1975; Folkman & Moskowitz, 2000; McMillen et al., 1995, 2001; O'Leary & Ickovics, 1995; Park et al., 1996; Ryff & Singer, 1998; Tennen et al., 1992; Yalom & Lieberman, 1991).

29.2 MEANING OF TRAUMA, POST-TRAUMATIC REACTIONS, AND POST-TRAUMATIC GROWTH

During the past decade, trauma has occupied a central place in the academic and professional spheres in the world. In the European parallel of the DSM, the International Statistical Classification of Diseases, and Related Health Problems 11th Revision (ICD-11), trauma results from an encounter and is likely to create distress with an overwhelming, negative, and extremely painful experience that is incongruent with the existing repertoire of perceptions, interpretations, and coping strategies.

Loss and trauma are different. Nevertheless, grief over the loss of people, objects, ideals, and beliefs is often central to the experience of trauma. The terms stressful events, trauma, and crisis are often interchangeably used to connote the same experience. Individual differences are observed in terms of adaptation to stressful events. The persons who adapt well to a stressful event in one situation may fail to adapt well in other situations. For example, persons who successfully handle stress in their workplaces may not become successful in managing stress in their personal life. However, traumatic events equally affect the individuals, groups, and communities, and through vicarious processes, to the whole nations and human society.

Post-traumatic reaction in its ordinary sense is likely to bring negative consequences in life. Be it terrorist attacks, natural disasters (e.g., tsunami and earthquakes), or wars (e.g., in the Middle East, Africa, and Eastern Europe), the aftermath leaves enduring negative impact. Reactions to trauma include the feeling of hopelessness, feeling of detachment or being unconcerned about others, having trouble in reasoning and decision-making process, feeling jumpy and getting startled easily at sudden noises, feeling on guard and constantly alert, and having frequently disturbing dreams and memories or flashbacks. These negative reactions increase the vulnerability of the person to more undesirable and harmful life consequences. Until the coining of the term "post-traumatic growth," the trauma was construed to bring negative orientation in life. Basing on the severity, it is assumed that the consequences of trauma are pathological that result in post-traumatic disorder.

Nevertheless, the genesis of the concept of post-traumatic growth has focused on the positive dimension of trauma—an extraordinary transformation. Calhoun and Tedeschi (1999) defined post-traumatic growth as the experience of positive change resulting from the struggle with major life crises. With the emergence of positive psychology, the foci of trauma research and intervention have shifted from negative consequences to potential positive outcomes (Calhoun & Tedeschi, 2006; Joseph & Linley, 2008). Therefore, major life events can influence a person's understanding of the world in a positive way. The assumptive world, as explained by Kuhn (1970), provides individuals with the general perspectives and paradigms within which they operate and guide their actions. Therefore, it is understood that growth, does not occur as a direct result of trauma, rather it is the individual's struggle that determines the extent to which post-traumatic growth takes place with a new reality in the aftermath (Richard & Lawrence, 2004). Hence, post-traumatic growth has a quality of transformation or a qualitative positive change in psychosocial functioning and an experience of improvement as it involves a variety of elements, such as personal characteristics, greater appreciation for life, a sense of personal strength, recognition of new possibilities, and self-disclosure.

29.3 RESILIENCE VERSUS POST-TRAUMATIC GROWTH

From the above discussion, it is understood that resilience may exist prior to exposure to trauma, crisis, or adverse situation in life. On the other hand, post-traumatic growth develops as a result of lessons learned from exposure to trauma or crisis (Tedeschi & Calhoun, 2004). According to Turner and Cox (2004), post-traumatic growth is manifested in behaviors and thought-forms and not necessarily exists prior to experience.

Resilience is a person's demonstration of survival in the face of overwhelming life circumstances. It is the ability to cope with negative emotions that arise from a stressful experience and function at normal or close to normal capacity. Research has shown that resilience is ordinary, not extraordinary. Society demonstrates resilience. One example is the September 11, 2001 terrorist attacks in America. The response of many Americans to the event can be witnessed by the efforts of the individuals to rebuild their lives. Several factors have been recognized that are associated with resiliency. The factors encompass attributes of a person, such as being sociable, having a sense of humor and being hopeful; as well as social dimensions such as experiences of parental warmth, nurturing in the family unit, support at school, and success in at least one area of life as explained by Fleming and Ledogar (2008).

Resiliency is thus explained as the ability to recover or return to the level of functioning present before the occurrence of adversity with only minimal reactions following the exposure to trauma or crisis (Bonanno et al., 2001). Though resilient persons may experience challenges in their normal functioning, yet they are assumed to have overall healthy functioning; the implication is that they have high level of coping after trauma or crisis in life.

Resilience in a specific area of an individual's life does not guarantee resilience across all functional domains. Indeed, research and theories of resilience indicate that resilience in one area of life can coexist with high-risk behaviors, social and emotional withdrawal, and maladaptive survival tactics. Further, it is observed that experience of childhood trauma is associated with lower levels of resilience and post-traumatic growth following a traumatic event in adulthood (Van Slyke, n.d.). Being resilient does not mean that a person doesn't experience difficulty or distress. Emotional pain and sadness are common in people who have suffered major adversity or trauma in their lives.

Nevertheless, people who suffer trauma can be viewed as having enormous potential to survive and the survival tactics used must be seen as normal reactions to abnormal circumstances. In this direction, people construct and develop new understandings of themselves, the world they live in, how to relate to other people, the kind of future they might have, and a better understanding of how to live life. These aspects have been emphasized by psychologists Richard Tedeschi and Lawrence Calhoun in the mid-1990s. Moreover, the theory behind post-traumatic growth explains an individual's transformation following trauma. Post-traumatic growth has been considered as the experiencing of positive post-event changes in one's functioning, attributed to the direct effects of the traumatic event. Far ahead, it has been conceptualized as positive changes experienced on different dimensions, as a result of the struggle with the negative event and its multilevel consequences as reported by Tedeschi and Calhoun (2004).

Reports and descriptions of post-traumatic growth may be found within existential and humanistic psychology as well (Jaffe, 1985). Within these approaches, mostly descriptions of identity and spiritual transformations, changes in interpersonal relationships, and meaning systems proliferate. The first studies that purposefully mention the possibility of experiencing growth in the aftermath of a negative event were those of Taylor (1977) on the effects of natural calamities, and Hamera and Shontz (1978) on life-threatening illness. Since then, the number of studies has exponentially grown covering a large range of traumatic events, and the methodological rigor has been improved as well. On the whole, post-traumatic growth is considered to be the experience of positive transformation that occurs as an outcome of the struggle with highly challenging life crises. The benefits of post-traumatic growth may bring an optimum level of enhancement in people's lives. This further emphasizes its consideration within psychotherapy, and prevention as well.

Hence, resilience is no doubt distinct from post-traumatic growth in the sense that post-traumatic growth has a higher connotation and deeper positive impact. Resilience accelerates the post-traumatic growth and prevents the fledgling victim to move toward post-traumatic disorder—the negative dimension of life. Empowerment, improvement, and strengthening are equally conceptualized in resilience and post-traumatic growth. However, in the case of post-traumatic growth, the trauma survivors bloom after traumatic experiences (Almedom, 2005; Hall, 2006).

29.4 POST-TRAUMATIC GROWTH AS A PROCESS AND AS AN OUTCOME

Post-traumatic growth is not only an outcome but also an ongoing positive transformation process. The outcomes of post-traumatic growth may be best considered as iterative or recursive. To trace the varied trajectories of the post-traumatic growth process, longitudinal or sequential research may be carried out. This process is likely to involve a powerful combination of demand for emotional relief and cognitive clarity, which is achieved through the construction of higher order schemas that allow for appreciation of paradox. Among trauma survivors, altruistic behavior was found to be an outcome of post-traumatic growth (Frazier et al., 2013; Staub & Vollhardt, 2008). Post-traumatic growth represents a pathway beyond resilience among trauma survivors and victims of natural disasters or personally experienced tragedy (Frazier et al., 2013). Individuals observed to have post-traumatic growth often engage in altruistic behavior. Linley and Joseph (2004) found that 30%–70% of survivors of a range of traumatic events including accidents, medical problems, assaults, and natural disasters experienced positive change following the traumatic event. Evidences have been reported in the literature of a very wide array of major life challenges that have acted as catalysts for post-traumatic growth. It appears that the phenomenon of post-traumatic growth occurs in a wide range of people facing a wide variety of traumatic circumstances. Research findings reveal that those experiencing post-traumatic growth do not return to the same baselines that existed prior to trauma, rather they emerge as transformed with a benchmark for future resilience trajectories. Therefore, post-traumatic growth has been viewed as both a process and an outcome (Zollner & Maercker, 2006) with distinctive cognitive abilities and emotional modules that result in behavioral changes.

29.5 CULTURAL MEANINGS ATTACHED TO POST-TRAUMATIC GROWTH

Culture is widely used as an explanatory concept for differences in attitudes, thoughts, emotions, and behaviors. There is no single definition of culture that is universally accepted. According to Fiske (2002), culture is a socially constructed constellation consisting of practices, competencies, ideas, schemas, symbols, values, norms, institutions, goals, constitutive rules, artifacts, and modifications of the physical environment. As suggested, we assume that culture is transmitted through environmental influences (Stuart, 2004), and interactions with other significant people (Brown, 2004). Nevertheless, the core components, like interpersonal relationships, values, beliefs, and view of self-persist in various combinations and variations across cultures (Weiss & Berger, 2010). People in collectivistic societies are highly committed to families, construct their identities in social interaction, and attribute events to external forces (Kirmayer, 1989). This indicates a culturally sensitive nature of post-traumatic growth, and also the importance of culturally competent practices. This can further be explained in terms of knowledge about populations, the characteristics and meaning of stress, crisis, and growth. Professionals equipped with this knowledge readily offer informed, differential, and culture-sensitive services.

There is an evidence that post-traumatic growth, both at the individual and communal level, is associated with communal coping and social processing (Paez et al., 2007) in the Spanish context. The findings relative to the individual post-traumatic growth emerged in the context of a unique cross-cultural comparison of expressive writing following S-11 in the United States and M-11 in Spain. Another study indicates strong similarities between American and Spanish participants in the emotional content of accounts about reactions to the terrorist attacks, including similar weight attached to support the idea that the concept of post-traumatic growth is applicable to Spain and has deep community components, specifically, as it relates to those vicariously traumatized by national trauma. Positive and negative emotions, communal coping, social sharing, coping by social support, and demonstrations are the main correlates and predictors of perceived individual and community benefits. These relationships could be explained, at least partially, by the egalitarian characteristic of the Spanish culture (Basabe & Ros, 2005) as well as by its secular nature.

It is quite convincing that across cultures, post-traumatic growth seems to be connected to coping mechanisms, cognitive functioning, social support, and religiosity or spirituality in its highest form. Coping strategies such as active cognitive coping and problem-focused coping are associated with post-traumatic growth (e.g., Japan, Turkey, and Kosova) whereas coping by avoidance is negatively correlated with post-traumatic growth (e.g., Turkey and Kosova). The type of coping is positively related to post-traumatic growth as evident from the literature. For instance, in Turkey, maintaining relationships and communication with others has shaped and supplemented growth in the interpersonal dimensions than in other facets of post-traumatic growth. In Germany, the Netherlands, Spain, Japan, and Australia studies have demonstrated that cognitive functioning is associated with post-traumatic growth. Owing to varying explicit characteristics, the importance of perceived social support for post-traumatic growth excels across cultures (Germany, Israel, Turkey, Spain, US Latinos, Australia, and Chinese in Australia). Sharing of feelings with close members like family, friends, and colleagues are exercised and encouraged toward post-traumatic growth across cultures (Israel, Spain, and Japan). Across the life span, the experience of post-traumatic growth is seemingly rooted in diverse cultures. It has been reported that children, adolescents, and adults from various cultures demonstrated post-traumatic growth (Meyerson et al., 2011; Splevins et al., 2010).

29.6 RELIGIOUS COPING AS A MODERATOR FOR POST-TRAUMATIC GROWTH

Due to the sudden onset of trauma or crisis, the fledgling victim may have four options to choose—flight, fright, fight, or faith. The first two choices (flight or fright) make the victim vulnerable to the negative outcome, that is, post-traumatic disorder, whereas the last two choices (fight or faith) orient the victim to transform in the form of post-traumatic growth. The victim constructively fights with the traumatic experience using positive resources, for example, resilience, to move toward transformation. This fight gains momentum and gets strengthened in the form of faith—faith toward internal resources and external agencies, like supernatural power in the form of God. In the process, faith gets reinforced through religious beliefs. Bert (2011) has recommended that religious beliefs are valuable for people who want to bring noticeable impacts in their lives. Mattoon (2011) suggested that religious beliefs influence post-traumatic growth. Religion enables post-traumatic growth in finding a meaning and benefit for the individual during traumatic event. Religion and spirituality assist the persons to direct their lives and decide what coping strategies are to be adopted to different events (Gall et al., 2009). Spirituality also serves as a significant factor affecting post-traumatic growth (Kimhi et al., 2010). It has been observed that spirituality, religiosity, and its predictors are linked with each other and contribute to post-traumatic growth in Christians, Jews, and Muslims in most cases (e.g., Israel, Palestinians, and Latinos in the United States) (Ransom et al., 2008; Williams et al., 2006). However, not all individuals (e.g., Australia) subscribe to religious beliefs as a coping strategy.

Moreover, the relationship between religious coping and post-traumatic growth is mediated by hope in some cultures (e.g., Kosovar). Laufer and Solomon (2006) reported that Jewish adolescents who had experienced terror attacks were found to have faith as a major factor in predicting their post-traumatic growth. Prati and Pietrantoni (2009) have conducted a meta-analysis and indicated that post-traumatic growth is influenced by religious coping and positive thinking in survivors of traumatic experiences. In the event of Hurricane Katrina, post-traumatic stress, psychological distress, and post-traumatic growth were studied in the framework of positive and negative religious coping strategies in female survivors. Results indicated that religious coping and positivity were related to post-traumatic growth. Further, post-traumatic stress and psychological distress were associated with negative religious coping (Chan & Rhodes, 2013).

Research also finds the role of religiosity in the post-traumatic growth of the caregivers of the persons suffering with chronic illness. In the area of caregiving, distinct dimensions of cancer-specific religious coping among the family caregivers have been identified. Post-traumatic growth in family caregivers of cancer patients at a similar phase of cancer treatment has been found to be associated with the increased use of positive religious coping strategies in the Indian context. Further, benevolent religious reappraisals and punishing God reappraisals have predicted the family caregivers' post-traumatic growth. The study has underlined that meaning-focused religious coping is crucial (Thombre et al., 2010). In the light of post-traumatic growth, evidence from a study documented that offering prayer was a possible coping behavior among trauma survivors belonging to Christianity (Harris et al., 2010). The study further illuminated that offering prayer to the deity by means of self-disclosure, reflection, and meditation fostered post-traumatic growth. It was also observed that survivors of traumatic experiences sought personal growth, wherein calm and focus prayer predicted post-traumatic growth. Therefore, as religious practice plays a significant role in coping during the post-traumatic phase, the therapists need to familiarize themselves with understanding various religious practices while working with clientele group in optimizing their religious coping for post-traumatic growth.

29.7 THERAPY FOR POST-TRAUMATIC GROWTH

Therapy for post-traumatic growth needs special attention as the therapist has to address a complex issue, which is both a process as well as the outcome. The therapist, who plays the role of a catalyst to post-traumatic growth, is considered as knowledgeable and mindful. For a therapist, post-traumatic growth is all about possibility and encourages clients in the wider spectrum of culture and religion. The more the person learns about post-traumatic growth, the higher would be the potency of the outcome. Moreover, Calhoun and Tedeschi (1998, 1999) have specified a set of procedures for integrating post-traumatic growth into therapeutic settings.

Post-traumatic growth emerges from a cognitive restructuring process for alleviating stress and grief. Trauma leaves a person with an emotionally dispirited belief system. Hence, cognitive processing is the key expression of post-traumatic growth (Janoff -Bulman, 1992). Hembree and Foa (2000) have emphasized that representation of the event through story-telling is the primary objective of the therapy intended for trauma survivors. According to Calhoun and Tedeschi (1998) by way of telling the story or narrating the event experienced by the client, the therapists can enable post-traumatic growth therapy. Support and encouragement are facilitated at the positive vicissitudes expressed by the client during the process of story telling. It is essential to simultaneously concede the client's pain and suffering, or help received during the traumatic event. There is a threat or danger involved in the usage of superficial statements used by the therapists, for example, "look at the positive side of life," "let's focus on the brighter side." Such statements attract the client toward refusal and cognitive escaping. In certain exceptional cases, it is possible that few trauma survivors with the worst types of traumatic experiences are unprepared due to their distorted thought process and fail to identify their personal growth. Foa (2006) has recommended that the primary objective of treating such individuals must be aimed at minimizing emotional distress through cognitive behavioral therapy.

In understanding post-traumatic growth, Cann et al. (2011) have proposed rumination thinking involving two styles—intrusive rumination and deliberate rumination. The thoughts related to events or experiences that appear in one's consciousness unexpectedly, without conscious intent is referred to as intrusive rumination. Deliberate ruminations are event-related thoughts that an individual invites, engages in purposefully, often with the intent of trying to understand or make sense of events and their implications. Both intrusive and deliberate rumination play distinct roles but deliberate rumination has been positively associated with post-traumatic growth (Lindstrom et al., 2013).

The post-traumatic growth therapy involves professionals who can display compassion, humility, and respect for survivors and bring the potential for strength and change (Tedeschi & McNally, 2011). These professionals are the expert companions as suggested by Calhoun and Tedeschi (2013), who display the elements as explained in the humanistic perspective. Hijazi (2012) observed that Iraqi refugees provided with three sessions of narrative exposure therapy showed increased post-traumatic growth at two and four-month follow-ups. Higher scores on cognitive emotion regulation strategies were found to be associated with higher post-traumatic growth (Hussain & Bhushan, 2011).

Through positive psychotherapy for Cancer Program, researchers have facilitated post-traumatic growth among cancer survivors (Casellas-Grau et al., 2017). Post-traumatic stress and emotional distress have significantly been reduced among the cancer survivors when compared with a control group. This specific program aims at enhancing patient's positive emotions, strengths, and personal meanings. It is assumed that growth, fulfillment, and happiness are inherent in individuals (Rashid & Seligman, 2013).

A new intervention developed by Xu et al. (2016) with an acronym Stress Management and Resiliency Training (SMART) has been crafted to enhance the quality of life and increase the quality of care in medical staff from China. The new intervention model applied positive psychological intervention based on the Chinese culture to improve medical workers' post-traumatic growth. The SMART was provided to the hospital staff to build coping skills, mindfulness-based stress reduction, communication, self-awareness, and wellness program and seminars on compassion fatigue (Sood et al., 2011).

Regular exercise and healthy eating habits result in positive coping mechanisms that further promote post-traumatic growth (Meadors & Lamson, 2008; Saadat et al., 2012). Research finding has revealed that cognitive-behavioral stress management (CBSM) is found to be effective in early-stage (0–II) breast cancer patients. The study has also mediated a strengthened immunity after a three-month follow-up (McGregor et al., 2004). It is observed that the CBSM promotes post-traumatic growth in women under treatment for early-stage breast cancer and elevates the high level of optimism (Antoni et al., 2001).

29.8 CONCLUSION

However, the evidence suggests that resilience is understood as an inherent personal trait of an individual, whereas post-traumatic growth represents an individual's experience of transformation by their struggles with adversity, trauma, or crisis. Hence, post-traumatic growth may represent a pathway beyond resilience among trauma survivors and victims of natural disasters or personally experienced tragedy through which the author proposes to address this conceptually. Therefore, post-traumatic growth is conceptualized as both a process as well as an outcome with distinct cognitive, emotional, and socio-cultural components that result in behavioral changes and transformation. Literature has shown the universality of post-traumatic growth as well as its culture-specific nature that explains the aspects and manifestations of cultural variation. There is also an evidence of knowledge about post-traumatic growth in cross-cultural contexts. Moreover, post-traumatic growth can be inculcated and fostered in the victims through interventions by the therapist using an integrated framework embracing positive psychosocial perspectives.

KEYWORDS

- **trauma**
- **post-traumatic reaction**
- **post-traumatic growth**
- **resilience**
- **trauma treatment and care**

REFERENCES

Almedom, A. M. (2005). Resilience, hardiness, sense of coherence, and post-traumatic growth: All paths leading to "light at the end of the tunnel"? *Journal of Loss and Trauma, 10*, 253–265.

Antoni, M. H., Lehman, J. M., Kilbourn, K. M., Boyers, A. E., Culver, J. L., Alferi, S. M., Carver, C. S. (2001). Cognitive behavioral stress management intervention decreases the prevalence of depression and enhances benefit finding among women under treatment for early-stage breast cancer. *Health Psychology, 20*, 20–32.

Basabe, N., & Ros, M. (2005). Cultural dimensions and social behavior correlates: Individualism-collectivism and power distance. *International Review of Social Psychology, 18*(1), 189–225.

Bert, S. C. (2011). The influence of religiosity and spirituality on adolescent mothers and their teenage children. *Journal Youth Adolescence 40*, 72–84

Bonanno, G. A., Papa, A., & O'Neill, K. (2001). Loss and human resilience. *Applied & Preventive Psychology, 10* (3), 193–206.

Bower, J. E., Kemeny, M. E., Taylor, S. E., & Fahey, J. L. (1998). Cognitive processing, discovery of meaning, CD4 decline and AIDS-related mortality among bereaved HIV seropositive men. *Journal of Consulting and Clinical Psychology, 66*, 979– 986.

Brown, D. K. (2004). *Social blueprints: Conceptual foundations of sociology*. New York: Oxford University Press.

Calhoun, L. G., & Tedeschi, R. G. (1991). Perceiving benefits in traumatic events: Some issues for practicing psychologists. *Journal of Training & Practice in Professional Psychology, 5*, 45–52.

Calhoun, L. G., & Tedeschi, R. G. (1998). Post-traumatic growth: Future directions. In R. G. Tedeschi, C. L. Park, & L. G. Calhoun (Eds.), *Post-traumatic growth: Positive change in the aftermath of crisis* (pp. 215–238). Mahwah, NJ: Erlbaum.

Calhoun, L. G., & Tedeschi, R. G. (1999). Facilitating post-traumatic growth*: A clinician's guide*. Mahwah, NJ: Erlbaum.

Calhoun, L. G., & Tedeschi, R. (2006). The foundations of post-traumatic growth: An expanded framework. In L. G. Calhoun, & R. G. Tedeschi (Eds), *Handbook of post-traumatic growth: Research and practice* (pp. 1–23). Mahwah, NJ: Erlbaum.

Calhoun, L. G., & Tedeschi, R. G. (2013). *Post-traumatic growth in clinical practice*. NY, USA: Routledge/Taylor & Francis Group.

Cann A., Calhoun L. G., Tedeschi R. G., Triplett K., Vishnevsky T., Lindstrom C. (2011). Assessing post-traumatic cognitive processes: The event related rumination inventory. *Anxiety, Stress, and Coping*. *24*(2), 137–156.

Casellas-Grau, A., Vives, J., Font, A., & Borràs, J. (2017). Positive psychotherapy for distressed cancer survivors: Post-traumatic growth facilitation reduces post-traumatic stress. *International Journal of Clinical and Health Psychology, 17*(1), 28–37.

Chan, C. S., & Rhodes, J. E. (2013). Religious coping, post-traumatic stress, psychological distress, and post-traumatic growth among female survivors four years after Hurricane Katrina. *Journal of Trauma Stress*. *26*(2):257–65. doi: 10.1002/jts.21801.

Finkel, N. J. (1974). Strens and traumas: An attempt at categorization. *American Journal of Community Psychology, 2*, 265–273.

Finkel, N. J. (1975). Strens, traumas, and trauma resolution. *American Journal of Community Psychology, 3*, 173–178.

Fiske, A. P. (2002). Using individualism and collectivism to compare cultures: A critique of the validity and measurement of the constructs: Comments on Oyserman, et al. (2002). *Psychological Bulletin, 128*(1), 78–88.

Fleming, J., & Ledogar, R. J. (2008). Resilience, an evolving concept: A review of literature relevant to aboriginal research. *Pimatisiwin, 6*(2), 7–23. Retrieved from https://www.ncbi.nlm.nih.gov/pmc/articles/PMC2956753/

Foa, E. B. (2006). Psychosocial therapy for post-traumatic stress disorder. *Journal of Clinical Psychiatry, 6*, 740–745.

Folkman, S., & Moskowitz, J. T. (2000). Stress, positive emotion, and coping. *Current Directions in Psychological Science*, *9*(4), 115–118.

Frazier, P., Greer, C., Gabrielsen, S., Tennen, H., Park, C., & Tomich, P. (2013). The relation between trauma exposure and prosocial behavior. *Psychological Trauma: Theory, Research, Practice, and Policy, 5*(3), 286–294.

Gall, T. L., Kristjansson, E., Charbonneau, C, & Florack, P. (2009). A longitudinal study on the role of spirituality in response to the diagnosis and treatment of breast cancer. *Journal of Behavioral Medicine, 32*, 174–186

Hall, M. J. (2006). Resilience and post-traumatic growth: The power of human attachment. Psychiatry: *Interpersonal & Biological Processes, 69*, 182–189.

Hamera, E. K. & Shontz, F. C. (1978). Perceived positive and negative effects of life-threatening illness. *Journal of Psychosomatic Medicine, 22*, 419–424.

Hariharan, M. & Rana, S. (2017). Conceptual complexity of resilience: Synergy approach to measurement. In U.

Kumar (Ed.), *The Routledge International Handbook of Psychosocial Resilience* (pp. 24–33). Oxon, UK: Routledge.

Harris, J. I., Christopher. R., Brian, E., Richard, G., Raymond, H., & Ann Marie, M. (2010). Coping functions of prayer and post-traumatic growth. *The International Journal for the Psychology of Religion, 20,* 26–38.

Hembree, E. A., & Foa, E. B. (2000). Post-traumatic stress disorder: Psychological factors and psychosocial interventions. *Journal of Clinical Psychiatry, 61*, 33–39.

Hijazi, A. M. (2012). Narrative exposure therapy to treat traumatic stress in Middle Eastern refugees: A clinical trial. Wayne State University Dissertations. Paper 543.

Hussain, D., & Bhushan, B. (2011). Post-traumatic stress and growth among Tibetan refugees: The mediating role of cognitive-emotional regulation strategies. *Journal of Clinical Psychology*, *67*(7), 720–735.

Jaffe, D. T. (1985). Self-Renewal: Personal transformation following extreme trauma. *Journal of Humanistic Psychology, 25*(4), 99–124. https://doi.org/10.1177/0022167885254012

Janoff-Bulman, R. (1992). *Shattered assumptions.* New York: The Free Press.

Joseph, S., & Linley, P. A. (2008). Positive psychological perspectives on post-traumatic stress: An integrative psychosocial framework. In S. Joseph & P. A. Linley (Eds.), *Trauma, recovery, and growth: Positive psychological perspectives on posttraumatic stress* (pp. 3–20). NJ: John Wiley & Sons Inc.

Kimhi, S., Eshel, Y., Zysberg, L., & Hantman, S. (2010). Post war winners and losers in the long run: determinants of war related stress symptoms and post-traumatic growth. *Community Mental Health Journal, 46*, 10–19.

Kirmayer, L. J. (1989). Cultural variations in the response to psychiatric disorders and emotional distress. *Social Science & Medicine*, *29*(3), 327–339.

Kuhn, T. S. (1970). *The structure of scientific revolutions*. Chicago: The University of Chicago Press.

Laufer, A. & Solomon, Z. (2006). Post-traumatic symptoms and post-traumatic growth among Israeli youth exposed to terror incidents. *Journal of Social and Clinical Psychology*, *25*, 429–447.

Lindstrom, C. M., Cann, A., Calhoun, L. G., & Tedeschi, R. G. (2013). The relationship pf core belief challenge, rumination, disclosure, and socio-cultural elements to post-traumatic growth. *Psychological Trauma: Theory, Research, Practice, and Policy, 5*(1), 50–55.

Linley, P. A., & Joseph, S. (2004). Positive change following trauma and adversity: A review. *Journal of Traumatic Stress: Official Publication of the International Society for Traumatic Stress Studies*, *17*(1), 11–21.

Masten, A. S., Best, K. M., & Garmezy, N. (1990). Resilience and development: Contributions from the study of children who overcome adversity. *Development and Psychopathology*, *2*(4), 425–444.

Mattoon, L. (2011). *The gift of trauma: Stories of post-traumatic growth and spiritual transformation in war survivors from Uganda and Vietnam* (Doctoral Dissertation). ProQuest Information & Learning, US. (71), Retrieved from http://ezproxy.ecu.edu.au

McGregor, B. A., Antoni, M. H., Boyers, A., Alferi, S. M., Blomberg, B. B., Carver, C. S. (2004). Cognitive-behavioral stress management increases benefit finding and immune function among women with early-stage breast cancer. *Journal of Psychosomatic Research, 56*, 1–8.

McMillen, J. C., Zuravin, S., & Rideout, G. B. (1995). Perceptions of benefit from child sexual abuse. *Journal of Consulting and Clinical Psychology*, 65, 733–739.

McMillen, C., Howard, M. O., Nower, L., Chung, S. (2001). Positive by-products of the struggle with chemical dependency. *Journal of Substance Abuse Treatment*, 20(1):69–79.

Meadors, P. & Lamson, A. (2008). Compassion fatigue and secondary traumatization: Provider self care on intensive care units for children. *Journal of Paediatric Health Care, 22*, 24–34.

Meyerson, D. A., Grant, K. E., Carter, J. S., & Kilmer, R. P. (2011). Post-traumatic growth among children and adolescents: A systematic review. *Clinical Psychology Review*, *31*(6), 949–964.

Norris, F. H., Stevens, S. P., Pfefferbaum, B., Wyche, K. F., & Pfefferbaum, R. L. (2008). Community resilience as a metaphor, theory, set of capacities, and strategy for disaster readiness. *American Journal of Community Psychology*, *41*(1–2), 127–150.

O'Leary, V. E., & Ickovics, J. R. (1995). Resilience and thriving in response to challenge: an opportunity for a paradigm shift in women's health. *Women's Health (Hillsdale, NJ)*, *1*(2), 121–142.

Paez, D., Basabe, N., Ubillos, S., & Gonzalez-Castro, J. L. (2007). Social sharing, participation in demonstrations, emotional climate, and coping with collective violence after the March 11th Madrid bombings. *Journal of Social Issues*, *63*, 323–337.

Park, C. L., Cohen, L. H., & Murch, R. (1996). Assessment and prediction of stress-related growth. *Journal of Personality, 64,* 71–105.

Prati, G., & Pietrantoni, L. (2009). Optimism, social support, and coping strategies as factors contributing to post-traumatic growth: A meta-analysis. *Journal of Loss and Trauma, 14*, 364–388.

Ransom, S., Sheldon, K. M., & Jacobsen, P. B. (2008). Actual change and inaccurate recall contribute to post-traumatic growth following radiotherapy. *Journal of Consulting and Clinical Psychology, 76(5),* 811–819.

Rashid, T., Seligman, M. E. P. (2013). Positive psychotherapy. In R. J. Corsini, D. Wedding (Eds.), *Current psychotherapies*. CA: Cengage, Belmont pp. 461–498,

Richard, G. Tedeschi and Lawrence, G. Calhoun. (2004). Post-traumatic growth: Conceptual foundations and empirical evidence. *Psychological Inquiry, 15*(1), 1–18.

Ryff, C. D., & Singer, B. (1998). The contours of positive human health. *Psychological Inquiry*, *9*(1), 1–28.

Saadat, H., Snow, D. L., Ottenheimer, S., Dai, F. & Kain, Z. N. (2012). Wellness program for anaesthesiology residents: A randomized, controlled trial. *Acta Anaesthesiologica Scandinavica, 56*, 1130–1138.

Sood, A., Prasad, K., Schroeder, D. & Varkey, P. (2011). Stress management and resilience training among Department of Medicine faculty: A pilot randomized clinical trial. *Journal of General Internal Medicine. 26*, 858–861.

Splevins, K., Cohen, K., Bowley, J., & Joseph, J. (2010). Theories of post-traumatic growth: Cross-cultural perspectives. *Journal of Loss and Trauma*, *15*(3), 259–277.

Staub, E., & Vollhardt, J. (2008). Altruism born of suffering: The roots of caring and helping after victimization and other trauma. *American Journal of Orthopsychiatry*, *78*(3), 267–80. doi: 10.1037/a0014223

Stuart, R. B. (2004). Twelve practical suggestions for achieving multicultural competence. *Professional Psychology: Research and Practice, 35,* 3–9.

Taylor, V. (1977). Good news about disaster. *Psychology Today, 11*, 124–126.

Tedeschi, R. G., & Calhoun, L. G. (1996). The post-traumatic growth inventory: Measuring the positive legacy of trauma. *Journal of Traumatic Stress*, *9*(3), 455–471.

Tedeschi, R. G., & Calhoun, L. G. (2004). Post-traumatic growth: Conceptual foundations and empirical evidence. *Psychological Inquiry, 15*(1), 1–18. doi: 10.1207/s15327965pli1501_01

Tedeschi, R. G., & Calhoun, L. G. (2006). Time of change? The spiritual challenges of bereavement and loss. *OMEGA—Journal of Death and Dying*, *53*(1), 105–116.

Tedeschi, R. G., & McNally, R. J. (2011). Can we facilitate post-traumatic growth in combat veterans? *American Psychologist, 66*(1), 19–24.

Tedeschi, R. G., & Kilmer, R. P. (2005). Assessing strengths, resilience, and growth to guide clinical interventions. *Professional Psychology: Research and Practice, 36*(3), 230–237.

Tennen, H., Affleck, G., Urrows, S., Higgins, P., & Mendola, R. (1992). Perceiving control, construing benefits, and daily processes in rheumatoid arthritis. *Canadian Journal of Behavioral Science, 24,* 186–203.

Thombre, A., Sherman, A. C., & Simonton, S. (2010). Religious coping and post-traumatic growth among family caregivers of cancer patients in India. *Journal of Psychosocial Oncology, 28*(2):173–88. *doi: 10.1080/07347330903570537.*

Turner, de S. & Cox, H. (2004). Facilitating post traumatic growth. *Health and Quality of Life Outcomes, 2*, 34. doi: 10.1186/1477-7525-2-34

Van Slyke, J. (n.d.) Post-traumatic growth. Combat & operational stress control. Retrieved January, 2015 from http://www.med.navy.mil/sites/nmcsd/nccosc/healthProfessionalsV2/reports/Documents/PTG_WhitePaper-Final.pdf

Walsh, F. (2003). Family resilience: A framework for clinical practice. *Family Process*, *42*(1), 1–18.

Weiss, T., & Berger, R. (2010). *Post-traumatic growth and culturally competent practice: Lessons learned from around the globe*. John Wiley & Sons.

Williams, M. S., Jerome, A., White, K., & Fisher, A. (2006). Making sense of suffering: A preliminary study of changes in religious women adjusting to severe adversity. *Counseling and Values*, *50*(2), 84–98.

Xu, X., Hu, M.-l., Song, Y., Lu, Z.-x., Chen, Y.-q., Wu, D.-x., and Xiao, T. (2016). Effect of positive psychological intervention on post-traumatic growth among primary healthcare workers in China: A preliminary prospective study. *Scientific Reports*, *6*, 39189.

Yalom, I. D., & Lieberman, M. A. (1991). Bereavement and heightened existential awareness. *Psychiatry*, *54*(4), 334–345.

Zollner, T., & Maercker, A. (2006). Post-traumatic growth in clinical psychology: A critical review and introduction of a two-component model. *Clinical Psychology Review, 26,* 626–653.

CHAPTER 30

Post-Traumatic Growth Following the Transition to Motherhood: Investigating the Role of Social Support[1]

JILLIAN M. MILLAR and JULIE ANN POOLEY*

School of Arts and Humanities, Edith Cowan University, Australia

*Corresponding author. E-mail: j.pooley@ecu.edu.au

ABSTRACT

Post-traumatic growth (PTG) or positive psychological outcomes can be experienced as a result of an individual's struggles with highly stressful or challenging life events, such as the birth of a child. The current investigation explores the relationship between PTG and the transition to motherhood. Two studies investigated women's retrospective perceptions of PTG regarding their experience of parenthood. The first study utilized a sample of 83 participants who completed four scales (Post-traumatic Growth Inventory, Impact of Events Scale-Revised, Mastery Self-efficacy Scale, and Multidimensional Scale of Perceived Social Support). Data were analyzed using a standard multiple regression to determine which variable(s) significantly and independently predicted the level of PTG. Results indicated that perceived social support was the only variable to achieve significance ($F(3,78) = 3.333$; $p < 0.05$). To gain a holistic perception of how and why social support played such a vital role in predicting PTG, a second qualitative ($N = 10$) study was conducted exploring perceptions of social support across the transition to motherhood. Thematic content analysis of transcripts revealed four main themes: reassurance, help seeking, self-efficacy, and changes in the self. This investigation has therapeutic implications for therapists who may be able to utilize these findings to encourage and enhance positive outcomes, coping, and adaptation in individuals experiencing a crisis and/or stressful life transitions. Additionally, this investigation adds to the evidence supporting the use of the PTGI as a general measure of growth, thereby allowing it to be applied to incidents and experiences outside of those traditionally defined as traumatic.

30.1 INTRODUCTION

Becoming a parent is often said to be one of the most life changing experiences and encompasses

[1] Authors' Note: This chapter is based on the doctoral thesis of the first author.

both stress and joy. Particularly for women, there is a higher risk for negative psychological effects and disorders around this period; much of which has been vigorously studied. However, little research has been conducted into the positive outcomes from the transition to motherhood. The concept of post-traumatic growth (PTG) has been an emerging area of study in the field of psychology generating interest, excitement, and hope for those who have experienced trauma and for the clinicians who work with traumatized populations. Generally, researchers have applied the concept of PTG to traumatic incidents, however there are now calls for it to be applied to events outside of those historically considered to be traumatic (e.g., the transition to motherhood). It is hoped that this might elucidate how the general process of human growth unfolds. Additionally it is anticipated that greater understanding of PTG and the factors, which facilitate or predict it, can then be generalized and applied to interventions with the aim of not only treating those in psychological distress; but also enhancing positive adaption to the variety of human experiences across the lifespan.

30.2 THE TRANSITION TO MOTHERHOOD

The onset of parenthood signifies a transition point in a person's life that requires adaptation to a variety of changes including new routines, responsibilities, challenges, and a new lifestyle (Bost et al., 2002). Parenthood is also associated with substantial reorganization of the new parents' lives in terms of social and occupational participation, restructuring of priorities and values, along with changes in psychological well-being and self-identity (Knoester & Eggebeen, 2006). Becoming a parent can be a source of joy and happiness, as well as a time of great strain and difficulty (Bondas & Erikkson, 2001). This life altering event has been described as being either a disorganizing incident or a developmental opportunity (Osofsky & Osofsky, 1984) a time for regression or an opportunity to grow (Antonucci & Mikus, 1988). It is possible that the transition to parenthood encompasses all of these descriptions as an individual gradually adapts to the new role of being a parent. Transition points in general are associated with increased levels of stress and pressure, which can render a person vulnerable to negative psychological outcomes such as depression and anxiety (Bost et al., 2002; Milward, 2006). The transition to parenthood in particular has the potential to have negative effects on individuals' mental health and marital satisfaction (Luo, 2006), which in turn can impact the quality of parenting provided to children (Bost et al., 2002).

In particular, women's lives are the most affected by the transition to parenthood, experiencing changes in all domains, physically, emotionally, occupational, and often financially (Bost et al., 2002; Luo, 2006). The postnatal period can be associated with increased psychiatric diagnoses such as PND, and increased hospital admissions (Mallikarjun & Oyebode, 2005). With the recognition that not all new mothers experience PND (Ayres et al., 2016), researchers have begun to investigate the factors associated with positive adaption to motherhood. A review of the existing literature indicates healthy life styles, religion, close families and supportive relationships are associated with more beneficial adjustment to motherhood (Bost et al., 2002; Hung, 2005; Luo, 2006; Page, 2004). A number of adjustments in women's identities, self-concepts, maternal efficacy, and priorities also seem to occur across the transition to motherhood (Milward, 2006). Mothers who have high parental efficacy beliefs reportedly experience less psychological distress than those with low efficacy beliefs (Halpern & McLean, 1997).

Social support appears to be especially salient and is believed to serve a cushioning affect or buffer against stressful life events; it is frequently associated with better psychological functioning, physical health, and social adjustment in general (Sarason et al., 1997). High levels of perceived social support for mothers also appear to have beneficial effects for their children (Dipietro et al., 2003; Huizink et al., 2003).

As the transition to motherhood necessitates immense changes in a woman's life, it is likely that her foundational assumptions (Parks, 1975) about the world and herself undergo radical change. Alterations in the meaning, purpose, focus, self-efficacy, and control of her life are likely to occur (Bandura, 1977; Beder, 2004, 2005), and thus the transition to motherhood acts as a central event for her "self" identity (Boal et al., 2010; Boals & Schuettler, 2011). Research suggests individual's assumptive worlds are linked to psychological health and adjustment (Rini et al., 2004), and when experiencing trauma or highly stressful periods in life the basic beliefs one holds may be shattered (Kaufmann, 2002). This led to the conception of PTG (also known as stress-related growth) whereby one's struggles with highly challenging crises or traumas may improve functioning and result in changes in the self, efficacy, interpersonal relationships, and life philosophies (Tedeschi & Calhoun, 1996; 2004). With the development of Tedeschi and Calhoun's (1996) PTGI researchers have supported the measurement of PTG to an array of traumatic and stressful events to ascertain whether growth occurs. In particular, Sawyer and Ayers' (2009) investigation into PTG after childbirth in the United Kingdom sought to understand how the process of PTG unfolds and thus provides insight into how mental health professionals and support systems may best facilitate adaptations across the transition to motherhood. However, the Sawyer and Ayers' (2009) study did not support a relationship between PTG and social support, which provides opportunity to clarify this space further.

Maternal efficacy beliefs have also been reported to mediate the effects of social support, depression, and infant temperament (Cutrona & Troutman, 1986). Additionally, high self-efficacy convictions have also been described as a psychological buffer against environmental stressors (Harwood et al., 2007). Despite the vast array of findings it remains unclear as to how and why social support and self-efficacy are associated with such beneficial adjustments to motherhood, and whether one or a combination of the two factors will predict more successful maternal adaptation; further research into the psychological impacts and processes are warranted.

Building on the positive psychology zeitgeist of the modern era (Seligman & Csikszentmihalyi, 2000), the current investigation sought to make an innovative and valuable contribution to the field of PTG, by exploring the experience of PTG across the transition to motherhood. Although it appears Sawyer and Ayers (2009) and McKenzie-McHarg et al. (2015) are some of the researchers to apply PTG to childbirth, further investigation of their contention that growth is experienced after childbirth is warranted. Therefore, the first study, of the current investigation, measured perceived PTG as an outcome via Tedeschi and Calhoun's (1996) PTGI. Based on the mixed findings that high levels of perceived social support, greater self-efficacy, and elevated levels of stress are associated with better scores on the PTGI, the current investigation aimed to determine which factor/factors (social support, self-efficacy and stress) best predicted the degree of PTG across the transition to motherhood. As the sole variable of perceived social support was found to be the only significant predictor of the PTGI score in the Australian sample and as suggested by Prati and Pietrantoni (2009), and Aspinwall and Tedeschi (2010) an investigation of social support (such as belonging, self-esteem, and tangible support) was undertaken. Hence, Study Two sought to qualitatively investigate the relationship of perceived social support to PTG across the transition to motherhood in order to illicit a rich, holistic, and comprehensive understanding of growth and adjustment to motherhood. Study Two explored women's discourses regarding their experience of PTG as it pertained to the birth of their first child along with their perceptions about the social support, they did or did not perceive around their transition to motherhood.

30.3 METHODOLOGY

30.3.1 STUDY ONE: PARTICIPANTS

A sample of 83 adult women participated. Descriptive statistics (see Table 30.1) revealed that the average age range of women at the time of sampling was 31–40 years old, and the average time elapsed since the birth of their first child was between 5 and 10 years. The women were also asked to provide a rating of the severity of distress they perceived experiencing around the birth, the mean rating was "a little distressing" however was only marginally short of reaching the "moderately distressing" category.

Participants completed a set of questionnaires containing an information letter, demographic information sheet, the Post-traumatic Growth Inventory (PTGI) (Tedeschi & Calhoun, 1996); the Multidimensional Scale of Perceived Social Support (MSPSS) (Zimet et al., 1988) the Mastery Self-efficacy Scale (MSES) (Pearlin & Schooler, 1978; Pearlin et al., 1981) and the Impact of Events Scale-Revised (IES-R) (Weiss & Marmar, 1997). All scales are well established with regard to validity and have been used in research extensively. Participants were recruited on a university campus and offered a questionnaire. Once completed, the data was entered and analyzed.

30.4 RESULTS OF STUDY ONE

Prior to the analysis, initial data screening was conducted, assumptions were found to be met. Initial reliability testing for the scales were conducted with the respective Cronbach's alpha

TABLE 30.1 Descriptive Statistics for Study One Sample

	N	Minimum	Maximum	Mean	Std Deviation
Participant age group	84	2	5	3.54	1.04
Time lapsed since birth	84	1	4	3.31	1.06
Severity of perceived distress rating	84	0	5	1.96	1.19
Perceived social	82			68.39	11.61
Support self-efficacy	82			27.24	4.50
Impact of event	82			8.12	10.26
PTGI	82			68.49	20.90

scores resulting: PTGI = 0.948; IES-R = 0.928; MSPSS = 0.919; and although the MSES=0.714.

Pearson's Product Moment Correlations were performed and are presented in Table 30.2. Both self-efficacy and social support were significantly correlated with the scores on the PTGI (r = 0.241, $p < 0.05$) and (r = 0.243, $p < 0.05$) respectively. Although IES-R was not significant with PTGI, IES-R did result in a significant negative correlation with social support (r = 0.250, $p < 0.05$). The mean rating for perceived degree of distress experienced during the adjustment to motherhood was 1.96 (standard deviation 1.19), coding values were as follows 1 = a little distressing, 2 = moderately distressing). The frequency distribution was bimodal with categories 1 and 2 being endorsed by 26 participants.

TABLE 30.2 Variable Correlations

	PTGI	IES-R	MSES	MSPSS
PTGI	–	0.040	0.241*	0.243*
IES-R		–	−0.056	−0.250*
MSES			–	0.124
MSPSS				–

*Correlation is significant at the 0.05 level (2-tailed).
**Correlation is significant at the 0.01 level (2-tailed).

After assumption testing was carried out, and met, Standard Multiple Regression analysis was then conducted to ascertain which independent variable/s best predicted the PTGI score (see Table 30.3).

The model accounted for 8% the variance in PTGI scores, although this was a relatively small figure the analysis of variance was significant ($F(3,78) = 3.333; p < 0.05$). Further examination of the coefficients revealed that perceived social support was the only independent factor to reach significance ($\underline{t}$ (78) = 2.148; $p < 0.05$).

TABLE 30.3 Predictors for Post-traumatic Growth (PTGI) Following the Transition to Motherhood

Predictors	Beta	B	R^2	Adjust.t R	Sig.
IES-R	0.114	0.233		1.040	
MSES	0.215	0.998		1.995	
MSPSS	0.238	0.429		2.148	
[$F(3,78) = 3.333$, $p < 0.05$]				114	0.80

$N = 83^*$, $p < 0.05$

30.5 SUMMARY AND DISCUSSION OF STUDY ONE

Becoming a parent can be a source of joy and happiness, as well as a time of great strain and difficulty (Bondas & Erikkson, 2001). This life altering event has been described as being either a disorganizing incident or a developmental opportunity (Osofsky & Osofsky, 1984) a time for regression or a chance to grow (Antonucci & Mikus, 1988). Analysis of the data from Study One revealed that women report experiencing PTG across the transition to parenthood. In trying to understand what best predicted PTG it was revealed that perceived social support significantly and solely predicted the amount of PTG reported. Participants who identified having greater levels of social support around the transition to motherhood, reported experiencing higher levels of PTG corroborating previous research (Bost et al., 2002; Hung, 2005; Hurdle, 2001; Luo, 2006; Page, 2004).

However, questions remain unanswered as to how this process of growth across the transition to motherhood unfolds? What is it about social support that seems to facilitate PTG? Consequently, Study Two employed a qualitative approach to explore women's experiences and perceptions regarding the social support they experienced during the transition to motherhood.

30.5.1 STUDY TWO: PARTICIPANTS

A sample of 10 women who volunteered from Study One consented to participating in semi structured interview process. The interviews commenced with a brief introduction and explanation of the study's purpose. The interview questions sought information about women's experiences of social support during the transition to motherhood (including the people involved in their support networks, the positive and negative aspects of the support, and how they feel the support could have been improved). At the conclusion of all interviews, participants were thanked for their time and reassured as to the confidentiality of their responses.

30.5.1.1 ANALYSIS

Following the completion of all interviews, the recordings were transcribed and then examined using interpretive phenomenological analysis (IPA) according to Smith (1995). This involved the transcripts being read and re-read in a cyclical manner in order to identify common themes and sub-themes in the participants' experiences.

Thematic content analysis of social support across the transition to motherhood revealed the following four major themes: reassurance, help seeking, self-efficacy, and changes in the women's identity and priorities. Each of these themes are described using participant anecdotes where appropriate.

30.5.1.2 THEME ONE: REASSURANCE

With the transition to motherhood all participants commented on their perceptions of embarking on an unknown experience in which they frequently grappled with both the unexpected and their own self-expectations. This experience seemed to leave most women feeling quite anxious about whether things were progressing "normally." Not surprisingly the women sought reassurance from a number of social support networks, including partners, family, friends, medical professionals (doctors, nurses, mid-wives, etc.), government parenting clinics, mothers' groups, and telephone help lines to name a few. The assortment of people that comprised the women's networks in the sample is consistent with those reported in the support network literature (Bost et al., 2002; Hopkins et al., 1984).

Existing evidence suggests social support can act as a buffer against stressful life events (like the transition to parenting), and is frequently associated with better psychological functioning, physical health, and social adjustment (Sarason et al., 1997). Additionally, the beneficial effects that social support provides for parenting have been demonstrated for infants (Crockenber, 1981), preschool children (Jennings et al., 1991), and throughout childhood from age 2 to 12 (Szykula et al., 1991). The value placed on support networks seems to link in with the second factor in Tedeschi and Calhoun's (1996) PTG model in which interpersonal relationships are perceived as more meaningful following a traumatic or stressful event. When trying to understand the impact that the support networks' reassurance had on mothers in the current study, a number of participants' quotes seemed to elucidate the process. The following quote illustrates how the reassurance seemed to prevent the mother from becoming extremely stressed and agitated over things by intervening in the anxiety cycle:

> (P3) "It was just having someone to listen to me; people who took me seriously. People who were able to give the reassurance and

> a bit of practical advice. I'm a person who worries.... So it was very nice to be able to have people come along and go 'you don't have to worry, that is fine.'."

Additionally, the reassurances appeared to calm new mothers' nerves, sooth concerns regarding their capabilities as a parent, inspire confidence, and self-efficacy in their new role. This finding appears to tie in with the third factor of PTG in which participants' felt an increased sense of personal strength (Tedeschi & Calhoun, 1996):

> (P1) "[Social support] just seemed to instill in me calmness and a belief that yes everything will be alright. So I was able to, when I was really worried and stressing out about whether I could fulfill my role as a parent, and be a mother, it just seemed to give me that little bit of hope that yes, I was going to be ok and do the job properly."

30.5.1.3 THEME TWO: HELP SEEKING

Many of the women described feeling initially uncertain and hesitant when it came to identifying what type of help would benefit them across the transition to motherhood. Furthermore, they often reported feeling lost or unconfident in trying to search for avenues which would provide the desired resources or information the women were seeking:

> (P1) "I probably didn't know where to source more help, or the sort of help that I knew would help me with breast feeding and the new things that were happening with my baby. I don't think I really asked a lot of questions, I just ... don't know, I was very inexperienced..."

A few of the mothers seemed to discuss a number of reasons why they felt unable to utilize their support networks for help at different times. Reported concerns included feeling they could not trust others to adequately care for their babies, their own separation issues, self-expectations about having/wanting to do it all on their own, anxieties related to the perceived burdening of their support networks, and negative experiences when seeking help in the past. Due to the crucial role social support plays in positive adjustment to motherhood and PTG, the perceived inability to access networks is disconcerting:

> (P4) "I tried to do it on my own... I was trying to be very strong and cope, and do it all by myself without relying too heavily on my husband. I felt hindered by that I guess, because I didn't want to put too much pressure on him, in case he melted down and couldn't cope with anything. He was already probably close to breaking point.... So I couldn't really share or moan about that too much. I kind of felt like I was doing it by myself... There was this community nurse that seemed like a tyrant! I didn't ask many questions or get much helpful feedback during those particular meetings. I also noticed that the other mothers in the group seemed to feel the same way; we just weren't as relaxed or game to ask questions. This nurse was so abrupt and her answers seemed inconsistent so you ended up feeling confused."

30.5.1.4 THEME THREE: SELF-EFFICACY

Self-efficacy is the belief in one's self regarding the necessary skills and abilities to successfully perform behaviors (Bandura, 1977) and appeared to be an important process in women's adaption to motherhood. The women's levels of self-efficacy varied greatly over time and between individuals. A number of factors were identified as impacting their beliefs including external and internal components. The external elements that seemed to build new mother's self-efficacy were often derived from members of the support network who were perceived as encouraged her confidence and persistence. Again, there seemed to be a link between helpful and meaningful support networks and mothers' perceptions of personal strength and confidence; the second and third factors of Tedeschi and Calhoun's (1996) model of PTG:

> (P3) "It was nice to have people go 'yes I've been there when your baby is screaming for like an hour and a half and there's seemingly nothing wrong and they're just screaming. You're not being a bad mum; that's just life.' It gave me more confidence that I was doing the right thing, because they were saying 'yes you're doing well, he looks great'. So yeah, the positive … um … feedback from them really built up my confidence."

Another aspect of self-efficacy seemed to be intrinsic to some mothers' personalities, like an internal sense of confidence. It is possible that some of the women in the current study exhibited high levels of perceived personal strength prior to the transition to motherhood, which may have placed them in a better position for adapting to their new role and the prospect of increased levels of PTG. As reported in the Bost et al. (2002) study, individuals with larger support networks tended to report elevated levels of adjustment however the authors also discovered a negative correlation between the frequency of contact with the support networks and adjustment. Similarly, research conducted by Wethington and Kessler (1986) indicated it was the *perception* of available social support, which was most important, and that individuals with high perceptions of support did not necessarily need to access the support:

> (P4) "I'm probably more resilient than I thought! A lot stronger than I thought. I can do a lot more on my own than I probably thought initially. I thought I was a strong person anyway, but this you know, proved oh yeah I can do it! I'm certainly more confident… like um, a growing sense of self-reliance. I was more confident in my self-reliance, more confident in my capabilities of looking after myself and a newborn baby, so even if I didn't do it really well, I can do it to survive until I can function better and then improve you know step by step as everything sort of comes together and you get more sleep and stuff like that… More personal strength, and confidence in what I can cope with."

30.5.1.5 THEME FOUR: CHANGES IN THE SELF

The women described noticing changes in their identity following the transition to motherhood. This finding seems to link in with Tedeschi and Calhoun's (1996) third and fourth areas of PTG in which people tend to perceive an increased sense of personal strength or assets, and changes in their priorities. Research indicates that the basic beliefs of an individual's assumptive world are linked to their physical and psychological health as well as adjustment (Rini et al., 2004). Examination of the transcript narratives regarding each woman's understanding of the impact to her self-identity as a result of the transition to motherhood was quite profound. The types of changes in the self that were reported by participants in the current investigation illuminated integral components in the process of PTG and adaptation across the transition to motherhood. The majority of mothers described increases in their tolerance levels and ability to put stressors and incidents into perspective. In line with previous findings from Smith (1999), women well into the transition to parenthood perceived themselves as calmer, more agreeable, tolerant, and content to allow others to handle things so they could focus on what they considered important; leading to the development of a less competitive, more relaxed, and cohesive self-conception:

> (P1) "I think the stresses that I went through have improved me, I have a higher tolerance level for other people who have problems... I think I learnt that I could probably cope with a lot more, well I learned to cope with a lot more things than I had ever done before; learnt to have a higher level of patience. Also you sort of understand that everybody copes in their own way and there's more than one way of coping with a problem to get to the positive end. There's more than one way of bringing up a child."

With their new found patience and perspective, the women also perceived themselves as being more relaxed, post-transition:

> (P9) "Yeah definitely flexibility, and yeah I think I've become a lot more relaxed since I've had kids... before I'd put all this pressure on myself to get done what I'd planned to get done. Whereas after it, if I didn't get it done then it was just like, oh well it doesn't matter; it's not going to be the end of the world.

The participants also spoke of fluctuations in their moods during the transition to motherhood, particularly in the earlier stages. Whilst some of the reported moods were likely attributable to hormonal changes during the final stages of pregnancy and the initial weeks post-birth some of the alterations in their moods appeared to relate more to the actual transition and acquisition of their new roles as mothers. As documented by Knoester and Eggebeen (2006), it is not uncommon to see changes in the psychological well-being and identities of new parents. Research has also shown the transition to parenthood can have potentially negative effects on individuals' mental health and marital satisfaction (Luo, 2006). Kaufmann (2002) equated the loss of one's assumptive world with the loss of predictable surroundings and a secure sense of self, potentially leaving the person without a framework to orientate the self or function. Due to the perceived loss of coherence, safety, and security provided by the old assumptive world, it may result in mood instability and leave the self in a state of panic and danger (Beder, 2004–2005; Kaufmann, 2002):

> (P9) "I think it's made me more grounded and it's made me more, I think for a while there I felt like I didn't' have a place where I belonged or like the kind of place I could call home. I think having this family now and I guess I do have a sense of purpose and a sense of belonging; and I think that's why I don't have the travel bug anymore... there's no where I want to go now, cause this is where I want to be."

30.6 SUMMARY OF STUDY TWO FINDINGS

In summary, the qualitative investigation of Study Two revealed four over-arching themes in participants' discourses regarding social support across the transition to motherhood: reassurance, help seeking, self-efficacy, and changes in the self. At times of high stress and worry the women's social support networks were able to intervene. Self-efficacy beliefs regarding adaption to the role of a first-time mother also seemed to weave throughout participants' discourses. Additionally, support networks were able to influence the women's self-efficacy and moderate their self-expectations and thus facilitate PTG.

30.7 DISCUSSION

The focus of the current chapter was to examine PTG across the transition to motherhood; initially via a quantitative investigation as to which factor(s) best predicted the experience of PTG, followed by a qualitative exploration of how this process unfolds. The current investigation attempted to build on the work of Sawyer and Ayers' (2009), by investigating PTG across the transition to motherhood in an Australian sample. Study One substantiated the existing literature regarding the beneficial effects of social support across the transition to motherhood and demonstrates its correlation to PTG (Bost et al., 2002; Hopkins, et al., 1984). The results of Study One lend further support to Lazarus et al. (1980) belief that social support might assist by sustaining emotional health and well-being of those experiencing stressful or traumatic changes, possibly acting as a buffer for the individual and aiding in adaptive adjustment.

Study Two sought to qualitatively explore the nuances of how perceived social support interacted with PTG and thereby provide some explanation and in-depth understanding of the relationship between the two variables. In particular, Study Two was able to discover how the process of growth unfolded across the transition to motherhood, why social support seemed to facilitate it, and which aspects of social support were considered to be beneficial or detrimental. The mixed method design utilized in the current chapter is one of its notable strengths and provides a valuable contribution to the PTG literature.

Many of the changes the women perceived across the transition to motherhood seemed to parallel those found in the Tedeschi and Calhoun's (1996) PTG model across all five factors. Perhaps the first three themes from Study Two highlight the processes by which PTG is achieved, and the final theme of changes in the self may be the location in which the growth is stored and understood. The majority of new mothers described increases in their tolerance levels and ability to put stressors and incidents into perspective, which is suggestive of an increase in their capacity to incorporate incidents into their world view.

30.7.1 LIMITATIONS

One of the greatest limitations lay in the sample selection, where there were no restrictions on whether the births examined in Study One or Two involved premature labors, birth complications and or defects, emergency procedures and deviations from the women's birth plans. Additionally, women in both studies were able to self-select into the convenience sample and this may have been indicative of greater motivation to discuss their experiences. Differential family types should also be a feature of future research. A further limitation of the current study was that the retrospective and cross-sectional nature of the research. Attempting a long-term study of transition to motherhood and fatherhood would be beneficial to further understanding PTG.

The regression model only accounted for 8% of the variance in PTGI scores, which although significant it seems there may be other factors which influence women across the transition to motherhood. Previous PTG research on childbirth (conducted by Sawyer & Ayers, 2009; Sawyer et al., 2012) may have arrived at different conclusions, particularly regarding social support due to their inclusion of any childbirth rather than a woman's initial experience only. Subsequent birthing experiences may result in the women constructing different meanings of the event.

30.8 CONCLUSION

Overall there is a great deal of change that occurs across the transition to parenthood, much of which parallels the areas of PTG identified by Tedeschi and Calhoun (1996). Specifically, in relation to the transition to motherhood, the findings of this investigation encourage professionals to use Tedeschi and Calhoun's (1996) PTGI as an assessment tool for measuring successful coping and adaption across the transition to mother.

The PTGI when applied outside of the typically defined areas of trauma seems to be a suitable instrument for measuring general human psychological growth processes in response to stressful life incidents and transition periods across the lifespan. Transitions such as returning to the work force after a period of absence, moving schools or towns, retirement, and relocating to aged care facilities etc. can be very distressing events, but may also provide opportunities for PTG and adaption. Furthermore, therapeutic interventions that incorporate (1) increasing clients' appreciation for life, (2) enhancing the meaningfulness of interpersonal relationships, (3) increasing personal strength, (4) awareness and flexible altering of priorities, and (5) exploration of existential and/or spiritual themes may help target and facilitate PTG and adjustment.

Any health professional working with a client or patient who is embarking on the transition to motherhood would be wise to routinely encourage women toward investing in and nurturing their support networks. Professionals would be wise in their interactions to provide reassurance, build confidence, and to normalize learning and the adjustment process across motherhood. Facilitating new mothers to comfortably express their worries and emotions while refraining from making judgments or being too directive when providing advice are also likely to assist new mothers.

KEYWORDS

- **post-traumatic growth**
- **transition**
- **motherhood**
- **social support**

REFERENCES

Antonucci, T., & Mikus, K. (1988). The power of parenthood: personality and attitudinal change during the transition to parenthood. In G. Michaels & W. Goldber (Eds.), *The transition to parenthood: Current theory and research.* Cambridge: Cambridge University Press.

Aspinwall, L., & Tedeschi, R. (2010). Of babies and bathwater: A reply to Coyne and Tennen's views on positive psychology and health. *Annals of Behavioral Medicine, 39,* 27–34.

Bandura, A. (1977). Self-efficacy: Toward a unifying theory of behavioral change. *Psychological Review, 84,* 191–215.

Beder, J. (2004–2005). Loss of the Assumptive world—How we deal with death and loss. *Omega, 50,* 255–265.

Boals, A., Steward, J., & Schuettler, D. (2010). Advancing our understanding of posttraumatic growth by considering event centrality. *Journal of Loss and Trauma, 15,* 518–533.

Bondas, T., & Eriksson, K., (2001). Women's lived experiences of pregnancy: A tapestry of job and suffering. *Qualitative Health Research, 11,* 824–840.

Bost, K., Cox, M., Burchinal, M., & Payne, C. (2002). Structural and supporting changes in couples' family and friendship networks across the transition to parenthood. *Journal of Marriage and Family, 64,* 517–536.

Crockenber, S. (1981). Infant irritability, mother responsiveness, and social support influences on the security of infant–mother attachment. *Child Development, 52,* 857–865.

Cutrona, C., & Troutman, B. (1986) Social support, infant temperament, and parenting self-efficacy: A mediational model of postpartum depression. *Child Development, 57*, 1507–1518.

DiPietro, J., Millet, S., Costigan, K., Gurewitsch, E., Caulfield, L. (2003). Psychological influences on weight gain attitudes and behaviors during pregnancy. *Journal of the American Dietetic Association, 103*, 1314–1320.

Halpern, L., & McLean, W. Jr. (1997) Hey mom, look at me! *Infant Behavior and Development, 20*, 515–529.

Harwood, K., McLean, N., & Durkin, K. (2007). First-time mothers' expectations of parenthood: What happens when optimistic expectations are not matched by later experiences? *Developmental Psychology, 43*(1), 1.

Hopkins, J., Marcus, M., & Campbell, S. (1984). Postpartum depression: A critical review. *Psychological Bulletin, 95*, 498–515.

Huizink, A., Robles de Medina, P., Mulder, E., Visser, G., & Buitelaar, J. (2003). Stress during pregnancy is associated with developmental outcome in infancy. *Journal of Child Psychology and Psychiatry and Allied Disciplines, 44*, 810–818.

Hung, C.-H. (2005). Women's postpartum stress, social support, and health status. *Western Journal of Nursing Research, 27*(2), 148–159.

Hurdle, D. (2001). Social support: A critical factor in women's health and health promotion. *Health & Social Work, 26*, 72–80.

Jennings, K., Stagg, V., & Connors, R. (1991). Social networks and mothers' interactions with their preschool children. *Child Development, 62*, 966–978.

Kaufmann, J. (Ed.). (2002). *Loss of the assumptive world*. New York: Brunner-Routledge.

Knoester, C., & Eggebeen, D. J. (2006). The effects of the transition to parenthood and subsequent children on men's well-being and social participation. *Journal of Family Issues, 27*, 1532–1559.

Lazarus, R., Kanner, A., & Folkman, S. (1980). Emotions: Cognitive phenomenological analysis. In R. Plutchik and H. Kellerman (Eds.), *Theories of emotions*. New York: Academic Press.

Luo, L. (2006). The transition to parenthood: Stress, resources, and gender differences in a Chinese society. *Journal of Community Psychology, 34*, 471–488.

Mallikarjun, P., & Oyebode, F. (2005). Prevention of postnatal depression. *The Journal of the Royal Society for the Promotion of Health, 125*, 221–227.

McKenzie-McHarg, K., Ayers, S., Ford, E., Horsch, A., Jomeen, J., Sawyer, A., ... Slade, P. (2015). Post-traumatic stress disorder following childbirth: an update of current issues and recommendations for future research. *Journal of Reproductive and Infant Psychology, 33*(3), 219–237.

Milward, L. (2006). The transition to motherhood in an organizational context: An interpretative phenomenological analysis. *Journal of Occupational and Organizational Psychology, 79*, 315–333.

Osofsky, J., & Osofsky, J. (1984). Psychological and developmental perspectives on expectant and new parenthood. In R. Parke (Ed.), *Review of child-development research, volume 7: The family*. Chicago: Chicago University Press.

Page, R. (2004). Positive outcomes in Mexican immigrants: What can we learn? *Journal of Obstetric, Gynecologic, and Neonatal Nursing: JOGNN, 33*, 783–790.

Parks, C. M. (1975). Determinants of outcome following bereavement. *Omega, 6*, 303–323.

Pearlin, L., Menaghan, E., Lieberman, M., & Mullan, J. (1981). The stress process. *Journal of Health and Social Behavior, 22*, 337–356.

Pearlin, L., & Schooler, C. (1978). The structure of coping. *Journal of Health and Social Behavior, 19*, 2–21.

Prati, G., & Pietrantoni, L. (2009). Optimism, social support, and coping strategies as factors contributing to post-traumatic growth: A meta-analysis. *Journal of Loss and Trauma, 14*, 364–388.

Rini, C., Manne, S., DuHamel, K. N., Austin, J., Ostroff, J., Boulad, F., ... Sexson, S. (2004). Changes in mothers' basic beliefs following a child's bone marrow transplantation: The role of prior trauma and negative life events. *Journal of Traumatic Stress: Official Publication of The International Society for Traumatic Stress Studies, 17*(4), 325–333.

Sarason, B., Sarason, I., & Gurung, R. (1997). Close personal relationships and health outcomes: A key to the role of social support. In S. Duck (Ed.), *Handbook of personal relationships: Theory, research and interventions* (2nd ed., 547–573). Hoboken, NJ: John Wiley & Sons Inc.

Sawyer, A., & Ayers, S. (2009). Posttraumatic growth in women after childbirth. *Psychology and Health, 24*, 457–471.

Sawyer, A., Ayers, S., Young, D., Bradley, R., & Smith, H. (2012). Posttraumatic growth after childbirth: A prospective study. *Psychology and Health, 27*, 362–377.

Seligman, M., & Csikszentmihalyi, M. (2000). Positive psychology: An introduction. *American Psychologist, 55*, 5–14.

Smith, J. (1995). Semi structured interviewing & qualitative analysis. In Smith, J., Harre, R., & Van Langenhove (Eds.), *Rethinking methods in psychology*. London: SAGE.

Smith, J. (1999). Identity development during the transition to motherhood: An interpretative phenomenological analysis. *Journal of Reproductive Infant Psychology*, *17*, 281–299.

Szykula, S., Mas, H., Turner, C., & Crowley, J. (1991). Maternal social support and prosocial mother-child interactions. *Journal of Family Psychology*, *5*, 82–92.

Tedeschi, R. G., & Calhoun, L. G. (1996). The posttraumatic growth inventory: Measuring the positive legacy of trauma. *Journal of Traumatic Stress*, *9*, 455–471.

Tedeschi, R., & Calhoun, L. (2004). Posttraumatic growth: Conceptual foundations and empirical evidence. *Psychological Inquiry*, *15*, 1–18.

Weiss, D., & Marmar, C. (1997). The impact of event scale–Revised. In J. Wilson & T. Keane (Eds.), *Assessing psychological trauma and PTSD* (pp. 399–411). New York, NY: Guilford Press.

Wethington, E., & Kessler, R. C. (1986). Perceived support, received support, and adjustment to stressful life events. *Journal of Health and Social behavior*, *27*(1), 78–89.

Zimet, G. D., Dahlem, N. W., Zimet, S. G., & Farley, G. K. (1988). Psychometric characteristics of the multidimensional scale of perceived social support. *Journal of Personality Assessment*, *55*, 610–17.

CHAPTER 31

The Utility of Adventure Programming for the Development of Resilience and Related Constructs: Examples from Adolescent, Corporate, and Clinical Contexts

CONRAD S. ZYGMONT[1*], A. VERNON NAIDOO[2], and PIETER SNYMAN[3]

[1]*Helderberg College of Higher Education & Stellenbosch University, Cape Town, 7135, South Africa*

[2]*Stellenbosch University, Stellenbosch, South Africa*

[3]*Tshwane University of Technology, Pretoria 0183, South Africa*

[*]*Corresponding author. E-mail: zygmontc@hche.ac.za*

ABSTRACT

Adventure programming is the purposeful use of adventure activities and/or wilderness for the purpose of cognitive, physical, social, and/or affective change in groups or individuals. It originated, in the form of Outward Bound, as a short-term intervention designed to help young seamen develop resilience. Contemporary applications of adventure programming are tailored to diverse clientele including young and old; individuals, families, groups, or organizations; with objectives ranging from recreation, education, development, adjunctive therapy to primary therapy. This chapter aims to provide a broad overview of the principles, theoretical frameworks, and types of adventure programming that can be used to foster the development of resilience and well-being. The types of outcomes described across studies, the mechanisms of change identified as important, and evidence of the effectiveness of adventure programs in the literature are reviewed. The authors also provide examples from their own experience and research of how adventure programming has been used for the development of resilience in clinical, corporate, and youth development contexts, providing insight into how adventure programming processes and contexts facilitate the development of resilience. The authors discuss models of resilience development in the context of adventure programming and make recommendations for program design to ensure optimal resilience development.

31.1 A HISTORY OF ADVENTURE PROGRAMMING

Many of the philosophical principles and applied practices used in adventure programming to facilitate the development of resilience have their roots in the Outward Bound School established by Kurt Hahn and Lawrence Holt on the 4th October 1941 in Wales. The school was established to provide a compelling demonstration of how a short-term program could fortify young seamen with the resolve and hardiness necessary for them to face the challenges in the Battle of the Atlantic (Veevers & Allison, 2011). Institutes fashioned after the Outward Bound model sprang up in Lumut, Malaysia in 1958, Marble Colorado in the United States in 1962, and continued on to over 35 countries all over the world through the 1970s and 1980s (Davis-Berman & Berman, 2008; Priest & Gass, 2005). The first training programs for outdoor leaders were established in 1944 in Great Britain, 1965 in the United States of America, 1969 in Australia, and 1977 in New Zealand (Priest, 1986). The early Outward Bound programs aimed to utilize the power of wilderness contexts, and pair this with targeted activities and challenges believed to foster prosocial values, resilience, and other valued character strengths. As adventure programs grew and diversified, the need for training, certification, safety standards, evidence of effectiveness, and a theoretical framework led to increased research and evaluation (Priest, 1999; Davis-Berman & Berman, 2008). Bandoroff and Newes (2004) describe the field as having matured since its "adolescence" of the 1990s to a stage where it is aware and willing to grapple with the key issues, making progress and yet comfortable with not having all the solutions. Adventure programming is now much more aware of its need to respond to the wider sociocultural, ecological, technological, and geopolitical circumstances in which it takes place (Beams et al., 2017).

31.2 TYPES OF ADVENTURE PROGRAMS

Adventure programming is broadly defined as the purposeful use of educational activities (e.g., games and trust activities), adventurous experiences (e.g., hiking and abseiling), and wilderness or other novel environments (e.g., high- or low-ropes courses) in order to foster cognitive (e.g., beliefs or wilderness knowledge), emotional (e.g., confidence and improved self-concept), physical (e.g., increased strength), and interpersonal (e.g., enhanced leadership or conflict resolution skills) growth within the safety of a supportive group system (Alvarez & Stauffer, 2001; Glass & Myers, 2001; Itin, 2001; Priest & Gass, 2005). There are many different types of adventure programs, and there is a lack of consistency in naming conventions used throughout the literature. Types of adventure programming can be very broadly distinguished by depth of intervention, setting characterization, and program duration. This typology of adventure programs, derived from Zygmont (2014), is presented in Figure 31.1. The depth of intervention ranges on a continuum from (a) recreation, which is activity orientated, allowing participants to experience a positive feeling and enjoyment; (b) education, which is task orientated, involving the use of adventure to enrich students understanding of important concepts, gain insight into problems and social behavior, form new attitudes and perspectives on life, and change thinking; (c) enrichment or development, which is process orientated, aiming to improve functional behaviors, and purposefully address common issues such as self-concept, trust, communication, and change functional behaving; (d) adjunctive therapy treatment blended with existing traditional therapy taking place before and after therapy, addressing primary therapeutic needs, and reducing resisting and denying; and (e) primary therapy that replaces traditional therapy in order to assess, diagnose, treat, and monitor therapeutic change (Gass, 1993a; Priest & Gass, 2005). The setting characterization reflects both the physical context and philosophical approach to place used in the program. For example, a view of "wilderness as therapy" can be differentiated from "wilderness as context for adventure" in that the former seeks a slower and more meaningful pace and experience, aims to provide for deeper and more meaningful connections to place, and uses wilderness as a central therapeutic construct. On the other hand using nature as a context for adventure aims to increase involvement and perceived risk in order to create disequilibrium, and places more emphasis on the *doing* of physical and often challenging activities in outdoor spaces than on the *being* in wilderness (Harper et al., 2018). Bowen and Neill's (2013) comprehensive meta-analysis of adventure programming outcome studies suggests that ropes/challenge/adventure-based programs are the most popular (42.7% of samples) type of adventure program being studied. Finally, adventure programs can be differentiated based on the duration of the intervention. In Russell's (2003) review of adventure programs that meet outdoor behavioral healthcare (OBH) criteria, he found that program participants can spend anywhere between 5 and 70 days in the wilderness. Continuous flow expedition type programs tend to spend the most time out in the wilderness, whereas indoor and outdoor high- and low-ropes courses tend to have the shortest duration.

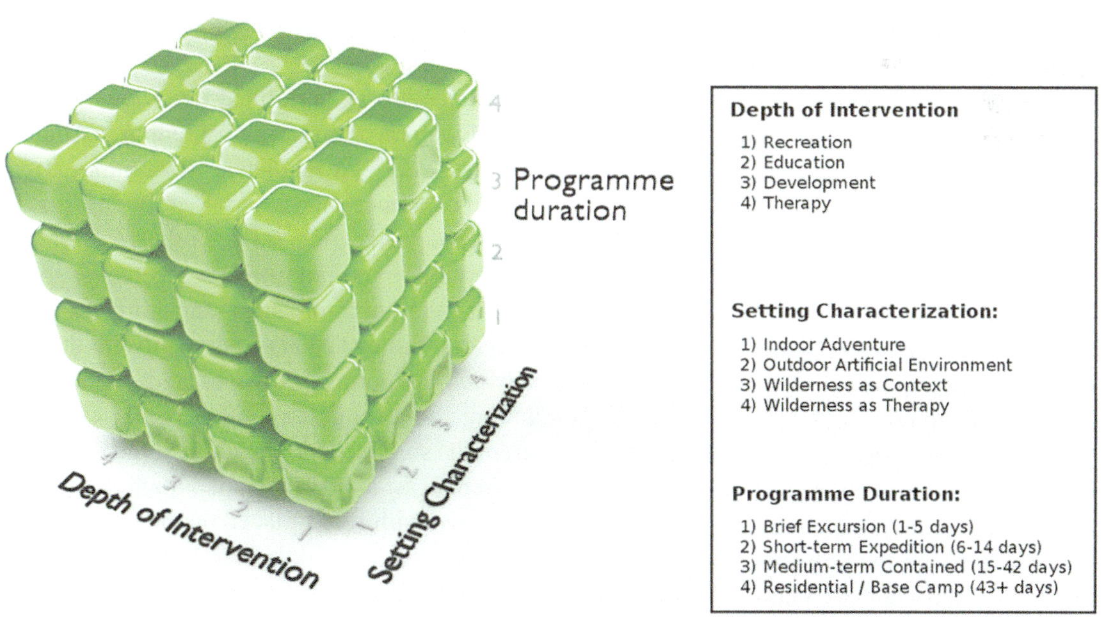

FIGURE 31.1 A typology of adventure programs (Zygmont, 2014).

31.3 THE IMPACT OF ADVENTURE PROGRAMMING ON WELLNESS

There is a burgeoning body of evidence of the effectiveness of adventure programs in addressing physical health enhancement, academic involvement and performance, family relationships, self-concept, body image and related constructs, social skills and psychosocial competencies, team building, spiritual or moral development, and reductions in clinical concerns (Naidoo et al., 2017). A perusal of the South African literature indicates that adventure programs have been applied with varying success with a wide range of diverse populations. Adventure program clients include youth with emotional and behavioral problems, demilitarized youth (Robertson & De Kiewit, 1998), physically and sexually abused children, children orphaned by HIV-AIDS, homeless people (Robertson, 2000), former military combatants (Schell-Faucon, 2001), youth referred by the court system (Botha, 2007), and even with elderly participants to facilitate successful ageing (Boyes, 2013). Adventure programs have also been utilized in a range of youth projects with community organizations (Benjamin, 2014; Fine & van Niekerk, 2014; Knoetze, 2003; van Heyden, 2004), gangs (Pinnock, 2016), and school groups (Naidoo & van Wyk, 2003; Nunez, 2010; Zygmont & Naidoo, 2018). Facilitators have also utilized adventure programs with corporate clients for the purpose of team building (Kriek, 2007; Theron, 2009).

In terms of program focus, some authors are diverting attention away from targeted individualistic therapeutic interventions and directing more emphasis on the prevention and development of positive health as a holistic concept that incorporates physical, social, emotional, and spiritual well-being (Neuman, 2010). There is a growing awareness that programs should include consideration of interpersonal relationships, community structures, intellectual development, emotional awareness and management, nutrition and other physical concerns, religiosity and spirituality, occupation, technological and ecological processes (Maller et al., 2002; Pryor et al., 2005; Van Lingen & De Jager, 2014). A growing body of research suggests that adventure programming is effective in facilitating people to engage physically, intellectually, emotionally, and spiritually with their own psyche, other people and the natural environment (Beil et al., 2014; Humberstone et al., 2016; Ryff & Singer, 2008). A summary of some of the dimensions of wellness being targeted as outcomes in adventure programming, as identified by various authors, is presented in Table 31.1.

31.4 THE EFFECTIVENESS OF ADVENTURE PROGRAMMING IN DEVELOPING RESILIENCE

Resilience has been defined in various ways in the adventure programming literature. Allan et al. (2012) defined resilience as "a multidimensional construct made up of individual assets and resources which reflect relative strengths in emotional and cognitive competence, social connectivity and physical capability" (p. 3). They cite Linley (2009) in showing that the link between adventure processes and resilience is among the strongest of 60 personal character strengths studied. Booth and Neill (2017) define resilience as a positive adaptation system to adversity, with focus placed on the way individuals deal with and think through risky situations by utilizing positive coping strategies. Neill and Dias (2001) define resilience as an individual's capacity for maintenance, recovery, or improvement in mental health following life challenges;

TABLE 31.1 Benefits of Outdoor Adventure (Darst & Armstrong, 1980)

Dimension	Benefit
Personal	New experience—adds a kick to participant's lives, allows them to experience something
	High-risk experience—facing perceived danger may help participants overcome fears, gain self-confidence, and enhance their ability to cope
	Escape—offers release from the tensions and complexities of modern life
	Success—allows participants to achieve a highly personal sense of accomplishment
	Knowledge—participants learn more about themselves and the environment
	Physical fitness—physical activity can help burn calories and increase strength, stamina, and flexibility
Economic	Minimal financial investment—provides interesting and pleasant activities at a reasonable cost
Social-psychological	Socializing—provides a chance to meet others who have similar interests
	Unity—promotes cohesiveness and doing things together without the distractions of everyday life
	Cooperation and trust—promotes better relationships through cooperation, appreciation of others, compassion, and respect
	Nature and outdoors —promotes aesthetic appreciation for nature and concern for vanishing wild places

TABLE 31.2 Benefits of Outdoor Adventure (Ewert, 1989)

Dimension	Benefit
Psychological	Benefits on a personal (versus group) basis—Self-concept (enhanced or strengthened view), self-efficacy (self-confidence), self-actualization (well-being, improved self-expression, and feelings of psychological health)
Sociological	Compassion, cooperation, respect for others, communication
Educational	Improved academic abilities, awareness of nature and problem solving, outdoor skills, and values clarification
Physical	Strength, coordination, balance, and cardiovascular endurance

TABLE 31.3 Goals and Benefits of Outdoor Adventure (Webb, 1999)

Dimension	Benefit
Recreational	Enjoyment, relaxation, entertainment, excitement, catharsis, and self-expression
Skill	Goal setting, decision-making, problem solving, responsibility, physical development, nature awareness, communication, and leadership
Character	Independence, interdependence, self-efficacy, willingness to take risks, tolerance, respect, trust, and compassion

successful adaptation following exposure to stressful life events; and the capacity for transformation and change. Resilience is likely to involve a number of interconnecting individuals, social and contextual factors. For example, when transitioning to university, adaptability depends on cognitive competence (e.g., problem solving skills), emotional intelligence (e.g., overcoming fears and growing from failure), and social connectedness (e.g., breaking down barriers, and knowing how to ask for help) (Allan et al., 2012). Adventure programming outcome studies have focused on different aspects and used different theoretical frameworks to define and assess resilience. For this reason, there is considerable variation in outcome study findings. Some studies have provided evidence of dramatic increases in self-perceived resilience. For example, Neill and Dias (2001) compared participants that completed a 22-day multielement adventure-based Outward Bound Australia program with a control group of undergraduate psychology students. They found the overall change in resilience for the experimental group was very large (ES = 1.10) with everyone who completed the program reporting increases in resilience. Some studies have reported only moderate gains in resilience. Beightol et al. (2012) compared an antibullying educational control group to a three-day high- and low-ropes course experimental group and found significant pre- to post-test increases in the treatment group for self-efficacy (ES = .26) and goals and aspirations (ES = .25) with continued gains in goals and aspirations from post-treatment to four month follow-up (ES = .21). In a study of Canadian aboriginal (First Nation) youth from Wikwemikong; Ritchie et al. (2014) found that participants experienced increased resilience (ES = 0.36). In a one year follow-up, they found that resilience scores had dropped below initial levels for younger participants (12–14) but had continued to increase for older participants (15–18). Hayhurst et al. (2015) describe two different studies of a 10-day developmental voyage on a 45-m, three-masted, Barquentine ship sailing around New Zealand that resulted in moderate increases of resilience (ES = 0.35 and ES = 0.3) that were maintained at 5-month follow-up. Other studies have failed to detect a significant change in self-reported resilience. For example, Boyers (2015) found no significant differences from pre- to post-test or follow-up resulting from participation in an Outward Bound program. She did however detect significant impacts on social and emotional well-being measured using the Mental Health Continuum Short Form (MHC-SF).

In order to explain the variability in findings researchers have begun to design studies to identify program or participant characteristics that could predict the impact of programs on resilience. For example, O'Brien and Lomas (2017) compared an adventure program that incorporated mindset theory in the design to a traditional program as a control, and found that resilience increased significantly from pretest to post-test for the experimental group (ES = 0.2), which was largely maintained at a one month follow-up. The control group failed to demonstrate significant gains. Similarly Passarelli et al. (2010) incorporated strengths-based education to a traditional adventure education program, and found personal growth was significantly correlated with strengths awareness ($r = 0.57$), strengths application ($r = 0.42$), and total strengths ($r = 0.5$). They found that being aware of one's strengths allows for (a) more meaningful and purposeful learning by focusing attention on opportunities for personal development, (b) providing tools students could use to enhance interpersonal relationships, and

(c) providing personal resources that could be activated in order to respond effectively to challenges. When looking at program context Tillmann et al. (2018) reviewed three studies that aimed to develop resilience using outdoor programming as well as other research on the impact of nature on youth development. They conclude that nature plays an important role in the development of resilience. Harper and colleagues (2018) argue that despite some evidence for improved outcomes in wilderness versus no-wilderness program settings, the relevance and rationalization of wilderness as a central component of adventure programming is sorely lacking in the literature. Other studies have argued that combining the inherent value offered in wilderness and artificial high- or low-ropes course environments may maximize on the strengths of each approach (Greffrath et al., 2013).

In terms of participant characteristics, culture, age, and gender may play a role in resilience development in adventure programming. A number of studies have found that males and females engage differently with different aspects of adventure programs and may develop resilience through different mechanisms. Beightol et al. (2012) and Overholt and Ewert (2015) found that females experienced greater overall gains than males. Norton and Watt (2014) found that male participants had significantly better outcomes in overall internal asset categories, specifically, social competencies and positive values compared to females. Bloemhoff (2012) found that male participants increased in value attached to achievement, whereas females did not, and females increased in interested and caring adults and levels of control against deviant behavior, whereas males did not. However, Hayhurst et al. (2012) found no significant differences by age or gender.

In summary, research suggests that adventure programs can and do provide opportunities for participants to develop resilience. Programs that are designed on the basis of established psychological theory, and delivered taking into account the individual, cultural, gender, and therapeutic needs of the client are most likely to be successful. Finally, researching resilience outcomes is challenging because of a lack of consistency in theoretical and measurement models, and the ethical, practical, and empirical issues that must be navigated carefully when conducting research on adventure programs (Gabrielsen et al., 2016).

31.5 THEORETICAL MODELS OF RESILIENCE DEVELOPMENT IN ADVENTURE PROGRAMMING

Various models of resilience have been developed that identify individual, social, cultural, and environmental factors that may promote resilience and offset risk. Van Rensburg et al. (2015) suggest that early conceptions of resilience defined resilience as a personality trait, skill, or genetic predisposition residing within the individual. In order to identify factors that could be leveraged to develop resilience, variable focused models sought to identify individual characteristics, and contextual protective and risk factors, which support access to, and use of, resilience promoting resources. Pathway models were developed to account for how adaptive and maladaptive pathways differed for individuals. Current research considers how pathways and variables interact within culturally aligned transactions facilitated by actions that social ecologies and the individual reciprocally take. In this way, contemporary models consider resilience to be the product of a complex interaction of risk, protective, developmental, contextual and individual factors (Zolkoski & Bullock, 2012). Protective factors believed to enhance resilience include biological or genetic factors (e.g., high intelligence); social environment (e.g., quality schools, cohesive family, neighborhood resources, interested and caring adults); perceived environment (e.g., sense of acceptance and belonging, models for conventional behavior, and high levels of control against deviant behavior); personality (e.g., value on health, positive attitude toward the future, intolerance of deviance, and value attached to achievement); and behavior (e.g., church attendance, involvement in school and voluntary activities, ability to work with others, ability to resolve problems in nonviolent ways, and enjoyment of and/or perceived competence in activity) (Bloemhoff, 2012; Jessor, 1993). Interventions based on these models suggest that resilience is more likely when protective factors are strengthened across all interactive levels of the socioecological model (i.e., individual, family, community, and society) (Zolkoski & Bullock, 2012).

31.6 DEVELOPMENT-BY-CHALLENGE OR STRESS INOCULATION MODEL

The most popular model of resilience underpinning adventure programming design is the development-by-challenge or stress inoculation model. This model proposes that if youth are presented with a series of stressors progressively increasing from low to moderate, they learn to mobilize resources necessary for adaptive coping with hardship (Zolkoski & Bullock, 2012). Adventure program curriculum typically consist of a series of activities, intentionally chosen so that they are incrementally sequenced in difficulty, and designed so that they are mastered only as groups persevere, exercise creativity, and develop a sense of interdependence leading to increased self-efficacy and competence (Kimball & Bacon, 1993). Challenges and perceived risk are believed to create the possibility for flow. This becomes possible when challenges are selected to push an individual's physical and psychological boundaries, when the person believes they have the ability to deal with the situation, and they feel deeply involved in the experience, highly motivated, cognitively efficient, and are truly enjoying the experience (Csikszentmihalyi & Csikszentmihalyi, 1999). Steyn (2012) suggests that, in line with Bandura's (1977) self-efficacy theory, adventure programming clients are encouraged to persevere in activities that are subjectively perceived as daunting despite the presence of contradictory information, existing barriers and absence of experience until they attain mastery across various circumstances. This newfound self-belief and successful strategies are likely to generalize to resilience in other contexts. Research supports the thesis that gains in resilience are largest among beginner outdoor pursuit participants and lowest among advanced participants for whom outdoor activities pose less of a challenge (Todd et al., 2018). In this way adventure programs seek to provide a generalized form of stress-inoculation training that fosters overall life effectiveness (Neill, 2008).

31.7 PROTECTIVE FACTOR MODELS

Alternative models of resilience emphasize the role of protective factors, such as peer support or positive role models in mediating, moderating, or regulating the impact of stressors on the development of resilience (Zolkoski & Bullock,

2012). Key factors identified include challenging activities, a caring and supportive social group, cognitive skills and attitudes developed through reflection, and living in a novel wilderness context in which new norms and patterns of behavior need to be established (Booth & Neill, 2017; Neill & Dias, 2001; Steyn, 2012). Activation and development of meaningful, positive relationships between participants and their program facilitators, who provide social support and encouragement to one another, has been identified as a central mechanism in adventure programming outcomes (Booth & Neill, 2017; Gass et al., 2012). Such models of adventure programming downplay the need for artificially increasing levels of perceived risk and emphasize the role of interpersonal relationships for ensuring a place of comfort, security, and acceptance (Davis-Berman & Berman, 2008). Neill and Dias (2001) suggest that social support accounts for around 24% of growth in resilience recorded in specific adventure programs. Research suggests that increased resilience in adventure programming is often attributable to increased social well-being (Boyers, 2015). Similarly, Zygmont and Naidoo (2018) found that adventure programs create a space in which previous stereotypes can be challenged, deep emotional bonds are formed, and communication can extend beyond the superficial boundaries of the comfortable. When adventure groups experience a strong sense of peer support, trust develops. This trust allows for intimate feelings and thoughts to be shared and strengthening social bonds, which then scaffolds as a support structure ensuring positive resolution of further difficulties and therapeutic growth in an iterative positively reinforcing cycle (Schoel & Maizell, 2002). Taken together these factors facilitate the development of problem focused coping strategies (e.g., positive reinterpretation, active coping, and suppression of competing activities), emotion focused coping strategies (e.g., humor, seeking emotional social support, and restraint), social support promotion (e.g., appropriate sharing, communication skills, trust, conflict resolution), and self-efficacy (self-concept, physical strength, self-esteem) that foster resilience across a range of contexts (Booth & Neill, 2017).

31.8 NEUROLOGICAL MODELS OF RESILIENCE THROUGH ADVENTURE PROGRAMMING

Some adventure programming researchers have recently highlighted the value of understanding neurological models of resilience based on contemporary neuroscience research in order to improve program design (Allen et al., 2012). From a neurological perspective resistance is defined as a buffering or protective ability allowing an individual to withstand harmful stressors that would otherwise compromise psychological functioning; whereas resilience is defined as the ability of an individual to recuperate and resume healthy functioning following a harmful stressor (Fleshner et al., 2011). These constructs likely have some shared and some dissociative neurological mechanisms but are often simply collapsed into the concept of neurological resilience (Maier & Watkins, 2010; Worley et al., 2018). A large body of animal and clinical research is accumulating providing evidence that behavioral interventions offering experiences of personal success or control, a sense of stressor temporality or limited severity, and habitual exercise can foster plasticity and strengthen cortical control over brainstem and limbic stress-responsive circuits (Christianson & Greenwood, 2014; Cooper et al., 2015;

Worley et al., 2018). The primary neurological mechanisms by which experience can modify resilience include neuroplasticity, neurogenesis, and epigenetics. Neuroplasticity includes changes in neural structures (e.g., dendritic and synaptic remodeling) and functions (e.g., rapid changes in synaptic activity or long-term potentiation processes). Neurogenesis involves the propagation and maturation of new glial cells or neurons in the subventricular zone, olfactory bulb, or subgranular zone of the dentate gyrus of the hippocampus. Epigenetic processes involve lasting but reversible changes in gene expression and play a significant role in both neuroplasticity and neurogenesis (Wakabayashi et al., 2014). Behavioral interventions can increase resilience through neurogenesis and neuroplasticity in several neural circuits including (a) modulation of the basolateral amygdala by the ventral hippocampus (Hill et al., 2015; Zannas & West, 2014); (b) mediation of the hypothalamic pituitary adrenal (HPA) axis through inhibition of the paraventricular nucleus (PVN) of the hypothalamus by a relatively circumscribed group of inhibitory and excitatory projections (Myers et al., 2014); (c) reduction of serotonin production in the dorsal raphe nuclei (DRN) and serotonin sensitivity in projection sites (Andolina et al., 2013; Calizo et al., 2011; Christianson & Greenwood, 2014; Cooper et al., 2015; Maier and Watkins, 2010); (d) increased activity in the ventral medial prefrontal cortex (vmPFC), which exercises top-down modulation of the DRN by means of the prelimbic (PL) neurons (Amat et al., 1998; Christianson & Greenwood, 2014; Worley et al., 2018), modulation of the basolateral amygdala (Andolina et al., 2013), bed nucleus of the stria terminalis (Kumar et al., 2014); and (e) increased circuit and intrinsic excitability of the dorsomedial striatum—vmPFC connection (Christianson & Greenwood, 2014; Shin et al., 2018). Studies of these mechanisms provide evidence that repeated exposure to stressors perceived as manageable, challenging, transient, and leading to positive outcomes does result in long-term resistance and resilience to stress at the neurological level.

31.9 MECHANISMS OF CHANGE IN ADVENTURE PROGRAMMING

Understanding the mechanisms by which any process is able to contribute to change is an important step in maximizing the effectiveness of any intervention. Rosenfeld (2009) identified eight elements, which in and of themselves are sufficient causes to account for change in psychotherapy. They include the following: (a) utilizing evidence-based techniques and avoiding discredited ones; (b) client characteristics (e.g., motivation to change, involvement in treatment); (c) therapeutic alliance; (d) therapist characteristics (e.g., warmth, genuineness, positive regard, empathy); (e) common processes present in most techniques (e.g., disconfirming irrational beliefs, a supportive environment); (f) extra-therapeutic forces (e.g., joining a supportive community, adopting alternative home and work patterns); (g) problem-related factors (e.g., whether disorders are more biologically based environmentally situated, the severity and duration of problem behaviors); and (h) the placebo effect explaining about 15% of the variance in treatment outcome. Zygmont (2014) reviewed the literature and identified the following important mechanisms facilitating improved adventure programming outcomes: (a) time for solitude in the wilderness allowing for attunement to nature and self; (b) helpful and caring facilitators that model positive, caring and involved adult roles; (c) activities that are intrinsically motivating and provide opportunities for peak experience; (d) a peer group that provides support, safety, positive social norms, and immediate, quality feedback; (e) motivation, receptivity, and desire for growth or change on the part of participants; (f) maintain moderate levels of perceived risk that creates optimum levels of physical and psychological eustress to allow for positive coping, while reducing actual risk; (g) a novel environment that provides a break from prevailing cultural norms and influences; (h) internalization and transfer of program experiences and metaphors to participants' daily lives; (i) the age of participants; (j) adopting a challenge-by-choice approach; (k) providing escalating exposure to difficult yet attainable goals; (l) direct assessment opportunities and reassessments of individual's coping methods; (m) utilizing a mode of operation that is more action centered than verbal; (n) tailoring program elements to clients' needs; addressing intellectual, emotional, social, and spiritual levels for change; (o) providing clear natural consequences; (p) adopting a salutogenic orientation; and (q) favoring a longer program duration, intensity, and higher level of intervention. Most of these, like client characteristics and therapeutic alliance are common to all interventions, while others like wilderness environment and physically strenuous but intrinsically motivating adventure activities are unique to adventure programming (Newes & Bandoroff, 2004).

When looking specifically at resilience as an outcome in adventure programming, key mechanisms contributing to program effectiveness include physical exercise, social and cognitive aspects of the program, attaining flow, and the wilderness context. It is likely that the physical, social and cognitive aspects of such programs are likely to reciprocally bolster one another. For example, Santos et al. (2017) found that self-report and teacher's evaluations of the emotional skills of adaptability, interpersonal relationships, positive mood, and stress management improved incrementally for participants of a soccer training school that included an outdoor education program compared with a regular soccer training program. Neill and Dias (2001) found that social support accounts for around a quarter of growth in resilience measured, with the support provided by the least supportive member of the group having the strongest impact on improvements in resilience across the program. Programs making extensive use of wilderness contexts have found sufficient support in the literature for the hypothesis that healing and growth occurs in wilderness contexts in ways that transcend traditional psychotherapy (Tillmann et al., 2018). Wilderness is seen to incorporate various physical, aesthetic, psychological, and spiritual properties that may be perceived differently by individual(s) within the specific cultural frameworks within which wilderness is defined and experienced (Harper et al., 2018).

31.10 CASE STUDIES

In this section, we present three brief case studies that provide examples of the use of adventure programming in a clinical, school, and corporate setting, respectively. In the first case study, the use of adventure programming in a youth diversion program under the auspices of Usiko Stellenbosch, a nonprofit organization in a peri-urban town in South Africa, is presented. The second case study illustrates the use of adventure programming as part of a school-based rites of passage program, whereas the third case study demonstrates the application of adventure programming with corporate clients, with the goal of improving various individual skills and group processes important for the company's strategic goals.

31.10.1 CASE STUDY ONE: THE USIKO DIVERSION PROGRAM AS EXAMPLE OF AP WITH A CLINICAL POPULATION

Usiko Stellenbosch is a nongovernmental organization providing a range of programs and services for at-risk youth in a peri-urban community in the district of Stellenbosch, 45 miles north-east of Cape Town. The development, implementation, evaluation and adaptation of what has evolved to be called the Usiko program has been documented and described in various studies (Anthonissen, 2011; Botha, 2007; Fabrik, 2004; Knoetze, 2003; Naidoo et al., 2007; Rousseau & Pinnock, 2014). Essentially, the Usiko project offers three youth development interventions: (a) a prevention focused program for at-risk youth participants at local high schools (Van Wyk & Naidoo, 2006); (b) a diversion program for youth referred from the court system (Botha, 2007); and (c) a livelihoods program for youth transitioning between school and the next stage in their lives. Use is made of adult volunteers from the community who are trained as mentors to support the staff and the youth participants in all three programs. Usiko's programs draw strongly from rites of passage work (Somes, 1998) and the Circle of Courage model of Brendtro et al. (2002). The four universal tasks of the circle of courage are as follows: fostering Belonging, Mastery, Independence, and Generosity that the adolescent needs to develop and integrate in transitioning to adulthood. Youth in conflict with the law are hypothesized to have not successfully integrated these tasks. Usiko adopts a restorative approach using the circle of change as the basis of its diversion program (Botha, 2007). The four tasks in diversion work using the cycle of change are as follows: review, restore, reconnect, and revision. Participants are guided within a restorative justice framework to engage with their offence. They review their current life situation, and seek to make amends for their behavior (restore their reputation with family, friends and the community), rekindle their vision for their life (reconnect), and plan for how they want to take forward their plans for their life (revision) (Botha, 2007).

With Usiko's diversion program, youths who are in conflict with the law are referred by the court system after appearing before a magistrate for minor offences. Although they may present with a range of diagnoses, the emphasis is on a group based restorative justice approach using activities that foster Usiko's circle of change. The diversion program is a 4-month program that involves regular one-on-one counselling sessions, group sessions, family sessions, mentor sessions, and wilderness excursions. Central to Usiko's diversion program is an emphasis on utilizing wilderness (both as a setting and as a therapeutic medium) as an integral part of the rehabilitation process (Anthonissen, 2011). Early in the program participants are taken on a wilderness excursion that is a key component of the program (Botha, 2007). The Usiko programs typically utilize two wilderness experiential program (WEP) variations: a base camp program with all activities located at a remote site, and a hiking expedition program following a mountain trail. In some programs a combination may be used with the base camp excursion occurring at the onset for the purpose of developing group cohesion and establishing rapport with staff and mentors. The hiking excursion may be utilized toward the latter part of the program for a deeper engagement with the restorative process linked to their criminal conviction. For the first wilderness activity, the youth participants are asked to prepare for a 3–4 days immersion into a remote wilderness setting. There is a build up before the excursion with information and instructions kept deliberately vague limited to a checklist of what kinds of clothing to bring and what not to bring with (cell phones, electronic devices, etc.) to create an aura of the unknown. The participants are taken out of their comfort zone, away from their "normal" environment into a natural setting, creating some dissonance and disruption (Pinnock, 2016). For some it is an escape from the routine, an opportunity to tune out of their home circumstances, social pressures and expectations to behave in certain ways. For others, it is being taken out of their comfort zone and being confronted with the unknown embodied in the wilderness—a mixture of excitement, uncertainty, risk and foreboding, an opportunity to be uninhibited, and engage with new experiences. With the view to harnessing the wilderness as therapeutic medium, natural settings are saturated with symbols and metaphors that can be applied effectively in guided activities (Gass, 1993b, 1993c).

At the wilderness base camp participants are introduced to the setting and are made aware of ecological principles and practices and program protocol. They engage in series of therapeutic and challenge-based experiential activities that focus raising awareness of their sense of self, their environment and their social connection. The central activity of the base camp program is the solo activity of approximately 5–8 h. Participants are given rations and are taken to an isolated setting, asked to find a place of solitude and to note their observations of nature around them, and to reflect on a designated aspect of their life. They may be asked to find an object that symbolizes their personal story. Participants in the diversion program are specifically asked to review the circumstances that led to their referral to the diversion program. This solo experience is thoroughly debriefed in a group feedback session with each participant given an opportunity to share their experience and reflections with the group. Invariably participants link aspects of their observation of nature and their symbolic object to their personal narrative (Botha, 2007). For many participants, this may be a unique opportunity to tell their story without being interrogated, judged or criticized. The first wilderness experience serves to connect the participants with nature, with one another and the facilitators, and can offer significant symbols, metaphors and reflection for the therapy process and for later discussions (Botha & Naidoo, 2006).

When the hiking excursion is utilized as part of the program, this allows for a combination of guided activities (at appropriate times in the hike) with ongoing internal reflection time during the hike and group debriefing in the evening around the campfire. Gass (1993a) suggests that the process of internal change in wilderness therapy is facilitated through the utilization of the physical body. Wilderness therapy engages the whole body in a completely interrelated physical and psychological experience linked to a conducive natural setting. Ecopsychology recognizes that there is an impact on cognition and well-being that results from the interaction between the body and the environment. For participants, the physical route of the hike and its physical obstacles, challenges, and achievements can provide a rich metaphor or parallel for contemplation and processing of their internal landscape (Naidoo et al., 2017).

The Usiko Project's wilderness-based activities have developed into a rites of passage practice that undergirds a strong sense of community values, identity and participation for all involved in the work of the project (Fabrik, 2004). When wilderness activities and their underlying goals are integrated into a broader community program, the benefits that accrue extend beyond the individual level of change and empowerment, offering impetus to form collective engagements and actions that are reinforcing of community initiatives (Naidoo et al., 2017). Qualitative feedback suggests that the majority of participants completing the diversion program are able to successfully divert from the juvenile justice system (Botha, 2007).

31.10.2 CASE STUDY TWO: SCHOOL-BASED WILDERNESS RITES OF PASSAGE PROGRAM

Zygmont (2014), Naidoo et al. (2017), and Zygmont and Naidoo (2018) provide a description of an adventure program that forms part of the outdoor education program at a private South African Secondary School. It is designed to provide grade-9 learners with health-promoting social and personal peak experiences that are aimed at helping learners transition into a senior role in the school, building social and school ties, breaking down cliques and discrimination, fostering self-discovery, interdependence, confidence, leadership, and resilience. The grade-9 learners were divided into small single-sex groups accompanied by two adult facilitators, with one group leaving from the school each day. The entire journey lasted 27 days traversing 360 km of mountainous wilderness areas and rural farmland on foot, bicycle, and canoe. During their expedition the groups participated in activities aimed at building individual and group competencies such as raft building, food groups, group debriefs, journaling, prescribed reading, and a 30-hour solo. The learners concluded their journey with a special symbolic ceremony to celebrate their achievement before returning by bus to the school where they were received by awaiting family and friends.

Program participants prepared their own meals, negotiated waking and sleeping times, pitched their own tents, washed their own clothes, and had to form a new, collaborative, cohesive social network. Structured readings, discussions, journaling, and debriefs beside a campfire provided participants with opportunities to reflect on their learning and discuss the significance of their experience to the broader context of their lives. Zygmont (2014) conducted interviews with participants in order to understand the different ways in which the program was experienced, and identify the key factors contributing to these different outcomes. Many of the learners spoke of learning to push through difficulties or finding inner strength they never knew they had. They suggested that resilience was gained by overcoming challenges that seemed insurmountable but were conquered through perseverance and social support. Examples of such utterances include:

"Um, like also like the biggest thing that changed for me was my mindset and how I've become more positive and mentally strong, like that was the biggest thing … I never thought that I could push myself, like just me saying to myself you can do it you can do it like I never thought that I could just listen to myself and do it … and then from then I realized OK well I actually can do it like my mind helps me a lot and I realized that your mind is stronger than like than you kind of and you can push yourself through" (Participant LC1FF).

"Um for me it's a huge like, success in my life it's like a huge accomp like a huge accomplishment, um there were so many times where, it was raining and all I wanted to do was go home but you know you literally you had to push through and like, it was like the hardest thing your mind's like no I can't do this can't do this but you have to push through and that was really nice for me" (participant TN1FF).

"Its its just such an amazing feeling, it it you feel like you can do anything like it's such a big accomplishment, um, like basically you feel like you can just, ah I don't know that's basically it like you feel like you can do anything" (participant ZE1FF) "my buddy had told me to give up ages ago, but and I had reached my limit and everyone you just pushed through that and you realize, well I'm actually really strong inside and physically and emotionally to get through that.... So you learn that about yourself you realize well I never ever thought I would, ever do something like that in my whole life and now I'm doing it" (participant OP9HF). Zygmont and Naidoo (2018) found that the (a) program characterization, (b) group processes, (c) adult leader alliance, (d) depth of engagement, (e) personal relevance, and (f) perceived outcomes are all important dimensions for ensuring positive program outcomes.

31.10.3 CASE STUDY THREE: DEVELOPMENTAL CORPORATE STRATEGIC INTERVENTION

Outdoor programs are perceived as being more effective than indoor or mixed contexts for team building interventions among South African organizations (Kriek & Venter, 2009). As part of a Company strategic corporate intervention, a group of 50 executives gathered at an Outdoor Centre next to the Magalies Mountains in South Africa (Snyman, 2017). The program objectives were agreed upon as independent decision-making, communication processes and dynamics, affirmation, resilience, celebration of successes, small group contributions, and big group success. On the adventure program application scale (recreation, education, development, therapy), this program was facilitated as a developmental application.

After strenuous theoretical sessions, the group was equipped with principles, models, and theories that they would typically need in the execution of their day-to-day work environment. It also reflected the models they would need to solve or navigate through the adventure related experiential learning process or challenge that was going to follow. As part of the theory, the facilitators conducted a variety of group and personal assessments including amongst others Whole Brain Thinking. Participants were made aware to include these dynamics in how they approach the diversity of participants as part of their "journey."

The group was divided into ten smaller groups of five each. Facilitators considered the various personality profiles and group analysis that was done earlier when putting small groups together. Each group was equipped with an activity kit containing material and resources for the upcoming activities and challenges. They were also equipped with a manual that contained the overall objective and specifications of the challenge as well as briefing sheets of each of the activities, a process that took them on a route of 20 km and lasted an intense 48 hours. Projects or objectives included; gather resources from the environment and follow a provided plan to build a shelter for the team to overnight in, complete an abseil at a specific site by following coordinates, build a stretcher when one of the team members was "injured," deal with the disappointment when the team was informed via cell-phones that they had to change plans and accommodate an additional two assignments. Technology was included in the scenario as participants had to refer to social media grid references, capture photographic evidence and complete electronic pictorial journaling. The briefing required that each project is conducted by a new group member. After each activity and subprocess, members were required to capture facts, findings, feelings, and potential future applications based on lessons learned in their journals (for later reflection).

The small groups functioned independently; however, they had to ensure each of their assignments had the necessary external scheduling of facilities and resources (e.g., ensuring a facilitator is available to assist in the 30 m abseil they had to complete as part of one of the outcomes). They had to be continually conscious of the diversity of team members as well as the fact that they were not the only group competing for resources, there were nine others. The group with the most meticulous planning and scheduling would be rewarded, but on the flipside, if they missed schedule times, they would be seriously penalized and could only be helped outside their schedule if these resources were available.

At the end of the process all groups had to come together at an agreed upon time and consolidate the individual and group lessons learned into a blueprint for the company's future endeavors and strategy. Other than company strategy, this plan included personal development strategies as well as team strategies.

Snyman (2017) concluded the program by debriefing the individuals and small groups via pictorial reflection (pictures triggering experiences), topical reflection (keywords triggerzing experiences), individual journals and group presentations. Although the learning was facilitated as an adventurous experiential learning program, with real-life physical, emotional and cognitive implications, the overall challenge (experience, project) was consciously set up to address agreed upon functional corporate variables and mirrored practices of employees, chasing objectives in silos, while being part of a macro-organization. This provided a mirror to a world or setting where participants are constantly confronted with a multitude day to day corporate challenges creating stress, fear, and fatigue while competing against time for available resources with a diverse team not always seeing eye to eye.

31.11 RECOMMENDATIONS

Sibthorpe and Morgan (2011) describe adventure programs as an exemplar of positive youth development, offering participants opportunities "to engage in healthy relationship building, feel acceptance, and develop socioemotional competence—assets that are particularly powerful in promoting positive outcomes and disrupting negative outcomes" (p. 116). Similarly, Kriek and Venter (2009) have found that corporate clients perceive outdoor programs to be very effective in nurturing team formation, enhancement, and resilience. Large-scale meta-analyses have established the effectiveness of adventure programs in supporting clinical and developmental gains among individuals, families, and organizations (Bowen & Neill, 2013), whereas neurobiological experiments have established the neuroscientific mechanisms by which challenges mastered in a supportive milieu can confer resilience at the molecular, cellular or neural network levels (Azadmarzabadi et al., 2018; Cooper et al., 2015; Majcher-Maślanka et al., 2018; Myers et al., 2014). In summary, the available literature suggests adventure programming can be effectively used in order to develop resilience for a variety of clients. More awareness, access and training in such programs should be encouraged. However, not all adventure programs will produce desired outcomes. In order to be effective programs should utilize established psychological theory to design physical and intellectual activities, challenge participants to extend their boundaries in a supportive manner that relies on progressive advancement, make use of activities that enhance engagement and personal meaning, allow space for reflection and transference of learning to other contexts, enact mechanisms of change common with other psychotherapeutic interventions, interchange wilderness and other contexts, be of a longer rather than shorter duration, and include follow-up interventions that maintain or build on gains made during the program.

KEYWORDS

- **resilience**
- **adventure programme**
- **adolescent**

REFERENCES

Allan, J. F., McKenna, J., & Hind, K. (2012). Brain resilience: Shedding light into the black box of adventure processes. *Australian Journal of Outdoor Education, 16*(1), 3–14.

Alvarez, A. G., & Stauffer, G. A. (2001). Musings on adventure therapy. *Journal of Experiential Education, 24*(2), 85–91.

Amat, J., Matus-Amat, P., Watkins, L. R., & Maier, S. F. (1998). Escapable and inescapable stress differentially alter extracellular levels of 5-HT in the basolateral amygdala of the rat. *Brain Research, 812*, 113–120.

Andolina, D., Maran, D., Valzania, A., Conversi, D., & Puglisi-Allegra, S. (2013). Prefrontal/amygdalar system determines stress coping behavior through 5-HT/GABA connection. *Neuropsychopharmacology, 38*, 2057–2067.

Anthonissen, L. (2011). *Exploring the influence of gender on how female facilitators of a wilderness programme view "wilderness"* (Unpublished master's thesis, Stellenbosch University). Retrieved from http://scholar.sun.ac.za/handle/10019.1/6462

Azadmarzabadi, E., Haghighatfard, A., & Mohammadi, A. (2018). Low resilience to stress is associated with candidate gene expression alterations in the dopaminergic signalling pathway. *Psychogeriatrics, 18*(3), 190–201.

Bandoroff, S., & Newes, S. (Eds.) (2004). *Coming of age: The evolving field of adventure therapy*. Boulder, CO: Association for Experiential Education.

Bandura, A. (1977). Self-efficacy: Toward a unifying theory of behavioral change. *Psychological Review, 84*, 191–215.

Beams, S., Humberstone, B., & Allin, L. (2017). Adventure revisited: critically examining the concept of adventure and its relations with contemporary outdoor education and learning. *Journal of Adventure Education and Outdoor Learning, 17*(4), 275–279.

Beightol, J., Jevertson, J., Carter, S., Gray, S., & Gass, M. (2012). Adventure education and resilience enhancement. *Journal of Experiential Education, 35*(2), 307–325.

Beil, K., Hanes, D., & Zwickey, H. (2014). Environmental influence on holistic health measures. *Explore (NY), 10*(2), 115–117.

Benjamin, L. (2014). *Community counsellors' experiences of trauma and resilience in a low-income community* (Unpublished doctoral dissertation, Stellenbosch University). Retrieved from http://scholar.sun.ac.za/handle/10019.1/86553

Bloemhoff, H. J. (2012). High-risk adolescent girls, resiliency and a ropes course. *African Journal for Physical, Health Education, Recreation and Dance (AJPHERD), Supplement 3*, 128–139.

Booth, J. W., & Neill, J. T. (2017). Coping strategies and the development of psychological resilience. *Journal of Outdoor and Environmental Education, 20*(1), 47–54.

Botha, M. (2007). *Harnessing wilderness in the rehabilitation of male adolescent offenders in a diversion programme* (Unpublished master's thesis, Stellenbosch University). Retrieved from http://scholar.sun.ac.za/handle/10019.1/3289

Botha, M.L., & Naidoo, A.V. (2006 April 12). *Unlocking the potential of wilderness for youth wellness and development in a local South African youth development project.* Paper presented at the International Conference on Sport and Development, University of the Western Cape.

Bowen, D. J., & Neill, J. T. (2013). A meta-analysis of adventure therapy outcomes and moderators. *The Open Psychology Journal, 6*, 1–38.

Boyes, M. (2013). Outdoor adventure and successful aging. *Aging and Society, 33*(4), 644–665.

Boyers, M. B. P. (2015). *Adventure based experiential learning and adolescents' self-reported levels of resilience and positive mental health* (Unpublished master's thesis, North-West University). Retrieved from https://repository.nwu.ac.za/handle/10394/17909

Brendtro, L., Brokenleg, M., & van Bockern, S. (2002). *Reclaiming youth at risk: Our hope for the future*. Bloomington, IN: Solution Tree Press.

Calizo, L. H., Akanwa, A., Ma, X., Pan, Y-Z., Lemos, J. C., Craige, C., Heemstra, L. A., & Beck, S. G. (2011). Raphe serotonin neurons are not homogenous: Electrophysical, morphological and neurochemical evidence. *Neuropharmacology, 61*(3), 524–543.

Christianson, J. P., & Greenwood, B.N. (2014). Stress-protective neural circuits: Not all roads lead through the prefrontal cortex. *Stress, 17*, 1–12. doi:10.3109/10253890.2013.794450

Cooper, M. A., Clinard, C. T., & Morrison, K. E. (2015). Neurobiological mechanisms supporting experience-dependent resistance to social stress. *Neuroscience, 291*, 1–14.

Csikszentmihalyi, M., & Csikszentmihalyi, I. (1999). Adventure and the flow experience. In J. Miles & S. Priest (Eds.), *Adventure programming* (pp. 153–158). State College, PA: Venture Publishing.

Darst, P. W., & Armstrong, G. P. (1980). *Outdoor adventure activities for school and recreation programs*. Minneapolis, MN: Burgess Publishing Company.

Davis-Berman, J., & Berman, D. (2008). *The promise of wilderness therapy*. Boulder, CO: Association for Experiential Education.

Ewert, A. W. (1989). *Outdoor adventure pursuits: Foundations, models, and theories*. Colombus, OH: Publishing Horizons.

Fabrik, C. (2004). *Reviving a forgotten custom: An evaluation of a community-based mentoring intervention—The Jamestown Usiko Youth Project* (Unpublished master's thesis, Stellenbosch University). Retrieved from http://scholar.sun.ac.za/handle/10019.1/3391

Fine, N., & van Niekerk, D. (2015). *In the hearts of men: Men changing lives*. Cape Town: Hearts of Men Institute.

Fleshner, M., Maier, S. F., Lyons, D. M., & Raskind, M. A. (2011). The neurobiology of the stress-resistant brain. *Stress, 14*(5), 498–502.

Gabrielsen, L. E., Fernee, C. R., Aasen, G. O., & Eskedal, L. T. (2016). Why randomized trials are challenging within adventure therapy research: Lessons learned in Norway. *Journal of Experiential Education, 39*(1), 5–14.

Gass, M. A. (Ed.) (1993a). *Adventure therapy: Therapeutic applications of adventure programming*. Dubuque, IO: Kendall/Hunt Publishing.

Gass, M. A. (1993b). Enhancing metaphor development in adventure therapy programs. In M. A. Gass (Ed.), *Adventure therapy: Therapeutic applications of adventure programming* (pp. 245–258). Dubuque, IO: Kendall/Hunt Publishing.

Gass, M. A. (1993c). Foundations of adventure therapy. In M. A. Gass (Ed.), *Adventure therapy: Therapeutic applications of adventure programming* (pp. 3–10). Dubuque, IO: Kendall/Hunt Publishing.

Gass, M. A., Gillis, H. L., & Russell, K. C. (2012). *Adventure therapy: Theory, research and practice*. New York, NY: Routledge.

Glass, J. S., & Myers, J. E. (2001). Combining the old and the new to help adolescents: Individual psychology and Adventure-Based Counseling. *Journal of Mental Health Counseling, 23*(2), 104–114.

Greffrath, G., Meyer, C. D., Strydom, H., & Ellis, S. (2013). A comparison between centre-based and expedition-based (wilderness) adventure experiential learning regarding group effectiveness: A mixed methodology. *South African Journal for Research in Sport, Physical Education and Recreation, 35*(1), 11–24.

Harper, N. J., Gabrielstein, L. E., & Carpenter, C. (2018). A cross-cultural exploration of the 'wild' in wilderness therapy: Canada, Norway and Australia. *Journal of Adventure Education and Outdoor Learning, 18*(2), 148–164.

Hayhurst, J., Hunter, J. A., Kafka, S., & Boyes, M. (2015). Enhancing resilience in youth through a 10-day developmental voyage. *Journal of Adventure Education & Outdoor Learning, 15*(1), 40–52.

Hill, A. S., Sahay, A., & Hen, R. (2015). Increasing adult hippocampal neurogenesis is sufficient to reduce anxiety and depression-like behaviors. *Neuropsychopharmacology, 40*, 2368–2378.

Humberstone, B., Prince, H., & Henderson, K. A. (Eds.) (2016). *Handbook of Outdoor Studies*. London: Routledge.

Itin, C. M. (2001). Adventure therapy—Critical questions. *Journal of Experiential Education, 24*(2), 80–84.

Jessor, R. (1993). Successful adolescent development among youth in high-risk settings. *American Psychologist, 48*(2), 117–126.

Kimball, R. O., & Bacon, S. B. (1993). The wilderness challenge model. In M. A. Gass (Ed.), *Adventure therapy: Therapeutic applications of adventure programming* (pp. 11-41). Dubuque, IO: Kendall/Hunt Publishing.

Knoetze, K. (2003). *Crossing the threshold: evaluation of a rites of passage programme in a peri-urban South African community* (Unpublished master's thesis, Stellenbosch University). Retrieved from http://scholar.sun.ac.za/handle/10019.1/97518

Kriek, H. S. (2007). A survey of the prevalence and nature of team building interventions in South African organisations. *South African Journal of Business Management*, 38(4), 1–7.

Kriek, H. S., & Venter, P. (2009). The perceived success of team building interventions in South African organisations. *South African Business Review*, *13*(1), 112–128.

Kumar, S., Hultman, R., Hughes, D., Michel, N., Katz, B. M., & Dzirasa, K. (2014). Prefrontal cortex reactivity underlies trait vulnerability to chronic social defeat stress. *Nature Communications*, *5*(4537), 1–9.

Linley, A. (2009). *Strengths measure, Realise 2* [Technical manual and statistical properties]. *Coventry*, *UK*: CAPP Press.

Maier, S. F., & Watkins, L. R. (2010). Role of the medial prefrontal cortex in coping and resilience. *Brain Research*, *1355*, 52–60.

Majcher-Maślanka, I., Solarz, A., Wędzdony, K., & Chocyk, A. (2018). Previous early-life stress modifies acute corticosterone-induced synaptic plasticity in the medial prefrontal cortex of adolescent rats. *Neuroscience*, *379*, 316–333.

Maller, C., Townsend, M., Brown, P. & St Leger, L. (2002). Healthy parks, healthy people: The health benefits of contact with nature in a park context: A review of current literature. *The George Wright Forum*, *26*(2), 51–83.

Myers, B., Dolgas, C. M., Kasckow, J., Cullinan, W. E., & Herman, J. P. (2014). Central stress-integrative circuits: Forebrain glutamatergic and GABAergic projections to the dorsomedial hypothalamus, medial optic area, and bed nucleus of the stria terminalis. *Brain Structure & Function*, *219*(4), 1287–1303.

Naidoo, A. V., & Van Wyk, S. B. (2003). Intervening in communities: Combining curative and preventive interventions. *Journal of Prevention and Intervention in the Community*, *25*(1), 65–80.

Naidoo, A.V., Zygmont, C. S., & Phillips, S. (2017). Harnessing the power of ecopsychology in community work. In M. Seedat, S. Suffla, & D. Christie (Eds.), *Emancipatory and participatory methodologies in peace, critical, and community psychology* (pp. 77–89). Peace Psychology Book Series. Cham, Switzerland: Springer.

Neill, J. T. (2008). *Enhancing life effectiveness: The impacts of outdoor education programmes: Volume 1* (Unpublished doctoral dissertation, University of Western Sydney). Retrieved from http://wilderdom.com/phd

Neill, J. T., & Dias, K. L. (2001). Adventure education and resilience: The double-edged sword. *Journal of Adventure Education and Outdoor Learning*, *1*(2), 35–42.

Neuman, J. (2010, November 18–21). *Outdoor activities and health.* Paper presented at the 5th International Mountain and Outdoor Sports Conference, Svatý Jan pod Skalou, Czech Republic. Retrieved from http://www.imosc.org/sbornik/sbornik2010_web.pdf

Newes, S., & Bandoroff, S. (2004). What is adventure therapy? In S. Bandoroff & S. Newes (Eds.), *Coming of age: The evolving field of adventure therapy* (pp. 1–30). Boulder, CO: Association for Experiential Education.

Nunez, C. (2010). *The outcomes of a wilderness experience programme on the resilience and psychological adjustment of South African adolescents* (Unpublished master's thesis, University of KwaZulu-Natal). Retrieved from https://researchspace.ukzn.ac.za/xmlui/handle/10413/6791

Norton, C. L., & Watt, T. T. (2014). Exploring the impact of a wilderness-based positive youth development program for urban youth. *Journal of Experiential Education, 37*(4), 335–350.

O'Brien, K., & Lomas, T. (2017). Developing a growth mindset through outdoor personal development: Can an intervention underpinned by psychology increase the impact of an outdoor learning course for young people? *Journal of Adventure Education & Outdoor Learning*, *17*(2), 133–147.

Overholt, J. R., & Ewert, A. (2015). Gender matters: Exploring the process of developing resilience through outdoor adventure. *Journal of Experiential Education*, *38*(1), 41–55.

Passarelli, A., Hall, E., & Anderson, M. (2010). A strengths-based approach to outdoor and adventure education: Possibilities for personal growth. *Journal of Experiential Education*, 33(2), 120–135.

Pinnock, D. (2016). *Gangtown.* Cape Town: NB Publishers.

Priest, S. (1986). *Outdoor leadership preparation in five nations* (Unpublished doctoral dissertation, University of Oregon). Retrieved from http://simonpriest.altervista.org/DOWNLOADS/Dissertation.pdf

Priest, S. (1999). National life cycles in outdoor adventure programming. *The Outdoor Network*, *10*(1), 16–17, 34–35.

Priest, S., & Gass, M.A. (2005). *Effective leadership in adventure programming* (2nd ed). Champaign, IL: Human Kinetics.

Pryor, A., Carpenter, C., & Townsend, M. (2005). Outdoor education and bush adventure therapy: A social-ecological approach to health and wellbeing. *Australian Journal of Outdoor Education, 9*(1), 3–13.

Ritchie, S. D., Wabano, M-J., Russell, K., Enosse, L., & Young, N. L. (2014). Promoting resilience and well-being through an outdoor intervention designed for Aboriginal adolescents. *Rural and Remote Health, 14*(2523), 1–19.

Robertson, G. (2000). *Wilderness therapy in South Africa: The development of community-based projects*. Paper presented at the Association for Experiential Education 28th International Conference, Tucson, Arizona.

Robertson, G., & De Kiewit, S. (1998). Wilderness therapy with militarised youths in traumatised communities. *Journal of Social Development in Africa, 13*(1), 53–58.

Rosenfeld, G. (2009). Beyond evidence-based psychotherapy: Fostering the eight sources of change in child and adolescent treatment. New York: Routledge.

Rousseau, E., & Pinnock, D. (2014). *Wild resilience: Working with high-risk adolescents using wilderness, ritual and mentorship*. Cape Town: African Sun Press.

Russell, K. C. (2003). A nation-wide survey of Outdoor Behavioral Healthcare Programs for adolescents with problem behaviors. *Journal of Experiential Education, 25*(3), 322–331.

Ryff, C. D., & Singer, B. H. (2008). Know thyself and become what you are: A eudemonic approach to psychological well-being. *Journal of Happiness Studies, 9*(1), 13–39.

Santos, M. F., Carrasco-Santos, M. J., Eça Costa Franco, M., Chinchilla, J. L. M., Reinoso-Caparrós, M., Bahia, S., Santos, N., & Beja, M. J. P. (2017). KidsTalentum: Outdoor training program to improve emotional intelligence in adolescents in Manchester Soccer Schools, Lisbon. *EPH – International Journal of Education, 1*(7), 1–25.

Schell-Faucon, S. (2001). *Journey into the inner self and encounter with the other: Transformation trails with militarised youths of opposing groups*. Research Report for the Centre for the Study of Violence and reconciliation.

Schoel, J., & Maizell, R. S. (2002). *Exploring islands of healing: New perspectives on adventure-based counseling*. Beverly, MA: Project Adventure, Inc.

Shin, J. H., Kim, D., & Jung, M. W. (2018). Differential coding of reward and movement information in the dorsomedial striatal direct and indirect pathways. *Nature Communications, 9*, 404.

Snyman, P. (2017). *Creative adventure related experiential learning programmes*. Paper presented at the meeting of the Uniting Christian Students Association (UCSA), Stellenbosch, South Africa.

Sibthorpe, J., & Morgan, C. (2011). Adventure-based programming: Exemplary youth development practice. *New Directions for Youth Development, 130*, 105–119.

Somes, M.P. (1998). *The healing wisdom of Africa: Finding life purpose through nature, ritual and community*. New York: Putnam Book.

Theron, J. P. (2009). *Group cohesion and team building through adventure: Experiences of a South African netball team in preparation for the world championships* (Unpublished master's thesis, University of Pretoria). Retrieved from http://nrfnexus.nrf.ac.za/handle/20.500.11892/92637

Tillman, S., Tobin, D., Avison, W., & Gilliland, J. (2018). Mental health benefits of interactions with nature in children and teenagers: A systematic review. *Journal of Epidemiology and Community Health, 72*(10), 958–966.

Todd, S., Breunig, M., O'Connell, T., Hutson, G., DiRezno, A., & Young, A. (2018, January 12–14). *Gains in outdoor pursuits program outcomes of sense of community, sense of place, nature relatedness, well-being and resilience by level of development*. Paper presented at the 14th Biennial Research Symposium of the Coalition for Education in the Outdoors, Martinsville, Indiana.

van Heyden, Y. (2004). *An exploratory study of the use of metaphor in the practice of ecotherapy* (Unpublished master's thesis, Stellenbosch University). Retrieved from http://scholar.sun.ac.za/handle/10019.1/49839

Van Lingen, J. M., & De Jager, A. C. (2014). Wellness and its dimensions—A South African student perspective. *Journal of Counselling and Development in Higher Education Southern Africa, 1*(1), 1–19.

Van Rensburg, A., Theron, L., & Rothmann, S. (2015). A review of quantitative studies of South African youth resilience: Some gaps. *South African Journal of Science, 111*(7/8). doi:10.17159/sajs.2015/20140164

Van Wyk, S.B., & Naidoo, A.V. (2006). Broadening mental health services to disadvantaged communities in South Africa. *Journal of Psychology in Africa, 2*, 273–282.

Veevers, N., & Allison, P. (2011). *Kurt Hahn: Inspirational, visionary, outdoor and experiential educator*. Rotterdam: Sense Publishers.

Wakabayashi, T., Hidaka, R., Fukimaki, S., Asashima, M., & Kuwabara, T. (2014). Chapter two—MicroRNAs and epigenetics in adult neurogenesis. *Advances in Genetics, 86*, 27–44.

Webb, D. J. (1999). Recreational outdoor adventure programs. In J. C. Miles & S. Priest (Eds.), *Adventure programming*. State College, PA: Venture.

Worley, N. B., Hill, M. N., & Christianson, J. P. (2018). Prefrontal endocannabinoids, stress controllability and resilience: A hypothesis. *Progress in Neuropsychopharmacology & Biological Psychiatry*, *85*, 180–188.

Zannas, A. S., & West, A. E. (2014). Epigenetics and the regulation of stress vulnerability. *Neuroscience*, *264*, 157–170.

Zolkoski, S. M., & Bullock, L. M. (2012). Resilience in children and youth: A review. *Children and Youth Services Review*, *34*, 2295–2303.

Zygmont, C. S. (2014). *A phenomenographical study of the qualitative variation of adventure /wilderness programme experiences among adolescent high school participants in the Western Cape* (Unpublished doctoral dissertation, University of Stellenbosch). Retrieved from http://scholar.sun.ac.za/handle/10019.1/86210

Zygmont, C. S., & Naidoo, A. V. (2018). A phenomenographic study of factors leading to variation in the experience of a school-based wilderness experiential programme. *South African Journal of Psychology*, *48*(1), 129–141.

Index

A

Activeness, 37
Actor–partner interdependence model (APIM), 324
ADRESSING model for multicultural counseling, 19
Adventure programming, 466
 case studies, 475–480
 in developing resilience, 468–471
 development-by-challenge or stress inoculation model, 472
 impact on wellness, 468–469t
 mechanisms of change in, 474–475
 neurological models, 473–474
 role of protective factors, 472–473
 theoretical models of, 471–472
 types, 466–467, 467f
Adversities, 162
Aging, 210–211
 retirement and worldview, 221–223
 successful, 214–215
 in technological era, 221
Aid-related support, 63
Al Qaeda, 433
Altruistic cherishing (ALTR), 54
American Indians/Alaskan Natives (AI/AN)
 risk for suicide, 24
Antecedent-focused emotion-regulation, 99
Appreciation-related support, 63
Ashby, W., 196
Asian Americans
 risk for suicide, 24
Association of Community Organizations for Reform Now (ACORN), 42
At-risk behaviors
 across racial/ethnic minorities, 23–25
Attachment, 36–37
 cherishing and, 52
Attention deployment, 100
Average lifespan of people, 210
Avoidance, 110–111, 123f

B

Belongingness support, 62–63
Bertalanffy, Ludwig von, 195
Bipolar disorder, 186
Body shaming, 230
Bogdanov, Alexander, 196–197
Borderline personality disorder (BPD), 18
Brief Illness Perception Questionnaire (BIPQ), 133
Brief Intervention of Thriving (BIT), 65–66
Brief Resilience Scale (BRS), 95, 343
Buffering Hypothesis, 184
Bullying, 230, 232–234
Burnout syndrome, 314

C

Cancer, 165–167
 negative effects of, 166
 resilience/coping skills, 166–167
Cardiovascular disease (CAD), 164–165, 313
 protective factors, 165
 psychological effects of, 165
 risk factors, 164
 sources of resilience, 165
Chandrasekhar, R., 360
Charismatic leadership, 290–291
Cherishing, 51–52, 57t
 attachment theories and, 52
 family, 52, 54
 friends, 52, 54
 measuring, 54
 nature of, 52
 personal, 52
 psychological well-being (PWB) and, 56
 structure, 55f
 well-being and, 52–54
Childhood trauma, 25
Chilean miners, case of, 33–34
Chronic fatigue syndrome, 312
Chronic illness in workplace, 167–169
 employer accommodations, 169–170
Chronic strains, 4
Chronic stressors, 3
Cognitive-behavioral stress management (CBSM), 447
Cognitive behavioral therapy (CBT), 39, 41, 179–180
Cognitive reappraisal, 63–64, 66–68, 70–73
Cognizant Technology Solutions, 360

Collective/social characteristics of resilience
accessibility to public amenities, 217
inclusivity, 218
safe neighborhood, 217
social services and policies, 218
social support system, 218
well-designed homes, 217
Combat-related trauma reactions, 38
Community resilience, 42–44, 255–256, 261–263
place identity and, 263–264
Compensatory model, 97
Connor-Davidson Resilience Scale (CD-RISC), 39–40, 150, 185, 342
COPD, 167
Coping, 8–9, 162–164
for chronic pain, 167
cognitive appraisal and, 162
contextual model of, 162
definition, 164
disengaged, 8
engaged, 8
personality traits and, 8
psychopathology and, 8
resilience and, 9
self-destructive behaviors and, 17
task- *versus* emotion-focused, 8
Coping Strategies Inventory, 111
Coping Strategy Indicator (CSI), 110–112
Coping Strategy Indicator (CSI), study of abridged version
classification agreement, 121
fit indices, 118–119, 119t
full version, 115
internal structure, 115
inter-version correlation, 119–121, 121t–122t
measurement parameters, 118
participants, 112
procedure, 113–115
psychometric analysis, 122–124
quantification of agreement, 121–122
Rasch analyses, 118–119
reliability, 115, 118
short version, 115–116, 118t
Counter-organizational activity, 203
Crossover, 324
Cultural aspects of resilience, 7
Cultural identities, 19
Cultural sanctions, 24
Culture, definition, 19
Culture, study of interplay between SWB and
comparison between Czechs and South Africans, 82, 87–89
Czech and South African participants, 81–82
dimensions, 82
measures, 84
methodology, 83–84
objectives, 83
procedure and data analysis, 84–85
result analysis, 85–87
Cyber bullying, 230, 234–235
Czech Republic (CZ), 81

D

Depression, 314
Developmental corporate strategic intervention, 479–480
Diabetes, 313
Diagnostic and Statistical Manual of Mental Disorders (DSM), 38
Dialectical behavioral therapy (DBT), 179
Dialectical Behavior Therapy group skills training (DBT-ST), 180
Difficulties in Emotion Regulation Scale (DERS), 95, 101
Direct harm, 16
Direct self-injury, 18
Disabilities, 150
Disengaged coping, 8
Doctor–patient communication
link between patient satisfaction and, 130–131
Doctor–patient communication, Indian psycho-oncology research in, 131–132
correlation analysis, 135–137, 136t
implications, 138–140
limitation of study, 140
measures, 132–133
participants, 132
patient characteristics, 134–135, 135t
predictors of, 137–138, 137t
procedure, 134

E

Edgy-creative thinkers, 305
Emotional intelligence, 289–290, 384
Emotional support, 62–63
Emotion-focused coping, 8
Emotion regulation (ER), 63–64, 70, 98–100
adaptive response tendencies and, 99
antecedent-focused, 99
process model of, 99
response-focused, 99
social support and, 64–65

Emotion Regulation Questionnaire (ERQ), 65–66
Emotions, understanding of, 284–285
  emotion types, 286
  emotion work, 285–286
  in terrorism, 425–426
Employee engagement, 292–293
Employee well-being (EWB), 380
  age and, 383
  antecedents of, 383–388, 386f
  benefits of, 382–383
  consequences of, 388–390
  elements of, 381–382, 382t
  emotional intelligence and, 384
  employee productivity and, 389–390
  employee retention and, 388–389
  flexibility at work and, 386–387
  job design and, 387–388
  job performance and, 389
  level of commitment and, 388
  link with employee productivity and, 380
  literature review, 381
  personality and, 383
  physical fitness and, 383–384
  wellness program, 387
  work environment and, 386
  work-related behaviors, 384–385
  work-supportive family members, 385
Employment, 309
Engaged coping, 8
Eritrean refugee, 36
Eudaimonic wellbeing, 380
Eudemonic theory of Aristotle, 62

F

Facebook, 43
Failure tolerance, 303–304
Family cohesion, 198
Family flexibility, 198
Family resilience, 197–198
  Bogdanov's theory, 201–204
  family resources as a source of, 198–201
Family ties, 198
Family values, 198
Flint water crisis, 34
Functional Assessment of Cancer Therapy (FACT-G), 133

G

General self-efficacy (GSE), 146, 148–149
God Locus Of Health Control (GLHC), 134
Gross stress reaction (GSR), 38

H

Happiness, 147–148
Happy person, characteristics of, 154
Harmony in life, 78–79
  correlates and predictors of, 86–87, 89–90
Health-related Quality of Life (HRQoL) questionnaire, 131
Healthy lifestyle, 216
Hedonic wellbeing, 380
Hidden ideators, 24
Hope, 317
  perceived, 79–80, 87
  transcendental conceptualizations of, 78
Hospital Anxiety Depression Scale (HADS), 133
Humor, 217
Hurricane Katrina, 34, 42–43
Hurricane Maria, 44

I

Idioms of distress, 24
Impact of Events Scale-Revised (IES-R), 454
Impulsive-aggressive behavior, 21
India, 54, 56
Indian IT Industry, 359–360
  job burnout, 361–365
Indirect harm, 16
Indirect self-injury, 18
Informational or appraisal support, 62
Information and communication technologies (ICTs), 360–361
Interactive models of resilience, 97
Interpersonal Support Evaluation List (ISEL), 65–66

J

Job rotation, 303

L

LatinX individual
  at-risk behavior among, 25
Leadership resilience, 298–299
  developing, 301–305
Life review, 217
Life satisfaction, 147, 149
Locus of control (LOC), 131
Lost leader syndrome, 304

M

Major depressive disorder (MDD), 21
Marine Operational Stress Control and Readiness program, 27

Marlowe–Crowne Social Desirability scale, 54
Mastery Self-efficacy Scale (MSES), 454
Measures of resilience, study of
  aim of study, 343
  analysis, 344–349
  confirmatory factor analysis (CFA), 345–349, 347f–348f
  descriptive statistics, 344–345, 345t
  discussion, 349–350
  instruments, 343–344
  limitations of study, 350–352
  principal component analysis (PCA), 345–346t
  sample and procedure, 343
Mediation modelling, 87
Medical Interview Satisfaction Scale (MISS), 130
Memorials, 43
Mental health-anxiety, 314
Mental healthcare workers, resiliency in, 40–42
Mental Health Continuum Short Form (MHC-SF), 470
Mental health disorders, resiliency and, 37–40
Mental skills in sports, 272–274
  goal setting, 273–274
  mental imagery, 273
  self-talk, 272–273
Metabolic syndrome, 313
Metatheory of resilience, 210
Meta-theory of resilience and resiliency, 96
Military sexual trauma, 27
Military trauma, 25
Mindfulness-based stress reduction, 41
Mindful sport performance enhancement (MSPE), 271
Minority stress, 24
MISS-21, 132–133
Motivation and Personality (Maslow), 146
Multicultural counseling, 19
Multidimensional Health Locus Of Control Cancer (MHLC-C), 133–134
Multidimensional Scale of Perceived Social Support (MSPSS), 454
Multiple intersecting identities, challenges of, 27
Musculoskeletal disorders, 313

N

Neemuchwala, Abid Ali, 360
Negative emotions, 80, 434–436
Negative events, 4
Neuroticism, 35
Nonsuicidal self-injury (NSSI), 180

O

Older age, 7
  employment and education, 213
  enabling environments, 213–214
  factors of resilience among elderly, 215–218, 220–221
  financial security, 211–212
  health status, 212–213
  risks and vulnerabilities, 211–214
Optimal stress, 97
Optimism, 216, 317

P

Parent–child relationship, 37
Parenthood
  transition to motherhood, 452–454
Patient dissatisfaction, 129
Patient satisfaction, 129–130
  link between doctor–patient communication and, 130–131
  locus of control (LOC) as predictor of, 131
Patient Satisfaction Questionnaire, 130
Perceived hope, 79–80, 87–88f
Personal cherishing (PERS), 54
Picker Patient Experience Questionnaire, 130
Place, definition, 256–257
Place identity, 257–258
  community and, 260–261
  functions of, 258–260
Positive emotions, 80, 433–434
Positive emotions, at workplace, 283–284, 286–288
  broaden-and-build theory of, 288
  fostering, 288–293
Positive Organizational Behavior (POB), 297–298, 309
Positive psychology, 146, 150
Positive psychology at work, 315
Post-traumatic Growth Inventory (PTGI), 454
Post-traumatic growth (PTG), 34, 44–45, 440–441, 452
  cultural meanings attached to, 443–444
  as process and outcome, 443
  religious coping as a moderator for, 444–446
  resilience *vs,* 441–443
  therapy for, 446–447
Post-traumatic growth (PTG), in transition to motherhood, study of
  changes in self, 458–459
  discussion, 460
  help seeking, 457
  limitations, 460
  participants, 454, 456

reassurance, 456–457
result analysis, 454–459
self-efficacy, 457–458
Post-traumatic reaction, 441
Post-traumatic stress disorder (PTSD), 4–5, 15, 27, 38–40, 42, 343
effect of stress on symptoms of, 6
military, 25
symptoms, 5
Problem solving, 110–111
Progressive muscular relaxation (PMR), 274
Project Liberty, 43
Protective factor model, 97
Protective factors, 410
Psychiatric illness, 20
Psychological aspects of resilience, 6–7
Psychological Capital (PsyCap), 297–298, 327
Psychological capital–WLB relationship, 325
Psychological flexibility, 397–398
as cognitive flexibility, 401–402
definition, 398–401
as emotional adaptability, 402–403
training for, 403–405
Psychological resilience, 96
Psychological skill training (PST), 270–271
Psychological stress
adverse health outcomes from, 4
environmental factors, 3
Psychological well-being (PWB), 53–54, 56, 146–147, 411
Psychosocial hazards, 311–312
impacts of, 312–315, 313t
Psychotherapy, 179–181

R

RAW Scale, 343, 349
Reactive approaches of resiliency, 301
Reappraisal, 71
Reappraising, 72
Reckless and self-destructive behavior (RSDB), 26–27
Reorganization, 43
Resilience attitude, 299
attitude of challenge, 300
attitude of commitment, 300
attitude of control, 300
social support, 300
transformational coping, 300
Resilience–emotion regulation relationship, study
correlation coefficients, 103t
findings, 102–105
measures, 101
procedure, 101–102
regression analysis, 102, 104t
sample, 100–101, 103t
therapeutic implications, 105–106
Resilience—proactive approaches
after action reviews (AARS), 304–305
asset-focused strategies, 301
developmental assignments, 303
experiences of trauma and successful resolution of trauma, 302
failure tolerance, 303–304
interaction with edgy-creative thinkers, 305
job rotation, 303
metaphorical problem solving, 305
process-focused strategies, 301
reflective inquiry and action plan (RIA), 304
risk–asset–process strategy of building resilience, 300–302
risk-focused strategies, 300–301
Resilience/resiliency, 5–7, 145–146, 149–151, 161, 195–196, 243, 315–316, 339–340, 439–440
attachment influence on, 36–37
in childhood, 17
coping strategies, 9
cultural aspects, 7
definition, 16, 340–342
as a developmental phenomenon across lifespan, 209–210
factors affecting, 316
individual's social support and, 35–36
measures of, 342–343. *See also* Measures of resilience, study of
as a personality factor, 182
personality features and level of, 35
positive emotion and, 291–292
as a protective factor, 34
PsyCap factors, 316–317
psychological aspects, 6–7
as psychological construct, 182–184
psychosocial risks and, 317–318
resources, 98
sexual and gender identities as risk factors, 23
social aspects, 5
social resources, 440
in sports, 269–270
theoretical perspectives on, 96–98, 244–245, 245f
at work, 318–319, 320t, 340
Resilient communities, 43
Resilient teachers, 244

Response-focused emotion-regulation, 99
Retirement planning, 222
Risk–asset–process strategy of building resilience, 300–302
Risk-taking or risk-seeking behavior, 17
Ryff Scale of Psychological Well-Being, 150

S

Scheier and Carver's Life Orientation Test, 150
Schematic Appraisals Model of Suicide, 187
School-based wilderness rites of passage program, 478–479
School–home relationships, 238
School life, 230
  academic pressure, 230–232
  bullying, 232–234
  cyber bullying, 234–235
  intervention to counter mental health problems, 236
  peer relations, 235, 238
  school–home relationships, 238
  social and emotional learning curriculum, 237
  student–teacher relationship, 237–238
Self-destructive behaviors, 17
  across racial/ethnic minorities, 23–25
  age and, 20–21
  age-specific protective factors, 22
  aggression and, 21
  among LGBTQ+, 23
  experience of trauma, 25–27
  impulsivity and, 21
  intersecting identities and, 27
  psychiatric illness and, 20
  risk factors, 18–19
  SES status and, 20
  sexual and gender identities as risk factors, 22–23
Self-efficacy, 7, 145–146, 148–149, 216, 317, 457–458
Self-harm ideation, 24
Self-harming behavior, 17–18
  direct *vs* indirect, 18
  risk factors, 18–19
Self-injury, 16
September 11th terrorist attacks, 34, 43
Sleep problems, 314
Smoking cessation programs, 167
Social and emotional learning curriculum, 237
Social aspects of resilience, 5
Social desirability, 54
Social discord, 24
Social interaction model, 130
Socialization, 7
Social network, 221
Social security system, 222
Social support, 35–36, 221, 300, 385
  association between depression and, 5
  emotion regulation (ER) and, 64–65
Social support, study of relationship between well-being and, 62–63
  analysis, 66–67
  hypothesis, 65
  limitation of study, 73–74
  measures, 65–66
  study sample and procedure, 65
South Africa (SA), 81–82
Spirituality, 220–221
Spiritual practice, 217
Sports psychology, 267
  global best practices, 274–276
  grounded theory of psychological resilience and performance, 271
  mental skills, 272–274
  models, 270–271
  quality of life model, 271
Stress, 109–110
  physiological effects of, 4
  psychological effects of, 4–5
  resilience to, 5–7
Stressful life circumstances, 163
Stress Management and Resiliency Training (SMART), 447
Stressors
  chronic, 3
  chronic strains, 4
  duration of, 3
  interpersonal impact of, 3
  nature of, 3–4
  negative events, 4
  proximity to, impact of, 4
  severity of, 3
  traumatic, 4
Stress-related appraisals, 9
Subjective well-being (SWB), 53, 61–62, 77–78, 147–148
Substance use disorder (SUD), 27
Successful aging, 214–215
  Indian perspective, 219
Suicidal behavior, 16
Suicidal ideation, 17–18, 184
Suicide, 15–16, 175–176, 314
  age as risk factor for, 186
  global differences in, 177
  individual differences in, 177–178

individual interventions for, 181
medication for, 181
preventative methods against, 179–181, 186–187
psychotherapy for, 179–181
as a public health concern, 176–178
resilience factors acting as moderators and protection against, 184–185
resilience scores, 185–186
risk and protective factors, 178–179
risk continuum and, 184
risk factors for, 16
in United States (US), 176–177
Suppression, 63–64
Symbolic convergence theory (SCT), 428–429
principles, 429–430
terrorism in terms of, 432–433
tools of, 430–432
System instability, 197

T

Tangible support, 62
Task-focused coping, 8
TCS, 360
Teachers' resilience, study of, 245–247
age and, 249–250
data analysis, 248
demographic characteristics, 247–248
gender and, 249
institutions and, 250–251
measures, 248
methodology, 248
recommendations, 251–252
results, 249
teaching experience and, 250
Teaching profession, 243
Technostress (TS) creators, 361–363
Terrorism, 426–427
emotions and, 427–428
Terrorist attacks, 43
Terrorist threat, 409–410
Terrorist threat–psychological well-being relationship, study
analysis of age differences, 413
analysis of gender differences, 413
correlation analysis, 415–416
data analysis, 413
descriptive statistics, 419–420t
features of life prospects, 418–421
intensity of terrorist threat experience, 417–418
method, 411–412
results, 413–421
r-Spearman correlation coefficients, 418t–417t
sample, 412–413
statistical analysis, 413
Transcendental conceptualizations of hope, 78
Trauma, 45, 440–441
as risk factor for self-destructive behavior, 25–27
Traumatic event, 38, 44
Traumatic stressors, 4
TS creators burnout relationship, study of
analysis and findings, 367–370
burnout as a mediator, 369–370
common method bias, 367
descriptive and correlation analysis, 368–369, 369t
discriminant validity, 368t
discussion, 370–373, 377t–378t
hypotheses, 365
limitations of study, 373–374
meaning work as a moderator, 370–371t, 372f
measures, 367
participants and procedure, 365–366
validity of measures, 367–368

U

Urban resilience, 263
Usiko diversion program, 476–478
Usiko Stellenbosch, 476

V

Virginia Tech school massacre, 45

W

Walsh, F., 198
Well-being, 52–54, 267, 379–380
emotion regulation (ER) and, 63–64
predictors of, 67–68, 68t
psychological (PWB), 53–54
social support and, 62–63
in sports, 268–269
subjective, 53, 61–62
Well-being of adolescents, study of
hearing disability and, 151–154t, 155
implications of, 155
methodology, 151
objectives, 151
qualitative analysis, 154–155
regression analysis, 153–154t
results and discussion, 151–153, 152t
Wiener, N., 196
Wipro Limited, 360
WLB–life satisfaction relationship, 325

Work–life balance (WLB), 324–326
crossover effect, 328
cultural aspects, 326
spillover effect, 326–328
Work–life balance (WLB), study
APIM results, 330–333t
descriptive statistics, 329–330, 330t
hypotheses, 328
implications of, 333–334
limitations and future directions, 334
measures, 329
sample and procedure, 328–329
Work–life interface, 324
Workplace bullying, 168
Workplace hazards
occupational/job stress, 311
occupational/workplace violence (WPV), 312
workplace bullying and harassment, 312
workplace harassment, 312
work-related fatigue, 311–312
Workplace health, 310–311
Work-related behaviors, 384–385
Work-supportive family members, 385

Y

Young onset dementia (YOD), 36

For Product Safety Concerns and Information please contact our EU representative GPSR@taylorandfrancis.com Taylor & Francis Verlag GmbH, Kaufingerstraße 24, 80331 München, Germany

Batch number: 10748878

Printed by Printforce, the Netherlands